AFFINITY AND EFFICACY

The Components of Drug-Receptor Interactions

AFFINITY AND EFFICACY

The Components of Drug-Receptor Interactions

FREDERICK J EHLERT

University of California, Irvine, USA

World Scientific

NEW JERSEY • LONDON • SINGAPORE • BEIJING • SHANGHAI • HONG KONG • TAIPEI • CHENNAI

Published by

World Scientific Publishing Co. Pte. Ltd.
5 Toh Tuck Link, Singapore 596224
USA office: 27 Warren Street, Suite 401-402, Hackensack, NJ 07601
UK office: 57 Shelton Street, Covent Garden, London WC2H 9HE

Library of Congress Cataloging-in-Publication Data
Ehlert, Frederick J., author.
Affinity and efficacy : the components of drug-receptor interactions / Frederick J. Ehlert.
p. ; cm.
Components of drug-receptor interactions
Includes bibliographical references.
ISBN 978-9814322560 (hardback : alk. paper)
I. Title. II. Title: Components of drug-receptor interactions.
[DNLM: 1. Receptors, Drug. 2. Ligand-Gated Ion Channels. 3. Protein Binding. 4. Radioligand Assay. 5. Receptors, G-Protein-Coupled. QV 36]
RM302.4
615.7'0452--dc23

2014024093

British Library Cataloguing-in-Publication Data
A catalogue record for this book is available from the British Library.

Typeset by Stallion Press
Email: enquiries@stallionpress.com

Printed in Singapore

For

Kelly and Molly

Contents

Acknowledgments

I am indebted to the University of California for enabling me to pursue my interests in drug-receptor interactions.

I owe thanks to many of my prior students and collaborators whose research is covered in this book including Beth Thomas, Evette Esqueda, Rennolds Ostrom, Greg Sawyer, Kate Figueroa, John Tran, Darakhshanda Shenaz and the late Les Candell. I would also like to thank Mike Griffin for his invaluable contributions to our research over the years. The same heartfelt thanks goes to Hinako Suga for her tireless and insightful contributions to our research during the past few years.

For having a look at a draft of Chapter 2, I would like thank Tony Auerbach.

For providing continual enthusiasm and support, I thank Larry Stein.

I thank Dennis Han for helping me with some of the figures and Christine Han for her continual support.

I especially thank my daughters, Kelly and Molly, for their support, enthusiasm and motivation. It is to them that I dedicate this book.

My research described in this book was supported by the National Institutes of Health (GM 69829).

Prologue — Drugs, Receptors and Therapeutic Messages

Most of us have personal experience with drug action, for we occasionally medicate ourselves with over-the-counter drugs and treatment by a physician often comes in the form of a prescription. This widespread use of drugs has led to common knowledge of their actions, and their molecular mechanisms have steadily come to light as reported in scientific journals and increasingly in lay publications. The ultimate explanation as to why drugs work often receives little attention, however. Simply stated, many drugs stimulate, inhibit or modulate natural signaling pathways.

Consider the use of the β_2 agonist, albuterol, in the treatment of symptomatic asthma. Following its absorption into the lung by inhalation, albuterol eventually binds to and turns on β_2-adrenoceptors expressed on the smooth muscle and mucosa of the airways, causing a decrease in both bronchial constriction and secretion of mucus.[1] Albuterol's chemical structure mimics that of the natural messenger molecule, epinephrine, which normally acts on this receptor (Figure P.1). Thus, albuterol does not generate a new response in the lungs; it simply acts in the way that epinephrine would under normal circumstances. The inflammatory conditions of asthma pose too much of an insult for normal homeostatic mechanisms to resolve, so albuterol provides a means of sending a therapeutic message through the natural signaling pathway of the β_2 adrenoceptor.

The idea that drugs act through natural mechanisms also applies to more complicated psychological and behavioral responses. When a highly addictive drug, like heroin or cocaine, is taken by a route that ensures rapid penetration to brain (e.g., intravenous), an intense feeling of pleasure is elicited. Heroin acts directly on receptors for natural opioid peptides, whereas cocaine acts indirectly by increasing the concentrations of the neurotransmitters dopamine and norepinephrine at synapses in the brain. These drugs also alter the brain so that the individual is a little more motivated to repeat the experience and engage in the behavior that caused the drug

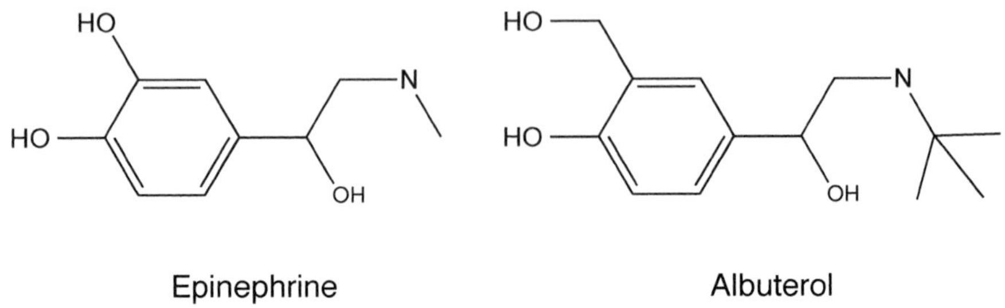

Figure P.1: Structures of epinephrine and albuterol.

effect (i.e., self administration of heroin or cocaine).[2–4] Over time with repetitive self-administration, the mind changes and the individual still feels intense satisfaction upon taking heroin or cocaine, but now feels compelled to continue taking these drugs long after their effects wear off.

If you ask a layperson why this happens you will probably get an answer like, "Of course addictive drugs make you feel good and of course if you continue to take them it will be difficult to stop." But why? How is it that heroin can elicit pleasurable feelings, and why do some people who enjoyed their first experience with the drug and may be considering taking it again, gradually turn into craving addicts while taking the drug, say, 60 times over the course of six weeks? Does heroin elicit a much stronger pleasurable effect after repetitive use such that the individual is willing to ignore personal health and legal consequences to pursue the drug? Probably not.

An adaptionist understands that this phenomenon requires a special explanation, one that addresses why a susceptibility to heroin addiction confers such a survival advantage that nearly all humans are likely have it. The explanation is complicated[2–4] so I will describe how the behavioral aspects of the system might function in a hypothetical creature whose time is spent mainly searching for food and occasionally reproducing. The elements of the reward pathway, after all, are present in very simple animals.

Imagine that the organism's nervous system is composed of a series of behavior and sensory modules overlaid with a reward system. When the animal mates or ingests food, the reward pathway releases dopamine, norepinephrine and opioid peptides (heroin-like substances) that act on neurons to reinforce the neuronal circuitry that caused the behavior. This mechanism increases the chances that the animal will engage in the same behavior in the future. Thus, the reward system is a type of behavior selector, and it reinforces behavior that increases the inclusive fitness of the organism. Natural selection has bequeathed humans with a reward system, and we have learned how to hijack its neuronal circuitry with heroin and cocaine to obtain immediate gratification.

The hypothesis of natural mechanisms for drug action also explains two essential features of how drugs work. First, the actions of a drug need to be amplified so that its effect can be seen, experienced or detected in some way. Second, the effect of a drug must be selective; that is, it must be focused on a specific aspect of physiology. Normal physiology depends on the concerted activity of the major organs including those of the cardiovascular, respiratory, digestive and musculoskeletal systems, all of which are controlled by the nervous and endocrine systems. Drugs produce effects by sending a therapeutic message through these control systems.

In thinking about how drugs do this, various targets come to mind including 1) receptors for neurotransmitters, peptides, eicosanoids and hormones, 2) ion channels that conduct electrical impulses through nerves, muscle, lymphocytes and other excitable cells, 3) transporters that accumulate chemical messengers in specialized compartments, and 4) enzymes that synthesize chemical messengers or other substrates. All of these proteins are essential elements of communication pathways, and drugs utilize them to send a therapeutic message. These targets are referred to as receptors, and many drugs elicit their effects by acting on receptors.

This is not to say that the actions of *all* drugs are mediated by receptors. For example, the antacid and cathartic, magnesium hydroxide, neutralizes gastric acid and remains in the gastrointestinal tract because it is not absorbed. This makes the contents of the gut salty, which draws in water through osmotic pressure to distend the colon and trigger reflex contractions and catharsis. Nonetheless, many other drugs elicit their effects by binding to specific targets called receptors.

Ultimately, the genome determines the origin and expression of the elements of communication pathways although genomic imprinting and environmental factors can influence the timing and pattern of expression. Throughout most of the prior century, the arrow of progress in understanding drug action has usually pointed from the top down. That is, the actions of drugs were first studied in the whole animal or in isolated tissues. As time progressed, the mechanisms of how drugs elicited responses inside the cell gradually came to light. The receptors upon which drugs act were initially purified with difficulty and ultimately their genes identified. Since many receptors share a similar genomic ancestry, the identification of related receptors was accelerated through gene cloning.

With the sequencing of the human genome, as well as those of other species, the arrow of progress now often points in the opposite direction. That is, putative receptors have been identified on the basis of the similarity of their DNA coding sequence with that of other known receptors. This is now accomplished using computer algorithms to search gene databases. The pattern of expression of a putative receptor can be determined in experiments mapping the distribution of its mRNA in the body, which yields clues to physiological function. With knowledge of its normal function, it is possible to speculate about how the receptor signaling system might be exploited to alleviate a specific disease condition. For example, the discovery of a receptor involved in maintaining the circadian sleep-wake cycle might be exploited to treat sleep disturbances. Thus, in this case, the progress of research began with the identification of DNA for the drug target and culminated with the identification of a disease whose condition may well be improved by the administration of a drug acting on the target receptor.

As mentioned, the term receptor can be used in a general sense to refer to any target to which a drug binds to elicit its effects. The term also has a more limited meaning relating to a member of one of the four families of physiological receptors. These receptors are targets for neurotransmitters, hormones and other signaling molecules, and include nuclear receptors, ligand-regulated enzymes, ligand-gated ion channels and *G* protein-coupled receptors. I will need to briefly summarize these before continuing with our discussion on how drugs work. Those already familiar with receptor families may want to skip the next section and continue reading under, "Endogenous ligands turn on their receptors".

Brief Survey of Receptor Families

Nuclear receptors mediate hormonal effects through gene transcription

Many hormones of the endocrine system act on nuclear receptors that alter gene expression and mediate relatively long-term changes in the cell. These receptors are essentially ligand-regulated transcription factors that bind to DNA and regulate the expression of specific genes in a hormone-dependent manner.[5] A number of hormones including vitamin D, thyroid hormone, retinoids, glucocorticoids, mineralocorticoids, androgens, estrogen and progesterone all act on nuclear receptors. Not all nuclear receptors respond to secreted signaling molecules. For example, intracellular metabolites act on some nuclear receptors (*PPARs*) to regulate lipid metabolism and fat cell differentiation, and various foreign compounds (xenobiotics), including some drugs, act on nuclear receptors to initiate the expression of enzymes (cytochrome P450) and transporters that affect the disposition of ingested xenobiotics and drugs. This regulatory system probably evolved to protect the organism against harmful

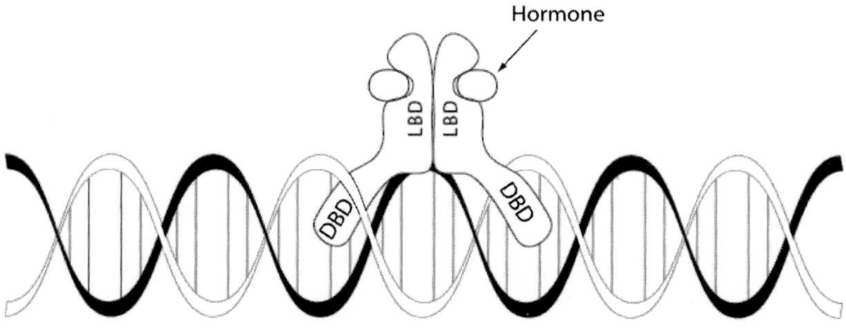

Figure P.2: The interaction of a nuclear receptor with DNA. The binding of hormone induces receptor dimerization, and ultimately, the interaction of the dimeric receptor complex with DNA. Abbreviations: LBD, ligand-binding domain; DBD, DNA-binding domain.

ingested xenobiotics. Some nuclear receptors are located in the cytosol and only travel to the nucleus after bound with their natural ligand or xenobiotic, whereas others are always in the nucleus bound to DNA, even in the absence of ligand.

Nuclear receptors are composed of three main parts, a ligand binding domain, a DNA binding domain and a transcriptional domain that activates or inhibits transcription (see Figure P.2). The ligand-binding domain is located on the carboxy-terminal side of the protein. When bound with hormone, this domain induces a conformational change in the receptor that ultimately influences gene transcription. Near the center of the receptor is a highly conserved sequence consisting of two polypeptide loops coordinated by a zinc ion interacting with cysteine residues on opposite loops of the receptor. This so-called "zinc finger" binds to DNA at specific regulatory sequences known as hormone response elements (HREs). Nuclear receptors bind to DNA as dimers, and this mode of interaction is manifest in the internal symmetry (palindromic) of the HRE. These dimers may be composed of the same (homodimers) or different (heterodimers) receptors. Each nuclear receptor has specific HREs with which it interacts to regulate the transcription of a nearby gene, although the two may be separated by several thousand base pairs. The third domain of the receptor is at the amino terminal end, and it plays a role in transcriptional activation. In many instances, the binding of hormone to a nuclear receptor results in the recruitment of the transcriptional apparatus to the regulator region of a gene resulting in transcription. In other instances, the hormone actually mediates an inhibition of transcription.

Hormones that act on nuclear receptors elicit effects after a typical lag period of approximately one-half to several hours. This delay corresponds to the time required for transcription of mRNA and the synthesis of new protein. Also, the effects of drugs acting on these receptors can persist for days after the drug has been eliminated because of the slow turnover of the proteins whose synthesis was turned on by the receptor.

Growth factors act on ligand-regulated enzymes on the cell surface

A group of proteins known as receptor tyrosine kinases mediate the actions of a number of other hormones and growth factors including insulin, fibroblast growth factor (FGF), epidermal growth factor (EGF) and platelet-derived growth factor (PDGF).[6,7] These receptors are essentially ligand-regulated enzymes located in the plasma membrane. Their structure includes a central, hydrophobic chain that passes through the plasma membrane once and connects the extracellular ligand binding domain with the intracellular kinase domain (Figure P.3). This kinase domain behaves as an enzyme that phosphorylates proteins on tyrosine residues. When bound by their natural ligand these receptors come together in pairs (i.e., dimerize),[8,9] and each partner phosphorylates the other triggering a signaling cascade often resulting in growth and differentiation. In the case of insulin, its receptor already exists as a dimer connected together by disulfide bonds.[10]

There is great diversity in the way receptor tyrosine kinases elicit their effects.[11,12] What follows is a general theme with one example. As noted above, when a receptor tyrosine kinase is activated by its ligand, it dimerizes and each kinase domain phosphorylates the other on tyrosine residues. One type of phosphorylation occurs on the substrate-binding site of the kinase domain, which increases the phosphorylating activity of the protein. The other type occurs in several other places on the receptor, most notably the C-terminal region. The latter act as docking sites for other proteins containing a structural feature known as a Src homology domain 2 (SH2 domain). SH2 domains bind to phosphotyrosine residues on proteins, which explain why proteins containing this domain bind to the phosphorylated receptor tyrosine kinase. There are two types of proteins that are recruited to activated receptor tyrosine kinases through their SH2 domains. The first is a simple adapter protein that has no activity by itself, but simply binds to the receptor and enables other signaling proteins to bind to it. The second type has enzymatic activity of its own or else increases the activity of another enzyme. The net result is that when a receptor tyrosine kinase is activated, it attracts signaling proteins to the cell membrane, bringing them in close proximity to each other so that they can initiate a signaling cascade.

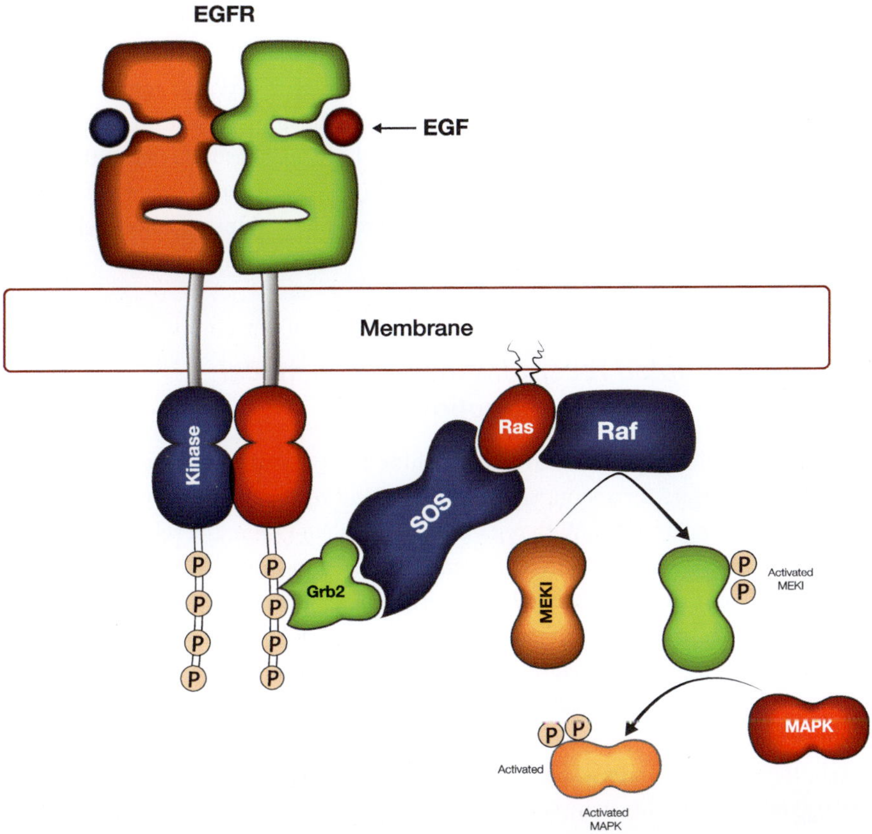

Figure P.3: Activation of MAP kinase (MAPK) by the epidermal growth factor receptor (EGFR).

One example is shown in Figure P.3.[13] When EGF binds to its receptor, it forms a dimer that undergoes transphosphorylation. Some of the resulting phosphotyrosine residues attract the docking proteins Grb2 and SHC. These in turn attract another protein called mSOS to the growth factor receptor-complex, enabling it to interact with a small *G* protein called Ras. Ras is normally inactive with GDP tightly bound to it. Its interaction with mSOS causes it to release GDP and bind GTP, which activates Ras. The activated Ras turns on another enzyme Raf, which phosphorylates yet another enzyme called mitogen activated protein kinase. Ultimately this kinase triggers a variety of effects necessary for growth and differentiation.

There are four other classes of receptors related to receptor tyrosine kinases in the sense that they function as a ligand-regulated enzyme. These are (1) *receptor guanylate cyclases*, which catalyze the formation of cGMP, (2) *receptor tyrosine phosphatases*, which cleave phosphate groups from phosphotyrosine on proteins, (3) *receptor tyrosine kinase associated receptors*, which have no kinase activity of their own but recruit tyrosine kinases in the cytosol, and (4) *receptor serine/threonine kinases*, which phosphorylate proteins on serine and threonine residues. Collectively, these receptors mediate diverse functions in the cell.

G protein-coupled receptors mediate chemical and sensory signal transduction

G protein-coupled receptors (GPCRs) include a large family of proteins that act as sensory transducers for vision, smell and taste and receptors for numerous natural chemical messengers.[12–15] With regard to mammalian vision, the relevant receptor, rhodopsin, already has a photo-activated ligand bound to it (11-*cis*-retinal). In the presence of light, retinal undergoes isomerization to the all *trans* form, which activates rhodopsin. In the case of taste, three types of receptors recognize compounds that are sensed as being sweet, bitter and umami (L-amino acids), respectively. The other two taste sensations, salty and sour, are mediated through the interactions of sodium and hydrogen ions, respectively, with ion channels. Finally, with regard to smell, there are hundreds of olfactory receptors that recognize volatile compounds, giving rise to the highly complex and sophisticated sense of smell in mammals.

There are also about 250 known *G* protein-coupled receptors for endogenous neurotransmitters, peptides and eicosanoids. The β_2 adrenoceptor described above in connection with asthma is a member of the *G* protein-coupled receptor family. These receptors mediate numerous physiological functions, making them important targets for therapeutic drugs and tools for research.

G protein-coupled receptors share a similar structure consisting of an extracellular amino terminus, an intracellular carboxy terminus and seven hydrophobic alpha helixes, each making a single pass through the plasma membrane. These are connected together by alternating intracellular and extracellular loops for a total of three each.

As their name implies, *G* protein-coupled receptors mediate physiological effects by interacting with *G* proteins.[16–18] These are heterotrimeric, consisting of three subunits, α, β and γ that function as a dimer of α and $\beta\gamma$ subunits. In the inactive condition, the α and $\beta\gamma$ subunits are bound together in an aggregate with GDP tightly bound to the α subunit. Activation occurs when an endogenous

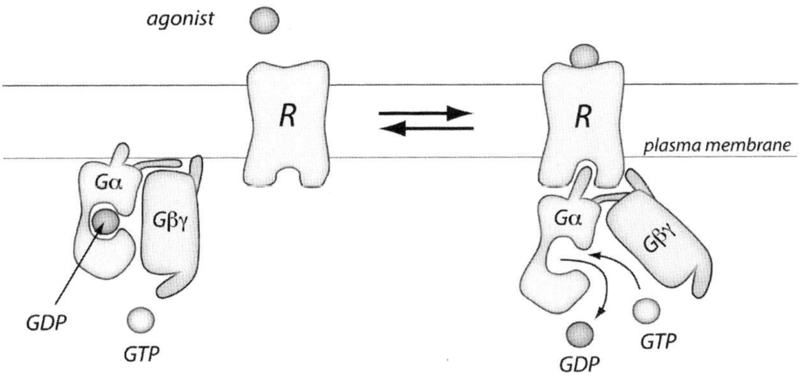

Figure P.4: Agonist-mediated guanine nucleotide exchange at a *G* protein-coupled receptor.

ligand binds to its receptor and persuades it to associate with its heterotrimeric *G* protein (Figure P.4). This transient association causes the *G* protein to give up GDP enabling GTP to bind, which is abundant in the cytosol. When this happens, the *G* protein splits apart into the GTP bound α and $\beta\gamma$ subunits. This pair of functional subunits activate a variety of enzymes and ion channels that mediate the cell's response to receptor activation. In total, there are about 20 different *G* protein α subunits interacting with a smaller number of $\beta\gamma$ subunits. These exhibit a variable selectivity for several hundred different *G* protein-coupled receptors. The α subunit of heterotrimeric *G* proteins shares a similar evolutionary ancestry with the Ras *G* protein mentioned above in connection with receptor tyrosine kinases.

Not all of the effects of *G* protein-coupled receptors are mediated through *G* proteins.[19] When the β_2 adrenoceptor is turned on, it initially activates a *G* protein called G_s. In time, the activated β_2 adrenoceptor recruits an enzyme that phosphorylates it, preventing it from interacting with G_s. In its phosphorylated form, the β_2-adrenoceptor interacts with another protein called β-arrestin, which ultimately enables it to travel inside the cell via endocytosis and activate a different signaling pathway resulting in the stimulation of MAP kinase. The internalized receptor can then recycle back to the plasma membrane or be degraded.

Ligand-gated ion channels mediate transmission at fast chemical synapses

Ligand-gated ion channels represent a large family of receptors that mediate fast neurotransmission at a variety of synapses and neuroeffector junctions throughout the central and peripheral nervous systems. This type of receptor is an ion channel that opens and carries an ionic current through the plasma membrane when bound by its natural ligand. Several neurotransmitters signal through ligand-gated ion channels including acetylcholine, GABA, glutamic acid, glycine, serotonin and ATP. Most of these also signal through *G* protein-coupled receptors.

At the neuromuscular junction, acetylcholine activates a ligand-gated ion channel known as the nicotinic acetylcholine receptor that conducts an inward current of mainly positive monovalent cations to cause depolarization and excitation of skeletal muscle (see Figure P.5). This receptor is a member of the Cys-loop family of ligand-gated ion channels that includes the $GABA_A$, $5HT_3$ and glycine receptors.[20] The muscle-type nicotinic acetylcholine receptor is composed of five protein subunits, two α and one each of β, γ and either δ or ε. These subunits line a central pore that forms an ion channel.[21] Two molecules of acetylcholine bind to the channel, each at the interface between an α subunit and an adjacent subunit. Different types of nicotinic acetylcholine receptors are expressed in the nervous system, and these vary in their subunit composition.[22] One type is composed of five α_7 subunits, and it conducts a Ca^{2+} current when activated by acetylcholine. This current rapidly inactivates or desensitizes in the continued presence of acetylcholine, presumably to protect the neuron from excessive accumulation of Ca^{2+}.

The major excitatory neurotransmitter in the brain is glutamic acid, and it acts on three broad classes of ligand-gated ion channels, whose members differ in their subunit composition.[23] Each is thought to be composed of four protein subunits. When activated by glutamic acid, these channels conduct an inward current of positive cations.

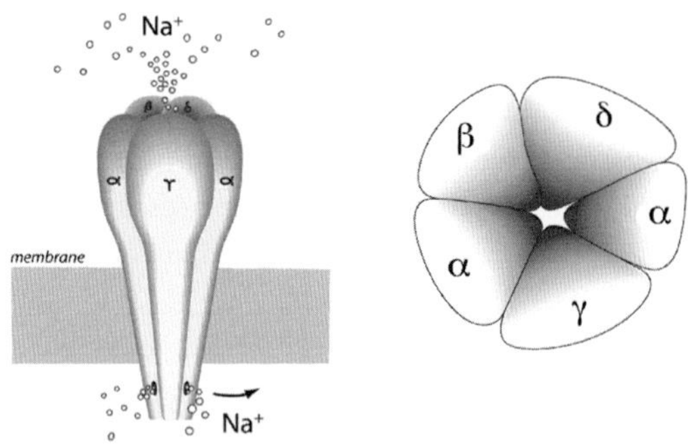

Figure P.5: The muscle-type nicotinic acetylcholine receptor. (a) Longitudinal view of the receptor. The small spheres represent individual hydrated sodium ions. (b) View looking down on the receptor from an extracellular perspective.

The major inhibitory neurotransmitter in the brain is GABA, and it acts on a pentameric ligand-gated ion channel that conducts an inward current of chloride ions when activated by GABA. This ligand-gated ion channel is known as the $GABA_A$ receptor, and there are several subtypes, varying in their subunit composition.[22–25] They are closely related to another pentameric ligand-gated ion channel for the inhibitory neurotransmitter, glycine. The glycine receptor is expressed in the spinal cord, and it also conducts a chloride current.

Endogenous Ligands Turn on their Receptors

Now that we know about different types of physiological receptors, how do their ligands trigger a response through them? Perhaps the clearest and most striking example of what happens when a neurotransmitter turns on its receptor comes from studies on ligand-gated ion channels. This picture was made possible by the development of the voltage clamp[26] and single channel recording techniques.[27] The voltage clamp technique uses a feedback circuit to maintain a constant voltage across the membrane. This is accomplished using a recording electrode to measure the electrical potential near the membrane and another electrode to deliver whatever current is necessary to maintain a constant voltage despite rapid changes in the permeability of the cell membrane. A high frequency amplifier with a feedback circuit continuously measures the difference between the intended voltage and the measured membrane potential and delivers a current to the cell to eliminate the difference between the two. When ion channels in the membrane open, an ionic current flows through them that, under normal circumstances, would elicit a voltage change across the membrane, but under voltage clamp conditions, the high frequency feedback amplifier delivers a current to maintain constant voltage. This delivered current is equivalent, though opposite in sign, to that which flows through ion channels in the membrane, so it is ultimately recorded as the channel current after changing its sign. Using fire polished electrodes it is possible to form a high resistance seal (gigaseal) on a small patch of membrane circumscribing only one or a few ion channels, enabling the measurement of single channel currents.

These techniques have been used to study the nicotinic acetylcholine receptor of the neuromuscular junction mentioned above, and the results are impressive (Figure P.6). In the absence of acetylcholine, the membrane current is essentially zero. When a relatively low concentration of acetylcholine is applied, the current occasionally jumps to a unitary conductance measured in picosiemens (pS). The transition between no conductance and the unitary value is abrupt, illustrating that the channel switches between closed and open states in the presence of a low

Figure P.6: Activation of single nicotinic acetylcholine receptor by acetylcholine (0.5 μM). The plot shows single channel recordings from *Xenopus* oocytes injected with mRNA for the α, β, γ and δ subunits of the calf nicotinic acetylcholine receptor. The plot has been prepared from that published by Mishina and coworkers.[50]

concentration of acetylcholine. As the concentration of acetylcholine increases, the frequency and duration of channel opening increases, and at a high concentration, the channel remains open nearly constantly over the short run. With continuous exposure to the agonist over a few seconds, the channel ultimately closes in time because of an inactivation process known as desensitization.

The behavior of the channel over the short run before desensitization is consistent with what is known about how a small ligand binds to a receptor. Acetylcholine binds in a reversible manner to the nicotinic acetylcholine receptor because its effects rapidly disappear after it is washed away. It follows that acetylcholine should bind to and dissociate from the receptor in a random manner at concentrations that are sufficient to cause a moderate response. If acetylcholine opens the ion channel when it is bound to it, then random openings in the channels should occur like those shown in Figure P.6. This explanation is overly simplified because channel opening occurs after two molecules of acetylcholine are bound to the channel. Regardless, at high concentrations, acetylcholine is expected to occupy single channels continuously resulting in near constant opening of the channel. Acetylcholine is not stuck to the channel under this condition, but rather, its rate of association at high concentrations is so fast that as soon as one molecule of acetylcholine dissociates from a channel, a nearby molecule rapidly replaces it resulting in continuous receptor occupation. Thus, the receptor is nearly continuously occupied by acetylcholine at high concentrations even though individual molecules of acetylcholine are constantly replacing each other on the receptor.

The behavior in Figure P.6 illustrates a fundamental principle about how neurotransmitters, drugs and other signaling molecules turn on their target receptor. The receptor itself already exists in at least two forms, active and inactive. In the case of a ligand-gated ion channel, these two forms correspond to the open and closed states of the channel. Usually the receptor is in the inactive form in the absence of ligands, and when the natural neurotransmitter binds to the receptor it greatly increases the probability that the receptor will turn on. An exogenous drug that turns on the receptor acts in the same manner; that is, it selects for the active state of the receptor. The output from the receptor is digital, corresponding to active and inactive states of the receptor. Ultimately, this behavior explains how many drugs that act on receptors do so through natural on and off states of a receptor to send a digital message.

Another way to consider the action of neurotransmitters on their receptors is from the perspective of a population of receptors. Imagine that it were possible to record the output of each nicotinic acetylcholine receptor in the cell simultaneously in the presence of acetylcholine. Although each single channel recording trace would exhibit random fluctuations in digital output, theoretically, the summation of the output from all of the channels should yield a constant level of current in time representing the whole cell current or ensemble average. In practice, the whole-cell current response is measured with a single set of voltage clamp recording electrodes that measure the total current across the plasma membrane. If this current is measured at various concentrations of acetylcholine, the resulting data can be plotted in the form of a concentration-response curve. The maximum of this plot represents the total current when acetylcholine is bound to all of the nicotinic acetylcholine receptors in the cell, and the half maximal current occurs when acetylcholine is occupying approximately half of these receptors (see Figure P.7). As shown in Chapter 2, the relationship between the whole-cell current response and receptor occupancy is roughly proportional, but not precisely, because each nicotinic receptor has two binding sites for acetylcholine.

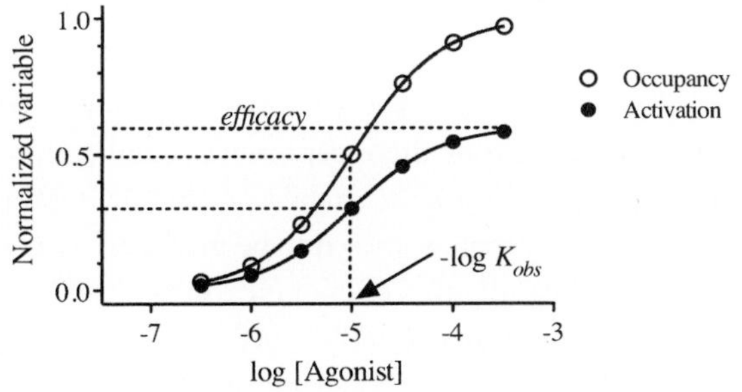

Figure P.7: Relationship between receptor occupancy and the activation state of a receptor population having a single ligand-binding site. Receptor activation is proportional to receptor occupancy, and the fraction of the population of ligand-receptor complexes in the active state is denoted as efficacy. In this case of one orthosteric binding site per receptor complex, the observed affinity constant (K_{obs}) is equivalent to the reciprocal of the agonist concentration eliciting both half-maximal receptor activation and occupancy.

Drugs that Act on Receptors can be Classified as Agonists, Antagonists, Inverse Agonists and Allosteric Modulators

Using ligand-gated ion channels as a frame of reference, we can now consider how other drugs besides the natural neurotransmitter act on receptors. Drugs that bind to receptors at the same site as the natural ligand are termed orthosteric ligands, and they come in the form of agonists and antagonists. Unless indicated otherwise, it is usually implicit that the terms agonist and antagonist refer to ligands that bind to the same site as the natural neurotransmitter without the need to use the adjective "orthosteric" to indicate this. The property that distinguishes agonists from antagonists is efficacy. An agonist activates a receptor when bound to it because it *has* efficacy, whereas an antagonist *lacks* efficacy. Almost all natural ligands are agonists. Although an antagonist usually has no effect on the receptor when bound to it, competitive antagonists have inhibitory effects *in vivo* because they block the actions of natural ligands.

Efficacy can be defined in two different ways depending upon the frame of reference. Let us begin with population thinking and consider the whole-cell current response to a nicotinic receptor agonist. As described above the concentration-response curve for the whole-cell current response to an agonist represents the summation of all of the unitary conductances of single channels plotted against the agonist concentration. Because the receptor is an ion channel, each unitary conductance represents a digital readout indicating that the channel is open. This concentration-response curve, therefore, represents receptor activation plotted against the agonist concentration. Efficacy represents the fraction of the occupied receptor population in the active state. If the receptor has a single orthosteric binding site, then efficacy is constant for any level of receptor occupancy, and hence, is proportional to the maximum of the receptor activation function (Figure P.7). Stephenson[28] and Furchgott[29] initially established this definition with regard to the analysis of responses that are now known to be mediated by *G* protein-coupled receptors. Although these investigators did not measure receptor activation directly at GPCRs, they postulated its existence and defined the term stimulus to describe it. Figure P.7 summarizes these concepts regarding affinity, efficacy and the receptor activation function.

If the receptor has two cooperatively linked sites, like the nicotinic acetylcholine receptor of the neuromuscular junction, then efficacy increases with receptor occupancy, and its value is no longer simply proportional the maximum current response. Nonetheless, efficacy is directly correlated with the maximum current response.

Another way to define efficacy is at a deeper level of analysis. The origin of this line of thinking is in the work of Changeux[30] and Koshland[31] and their respective colleagues. These investigators were interested in understanding how small ligands selected or induced conformational changes in enzymes. They suggested that if a ligand caused a conformational change in the shape of a protein when bound to it, then the ligand must have higher affinity for the conformation that it induces. The implications for receptor theory were obvious. Using single channel analysis and mathematical modeling, Colquhoun showed that the ability of an agonist to open the nicotinic acetylcholine receptor-ion channel is proportional to its affinity for the open state relative to that of the closed state.[32,33]

It follows that as a ligand-gated ion channel undergoes quantal changes in affinity as it opens and closes as illustrated by the theoretical example in Figure P.8. Acetylcholine is thought to have affinity constants of 2.4×10^7 and 6.2×10^3 M^{-1} for the open and closed states of the muscle type nicotinic acetylcholine receptor, respectively, whereas the corresponding values for the less efficacious agonist, nicotine, are 2×10^5 and 4.8×10^2 M^{-1}. Thus, acetylcholine exhibits about 4,000-fold selectivity for the open state of the channel, whereas nicotine only exhibits 400– fold selectivity.

Colquhoun's concept of efficacy is a more fundamental definition expressed in terms of first causes as compared to that of Stephenson and Furchgott. To appreciate the difference between the two, consider two agonists, one having 10^3-fold higher affinity for the open state relative to the closed (agonist A) and the other having a corresponding ratio of 10^4-fold (agonist B). For this example, assume that the agonist must have 10^3-fold higher affinity for the active state relative to the inactive to cause a near continuous, 95% opening of all of the ion channels at 100% receptor occupancy. Thus, both A and B have sufficient selectivity for the open state to cause near complete activation of ion channels, and their $E_{\max}$ values for the whole cell current response are both approximately equal to the maximal current response of

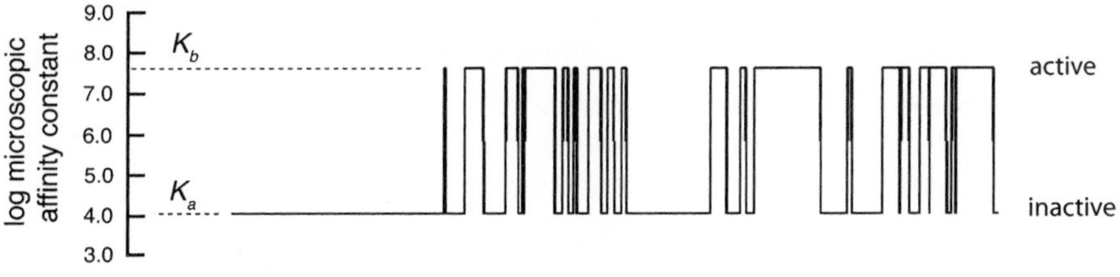

Figure P.8: Theoretical plot of single receptor activity showing the quantal changes in affinity that occur as the receptor switches from the active state (high affinity) to inactive (low affinity) states in the presence of agonist.

the cell when all of the channels are open. Because the agonists' E_{max} values for the receptor activation function (current response) are nearly the same, they will have similar efficacy values according to Stephenson's and Furchgott's definition. From the perspective of receptor states, however, agonist B has 10-fold higher capacity to induce the active state. These two definitions of efficacy do not conflict; they simply represent different types of information based on different levels of analysis. Stephenson's and Furchgott's more superficial definition enables us to estimate the fraction of the receptor population in the active state at any level of receptor occupancy, and Colquhoun's more ultimate definition measures the capacity of the agonist to induce the active state. Hereafter, I will refer to Stephenson's and Furchgott's definition as observed efficacy.

The property of antagonism can also be defined at the level of receptor states.[34,35] Two situations are possible. An antagonist can have either have the same affinity for ground and active states or higher affinity for the ground state. An antagonist with the same affinities for the ground and active states (neutral antagonist) will not induce a conformational change in the receptor. If the receptor is inactive in the absence of antagonist, it will remain so when bound with the antagonist. An antagonist with selectivity for the ground state will tend to induce the ground state. Because the receptor is usually already in the ground state, this type of antagonist usually has no effect, but it was predicted to inhibit basal receptor activity.[34,35] Since these definitions were established, several examples of *G* protein-coupled receptors exhibiting spontaneous basal activity have been described. In these systems, some well-known antagonists inhibit basal activity, presumably by binding with higher affinity to the ground state. The term "inverse agonist" is now established in our lexicon for reference to these agents even though this behavior was designated previously as antagonism by Colquhoun and myself.

Constitutive activity is often detected at *G* protein-coupled receptors because the constitutive response is usually measured at a point downstream in the signaling pathway, and the signaling pathway amplifies the initial output of the receptor, particularly if the receptor or some component of its signaling pathway (e.g., *G* protein) is overexpressed. Under more common physiological conditions, neutral antagonists and inverse agonists both behave like antagonists.

What has come as a surprise to some is that many common highly potent competitive antagonists behave as inverse agonists in systems with constitutive activity. This situation is predictable on the basis of an analogous phenomenon noted by Carl Pfeiffer regarding the potency and relative pharmacological activity of stereoisomers of receptor active drugs. Pfeiffer found that the ratio of activities of bioactive stereoisomers is generally proportional to the potency of the active enantiomer[36]

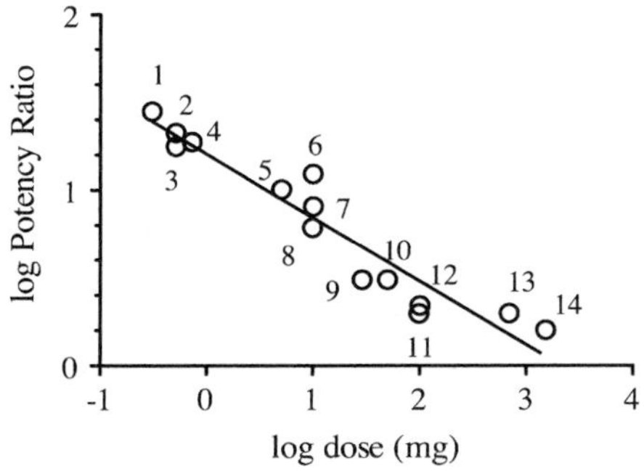

Figure P.9: Relationship between potency and the ratio of activities of the enantiomers of a drug. The plot shows the ratio of activities of the enantiomers of a series of drugs plotted against the dose of the racemate used clinically. The data have been plotted from that described by Pfeiffer. 1, norepinephrine; 2, atropine; 3, epinephrine; 4, scopolamine; 5, Dromoran; 6, methadone; 7, amphetamine; 8, methamphetamine; 9, isomethadone; 10, ephedrine; 11, Nervanol; 12, Mesantoin; 13, quinine and 14, atrolactamide.

(see Figure P.9). It follows that if the active enantiomer is highly potent, then its three-dimensional structure must match that of its binding pocket on the receptor very well. This correspondence would be lost in its mirror image, giving rise to a large loss in potency of the less active enantiomer and a large enantiomeric potency ratio. If neither isomer is very potent, then the match between the binding site and either enantiomer is not very good and replacing one enantiomer with the other would not be expected to disrupt or improve binding forces much.

If we change the analogy from two ligands and one receptor conformation to two receptor conformations and one ligand, we can understand the abundance of highly potent, inverse agonists. Any antagonist that exhibits high potency must have high affinity for the inactive state, and its structure must match its binding pocket on the inactive conformation very well because of its high potency. Consequently, this close match in structure would be lost in a different conformation of the receptor (i.e., the active state), and the antagonist would exhibit greater selectivity for the ground state relative to the active state and behave as an inverse agonist in systems with basal activity.

The relationship shown in Figure P.9 is not a rule in the strict sense. It simply predicts that within a random sample of a large number of enantiomers, there should be a positive correlation between the potency of the active enantiomer and the ratio of activities of the two enantiomers. Obviously if the chiral center is located on an unimportant part of a highly potent drug molecule, one might expect little difference in the activity of the enantiomers.

Not all receptor-active drugs bind at the same site on the receptor as the natural ligand (orthosteric site). Some drugs, known as allosteric modulators, bind at a distinct allosteric site and tune up or tune down the actions of the natural ligand or other agonists.[37] The nature of the modulation on the actions of agonists can be measured in terms of their effect on the ability of the agonist to turn on the receptor (efficacy modulation) or to bind to the receptor (affinity modulation). Many allosteric modulators have no effect on the receptor in the absence of an orthosteric agonist, whereas some positive modulators are actually able to turn on the receptor through the allosteric site in the absence of an orthosteric agonist. Ultimately, the nature of the allosteric effect depends on the selectivity of the allosteric modulator for the ground and active states of the receptor (i.e., open and closed states of a ligand-gated ion channel).

Drug Action in Isolated *in vitro* Assays

So far, we have considered how drugs bind to and turn on receptors mainly by considering ionic currents at ligand-gated ion channels. The basic principles of drug-receptor interactions are easier to describe and to understand from this perspective, because the receptor *is* a channel and the current response *is* a direct measure of receptor activation. At *G* protein-coupled receptors, it is not a simple matter to measure receptor activation directly. Usually a variety of downstream responses are measured, depending upon the goals of the experiment. If direct information about the role of a receptor in a physiological response is the topic of investigation, then the experiment might be performed on an isolated tissue, like a blood vessel or gland for example, and the measured response might be contraction of smooth muscle or secretion of electrolyte and protein, respectively. If the focus is to screen compounds to discover novel agonists and antagonists, then the target human receptor might be stably expressed in a cell line, and the measured response could be some endpoint in the intracellular signaling pathway.

One common approach is to measure Ca^{2+} mobilization using a fluorescent probe, like Fura 2. This strategy works well with *G* protein-coupled receptors that couple to G_q to activate phospholipase C-β and ultimately cause Ca^{2+} release from intracellular stores.[38] Even if the receptor does not couple to G_q, Ca^{2+} mobilization can still be measured as an endpoint by employing promiscuous *G* proteins, like G_{15}, that couples G_i and G_s-linked receptors to phospholipase C-β,[39] or by using a chimeric *G* protein composed mostly of G_q and the C terminal tail G_i or G_s. These chimeras activate phospholipase C-β and sometimes interact with G_i- and G_s-linked receptors because the C-terminal tail of the Gα subunit often has a prominent role in receptor-*G* protein interactions.[40,41] By measuring the increase in fluorescence as a function of the agonist concentration, a graded concentration-response curve to the agonist is obtained.

There are hundreds of different assays that could be performed; those just mentioned represent but a few examples. It is possible to apply an agonist at known concentrations and estimate how potently and effectively it turns on a receptor through the analysis of these types of *in vitro* concentration-response curves. One can also use *in vitro* assays to identify antagonists by measuring the ability of compounds to inhibit the response to an agonist. This approach is used because the concentration of the agonist or antagonist or both can be carefully adjusted, and a fairly uncomplicated response can be measured to give an unambiguous measure of receptor activity. In contrast, it would be difficult to obtain the same information with precision from an *in vivo* experiment because of variation in the absorption, distribution, metabolism and excretion of drugs.

One general feature of downstream physiological responses is that they tend to be highly amplified and have some redundancy at various steps in the pathway. That is, the output from one particular step is amplified, but it might reach a maximal level before the input to that step has reached its maximum. The increase in amplification makes it easier to detect a signal, and the redundancy makes the system reach a maximal response at a very low level of receptor occupancy. The net result is that the response is not proportional to receptor occupancy, and the EC_{50} value for the response can be as little as one hundredth the concentration of agonist required for half-maximal receptor occupancy if the agonist has high efficacy.

An example of this behavior is illustrated in Figure P.10, which shows the concentration-response curves for the highly efficacious muscarinic agonist,

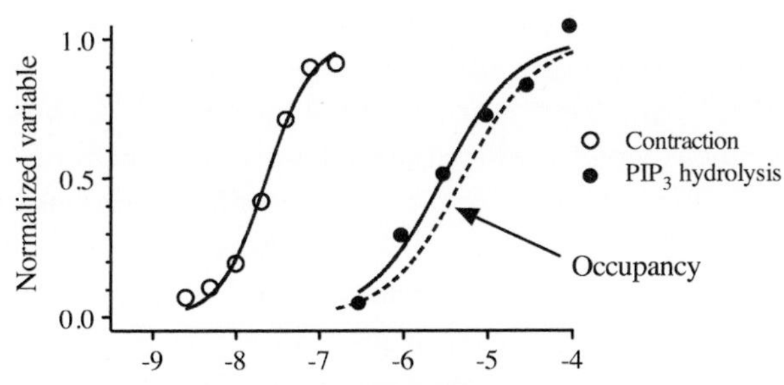

Figure P.10: Relationships among contraction, phosphoinositide hydrolysis and receptor occupancy for the highly efficacious muscarinic agonist, oxotremorine-M, in smooth muscle. The plot has been adapted from that described by Ehlert.[51]

oxotremorine-M, for eliciting smooth muscle contraction and phosphoinositide hydrolysis in the guinea pig ileum. Both of these effects are initiated by the M_3 muscarinic receptor. Also shown in the plot is a theoretical curve for receptor occupancy by oxotremorine-M. In this case the EC_{50} value of the agonist for eliciting contraction (20 nM) is approximately on fiftieth that for eliciting the phosphoinositide response, which in turn is about one fifth that of the observed dissociation constant for binding to the receptor. Overall, the agonist is 100-fold more potent at eliciting contraction then it is at binding to the receptor. This means that it only requires a very low level of receptor activation to elicit a maximal contractile response. So a wide range of agonists with varying degrees of observed efficacy should be able to elicit a maximal contractile response. Only those agonists with very low intrinsic efficacy would have a maximal contractile response less than that of oxotremorine-M. It is clear, therefore, that the EC_{50} and E_{max} values of agonists from these types of contractile experiments do not represent the observed affinity and observed efficacy of the agonist. That is, they do not have the same meaning as the corresponding values for the whole cell current response to an agonist at a ligand-gated ion channel.

As described in greater detail in Chapter 5, *G* protein-coupled receptors that mobilize Ca^{2+} and elicit contraction in smooth muscle often signal through the *G* protein, G_q.[42] This *G* protein activates the enzyme phospholipase C-β, which cleaves a phospholipid in the membrane called, phosphatidylinositol-4, 5-bisphosphate (PIP_2).[43] The cleavage products are inositol-1,4,5-trisphosphate (IP_3) and diacylglycerol. IP_3 releases Ca^{2+} from intracellular stores and diacylglycerol activates an enzyme called protein kinase C. The Ca^{2+} for contraction, however, comes from an extracellular source making it unlikely that IP_3 mediated Ca^{2+} release initiates contraction.[44] How the muscarinic receptor mobilizes extracellular Ca^{2+} is unclear, but there are some promising possibilities. Several potassium channels are activated by PIP_2 and protein kinase C and some transient receptor potential (Trp) channels are inhibited by PIP_2.[43–46] Thus, it is possible that the muscarinic receptor mediated breakdown in PIP_2 activates Trp channels and inhibits potassium channels, resulting in an influx of cations and an inhibition of potassium efflux, respectively. Both of these effects should cause depolarization of smooth muscle and an influx of Ca^{2+} through voltage dependent Ca^{2+} channels. Regardless of the mechanism for contraction, the data in Figure P.10 illustrate an important point. The activation of a *G* protein-coupled receptor in a cell often triggers a variety of responses. Although a particular second messenger, like IP_3, may not be involved in mediating the response of interest (e.g., contraction), it does nonetheless provide a useful reporter assay for measuring agonist activity.

In discussing receptor activation at ligand-gated ion channels, it was shown that the agonist switches the channel from a closed state into an open state having a characteristic unitary conductance. *G* protein-coupled receptors probably also behave in a similar manner. When an agonist occupies a *G* protein-coupled receptor, it increases the probability that the receptor will assume the active conformation and interact with a *G* protein. The active state of the agonist-receptor complex coupled to the *G* protein behaves like a quantum of activity, analogous to the open state of a ligand-gated ion channel.[47] But the situation is more complex, because a given *G* protein-coupled receptor may interact with more than one type of *G* protein or other coupling protein. If the same active state of the receptor is involved in the coupling, then the rank order of the relative efficacies of a series of agonist for eliciting responses through the different *G* proteins should be the same. There is now substantial evidence, however, that the efficacy profiles of some agonists for eliciting responses through different *G* proteins change.[48] These results suggest that there may be more than one active conformation of a *G* protein coupled receptor, and that the different active conformations exhibit selectivity for different *G* proteins. Some agonists may be unique in having the ability to stabilize the receptor in a confirmation that signals preferentially through one *G* protein relative to another. This situation raises the possibility that if the activity of agonists at a *G* protein-coupled receptor are characterized in an *in vitro* assay based on Ca^{2+} mobilization through G_{15}, for example, but mediate their physiological response through G_i, then there may be a difference in the behavior of agonists in the two systems. More details on the signaling of GPCRs and the analysis of their responses are described in Chapters 5–9.

At the beginning of this section on *in vitro* assays, it was mentioned that assays on cell lines expressing a target receptor are useful for screening the activity of compounds. If the endpoint can be measured at a distance (e.g., light detection) without invasive manipulation of the cells, then the assay can be automated so that the activity of thousands of compounds can be tested in a single day. This process is known as high throughput screening, and it is widely used in the pharmaceutical industry. One approach mentioned above is to express a human *G* protein-coupled receptor in a cell line also expressing a *G* protein or chimeric *G* protein that couples the receptor to phospholipase C-β. Upon stimulation of the receptor by a test compound, phospholipase C-β is activated resulting in an increase in Ca^{2+} levels in the cell. The Ca^{2+} response can be measured as light if the cells are first treated with a fluorescent Ca^{2+} dye or if the cells express a bioluminescent protein. To adapt the assay for high throughput screening, cells are grown in individual wells containing media on plates having 96 to 2000 wells. Test drugs are added to each well using a robotic fluid handling device, and systems have been developed to catalog a large number of test compounds, enabling the screening of large compound libraries often of synthetic origin.

As the naturalist, Edward O. Wilson, pointed out in 1986 in his book, *Biodiversity*, 40% of prescriptions filled in the US at the time were for drugs derived from natural origins, mainly from plants but also from animals and bacteria. He suggested that high throughput screening should be useful in discovering new therapeutic agents from natural sources. So far, therapeutic drugs have been discovered from a very small number of known species. Systematists place the total number of identified species on Earth somewhere between 1.6 and 1.8 million, but most agree that the actual number is 10 to 100 times larger. The relatively recent discovery of a subterranean ecosystem composed of bacteria and microscopic fungi (subterranean lithoautotrophic microbial ecosystems, SLIMES) that may outweigh all living matter on the surface of our planet illustrates how little we know about biological diversity. This diversity represents a large, untapped resource likely to yield many promising drugs through advances in high throughput screening.

More recently new lead compounds have been identified by virtual ligand screening without ever doing an experiment on a real receptor. This approach can be done if there is a quality X-ray structure of the receptor. The binding pocket can be represented on the computer as a three-dimensional grid of receptor interaction potentials. The computer algorithm docks ligands from a large compound database (10^5–10^6 compounds) into the binding pocket and evaluates the quality of the fit *in silico*. During the docking process the ligand is transposed within the binding pocket, its orientation altered and the bond angles of its constituent atoms are rotated to achieve global energy optimization. The docking procedure also incorporates random steps to displace the ligand form its current optimized binding position and repeats the energy optimization process to determine if a better fit can be obtained. Some algorithms incorporate minor changes in the structure of the receptor. So far it is impractical to obtain good estimates of the change in free energy of binding, and hence of binding affinity, but with advances in computer power, this may be possible in the future.

Having identified lead compounds through virtual screening, these and related compounds can be tested in biological screening assays, like those mentioned above, to choose optimal compounds for clinical development.

Hierarchical Levels of Analysis of Drug Action and Receptor Theory

The different methods for measuring drug-receptor interactions described earlier in this chapter can be categorized into three hierarchical levels (Figure P.11). The deepest level of analysis focuses exclusively on the drug-receptor interaction and how the drug interacts with active and inactive states of the receptor (level 3). The second level of analysis considers drug action from the perspective of a population of receptors. How the drug activates the population is determined mainly by the properties of the drug-receptor complex, yet for *G* protein coupled receptors, the receptor activation function depends on the *G* protein with which the receptor interacts and the concentration of GTP. The most superficial level of analysis (level 1) considers the concentration-response curve of the drug at a point in the signaling cascade downstream from receptor activation, and hence, the parameters of the concentration-response curve may differ greatly from those of the initial stimulus.

Each level of analysis has its own set of advantages and disadvantages. For example, the experimenter can get a better picture of how a drug will behave at

Hierarchical Levels of Analysis

Level 1: Downstream response
Parameters: EC_{50}, E_{max}

↓

Level 2: Receptor population
Parameters: observed affinity (K_{obs}) and efficacy (ε)

↓

Level 3: Single receptor
Parameters: microscopic constants, K_b and K_a

Figure P.11: Hierarchical levels of analysis of drug-receptor interactions. Agonist-receptor interactions can be analyzed at three hierarchical levels relating to the response downstream from receptor activation (Level 1), the activation state of the receptor population (Level 2), and the behavior of single receptors (Level 3). The parameters that are usually used to described these levels of analysis are potency (pEC_{50}) and maximal response ($E_{\max}$) (Level 1), observed affinity (K_{obs}) and efficacy (ε) (Level 2), and the microscopic affinity constants for active (K_b) and inactive (K_a) states of the receptor (Level 3).

the organ or system level if its effects are measured at this level (level 1). However, it is harder to determine how the drug interacts with active and inactive states of the receptor if the response is measured at the tissue or organ level. Understanding drug action at all three levels of analysis explains how the affinity and efficacy of a drug ultimately determine its effects on the tissue. This information is useful in addressing a host of questions including how structural modifications within a series of the drugs modify their actions. Often, the goal of pharmacological analysis is to estimate the more fundamental parameters of the drug-receptor interaction through the analysis of a response measured downstream from receptor activation. Simply stated, the process involves the estimation of level 2 and level 3 parameters from experimental data obtained at a more superficial level (level 1).

This analytical process falls within the purview of what is commonly known as receptor theory, which includes not only methods for measuring the activity of drug-receptor complexes, but also a set of tacit assumptions held by pharmacologists. Often the latter conflict with the former, and it requires a digression on the origins of receptor theory to explain this dichotomy.

Around the end of the 19th century and the beginning of the 20th century, several investigators were impressed by the remarkable potency and specificity of drugs and suggested that some drugs must interact specifically with receptors to elicit their unique effects. The novelty of this line of thinking is apparent in the cautious writings of A. J. Clark, one of the early proponents of receptor theory. In discussing the actions of acetylcholine on heart muscle[49] in 1926, he wrote:

> "The relation between the concentration and action of acetyl choline in most cases follows the formula $Kx = y/(100 - y)$ and the simplest explanation of this fact is to suppose that a reversible monomolecular reaction occurs between the drug and some receptor in the cells."

With little direct information about receptors, early scientists used simple rules governing the interaction of small ligands with a protein (law of mass-action) to predict how drugs bind to or occupy a receptor. The next hurdle to address was how occupancy of the receptor was translated into a downstream response (level 1 response). Stephenson and Furchgott circumvented this unknown relationship using an approach known as a "null method" in which the experimenter estimates equi-active concentrations of agonist before and after perturbing the amount of available receptors. Another variation involves measuring equiactive concentrations of different agonists. These approaches can be described mathematically by an equality (null equation) in which the unknown transducer function appears on both sides of the equation, effectively nullifying its influence on the data and enabling it to be cancelled out of the equation. The analysis essentially represents a "response-clamp analysis" even though the response is not measured under response-clamp conditions. This approach enabled scientists to estimate the observed dissociation constants of agonists and antagonists and the relative efficacies of agonists. From the perspective of hierarchical levels of analysis, the method involves measuring a superficial level 1 response under different conditions and analyzing the data to deduce how drugs activate the receptor population (level 2 analysis). These strategies were developed in the 1950s and 1960s and were applied mainly to the analysis of responses now known to be mediated by *G* protein coupled receptors.

At a later time, electrophysiologists' using whole-cell recording techniques and single channel analysis measured receptor activation directly (level 2 measurement) as described earlier in this chapter in connection with ligand-gated ion channels. From these measurements, it was possible to determine how agonists interact with open and closed states of ligand-gated ion channels (level 3 analysis). This type of analysis is distinct from the developments in classic receptor theory just mentioned. That is, the estimation of level 2 parameters from a level 1 response (classic receptor theory) is distinct from the estimation of level 3 parameters from single receptor activity (level 3 response) or a population of receptors (level 2 response). The methods are distinct because they were intentionally designed by scientists to measure different things.

Despite the lack of overlap in the two methods of analysis contemporary authors often claim that receptor theory developed by Stephenson and Furchgott provides a set of prescriptive rules regarding the behavior of receptors at the third level of analysis; that is, at the level of receptor states. Here is a short list of some of these rules:

1. Receptor theory is based on the assumption that there is only one active state of the receptor.
2. Receptor theory is based on the assumption that the efficacy of an agonist at a *G* protein-coupled receptor should be constant regardless of the *G* protein mediating the response.
3. Affinity and efficacy are independent parameters.

Because these rules sometimes conflict with the behavior of receptors, classic receptor theory is often considered flawed in contemporary writing.

But hold on. The methods for determining the stimulus through the analysis of a downstream response do not depend on the existence of only one active state; the strategy works satisfactorily even if there is more than one active state, which contradicts rule #1. If there is more than one active state, then it is possible that some agonists may exhibit different relative efficacy values depending upon the *G* protein through which a *G* protein-coupled receptor signals. The methods of receptor theory *do* provide an accurate measurement of observed affinity and relative efficacy under these conditions, which contradicts rule #2.

With regard to rule #3, it is often stated that receptor theory is based on the assumption that observed affinity and intrinsic efficacy are independent parameters because they appear as distinct variables in the stimulus function. This has led to the opinion that receptor theory is flawed if it can be shown that these parameters are interdependent. In fact, each of these parameters can be defined at a deeper level of analysis in terms of the same active and inactive states of the receptor showing that observed affinity and efficacy are not independent. But the presumed flaw in receptor theory mentioned above is inaccurate because the use of distinct variables for parameters does not require the parameters to be independent or incapable of being described in terms of more fundamental constants.

This type of hierarchical analysis is common in other disciplines, like chemistry, physics and philosophy, for example. Thus, while an atom is a useful fundamental unit in a variety of chemical reactions, it can be split into more fundamental particles, and it is the different combinations of these same fundamental particles that give rise to different atoms. But no scientifically literate individual would criticize a chemist for identifying atoms as autonomous structures in chemical reactions. Similarly, there is no reason to dismiss the population concepts of observed affinity and efficacy just because they can be defined in terms of more fundamental constants.

Throughout this book the internal consistency of different hierarchical levels of pharmacological analysis is explained.

Summary of this Book

The thesis of this book is that the interaction of a drug with a receptor generates a code of information having two components, affinity and efficacy. How this information is translated into a response depends on the unique cells, tissue, organ or system in which the particular response is measured. This book describes how to analyze various responses to estimate the affinity and efficacy components of the initial drug-receptor interaction and how these population parameters are related to active and inactive states of the receptor.

This approach might seem like an academic exercise in reducing a train of remarkable biological events into two abstract mathematical quantities, affinity and efficacy. Modern biological techniques have provided an exquisite picture of the structure of receptors, and how their constituent domains mediate function.

Advances in light microscopy, fluorescence techniques and computing power have been combined to give us a more detailed look at receptors in their natural context. Scientists can now see where receptors are located in a variety of cell types including complicated ones, like neurons. These advances have shown that receptors are often located in receptosomes with other neighboring proteins that regulate and transduce their output. It is now possible to track the movement of a receptor or some aspect of its molecular activation in real time, even at the level of single receptor molecules. So, why write a book on estimating affinity and efficacy when more interesting and dynamic molecular details about receptors can be gleaned from biochemical and biophysical approaches? The answer is that without a way of quantifying the activity of receptor structures, we have no way of assessing their significance however exquisitely defined they may be.

Another way to look at this question is to consider a hypothetical physiological response mediated by a known receptor through a mechanism that is completely understood, not just in terms of the proteins involved, but also in terms of their location in the cell and their behavior in time. There may be many reasons for measuring the affinity and efficacy of drugs in this well understood system. Suppose there is an agonist with optimal properties that acts on this receptor to produce a useful therapeutic action. What is the unique combination of affinity and efficacy that allows this drug to produce its optimal effects? Suppose it is a goal to develop a more potent analog that has higher affinity, but retains the same amount of intrinsic efficacy as the initial drug. How does one measure these properties in a series of new candidate drugs using an *in vitro* assay system? This book illustrates how to answer these questions, and the methods do not require an in depth understanding of the intermediate steps in the signaling pathway of the final response.

These methods also apply to a host of other questions. If the drug's affinity and efficacy are sufficiently distinctive at one receptor subtype compared to others, it may be possible to use it as a selective probe to identify the function of receptors *in vitro*. In this application, the drug could be an agonist, antagonist, or allosteric ligand. This is an important question in tissue preparations from animals other than mice in which it is unfeasible to make null mutations in the gene for the receptor in question. The methods described here are also useful for determining whether a G protein-coupled receptor can signal through alternative pathways preferentially depending upon the agonist turning on the receptor or the allosteric ligand modulating the receptor. Other applications are the study of interactions between receptors and the determination of whether a particular signaling protein or enzyme is involved in transducing the response to an agonist.

I use a three-level framework of different perspectives to consider drug action and the explanation of affinity and efficacy. These are the three hierarchical levels of analysis mentioned previously — the interaction of a drug with states of a single receptor, the effect of a drug on a population of receptors, and the consequences of drug action at the level of a downstream response. This approach tracks the chain of causation from the initial drug-receptor interaction to the downstream response. The final outcome depends not only on the initial step, but also on other factors related to the tissue and the response being measured. Thus, with knowledge of the nature of the signaling pathway, one can predict that self administration of a β_2 agonist should produce relief from asthma, whereas an opiate agonist having correspondingly similar affinity and efficacy for its receptor should cause pleasure and predispose the individual to compulsive seeking of opiates.

In describing the relationship between the three levels of analysis, mathematics is required. However, do not put this book down just yet. One can get an intuitive understanding of the concepts of affinity and efficacy at each level of analysis, and how the selectivity of an agonist for ground and active states of the receptor determines the activation of a population of receptors and how the latter triggers a downstream response. But it requires healthy doses of algebra to illustrate this in an unambiguous, quantitative fashion and to derive equations for the analysis of real data and the estimation of affinity and efficacy by nonlinear regression analysis. In these situations, I give a pictorial description of the model, the requisite assumptions, and the strategy for solving the model algebraically. The reader can get an intuitive understanding of the basis of the model without going through the details of the algebra. Nonetheless, the individual steps are shown in an abbreviated format. The final equations are given, and simulations of the model are shown to foster an intuitive understanding of the various concepts.

To avoid conveying the idea that all we know about drug action can be reduced to affinity and efficacy, I begin with a description of the natural history of a receptor class including the structure, function, location, regulation and the biochemical events triggered by members of the class. Having completed this description a model is developed to account for the known behavior of the receptor at the level of receptor states. Then, methods for analyzing receptor activation to estimate the state parameters related to affinity and intrinsic efficacy are described. Through a process of reverse engineering, the model is used as input into a generic transducer function (operational model) to generate theoretical downstream responses of the sort typically measured in a variety of *in vitro* assays. This output is then analyzed using pharmacological methods to determine how estimates of *observed* affinity and intrinsic efficacy are related to more fundamental state parameters. Finally, methods are described for the estimation of some of the receptor state constants through the analysis of downstream responses. In addition, examples of how to estimate these parameters from real data are described. In each case, step-by-step instructions are given so that the scientist or student can easily carry out the same analyses on their own data.

Let me give a brief summary of how this book is organized. My analysis of drug-receptor interactions is focused on two classes of receptors, ligand-gated ion channels and *G* protein-coupled receptors. Part 1 of this book covers ligand-gated ion channels. If we can understand how to measure drug-receptor interactions at ligand-gated ion channels, then we can apply these concepts to study more complicated receptors, like G protein-coupled receptors, which depend on an interaction with other proteins to mediate a response. Fortunately, much work has been accomplished in the area of ligand-gated ion channels already. Chapter 1 describes ligand-gated ion channels, focusing on nicotinic acetylcholine and $GABA_A$ receptors as examples. Their overall structure, how their natural ligands bind to them and how electrophysiological techniques are used to study them are all described.

Chapter 2 describes the analysis of how agonists bind to and activate ligand-gated ion channels. Almost two-thirds of this chapter covers how to analyze the behavior of a hypothetical ligand-gated ion channel having only one agonist-binding site per channel. Of what value is a theoretical discussion about a receptor that does not exist you might ask? In understanding ligand-gated ion channels or the function of any protein, it is helpful to consider how its components work, and thus, I begin my discussion with how a one-site ligand-gated ion channel works. It is much easier to describe the process of ligand activation of a receptor if there is only one ligand site, and how the behavior of such a receptor, at both the single-receptor and population level, relates to affinity and efficacy. It not difficult to add the component principles mentally to understand the behavior of realistic ligand-gated ion channels having two or more ligand-activation sites. Thus, my strategy of first explaining the behavior of a one-site ligand gated ion channel is simply the first step in explaining a multi-site ligand-gated ion channel. The underlying principles will go a long way in helping us understand cooperative interactions between binding sites of any type. Chapters 3 and 4 cover the analysis of competitive antagonist and allosteric ligand interactions with ligand-gated ion channels respectively.

Part 2 covers *G* protein-coupled receptors and consists of Chapters 5–11. Chapter 5 describes the natural history of *G* proteins and *G* protein-coupled receptors, including those involved in vision, taste, smell and adhesion and the more commonly known type involved in chemical transmission between cells. We will see that the recruitment of heterotrimeric *G* proteins by *G* protein-coupled receptors represents a specialization of a generalized mechanism for transferring information between two destinations using the hydrolysis of GTP as the source of energy. This mechanism involves allosteric interactions among receptors, *G* proteins, their subunits and attendant ligands, and these interactions are easier to understand after having established the rules for allosteric interactions in Chapter 4.

Chapter 6 describes these allosteric interactions and explains how they relate to affinity and efficacy and how receptor activation is consistent with a simple stimulus function. We will use this function in Chapter 7 to apply reverse engineering (operational model) and response-clamp analysis (null method) to infer the activation state of the receptor population from the behavior of a response downstream in the signaling pathway. Ultimately these methods will enable us to estimate the observed affinity and efficacy of agonists from their ability to elicit downstream responses. Knowing affinity and efficacy, it is possible to calculate relative estimates of the affinity constants of agonists for active and inactive states of the receptor using an approach described in Chapter 2 for ligand-gated ion channels.

Most receptors exhibit little or no measureable constitutive activity in the absence of agonist, but because the signaling pathways of *G* protein-coupled receptors are highly amplified, it is sometimes possible to measure the signal elicited by the unoccupied receptor when it or its *G* protein is expressed in high abundance. The endogenous signaling pathway can be used to measure constitutive activity in the same way that a scientist uses an enzyme-linked immunosorbent assay to detect a small amount of protein. Both approaches rely on the catalytic activity of an enzyme or signaling protein to amplify a small signal. Chapter 8 describes how to analyze agonist and inverse agonist activity for constitutively active receptors. Having a measure of the constitutive activity of a receptor-signaling pathway, it is possible to estimate the microscopic affinity constants of orthosteric ligands for active and inactive states of the receptor in actual units (e.g., M^{-1}) and not just relative units.

Chapter 9 illustrates how it is possible to estimate the affinity constants of agonists for different active states of the receptor that elicit different signaling pathways. Finally, the last two chapters of Part 2 cover competitive (Chapter 10) and allosteric (Chapter 11) interactions at *G* protein coupled receptors.

Part 3 consists of a single chapter on radioligand-binding analysis (Chapter 12). While it might seem that a book on affinity and efficacy would devote more effort to describing how to measure ligand-binding affinity directly, it is often less than obvious what component of an agonist's binding curve should be used to estimate observed affinity. If the goal is to determine the observed affinity of an agonist for that component of a population of *G* protein-coupled receptors that interacts with a *G* protein to elicit a response of interest, then problems arise because not all of the receptors in the cell may interact with the *G* protein, and observed affinity and efficacy are affected by GTP and GDP. If the receptor is expressed in a CHO cell at high density, for example, perhaps only a small fraction of the total population may be interacting with its *G* protein, and this component of binding may have an observed affinity different from that of the rest of the pool when measured in an intact cell. Problems also arise if binding is measured using cellular homogenate. The appropriate concentration of GTP needs to be included in the assay because intracellular GTP affects observed affinity.

These issues are less of a concern when the binding of antagonists is measured because the latter usually do not affect *G* protein coupling provided that the assay buffer contains physiological salts at the appropriate concentrations. The same is true for allosteric modulators. Measuring the binding affinity of these types of ligands is useful in identifying the receptor subtype mediating a particular response and in determining both the affinity and efficacy components of allosteric regulation. Examples of these approaches are described in Chapter 12.

Pharmacologists often debate the validity of the parameters, affinity and efficacy, or prescribe rules about what they should mean often without concern for different hierarchical levels of analysis. The logical basis for measuring observed affinity and efficacy enjoys a long-term history of acceptance in other disciplines, like the cognitive sciences, where reverse engineering is often employed to identify brain mechanisms, or in electrophysiology, where a null approach is used to deduce channel current (voltage clamp). The validity of the corresponding methods in pharmacology (operational model and null methods) are often debated, however. Efficacy itself is sometimes viewed as a mysterious parameter in search of a meaning. These are among the many reasons why I have written this book.

References

1. Kelly, H. W. New beta 2-adrenergic agonist aerosols. *Clin Pharm* **4**, 393–403 (1985).
2. Wise, R. A. The role of reward pathways in the development of drug dependence. *Pharmacol Ther* **35**, 227–263 (1987).
3. Stein, L. and Belluzzi, J. D. in *Encyclopedia of Neuroscience Volume II, Boston, Birkhauser* Vol. 2 (ed. G. Adelman) 1043–1045 (Birkhauser, 1987).
4. Stein, L. in *International Encyclopedia of Social and Behavioral Science*, 2nd edn. (ed. JD Wright) (Elsevier, 2014).
5. Laudet, V. Evolution of the nuclear receptor superfamily: early diversification from an ancestral orphan receptor. *J Mol Endocrinol* **19**, 207–226 (1997).
6. Robinson, D. R., Wu, Y. M. and Lin, S. F. The protein tyrosine kinase family of the human genome. *Oncogene* **19**, 5548–5557 (2000).

7. Carpenter, G., King, L., Jr. and Cohen, S. Epidermal growth factor stimulates phosphorylation in membrane preparations in vitro. *Nature* **276**, 409–410 (1978).
8. Defize, L. H., Moolenaar, W. H., van der Saag, P. T. and de Laat, S. W. Dissociation of cellular responses to epidermal growth factor using anti-receptor monoclonal antibodies. *Embo J* **5**, 1187–1192 (1986).
9. Ogiso, H., Ishitani, R., Nureki, O., Fukai, S., Yamanaka, M., Kim, J. H., Saito, K., Sakamoto, A., Inoue, M., Shirouzu, M. and Yokoyama, S. Crystal structure of the complex of human epidermal growth factor and receptor extracellular domains. *Cell* **110**, 775–787 (2002).
10. Ottensmeyer, F. P., Beniac, D. R., Luo, R. Z. and Yip, C. C. Mechanism of transmembrane signaling: insulin binding and the insulin receptor. *Biochemistry* **39**, 12103–12112 (2000).
11. Lemmon, M. A. and Schlessinger, J. Cell signaling by receptor tyrosine kinases. *Cell* **141**, 1117–1134 (2010).
12. Schlessinger, J. Cell signaling by receptor tyrosine kinases. *Cell* **103**, 211–225 (2000).
13. Qi, M. and Elion, E. A. MAP kinase pathways. *J Cell Sci* **118**, 3569–3572 (2005).
14. Fredriksson, R., Lagerstrom, M. C., Lundin, L. G. and Schioth, H. B. The *G*-protein-coupled receptors in the human genome form five main families. Phylogenetic analysis, paralogon groups, and fingerprints. *Mol Pharmacol* **63**, 1256–1272 (2003).
15. Lagerstrom, M. C. and Schioth, H. B. Structural diversity of *G* protein-coupled receptors and significance for drug discovery. *Nat Rev Drug Discov* **7**, 339–357 (2008).
16. Oldham, W. M. and Hamm, H. E. Structural basis of function in heterotrimeric *G* proteins. *Q Rev Biophys* **39**, 117–166 (2006).
17. Oldham, W. M. and Hamm, H. E. Heterotrimeric *G* protein activation by *G*-protein-coupled receptors. *Nat Rev Mol Cell Biol* **9**, 60–71 (2008).
18. Johnston, C. A. and Siderovski, D. P. Receptor-mediated activation of heterotrimeric *G*-proteins: Current structural insights. *Mol Pharmacol* **72**, 219–230 (2007).
19. Daaka, Y., Luttrell, L. M. and Lefkowitz, R. J. Switching of the coupling of the beta2-adrenergic receptor to different *G* proteins by protein kinase A. *Nature* **390**, 88–91 (1997).
20. Thompson, A. J., Lester, H. A. and Lummis, S. C. The structural basis of function in Cys-loop receptors. *Q Rev Biophys* **43**, 449–499 (2010).
21. Unwin, N. Refined structure of the nicotinic acetylcholine receptor at 4A resolution. *J Mol Biol* **346**, 967–989 (2005).
22. Buccafusco, J. J. Neuronal nicotinic receptor subtypes: defining therapeutic targets. *Mol Interv* **4**, 285–295 (2004).
23. Hollmann, M. and Heinemann, S. Cloned glutamate receptors. *Annu Rev Neurosci* **17**, 31–108 (1994).
24. Olsen, R. W. and Sieghart, W. GABA A receptors: subtypes provide diversity of function and pharmacology. *Neuropharmacology* **56**, 141–148 (2009).
25. Olsen, R. W. and Sieghart, W. International Union of Pharmacology. LXX. Subtypes of gamma-aminobutyric acid(A) receptors: classification on the basis of subunit composition, pharmacology, and function. Update. *Pharmacol Rev* **60**, 243–260 (2008).
26. Hodgkin, A., Huxley, A. and Katz, B. Ionic currents underlying activity in the gian axon of the squid. *Arch Sci Physiol* **3**, 129–150 (1949).
27. Neher, E. and Sakmann, B. Single-channel currents recorded from membrane of denervated frog muscle fibres. *Nature* **260**, 799–802 (1976).

28. Stephenson, R. P. A modification of receptor theory. *Br J Pharmacol* **11**, 379–393 (1956).
29. Furchgott, R. F. The use of β-haloalkylamines in the differentiation of receptors and in the determination of dissociation constants of receptor-agonist complexes. *Adv Drug Res* **3**, 21–55 (1966).
30. Monod, J., Wyman, J. and Changeux, J. P. On the Nature of Allosteric Transitions: A Plausible Model. *J Mol Biol* **12**, 88–118 (1965).
31. Koshland, D. E., Nemethy, G. and Filmer, D. Comparison of experimental binding data and theoretical models in proteins containing subunits. *Biochemistry* **5**, 365–385 (1966).
32. Colquhoun, D. and Sakmann, B. Fast events in single-channel currents activated by acetylcholine and its analogues at the frog muscle end-plate. *J Physiol* **369**, 501–557 (1985).
33. Colquhoun, D. Binding, gating, affinity and efficacy: The interpretation of structure-activity relationships for agonists and of the effects of mutating receptors. *Br J Pharmacol* **125**, 924–947 (1998).
34. Colquhoun, D. in *Drug Receptors* (ed. H. P. Rang) 149–182 (University Park, 1973).
35. Ehlert, F. J. 'Inverse agonists', cooperativity and drug action at benzodiazepine receptors. *Trends in Pharmacol Sci* **7**, 28–32 (1986).
36. Pfeiffer, C. C. Optical isomerism and pharmacological action, a generalization. *Science* **124**, 29–31 (1956).
37. Christopoulos, A. Allosteric binding sites on cell-surface receptors: novel targets for drug discovery. *Nat Rev Drug Discov* **1**, 198–210 (2002).
38. Berridge, M. J. Inositol trisphosphate and diacylglycerol as second messengers. *Biochem J* **220**, 345–360 (1984).
39. Milligan, G., Marshall, F. and Rees, S. G16 as a universal G protein adapter: implications for agonist screening strategies. *Trends Pharmacol Sci* **17**, 235–237 (1996).
40. Conklin, B. R., Farfel, Z., Lustig, K. D., Julius, D. and Bourne, H. R. Substitution of three amino acids switches receptor specificity of Gq alpha to that of Gi alpha. *Nature* **363**, 274–276 (1993).
41. Conklin, B. R., Herzmark, P., Ishida, S., Voyno-Yasenetskaya, T. A., Sun, Y., Farfel, Z. and Bourne, H. R. Carboxyl-terminal mutations of G_q alpha and $G_s$alpha that alter the fidelity of receptor activation. *Mol Pharmacol* **50**, 885–890 (1996).
42. Michell, R. H. Inositol phospholipids and cell surface receptor function. *Biochim Biophys Acta* **415**, 81–147 (1975).
43. Young, R. C. and Downes, C. P. Inositol phospholipid-dependent cellular signalling: opportunities for drug discovery. *Drug Des Deliv* **6**, 1–13 (1990).
44. Bolger, G. T., Gengo, P., Klockowski, R., Luchowski, E., Siegel, H., Janis, R. A., Triggle, A. M. and Triggle, D. J. Characterization of binding of the Ca++ channel antagonist, [3H]nitrendipine, to guinea-pig ileal smooth muscle. *J Pharmacol Exp Ther* **225**, 291–309 (1983).
45. Hardie, R. C. TRP channels and lipids: from Drosophila to mammalian physiology. *J Physiol* **578**, 9–24 (2007).
46. Hilgemann, D. W., Feng, S. and Nasuhoglu, C. The complex and intriguing lives of PIP2 with ion channels and transporters. *Sci STKE* **2001**, re19 (2001).
47. Rasmussen, S. G., DeVree, B. T., Zou, Y., Kruse, A. C., Chung, K. Y., Kobilka, T. S., Thian, F. S., Chae, P. S., Pardon, E., Calinski, D., Mathiesen, J. M., Shah, S. T., Lyons, J. A., Caffrey, M., Gellman, S. H., Steyaert, J., Skiniotis, G., Weis, W. I.,

Sunahara, R. K. and Kobilka, B. K. Crystal structure of the beta2 adrenergic receptor-Gs protein complex. *Nature* **477**, 549–555 (2011).
48. Kenakin, T. and Miller, L. J. Seven transmembrane receptors as shapeshifting proteins: the impact of allosteric modulation and functional selectivity on new drug discovery. *Pharmacol Rev* **62**, 265–304 (2010).
49. Clark, A. J. The reaction between acetyl choline and muscle cells. *J Physiol* **61**, 530–546 (1926).
50. Mishina, M., Takai, T., Imoto, K., Noda, M., Takahashi, T., Numa, S., Methfessel, C. and Sakmann, B. Molecular distinction between fetal and adult forms of muscle acetylcholine receptor. *Nature* **321**, 406–411 (1986).
51. Ehlert, F. J. Pharmacological analysis of the contractile role of M_2 and M_3 muscarinic receptor in smooth muscle. *Receptors and Channels* **9**, 261–277 (2003).

PART 1

Ligand-Gated Ion Channels

Ligand-Gated Ion Channels 1

Ligand-gated ion channels mediate the rapid transfer of information across nerve synapses in the brain and peripheral nervous system. They are also located in extrasynaptic regions of neurons as well as on glia and non-neuronal tissue. These receptors are composed of four to five protein subunits that span the plasma membrane and circumscribe a central ion pore.[1] When the neurotransmitter binds to its pocket on the extracellular domain of the receptor it causes the channel to open, allowing specific ions to flow in or out of the postsynaptic cell depending upon their electrochemical gradients and the selectivity of the channel. Channels that conduct an influx of sodium or calcium depolarize the cell, whereas those that conduct an influx of Cl^- hyperpolarize the cell or make it more difficult to be depolarized. These channel currents instantaneously generate electrotonic currents in the region of the postsynaptic membrane. The sum of the depolarizing and hyperpolarizing currents ultimately determines the excitability of the cell. Without this rapid trans-synaptic communication, the information processing capabilities of neural networks would be severely limited.

Because ligand-gated ion channels incorporate both the ion pore and binding pocket for the chemical messenger into a single structure, it is possible to measure the activity of the ligand-receptor complex directly. This enables an unambiguous measurement of how well a drug binds to and activates the receptor. Thus, ligand-gated ion channels provide a unique opportunity to examine the properties of affinity and efficacy of the ligand-receptor complex. Any book on affinity and efficacy, therefore, must clearly explain these principles for ligand-gated ion channels. We can then apply these concepts to *G* protein-coupled receptors, whose activation state is not usually measured directly, but rather, inferred from measurements made at different points downstream in the receptor-signaling pathway.

This chapter focuses on the structure of ligand-gated ion channels so that we will have a better understanding of what physical properties our estimates of affinity and efficacy represent. Chapters 2–4 will describe how to measure affinity, efficacy and allosteric modulation at ligand-gated ion channels.

Ligand-gated ion channels represent a large group of receptors that can be divided into two evolutionarily related groups. One group includes receptors for acetylcholine (nicotinic acetylcholine receptors), γ-aminobutyric acid ($GABA_A$ receptors), glycine and serotonin ($5\text{-}HT_3$ receptor).[2] The other includes many types of receptors for glutamate.[3] This chapter will focus on nicotinic acetylcholine and $GABA_A$ receptors. The reasons are that much has been known about the native structure of the nicotinic acetylcholine receptor for quite some time, and $GABA_A$ receptors exhibit abundant allosteric interactions. Thus, having a close look at these receptors will provide the structural basis for understanding the principles of affinity, efficacy and allosterism at ligand-gated ion channels.

Nicotinic Acetylcholine Receptors

The nicotinic receptor is so-named because it represents one of the two subgroups of acetylcholine receptors that is activated by the plant alkaloid, nicotine. The other group, muscarinic receptors, is activated by muscarine, which accumulates in the fruiting bodies (basidiocarps or mushrooms) of some members of the *Inocybe* and *Clitocybe* genera of fungi. Muscarinic receptors are members of the *G* protein-coupled receptor family and are described in Chapter 5. Because acetylcholine is a linear flexible molecule, it can fit into the orthosteric binding pockets on both classes of receptors, whereas the more rigid alkaloids have a preference for only one type (Figure 1.1). Both nicotinic and muscarinic receptors represent groups of receptors with several subtypes.

Our knowledge of the structure of the nicotinic receptor would not have progressed the way it did without a gift from nature in the form of the electric organ, particularly those from two marine rays, *Torpedo californica* and *Torpedo marmorata*.[4]

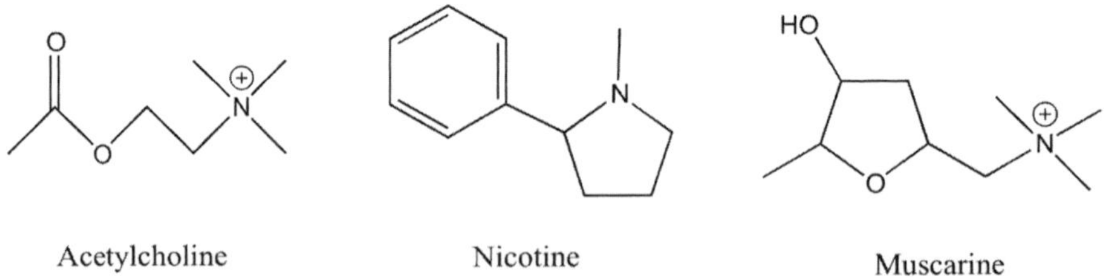

Acetylcholine Nicotine Muscarine

Figure 1.1: Structures of acetylcholine, nicotine and muscarine.

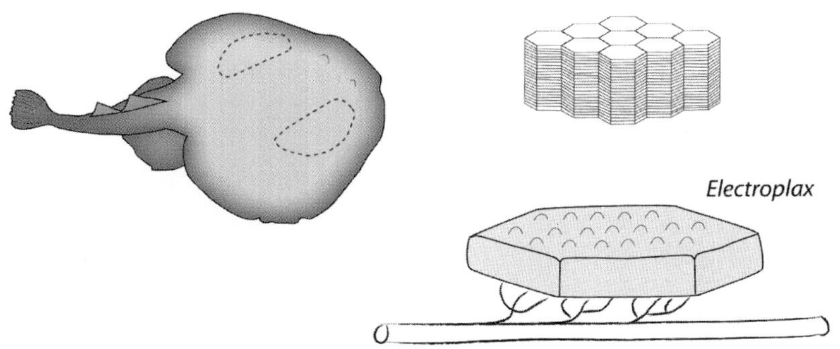

Figure 1.2: The electric organ from *Torpedo californica*. (a) The location of the bilateral electric organs in *Torpedo californica* is indicated by the dashed lines. (b) Schematic drawing of the columns of electroplax within the electric organ. (c) A schematic drawing of a single electroplax showing its innervation.

These fishes have large bilateral electric organs occupying a substantial portion of the flattened pectoral fins that define the skate and ray body plan. Each electric organ consists of a few hundred closely packed columns of cells traversing the ventral to dorsal sides of the ray (Figure 1.2). Each column contains a stack of about 1000 flattened cells (electroplaxs) having a diameter and thickness of approximately 5 mm by 0.02 mm. Cholinergic nerves innervate the ventral surface of each electroplax.[5] When stimulated, these nerves release acetylcholine, which activates nicotinic receptors on the ventral surface of the electroplax, causing an influx of positive monovalent cations. This depolarizes the ventral side of the electroplax relative to the dorsal side causing a potential difference of about 50–100 mV. Because the 1000 or so electroplaxs in each column are linked in series and coupled in parallel with those of other columns, the total electromotive force from the discharge of the electric organ can be 50–100 V, depending upon how it is measured.

This large potential difference and its associated current would not be possible without the contribution of a colossal population of nicotinic acetylcholine receptors, each having a conductance of about 50 pS. This great abundance of receptors enabled Michael Raftery and colleagues[6] to purify the native receptor from the electric organs of *Torpedo californica*, determine its pentameric subunit composition (two α_1 subunits and one each of β_1, γ_1 and δ_1), and sequence the first 54–56 amino acids of the amino terminus of the four subunits. Knowing the partial sequences, Shosaku Numa and colleagues[7] determined the complete sequences by gene cloning. When

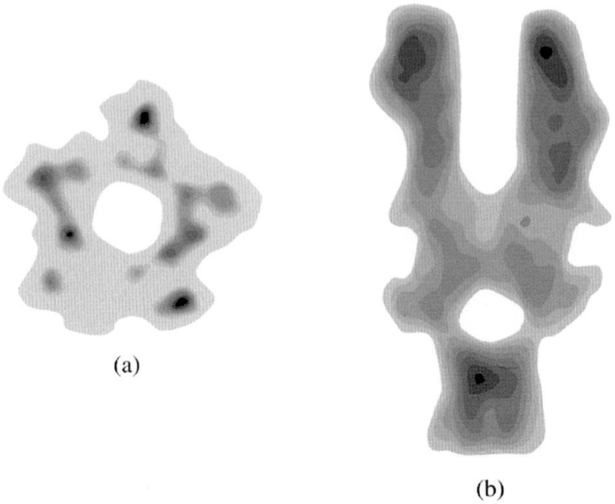

Figure 1.3: Structure of single nicotinic acetylcholine receptors from *Torpedo marmorata*. Crude replicas of electron density maps of cross-sectional (a) and longitudinal (b) views of the receptor calculated by averaging data from separate images using helical diffraction techniques. The view in (a) is at the level of the outer vestibule approximately 30 Å above the lipid bilayer. In (b) the extracellular side of the channel is on top. The figure has been constructed from the electron density contours published by Unwin.[11]

injected into *Xenopus* oocytes, the mRNAs for the four subunits are translated into functional nicotinic acetylcholine receptors.[8]

One convenient property of the electric organ from *Torpedo* is that vesicle preparations containing only nicotinic acetylcholine receptors can be prepared from it without great difficulty. After incubation in aqueous buffer at low temperature for a few weeks, these vesicles gradually form elongated tubes composed of a densely packed crystalline array of nicotinic acetylcholine receptors within the plane of the membrane.[9] Electron microscopic images taken normal to the surface of these tubes show clear pore-like structures representing single nicotinic acetylcholine receptors (Figure 1.3a).[10,11] Nigel Unwin and coworkers[12] developed a refined model of the nicotinic acetylcholine receptor at 4 Å resolution by averaging separate electron microscopic images using helical diffraction techniques. A model of the open state of the receptor was obtained by imaging vesicular tubes that were sprayed with a solution of acetylcholine 5 ms prior to freezing.[13] By integrating both structural and sequence data and applying computer modeling methods, Unwin and coworkers developed a striking model of the nicotinic acetylcholine receptor (Figure 1.4).

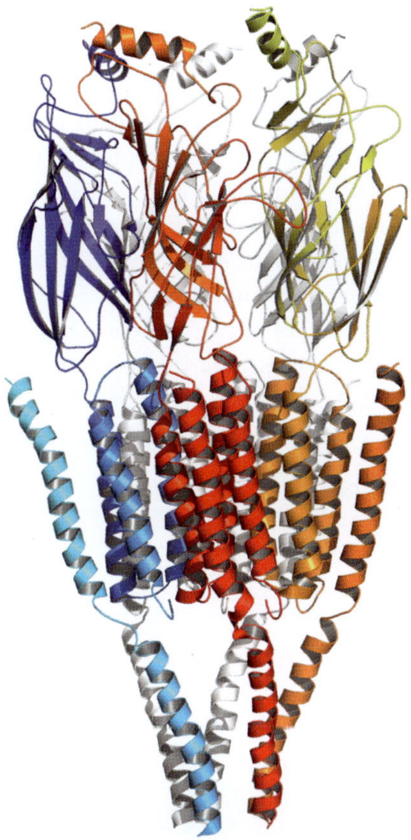

Figure 1.4: (Color online) Refined model of the nicotinic acetylcholine receptor from *Torpedo marmorata*. In this longitudinal view, the γ subunit in the center (red) is flanked by α subunits (blue and orange-yellow). The model was developed by Unwin[12] and is deposited in the protein data bank (PDB code 2bg9).

Subunit composition and transmembrane topology of the nicotinic acetylcholine receptor from Torpedo

When viewed along an axis normal to the surface of the plasma membrane, average electron-density images of the *Torpedo* nicotinic acetylcholine receptor show a ring having radial symmetry with five distinguishable lobes surrounding a central pore (Figure 1.3a). The lobes represent the five subunits of the receptor. The central pore is maintained in cross-sectional images throughout the length of the receptor. It is widest at the extracellular end of the receptor where it forms a vestibule with a diameter of 20 Å. The central pore narrows to a minimum diameter of about 8 Å at a locus near the center of the plasma membrane, which is thought to represent the ion gate. It continues to the cytosolic end of the receptor where it forms an inner vestibule with transverse openings to the cytosol. Longitudinal images of the receptor resemble a goblet with an open vestibule protruding into the synaptic space a distance almost half the length of the receptor[10,11] (Figure 1.3b). The outside diameter of the receptor narrows at the level of the plasma membrane and narrows further as it projects into the cytosol. Approximately 30% of the length of the receptor projects into the cytosol. The overall length of the receptor is about 160 Å, and about 70 Å project into the synaptic space.

The primary sequences of the four homologous subunits of the *Torpedo* nicotinic receptor (α_1, β_1, γ_1 and δ_1) consist of a large amino terminus, four membrane-spanning domains (M1–M4), a short loop joining M1 and M2, another joining M2 and M3, and a long cytosolic loop joining M3 and M4[7,12] (Figure 1.5). The

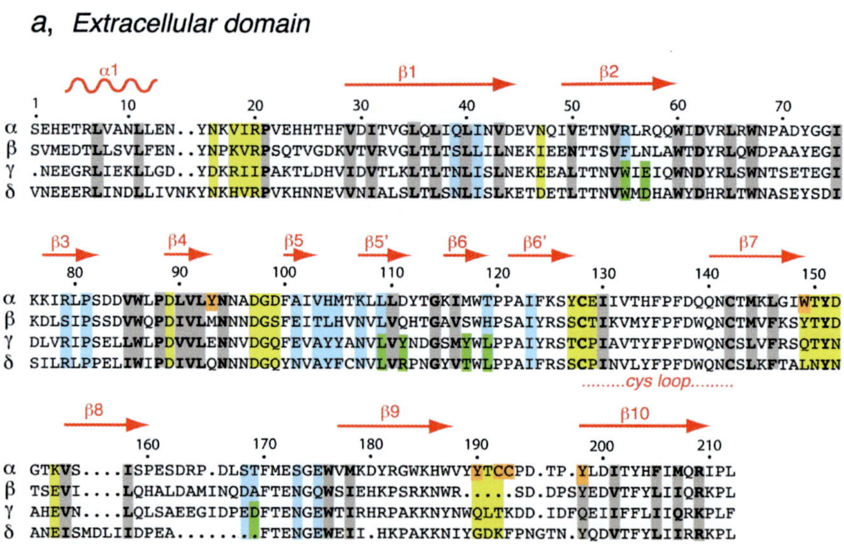

Figure 1.5: (Color online) Sequence alignment of the nicotinic acetylcholine receptor subunits from *Torpedo Marmorata*. The numbering corresponds to the sequence of the α subunit. Secondary structure (α helix or β strand) is indicated above the sequence of the α subunits. Residues indicated in bold with gray background are conserved throughout the Cys-loop family. (a) Extracellular ligand-binding domain. The location of the Cys-loop is identified in red letters and dots beneath the second group of sequences. Residues located on the principal (+) and complementary (−) sides of the subunit interfaces are indicated in yellow and light-blue background, respectively. Residues critical for the binding of acetylcholine and other othosteric ligands are indicated on the α (orange) and γ and δ (green) subunits. (b) Intracellular and membrane spanning domains. Hydrophilic (yellow) and hydrophobic (light-blue) residues lining the channel pore are indicated. The intracellular amphipathic helix is denoted as "MA".

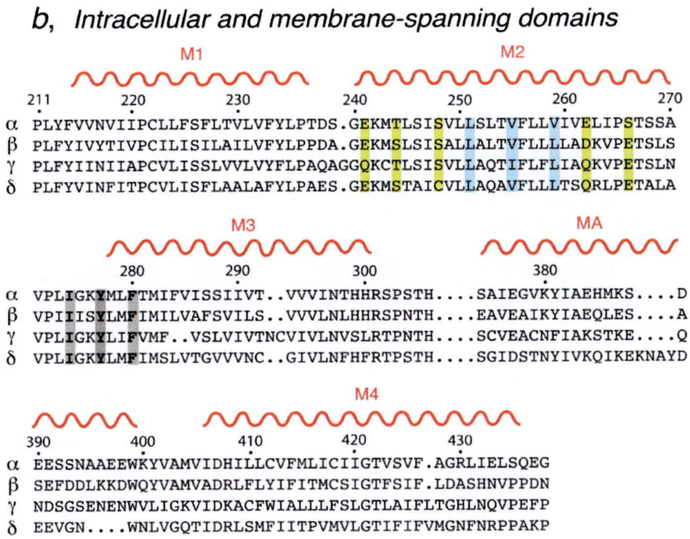

Figure 1.5: (*Continued*)

extracellular part of the receptor is formed primarily from the amino terminal domains, but includes a 10 Å extension of the M1–M4 membrane-spanning domains. The inner vestibule is formed from the cytosolic M3–M4 loop. Near its amino terminal end, close to M4, this loop assumes an alpha helical structure that extends M4 into the cytosol.

The sequences of the *Torpedo* nicotinic acetylcholine receptor subunits place them within the homologous Cys-loop family of ligand-gated ion channels, which includes other nicotinic receptors and the $GABA_A$, glycine and $5HT_3$ receptors.[1] The receptor from the muscle-derived electric organ of *Torpedo* is analogous to that of the mammalian embryonic neuromuscular junction. Both are formed from a ring of five receptor subunits — two α_1 and one each of β_1, γ and δ. In the adult muscle-type receptor, the γ subunit is replaced with an ε subunit, which causes a reduction in the mean open time of the channel and an increase in the single channel conductance from 40 pS to about 60 pS.[14]

The results of experiments with labeled irreversible receptor ligands are consistent with the hypothesis that acetylcholine binds to each of the two α subunits in a pocket between the α–γ and α–δ subunit interfaces on the amino terminal domain of the α subunit. Studies by Raftery coworkers[15] on the interaction of [^{3}H]bromoacetylcholine ([^{3}H]BrACh) with the *Torpedo* receptor serve as a useful example. This labeled acetylcholine derivative binds in a reversible fashion to the native receptor in *Torpedo* membranes with high affinity (K_D = 17 nM). If the receptor is first treated with the sulfhydryl group reducing agent, dithiothreitol, [^{3}H]BrACh binds irreversibly. Both the reversible and irreversible binding of [^{3}H]BrACh to the receptor are competitively inhibited by acetylcholine and the selective nicotinic receptor antagonist, *d*-tubocurarine, showing that [^{3}H]BrACh binds to the same site on the receptor as acetylcholine. When the [^{3}H]BrACh-labeled membranes from *Torpedo* were solubilized and examined by electrophoresis, it was found that [^{3}H]BrACh was incorporated into the α subunit only. Presumably, it reacts covalently with adjacent cysteine residues 192 and 193 in the amino terminal domain of the α subunit. Dithiothreitol is expected to reduce the vicinal disulfide linkage between these cysteines, thereby exposing two free sulfhydryl groups for reaction with [^{3}H]BrACh. These cysteine residues are absent in the β, γ and δ subunits.

Another example of a related experiment involves the competitive antagonist [^{3}H]DDF (N,N-dimethylaminophenyldiazonium fluoroborate). This photoactivated ligand has been used to label the native nicotinic receptor in *Torpedo* membranes irreversibly. Changeux and coworkers[16] solubilized *Torpedo* membranes labeled in this manner and sequenced fragments of the cleaved receptor to identify the amino acids covalently linked with [^{3}H]DDF. It was shown that [^{3}H]DDF labeled tryptophan 149, tyrosines 93 and 190 and cysteines 192 and 193 in the α subunit. Using a similar peptide mapping strategy, but with [^{3}H]nicotine as the photoaffinity label, Jonathan Cohen and coworkers[17] found that tyrosine 198 and cysteine 192 were labeled in the α subunit. A number of other investigators employing reversible and irreversible ligands and the technique of site-directed mutagenesis have obtained evidence consistent with the location of the acetylcholine binding sites on the α subunits at the interfaces between adjacent γ and δ subunits.

A similar type of approach involving reactive ligands that penetrate the ion pore has been used to identify the pore lining amino acids on the membrane-spanning regions of the receptor. Both phencyclidine (PCP)[18] and chlorpromazine (CPZ)[19] behave as noncompetitive inhibitors of the current response to acetylcholine at the nicotinic receptor from *Torpedo*. These agents bind to the receptor at a locus distinct from that of acetylcholine and inhibit the current response. The binding of [^{3}H]CPZ is inhibited by PCP suggesting that both ligands occupy the same site in the channel pore.[20] When bound to its site, [^{3}H]CPZ photolabels all four receptor subunits of the *Torpedo* receptor. Photolabeling is enhanced with simultaneous addition of an agonist, but is inhibited by prior treatment with the same agonist, suggesting that [^{3}H]CPZ labels a site in the channel pore in the open state, but not in the closed and desensitized states.[21] After labeling the native receptor in *Torpedo* membranes with [^{3}H]CPZ, Changeux and coworkers identified that the amino acids covalently bound with [^{3}H]CPZ by sequencing purified fragments of the solubilized receptor and counting the radioactivity released after each cycle of the analysis. The results showed that [^{3}H]CPZ labels serine 248 in the α subunit, serine 254 and leucine 257 in the β subunit, threonine 253, serine 257 and leucine 260 in the γ subunit, and serine 262 in the δ subunit.[22–24] All of these residues are in the M2 domain of each subunit approximately two-thirds of the way through the membrane-spanning central pore from the extracellular side, just past a hydrophobic region (Figure 1.5). In the case of the β and γ subunits, labeling occurs on more than one amino acid, each spaced by three to four residues. This pattern is consistent with an alpha helical secondary structure of the pore-lining M2 segment because the peptide backbone of the alpha helix completes a full rotation for every 3.6 residues.

The electron density maps of the open state of the *Torpedo* nicotinic receptor are consistent with these peptide-mapping results. The location of the residues forming the acetylcholine-binding pocket is thought to lie approximately 40 Å above the plasma membrane. When taken 5 ms after treatment with acetylcholine, electron density images in the regions of the two acetylcholine-binding pockets show a change in the α subunits.[13] The change could be explained as a small clockwise rotation (about 4°) in the two α subunits, but evidence described below suggests a larger (11°) rotation in only the inner pore-lining section of the α subunits. This clockwise rotation was transmitted to the alpha helices lining the membrane-spanning ion pore, causing the pore to widen. Images focused on the membrane-spanning portion of the receptor show five rods of high electron density surrounding the central pore, which undoubtedly represent the five M2 domains.[11]

The extracellular ligand binding domain

The extracellular domain of the nicotinic acetylcholine receptor is homologous to that of other Cys-loop family members as well as to a soluble acetylcholine-binding protein from the snail, *Lymnaea stagnalis* (20–23% sequence identity). This protein is secreted from glial cells in an acetylcholine-dependent manner and is involved in modulating synaptic transmission.[25] It also binds other ligands including nicotine and the nicotinic antagonists, *d*-tubocurarine and α-bungarotoxin. The crystal structure of a pentameric ring of this protein was solved, and it represents a useful model for the extracellular ligand-binding domain of the nicotinic receptor from *Torpedo* (Figure 1.6).[26]

The acetylcholine binding protein forms a homomeric pentamer similar in size to the extracellular domain of the nicotinic receptor (Figure 1.6). When viewed parallel to an axis through the central pore, the pentamer forms a ring with five lobes remarkably similar to the extracellular portion of the nicotinic receptor from *Torpedo*. The amino terminus is on the top surface of the ring and the carboxyl

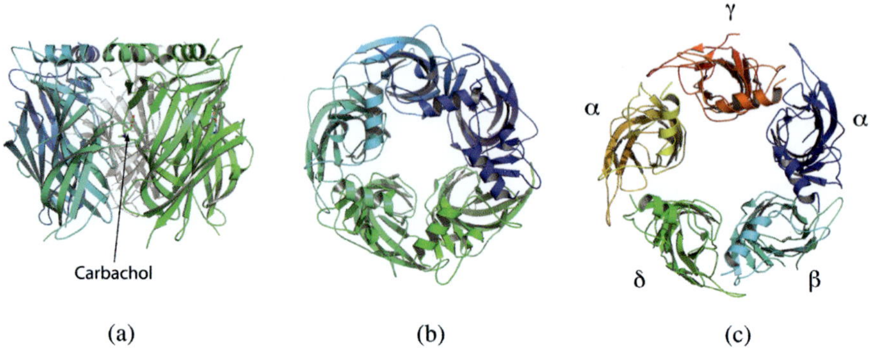

Figure 1.6: Comparison of the structure of the acetylcholine binding protein from *Lymnaea stagnalis* with the extracellular domain of the nicotinic acetylcholine receptor from *Torpedo marmorata*. The N-terminal α helix is visible on the top side of the receptor. (a) The crystal structure of the pentameric acetylcholine binding protein with bound carbachol is shown. (b) Top view of the acetylcholine binding protein. (c) Top view of the extracellular domain of the nicotinic acetylcholine receptor. The structures were obtained by Celie *et al.*[83] (PDB code 1uv6) and Unwin[12] (PDB code 2bg9).

terminus on the bottom. These two surfaces correspond to the synaptic and membrane-proximal sides of the extracellular domain of the nicotinic receptor, respectively. The primary sequences of the acetylcholine binding protein and the amino terminal domains of the α_1, β_1, γ and δ subunits from *Torpedo* include 10 β strands ($\beta1-\beta10$). In the crystal structure of the acetylcholine binding protein, these are arranged in two antiparallel β sheets that form a sandwich with a one-quarter twist. These represent inner ($\beta1$, $\beta2$, $\beta3$, $\beta5$ and $\beta6$) and outer ($\beta4$, $\beta7$, $\beta8$, $\beta9$ and $\beta10$) β sheets lining the central pore and the outer surface of the pentamer, respectively. The two sheets are held together, in part, by a disulfide bond between two conserved cysteine residues located on the underside of each monomer. In the primary sequence, these cysteines occur at the C terminal end of $\beta6$ and the amino terminal beginning of $\beta7$ and delineate a 15-residue Cys-loop, which is conserved in all subunits of the nicotinic receptor as well as in other members of the Cys-loop family of ligand-gated ion channels. The secondary structure also contains a short alpha helix near the beginning of the amino terminus and two more short 3_{10} helices within the $\beta2-\beta3$ and $\beta3-\beta4$ loops, respectively.

The structure of the acetylcholine binding protein is consistent with the postulate that the ligand-binding sites on the nicotinic receptor are located in the $\alpha-\gamma$ and $\alpha-\delta$ subunit interfaces (Figure 1.7). Recall that tyrosines 190 and 198 and cysteines 192 and 193 of the alpha subunit constitute part of the acetylcholine binding pocket

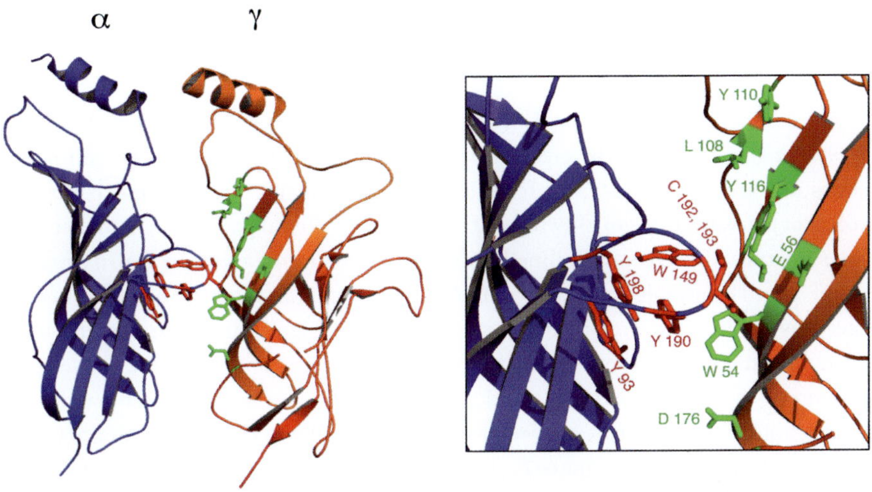

Figure 1.7: (Color online) The location of critical residues for orthosteric ligand binding on the $\alpha-\gamma$ subunit interface of the nicotinic acetylcholine receptor from *Torpedo marmorata*. The point of view in (a) and (b) is from outside the channel. (b) An expanded view (20 Å × 20 Å) of the ligand binding pocket. The α and γ subunits are shown in blue and orange, respectively.

because they react covalently with photoaffinity ligands. The cognate residues of the acetylcholine binding protein (Y185, Y192, C187 and C188) are located on the outer β sheet on the loop between $\beta 9$ and $\beta 10$ (C Loop). From a downward view of the top of the homopentamer, this loop projects in a counterclockwise direction into the interface between adjacent subunits. Immediately above it is a loop between $\beta 7$ and $\beta 8$ (B Loop) that contains tryptophan 149, a critical residue for acetylcholine binding. Acetylcholine is thought to bind between these B and C Loops, and the resulting interaction is thought to bring these loops closer together. Other residues important for the binding of *d*-tubocurarine have been identified on the γ and δ subunits of the nicotinic receptor by site-directed mutagenesis. The cognate residues on the acetylcholine binding protein are located on $\beta 2$ and $\beta 6$ of the inner β sheet. These residues are on the subunit interface opposite to that bearing the B and C loops. Thus, the residues on the acetylcholine binding protein that define the ligand-binding pockets of the nicotinic receptor are on complementary sides of the subunit interfaces in the homomeric crystal structure, depending upon whether they are cognate to residues in the α subunit (counterclockwise face; principal or positive side) or the γ and δ subunits (clockwise face; complementary or negative side). This organization is consistent with the binding sites for acetylcholine being located on the $\alpha-\gamma$ and $\alpha-\delta$ subunit interfaces and with the arrangement of the $\alpha_1-\gamma-\alpha_1-\delta-\beta_1$ ring of subunits being counterclockwise in the *Torpedo* nicotinic receptor.

A comparison of the crystal structure of the acetylcholine binding protein from *Lymnaea stagnalis* with the electron microscopic images of the resting and active states of the nicotinic receptor from *Torpedo* yields insight into the mechanism of receptor activation. Nigel Unwin and coworkers[12,27] found good agreement between the structure of the acetylcholine binding protein and the electron density map of the β, γ and δ subunits in images of the *Torpedo* receptor, regardless of whether the receptor was in the active or inactive state. There was significant disagreement, however, with the map of the α subunit in the inactive state, but not in the active state. A good fit to the inactive state could be obtained if the inner and outer β-sheets of acetylcholine binding protein were fitted independently to the electron density map. This involved rotating the inner β-sheet of the α subunit counterclockwise 11° from its position in the acetylcholine binding protein. Thus, there appear to be two conformations of the α subunit. In the active conformation, the orientation of the inner and outer β sheets resembles those of the non-α subunits and the acetylcholine binding protein. In the resting state, however, the position of the inner β-sheet is rotated counterclockwise 11° away from the outer β-sheet. In this inactive conformation, the binding pocket for acetylcholine between the B loop on the inner β sheet and the C loop on the outer β sheet is open. The binding of acetylcholine between the two loops may bring them closer together causing a clockwise rotation of the inner β-sheet by 11°. This rotation could be transmitted to the M2 pore lining helix via the short loop between M2 and M3, causing an opening of the channel gate as described below.

The ion conduction pathway

Under physiological conditions, the muscle-type nicotinic acetylcholine receptor conducts an inward current of mainly positive monovalent cations. The channel also conducts calcium and Mg^{2+}, but their permeabilities are much lower than those of sodium and potassium. At a negative resting-membrane potential (e.g., −80 mV), the electrochemical force driving sodium into the cell is much greater than that driving potassium out. This is because the equilibrium potential of potassium is near the resting membrane potential, whereas that for sodium is around +40 mV. Thus, under physiological conditions, the muscle-type nicotinic receptor conducts an inward current of mainly sodium, whereas at positive holding potentials, the current is mainly an outward flux of potassium.

The surface charge on the inner lining of the 20 Å-diameter intra- and extracellular vestibules is consistent with the nicotinic receptor being a nonselective cation channel. The inner β sheet of the extracellular domain has an excess of negative charge on the inside surface of the vestibule and is dotted with a combination of about four or five negatively charged glutamate and aspartate residues per subunit.[12] The alpha helical extensions of membrane spanning domain M4 that are formed by the carboxyl terminal end of the cytosolic loop between M3 and M4 also have several glutamate and aspartate residues on the inside of the vestibule and the outside perimeter of the lateral openings into the cytosol.[12] This abundance of negative charge at each end of the channel increases the concentration of cations near the channel openings, thereby facilitating their flux through the channel.

As described above, four α helical segments (M1–M4), arranged in an anti-parallel fashion, form the membrane-spanning domain of each subunit. The M2 segment from each of the five subunits lines the 40 Å-long ion pore, whereas the other segments (M1, M3 and M4) form an outer shell around the M2 helix of the same subunit. Looking down on the channel from the synaptic space, the M2 helices form the vertices of a pentagon surrounding the central pore (Figure 1.8). They are close together at a point near the center and intracellular side of the plasma membrane. As they traverse upward towards the extracellular side, they bend outwardly along a radial axis in a gradual arch punctuated by slight kinks near

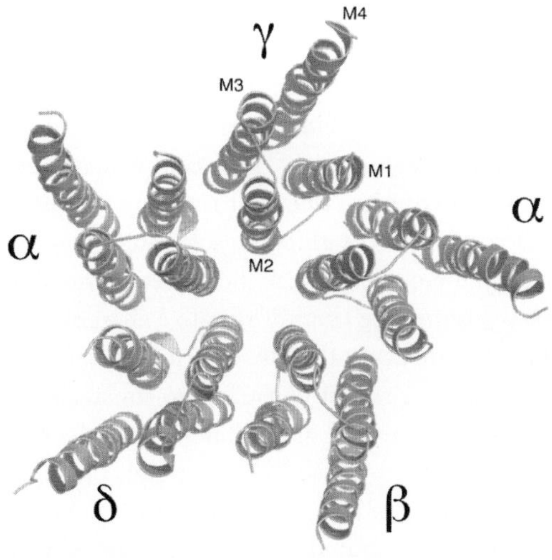

Figure 1.8: Membrane spanning domains of the nicotinic acetylcholine receptor from *Torpedo marmorata*. The point view is from the extracellular space looking down on the channel. The membrane spanning domains (M1–M4) for each receptor subunit are shown. Their orientation is the same in all subunits and is indicated for the γ subunit. The structure is from Unwin[12] (PDB code 2bg9).

the center of the membrane and at a proline near the extracellular end of the helix. The result is a funnel-shaped channel pore with the wider end on the extracellular side of the channel. The surrounding M1, M3 and M4 helices are also splayed outward on the extracellular side, but their bends follow both radial and tangential paths resulting in a slight clockwise turn as they rise from the inner side of the membrane. The M1, M3 and M4 helices make substantial contacts with each other whereas the M2 helix is separated from them.

The primary sequence of the M2 segment of each subunit is consistent with its functional role of lining a cation channel and forming a ligand-controlled hydrophobic plug near the center of the membrane (see Figure 1.5). The M2 domain of the α subunit shown in Figure 1.9a illustrates these attributes. The residues lining the central axis of the pore are highlighted. Most of the cognate residues on the M2 helices of the other subunits are conserved or represent conservative replacements (Figure 1.9b). Each residue on the pore-lining side of the α subunit, therefore, designates one residue of a five-membered ring of similar residues surrounding the pore. Beginning at the intracellular end of the channel, there is a ring of negative charge composed of four glutamic acids and a glutamine at position E241 in the α subunit. One helical turn upward is a ring containing two serines and three threonines. The next helical turn upward contains four serines and a cysteine. The next three helical turns contain three hydrophobic rings of amino acids corresponding to leucine 251, valine 255 and valine 259 on the α subunits. These rings are composed mainly of leucine and valine residues but also include an isoleucine. Following this hydrophobic stretch of amino acids, there is a hydrophilic ring corresponding to glutamine 262 in the alpha sequence. This ring contains an aspartic acid, two glutamic acids and two glutamines. The next ring upward corresponding to serine 266 on the α subunit contains three glutamic acids and two serines. Thus, with the exception of the hydrophobic stretch near the center of the plasma membrane, the ion channel consists of alternating rings of negative charge or hydrophilic hydroxyl groups — an ideal environment for the flux of hydrated cations.

In the closed state, the channel constricts to its smallest diameter of 3–6 Å at a locus aligning with leucine 251 and valine 255 in the α subunit. It remains at a diameter of less than 7 Å at valine 259. Although this opening may be larger than the diameter of sodium (2.32 Å) or potassium (3.04 Å), it is smaller than the diameters of these ions when hydrated with a single shell of water molecules (about 8 Å). The hydration shells of sodium and potassium are strongly bonded (heats of hydration of –105 and –85 kcal mol^{-1}, respectively) and cannot be removed unless replaced by polar residues lining the channel. Because there are none, sodium and potassium remain hydrated and cannot penetrate the hydrophobic stretch of the channel. In the closed state, therefore, the hydrophobic gate aligning with leucine

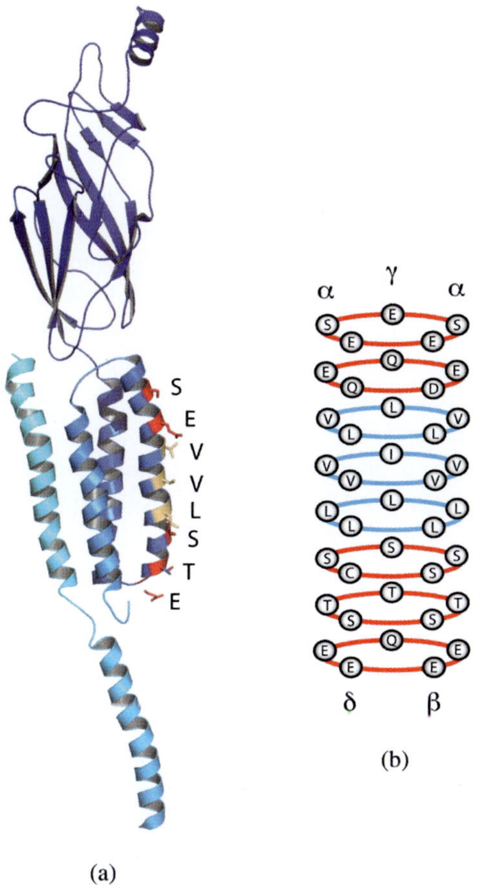

Figure 1.9: Pore-lining residues of the nicotinic acetylcholine receptor from *Torpedo marmorata*. (a) The pore-lining residues on the M2 segment of the α subunit are indicated in red (hydrophilic) and tan (hydrophobic). (b) The pore-lining residues of the α, β, γ and δ subunits are shown. The structure is from Unwin[12] (PDB code 2bg9).

251 to valine 259 in the α subunit forms a hydrophobic plug that prevents the flow of ions.

Ligand-induced channel opening

The binding of acetylcholine to the extracellular domains of the two α subunits transmits a signal to the membrane spanning M2 helix to open the hydrophobic channel plug. This information is most likely communicated through physical connections between these two domains of the alpha subunit. There are three possible routes of communication. Two involve interactions of the $\beta1-\beta2$ and Cys-loops of the ligand-binding domain with the M2–M3 loop of the channel domain. The third is the covalent connection of the peptide chain between $\beta10$ and M1.

The Cys-loop contains a phenylalanine-proline-phenylalanine (FPF) motif that is thought to interact with three residues near the top of M3 (isoleucine, tyrosine and phenylalanine), numbered 273, 277 and 280 in the alpha subunit.[12] These two hydrophobic motifs are conserved in the β, γ and δ subunits, and their hydrophobic and aromatic character is also conserved throughout the Cys-loop family of ligand-gated ion channels. The FPF motif is near the center of the Cys-loop, and it projects downward to anchor the top of helix M3 to the ligand-binding domain.

The $\beta1-\beta2$ loop of all nicotinic receptor subunits contains a conserved glutamate residue at position 43 in the α subunit. It too is conserved in members of the Cys-loop family. In the α subunit, glutamate 43 and valine 44 clasp the backbone of the M2–M3 loop near the top of M2 in the α subunit.[12] At this level, M2 lacks contact with the other helices of the channel, and its interaction with the $\beta1-\beta2$ loop determines its position. In the non-α subunits, this interaction is displaced along the M2-M3 loop in a C-terminal direction away from the top of M2.[12]

As described above, the binding of acetylcholine is thought to cause an 11° clockwise rotation in the inner β-sheet of the ligand-binding domain on the α subunit. On the basis of their model of the *Torpedo* receptor, Nigel Unwin and colleagues[12,28] have postulated how this rotation might open the hydrophobic channel plug. The inner β-sheet is connected to the top of the pore-lining M2 helix via the downward projecting $\beta1-\beta2$ loop. A rotation of the inner β-sheet is expected, therefore, to cause the top of M2 to track an 11° arch pointing away from the center of the channel and having an axis of rotation near the top of M3, which acts as a pivot. The connection between M2 and its axis of rotation is the M2–M3 loop, which has the requisite flexibility because of a highly conserved glycine residue at position 275 in the alpha subunit near the top of M3. The lack of a side chain on glycine imparts flexibility to the loop. Because the M2 helix lacks interactions with other helices at this level, movement of the inner β-sheet should influence its position. Thus, a clockwise rotation of the inner β-sheet could cause the pore lining M2 helix to bend away from the central axis of the channel and break its hydrophobic interactions with the other M2 helices at the channel plug. This increases the aperture of the ion pore allowing a flux of ions.

In the open state, the narrowest part of the channel is at the bottom of M2 near the ring of negative charge aligning with glutamate 241 in the alpha subunit (Figure 1.9). In other members of the Cys-loop family, this ring is conserved and its charge has a large effect on the selectivity of the ion pore. In cation-conducting receptors, like the nicotinic and $5HT_3$ receptors, a negatively charged glutamate is usually present, whereas in the anion conducting receptors, like the glycine and $GABA_A$ receptors, the residue is uncharged.[1,29] The lateral openings of the inner vestibule are of approximately the same size as the selectivity filter at the bottom of M2, and these may also act as a filter, given the abundance of negative charge around them.

Nicotinic receptor subtypes

A total of 16 subunits have been identified for nicotinic acetylcholine receptors. These include nine α subunits ($\alpha_1-\alpha_7$, α_9 and α_{10}), four β subunits ($\beta_1-\beta_4$), and one each of γ, δ and ε.[30,31] The subunits of the muscle-type nicotinic receptor, which are most closely homologous to the *Torpedo* receptor, are designated with the subscript one (i.e., α_1, and β_1). All of the α subunits contain adjacent cysteine residues aligning with cysteines 192 and 193 on the C loop of the α_1 subunit. The non-α subunits lack these adjacent cysteine residues. The predicted secondary structures of all of the subunit sequences are similar to that of the α_1 subunit from *Torpedo*. That is, they have a large amino terminus that aligns with the β strands, loops and helices of the acetylcholine binding protein from *Lymnaea stagnalis*, a set of four, antiparallel, membrane spanning domains (M1–M4), and a large cytosolic loop between M3 and M4.

A variety of measures have been used to assess nicotinic receptor subtypes including the anatomical distribution of subunit mRNA and immunoreactivity, immuno-characterization of oligomeric receptor complexes, characterization of receptors from knockout mice, and the functional and pharmacological properties of heterologously expressed channels. Each approach has its advantages and disadvantages. One can never be certain, for example, that a given type of mRNA in a tissue expresses a subunit that is incorporated into the channel whose functional responses are being measured, particularly when a mixture of mRNAs is present. Similarly, when a mixture of subunits is expressed heterologously, it is unclear whether the resulting receptor population is homogeneous or whether all of the transfected subunits contribute to the functional channels. The measurement of single-channel properties can alleviate some issues associated with the former problem.

Despite these limitations, three general concepts have emerged from a large group of experimental findings: Acetylcholine-gated channels require at least two α subunits; many nicotinic receptor subtypes exhibit different pharmacological properties depending on their subunit combination; and while the total possible number of combinations of five nicotinic receptor subunits — each containing two α — is quite large, many combinations do not form functional channels or do not appear to exist in nature.[30,31] In addition to these general principles, there is now specific information about the subunit composition and pharmacological properties of the major subtypes of the nicotinic receptor expressed in some prototypical tissues.

As mentioned, the neuromuscular junction expresses nicotinic receptors composed of $(\alpha_1)_2\beta_1\gamma\delta$ or $(\alpha_1)_2\beta_1\varepsilon\delta$ subunit stoichiometry. These two receptors have high permeability to sodium and potassium and about one-tenth the permeability to calcium.[32] They are weakly activated by nicotine and choline and are blocked by α-bungarotoxin with high potency.[33–35] How α-bungarotoxin interacts with the receptor is known because the crystal structure of the monomeric N-terminal domain of the α_1 subunit bound to bungarotoxin was solved.[36] α-Bungarotoxin is a relatively large peptide ligand having a molecular weight of about 8,000. Its primary sequence consists of 74 residues including 10 cysteines that form five disulfide bonds. The secondary structure includes five β-strands that form two antiparallel β-sheets. In the crystal structure of the α-bungarotoxin-α_1 subunit complex, the C loop of the α_1 subunit is sandwiched between the two β-sheets of α-bungarotoxin. Recall that the C loop contains critical residues in the acetylcholine-binding pocket including adjacent cysteines, 192 and 193. Thus, the toxin interacts competitively with acetylcholine at the ligand-binding site.

Autonomic ganglia expresses an abundance of α_3 and β_4 subunits, and these interact with additional subunits to form channels containing $\alpha_3\beta_2$, $\alpha_3\beta_4$ and $\alpha_3\beta_4\alpha_5$ subunit combinations.[37] These receptors having similar permeability to sodium and calcium are relatively insensitive to inhibition by α-bungarotoxin. Nicotine potently activates the α_3 and β_4 subunit containing channels, but only weakly activates the $\alpha_3\beta_2$ receptor. In the central nervous system, an abundance of α_4 and β_2 subunits are expressed,[38] and these form channels with subunit stoichiometries of $(\alpha_4)_2(\beta_2)_3$, $(\alpha_4)_3(\beta_2)_2$ and $(\alpha_4)_2(\beta_2)_2(\alpha_5)$.[37] These receptors are potently activated by nicotine and are relatively insensitive to blockade by α-bungarotoxin. Their permeability to monovalent cations is greater than that to calcium.

Another type of neuronal nicotinic receptor is the homomeric α_7 channel.[37] This receptor is potently activated by nicotine and blocked by α-bungarotoxin with high potency. It exhibits much higher permeability to calcium than to sodium and potassium (10- to 20-fold), and it desensitizes rapidly.

Electrical responses of nicotinic receptors

Responses to nicotinic receptor activation can be measured in a variety of ways depending on the type of experimental preparation and the goals of the investigator. Extracellular recordings of single neurons (single unit analysis) or groups of neurons can be measured following activation of nicotinic receptors *in situ*. The resulting potentials for single neurons are small, whereas field potentials are larger and measure the activity of many nearby cells.

Intracellular recordings of membrane voltage provide more specific information about the receptors on the membrane of a given cell. The application of a nicotinic agonist causes depolarization, which may trigger a volley of action potentials depending on the concentration of applied agonist. The current flow through ion channels depends on the voltage difference and the concentration gradient of conducted ions across the membrane. For a given ionic gradient, a change in voltage will change the current flow induced by a given concentration of agonist. Because agonist activation of the channel causes a voltage change, the agonist inhibits its own effect by bringing membrane potential closer to the reversal potential and reducing the drive for ion flux. Thus, the current induced by a nicotinic agonist decreases in time as the voltage response increases and comes closer to the reversal potential.

The voltage-clamp technique avoids this problem by maintaining a constant voltage across the membrane. This is achieved by applying whatever current is necessary to counteract the change in voltage caused by agonist-induced channel activation (see Prologue). The applied current is equivalent, though opposite in sign, to that elicited by the channels. It is automatically determined and displayed by the feedback circuitry of the voltage clamp apparatus. Figure 1.10 shows the results of experiments in which the whole cell current response to various concentrations of nicotine are plotted against the log concentration of agonist.[39] As acetylcholine occupies more of the population of nicotinic receptors, the current response approaches a maximum, which represents the current elicited when nicotine occupies nearly 100% of the receptor population.

The voltage clamp technique can also be used to monitor the activity of single channels. In this type of experiment, the recording electrode circumscribes a small patch of membrane encompassing one or a few ion channels. Depending upon the configuration of the patch, it is possible to investigate the properties of individual channels in their native context (cell attached mode) or to manipulate the ionic gradient and other constituents (inside out and outside out configurations). Figure 1.11 shows a single channel recording of a nicotinic receptor in the presence of a fixed concentration of carbachol.[40] As described in more detail in Chapter 2, channel openings occur in bursts, corresponding to the time when two molecules of agonist are occupying the receptor. The inter-burst interval represents a receptor bound with a single molecule of agonist or none at all. Long periods of quiescence are attributed to desensitization, and this is common at high concentrations of agonist.

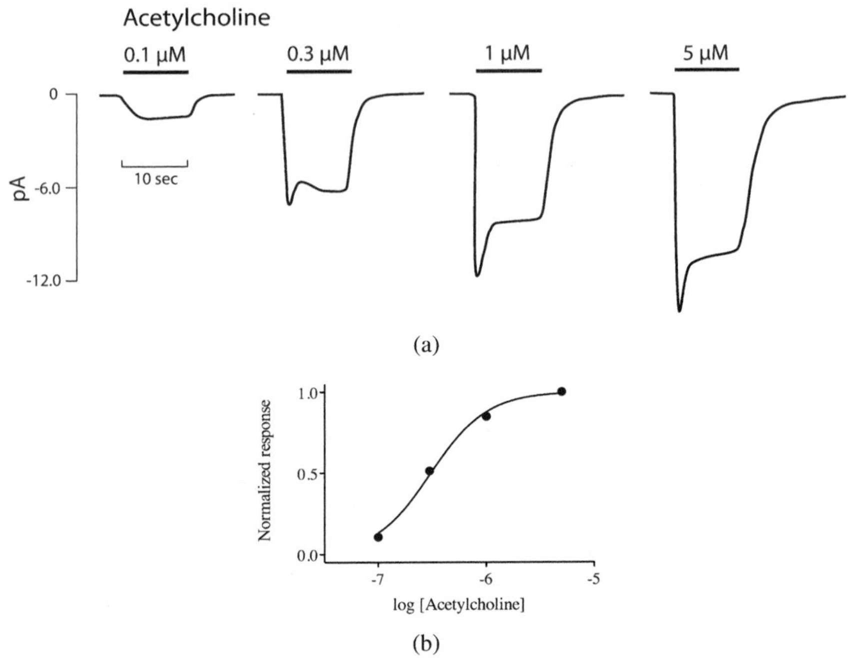

Figure 1.10: The current response of a population of nicotinic acetylcholine receptors to acetylcholine in *Xenopus* oocytes expressing chicken α and non-α subunits. (a) The individual current responses to various concentrations of acetylcholine are shown. (b) The current response is plotted against the log molar concentration of acetylcholine. The data are from Bertrand *et al.*[39]

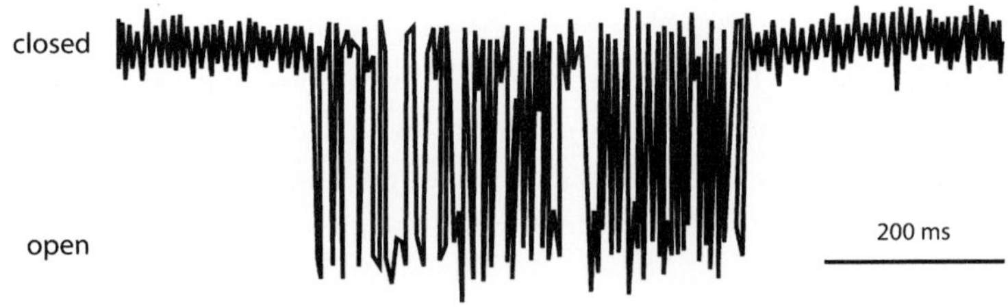

Figure 1.11: Single channel recording of nicotinic acetylcholine receptors from an outside-out patch of rat embryonic muscle. The recording was made in the presence of carbachol (0.5 mM). The data are from Hamill *et al.*[40]

$GABA_A$ Receptors

The $GABA_A$ receptor is one of the two classes of receptors activated by the neurotransmitter, GABA. The other class, the $GABA_B$ receptor, is a dimeric *G* protein-coupled receptor described in Chapter 5. The $GABA_A$ receptor is abundantly expressed in the nervous systems of many types of animals, and it is the major ligand-gated ion channel mediating inhibitory neurotransmission.[41] Drugs that block $GABA_A$ receptors cause seizures, whereas those that activate the receptor inhibit seizures and cause sedation and tranquility. The $GABA_A$ receptor is the target for many drugs used in medicine (e.g., benzodiazepines, barbiturates, anesthetics and ethanol) that produce their therapeutic effects by enhancing $GABA_A$ receptor-mediated neurotransmission in the brain.[42]

When activated, the $GABA_A$ receptor conducts a current of mainly chloride.[43,44] The concentration of chloride is usually much higher extracellularly (140 mM) than intracellularly (7 mM), resulting in a chloride-equilibrium potential (−70 mV) close to the resting membrane potential. Thus, activation of the $GABA_A$ receptor causes an increase in membrane conductance and a decrease in capacitance such that the membrane potential is stabilized near the resting potential. This makes it more difficult for excitatory inputs to depolarize the cell. The actual changes in membrane potential caused by activation of the $GABA_A$ receptor at inhibitory synapses vary from hyperpolarizing to moderately depolarizing, depending on the difference between the resting membrane potential and the chloride equilibrium potential.

In some cells, particularly during development, the expression of chloride transporters causes the concentration of chloride in the cytosol to increase to about 40 mM.[45] This reduces the chloride gradient across the plasma membrane and sets the chloride-equilibrium potential to a value (−40 mV) much more positive than the resting membrane potential. Under this condition, activation of the $GABA_A$ receptor causes chloride efflux and a large depolarization that can elicit a discharge of action potentials in the postsynaptic cell.[46]

$GABA_A$ receptors represent a group of pentameric ligand-gated ion channels of varying subunit combinations. A total of 19 human $GABA_A$ receptor genes have been identified, including six α-subunits (α_{1-6}), three β-subunits (β_{1-3}), three γ-subunits ($\gamma_1-\gamma_3$), one each of δ, ε, θ and π and three ρ-subunits ($\rho_1-\rho_3$).[47] In the past, responses to GABA were classified as $GABA_C$ receptor-mediated if they were not blocked by the $GABA_A$ receptor antagonist, bicuculline, nor mimicked by the $GABA_B$ receptor agonist, baclophen. Many of these responses are now known to be mediated by channels composed of ρ subunits. Given the homology between ρ subunits and other $GABA_A$ receptor subunits, the $GABA_C$ receptor nomenclature is no longer recommended.[48]

The primary sequences of the $GABA_A$ receptor subunits share homology (about 30%) with other the Cys-loop family members including the nicotinic receptors described earlier.[1] They are expected to have similar secondary, tertiary and quaternary structure as well, and several homology-modeling studies show this to be the case. That is, $GABA_A$ receptors have a large amino-terminal extracellular domain, four membrane-spanning domains (M1–M4), and an intracellular domain on the loop between M3 and M4 that assumes an α helical structure near its C-terminal end. The secondary structure of the amino terminal domain is primarily composed of 10 β-strands and their adjoining loops that together form two anti-parallel β-sheets lining the inside (inner β-sheet) and outside (outer β-sheet) of the extracellular vestibule. The membrane spanning M1, M3 and M4 segments surround the inner pore-lining M2 segment, and the intracellular loop between M3 and M4 is thought to form an inner vestibule that, together with M2, contributes to ion selectivity.

When $GABA_A$ receptor subunits from the group α, β, γ, δ and ε are expressed in heterologous cells, one subunit at a time, there is little reconstitution of functional GABA-mediated chloride flux as judged by the whole-cell current response.[49] The reconstitution of robust $GABA_A$-mediated chloride currents usually, but not always, requires the coexpression of both α and β subunits, and the complete pharmacological properties of a given channel usually requires co-expression of an additional subunit, like the γ subunit for benzodiazepine modulation, for example. In contrast, functional homomeric channels can be formed from the individual ρ subunits.

In the following sections, I describe the structural features of the channel that determine the ligand-binding pockets and ion selectivity and give a brief description of $GABA_A$ receptor heterogeneity and the allosteric modulation of chloride currents.

The orthosteric ligand-binding site

To identify critical amino acid residues of the ligand-binding site, many investigators have systematically introduced point mutations in the $GABA_A$ receptor and measured the consequences on radioligand binding or GABA-induced chloride currents. A substantial loss of function can be interpreted as evidence that the side chain of the amino acid contributes to the GABA binding site. A few examples of this work are described next.

Irwin Sigel and coworkers found that replacement of phenylalanine 64 in the α_1 subunit with leucine caused a large decrease in the potency of GABA for eliciting chloride currents in *Xenopus* oocytes expressing a combination of α_1, β_2 and γ_2 subunits.[50] Using a different approach, Geoffrey Smith and Richard Olsen purified the $GABA_A$ receptor from bovine brain, photolabeled the intact receptor with the

GABA agonist, [^{3}H]muscimol, and then sequenced a chymotrypsin fragment of it to identify the labeled amino acid.[51] These investigators found a release of radioactivity on the fifth cycle of sequencing corresponding to phenylalanine 65 in the bovine α_1 sequence. This residue is homologous to phenylalanine 64 in the rat α_1 sequence and is present in all α subunits and the β_2 and γ subunits. Thus, it seems likely that the GABA binding site includes part of the α subunit that aligns with phenylalanine 64 in the rat α_1 subunit.

Jahanshah Amin and David Weiss identified two loci on the β subunit that are critical for functional $GABA_A$ receptors.[52] Conservative point mutations in tyrosine 157 and threonine 160 of the β_2 subunit caused a large reduction in the potency of GABA for eliciting chloride currents in *Xenopus* oocytes expressing α_1, β_2 and γ_2 subunits. Another locus was identified on the basis of the large loss in GABA potency associated with conservative point mutations in threonine 202 and tyrosine 205.

Given the sequence homology within the Cys-loop family of ligand-gated ion channels, it is possible to map the residues identified by mutagenesis onto the structure of the acetylcholine binding protein from *Lymnaea stagnalis* to identify the probable locus of the binding site for GABA on the $GABA_A$ receptor. The phenylalanine 64 residue in the rat α_1 sequence required for GABA-mediated chloride currents aligns within the $\beta2$ strand. With regard to the residues on the β_2 subunit required for GABA-mediated chloride current, tyrosine 157 and threonine 160 are located on the loop between β strands $\beta7$ and $\beta8$ (Loop B), whereas threonine 202 and tyrosine 205 are located on the loop between β strands $\beta9$ and $\beta10$ (Loop C). Figure 1.12 shows the predicted location of these residues in the extracellular domain of the $GABA_A$ receptor structure based on homology with the acetylcholine binding protein. These residues are on the interfaces between adjacent α and β subunits (panel b). Recall from Figure 1.7 that the binding sites for acetylcholine on the muscle-type nicotinic acetylcholine receptor are between $\alpha-\gamma$ and $\alpha-\delta$ interfaces. The acetylcholine-binding pocket is formed from Loops B and C on the α subunit (principal side) and from the $\beta2$ strand on the γ or δ subunits (complementary side). In the $GABA_A$ receptor the orientation is reversed. That is, the B and C Loops for the GABA binding pocket are from the β subunit (principal side), whereas the $\beta2$ strand is from the α subunit (complementary side).

Allosteric binding sites

Different types of drugs are known to modulate the $GABA_A$ receptor, allosterically, including endocannabinoids, benzodiazepines, barbiturates, neurosteroids and general anesthetics.[42] The location of their allosteric binding sites have been identified

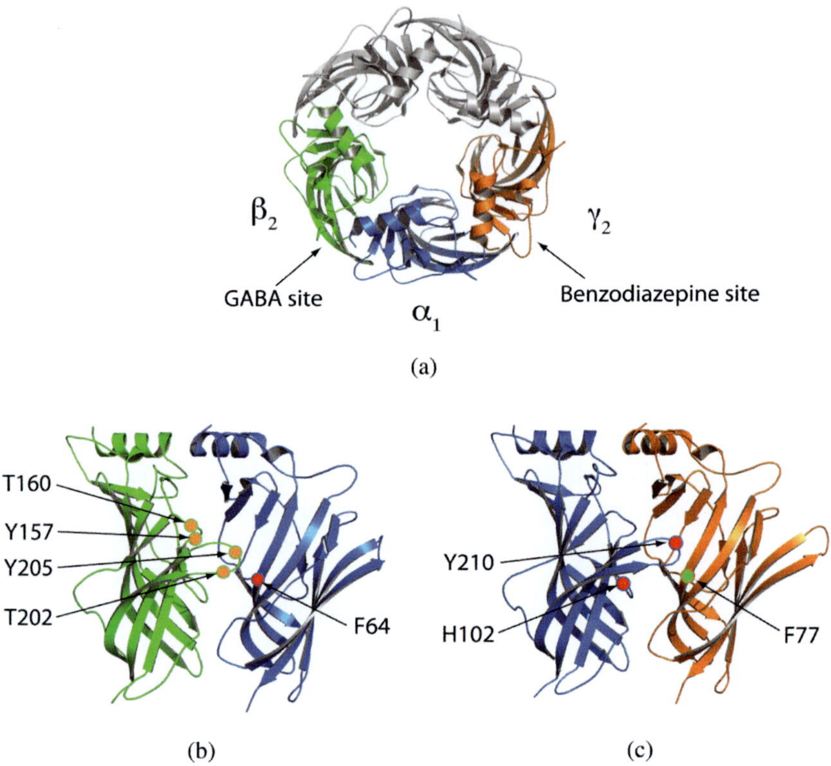

Figure 1.12: The location of critical residues within the orthosteric site and the allosteric site for benzodiazepines on the $GABA_A$ receptor. The crystal structure of the acetylcholine binding protein[83] (PDB code 1uv6) was used as a scaffold to map the location of cognate residues on $GABA_A$ receptor subunits that are critical for ligand binding. The numbering of the residues and the labeling of the subunits corresponds to the appropriate $GABA_A$ receptor subunits. (a) Location of the GABA and benzodiazepine sites at the subunit interfaces of the $GABA_A$ receptor. The point of view is from the extracellular space above the channel. (b) Critical residues for ligand binding to the orthosteric site at the $\alpha_1-\beta_2$ subunit interface. (c) Critical residues for ligand binding to the benzodiazepine site at the $\alpha_1-\gamma_2$ subunit interface. The point of view in panels (b) and (c) is from outside the channel.

through site-directed mutagenesis, cysteine substitution mutagenesis, and more unequivocally, by sequencing fragments of the native receptor previously labeled with covalently bound allosteric ligand. A few examples of these studies on the allosteric sites for benzodiazepines and anesthetics are described below.

The benzodiazepine analogs in clinical use for their sedative, anti-anxiety and anti-seizure properties have positive modulatory effects, and thus, enhance

GABA-mediated chloride currents. At the level of the whole cell, this modulation is manifest as a two- to three-fold increase in the potency of GABA for eliciting chloride currents,[53,54] whereas at the level of single $GABA_A$ receptors, the effect represents an increase in the frequency of opening.[55] Because the modulation by benzodiazepines is relatively small and channel opening is so improbable in the absence of GABA, benzodiazepines have essentially no effect on chloride conductance by themselves. Many benzodiazepines bind to the $GABA_A$ receptor with nanomolar affinity, and hence, are useful radioligands.[56,57] One such example is the positive allosteric modulator, [^{3}H]flunitrazepam, which can also be used as a photoaffinity label.[58] Other ligands bind to the benzodiazepine site on the $GABA_A$ receptor and have negative modulatory effects (e.g., β-carbolines; Ro15-4513),[59,60] while still others have no effect, but competitively antagonize the modulatory action of positive and negative allosteric modulators (Ro15-1788).[61,62]

Some of the strongest evidence identifying the location of the benzodiazepine-binding site has come from experiments in which the $GABA_A$ receptor was first purified from bovine brain, photolabeled with [^{3}H]flunitrazepam, and partially digested with specific proteases. The photolabeled peptide fragments were then purified and sequenced. Although the yield of receptor protein from bovine brain is often too low to identify a given amino acid after each cycle of Edman degradation, it has been possible to identify a substantial peak of radioactivity at a specific cycle. Given the specificity of the protease employed and the size of the peptide fragment, it is usually possible to identify the peptide within the known sequence of the $GABA_A$ receptor, and ultimately, the amino acid in the peptide fragment corresponding to a specific cycle in the Edman degradation. Using this approach Susan Dunn and coworkers identified histidine 102 in the bovine α_1 subunit as the substrate for [^{3}H]flunitrazepam.[63] Using a similar strategy but with different proteases Geoffrey Smith and Richard Olsen identified the same residue as well as secondary labeling of proline 97.[64]

In other studies, Olsen and coworkers[65] purified the $GABA_A$ receptor from bovine brain and photolabeled it with the benzodiazepine analog, [^{3}H]Ro15-4513. Jonathan Cohen's laboratory used a peptide-mapping strategy to identify the labeled residues in the purified receptor preparation. It was found that [^{3}H]Ro15-4513 labeled homologous tyrosine residues in the α_1 (tyrosine 210), α_2 (tyrosine 209) and α_3 (tyrosine 234) subunits.

While the residue labeled by [^{3}H]Flunitrazepam (α_1, histidine 102) is different from that labeled by [^{3}H]Ro15-4513 (α_1, tyrosine 210), both occur on the counterclockwise face of the α_1 subunit when mapped onto the structure of the acetylcholine binding protein as shown schematically in Figure 1.12c. These tyrosines occur on the loops between β strands β4 and β5 (tyrosine 102) and β9 and β10 (tyrosine 210). Cysteine mutagenesis has shown that tyrosine 102 is essential for the covalent binding of reactive benzodiazepine ligands and that tyrosine 210 is required for benzodiazepine binding and modulation of chloride currents in cells transfected with α_1, β_2 and γ_2 $GABA_A$ receptor subunits.[66,67] Similar mutagenesis studies have shown that phenylalanine 77 in the γ_2 subunit is also important for benzodiazepine binding.[68] Using a homology model of the $GABA_A$ receptor, Olsen, Cohen and coworkers[65] showed that this residue occurs on the $\beta 2$ strand of the γ_2 subunit within 4–6 Å of the reactive azide group of Ro15-4513. Thus, the benzodiazepine site is located at the interface between α (principal side) and γ (complementary side) subunits with the orientation of the subunits the same as that of acetylcholine binding sites between the $\alpha - \gamma$ and $\alpha - \delta$ interfaces of the muscle type nicotinic receptor.

A similar type of microsequencing strategy has been used to identify the binding site for the anesthetic etomidate. Following photoaffinity labeling of purified $GABA_A$ receptors from bovine brain with an azido-derivative of labeled etomidate, Olsen and coworkers[69] found that [^{3}H]azietomidate covalently reacted with methionine 236 in the α_1 subunit and methionine 286 in the β_3 subunit. The labeling of homologous residues on the other α (α_2, α_3 and α_5) and β (β_1 and β_2) subunits of the $GABA_A$ receptor in bovine brain was also detected. These residues occur on the membrane spanning α helices of the α (M1) and β (M3) subunits approximately one-third of the distance down the α helix from its junction with the extracellular domain. This location is directly beneath the GABA binding site, which is at the interface between the extracellular domains of the same subunits. The irreversible labeling of the $GABA_A$ receptor by [^{3}H]azietomidate appears to be competitively inhibited by low concentrations of barbiturates, propofol and isoflurane, suggesting that these general anesthetics act at the same site.[70] The location of the binding site for [^{3}H]azietomidate is shown schematically in Figure 1.13. The M2 helix lines the channel pore, whereas the M1, M4 and M3 helices form a membrane barrier around the M2 helix resembling a semicircle in a cross-sectional view of the receptor. Etomidate and the other anesthetics probably bind in an aqueous pocket between the M1 and M3 α helices of adjacent α and β subunits, respectively.[70,71] Presumably, this causes a disruption of the hydrophobic interaction among the five M2 segments resulting in an opening of the channel pore.

$GABA_A$ receptor subtypes

As mentioned above, 19 human $GABA_A$ receptor genes have been identified, including six α subunits (α_{1-6}), three β subunits (β_{1-3}), three γ subunits ($\gamma_1 - \gamma_3$), one

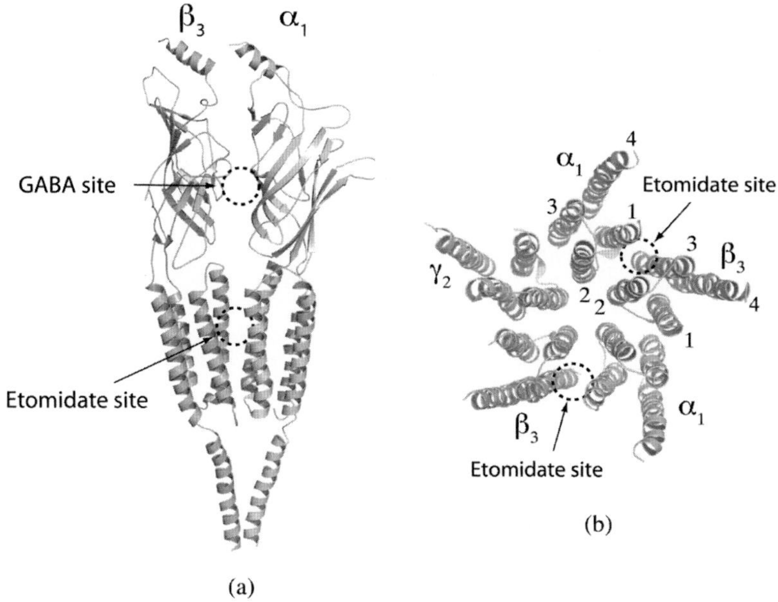

Figure 1.13: The general location of the orthosteric site and the allosteric site for etomidate on the $GABA_A$ receptor. Unwin's model[12] (PDB code 2bg9) of the nicotinic acetylcholine receptor was used as a scaffold to map the location of the binding sites based on the peptide mapping and modeling of studies Li *et al.*[69] The labeling of the subunits corresponds to the appropriate $GABA_A$ receptor subunits. (a) Location of the GABA and etomidate sites (dashed circles). The point of view is from outside the channel. (b) The location of the etomidate site (dashed circle) within the membrane spanning domains (M1–M4). These are labeled 1–4 for two of the subunits. The point of view is from the extracellular space above the channel.

each of δ, ε, θ and π, and three ρ-subunits ($\rho_1-\rho_3$).[47] These could combine in a large number of different ways to give rise to hundreds of different types of receptor. The probable existence of some specific subunit combinations has been gleaned from studies on the co-localization of receptor subunits within individual neurons, the expression of receptor subunits in heterologous cells, the co-immunoprecipitation of multiple subunits by a single subunit-specific antibody, and the functional and pharmacological consequences of genetic manipulation of specific receptor subunits. The reader is directed to excellent reviews by Olsen and Sieghart on this topic.[42,48,72] Below, I quickly describe some of the $GABA_A$ receptor subunit combinations that are likely to form native $GABA_A$ receptors in the mammalian brain.

Studies using *in situ* hybridization have shown that mRNAs for the α_1, β_2, β_3 and γ_2 subunits are widely distributed throughout the brain with some differences in their individual distribution.[73–75] By far, the most abundant subunits are α_1, β_2 and γ_2. It has been estimated that 75%–80% of $GABA_A$ receptors in the brain contain the γ_2 subunit, and knockout of either the α_1 or β_2 gene causes a 50% reduction in $GABA_A$ receptors in mouse brain.[76] The α_2, α_3, α_4, α_5, α_6, β_1, γ_1 and δ subunits show a more regional distribution in brain, whereas the ρ subunits are expressed preferentially in the retina in a pattern that does not overlap with α subunits.[77,78] The ρ subunits, particularly ρ_1, are also present in a few discrete brain regions.

The distribution of $GABA_A$ receptor subunits is consistent with more specific evidence for formation of receptors from the following subunit combinations — $\alpha_1\beta_2\gamma_2$, $\alpha_4\beta_2\delta$, $\alpha_4\beta_3\delta$, $\alpha_6\beta_2\delta$ and $\alpha_6\beta_3\delta$.[48,72] There is also strong evidence for the formation of receptors from the γ_2 subunit, a specific α subunit and an unspecified β subunit — $\alpha_2\beta\gamma_2$, $\alpha_3\beta\gamma_2$, $\alpha_4\beta\gamma_2$, $\alpha_5\beta\gamma_2$ and $\alpha_6\beta\gamma_2$. Finally, there is strong evidence for the formation of receptors from ρ subunits. It is also highly likely that other subunit combinations form functional channels — $\alpha_1\beta_3\gamma_2$, $\alpha_1\beta\delta$, $\alpha_5\beta_3\gamma_2$, $\alpha\beta_1\gamma/\alpha\beta_1\delta$, $\alpha\beta$ and $\alpha_1\alpha_6\beta_3\gamma/\alpha_1\alpha_6\beta_3\delta$. No doubt, this list will steadily increase and be refined in terms of the unspecified subunits as time progresses.

Benzodiazepine modulation is expected to occur on receptors containing α and γ subunits because benzodiazepines bind at the interface between these two subunits. Among the subtypes verified so far, receptors containing the γ_2 and one of the α_1, α_2, α_3 or α_5 subunits ($\alpha_1\beta\gamma_2$, $\alpha_2\beta\gamma_2$, $\alpha_3\beta\gamma_2$ and $\alpha_5\beta\gamma_2$) exhibit a robust modulation by the more traditional benzodiazepines used in medicine.[72] Benzodiazepines exhibit no activity at receptors containing the α_4 or α_6 subunits[79,80] and reduced activity at γ_1- and γ_3-containing receptors.[72] Benzodiazepine modulation of receptors composed of $\alpha_1\beta\gamma_2$ subunits is slightly affected by the subtype of β subunit.[81]

Allosteric modulation of electrical responses of $GABA_A$ receptors

The effects of allosteric modulators on the current response elicited by GABA and GABA agonists can be studied in different ways depending on the goal of the experiment as described above in the section on nicotinic receptors. Figure 1.14 shows the effect of the benzodiazepine, diazepam (50 nM) on the current response of single $GABA_A$ receptors under voltage-clamp conditions.[55] In the absence of GABA, there is no measureable response, but upon the addition of GABA (2 μM),

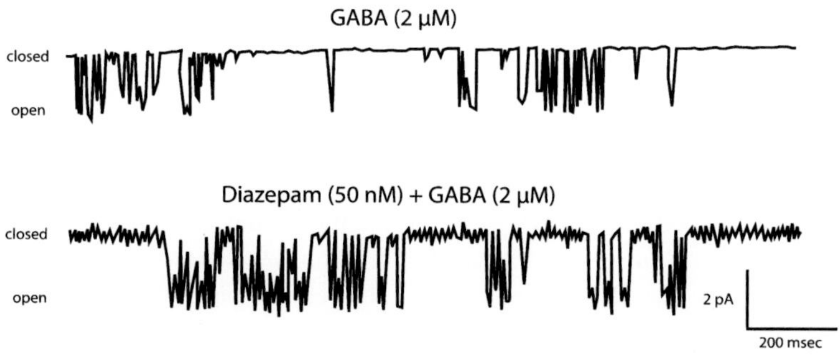

Figure 1.14: Effects of diazepam on single channel currents elicited by GABA from an outside-out patch of cultured mouse spinal cord neurons. The plot is a representation of data published by Rogers *et al.*[55]

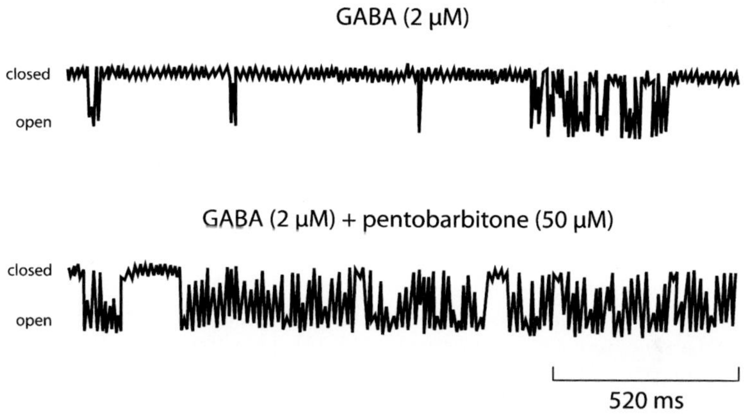

Figure 1.15: Effects of pentobarbitone on single channel currents elicited by GABA from an outside-out patch of cultured mouse spinal cord neurons. The plot is a representation of data published by MacDonald *et al.*[82]

spontaneous single channel currents occur. The addition of diazepam has no effect by itself, but causes an increase in the frequency of single channel openings elicited by GABA.

Figure 1.15 shows the effects of pentobarbitone on single channel openings elicited by GABA at $GABA_A$ receptors.[82] By itself, pentobarbitone elicits a concentration-dependent increase in channel opening similar to that caused by GABA. In the presence of a low concentration of pentobarbitone (50 μM) that has little effect by itself, the rate of channel opening by GABA is greatly increased.

Structural Correlates of Affinity and Efficacy at Ligand-Gated Ion Channels

We now see that acetylcholine binds to two sites on the extracellular domain of the nicotinic receptor to cause the five pore-lining M2 segments to disrupt their hydrophobic interaction near the center of the membrane spanning pore and allow a flux of positive cations. Each acetylcholine-binding site exists in two shapes, corresponding to the closed and open states of the channel pore. When the ion pore is closed, the inner β sheet of the ligand-binding site is rotated away from the outer β sheet in a counterclockwise direction. The binding of acetylcholine causes the inner β sheet to rotate in the opposite direction, bringing the two sheets closer together and opening the channel pore. Presumably a related mechanism exists for the $GABA_A$ receptor. Thus, our definition of the affinity of an orthosteric ligand for the receptor needs to account for these two shapes of the ligand-binding site, and our definition of efficacy needs to address how the differential affinity of a ligand for these two shapes is related to its ability to open the ion pore. In the next chapter, I explain these concepts in detail.

References

1. Thompson, A. J., Lester, H. A. and Lummis, S. C., The structural basis of function in Cys-loop receptors. *Q Rev Biophys* **43**, 449–499 (2010).
2. Ortells, M. O. and Lunt, G. G., Evolutionary history of the ligand-gated ion-channel superfamily of receptors. *Trends Neurosci* **18**, 121–127 (1995).
3. Hollmann, M. and Heinemann, S., Cloned glutamate receptors. *Annu Rev Neurosci* **17**, 31–108 (1994).
4. Keesey, J., How electric fish became sources of acetylcholine receptor. *J Hist Neurosci* **14**, 149–164 (2005).
5. Sheridan, M. N., The fine structure of the electric organ of Torpedo Marmorata. *J Cell Biol* **24**, 129–141 (1965).
6. Raftery, M. A., Hunkapiller, M. W., Strader, C. D. and Hood, L. E., Acetylcholine receptor: complex of homologous subunits. *Science* **208**, 1454–1456 (1980).
7. Noda, M. *et al.*, Structural homology of Torpedo californica acetylcholine receptor subunits. *Nature* **302**, 528–532 (1983).
8. Takahashi, T., Kuno, M., Mishina, M. and Numa, S., A physiological study on acetylcholine receptor expressed in Xenopus oocytes from cloned cDNAs. *J Physiol (Paris)* **80**, 229–232 (1985).
9. Brisson, A. and Unwin, P. N., Tubular crystals of acetylcholine receptor. *J Cell Biol* **99**, 1202–1211 (1984).
10. Toyoshima, C. and Unwin, N. Ion channel of acetylcholine receptor reconstructed from images of postsynaptic membranes. *Nature* **336**, 247–250 (1988).

11. Unwin, N., Nicotinic acetylcholine receptor at 9 Å resolution. *J Mol Biol* **229**, 1101–1124 (1993).
12. Unwin, N., Refined structure of the nicotinic acetylcholine receptor at 4 Å resolution. *J Mol Biol* **346**, 967–989 (2005).
13. Unwin, N., Acetylcholine receptor channel imaged in the open state. *Nature* **373**, 37–43 (1995).
14. Mishina, M. *et al.*, Molecular distinction between fetal and adult forms of muscle acetylcholine receptor. *Nature* **321**, 406–411 (1986).
15. Moore, H. P. and Raftery, M. A., Studies of reversible and irreversible interactions of an alkylating agonist with Torpedo californica acetylcholine receptor in membrane-bound and purified states. *Biochemistry* **18**, 1862–1867 (1979).
16. Dennis, M. *et al.*, Amino acids of the Torpedo marmorata acetylcholine receptor alpha subunit labeled by a photoaffinity ligand for the acetylcholine binding site. *Biochemistry* **27**, 2346–2357 (1988).
17. Middleton, R. E. and Cohen, J. B., Mapping of the acetylcholine binding site of the nicotinic acetylcholine receptor: [3H]nicotine as an agonist photoaffinity label. *Biochemistry* **30**, 6987–6997 (1991).
18. Tsai, M. C. *et al.*, Sites of action of phencyclidine. I. Effects on the electrical excitability and chemosensitive properties of the neuromuscular junction of skeletal muscle. *Mol Pharmacol* **18**, 159–166 (1980).
19. Koblin, D. D. and Lester, H. A., Voltage-dependent and voltage-independent blockade of acetylcholine receptors by local anesthetics in Electrophorus electroplaques. *Mol Pharmacol* **15**, 559–580 (1979).
20. Heidmann, T., Oswald, R. E. and Changeux, J. P., Multiple sites of action for noncompetitive blockers on acetylcholine receptor rich membrane fragments from torpedo marmorata. *Biochemistry* **22**, 3112–3127 (1983).
21. Heidmann, T. and Changeux, J. P., Characterization of the transient agonist-triggered state of the acetylcholine receptor rapidly labeled by the noncompetitive blocker [3H]chlorpromazine: additional evidence for the open channel conformation. *Biochemistry* **25**, 6109–6113 (1986).
22. Giraudat, J. *et al.*, Structure of the high-affinity binding site for noncompetitive blockers of the acetylcholine receptor: [3H]chlorpromazine labels homologous residues in the beta and delta chains. *Biochemistry* **26**, 2410–2418 (1987).
23. Giraudat, J., Dennis, M., Heidmann, T., Chang, J. Y. and Changeux, J. P., Structure of the high-affinity binding site for noncompetitive blockers of the acetylcholine receptor: serine-262 of the delta subunit is labeled by [3H]chlorpromazine. *Proc Natl Acad Sci USA* **83**, 2719–2723 (1986).
24. Revah, F. *et al.*, The noncompetitive blocker [3H]chlorpromazine labels three amino acids of the acetylcholine receptor gamma subunit: implications for the alpha-helical organization of regions MII and for the structure of the ion channel. *Proc Natl Acad Sci USA* **87**, 4675–4679 (1990).
25. Smit, A. B. *et al.*, A glia-derived acetylcholine-binding protein that modulates synaptic transmission. *Nature* **411**, 261–268 (2001).
26. Brejc, K. *et al.*, Crystal structure of an ACh-binding protein reveals the ligand-binding domain of nicotinic receptors. *Nature* **411**, 269–276 (2001).

27. Unwin, N., Miyazawa, A., Li, J. and Fujiyoshi, Y., Activation of the nicotinic acetylcholine receptor involves a switch in conformation of the alpha subunits. *J Mol Biol* **319**, 1165–1176 (2002).
28. Miyazawa, A., Fujiyoshi, Y. and Unwin, N., Structure and gating mechanism of the acetylcholine receptor pore. *Nature* **423**, 949–955 (2003).
29. Sine, S. M., Wang, H. L., Hansen, S. and Taylor, P., On the origin of ion selectivity in the Cys-loop receptor family. *J Mol Neurosci* **40**, 70–76 (2010).
30. Millar, N. S. and Gotti, C., Diversity of vertebrate nicotinic acetylcholine receptors. *Neuropharmacology* **56**, 237–246 (2009).
31. Albuquerque, E. X., Pereira, E. F., Alkondon, M. and Rogers, S. W., Mammalian nicotinic acetylcholine receptors: from structure to function. *Physiol Rev* **89**, 73–120 (2009).
32. Huang, L. Y., Catterall, W. A. and Ehrenstein, G., Selectivity of cations and nonelectrolytes for acetylcholine-activated channels in cultured muscle cells. *J Gen Physiol* **71**, 397–410 (1978).
33. Luetje, C. W. and Patrick, J., Both alpha- and beta-subunits contribute to the agonist sensitivity of neuronal nicotinic acetylcholine receptors. *J Neurosci* **11**, 837–845 (1991).
34. Akk, G. and Auerbach, A., Activation of muscle nicotinic acetylcholine receptor channels by nicotinic and muscarinic agonists. *Br J Pharmacol* **128**, 1467–1476 (1999).
35. Chang, C. C., Chen, T. F. and Lee, C. Y., Studies of the presynaptic effect of — bungarotoxin on neuromuscular transmission. *J Pharmacol Exp Ther* **184**, 339–345 (1973).
36. Dellisanti, C. D., Yao, Y., Stroud, J. C., Wang, Z. Z. and Chen, L., Crystal structure of the extracellular domain of nAChR alpha1 bound to alpha-bungarotoxin at 1.94 A resolution. *Nat Neurosci* **10**, 953–962 (2007).
37. McGehee, D. S. and Role, L. W., Physiological diversity of nicotinic acetylcholine receptors expressed by vertebrate neurons. *Annu Rev Physiol* **57**, 521–546 (1995).
38. Schoepfer, R. *et al.*, cDNA clones coding for the structural subunit of a chicken brain nicotinic acetylcholine receptor. *Neuron* **1**, 241–248 (1988).
39. Bertrand, D., Ballivet, M. and Rungger, D., Activation and blocking of neuronal nicotinic acetylcholine receptor reconstituted in Xenopus oocytes. *Proc Natl Acad Sci USA* **87**, 1993–1997 (1990).
40. Hamill, O. P., Marty, A., Neher, E., Sakmann, B. and Sigworth, F. J., Improved patch-clamp techniques for high-resolution current recording from cells and cell-free membrane patches. *Pflugers Arch* **391**, 85–100 (1981).
41. Macdonald, R. L. and Olsen, R. W., GABAA receptor channels. *Annu Rev Neurosci* **17**, 569–602 (1994).
42. Olsen, R. W. and Sieghart, W., GABA A receptors: subtypes provide diversity of function and pharmacology. *Neuropharmacology* **56**, 141–148 (2009).
43. Sakmann, B., Hamill, O. P. and Bormann, J., Patch-clamp measurements of elementary chloride currents activated by the putative inhibitory transmitter GABA and glycine in mammalian spinal neurons. *J Neural Transm Suppl* **18**, 83–95 (1983).
44. Macdonald, R. L., Rogers, C. J. and Twyman, R. E., Kinetic properties of the GABAA receptor main conductance state of mouse spinal cord neurones in culture. *J Physiol* **410**, 479–499 (1989).
45. Rivera, C. *et al.*, The K+/Cl− co-transporter KCC2 renders GABA hyperpolarizing during neuronal maturation. *Nature* **397**, 251–255 (1999).

46. Ben-Ari, Y., Excitatory actions of gaba during development: the nature of the nurture. *Nat Rev Neurosci* **3**, 728–739 (2002).
47. Simon, J., Wakimoto, H., Fujita, N., Lalande, M. and Barnard, E. A., Analysis of the set of GABA(A) receptor genes in the human genome. *J Biol Chem* **279**, 41422–41435 (2004).
48. Olsen, R. W. and Sieghart, W., International Union of Pharmacology. LXX. Subtypes of gamma-aminobutyric acid (A) receptors: classification on the basis of subunit composition, pharmacology, and function. Update. *Pharmacol Rev* **60**, 243–260 (2008).
49. Sigel, E., Baur, R., Trube, G., Mohler, H. and Malherbe, P., The effect of subunit composition of rat brain GABAA receptors on channel function. *Neuron* **5**, 703–711 (1990).
50. Sigel, E., Baur, R., Kellenberger, S. and Malherbe, P., Point mutations affecting antagonist affinity and agonist dependent gating of GABAA receptor channels. *Embo J* **11**, 2017–2023 (1992).
51. Smith, G. B. and Olsen, R. W., Identification of a [3H]muscimol photoaffinity substrate in the bovine gamma-aminobutyric acid A receptor alpha subunit. *J Biol Chem* **269**, 20380–20387 (1994).
52. Amin, J. and Weiss, D. S., GABAA receptor needs two homologous domains of the beta-subunit for activation by GABA but not by pentobarbital. *Nature* **366**, 565–569 (1993).
53. Sigel, E. and Baur, R., Allosteric modulation by benzodiazepine receptor ligands of the GABAA receptor channel expressed in Xenopus oocytes. *J Neurosci* **8**, 289–295 (1988).
54. Maksay, G., Thompson, S. A. and Wafford, K. A. Allosteric modulators affect the efficacy of partial agonists for recombinant GABA(A) receptors. *British J Pharmacol* **129**, 1794–1800 (2000).
55. Rogers, C. J., Twyman, R. E. and Macdonald, R. L. Benzodiazepine and beta-carboline regulation of single GABAA receptor channels of mouse spinal neurones in culture. *J Physiol* **475**, 69–82 (1994).
56. Mohler, H. and Okada, T. Benzodiazepine receptor: demonstration in the central nervous system. *Science* **198**, 849–851 (1977).
57. Squires, R. F. and Brastrup, C. Benzodiazepine receptors in rat brain. *Nature* **266**, 732–734 (1977).
58. Battersby, M. K., Richards, J. G. and Mohler, H. Benzodiazepine receptor: photoaffinity labeling and localization. *Eur J Pharmacol* **57**, 277–278 (1979).
59. Oakley, N. R. and Jones, B. J. The proconvulsant and diazepam-reversing effects of ethyl-beta-carboline-3-carboxylate. *Eur J Pharmacol* **68**, 381–382 (1980).
60. Nutt, D. J. and Lister, R. G. The effect of the imidazodiazepine Ro 15-4513 on the anticonvulsant effects of diazepam, sodium pentobarbital and ethanol. *Brain Res* **413**, 193–196 (1987).
61. Hunkeler, W. *et al.* Selective antagonists of benzodiazepines. *Nature* **290**, 514–516 (1981).
62. Ehlert, F. J. Inverse agonists, cooperativity and drug action at benzodiazepine receptors. *Trends in Pharmacol Sci* **7**, 28–32 (1986).
63. Duncalfe, L. L., Carpenter, M. R., Smillie, L. B., Martin, I. L. and Dunn, S. M. The major site of photoaffinity labeling of the gamma-aminobutyric acid type A receptor by [3H]flunitrazepam is histidine 102 of the alpha subunit. *J Biol Chem* **271**, 9209–9214 (1996).

64. Smith, G. B. and Olsen, R. W. Deduction of amino acid residues in the GABA(A) receptor alpha subunits photoaffinity labeled with the benzodiazepine flunitrazepam. *Neuropharmacology* **39**, 55–64 (2000).
65. Sawyer, G. W., Chiara, D. C., Olsen, R. W. and Cohen, J. B. Identification of the bovine gamma-aminobutyric acid type A receptor alpha subunit residues photolabeled by the imidazobenzodiazepine [3H]Ro15-4513. *J Biol Chem* **277**, 50036–50045 (2002).
66. Buhr, A., Schaerer, M. T., Baur, R. and Sigel, E. Residues at positions 206 and 209 of the alpha1 subunit of gamma-aminobutyric Acid A receptors influence affinities for benzodiazepine binding site ligands. *Mol Pharmacol* **52**, 676–682 (1997).
67. Berezhnoy, D. *et al.* On the benzodiazepine binding pocket in GABAA receptors. *J Biol Chem* **279**, 3160–3168 (2004).
68. Buhr, A., Baur, R. and Sigel, E. Subtle changes in residue 77 of the gamma subunit of alpha1beta2gamma2 GABAA receptors drastically alter the affinity for ligands of the benzodiazepine binding site. *J Biol Chem* **272**, 11799–11804 (1997).
69. Li, G. D. *et al.* Identification of a GABAA receptor anesthetic binding site at subunit interfaces by photolabeling with an etomidate analog. *J Neurosci: The official J Soc Neurosci* **26**, 11599–11605 (2006).
70. Li, G. D., Chiara, D. C., Cohen, J. B. and Olsen, R. W. Numerous classes of general anesthetics inhibit etomidate binding to gamma-aminobutyric acid type A (GABAA) receptors. *J Biol Chem* **285**, 8615–8620 (2010).
71. Olsen, R. W. and Li, G. D. GABA(A) receptors as molecular targets of general anesthetics: identification of binding sites provides clues to allosteric modulation. *Can J Anaesth; J Can Anesth (French)* **58**, 206–215 (2011).
72. Sieghart, W. Structure and pharmacology of gamma-aminobutyric acid A receptor subtypes. *Pharmacol Rev* **47**, 181–234 (1995).
73. Laurie, D. J., Seeburg, P. H. and Wisden, W. The distribution of 13 GABAA receptor subunit mRNAs in the rat brain. II. Olfactory bulb and cerebellum. *J Neurosci* **12**, 1063–1076 (1992).
74. Wisden, W., Laurie, D. J., Monyer, H. and Seeburg, P. H. The distribution of 13 GABAA receptor subunit mRNAs in the rat brain. I. Telencephalon, diencephalon, mesencephalon. *J Neurosci* **12**, 1040–1062 (1992).
75. Persohn, E., Malherbe, P. and Richards, J. G. Comparative molecular neuroanatomy of cloned GABAA receptor subunits in the rat CNS. *J Comp Neurol* **326**, 193–216 (1992).
76. Sur, C. *et al.* Loss of the major GABA(A) receptor subtype in the brain is not lethal in mice. *J Neurosci* **21**, 3409–3418 (2001).
77. Enz, R. and Cutting, G. R. GABAC receptor rho subunits are heterogeneously expressed in the human CNS and form homo- and heterooligomers with distinct physical properties. *Eur J Neurosci* **11**, 41–50 (1999).
78. Greka, A., Lipton, S. A. and Zhang, D. Expression of GABA(C) receptor rho1 and rho2 subunits during development of the mouse retina. *Eur J Neurosci* **12**, 3575–3582 (2000).
79. Wieland, H. A. and Luddens, H. Four amino acid exchanges convert a diazepam-insensitive, inverse agonist-preferring GABAA receptor into a diazepam-preferring GABAA receptor. *J Med Chem* **37**, 4576–4580 (1994).
80. Wieland, H. A., Luddens, H. and Seeburg, P. H. A single histidine in GABAA receptors is essential for benzodiazepine agonist binding. *J Biol Chem* **267**, 1426–1429 (1992).

81. Hadingham, K. L. *et al.* Role of the beta subunit in determining the pharmacology of human gamma-aminobutyric acid type A receptors. *Mol Pharmacol* **44**, 1211–1218 (1993).
82. MacDonald, R. L., Rogers, C. J. and Twyman, R. E. Barbiturate regulation of kinetic properties of the GABAA receptor channel of mouse spinal neurones in culture. *J Physiol* **417**, 483–500 (1989).
83. Celie, P. H. *et al.* Nicotine and carbamylcholine binding to nicotinic acetylcholine receptors as studied in AChBP crystal structures. *Neuron* **41**, 907–914 (2004).

Affinity and Efficacy of Orthosteric Ligands at Ligand-Gated Ion Channels

2

Ligand-gated ion channels provide a unique opportunity to study affinity and efficacy. This is because the receptor *is* an ion channel, and its activation state can be measured directly using electrophysiological methods. The measured current can be analyzed to determine how occupancy of a single receptor increases the probability of channel opening or how occupancy of the receptor population causes a graded increase in the current response of the cell. From this approach, we should be able to develop a quantitative measure of affinity and efficacy.

The process of measuring affinity and efficacy has progressed along different paths for scientists investigating ligand-gated ion channels and *G* protein-coupled receptors because of inherent differences in the receptors and the methods used to study them. Equipped with sensitive recording techniques and evidence that channel opening requires the binding of two molecules of neurotransmitter, electrophysiologists developed a two-site cooperative model for the muscle-type nicotinic acetylcholine receptor.[1] In contrast, scientists studying *G* protein-coupled receptors developed a method for inferring ligand-receptor occupancy based on responses measured downstream in the signaling pathway[2] because it was unfeasible to measure receptor activation directly (see Chapter 7). A conservative one-site model was used in the analysis because the number of binding sites per receptor was unknown at the time. Now, there is clear evidence that some *G* protein-coupled receptors of the rhodopsin family behave as monomers[3] regardless of their oligomeric association with adjacent receptor neighbors, suggesting that the simple one-site model is appropriate for measuring the function of many types of *G* protein-coupled receptors.

While the models for drug action at ligand-gated ion channels and *G* protein-coupled receptors are different, the fundamental principles of affinity and efficacy are similar for both receptor types. This congruence would be obvious in the case of a ligand-gated ion channel having only one binding-site per receptor complex. It is also easier to describe and understand how two linked binding sites interact with one another if the behavior of a single-site receptor is understood. Hence, I begin this chapter with a description of such a hypothetical ligand-gated ion channel and address the properties of affinity and efficacy at both the level of single receptors and the receptor population. Following this hypothetical example, I proceed with the more realistic case of two binding sites per channel, as represented by the muscle-type nicotinic acetylcholine and $GABA_A$ receptors. My hope is that by using this approach, the similarity in the principles of affinity and efficacy for ligand-gated ion channels and *G* protein-coupled receptor will be more apparent.

Hypothetical One-Site Model

Population analysis

I begin by considering how the ligand (D) binds to and activates a population of receptors (R). The simplest model to represent this interaction incorporates a single ligand-binding site per channel as shown in Figure 2.1a. This model gives rise to four types of receptor complexes defined by the state of the channel (open or closed) and the ligand-binding site (occupied or unoccupied). Although the receptor may make hundreds of transitions between the active and inactive states when occupied by an agonist for a few seconds, the lifetime of the functional receptor can be attributed almost entirely to its time spent in either the open (R_s^*, DR_s^*) or closed (R_s, DR_s) states. The time spent in transition states is very brief, and these are denoted by the double arrows ($\leftrightarrow$) in Figure 2.1. The affinity constants of the agonist for the open and closed states of the receptor are denoted by K_b and K_a, respectively, and are defined in the usual manner:

$$K_b = \frac{[DR_s^*]}{[D][R_s^*]}, \tag{2.1}$$

$$K_a = \frac{[DR_s]}{[D][R_s]}. \tag{2.2}$$

These two constants have units of inverse molarity (M^{-1}) and are known as the microscopic affinity constants for the open (K_b) and closed (K_a) states of the receptor.

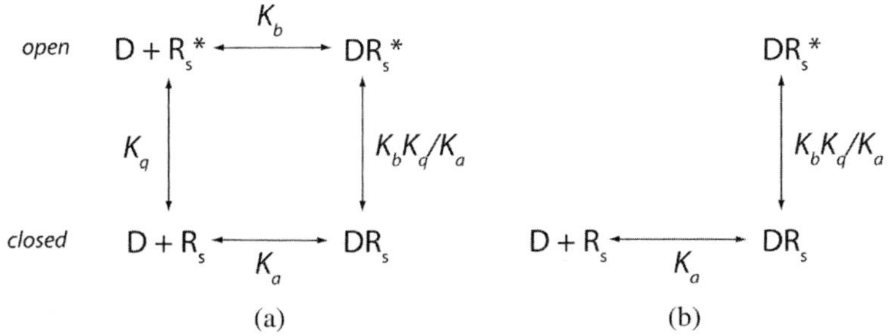

Figure 2.1: Simple one-site model for a hypothetical ligand-gated ion channel. (a) The interaction of a ligand with its binding site on the open (R_s^*) and closed (R_s) states of the channel. (b) The component of the model in (a) that represents the process of conformational induction.

The square brackets denote molar concentration units. The constant describing the isomerization of the free receptor (K_q) is defined as:

$$K_q = \frac{[R_s^*]}{[R_s]}. \tag{2.3}$$

The three constants, K_b, K_a and K_q, represent fundamental properties of the ligand-receptor complex.

Because the receptor shown in Figure 2.1 exhibits only reversible transitions, with no input of energy or the breaking of covalent bonds, it must achieve equilibrium in time, provided that the concentration of drug is constant. That is, at equilibrium, there is no net flux across each of the four transitions, nor cycling around the system. The behavior of each receptor is random so it is possible that some receptors might actually go through a complete cycle of transitions (i.e., $R \rightarrow DR_s \rightarrow DR_s^* \rightarrow R_s^* \rightarrow R$) at equilibrium even though the collective behavior of all of the receptors in the population at equilibrium exhibits no net flux. Ultimately, this means that any one of the four equilibrium constants can be described in terms of the other three.[4] Given the uniqueness of the three constants, K_b, K_a and K_q, the fourth equilibrium constant describing the induction of the active state ($DR_s \leftrightarrow DR_s^*$) has been defined as:

$$K_g = \frac{K_b K_q}{K_a} = \frac{[DR_s^*]}{[DR_s]}. \tag{2.4}$$

Equation (2.4) can be checked by substituting the expressions for K_b, K_a and K_q from Equations (2.1)–(2.3). The result simplifies to the equality, $DR_s^*/DR_s = DR_s^*/DR_s$, which proves the relationship. This property is commonly known as the "law of microscopic reversibility".

There are two processes by which an agonist can increase the amount of receptors in the open state. The agonist can bind to the closed state and cause it to open by a process known as "conformational induction":

$$D + R_s \overset{K_a}{\longleftrightarrow} DR_s \overset{K_b K_q / K_a}{\longleftrightarrow} DR_s^*, \tag{2.5}$$

or it can wait for the receptor to open spontaneously and then bind selectively to the open state by a process known as "conformational selection":

$$D + R_s \overset{K_q}{\longleftrightarrow} D + R_s^* \overset{K_b}{\longleftrightarrow} DR_s^*, \tag{2.6}$$

either mechanism yields the same behavior at equilibrium. Given that most ligand-gated ion channels evolved to be closed in the absence of ligand, conformational induction is probably the most common mechanism for channel opening. Nonetheless, both processes require some spontaneous formation of the active receptor complex in the absence of ligand (i.e., $K_q > 0$). Castillo and Katz[5] were the first to apply the conformational induction model to ligand-gated ion channels.

Conformational induction may be the only possible mechanism for glycine to activate its receptor. There is evidence that glycine becomes trapped on its receptor during channel opening, perhaps due to the closure of the subunit interfaces around glycine.[6] Hence, the direct binding of glycine to an unoccupied open channel may be impossible. The analogous condition for our simple model is represented by the scheme in Figure 2.1b. This model does not imply that glycine has infinite affinity for the active state (i.e., $K_b = \infty$), for if it did, it would be impossible to wash glycine off the receptor. One could ask, what value could be assigned to K_b in the conformational selection process described by scheme 2.6 that would yield the same amount of DR_s^* as the conformational induction process shown in Figure 2.1b? The constant K_b in Figure 2.1a is intended to represent this value. Another way to consider this question begins with the assumption that the expression $K_b K_q / K_a$ accurately describes the induction or gating of the active drug receptor complex ($DR_s \leftrightarrow DR_s^*$) in Figure 2.1b. The value of K_b in this expression is equivalent to the hypothetical microscopic affinity constant for the active state as defined in Figure 2.1a. Thus, the scheme in Figure 2.1a enables us to address the affinity of the ligand for the open state quantitatively even though the ligand may not bind directly to the open state by a process of conformational selection.

Derivation of equilibrium equations

Having defined the model for ligand interactions, we can now derive equations for observed affinity (K_{obs}), observed efficacy (ε) and constitutive activity. We can also derive functions for receptor occupancy, the fraction of the population of ligand-receptor complexes in the active state, the fraction of the free receptor population in the active state (constitutive activity), and the total fraction of receptor population in the active state as a function of the ligand concentration. This will be done for equilibrium conditions because many agonists equilibrate rapidly with ligand-gated ion channels.

To begin, we express each receptor species in terms of equilibrium constants, the free receptor concentration, and if necessary, the ligand concentration using the definitions of the various constants described above (i.e., Equations (2.1)–(2.3)). Thus,

$$[R_s^*] = K_q[R_s] \tag{2.7}$$

$$[DR_s] = DK_a[R_s] \tag{2.8}$$

$$[DR_s^*] = [D]K_bK_q[R_s]. \tag{2.9}$$

Then, equations are derived for expressing the different types of receptor complexes (e.g., $[DR_s^*]$) or sums of receptor complexes ($[DR_s] + [DR_s^*]$) as a function of the agonist concentration using Equations (2.7)–(2.9). Some of these equations are used to define other parameters, like K_{obs} and ε.

The derivation of the equation for fractional receptor occupancy (occupancy) begins with the expression:

$$\text{occupancy} = \frac{[DR_s] + [DR_s^*]}{[R_T]} = \frac{[DR_s] + [DR_s^*]}{[DR_s] + [DR_s^*] + [R_s^*] + [R_s]} \tag{2.10}$$

in which $[R_T]$ denotes the total receptor concentration. Substituting Equations (2.7)–(2.9) for R_s^*, DR_s and DR_s^*, respectively, and simplifying yields:

$$\text{occupancy} = \frac{[D]K_a + [D]K_bK_q}{[D]K_a + [D]K_bK_q + K_q + 1} \tag{2.11}$$

This equation can be rearranged into the following form,

$$\text{occupancy} = \frac{[D]}{[D] + \frac{1}{K_{\text{obs}}}} \tag{2.12}$$

in which,

$$K_{\text{obs}} = \frac{K_qK_b + K_a}{1 + K_q}. \tag{2.13}$$

The K_{obs} value is known as the observed affinity constant of the ligand, and it has units of inverse concentration (e.g., M^{-1}). The reciprocal of K_{obs} is known as the observed dissociation constant (K_D), and its value is equivalent to the concentration of ligand required for half-maximal receptor occupancy.

This is a good time to raise the point that Equation (2.12) represents a description of drug-receptor interactions at the population level of analysis without concern for what state the receptor is in:

$$D + R \overset{K_{obs}}{\longleftrightarrow} DR. \tag{2.14}$$

In this scheme, R and DR each represent the summation of the two states of the free ($R_s + R_s^*$) and bound ($DR_s + DR_s^*$) forms of the receptor. This is the more common way of analyzing drug-receptor interactions (level 2, population analysis), and it is entirely consistent with the more detailed level of analysis that addresses receptor states (level 3). A more complete description of the relationship between the microscopic constants (K_b and K_a) and K_{obs} is given below.

Getting back to the derivation of equations, we define the function for the amount of ligand-receptor complex in the active state by:

$$\frac{[DR_s^*]}{[R_T]} = \frac{[DR_s^*]}{[DR_s] + [DR_s^*] + [R_s^*] + [R_s]}. \tag{2.15}$$

Substitution of the appropriate equations for the receptor complexes and simplification yields:

$$\frac{[DR_s^*]}{[R_T]} = \frac{[D]K_bK_q}{[D]K_a + [D]K_bK_q + K_b + 1}. \tag{2.16}$$

Further simplification yields:

$$\frac{[DR_s^*]}{[R_T]} = \frac{1}{1 + \frac{[D]K_a + K_q + 1}{[D]K_bK_q}}. \tag{2.17}$$

At the population level of analysis, observed intrinsic efficacy (ε) is defined as the fraction of the population of ligand-receptor complexes in the active state. For the simple one-site model shown in Figure 2.1, this fraction is constant for any level of receptor occupancy. When fractional receptor occupancy is at its maximum (i.e., 1), then the fractional amount of active receptor complex is equivalent to observed intrinsic efficacy. Thus, the limit of Equation (2.17) as D approaches infinity yields a value of receptor activation that is equivalent to observed efficacy:

$$\varepsilon = \lim_{D \to \infty} \frac{1}{1 + \frac{[D]K_a + K_q + 1}{[D]K_b K_q}}. \tag{2.18}$$

Taking this limit yields:

$$\varepsilon = \frac{1}{1 + \frac{K_a}{K_b K_q}}. \tag{2.19}$$

The function for the amount of free receptor complex in the active state is given by:

$$\frac{[R_s^*]}{[R_T]} = \frac{[R_s^*]}{[DR_s] + [DR_s^*] + [R_s^*] + [R_s]}. \tag{2.20}$$

Substituting the appropriate equations for the receptor complexes and simplifying yields:

$$\frac{[R_s^*]}{[R_T]} = \frac{K_q}{[D]K_b K_q + [D]K_a + K_q + 1}. \tag{2.21}$$

The fractional amount of the free receptor complex in the active state (R_s^*) in the absence of agonist is defined as constitutive activity, and its value can be determined by substituting 0 for $[D]$ in Equation (2.21):

$$\varepsilon_{\text{sys}} = \frac{K_q}{K_q + 1}. \tag{2.22}$$

In this equation, the efficacy of the system (ε_{sys}) represents the fractional amount of the receptor population in the active state in the absence of ligand.

Finally, the derivation of the function for the total amount of receptor in the active state begins with:

$$\frac{[R_s^*] + [DR_s^*]}{[R_T]} = \frac{[R_s^*] + [DR_s^*]}{[DR_s] + [DR_s^*] + [R_s] + [R_s^*]}. \tag{2.23}$$

Substituting the appropriate equations for the receptor complexes and simplifying yields:

$$\frac{[R_s^*] + [DR_s^*]}{[R_T]} = \frac{[D]K_b K_q + K_q}{[D]K_b K_q + [D]K_a + K_q + 1}. \tag{2.24}$$

Further simplification yields:

$$\frac{[R_s^*] + [DR_s^*]}{[R_T]} = \frac{1}{1 + \frac{[D]K_a + 1}{K_q([D]K_b + 1)}}. \tag{2.25}$$

We will use these equations to estimate observed affinity (K_{obs}) and efficacy (ε) of the agonist from its concentration-response curve for eliciting a current through a population of receptors.

Measurement of ligand-gated ion channel currents

The current response of a population of receptors can be measured in a whole-cell voltage-clamp experiment as illustrated in Figures 2.2a and 2.2b. A patch clamp electrode is applied to the surface of the plasma membrane, and a small suction is applied through the recording electrode so that it firmly attaches to the cell and forms a high resistance seal (gigaseal). Then a greater suction is applied to remove a small patch of membrane circumscribed by the recording electrode. This allows the inside of the recording electrode to communicate with the intracellular fluid and

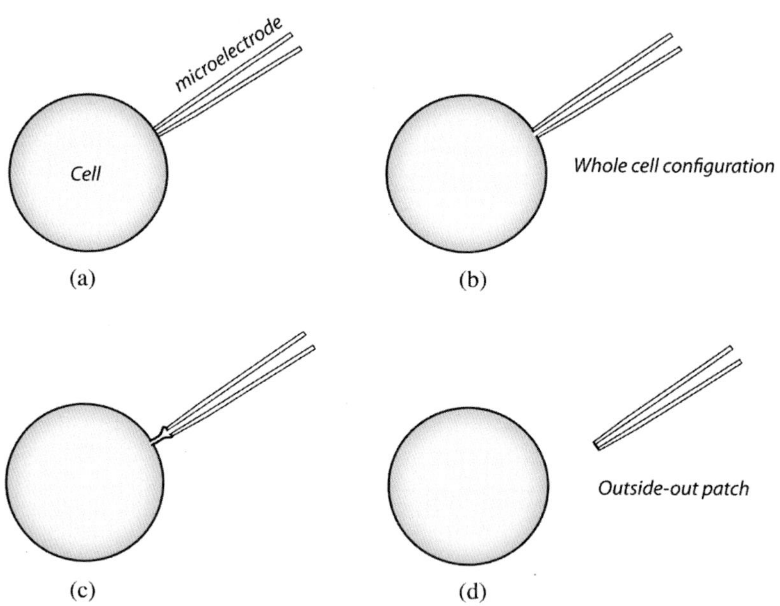

Figure 2.2: Process for establishing whole-cell and outside-out patch configurations for measuring ligand-gated ion channel currents. (a) Initially, a high resistance gigaseal is established between the recording electrode and the cell by applying a small suction. (b) A greater suction is subsequently applied to remove a patch of membrane and establish the whole-cell recording mode. (c) Alternatively, after establishing a gigaseal as shown in (a), the electrode is withdrawn, and ultimately, the connection between the cell and electrode is broken resulting in the outside-out patch configuration (d).

enable the potential difference across the plasma membrane to be measured. An agonist can be applied to the bathing medium, and the current elicited by a population of ligand-gated ion channels expressed on the cell can be measured under voltage clamp conditions. This is done by holding the membrane potential at a target value, like that approximating the resting membrane potential (−70 mV). Any deviation from −70 mV is detected by the recording electrode, and a feedback amplifier applies current to the cell to offset the difference. When an agonist is applied, it binds to the ligand-gated ion channels and reaches equilibrium quickly. The agonist-induced current tends to change the membrane potential. This potential change is quickly detected by the feedback amplifier, and a current is applied to the cell to maintain the membrane voltage at −70 mV. The applied current is equal but opposite in sign to that flowing through the ion channels. It is automatically recorded as the response to the agonist. Because the current response is measured under constant voltage, the electrochemical gradient is constant throughout the measurement. Various concentrations of agonist can be applied to the cell to obtain the requisite data for a concentration-response curve (Figures 2.3a and 2.3b). These data are then

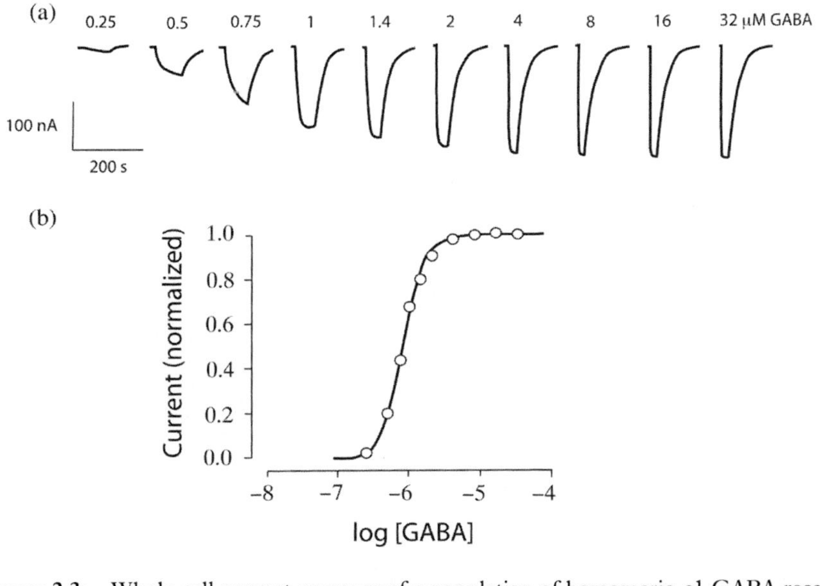

Figure 2.3: Whole-cell current response of a population of homomeric $\rho 1$ GABA receptors expressed in Xenopus oocytes. (a) Whole cell current response elicited by increasing concentration of GABA. (b) The concentration-response curve represents the peak current response to GABA plotted against the log molar concentration of GABA.

Sources: The illustration has been redrawn from Figure 1 of Chang and Weiss.[28]

analyzed as described in the next section to estimate the affinity and efficacy of the ligand-receptor complex.

Estimation of observed affinity and efficacy

If our hypothetical one-site ligand-gated ion channel does not desensitize rapidly, the observed affinity and relative efficacy of an agonist for the receptor could be determined by analysis of its concentration-response curve for activating the receptor population. The basis for this method is illustrated in Figure 2.4a for a receptor lacking constitutive activity. The plot shows total receptor occupancy ($DR_s + DR_s^*$) and the fraction of the agonist-receptor complex in the active state (DR_s^*) as a function of the agonist concentration. The theoretical curves for occupancy and the active state of the agonist-receptor complex were generated using Equations (2.11) and (2.17), respectively, assuming very little constitutive activity ($K_q = 0.001$) and log K_b and log K_a values of 9.83 and 6.52, respectively. Agonist receptor occupancy increases with the agonist concentration in a manner consistent with a one-site model (i.e., Equation (2.11)). That is, half-maximal receptor occupancy occurs at an agonist concentration equivalent to the reciprocal of K_{obs}. The component of the agonist occupancy curve representing the active state of the receptor also resembles a one-site model having the same K_{obs} value as the occupancy curve (i.e., log $K_{\mathrm{obs}} = 7.0$). The fraction of the population of agonist-receptor complexes in the active state is defined as observed efficacy (ε), and it can be estimated from the maximum of the agonist concentration-response curve for receptor activation as shown in the plot. A related parameter termed inactivity (ε_i)

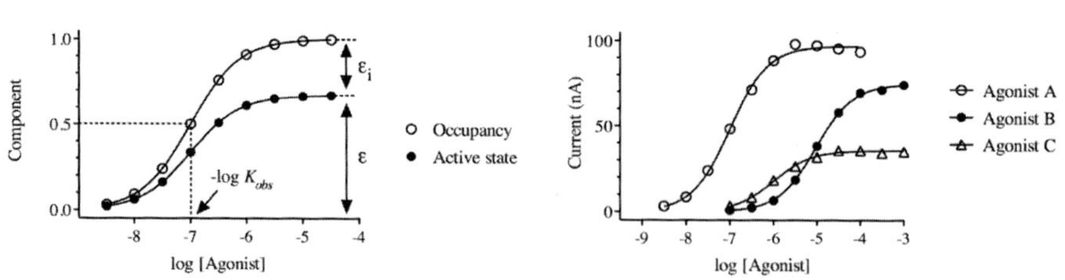

(a) (b)

Figure 2.4: Behavior of the simple one-site model illustrated in Figure 2.1 as a function of the agonist concentration. (a) Relationship between receptor occupancy and the fraction of the population of ligand-receptor complexes in the active (ε) and inactive (ε_i) states. (b) Receptor activation functions of three agonists have different affinities ($A > C > B$) and efficacies ($A > B > C$).

represents the fraction of the agonist-receptor complex in the inactive state, and it is defined as:

$$\varepsilon_i = 1 - \varepsilon \tag{2.26}$$

The method for estimating K_{obs} and ε rests on the assumption that both half-maximal receptor occupancy and activation occur at the same concentration of agonist (i.e., $1/K_{\text{obs}}$) as proven graphically in Figure 2.4a. This useful relationship can also be proven analytically using the following rationale. If we solve the receptor activation function (Equation (2.17)) for D (agonist concentration) when the value of the function is half-maximal, then the result should be equivalent to the reciprocal of K_{obs}. The following equation illustrates this reasoning:

$$\frac{1}{2} \times \frac{1}{1 + \frac{K_a}{K_b K_q}} = \frac{1}{1 + \frac{[EC_{50}]K_a + K_q + 1}{[EC_{50}]K_b K_q}}. \tag{2.27}$$

The right-hand side of this equation represents receptor activation with the independent variable D replaced with EC_{50} (Equation (2.17)) and the left-hand side represents half of its maximal value (i.e., the product of $\frac{1}{2}$ and ε, Equation (2.19)). Cross multiplying and simplifying yields:

$$\frac{1 + K_q}{K_b K_q + K_a} = EC_{50}. \tag{2.28}$$

The left-hand side of this equation is equivalent to $1/K_{\text{obs}}$ as shown by Equation (2.13). Thus, Equation (2.26) can be rearranged into the form:

$$\frac{1}{K_{\text{obs}}} = EC_{50}. \tag{2.29}$$

Thus, it follows that half-maximal receptor activation occurs when the concentration of agonist is equivalent to the reciprocal of the K_{obs}. It follows that the receptor activation function can be expressed as the product of receptor occupancy (Equation (2.11)) and the maximum of the receptor activation function (ε):

$$\frac{[DR_s^*]}{[R_T]} = \frac{\varepsilon[D]}{[D] + \frac{1}{K_{\text{obs}}}}. \tag{2.30}$$

This function is equivalent to that described by Stephenson's[2] and Furchgott's[7] equations for agonist-induced receptor activation:

$$\text{stimulus} = \frac{\varepsilon[D]}{[D] + K_D}. \tag{2.31}$$

These investigators referred to receptor activation as the stimulus and used the dissociation constant (K_D) in their equation instead of the affinity constant, K_{obs}.

It follows that the observed affinity (K_{obs}) and relative efficacy of an agonist for our hypothetical one-site, ligand-gated ion channel can be estimated from its concentration-response curve for eliciting currents in a cell expressing a population of these receptors. The plot in Figure 2.4b shows theoretical examples of these curves for three agonists (A–C). The difference between the plots in Figure 2.4b and the theoretical receptor activation plot in Figure 2.4a is that the latter represents the proportion of receptors in the active state, whereas the former represents the actual current response. The maximum response of each agonist (I_{max}) has units of current (e.g., nA), and is not equivalent to the parameter, ε, which has no units. Nonetheless, the I_{max} of each agonist is proportional to its ε value, and therefore, it can be used to estimate a relative measure of ε. A rough value of K_{obs} can be obtained by interpolating the molar concentration of the agonist that yields a half-maximal current response (EC_{50}). The reciprocal of EC_{50} is equivalent to K_{obs} ($1/EC_{50} = K_{obs}$). Similarly, an estimate of I_{max} can be obtained by extrapolating the maximal current response.

A more unbiased approach is to fit the following equation to the data by non-linear regression analysis:

$$i = \frac{[D]i_{max}}{[D] + \frac{1}{K_{obs}}}, \tag{2.32}$$

in which, i denotes the current response and i_{max} the maximum current response to the agonist. This equation is equivalent to the simple one-site model (Equation (2.11)) except that the measured, independent variable is current (i), and the right side of the equation has been multiplied by i_{max}. Regression analysis yielded log K_{obs} values of 6.99, 5.03 and 5.99 and i_{max} values of 97, 75 and 36 nA for agonists A, B and C, respectively. The relationship between i_{max} and ε is given by the following equation:

$$\varepsilon = \frac{i_{max}}{i_{max-sys}} \tag{2.33}$$

in which, $i_{max\text{-}sys}$ represents the current when all of the ligand-gated ion channels in the population are open. There is no simple way to estimate $i_{max\text{-}sys}$ from the whole cell current response. If a group of highly efficacious agonists all exhibit the same i_{max} value, then it would seem that their i_{max} values are approximately equal to $i_{max\text{-}sys}$. We can estimate the relative efficacies of the agonists as shown by the relationship:

$$\frac{\varepsilon_{B}}{\varepsilon_{A}} = \frac{\frac{i_{max-B}}{i_{max-sys}}}{\frac{i_{max-A}}{i_{max-sys}}} = \frac{i_{max-B}}{i_{max-A}} \tag{2.34}$$

In this equation, the subscripts A and B are used to denote the ε and I_{max} values of agonists A and B, respectively. Equation (2.34) simply shows that the efficacy of one agonist, expressed relative to that of another, is equivalent to the corresponding ratio of i_{max} values. When expressed relative to that of agonist A, the relative efficacy values of the agonists B and C are 0.77 and 0.37, respectively.

We can also consider affinity and efficacy for ligand-gated ion channels that exhibit constitutive activity. This may seem like an academic exercise because most native ligand-gated ion channels lack constitutive activity. Nonetheless, scientists often introduce constitutively activating point mutations into ligand-gated ion channels for a variety of reasons, so the methods for analyzing these mutants are useful. Figure 2.5a shows such a plot for a hypothetical receptor exhibiting an unusually high level constitutive activity. The latter was adjusted to a high value so that it would be easier to visualize the relationship between the population parameters and the various receptor components more easily. The various components of the plot ($[DR_s^*]$, $[R_s^*] + [DR_s^*$ and $R_s^*]$) were simulated using Equations (2.17), (2.21) and (2.25), respectively.

The plot in Figure 2.5a shows that total receptor occupancy ($DR_s + DR_s^*$) of a constitutively active receptor increases with the agonist concentration in a manner consistent with a simple one-site model. The fractional amount of agonist-receptor complex in the active state (DR_s^*) is equivalent to the observed efficacy of the agonist, and its value can be estimated from the maximum of the receptor activation function. In the absence of agonist, the fractional amount of receptor in the active state is equivalent to the observed efficacy of the system (ε_{sys}). Increasing concentrations of agonist cause an increase in receptor occupancy, and hence, a decrease in both the free active (R_s^*; constitutive activity) and inactive (R) states of the receptor. The total amount of receptor in the active state ($R_s^* + DR_s^*$) increases from a low value of ε_{sys} in the absence of agonist, to a high value of ε in the presence of a maximally effective concentration of agonist. Two additional parameters are equivalent to the fraction of the receptor population in the free (R_s) and ligand-occupied receptor (DR_s) inactive states, and are denoted with the symbols, ε_{i-sys} (inactivity of the system) and ε_i, respectively. The relationship between ε_{sys} and ε_{i-sys} is given by:

$$\varepsilon_{i-sys} = 1 - \varepsilon_{sys}. \tag{2.35}$$

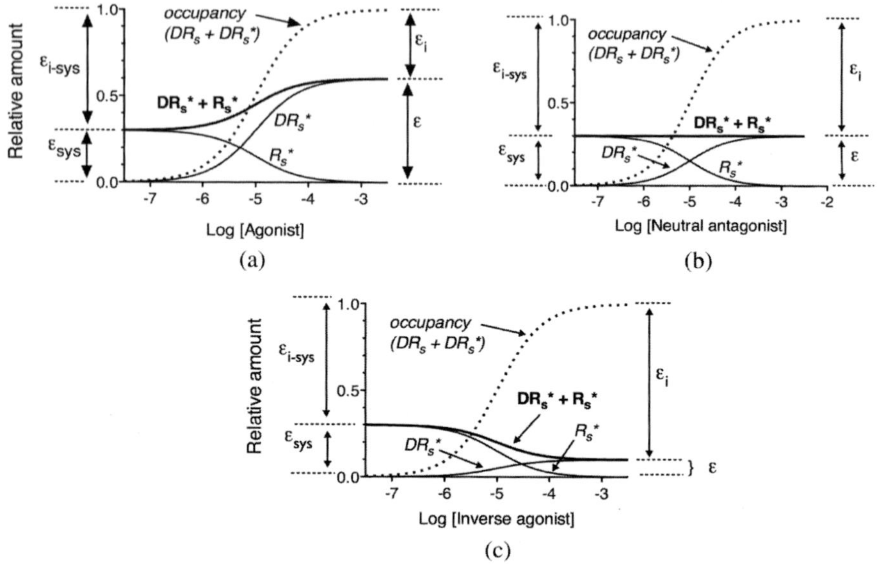

Figure 2.5: Behavior of the simple one-site model illustrated in Figure 2.1 when the receptor is constitutively active. The pharmacological parameters and relative amounts of various components of the receptor population are plotted against the concentration of agonist (a), neutral antagonist (b) and inverse agonist (c).

An analogous plot for the behavior of a neutral antagonist ($K_b/K_a = 1$) is shown in Figure 2.5b. In this case, the total amount of receptor in the active state ($R_s^* + DR_s^*$) is unaffected by the neutral antagonist. Because receptor occupancy by any ligand decreases the amount of free receptor, a neutral antagonist reduces constitutive activity and replaces it with an equivalent amount of ligand-receptor complex in the active state (DR_s^*). Thus, the observed efficacy of a neutral antagonist is equivalent to the efficacy of the system ($\varepsilon = \varepsilon_{\text{sys}}$).

The behavior of an inverse agonist ($K_b/K_a < 1$) is shown in Figure 2.5c. In this case, the total amount of receptor in the active state ($DR_s + DR_s^*$) decreases with an increase in receptor occupancy by the inverse agonist. At maximal receptor occupancy, the fraction of the ligand-receptor complex in the active state (ε) is less than the efficacy of the system ($\varepsilon < \varepsilon_{\text{sys}}$).

It may seem odd that neutral antagonists and inverse agonists have efficacy, for these ligands are usually considered to lack efficacy (neutral antagonist) or have negative efficacy (inverse agonist). Efficacy was defined by Stephenson[2] and Furchgott[7] at the population level, however, and at this level intrinsic efficacy is equivalent to the fraction of the ligand-receptor complex in the active state. When there is constitutive activity, this fraction is greater than zero for neutral antagonists and inverse agonists as shown by the plots in Figures 2.4b and 2.4c. At a deeper level of analysis, the efficacy of an agonist is related to its selectivity for the active state of the receptor. That is, the ratio, K_b/K_a is greater than one for agonists ($\log K_b/K_a > 0$), equal to one for neutral antagonists ($\log K_b/K_a = 0$), and less than one for inverse agonists ($\log K_b/K_a < 0$).

A useful property of all the plots in both Figures 2.3 and 2.4 is that the EC_{50} value of the ligand for causing a change in the amount of a receptor complex (e.g., R_s^*), or sum of receptor complexes (e.g., $DR_s + DR_s^*$), is equivalent to the reciprocal of K_{obs}. Thus, the K_{obs} value can be estimated from the concentration-response curve of the ligand for changing the amount of any one complex or the sum of several complexes. We can show mathematically that this is the case for the sum of the constitutive and the ligand-induced current ($R_s^* + DR_s^*$). For an agonist, the range of its concentration-response curve for receptor activation ($\varepsilon - \varepsilon_{\text{sys}}$) spans from a low value of ε_{sys} to a high value of ε, with half of its maximal effect on receptor activation occurring when half of this range is added to ε_{sys}:

$$i_{50} = \varepsilon_{\text{sys}} + \frac{1}{2}(\varepsilon - \varepsilon_{\text{sys}}) \tag{2.36}$$

In this equation, i_{50} denotes the total current response at the EC_{50} concentration of the ligand. For an inverse agonist, the range ($\varepsilon_{\text{sys}} - \varepsilon$) spans from a high value of ε_{sys} to a low value of ε, with half of its maximal effect on receptor activation occurring when half of this range is subtracted from ε_{sys}:

$$i_{50} = \varepsilon_{\text{sys}} + \frac{1}{2}(\varepsilon_{\text{sys}} - \varepsilon) \tag{2.37}$$

It can be shown that Equation (2.37) is equivalent to Equation (2.36). To derive an equation for the EC_{50} value of a ligand for affecting receptor activation, we can substitute EC_{50} for D in the equation for receptor activation in the presence of constitutive activity (Equation (2.22)) and set it equal to Equation (2.36):

$$\varepsilon_{\text{sys}} + \frac{1}{2}(\varepsilon - \varepsilon_{\text{sys}}) = \frac{[EC_{50}]K_bK_q + K_q}{[EC_{50}]K_bK_q + K_q + [EC_{50}]K_a + K_q + 1}. \tag{2.38}$$

Substituting in Equations (2.19) and (2.22) for ε and ε_{sys} yields:

$$\frac{K_q}{K_q+1} + \frac{1}{2}\left(\frac{1}{1+\frac{K_q}{K_bK_q}} - \frac{K_q}{K_q+1}\right) = \frac{[EC_{50}]K_bK_q + K_q}{[EC_{50}]K_bK_q + [EC_{50}]K_a + K_q + 1}. \tag{2.39}$$

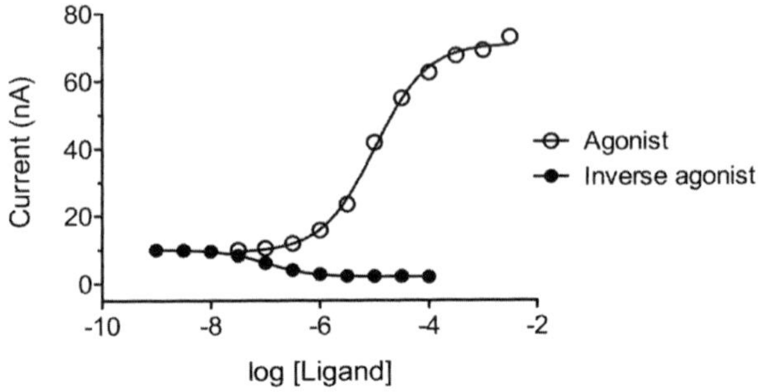

Figure 2.6: Hypothetical current responses elicited by an agonist and an inverse agonist at a ligand-gated ion channel exhibiting constitutive activity. The theoretical curve represents the least-squares fit of Equation (2.40) to the data.

Ultimately, this equation can be simplified to Equation (2.26), illustrating that the EC_{50} value of the ligand is equivalent to the reciprocal of K_{obs}.

Because the EC_{50} value for affecting the current response is equivalent to K_{obs}, we can use an empirical one-site equation to describe the effect of the ligand on the current response in the presence of constitutive activity:

$$i = i_{\text{sys}} + \frac{[D](i_{\max} - i_{\text{sys}})}{[D] + \frac{1}{K_{\text{obs}}}}. \tag{2.40}$$

In this equation, i_{sys} represents the current caused by constitutive activity.

An example of how to estimate the observed affinity and efficacy of an agonist and inverse agonist from the current response of a population of one-site ligand-gated ion channels exhibiting constitutive activity is illustrated in Figure 2.6. Equation (2.40) was fitted to the theoretical data using nonlinear regression analysis and the following parameter estimates were obtained: agonist, log K_{obs} 4.99; $i_{\max}$, 71 nA; inverse agonist, K_{obs}, 6.98, $i_{\max}$, 2.01 nA. As described above, the $i_{\max}$ value is proportional to ε. Thus, the efficacy of the inverse agonist can be expressed relative to that of the agonist using Equation (2.34). When this was done, a value of 0.028 was estimated for the efficacy of the inverse agonist.

Relationship between population parameters and microscopic constants

The population parameters (K_{obs} and ε) describe the average behavior of the receptor population, and these macroscopic constants are ultimately determined by the more fundamental microscopic constants, K_a, K_b and K_q. Before describing the relationship between population and state parameters, we need to address the fundamental difference between the two. This difference is related to the two properties that the affinity constant is a measure of: 1) with what potency the ligand occupies the receptor population and 2) the strength of the binding interaction. The population affinity constant, K_{obs}, defines receptor occupancy (property 1), whereas the microscopic affinity constants (K_b and K_a) describe the strength of the binding interaction of the ligand with the two shapes of the receptor (property 2).

Consider the theoretical occupancy plot shown by the solid line in Figure 2.7. This curve represents receptor occupancy by a ligand having log affinity constants of 8.0 and 3.0 for the active (K_b) and inactive (K_a) states, respectively. The gating constant of the unoccupied receptor (K_q) is equivalent to 10^{-4}, indicating little constitutive activity in the absence of agonist. The K_{obs} value of the ligand for this receptor is about 10^{-5} (see Equation (2.11)). Thus, at a concentration of 10^{-5} M, the agonist occupies 50% of the receptor population at equilibrium, and K_{obs} accurately describes the potency of the ligand–receptor interaction (property 1). There is no receptor structure that exhibits an affinity constant of 10^5 M^{-1}, however. This value represents a weighted average of the microscopic affinity constants for the active and inactive states (see Equation (2.11)). If we could lock the receptor in the active state, agonist occupancy would resemble the dashed curve on the left, labeled with K_b. Thus, K_b is an accurate measure of the strength of the interaction of the agonist with the active state of the receptor (property 2). However, there is no native receptor structure exhibiting 50% receptor occupancy when the ligand concentration is equivalent to $1/K_b$. In other words, K_b is not a measure of the potency of the

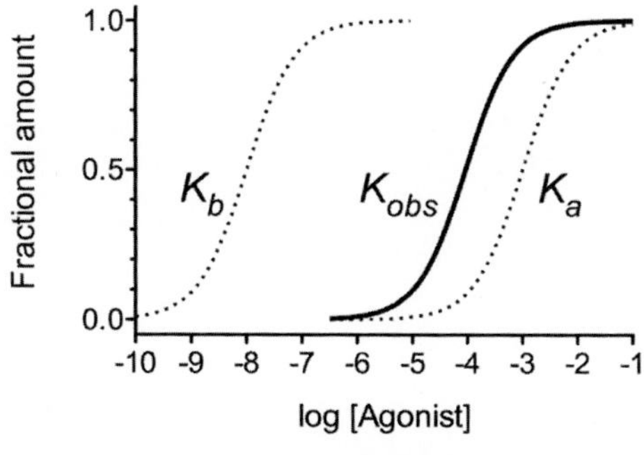

Figure 2.7: Relationship between the observed affinity constant (K_{obs}) and the microscopic affinity constants for the active (K_b) and inactive (K_a) states of the receptor. The plot shows the real occupancy function of the receptor (solid line, Equation (2.12)) and the imaginary occupancy functions (dashed lines) if the receptor were locked in either the active (left curve) or inactive (right curve) state. $K_q = 10^{-4}$.

agonist-receptor interaction. An analogous type of argument applies to the affinity constant of the inactive state as illustrated by dashed curve on the right, although there may be little difference between K_{obs} and K_a for ligands having low efficacy. To summarize, the population parameter, K_{obs}, determines receptor occupancy of both the active and inactive states of the receptor, whereas the microscopic affinity constants describe the strength of the interaction of the ligand with the two shapes of the receptor. Even though there are two states, the agonist occupancy curve resembles a simple one-site model because the two states interconvert.

The efficacy of an agonist is related to its selectivity for the active state of the receptor (K_b/K_a). Figure 2.8a shows the efficacy of the agonist plotted against the log ratio of K_b/K_a for a hypothetical receptor lacking constitutive activity ($K_q = 10^{-3}$). The theoretical ε values were calculated using Equation (2.19). Agonist efficacy increases with an increase in K_b/K_a until efficacy approaches a maximum value of one. A further increase in K_b/K_a has little effect on receptor activation because the receptor is already maximally activated. When the ratio, K_b/K_a, is equal to the reciprocal of K_q, efficacy is equivalent to 0.5. Thus, at the population level, efficacy ranges from 0–1, whereas at the level of receptor states, the ability of the agonist to induce the active state is determined by the ratio of K_b/K_a, which can vary over a broad range. The latter is the most fundamental measure of an agonist's ability to activate a receptor because it is proportional to the gating constant (K_bK_q/K_a) that describes the agonist-induce activation (i.e., $DR_s \leftrightarrow DR_s^*$, see Figure 2.1).[8]

If constitutive activity is lacking ($K_q \ll 1$), then nearly all of the receptors are in the inactive state in the absence of ligands. Because inverse agonists and neutral antagonists lack selectivity for the active state, these types of ligands will not

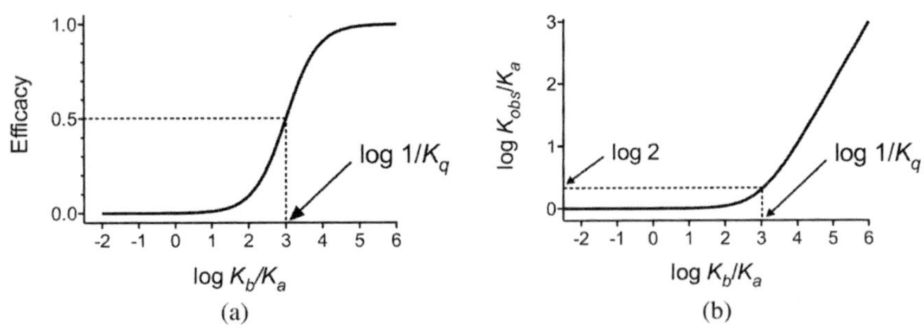

Figure 2.8: How the selectivity of a ligand for the active and inactive states of the receptor influences observed efficacy (ε) and affinity (K_{obs}). Observed efficacy (a) and affinity (b) are plotted against the ratio of microscopic affinity constants of the active and inactive states (K_b/K_a).

induce a conformational change in the receptor, and hence, their respective K_{obs} values will be equivalent to K_a. The same is essentially true of agonists with very low K_b/K_a ratios. This principle is illustrated in Figure 2.8b for a receptor having little constitutive activity ($K_q = 10^{-3}$). The graph shows the log ratio of K_{obs}/K_a plotted against log K_b/K_a. When K_b/K_a is less than one (log $K_b/K_a < 0$), K_{obs} is essentially equal to K_a as shown by the K_{obs}/K_a ratio of 1.0 (log $K_{obs}/K_a = 0$). As K_b/K_a increases above 1, K_{obs} slowly increases. When K_b/K_a is equivalent to the reciprocal of K_q (10^3), the K_{obs} value is twice the value of K_a. A further increase in K_b/K_a ultimately causes a proportional increase in K_{obs}/K_a. Because most native receptors have little or no constitutive activity, ligands that either lack or have low selectivity for the active state will exhibit K_{obs} values similar to K_a. It is only the most efficacious agonists that exhibit a K_{obs} value substantially greater than K_a ($K_bK_q/K_a > 1$).

Estimation of K_b and K_a

When a receptor lacks constitutive activity, it is impossible to estimate the K_b value of an agonist from its concentration-response curve for eliciting current through a population of receptors, but it is possible to estimate the K_b of one agonist (A) relative to that of another (B) based the following principle:

$$RA_i = \frac{K_{b-\mathrm{B}}}{K_{b-\mathrm{A}}} = \frac{\varepsilon_{\mathrm{B}} K_{\mathrm{obs\text{-}B}}}{\varepsilon_{\mathrm{A}} K_{\mathrm{obs\text{-}A}}}. \tag{2.41}$$

In this equation, the subscripts "A" and "B" are used to denote the parameters of the agonists (A and B, respectively). The variable RA_i is known as *intrinsic relative activity*. As shown by the equation, this parameter is a relative measure of 1) the affinity constant of an agonist for the active state of the receptor and 2) the product of observed affinity and efficacy.[9] This equation can be proved by substituting in Equations (2.13) and (2.19) for K_{obs} and ε, respectively:

$$\frac{K_{b-\mathrm{B}}}{K_{b-\mathrm{A}}} = \frac{\dfrac{1}{1+\frac{K_{a-\mathrm{B}}}{K_{b-\mathrm{B}}K_q}} \times \dfrac{K_{b-\mathrm{B}}K_q + K_{a-\mathrm{B}}}{1+K_q}}{\dfrac{1}{1+\frac{K_{a-\mathrm{A}}}{K_{b-\mathrm{A}}K_q}} \times \dfrac{K_{b-\mathrm{A}}K_q + K_{a-\mathrm{A}}}{1+K_q}}. \tag{2.42}$$

Again, the subscripts "A" and "B" are used to discriminate between the parameters of the two agonists. This equation reduces to the equality, $K_{b-B}/K_{b-A} = K_{b-B}/K_{b-A}$, which proves the equation from which it was derived (Equation (2.41)). Figure 2.9 illustrates the consequences of Equation (2.41).

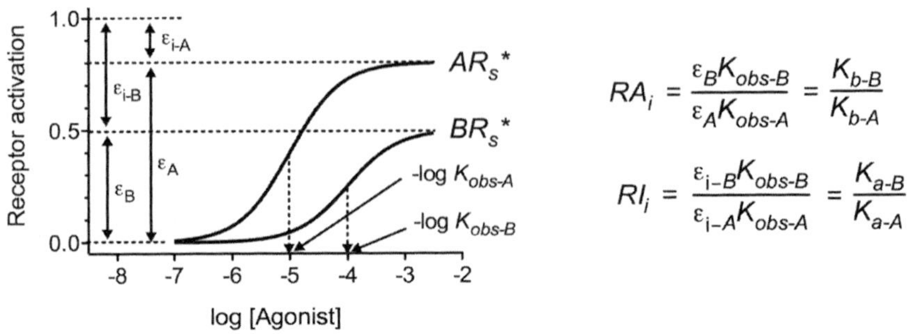

Figure 2.9: Relationship between the empirical parameters of the receptor-activation functions of two agonists (A and B) and relative estimates of the microscopic affinity constants of agonist B for the active (RA_i) and inactive (RI_i) states of the receptor, expressed relative to those of agonist A. The parameters of the two agonists are denoted with the corresponding subscripts, A and B.

Because the maximum current response elicited by an agonist acting on a population of receptors (I_{max}) is proportional to ε, i_{max} can be substituted for ε in Equation (2.41) to yield:

$$RA_i = \frac{K_{b-B}}{K_{b-A}} = \frac{i_{max\,-B}K_{obs-B}}{i_{max\,-A}K_{obs-A}}. \tag{2.43}$$

This equation can be used to estimate the log RA_i values of the agonist in Figure 2.4b. Substituting in the appropriate K_{obs} and I_{max} values yields log RA_i estimates of -2.11 and -1.43 for agonists B and C, respectively, when expressed relative to agonist A. The log K_b values that were used to generate the theoretical curves for agonists A–C were 10, 7.88 and 8.56, respectively. Thus, the log RA_i values of agonists B and C are accurate estimates of the log K_b value expressed relative to that of agonist A.

There is an analogous relationship for the ratio of the K_a value of one agonist (B) relative to that of another (A):

$$RI_i = \frac{K_{a-B}}{K_{a-A}} = \frac{\varepsilon_{i-B}K_{obs-B}}{\varepsilon_{i-A}K_{obs-A}}. \tag{2.44}$$

This ratio is termed *intrinsic relative inactivity* (RI_i).[10] The relationship between the population parameters and RI_i is illustrated in Figure 2.9. Equation (2.44) can be proven by substituting in Equations (2.35), (2.19) and (2.13) for ε_i, ε and K_{obs}, respectively, and then simplifying the equation, which yields the equality, $K_{a-B}/K_{a-A} = K_{a-B}/K_{a-A}$. Because the ε_i values of agonists are proportional to the maximum current response of the system ($i_{max\text{-}sys}$) minus the maximum current response of the agonist ($i_{\max}$), Equation (2.44) can be expressed in the following form for the estimation of agonist parameters from the current response of a population of receptors:

$$RI_i = \frac{K_{a-\mathrm{B}}}{K_{a-\mathrm{A}}} = \frac{(i_{\max\text{-sys}} - i_{\max\text{-B}})K_{\mathrm{obs-B}}}{(i_{\max\text{-sys}} - i_{\max\text{-A}})K_{\mathrm{obs-A}}}. \tag{2.45}$$

In most instances, it would be difficult to determine the value of $i_{\max-\mathrm{sys}}$ for a population of ligand-gated ion channels as described above. Perhaps in some instances, the $i_{\max}$ value of a highly efficacious agonist may be very similar to $i_{\max-\mathrm{sys}}$, in which case Equation (2.45) could be used in the estimation of the relative K_a values (RI_i values) of partial agonists. Because the difference between $i_{\max-\mathrm{sys}}$ and $i_{\max}$ is very small for highly efficacious agonist, it would be difficult to estimate the relative K_a values of highly efficacious agonists accurately, even if $i_{\max-\mathrm{sys}}$ were known. As described below, a different form of Equation (2.45) is useful for estimating the relative K_a values of agonists from single channel recordings.

It is possible to derive equations expressing K_b and K_a in terms of the population parameters for the condition of constitutive activity. The equations for the estimation of K_a and K_q, however, require accurate estimation of $i_{\max-\mathrm{sys}}$, which may or may not be possible. Regardless, these equations are useful in the analysis of single channel recordings as described in the next section.

The equations describing the basis for the estimation of the microscopic affinity constants for the active (K_b) and inactive (K_a) states of the receptor are:

$$K_b = \frac{\varepsilon K_{\mathrm{obs}}}{\varepsilon_{\mathrm{sys}}}, \tag{2.46}$$

$$K_a = \frac{\varepsilon_i K_{\mathrm{obs}}}{\varepsilon_{i-\mathrm{sys}}}, \tag{2.47}$$

These can be proved by substituting in Equations (2.13), (2.19), (2.26) and (2.35) for K_{obs}, ε, ε_i and $\varepsilon_{i-\mathrm{sys}}$, and then simplifying the equations, which yield the two equalities $K_b = K_b$ and $K_a = K_a$. An illustration of these relationships is shown in Figure 2.10. The isomerization constant (K_q) is described by:

$$K_q = \frac{\varepsilon_{\mathrm{sys}}}{\varepsilon_{i-\mathrm{sys}}}. \tag{2.48}$$

These equations can be easily converted into a form for analyzing the current response of a population of receptors by making the following substitutions for

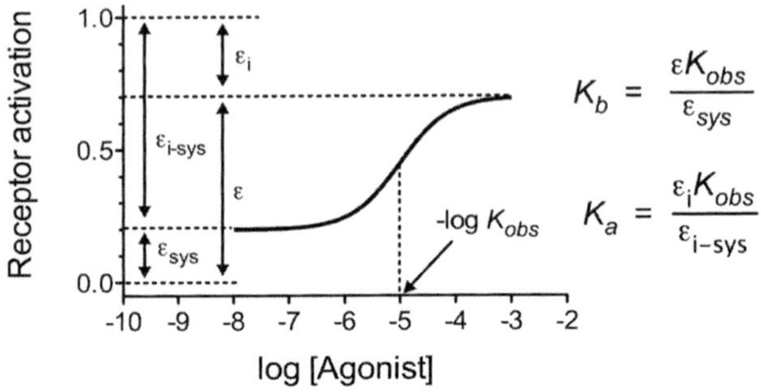

Figure 2.10: Relationship between the microscopic affinity constants of an agonist for the active (K_a) and inactive states (K_b) of a receptor and the empirical parameters of the receptor-activation function of an agonist at a constitutively active receptor.

ε, ε_i, ε_{sys} and $\varepsilon_{i\text{-sys}}$:

$$\varepsilon = \frac{\imath_{\max}}{\imath_{\max\text{-sys}}}, \tag{2.49}$$

$$\varepsilon_i = \frac{\imath_{\max\text{-sys}} - \imath_{\max}}{\imath_{\max\text{-sys}}}, \tag{2.50}$$

$$\varepsilon_{\text{sys}} = \frac{\imath_{\text{sys}}}{\imath_{\max\text{-sys}}}, \tag{2.51}$$

$$\varepsilon_{i\text{-sys}} = \frac{\imath_{\max\text{-sys}} - \imath_{\text{sys}}}{\imath_{\max\text{-sys}}}. \tag{2.52}$$

Finally, making these substitutions into Equations (2.46)–(2.48) yields equations for the microscopic affinity constants in terms of the population currents:

$$K_b = \frac{\imath_{\max} K_{\text{obs}}}{\imath_{\text{sys}}}, \tag{2.53}$$

$$K_a = \frac{(\imath_{\max\text{-sys}} - \imath_{\max}) K_{\text{obs}}}{\imath_{\max\text{-sys}} - \imath_{\text{sys}}}, \tag{2.54}$$

$$K_q = \frac{\imath_{\text{sys}}}{\imath_{\max\text{-sys}} - \imath_{\text{sys}}}. \tag{2.55}$$

Using Equation (2.53) and the estimates of $\imath_{\text{sys}}$ and the $\imath_{\max}$ values of agonist A and inverse agonist B in Figure 2.6, it is possible to estimate their respective log K_b values (5.78 and 6.30). These values are equivalent to those used to simulate the data in the first place.

Summary of population analysis

If there is little desensitization and no measureable constitutive activity, the one-site model predicts that receptor activation is equivalent to the product of receptor occupancy and intrinsic efficacy as initially postulated by Stephenson[2] and Furchgott.[7] It follows that the concentration of agonist eliciting a half-maximal increase in current is equivalent to $1/K_{obs}$ and that the maximum current response is proportional to ε. The affinity constant of an agonist for the active state of the receptor (K_b) is proportional to the product, εK_{obs}. Hence, the K_b value of one agonist, expressed relative to that of another can be estimated from the corresponding ratios of the products of affinity and maximum current response ($i_{max}K_{obs}$).

When there is constitutive activity, the current response is equivalent to the sum of the constitutive and ligand-induced current. The concentration of ligand causing a half-maximal change in the current response is equivalent to $1/K_{obs}$. When the ligand concentration is high enough to cause nearly complete receptor saturation, there is no constitutive activity and the maximally effective concentration of ligand is proportional to ε. Thus, the intrinsic efficacy of a neutral antagonist is equivalent to the constitutive activity of the receptor in the absence of ligand and the constitutive activity of an inverse agonist is less than this value. The affinity constant of a ligand for the active state of the receptor is equivalent to the product of observed affinity and efficacy divided by the efficacy of the free receptor complex ($i_{max}K_{obs}/\varepsilon_{sys}$). Hence, K_b can be calculated as the product of the affinity constant and maximum current response divided by the constitutive current ($i_{max}K_{obs}I/i_{sys}$).

A limitation in the whole cell population analysis is that it may not be possible to obtain an estimate of the maximal current when all of the channels are open ($i_{max-sys}$), although the maximal current response to a highly efficacious agonist may approach this value provided there is little desensitization. Without an estimate of $i_{max-sys}$, the efficacy of an agonist cannot be estimated, but it can be expressed relative to that of another agonist. It is also impossible to obtain a precise estimate or relative estimate of the affinity constant of an agonist for the inactive state of the receptor (K_a) unless the ligand has low efficacy. For partial agonists with ε values less than 0.5, the K_a value is never less than one-half K_{obs}. An advantage of single channel analysis is that ε can be estimated accurately, which enables the estimation of K_a.

Single Channel Analysis

In this section, I describe how affinity and efficacy are related to estimates of the open probability of our hypothetical, one-site ligand-gated ion channel. The emphasis is on how to relate population parameters to the behavior of single channels and how to estimate the microscopic constants, K_b and K_a, given our hypothetical one-site reaction scheme. Ultimately, it is shown that the open probability of a channel is proportional to the fractional amount of receptor in the $R_s^* + DR_s^*$ states or simply the DR_s^* state if the receptor lacks constitutive activity. One can apply, therefore, the population analysis to estimate the population (K_{obs} and ε) and state (RA_i and RI_i or K_b and K_a) parameters through analysis of the open probability estimates from single channel recordings. I begin with the theoretical basis for these calculations and a brief description of the single-channel recording technique, its advantages and disadvantages.

Measurement of single channel currents

The effects of agonists on the behavior of single ligand-gated ion channels have been measured using patch-clamp recording electrodes filled with agonist in the cell-attached mode or by adding agonist to the bathing medium in the outside-out recoding mode. In the latter case, the whole-cell recording mode is first established as described in Figure 2.2b, and then the electrode is pulled away from the cell, causing the attached plasma membrane to stretch into a short tube connecting the cell to the tip of the electrode. This connection ultimately breaks, and remarkably, the plasma membrane reseals the opening of the electrode tip in an outside-out configuration (Figures 2.2c and 2.2d).

Agonist-occupied ligand-gated ion channels undergo transitions between open and closed states at a speed that exceeds the detection limit or dead time ($< 10\ \mu$sec) of recording techniques. Thus, brief channel opening or closing events are missed, which can lead to error in the estimation of open and closed times. Two adjacent open events punctuated by a brief closing event, for example, would appear as one long open event. Several investigators have developed analytical methods that correct these missed events.[11] Another source of error is related to potential ambiguity in the number of channels in the membrane patch. If the patch contains a population of receptors and the current output is graded, fluctuation analysis can be used to estimate the number of channels, based on the principle that the variance of the mean current is related to the square root of the number of channels.[12,13] In the case of records containing only single unitary-conductance events, however, there is always the question of whether the events are caused by more than one channel. The probability of the latter is very low for high-frequency records containing only unitary conductance events.

Agonists have both positive and negative effects on ligand-gated ion channels. For those interested in the positive mechanism of ligand-induced channel opening, the negative effect of ligand-induced desensitization can have a confounding influence. Many investigators simply omit those regions of the record that exhibit desensitization (long closed-channel periods) in their analysis, so that it does not adversely affect the estimation of agonist gating properties.

Because my interest is in describing how to relate the effects of agonists on single channel activity to their influence on a population of ligand-gated ion channels, I will focus on the analysis of records that lack desensitization and represent the activity of a single channel. The overall goal is to explain how open probability is related to microscopic constants and the population parameters of affinity and efficacy.

The population model applies to single receptor events

If the microscopic affinity constants for the model in Figure 1a are replaced with the corresponding rate constants, the resulting scheme can be used to predict the probability of single channel events. Figure 2.11 shows the revised scheme with rate constants defined as:

$$\frac{k_{a+}}{k_{a-}} = K_a, \tag{2.56}$$

$$\frac{k_{b+}}{k_{b-}} = K_b, \tag{2.57}$$

$$\frac{k_{q+}}{k_{q-}} = K_q, \tag{2.58}$$

$$\frac{k_{g+}}{k_{g-}} = K_g. \tag{2.59}$$

As described above, equilibria of the type shown in Figure 2.1a can be described by a minimum of three constants (e.g., K_a, K_b and K_q); the fourth can always be described in terms of the other three (e.g., $K_g = K_b K_q / K_a$). It follows that one of

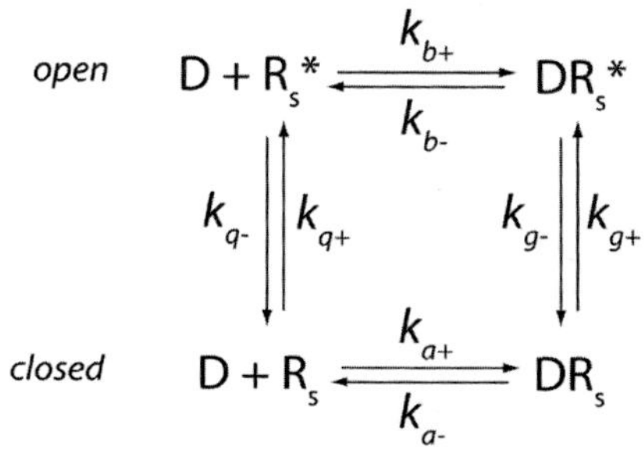

Figure 2.11: Simple one-site model illustrated in Figure 2.1 but with the microscopic affinity constants replaced with the corresponding rate constants.

the eight rate constants shown in Figure 2.11 can be described in terms of the other seven. Given the uniqueness of the rate constants related to K_a, K_b and K_q, the constant k_{g-} has been defined as, $k_{g-} = k_{g+}k_{q-}k_{a+}k_{b-}/k_{q+}k_{a-}k_{b+}$.

At the population level, the rate constants specify the fraction of the receptor population undergoing a specific change per unit time. For example, the binding of the agonist to the closed state of the channel is defined by the following reaction:

$$[D] + [R_s] \xrightarrow{k_{a+}} [DR_s]. \tag{2.60}$$

The second order rate constant, k_{a+}, has units of both molarity and inverse time. If we multiply this constant by the molar concentration of agonist, the rate constant for the reaction is obtained in units of inverse time. For example, if the agonist concentration is 10^{-5} M and the value of k_{a+} is 10^6 M^{-1} min^{-1}, then a pseudo first order rate constant of 10 min^{-1} is obtained (i.e., $Dk_{a+} = 10$). This rate constant corresponds to the equivalent of 10 receptor populations being bound with agonist per minute at the start of the reaction. Obviously this initial starting rate slows down as the reaction progresses because the amount of agonist bound at equilibrium never reaches a value greater than one unit (i.e., the receptor population). To calculate the actual rate of drug binding near the start of the reaction, the psuedo first order rate constant (Dk_{a+}) is multiplied by the fractional amount of free receptor (i.e., $[R] = 1$) and a small time increment (Δt). The result ($Dk_a + [R_s]\Delta t$) represents the fractional amount of DR_s formed during the time interval, Δt. For example, using a small value of 0.005 min for Δt yields an estimate of 0.05 for the fractional amount of DR_s formed. In other words, in the first 0.3 s (i.e., 0.005 min), 5% of the receptors in the population become bound with agonist provided that none were bound at the beginning of the reaction (i.e., $[R_s] = 1$). This rate decreases in time as occupancy increases because the amount of free receptor decreases from its starting amount of 1.0. At equilibrium the rate of formation of the DR_s complex is equivalent to its rate of dissociation.

The constant (k_{a-}) describing the dissociation of the $[DR_s]$ complex is an example of a different type of rate constant:

$$[DR_s] \xrightarrow{k_{a-}} [D] + [R_s]. \tag{2.61}$$

It represents the fraction of the population of drug-receptor complexes that dissociates per unit time. Unlike the second order rate constant, k_{a+}, k_{a-} is a first order rate constant having units of inverse time only. For example, if the value of k_{a-} is 0.1 min^{-1}, then the initial rate of dissociation corresponds to 10% of the population of $[DR_s]$ complexes dissociating per min. Again, this is an initial rate and the actual rate changes in time depending on the amount of the $[DR_s]$ complex remaining in the population. Applying the same small time increment used above (0.005 min), the initial amount of the population of bound receptors that dissociates in 0.3 s is 0.05%. At a starting condition of no receptor occupancy, the initial rate of dissociation would be zero, and it would increase to a maximum value when equilibrium is achieved.

These rate constants can be used to estimate the probability of single channel events. For example, if the pseudo first order rate constant (Dk_{a+}) describing the binding of agonist to the closed receptor is multiplied by a small time increment (0.005 min), the result is the unitless probability that the agonist will bind to the receptor during the time increment, provided that the receptor is unoccupied at the start of the time interval.[14] This type of probability is known as a conditional probability because it is contingent upon the receptor being unoccupied at the beginning of the time interval. It can be denoted in the following way:

$$Dk_{a+}\Delta t = \text{Prob}\{DR_s, t + \Delta t | R_s, t\}. \tag{2.62}$$

This expression means that $Dk_{a+}\Delta t$ is equivalent to the probability that the unoccupied, closed channel (R_s) becomes bound with agonist (DR_s) during the time interval spanning from t to $t+\Delta t$, provided that the channel is unoccupied and closed at time t. The scheme in Figure 2.11 shows that there are two other possibilities for the R_s state during the time interval, t to $t + 1$. It could undergo a rare constitutive transition to the active state, R_s^*, or it could remain in the closed, unoccupied state. These conditional probabilities are given by Equations (2.63) and (2.64), respectively:

$$k_{q+}\Delta t = \text{Prob}\{R_s^*, t + \Delta t | R_s, t\}, \tag{2.63}$$

$$1 - \Delta t(Dk_{a+} + k_{q+}) = \text{Prob}\{R_s, t + \Delta t | R_s, t\}. \tag{2.64}$$

Equation (2.63) indicates that $k_{q+}\Delta t$ is equivalent to the probability that the receptor will undergo constitutive activation (R^*) during the time interval from t to $t + \Delta t$, provided that the receptor is in the R state at time t. Lastly, Equation (2.64) indicates that the corresponding conditional probability for the receptor remaining in the R_s state over the same time interval is equivalent to one minus the two other conditional probabilities. Thus, given that the receptor is in the R_s state at the start of the time interval, the sum of the probabilities that the receptor transitions to the DR_s or R_s^* states or remains in the R_s state is equal to one.

Using the same strategy, it is possible to define three conditional probabilities for each of the three other states of the receptor (i.e., DR_s, DR_s^* and R_s^*). For example, given that the receptor is in the DR_s state at the start of the time interval, the conditional probabilities for transitioning to the DR_s^* or R_s states or remaining in the DR_s state could be derived using the appropriate rate constants (k_{g+} and k_{a-}) and the axiom that the three conditional probabilities sum to one. Analogous reasoning applies to the receptor when it is in the either the DR_s^* or R_s^* states at the start of the time interval. These 12 conditional probabilities are listed in a transition-matrix (**T**) described in the next section.

Using this type of thinking, one can imagine that the receptor jumps randomly, one equilibrium step at a time, among the four different states of the receptor shown in Figure 2.11 in a clockwise or counterclockwise manner, but not across diagonals (e.g., not from R_s to DR_s^*). In a given state, the receptor is continuously submitted to three conditional probabilities unique to that state over a time interval measured in quantal units (Δt), until it randomly jumps to a new state, whereupon it is submitted to a new set of three conditional probabilities. This type of system is known as a continuous Markov process, and its mathematics are described next.

Single channel behavior resembles a continuous Markov process

A Markov process refers to a system that can exist in discrete states. The probability of being in a given state at a particular time depends only on the state of the system at the prior time interval. A Markov process can have a discrete time interval, like a daily weather forecast, or it may evolve continuously in time, like the random behavior of single channels (continuous Markov process). The scheme in Figure 2.11 can be modeled as a continuous Markov process, and this will enable us to determine the probability of channel opening at various times following addition of agonist.

The set of conditional probabilities mentioned above govern the Markov process, and these are entered into a two-dimensional square array (matrix) having dimensions equivalent to the number of states. The transition matrix (**T**) for the ion channel scheme shown in Figure 2.11 is:

$$\mathrm{T} = \begin{array}{c} \begin{array}{cccc} (R_s) & (R_s^*) & (DR_s) & (DR_s^*) \end{array} \\ \begin{bmatrix} 1-\Delta t(k_{q+}+Dk_{a+}) & k_{q-}\Delta t & k_{a-}\Delta t & 0 \\ k_{q+}\Delta t & 1-\Delta t(k_{q-}+Dk_{b+}) & 0 & k_{b-}\Delta t \\ Dk_{a+}\Delta t & 0 & 1-\Delta t(k_{a-}+Dk_{g+}) & k_{g-}\Delta t \\ 0 & k_{b+}\Delta t & k_{g+}\Delta t & 1-\Delta t(k_{b-}+Dk_{g-}) \end{bmatrix} \begin{array}{c} (R_s) \\ (R_s^*) \\ (DR_s) \\ (DR_s^*) \end{array} \end{array} \quad (2.65)$$

This transition matrix consists of a set of conditional probabilities ($P_{i,j}$) indexed by row (i) and column (j) to identify the state of the receptor at the start (column) and end (row) of the time interval. These are also denoted by the parenthetical expressions on the top and right side of the matrix. For example, the entry ($P_{1,1}$) in row one column one gives the probability that the receptor will remain in the R state over the time interval t to $t + \Delta t$, given that it is in the R state at time t.

The entry, $P_{1,3}$, gives the probability that the receptor will transition to the R state, given that it is in the DR_s state at time t. Finally, the entry, $P_{4,3}$, gives the probability that the receptor will transition to the DR_s^* state, given that it is in the DR_s state at time t. Transitions that cannot occur in one step (e.g., $R \to DR_s^*$) are assigned a probability of zero. The sum of the probabilities in each column equals 1.0 because they account for all possible routes to a given state in one time interval.

Before we can use transition matrix **T**, we need to define a one-dimensional column array (column vector) that identifies the state of the Markov process. This state vector (**P**) for the condition of the receptor being in the R state is:

$$\mathbf{P} = \begin{bmatrix} 1 \\ 0 \\ 0 \\ 0 \end{bmatrix} \begin{matrix} (R_s) \\ (R_s^*) \\ (DR_s) \\ (DR_s^*) \end{matrix}. \tag{2.66}$$

The entries denote the probability that the receptor is in a given state, and hence, they represent ordinary probabilities, and not the conditional type that require a two-dimensional matrix. Because we know that the receptor is in the R state, the probability of being in the R state is assigned a value of 1 in the first row of the vector. The other rows are assigned values of zero because we know that the receptor is not in states R_s^*, DR_s or DR_s^* (rows 2–4, respectively).

To determine the state of the system $\mathbf{P}^{(n)}$ after a number (n) of time intervals the following matrix calculation is performed:

$$\mathbf{P}^{(n)} = \mathbf{T}^{(n)}\mathbf{P}^{(0)}, \tag{2.67}$$

in which $\mathbf{P}^{(0)}$ denotes the state of the system at the start of the Markov process. For example, after one time interval the state of the system ($\mathbf{P}^{(1)}$) is given by:

$$\mathbf{P}^{(1)} = \mathbf{T}\mathbf{P}^{(0)}. \tag{2.68}$$

After six time intervals, the state of the system $\mathrm{P}^{(6)}$ is:

$$\mathbf{P}^{(6)} = \mathbf{T}^{6}\mathbf{P}^{(0)}. \tag{2.69}$$

The state of the system as a function of time can be calculated by evaluating Equation (2.67) over a series of time intervals, with each time interval (Δt) being small so that the fastest process only involves a very small change in the probability of a given receptor state. This can be easily done on a computer using an application, like MATLAB (MathWorks, Inc.), for example. Figure 2.12 shows a plot of the state of the scheme in Figure 2.11 as a function of time when the agonist concentration is 10^{-3} M and the system is in the R state at the start. That is, the probability

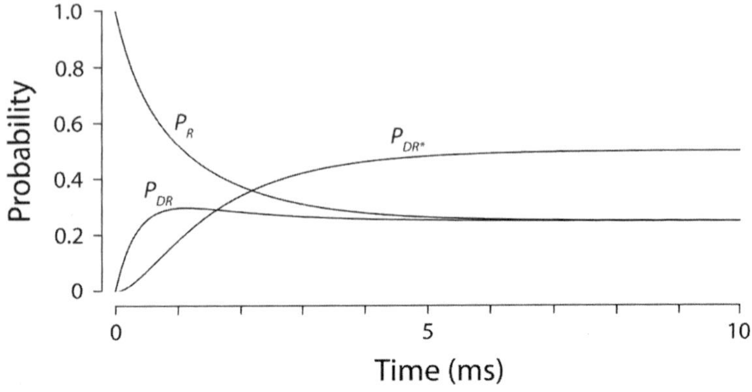

Figure 2.12: The probability of a single receptor being in the R_s, (P_R), $DR_s(P_{DR})$ and DR_s^* (P_{DR}^*) states as a function of time for the case where the agonist concentration is 10^{-3} M and the receptor is in the R_s state at time zero. Parameters of the simulation: k_{a+}, 10^6 M^{-1} s^{-1}; k_{a-} 10^3 s^{-1}, k_{b+}, 2×10^6 M^{-1} s^{-1}; k_{b-}, 0.1 s^{-1}; k_{q+}, 10^2 s^{-1}; k_{q-}, 10^6 s^{-1}; k_{g+}, 10^3 s^{-1} and k_{g-}, 500 s^{-1}.

of the receptor being in the R, R_s^*, DR_s and DR_s^* states (P_R, P_{R*}, P_{DR} and P_{DR*}, respectively) is plotted against time. The values of the parameters for this simulation are given in the legend to the figure. The plot shows that P_R decreases in time as P_{DR} and P_{DR*} increase. The P_{R*} value is so small that it would be nearly undetectable above the abscissa because the probability of constitutive activity is low for this simulation (i.e., $k_{q+}/k_{q-} = 10^{-3}$). Thus, P_{R*} is not shown.

Figure 2.12 shows that equilibrium is achieved after a large number of time intervals, corresponding to a total time of about 5 ms. **P** and **T** have interesting properties, and I mention just a few here before getting back to our discussion of how single receptor events relate to population behavior. If the number of time intervals required to reach equilibrium is equivalent to m, then the state of the system at equilibrium can be designated as $\mathbf{P}^{(m)}$. It follows from Equation (2.67) that:

$$\mathbf{P}^{(m)} = \mathbf{T}^m \mathbf{P}^{(0)}. \tag{2.70}$$

It can also be shown that:

$$\mathbf{P}^{(m)} = \mathbf{T}\mathbf{P}^{(m)}. \tag{2.71}$$

Thus, $\mathbf{P}^{(m)}$ is an eigenvector of **T** having an eigenvalue of 1. It can be shown that each column vector in $\mathbf{T}^m$ is identical to $\mathbf{P}^{(m)}$.

Getting back to the relation between population behavior and single receptor events, the plot in Figure 2.12 resembles that of the fractional amount of the receptor population in the R, R_s^*, DR_s and DR_s^* states following addition of the agonist. If so, then the probabilities of the receptor being in the various states (P_R, P_{R^*}, P_{DR} and P_{DR^*}) at equilibrium should be equivalent to the fractional amounts of the receptor population in the various receptor complexes (R, R^*, DR_s and DR_s^*, respectively) at equilibrium. From the rate constants used in this simulation (see legend of Figure 2.12), the following values can be calculated for the microscopic affinity constants: K_b, $10^8\ \mathrm{M}^{-1}$; K_a, $10^3\ \mathrm{M}^{-1}$; and K_q, 10^{-4}, using the relationships in Equations (2.56)–(2.58). Using these constants and Equation (2.17), the fractional amount of receptor in the DR_s^* state at equilibrium can be calculated (i.e., $DR_s^*/R_T = 0.50$). The amount of free receptor in the active state at equilibrium can be calculated using Equation (2.21), which yields, $R^*/R_T = 2.5 \times 10^{-5}$. Total receptor occupancy can be calculated using Equation (2.11) ($(DR_s + DR_s^*)/R_T = 0.75$). Subtracting the value of DR_s^*/R_T from total receptor occupancy yields the fractional amount of receptor in the DR_s state at equilibrium ($0.75 - 0.50 = 0.25$). To summarize, the fractional amounts of the receptor population in the R, R^*, DR_s and DR_s^* states at equilibrium are equivalent to about 0.25, 0.000025, 0.25 and 0.50, respectively. These are essentially equal to the values of P_R, P_{R^*}, P_{DR} and P_{DR^*} at equilibrium for the simulation shown in Figure 2.12 (0.25, 0.25×10^{-5}, 0.25 and 0.50, respectively). Thus, the equilibrium probabilities of single receptor events resemble the behavior of the receptor population at equilibrium.

There is another way to think about the probability of single channel events that illustrates their similarity to population behavior. I briefly describe this approach without going into much mathematical detail. One common method for developing a time function for a system consisting of discrete states is to derive a series of differential equations that describe the rate of change of each state. These are then solved analytically or numerically to identify the time function of each species. The corresponding differential equations for the channel probabilities are:

$$\frac{dP_R}{dt} = -(Dk_{a+}P_R + k_{q+}P_R) + (k_{a-}P_{DR} + k_{q-}P_{R^*}), \tag{2.72}$$

$$\frac{dP_{R^*}}{dt} = -(Dk_{b+}P_{R^*} + k_{q-}P_{R^*}) + (k_{q+}P_R + k_{b-}P_{DR^*}), \tag{2.73}$$

$$\frac{dP_{DR}}{dt} = -(k_{g+}P_{DR} + k_{a-}P_{DR}) + (Dk_{a+}P_R + k_{g-}P_{DR^*}), \tag{2.74}$$

$$\frac{dP_{DR^*}}{dt} = -(k_{g-}P_{DR^*} + k_{b-}P_{DR^*}) + (Dk_{b+}P_{R^*} + k_{g+}P_{DR}). \tag{2.75}$$

Each differential equation is derived in the following manner, using the rate of change in P_R as an example. The probability that the channel is in the R state at time $t + \Delta t(P_R(t + \Delta t))$ is:

$$P_R(t + \Delta t) = P_R(t) - \Delta t(Dk_{a+}P_R(t) + k_{q+}P_R(t)) + \Delta t(k_{a-}P_{DR}(t) + k_{q-}P_{R^*}(t)). \tag{2.76}$$

In other words, this equation states that $P_R(t + \Delta t)$ is equal to the probability that the receptor is in in the R_s state at time $t(P_R(t))$ minus [the probabilities that the receptor leaves the R_s state] plus [the probabilities that the receptor enters the R_s state]. Figure 2.11 shows that there are two ways the receptor can leave the R_s state. It can transit to either DR_s or R_s^*. The probability of the former is equal to $Dk_{a+}\Delta t$ times the probability that the receptor is indeed in the R_s state at time $t(P_R(t))$. The probability of the latter is equal to $k_{q+}\Delta t$ times $P_R(t)$. The sum of the probabilities of the receptor leaving the R_s state is, therefore, $\Delta t(Dk_{a+}P_R(t) + k_{q+}P_R(t))$. Using analogous reasoning, the sum of the probabilities of the receptor entering the R_s state is $\Delta t(k_{a-}P_{DR}(t) + k_{q-}P_{R*}(t))$. Thus, Equation (2.76) accurately describes the probabilities of the receptor being in the R_s state at time $t + \Delta t$. Taking the limit as t approaches 0 yields differential equation (2.72). The other differential equations ((2.73)–(2.75)) are derived in an analogous manner. These equations can be solved numerically on the computer using a technique known as Euler's method. This technique is equivalent to solving Equation (2.67) iteratively on the computer over a number of time intervals until equilibrium is achieved. Hence, solving differential equations (2.72)–(2.75) yields results identical to those shown in Figure 2.12.

The point of this digression is that differential equations (2.72)–(2.75) are analogous to those describing the change in the various states of the receptor population:

$$\frac{dR}{dt} = -(Dk_{a+}R_s + k_{q+}R_s) + (k_{a-}DR_s + k_{q-}R_s^*), \tag{2.77}$$

$$\frac{dR^*}{dt} = -(Dk_{b+}R_{s^*} + k_{q-}R_{s^*}) + (k_{q-}R_s + k_{b-}DR_s^*), \tag{2.78}$$

$$\frac{dDR}{dt} = -(k_{g+}DR_s + k_{a-}DR_s) + (Dk_{a+}R_s + k_{g-}DR_s^*), \tag{2.79}$$

$$\frac{dDR_s^*}{dt} = -(k_{g-}DR_s^* + k_{b-}DR_s^*) + (Dk_{b+}R_s^* + k_{g+}DR_s). \tag{2.80}$$

The only difference is that each probability term in Equations (2.72)–(2.75) has been replaced with a receptor species in Equations (2.77)–(2.80). That is R_s, R_s^*,

DR_s and DR_s^* replace P_R, P_{R*}, P_{DR} and P_{DR*}, respectively, in Equations (2.72)–(2.75) to yield Equations (2.77)–(2.80). The value of each receptor species term is a real number bounded by 1 and 0 representing the fraction of the receptor population in a given state. The solutions to both sets of differential equations are identical, therefore, and the probability of a receptor being in a given state is equivalent to the fraction of the receptor population in that state. Thus, Figure 2.12 accurately describes the fractions of the receptor population in the various states as a function of time.

We will use this relationship to estimate affinity, efficacy and the microscopic constants, K_b and K_a, from measurements of open probability in the presence of different concentrations of agonist. Before describing the details of the method, I describe the behavior of single channels using a simulation technique.

Simulation of single channel events

The behavior of a single channel in time can be simulated using the conditional probabilities of transition matrix **T**. First, we choose a starting state (e.g., R_s) and a small time increment (Δt) and then estimate the three probabilities associated with the three possible events, given that the receptor is in state R_s: remain in state $R_s(1 \quad \Delta t(k_{q+} + Ak_{a+})$, transition to state R_s^* (Δtk_{q+}) or transition to state DR_s (ΔtAk_{a+}) (see the first column of **T** in Equation (2.65)). These conditional probabilities are used to calculate three probability intervals over a real number scale from 0 to 1.0. If Δt is small, then the probability for remaining in the R_s state is usually very close to 1.0 unless the agonist concentration is very high. For the sake of argument, let's assume the probability is 0.90, and the probabilities of transitioning to either the R_s^* or DR_s states are 0.01 and 0.09, respectively. We use these to establish three probability intervals from 0 to 1.0. These are 0–0.8999, 0.900–0.9099 and 0.910–1.0, for the R_s, R_s^* and DR_s states, respectively. Then, we pick a random number between 0 and 1 and determine in which probability interval the number falls. This interval corresponds to the state of the receptor during the time interval from t to $t + \Delta t$. If the starting condition is state R_s and Δt is small and the agonist concentration is not too high, the most likely event is that the receptor remains in the R_s state over the first time interval, 0 to Δt. This process is repeated over subsequent time intervals (e.g., Δt to $2\Delta t$, $2\Delta t$ to $3\Delta t$, $3\Delta t$ to $4\Delta t, \ldots$) until a random number is generated that falls within the probabilities intervals of either the R_s^* or DR_s states. The receptor has now transitioned to a new state; let us assume that it is DR_s. A new set of three probability intervals between 0 and 1.0 are calculated, based on the three conditional probabilities associated with the receptor being in state DR_s (see the third column of **T** in Equation (2.65)). The process just described for the prior state of the receptor is repeated until the receptor transitions to another state. This process is repeated iteratively to define the state of the receptor in time and to generate a series of open and closed times for the receptor. The entire simulation can be completed in a few seconds on the computer even though it involves innumerable steps.

Figure 2.13 shows the output from this simulation technique using an agonist concentration of 10^{-4} M and rate constants similar to those used in Figure 2.12 except that k_{g+} and k_{g-} are three-fold larger (see legend). The lower graph shows the state of the receptor (R_s, DR_s, DR_s^* and R_s^*). The upper plot shows the open (DR_s^* and R_s^*) and closed (DR_s and R_s) states of the receptor. It is apparent that, at this concentration of agonist, channel opening occurs in bursts, with the majority of the closed time being accounted for by the R_s state. Periods of channel opening occur as a series of rapid transitions between the DR_s and DR_s^* states because the gating constant (k_{g+}) is larger than the rate constant for the dissociation of the DR_s complex (k_{a-}). There are four constitutive receptor activations in the simulated

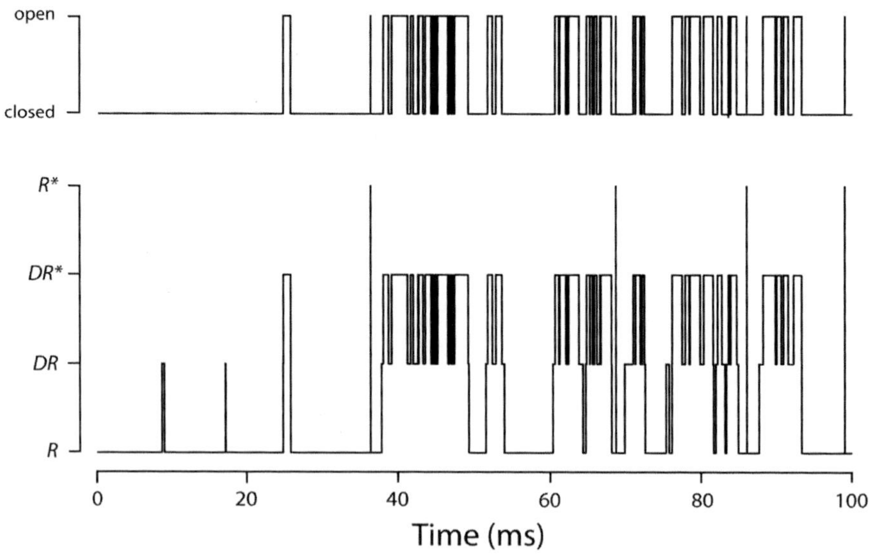

Figure 2.13: Simulation of single receptor events based on the one-site model shown in Figure 2.11. The agonist concentration is 10^{-4} M. Parameters of the simulation: k_{a+}, 10^6 M^{-1} s^{-1}; k_{a-} 10^3 s^{-1}, k_{b+}, 2×10^6 M^{-1} s^{-1}; k_{b-}, 0.1 s^{-1}; k_{q+}, 10^2 s^{-1}; k_{q-}, 10^6 s^{-1}; k_{g+}, 3×10^3 s^{-1} and k_{g-}, 1.5×10^3 s^{-1}.

100 ms record. On average, these have a mean open time (1 μs) smaller than the thickness of the line used to draw them. The simulated trace also shows three brief transitions to the DR_s state that did not result in channel opening. Two occur before 20 ms and the third occurs before the burst centered around 80 ms. The channel opening that occurs at approximately 25 ms appears to be a single transition from R_s to DR_s^*, but this is an illusion. When the time scale is expanded, the trace shows that the receptor remained in the DR_s state for about 31 μs before transitioning to DR_s^*, and upon channel closing, the receptor remained in the DR_s state for about 4 μs before transitioning to R_s.

Figure 2.14 shows seven single channel recordings that were simulated using the same values for the rate constants and agonist concentration (10^{-3} M) as in Figure 2.12. The average of these seven simulated recordings (lower trace in Figure 2.14) rises and approaches equilibrium with considerable fluctuation in the average current. An ensemble of 100 channels having the same microscopic constants is simulated in Figure 2.15 for the same concentration of agonist. The ensemble current rises and approaches equilibrium with the same kinetics as the plot of P_{DR*} (see Figure 2.12). That is, the time to reach equilibrium is approximately 5 ms, and at equilibrium, the current is half-maximal. It follows that the summation of many random quantal currents gives rise to a predictable graded current at the level of the receptor population.

The effect of the concentration of agonist on single channel recordings is simulated in Figure 2.16. Again, the rate constants are the same as those used in Figure 2.12. As the concentration of agonist increases, the time between bursts decreases. Constitutive channel openings are apparent at the low agonist concentrations (10^{-5}–10^{-4} M), and these appear as vertical lines in the simulated trace. The probability of channel opening can be calculated as the total open time divided by the total time of the simulation. Figure 2.17 shows a plot of this open probability (P_{DR^*}) as a function of the agonist concentration. The data can be fitted to the one-site model:

$$P_{DR^*} = \frac{[D]\varepsilon}{[D] + \frac{1}{K_{obs}}}. \tag{2.81}$$

This equation is analogous to Equation (2.30), but with the fraction of the agonist receptor complex in the open state (DR_s^*/R_T) replaced with P_{DR^*}. Nonlinear regression analysis yielded estimates of 0.65 ± 0.007 and 3.47 ± 0.02, for ε and log K_{obs}, respectively. These are essentially equivalent to the theoretical values (0.667 and 3.48, respectively), which can be calculated using Equations (2.19) and (2.13) and the theoretical microscopic constants (K_a, K_b and K_q), which in turn

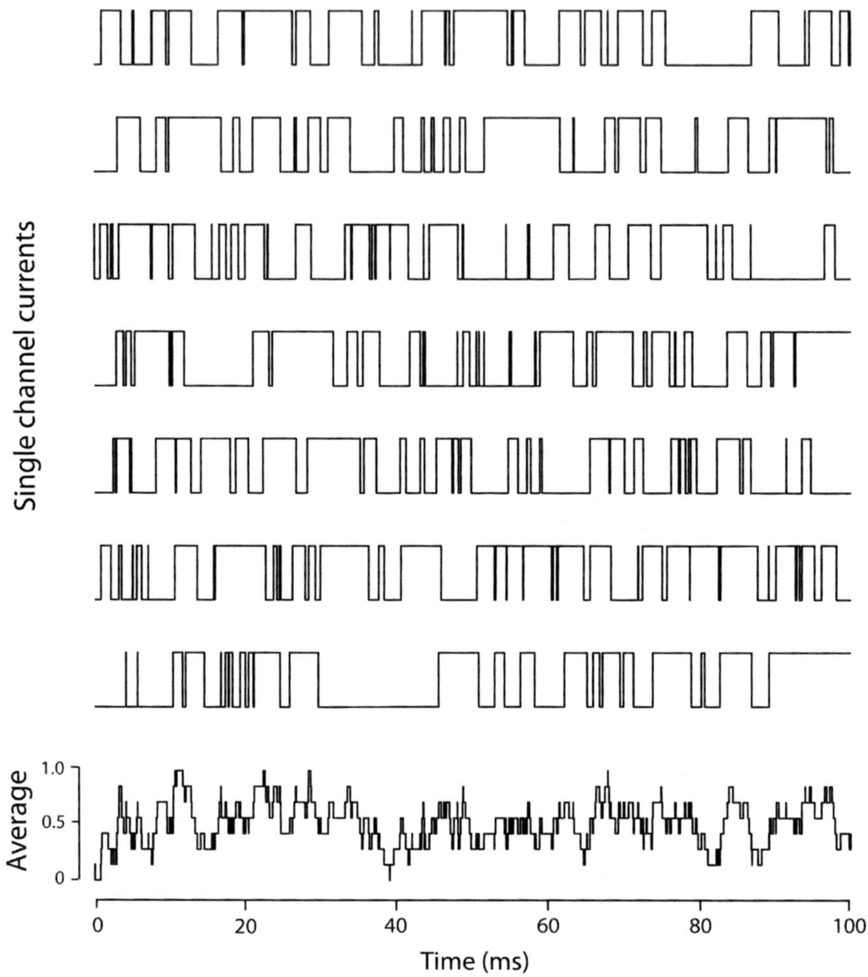

Figure 2.14: Simulation of seven single-channel recordings and their ensemble average (bottom trace). The agonist concentration is 10^{-3} M, and the rate constants are given in the legend to Figure 2.12.

can be calculated from the rate constants using Equations (2.56)–(2.58). Thus, open probability is proportional to the fraction of the receptor population in the active state. When there is little constitutive activity, open probability is equivalent to P_{DR^*}, whereas in the case of constitutive activity, it is equivalent to $P_{R^*} + P_{DR^*}$.

Starting conditions other than the case of the receptor being in the R_s state can also be simulated. Figure 2.18 shows the example where the receptor is in the DR_s^*

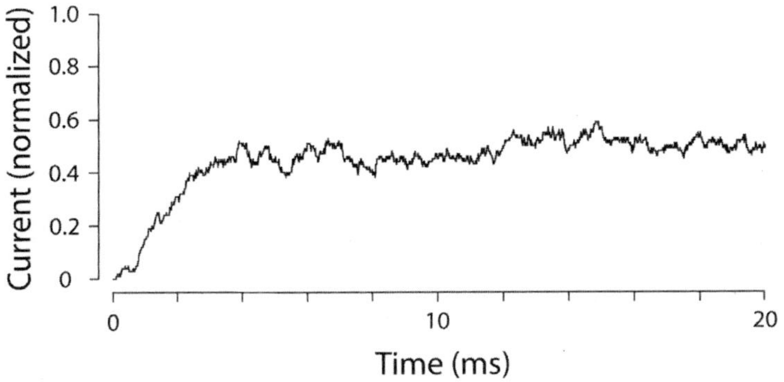

Figure 2.15: Simulation of an ensemble of 100 single-channel recordings. The agonist concentration is 10^{-3} M, and the rate constants are given in the legend to Figure 2.12.

state at the start of the time interval, and the agonist concentration is equal to 0. This condition corresponds to removing the agonist from the medium surrounding a receptor in the open state. Upon removal of the agonist, bursting continues until the agonist dissociates from the receptor. Figure 2.19 shows a series of nine single channel simulations for the case of a receptor that had been previously saturated with agonist, and then, the agonist removed at time = 0. Because the microscopic constants are the same as those used in Figure 2.17, the probabilities of the receptor being in the DR_s^* and DR_s states at the start of the time interval are 0.67 and 0.33, respectively. Each receptor finishes its burst and then remains silent except for rare constitutive activations. The third channel in Figure 2.19 never opened because it was in the DR_s state at the start of the time interval and subsequently transitioned to the R_s state. Channels 6 and 9 began the interval in the DR_s state, but each made a few transitions to the DR_s^* state before closing. The other channels were all in the DR_s^* state at time = 0. The average plot (bottom trace in Figure 2.19) shows that the decay in population current is determined by the random endings of the bursts of the nine individual channels. The behavior of an ensemble of 100 channels under the same condition is illustrated in Figure 2.20. The fractional population current at the start of the interval is approximately equal to efficacy (0.67) and this decays according to an exponential equation having a rate constant equivalent to the reciprocal of the mean apparent burst length given by Equation (2.92). The explanation for this behavior is described under "Distribution of open and closed channel times".

One can also use transition matrix **T** with Equation (2.70) to determine the probabilities of the various states as a function of time following removal of the agonist. In this case, the starting state vector ($\mathbf{P}^{(0)}$) is assigned probabilities of about

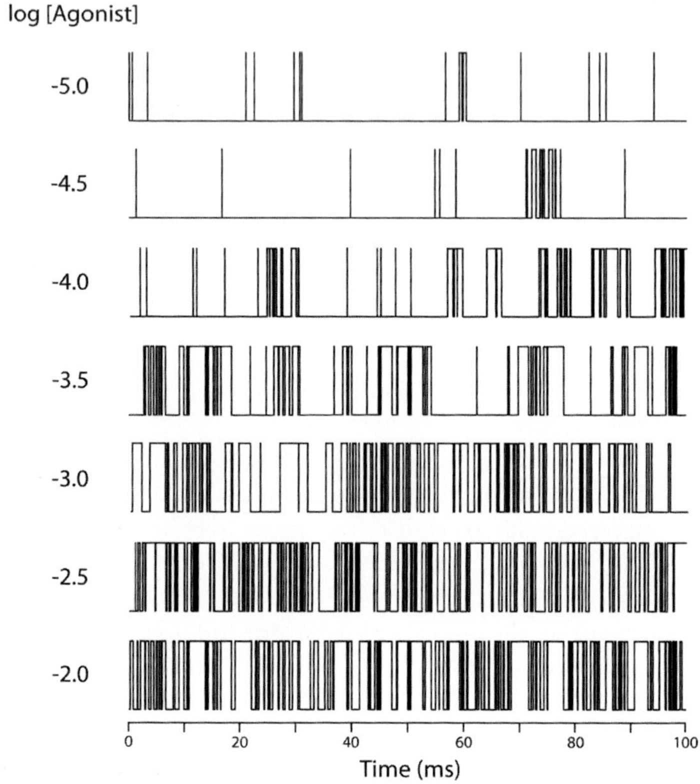

Figure 2.16: Simulation of single-channel recordings in the presence of different concentrations of agonist. Representative 100 ms simulations of longer one-second simulations are shown. Parameters of the simulation: k_{a+}, 10^6 M^{-1} s^{-1}; k_{a-} 10^3 s^{-1}, k_{b+}, 2 x 10^6 M^{-1} s^{-1}; k_{b-}, 0.1 s^{-1}; k_{q+}, 10^2 s^{-1}; k_{q-}, 10^6 s^{-1}; k_{g+}, 3 x 10^3 s^{-1} and k_{g-}, 1.5×10^3 s^{-1}.

0, 0, 0.33 and 0.67 for the R_s, R_s^*, DR_s and DR_s^* states, respectively, assuming the condition illustrated in Figure 2.20 (concentration jump to zero following prior saturation of the receptor population with agonist). Determining the probability of the receptor being in the DR_s^* state as a function of time using Equation (2.70) yields a smooth curve that agrees with the ensemble average in Figure 2.20 despite the random fluctuations in the latter.

Distribution of channel open times

If the time scale on the abscissa is divided into a series of intervals or bins and the number of open times that fall within a bin are plotted on the ordinate, the

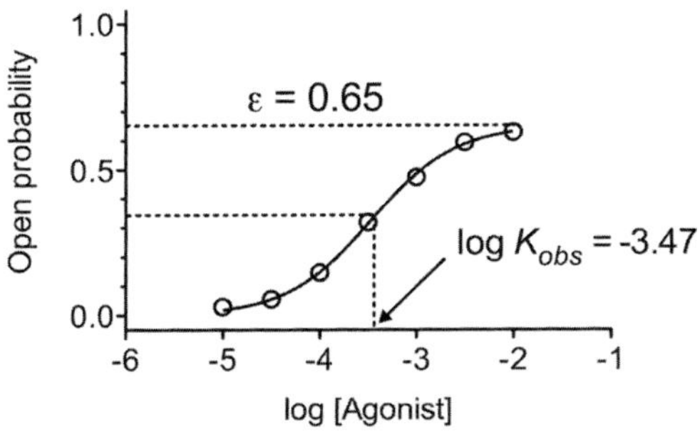

Figure 2.17: Relationship between the population parameters and the behavior of the open-probability plot. Open probability is plotted against the log agonist concentration. Open probability was estimated from the simulated data in Figure 2.16 over a 1 s time interval.

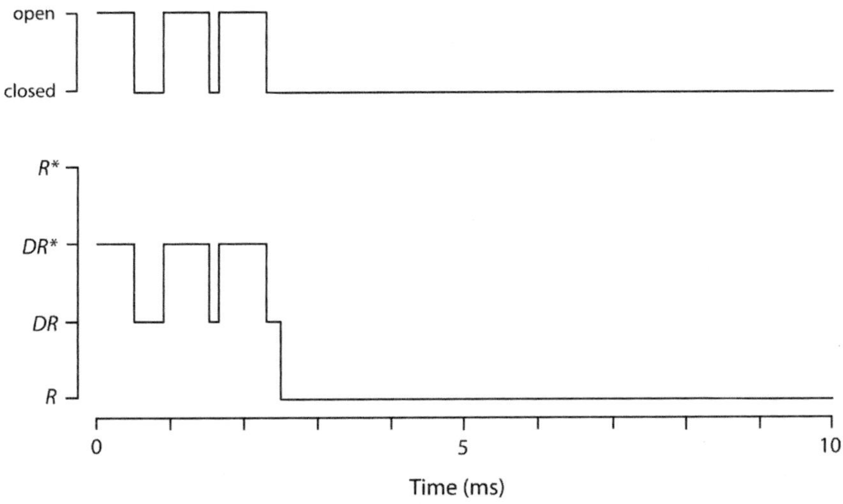

Figure 2.18: Simulation of a single channel recording following removal of agonist for a ligand-gated ion channel in the DR_s^* state at time = 0. The rate constants are given in the legend to Figure 2.16.

result is the distribution of channel-open times. An analogous procedure can be used to determine the distribution of channel-closed times. In the case of single channel events, these distributions are exponential, and the transition rate constants (e.g., k_{a+}, k_{a-}, k_{a+} and k_{a-}) are related to the observed rate constant (λ) of the distribution.

One way to think about this distribution is to consider a population of channels that are all open at a given time, t_0. The fraction of the population remaining in this

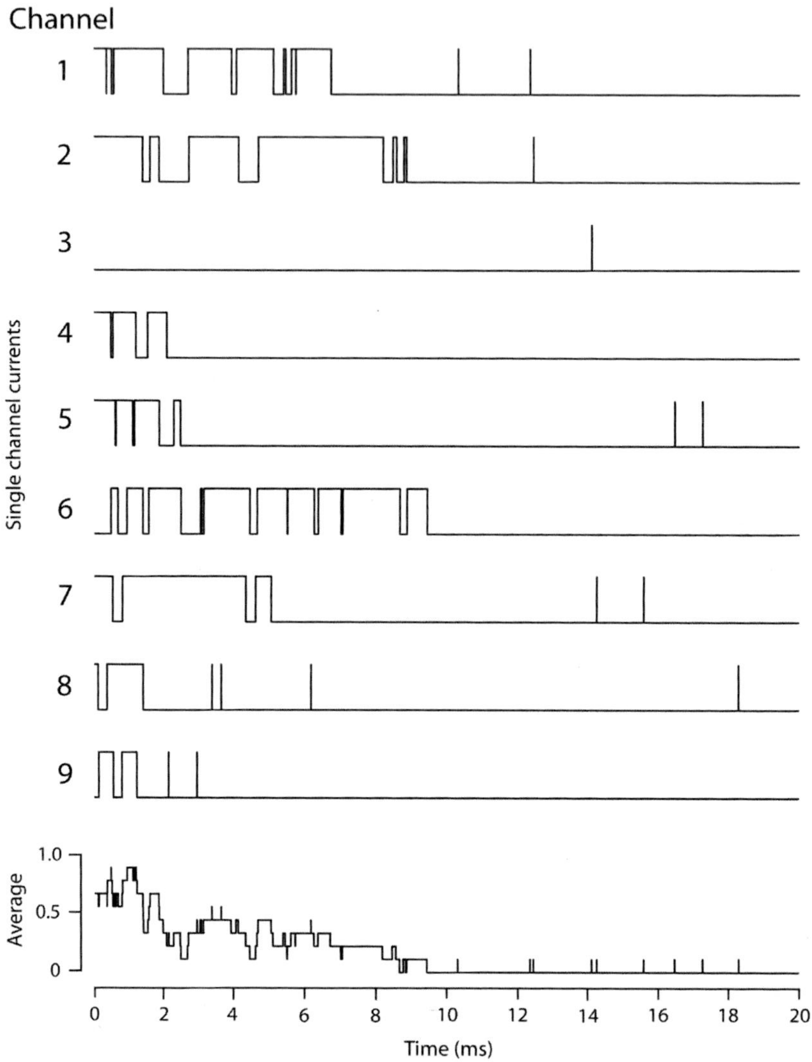

Figure 2.19: Simulation of nine single-channel recordings and their ensemble average (bottom trace) for the condition of a jump in the agonist concentration to zero following prior equilibration of a ligand-gated ion channel with a saturating concentration of agonist. The rate constants are given in the legend to Figure 2.16.

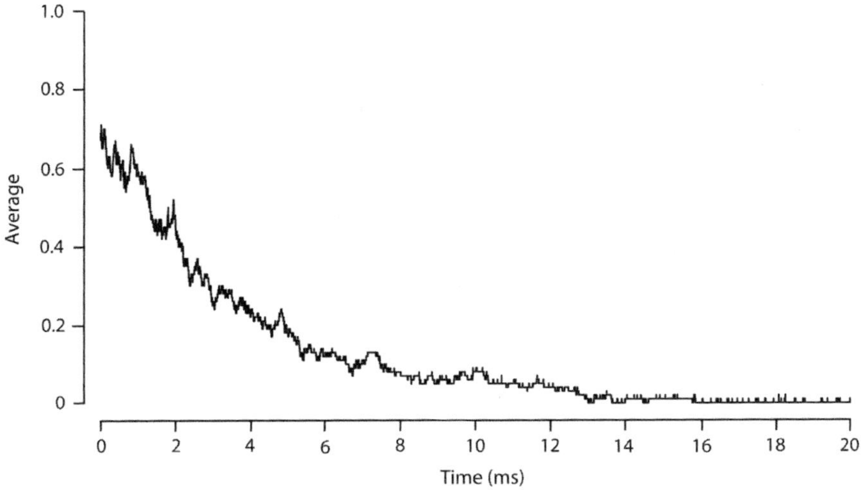

Figure 2.20: Simulation of an ensemble of 100 single-channel recordings for the condition of a jump in the agonist concentration to zero following prior equilibration of a ligand-gated ion channel with a saturating concentration of agonist. The rate constants are given in the legend to Figure 2.16.

first open state over time is described by an exponential decay equation ($y = e^{-\lambda t}$) essentially the same as that used to describe the decay of radioactivity, for example. The observed rate constant (λ) represents the transition rate leaving the open state. For the simulations in Figures 2.13–2.16, 2.18 and 2.20, there is one main open state (DR_s^*) and one main transition constant leaving this state (k_{g-}). Thus, λ is equivalent to k_{g-}:

$$\lambda = k_{g-}. \tag{2.82}$$

Thus, for our population of channels that were open at the starting time $t = 0$, the fraction continuously remaining open (y) at a later time is equivalent to:

$$y = e^{-k_{g-}t}. \tag{2.83}$$

Figure 2.21a shows an idealized plot of the fraction of a population of 10 channels undergoing their first closing event as a function of time. For this example, the value of k_{g-} is 1.5 ms^{-1}. Each horizontal bar represents a channel open time. The bars have been aligned to represent the condition of being open at time = 0, and the bars are stacked in rank order with the longest open time on the bottom. Obviously, a sample of 10 open times would have more random lengths; those shown in the figure would represent the average lengths of the shortest 10% of a large population, the

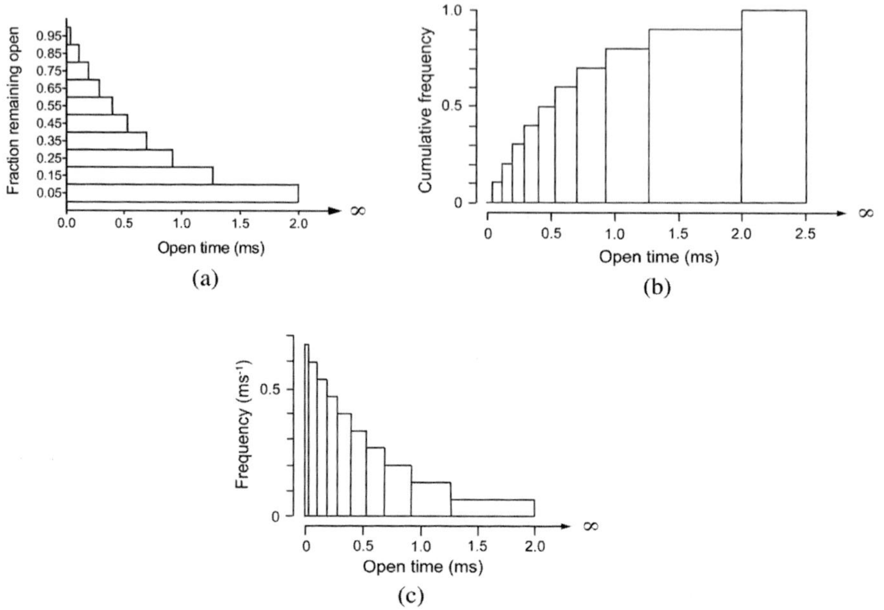

Figure 2.21: Relationships among the rank order of open times (a), the cumulative frequency distribution of open times (b) and the probability density function of open times (c).

next shortest 10% of the population, and so on. The figure shows the fraction of the population of channels remaining open as a function of time within the context of a single open event. If the channel opens again after its first closing, this is not shown on the plot.

The information in Figure 2.21a can be arranged in the form of a cumulative frequency distribution. That is, the fraction of the receptor population having an open time equal to or less than a given value is plotted against open time as shown in Figure 2.21b. This plot represents an increasing exponential equation of the form $y = 1 - e^{-\lambda t}$. It can be derived by subtracting the ordinate values in Figure 2.21a from one. To obtain the equation for the frequency distribution of open times ($f_o(t)$), the derivative of the equation for the cumulative frequency distribution is taken:

$$f_o(t) = \frac{d(1 - e^{-\lambda_o t})}{dt} = \lambda_o e^{-\lambda_o t}. \tag{2.84}$$

In this equation, λ_o denotes the rate constant for the exponential. It is equivalent to the sum of the two rate constants describing transitions away from DR_s^*

$(k_{g-} + k_{b-})$.[14] But because k_{g-} is so much larger than k_{b-} in this case ($k_{g-}/k_{b-} = 10^4$), the value of λ_o is essentially equivalent to k_{g-}. The distribution described by Equation (2.84) is proportional to that mentioned at the beginning of this section. That is, if we section the time scale beginning at time = 0 into bins and plot the fraction of the population of open times that occur in each bin, a histogram is obtained (Figure 2.21c). The height of each bar is proportional to e to the power of $-k_{g-}t$. Equation (2.84) denotes a special type of frequency distribution, however, for it has a total area of one and is known as a probability density function (pdf). The reciprocal of the observed rate constant ($1/\lambda_o$) is known as the time constant (τ_o), and it is equivalent to the mean open time:[14]

$$\tau_o = \frac{1}{\lambda_o} = \frac{1}{k_{g-}}. \tag{2.85}$$

Exponential distributions have some unique properties. If we consider a very large population of open times and rank each open time according to size, the middle-sized open time (median) is equivalent to the negative of the natural log of 0.5 times τ_o ($0.693\tau_o$). The median for the example shown in Figure 21a is approximately 0.47 ms (average of the fifth and sixth bars). The mean open time, however, is 0.67 ms. Hence, an exponential distribution is skewed to the right because its mean (τ_o) is greater than its median. The median divides an exponential distribution in half, and the frequency value of a bin at the median is about half that of an equivalent-sized bin for the shortest open times. An exponential distribution describes a memoryless process. For no matter how long a channel has already been open, its probability of remaining open still obeys the same probability density function.

An example of a frequency distribution of open times is shown in Figure 2.22a. This distribution was obtained by simulating single channel activity and plotting the number of open events for each 0.2 ms bin over the time interval from 0–6 ms. The agonist concentration (10^{-3} M) and rate constants are the same as those used for the simulation in Figure 2.12. There were 25 constitutive (R_s^*) and 741 agonist-induced (DR_s^*) channel openings. The behavior of the distribution is determined mainly by the agonist-induced channel openings because the proportion of constitutive openings is small. The mean open time is 0.64 ms, which is close to the predicted value (i.e., $1/k_{g-} = 0.67$ ms). The distribution is consistent with an exponential decay curve having a rate constant, λ, approximately equal to k_{g-}.

Because of the large difference in the predicted open times of spontaneous (R_s^*) and agonist-induced (DR_s^*) channel openings, it is difficult to resolve these on a single plot having a constant bin size. To alleviate this problem, investigators often plot the distribution using logarithmic bin sizes as shown in Figure 2.22b for the

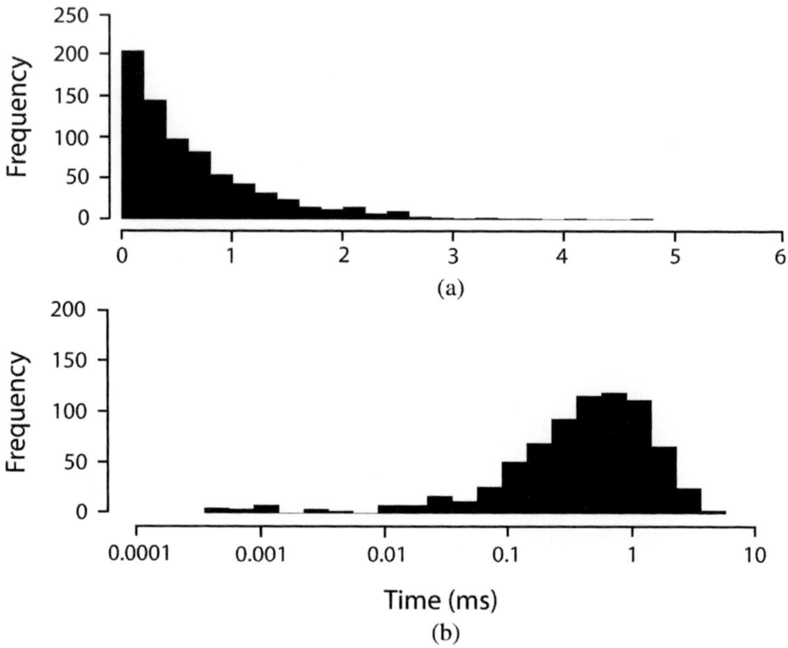

Figure 2.22: Simulation of the distribution of channel open times. The number of open events are plotted against open time using 0.2 ms bins (a) or bins of 0.2 log units (b). The rate constants are given in the legend to Figure 2.16.

same data plotted in Figure 2.22a. The peak of the main distribution falls within the bin containing the mean open time (0.64 ms). There are two small groups of events clustered around a mean open time of 1 μs, which is the predicted value for constitutive channel openings ($1/k_{q-} = 1\ \mu$s).

To convert the distribution into a PDF, each bar of the histogram in Figure 2.22a is divided by the total number of channel openings (766). When calculated in this manner, the combined areas of all of the bars sums to one, and the area of a given bar represents the probability of observing an open time within its bin interval.

This brings us to a fundamental corollary — the mean life of a state is equal to the reciprocal of the sum of the transition rates leaving the state.[14] Thus, when there is little constitutive activity, the mean life of the DR_s^* state (mean open time, τ_o), is equal to:

$$\tau_o = \frac{1}{k_{g-}}. \tag{2.86}$$

And the distribution of all open times resembles a single exponential function as described above in relation to Figure 2.22a. From the scheme in Figure 2.11 and the foregoing corollary, it follows that the mean life of the DR_s state is:

$$\tau_{DR} = \frac{1}{k_{g+} + k_{a-}}. \tag{2.87}$$

As mentioned above, the distribution shown in Figure 2.22 is actually more complicated than a single exponential because of the very small contribution of the R_s^* state. If constitutive activity were more prominent, one could obtain a more reliable estimate of k_{g-} by examining only those open times that occur within a burst and not the constitutive events ($R_s \rightarrow R_s^*$) that occur mainly during periods between bursts. Alternatively, one could examine the entire distribution with a bi-exponential equation ($y = a_1\lambda_1 e^{-\lambda_1 t} + a_2\lambda_2 e^{-\lambda_2^t}$) to dissect out the individual components (i.e., DR_s^* and R_s^*). The observed rate constant for each component of the distribution is usually a function of more than one transition constant, however, and the derivation of the areas of each component (a_1 and a_2) and their observed rate constants (λ_1 and λ_2) in terms of the transition rate constants (e.g., k_{g+}, k_{g-}, k_{a-}, etc.) is not trivial.

Distribution of burst lengths

An apparent burst is defined as a series of transitions between the DR_s and DR_s^* states that is separated from an adjacent burst by a closed time that is much longer than the small closed times within a burst. As shown in Figure 2.13, each burst begins with an $R_s \rightarrow DR_s$ transition followed by a $DR_s \rightarrow DR_s^*$ transition that marks the beginning of the apparent burst. Each apparent burst ends with a $DR_s^* \rightarrow DR_s$ transition followed by a $DR_s \rightarrow R_s$ transition that marks the end of the burst. Thus, the length of a burst is equivalent to the corresponding apparent burst measured by recording single channel activity plus the lengths of times of two DR_s states, one before and one after the apparent burst. Like many other aspects of single channel behavior, channel bursts have been described in detail by David Colquhoun and Alan Hawkes,[1,14,15] including the mean number of openings within a burst, and the distribution of burst times. The number of open times within an apparent burst (n_{ob}) for our hypothetical one-site ligand-gated ion channel is given by:

$$n_{ob} = 1 + \frac{k_{g+}}{k_{a-}}. \tag{2.88}$$

The mean open time per apparent burst ($mean_{oab}$) is equal to the product of n_o and the mean open time from Equation (2.84), $1/k_{g-}$:

$$mean_{oab} = \frac{k_{a-} + k_{g+}}{k_{g-}k_{a-}}. \tag{2.89}$$

Because there is always one more open time per burst than the number of closed times (n_{cb}), the latter is equal to:

$$n_{cb} = \frac{k_{g+}}{k_{a-}}. \tag{2.90}$$

The mean closed time per apparent burst ($mean_{cab}$) is equal to the product of n_c and the mean closed time (τ_{DR}) given by Equation (2.87):

$$mean_{cab} = \frac{k_{g+}}{k_{a-}(k_{g+} + k_{a-})}. \tag{2.91}$$

The mean length of an apparent burst ($mean_{ab}$), therefore, is equal to the sum of the mean open $mean_{oab}$ and closed $mean_{cab}$ times within the burst,

$$mean_{ab} = \tau_{ab} = \frac{k_{g+} + k_{a-}}{k_{g-}k_{a-}} + \frac{k_{g+}}{k_{a-}(k_{g+} + k_{a-})}, \tag{2.92}$$

and it is unaffected by the concentration of agonist.

The distribution of apparent burst lengths is described by the summation of two exponentials, although the two components may not correspond to channel states. In many instances, the distribution is consistent with a single exponential, like that shown for the simulation of the decay in the population current following removal of agonist (Figure 2.20).

This gives us another way to think about the distribution of burst lengths, which involves considering the decay in the population current following removal of the agonist. Immediately before removal of agonist, only those receptors that are occupied and currently undergoing a burst will contribute to the population current after removing the agonist from the incubation medium. Even though the agonist has been removed from the incubation medium, receptors undergoing a burst still remain bound with agonist. Thus, the decay in the population current is caused by the random endings of single channel burst events, and the decay in the population current is equivalent to the summation of these burst events. As described in connection with Figure 2.18, the time constant for this decay is equivalent to the mean apparent burst time given by Equation (2.92). Thus, the decay in the population current is proportional to the distribution of burst times.

This relationship may seem intuitively correct at this point, but recall that the decay in the population current is not a frequency distribution. When its maximum is normalized to one ($e^{-t/\tau}$), it represents the complement of the cumulative frequency distribution of burst times ($1 - e^{-t/\tau}$). Taking the derivative of the cumulative frequency distribution yields the frequency distribution as a probability density function ($e^{-t/\tau}/\tau$) (see derivation of Equation (2.84)). Because both the decay in the population current and the distribution of burst times are described by exponentials having the same time constant (τ), there is agreement in the estimates of τ_b by both methods.

Finally it may seem that measuring the mean apparent burst length from the decay of the population current would yield an estimate less than the true value because bursts are being sampled after they have already begun, and surely the time left in a burst that has already begun is less than its entire burst time. There is no error, however, because Markov processes are memoryless when viewed from the perspective of the population. This issue is sufficiently confusing that it has been given a name — the Waiting Time Paradox. The basis for the explanation lies in the skewed nature of the exponential distribution as described by Colquhoun.[14] At any given instant in time, those receptors undergoing a burst are more likely to have a mean apparent burst length longer than the average apparent burst length of the population. Removing the agonist at a given instant in time, therefore, selects for longer bursts, and the distribution of their remaining burst times is equivalent to that of the average value of the population.

Distribution of closed times

The principles described above for open times also apply to the distribution of closed times, which are shown in Figure 2.23 for our hypothetical one-site, ligand-gated ion channel. The conditions of this simulation are the same as those of the prior figure except that the concentration of agonist was reduced to 3×10^{-4} M. The frequency distribution in Figure 2.23a shows one main distribution with a spread of much longer closed times to the right of it. Most of these longer events represent times between bursts spent in the R_s and DR_s states. These closed intervals are occasionally bisected by brief constitutive transitions to R_s^*. In this simulation, the rate of gating ($DR_s \rightarrow DR_s^*$; $k_{g+} = 3\ \text{ms}^{-1}$) is three-fold greater than the rate of dissociation of agonist from the receptor ($DR_s \rightarrow R_s$; $k_{a-} = 1\ \text{ms}^{-1}$), and hence, the number of closed times within a burst is greater than the number of intervals between bursts. Figure 2.23b shows the frequency distribution of closed events plotted with logarithmic bin intervals. There appear to be two main distributions. The larger

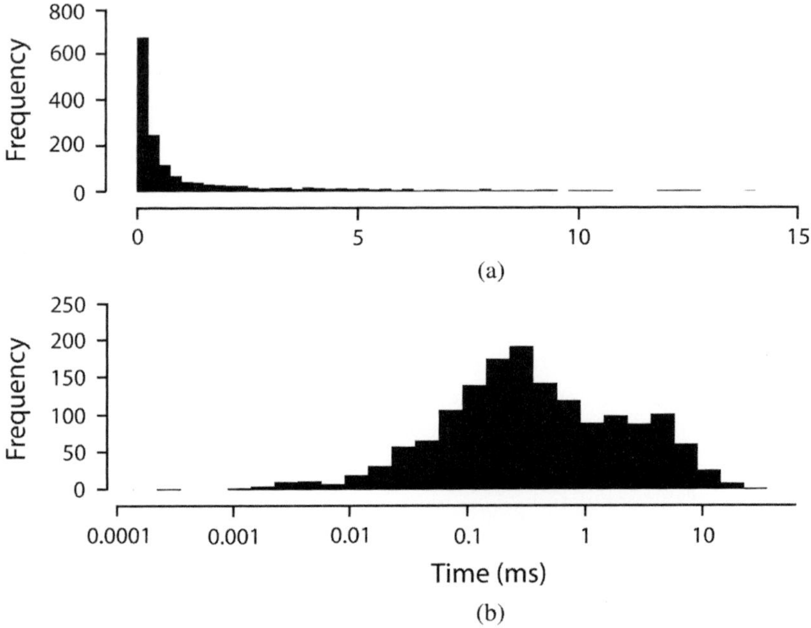

Figure 2.23: Simulation of the distribution of channel closed times. The number of closed events are plotted against closed time using 0.2 ms bins (a) or bins of 0.2 log units (b). The rate constants are given in the legend to Figure 2.16.

is composed mainly of closed times within bursts and has a peak frequency at about 0.3 ms. These closed times are attributed to the DR_s state, and this is predicted to have a mean duration of 0.25 ms (see Equation (2.87); $\tau_{DR} = 1/(k_{g+} + k_{a-}) = 0.25$ ms). Hence, is it not surprising that the compound distribution in Figure 2.23b has a major peak near 0.25 ms. The smaller distribution having a broader peak at approximately 3 ms is attributed to the closed times between bursts.

One can eliminate the contribution of closed times between bursts by sampling only within bursts. As inferred from Figure 2.23, the probability density function for these closed times ($f_{cb}(t)$) is:

$$f_{cb}(t) = \lambda_{cb} e^{-\lambda_{cb} t}, \tag{2.93}$$

where,

$$\lambda_{cb} = k_{g+} + k_{a-} = \frac{1}{\tau_{cb}}. \tag{2.94}$$

In this latter expression, τ_{cb} denotes the mean closed time within a burst.

Alternatively, if the receptor lacks constitutive activity the distribution of all closed events can be fitted to a bi-exponential equation as described by Colquhoun and Hawkes.[16] The latter approach can provide estimates of k_{g+}, k_{a+} and k_{a-}, and therefore, the microscopic affinity constant for the DR_s state (i.e., $K_A = k_{a+}/k_{a-}$).

Finally, the distribution of apparent closed times between bursts can be analyzed separately using a compound exponential equation as described by Colquhoun and Hawkes,[1] and these results also provide estimates of k_{g+}, k_{a-} and k_{a+}. As mentioned above, the closed time between bursts decreases with an increase in agonist concentration. At receptor-saturating concentrations of agonist, the receptor would appear to be in a continuous burst if it were not for desensitization and potential blockage of the channel pore by the agonist.

Estimation of microscopic constants and the macroscopic constants, affinity and efficacy, from the distributions of single channel dwell times

Using the time constants for the various distributions of single channel events, it is possible to estimate observed affinity ($K_{\rm obs}$) and efficacy (ε) and some of the microscopic constants (K_a and K_g) of the model shown in Figure 2.1.

The equation for mean apparent burst length (Equation (2.92)) can be used to estimate k_{a-} and k_{g+}. This equation contains three unknowns (k_{a-}, k_{g+} and k_{g-}) so it is necessary to make some substitutions to solve for k_{a-}, for example. As mentioned above, τ_o can be estimated from the distribution of open times, and its reciprocal yields an estimate of k_{g-}. From Equation (2.90), it follows that the product of n_{cb} and k_{a-} can substitute for k_{g+}. Making these substitutions into Equation (2.92) yields:

$$mean_{ab} = \frac{n_{cb}k_{a-} + k_{a-}}{\frac{1}{t_o}k_{a-}} + \frac{n_{cb}k_{a-}}{k_{a-}(n_{cb}k_{a-} + k_{a-})}. \tag{2.95}$$

Solving for the reciprocal of k_{a-} yields:

$$\left(\frac{n_{cb}+1}{n_{cb}}\right)[mean_{ab} - (n_{cb}+1)\tau_o] = \frac{1}{k_{a-}}. \tag{2.96}$$

After calculating k_{a-} using Equation (2.96), k_{g+} can be calculated from Equation (2.90) using the estimate of n_{cb}. We will use these estimates in developing a model for conformational selection.

Before proceeding though, let us briefly examine what Equation (2.96) tells us about the mean closed time of a burst. The product of $n_{cb} + 1$ (number of openings per apparent burst) and τ_o (length of a brief opening within an apparent burst) is equivalent to the mean open time per apparent burst (m_{oab}). Making this substitution in Equation (2.96) yields:

$$\left(\frac{n_{cb}+1}{n_{cb}}\right)[mean_{ab} - m_{aob}] = \frac{1}{k_{a-}}. \tag{2.97}$$

The quantity in square brackets represent the total time of an apparent burst minus the mean open time of the apparent burst, which yields the mean closed time of the apparent burst. Making this substitution yields:

$$\left(\frac{n_{cb}+1}{n_{cb}}\right) mean_{cab} = \frac{1}{k_{a-}}. \tag{2.98}$$

Multiplying the mean closed time of an apparent burst by the ratio of number of openings ($n_{cb} + 1$) to closings (n_{cb}) within an apparent burst has the effect of adding one closed time to the mean closed time of the *apparent* burst. If one closed time is added to an apparent burst the result is the time of the burst. Hence the left side of Equation (2.95) is equivalent to the mean closed time of a burst ($mean_{cb}$):

$$mean_{cb} = \frac{1}{k_{a-}}. \tag{2.99}$$

Thus, the total time spent in the closed state during a burst is equivalent to $1/k_{a-}$. Based on the rule that the reciprocal of the sum of the rates leaving a state is equivalent to the mean life of a state, $1/k_{a-}$ would be equivalent to the mean time spent in the DR_s state if the receptor were incapable of opening.[1] In a sense, the burst represents a sojourn in the DR_s state that is punctuated by a series of openings. If all of the brief open times within a burst are subtracted from total burst time, the remaining time is analogous to that predicted for the DR_s state if it were incapable of opening.

Now let us return to the estimation of rate constants for the distributions of single receptor events. So far we have estimated k_{a-}, k_{g+} and k_{g-}. Let us assume that we can estimate k_{a+} through analysis of the distribution of closed times between bursts as described by Colquhoun and Hawkes.[1] With these four rate constants, we can calculate the microscopic constants for gating ($K_g = k_{g+}/k_{g-}$) and the binding of the agonist to the inactive state of the receptor ($K_a = k_{a+}/k_{a-}$). Now, the model for conformational selection can be derived as described by the scheme in 2.1, but with the isomerization step described by K_g instead of K_bK_q/K_a

(see Equation (2.4)):

$$D + R_s \overset{K_a}{\longleftrightarrow} DR_s \overset{K_g}{\longleftrightarrow} DR_s^*. \tag{2.100}$$

Using this model, we can relate affinity and efficacy to the microscopic constants (K_a and K_g) estimated from the distributions of single channel events. To begin, we solve the model to obtain equations for occupancy, observed affinity (K_{obs}) and efficacy (ε). Fractional receptor occupancy is given by:

$$\text{occupancy} = \frac{[DR_s] + [DR_s^*]}{[DR_s] + [DR_s^*] + [R_s]}. \tag{2.101}$$

Substituting Equations (2.8) and (2.9) for $[DR_s]$ and $[DR_s^*]$ and simplifying yields the standard equation for receptor occupancy described above (Equation (2.12)) but with the K_{obs} replaced with:

$$K_{\text{obs}} = K_a(1 + K_g). \tag{2.102}$$

Thus, Equation (2.102) can be used to estimate observed affinity from the estimates of K_a and K_g, provided that there is little constitutive activity (i.e., $K_q \ll 1$). When there is constitutive activity, the appropriate equation for K_{obs} is:

$$K_{\text{obs}} = \frac{K_a(1 + K_g)}{1 + K_q}. \tag{2.103}$$

It is easy to see that there is little difference between the value of K_{obs} from Equations (2.102) and (2.103) provided that $K_q \ll 1$. Equation (2.103) is equivalent to the previously described Equation (2.13). The latter expresses K_{obs} in terms of K_a, K_b and K_q, whereas the former uses K_g instead of K_q.

To derive an equation for efficacy, we begin with the definition of efficacy — the fraction of the population of agonist-receptor complexes in the active state:

$$\varepsilon = \frac{[DR_s^*]}{[DR_s] + [DR_s^*]}. \tag{2.104}$$

Substituting in $[DR_s^*]/K_g$ for $[DR_s]$ (see Equation (2.4)) and simplifying yields:

$$\varepsilon = \frac{K_g}{K_g + 1}. \tag{2.105}$$

Thus, efficacy can be estimated from Equation (2.105) using the estimate of K_g calculated from the estimates of k_{g+} and k_{g-} (Equation (2.59)).

To summarize, it is possible to estimate some of the microscopic constants in the simple, one-site, conformational induction model from the distribution of single receptor events.

With a more realistic and complicated model, however, this approach is unfeasible. A much more robust method is to compute the likelihood of obtaining the entire sequence of open and closed times given a set of rate constants. This has been made possible because the distributions of observed open and closed times have been worked out in a general way for a variety of relevant reaction schemes.[15] Being able to calculate the likelihood of a given sequence of alternating open and closed times enables one to adjust the parameters and identify a vector of parameter estimates that maximize the likelihood of the observed record. Different models can be compared based on the maximum likelihood estimate. Colquhoun and Hawkes[14] and other investigators[17] have developed these methods, and two computer programs for this analysis are available (HJCFIT, http://www.ucl.ac.uk/Pharmacology/dcpr95.html#hjcfit and QUB, http://www.qub.buffalo.edu/).

Estimation of observed affinity and efficacy from measurements of open probability

In this and the following sections, I describe the approach for estimating microscopic constants (K_b and K_a) and observed affinity (K_{obs}) and efficacy (ε) from measurements of open probability. The method is based on the assumption that the simple one-site model accounts for the underlying mechanism (Figure 2.11). The method takes advantage of the proportionality between the probability of single receptor events and the average behavior of the population. Essentially, open probability is used as a fractional measure of DR_s^* or the sum of DR_s^* and R_s^* for the case of constitutive activity.

As described earlier, open probability (P_{open}) is equivalent to the time spent in the open state divided by the total time of the observation, which can be determined from an idealized record of open and closed channel times. A simpler method of estimation is based on the principle that open probability is equivalent to the charge passed during the observation time (C_o) divided by the theoretical amount of charge passed if the channel or channels were open continuously during the same period. The latter is equal to the product of the single channel current (I_{unit}), the number of channels and observation time (t), whereas the former is equivalent to the integral of the measured current trace. Both methods, of course, underestimate open probability whenever desensitization occurs, which is likely at high agonist concentrations. In this analysis, it is assumed that appropriate measures have been taken to eliminate the contribution of desensitization, missed events and potential errors associated with ambiguity in the number of channels in the record. This might be difficult, and perhaps impossible, at some ligand-gated ion channels. Nonetheless, my goal is to explain the relationships among open probability, the microscopic constants, and observed affinity and efficacy.

The approach for estimating affinity and efficacy simply involves fitting the appropriate equation based on population analysis to the open probability plot using nonlinear regression analysis. A trivial change is required, and this simply involves replacing variables related to population current with the corresponding open probability variables. Recall that Equation (2.32) expresses the population current as a function of the agonist concentration. Substituting open probability (P_{open}) and open probability at 100% receptor occupancy by the ligand ($P_{open\text{-}max}$) for current (i) and maximum current (I_{max}) in Equation (2.32) yields:

$$P_{open} = \frac{[D]P_{open\text{-}max}}{[D] + \frac{1}{K_{obs}}}. \tag{2.106}$$

This equation represents a simple one-site model having an EC_{50} of $1/K_{obs}$ and a maximum of $P_{open\text{-}max}$, which is equivalent to observed efficacy:

$$P_{open\text{-}max} = \varepsilon. \tag{2.107}$$

Figure 2.24a shows an example of this type of analysis using theoretical data. The best fit of Equation (2.106) to the plot of open probability against the agonist concentration yielded estimates of 4.58 and 0.77 for log K_{obs} and ε, respectively. These are consistent with the microscopic constants used to simulate the data, which are given in the Figure 2.24a legend.

For a receptor exhibiting constitutive activity, the effects of various agonist concentrations on open probability is measured and the resulting data are fitted to a modified form of Equation (2.40), which is given below with the appropriate variables replaced to indicate that the independent variable is open probability:

$$P_{open} = P_{open\text{-}sys} + \frac{[D](P_{open\text{-}max} - P_{open\text{-}sys})}{[D] + \frac{1}{K_{obs}}}. \tag{2.108}$$

In this equation, $P_{open\text{-}max}$ is defined as in Equation (2.106) and $P_{open\text{-}sys}$ represents the probability of opening in the absence of agonist, which is equivalent to the efficacy of the system:

$$P_{open\text{-}sys} = \varepsilon_{sys}. \tag{2.109}$$

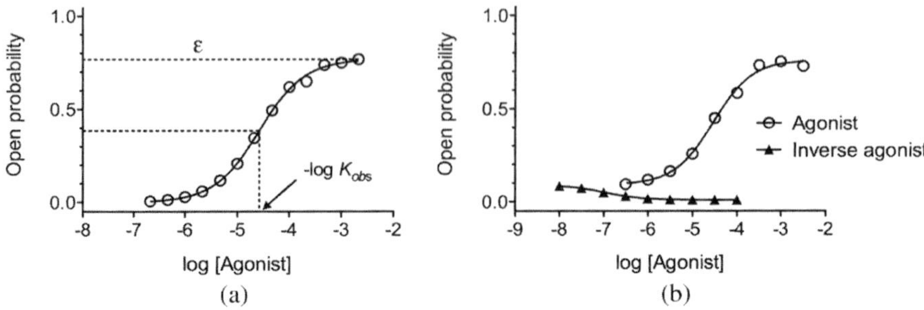

Figure 2.24: Simulation of open-probability plots. (a) Relationship between the population parameters (ε and K_{obs}) and the behavior of the open-probability plot. The theoretical data were generated using Equation (2.17) plus the addition of a random error having a range of $\pm$ 5%. The theoretical curve represents the least-squares fit of Equation (2.106) to the data. The microscopic constants for the simulation are: log K_a, 4.0; log K_b, 8.48 and log K_q, -4. These microscopic constants yield population parameters of log K_{obs}, 4.60 (Equation (2.13)) and ε, 0.75 (Equation (2.19)). (b) Simulated open-probability plots for an agonist (log K_a, 4.0; log K_b, 5.48) and inverse agonist (log K_a, 7.0; log K_b, 6.0) at a ligand-gated ion channel exhibiting constitutive activity (log $K_q = -1.0$). The simulated data were generated using Equation (2.25) with the addition of a $\pm$5% random error.

Analysis of the data from single channel records exhibiting constitutive activity is analogous to that shown for the population current in the presence of constitutive activity (see Figure 2.5). The principle difference is that the dependent variable is open probability, and the estimated parameters are those of Equation (2.108) instead of the corresponding population parameters in Equation (2.40). From this type of analysis, it is possible to estimate the K_{obs} and efficacy values of both agonists and inverse agonists and the efficacy of the system. Recall that efficacy is a population term whose value ranges from 0 to 1 for orthosteric ligands. Inverse agonists have efficacy values greater than zero, but less than that of ε_{sys}.

Estimation of microscopic affinity constants from measurements open probability

The estimation of microscopic constants from measurements of ligand-induced changes in open probability is analogous to the procedure used in the analysis of population currents. That is, the estimates of K_{obs} and ε are used to calculate relative or actual values of K_b and K_a, depending upon the presence of constitutive activity. A limitation in the population analysis is that it is never possible to obtain an actual value of ε for a ligand, only a value expressed relative to that of another ligand. Without a value for ε, it is impossible to estimate ε_i, and hence, the RI_i or

K_a values of efficacious agonists. This limitation does not exist in the analysis of open probability, because the open probability of an occupied receptor is equivalent to ε.

The requisite equations for estimating RA_i, RI_i, K_b and K_a are analogous to Equations (2.41), (2.44), (2.53) and (2.54) of the population analysis, but with the replacement of population parameters with open-probability parameters:

$$RA_i = \frac{K_{b-\mathrm{B}}}{K_{b-\mathrm{A}}} = \frac{\varepsilon_{\mathrm{B}} K_{\mathrm{obs-B}}}{\varepsilon_{\mathrm{A}} K_{\mathrm{obs-A}}} = \frac{P_{\mathrm{open-max-B}} K_{\mathrm{obs-B}}}{P_{\mathrm{open-max-A}} K_{\mathrm{obs-A}}}, \tag{2.110}$$

$$RI_i = \frac{K_{a-\mathrm{B}}}{K_{a-\mathrm{A}}} = \frac{\varepsilon_{i-\mathrm{B}} K_{\mathrm{obs-B}}}{\varepsilon_{i-\mathrm{A}} K_{\mathrm{obs-A}}} = \frac{(1 - P_{\mathrm{open-max-B}}) K_{\mathrm{obs-B}}}{(1 - P_{\mathrm{open-max-A}}) K_{\mathrm{obs-A}}}, \tag{2.111}$$

$$K_b = \frac{\varepsilon K_{\mathrm{obs}}}{\varepsilon_{\mathrm{sys}}} = \frac{P_{\mathrm{open-max}} K_{\mathrm{obs}}}{P_{\mathrm{open-sys}}}, \tag{2.112}$$

$$K_a = \frac{\varepsilon_i K_{\mathrm{obs}}}{\varepsilon_{i-\mathrm{sys}}} = \frac{(1 - P_{\mathrm{open-max}}) K_{\mathrm{obs}}}{1 - P_{\mathrm{open-sys}}}. \tag{2.113}$$

The equation describing the constant for the spontaneous activation of the receptor (constitutive activity) is:

$$K_q = \frac{\varepsilon_{\mathrm{sys}}}{\varepsilon_{i-\mathrm{sys}}} = \frac{P_{\mathrm{open-sys}}}{1 - P_{\mathrm{open-sys}}}. \tag{2.114}$$

An example of the analysis of the effects of an agonist and inverse agonist on open probability for a receptor exhibiting constitutive activity is shown in Figure 2.24b. Fitting Equation (2.108) to the data yields the following parameter estimates for the agonist (log K_{obs}, 4.54 ± 0.031; $P_{\mathrm{open-max}}$ (ε), 0.76 ± 0.008; log K_b, 5.46; and log K_a, 3.97) and inverse agonist (log K_{obs}, 6.96 ± 0.23; $P_{\mathrm{open-max}}$ (ε), 0.0097 ± 0.0063; K_b, 5.99; and K_a, 7.00). The system parameters are $\varepsilon_{\mathrm{sys}}$, 0.090 ± 0.006; $\varepsilon_{i-\mathrm{sys}}$, 0.90 ± 0.006 and log K_q, -1. The estimates of the microscopic constants are practically the same as those used in the simulation.

Note that the parameter, $P_{\mathrm{open-max}}$, represents the asymptotic value of open probability at maximal occupancy. For an agonist, therefore, $P_{\mathrm{open-max}}$ corresponds to the maximum open probability, but for an inverse agonist, it represents the minimum open probability.

One final point on microscopic constants and the one-site model; it is unnecessary to estimate these constants from the population parameters. One can fit an equation to the data directly that expresses open probability as a function of the microscopic constants and the ligand concentration. The appropriate equations are

(2.17) and (2.25), which express fractional receptor activation in terms of microscopic constants for the condition of measureable (Equation (2.25)) and immeasurable (Equation (2.17)) constitutive activity. Although the latter equations were developed for population analysis, the dependent variable, open probability, can be swapped for fractional activation of the receptor population because the equivalence of these measured as described above (see discussion of figure).

In the case of no constitutive activity, it is impossible to estimate unique values of K_b and K_q using this approach, but it is possible to estimate their product, and hence, the K_b value of an agonist is expressed relative to that of another agonist (RA_i value). For the case of constitutive activity, it is possible to estimate both K_b and K_q. In all cases, it is difficult to obtain an accurate estimate of K_a when $P_{\text{open-max}}$ approaches one. Also, it is difficult to estimate the K_b of an inverse agonist in a system with constitutive activity if $P_{\text{open-max}}$ is approximately equal to zero.

Summary of single channel analysis

The analysis of single channel recordings provides a direct measure of open probability in the presence of an agonist. Open probability is equivalent to the fraction of the receptor population in the active state, and hence, observed affinity (K_{obs}) and efficacy (ε) can be estimated by fitting a one-site model to a plot of open probability vs. agonist concentration. Knowing affinity and efficacy, it is possible to estimate a relative measure of the affinity constant of the agonist for the active state of the receptor. Measuring the effects ligands on the open probability of a receptor exhibiting constitutive activity enables the estimates of the observed affinities and efficacies of both agonists and inverse agonists, as well as the efficacy of the free receptor (ε_{sys}). With these population estimates, it is a simple matter to calculate the microscopic affinity constants of an orthosteric ligand for active K_b and inactive K_a states of the receptor. These are the fundamental parameters that determine drug-receptor interactions.

Cooperative Two-Site Model

Population analysis

Now let us consider a more realistic model that explains the behavior of ligand-gated ion channels, like the nicotinic and $GABA_A$ receptors described in Chapter 1. These receptors have two ligand-binding sites per receptor-channel complex, so we will need a model that incorporates two linked sites. Figure 2.25b shows such

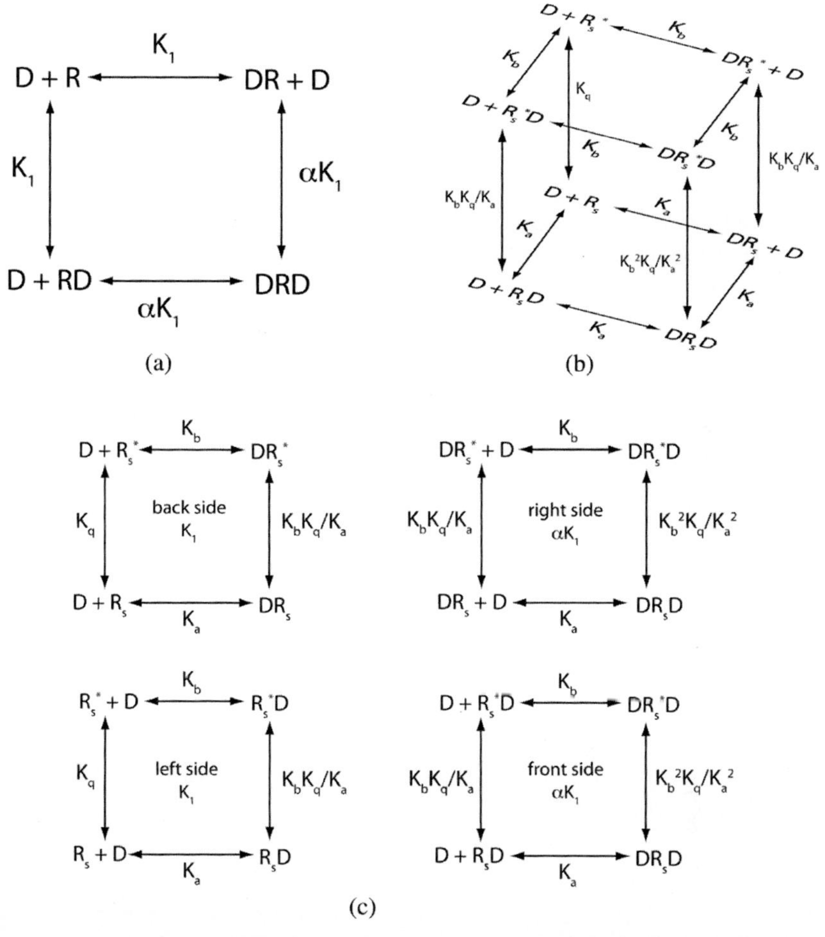

Figure 2.25: Two-site cooperative model for ligand-gated ion channels. (a) Cooperative binding of a ligand to two linked sites on a receptor. (b) The model in (a) is expanded to identify the active and inactive states of each receptor complex. Their properties are defined by the microscopic constants, K_a, K_b and K_q. (c) Each side of the cubic model in (b) is illustrated to show how the microscopic constants relate to the more superficial constants, K_1 and α.

a model at the level of receptor states. Each receptor complex exists in active (R_s^*) and inactive (R_s) states, representing the open and closed states of the channel. The microscopic affinity constants of the ligand for these states are denoted by K_b and K_a, respectively, and the constant describing the isomerization of the free receptor complex into the active state (constitutive activity) is denoted by K_q.

When a receptor is modeled at the level of receptor states, ligand affinity is determined by the state of the receptor. Thus, an agonist has the same high affinity (K_b) for all of the active receptor complexes on the top side of the cube (R_s^*, DR_s^*, R_s^*D), and also, the same low affinity (K_a) for all of the inactive receptor complexes on the bottom side (R_s, DR_s and R_sD).

The model can also be described by indicating how many ligands are bound to the receptor without specifying if the receptor is open or closed (Figure 2.24a). At this population level of analysis, the observed affinity constant of the binary complexes (DR and RD) is denoted by K_1 and that of the ternary complex by αK_1. Thus, the binding of one molecule of agonist enhances the binding of the second by the scalar, α, which is known as the cooperativity constant. Changeaux and colleagues[18] defined this type of cooperativity as homotropic because it represents an interaction between binding sites for the same ligand. In the case of agonists, which induce or select the active state, the cooperativity is positive, or more completely, positive homotropic cooperatively. The value of α determines whether the cooperativity is positive ($\alpha > 1$) or negative ($0, \alpha < 1$). In this case, where the same agonist binds to each homologous site, the most likely event is that each ligand will induce the same type of conformational change. This can only result in positive cooperativity. To complete the model in Figure 2.24a, we need to ascribe an efficacy value to each type of complex. This is complicated, however, because the efficacy of a given type of agonist-receptor complex (i.e., DRD, DR and RD) changes as a function of receptor occupancy, and hence, efficacy is no longer a simple scalar as in the case of the one-site model described earlier in this chapter. The equation that expresses efficacy as a function of receptor occupancy is given below.

Each equilibrium step or side of the population model shown in Figure 2.24a can be expanded into a square of four equilibria that describes the interaction of the ligand with the active and inactive states of the various receptor complexes (Figure 2.24c). Each of these four squares represents a side of the cube of equilibria shown in Figure 2.24b. Because there is neither input of energy nor breaking of covalent bonds, the system reaches equilibrium in time, and consequently, there is no net flux across each of the steps at equilibrium. Therefore, the four equilibria on any face of the cube can always be described in terms of three constants, K_b, K_a and K_q as shown in Figure 2.24c.[4]

The population constants, K_1 and αK_1, can be expressed in terms of the microscopic constants using the equilibrium schemes shown in Figure 2.24c. For example, from the upper two squares of equilibria in Figure 2.24c, it follows that

$$K_1 = \frac{[DR]}{[D][R]} = \frac{[DR_s^*] + [DR_s]}{[D]([R_s^*] + [R_s])}, \tag{2.115}$$

$$\alpha K_1 = \frac{[DRD]}{[D][DR]} = \frac{[DR_s^*D] + [DR_sD]}{[D]([DR_s^*] + [DR_s])}. \tag{2.116}$$

Each receptor term on the right sides of these equations (R_s^*, DR_s, DR_s^*, DR_sD and DR_s^*D) can be expressed in terms of R_s, the microscopic constants and usually the ligand concentration as prescribed by Equations (2.3), (2.119), (2.121), (2.123) and (2.125). Making these substitutions and simplifying yields:

$$K_1 = \frac{K_a + K_bK_q}{1 + K_q}, \tag{2.117}$$

$$\alpha K_1 = \frac{K_a^2 + K_b^2K_q}{K_a + K_bK_q}. \tag{2.118}$$

The more complete model in Figure 2.25b enables one to predict how the receptor behaves given a set of microscopic constants that describe the affinity of the ligand for the active and inactive states of the receptor. These are defined as described for the simple one site model, but there are additional complexities because two agonists can bind to each state of the receptor:

$$K_b = \frac{[DR_s^*]}{[D][R^*]}, \tag{2.119}$$

$$K_b = \frac{[R_s^*D]}{[D][R_s^*]}, \tag{2.120}$$

$$K_a = \frac{[DR_s]}{[D][R_s]}, \tag{2.121}$$

$$K_a = \frac{[R_sD]}{[D][R_s]}. \tag{2.122}$$

To simplify the analysis, the two types of binary complexes (DR_s and R_sD) have been given the same microscopic affinity constants. In practice, these affinity constants may differ depending on the structure and size of the ligand. It is clear that large antagonists of the muscle-type nicotinic acetylcholine receptor (e.g., D-tubocurarine) exhibit different affinities for the two sites.[19] These antagonists probably detect differences between the $\alpha - \gamma$ and $\alpha - \delta$ interfaces of the two binding pockets.

The analysis of single channel recordings of the muscle-type nicotinic acetylcholine receptor, however, reveals no evidence that acetylcholine and other small agonist ligands exhibit a difference in their microscopic affinity constants for the two sites.[20–22] Thus, the assumption that the microscopic constants are similar for the two sites is reasonable.

The binding of the second molecule of ligand to the active and inactive states of the binary complexes is characterized by the same affinity constant that described the binding of the first molecule:

$$K_b = \frac{[DR_s^* D]}{[D][DR_s^*]}, \tag{2.123}$$

$$K_b = \frac{[DR_s^* D]}{[D][R_s^* D]}, \tag{2.124}$$

$$K_a = \frac{[DR_s D]}{[D][DR_s]}, \tag{2.125}$$

$$K_a = \frac{[DR_s D]}{[D][R_s D]}. \tag{2.126}$$

The constant (K_q) describing the equilibrium between the active and inactive states is defined according to Equation (2.3) as mentioned previously in relation to the simple one-site model (i.e., $K_q = R_s^*/R_s$).

Derivation of equilibrium equations for the cooperative model

Having developed a realistic model for ligand-receptor interactions, we need to derive equations that describe the behavior of the receptor so that it will be possible to estimate relevant parameters related to affinity, efficacy and the open and closed states of the channel. Below, I describe the derivation of equations for occupancy, the fraction of the receptor population in the binary and ternary complexes, and the components of these complexes in the active state. The approach is the same as that used in the derivation of the equilibrium equations for the one-site model described earlier.

To derive the function that expresses receptor occupancy as a function of the ligand concentration, we begin with the following expression:

$$\text{occupancy} = \frac{[DR_s] + [R_s D] + [DR_s^*] + [R_s^* D] + [DR_s^* D] + [DR_s D]}{[DR_s] + [R_s D] + [DR_s^*] + [R_s^* D] + [DR_s^* D] + [DR_s D] + [R_s^*] + [R_s]}. \tag{2.127}$$

This equation defines fractional receptor occupancy as the sum of the receptor complexes bound with either one or two ligand molecules (numerator) divided by the sum of all of the receptor complexes (denominator). Next, each receptor species is substituted with an expression always containing R_s and usually other variables including D and the microscopic constants (K_a, K_b and K_q). These substitution equations are derived from Equations (2.119)–(2.126) that define the states of the various receptor complexes. Making these substitutions and canceling out the R_s variable yields:

$$\text{occupancy} = \frac{[D]^2K_b^2K_q + 2[D]K_bK_q + [D]^2K_a^2 + 2[D]K_a}{[D]^2K_b^2K_q + 2[D]K_bK_q + [D]^2K_a^2 + 2[D]K_a + K_q + 1}. \tag{2.128}$$

This equation can be rearranged into the following form, which gives some insight into the behavior of the occupancy curve:

$$\text{occupancy} = \frac{[D]}{[D] + \frac{K_q+1}{[D](K_b^2K_q+K_a^2)+2(K_bK_q+K_a)}}. \tag{2.129}$$

Finally, one more simplification can be made to illustrate the relationship between the cooperative model and the one-site model:

$$\text{occupancy} = \frac{[D]}{[D] + K_{\text{occ}}}. \tag{2.130}$$

In Equation (2.130), K_{occ} is defined as the concentration of ligand that yields half-maximal receptor occupancy, and its value depends on the ligand concentration and microscopic constants:

$$K_{\text{occ}} = \frac{K_q + 1}{[D](K_b^2K_q + K_a^2) + 2(K_bK_q + K_a)}. \tag{2.131}$$

On the surface, Equation (2.130) resembles that for the simple one-site model described earlier (Equation (2.12)) except that the empirical term, K_{occ}, replaces the reciprocal of the affinity constant (K_{obs}) in Equation (2.12). A more empirical term is used here because the value of K_{occ} depends on the ligand concentration as shown by Equation (2.131). That is, as the concentration of ligand increases, the affinity of the receptor changes. An increase in the concentration of an agonist, for example, causes the affinity of the receptor for the agonist to increase (i.e., decrease in K_{occ}). In contrast, the value of K_{obs} in the simple one-site model is constant and unaffected by the concentration of ligand. The change in K_{occ} with the ligand concentration is ultimately caused by the cooperative interaction between the two binding sites on the receptor.

An alternative way to describe receptor occupancy is with the population model shown in Figure 2.25a:

$$\text{occupancy} = \frac{[DR] + [RD] + [DRD]}{[DR] + [RD] + [DRD] + [R]}. \tag{2.132}$$

Using the same strategy to solve this model yields Equation (2.130) above, but with K_{occ} defined as:

$$K_{\mathrm{occ}} = \frac{1}{K_1([D]\alpha K_1 + 2)}. \tag{2.133}$$

Substituting from Equations (2.117) and (2.118) the values of K_1 and αK_1 in Equation (2.133) yields Equation (2.131).

We can also use the model in Figure 2.25b to define the various components of the receptor occupancy curve, and the definitions and solutions to these are given in the next series of Equations (2.134)–(2.141), beginning with the proportion of receptors in the ternary complex (DRD)/R_T:

$$\frac{[DR_s^*D] + [DR_sD]}{[R_T]} = \frac{[DR_s^*D] + [DR_sD]}{[DR_s] + [R_sD] + [DR_s^*] + [R_s^*D] + [DR_s^*D] + [DR_sD] + [R_s^*] + [R_s]}. \tag{2.134}$$

In this equation, $[R_T]$ denotes the total amount of receptors. Substituting each receptor complex in Equation (2.134) with the appropriate expression and simplifying yields:

$$\frac{[DRD]}{R_T} = \frac{D}{D + \frac{2[D](K_bK_q+K_a)+K_q+1}{[D](K_b^2K_q+K_a^2)}}. \tag{2.135}$$

The fraction of the receptor population in the binary complex is defined as:

$$\frac{[DR] + [RD]}{[R_T]} = \frac{[DR_s] + [R_sD] + [DR_s^*] + [R_s^*D]}{[DR_s] + [R_sD] + [DR_s^*] + [R_s^*D] + [DR_s^*D] + [DR_sD] + [R_s^*] + [R_s]}. \tag{2.136}$$

Following substitution and simplification, Equation (2.136) becomes:

$$\frac{[DR] + [RD]}{[R_T]} = \frac{[D]}{[D] + \frac{[D]^2(K_b^2K_q+K_a^2)+K_q+1}{2(K_bK_q+K_a)}}. \tag{2.137}$$

The fractional amount of ligand-receptor complex in the active state is defined as:

$$\frac{[DR_s^*]+[R_s^*D]+[DR_s^*D]}{[R_T]} = \frac{[DR_s^*]+[R_s^*D]+[DR_s^*D]}{[DR_s]+[R_sD]+[DR_s^*]+[R_s^*D]+[DR_s^*D]+[DR_sD]+[R_s]+[R_s^*]}. \tag{2.138}$$

Making the appropriate substitutions and simplifying yields:

$$\frac{[DR_s^*]+[R_s^*D]+[DR_s^*D]}{[R_T]} = \frac{1}{1+\frac{([D]K_a+1)^2}{[D]K_bK_q([D]K_b+2)}}. \tag{2.139}$$

The fractional amount of receptor complex in the active state is defined as:

$$\frac{[DR_s^*]+[R_s^*D]+[DR_s^*D]+[R_s^*]}{[R_T]} = \frac{[DR_s^*]+[R_s^*D]+[DR_s^*D]+[R_s^*]}{[DR_s]+[R_sD]+[DR_s^*]+[R_s^*D]+[DR_s^*D]+[DR_sD]+[R_s]+[R_s^*]}. \tag{2.140}$$

Making the appropriate substitutions and simplifying yields:

$$\frac{[DR_s^*]+[R_s^*D]+[DR_s^*D]+[R_s^*]}{[R_T]} = \frac{1}{1+\frac{([D]K_a+1)^2}{K_q([D]K_b+1)^2}}. \tag{2.141}$$

We will use these equations in the next two sections to describe the behavior of the model and determine the meaning of observed affinity and efficacy at the population level of analysis.

The meaning of observed affinity

In the case of the one-site model, a quantitative measure of observed affinity is straightforward. It represents the observed affinity constant (K_{obs}) that describes the binding of the ligand to the receptor in the simplest context, without concern for different states of the receptor as shown in scheme 2.14. The occupancy curves of the different states of the receptor are also defined by the same observed affinity constant, and these are consistent with a one-site model having a Hill slope of one.

In the case of the cooperative model, however, occupancy can be defined in different ways, for example, depending upon whether one normalizes the binding curve relative to 1) the amount of ligand bound, 2) the amount of receptor bound or 3) the amount of the ternary complex (DRD). The occupancy curves for the second

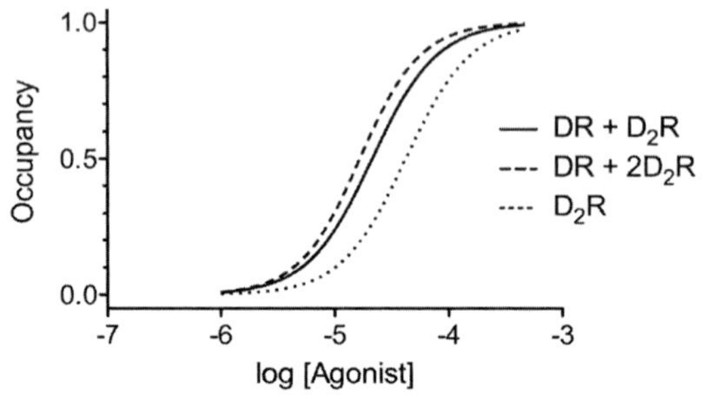

Figure 2.26: Some different ways of defining receptor occupancy for the cooperative two site model. The theoretical curves were generated using the appropriate equations in the text ((2.142), (2.129) and (2.135) for the dashed, solid and dotted curves, respectively) and the following values for microscopic constants of K_a, 10^4 M^{-1}; K_b 10^8 M^{-1} and K_q, 10^{-7}.

and third methods are described by Equations (2.129) and (2.135), respectively. The occupancy curve for the first method can be derived from Equation (2.127) by multiplying the ternary complex by two to account for two equivalents of ligand. Further simplification yields the following equation:

$$\frac{\text{Bound}_{\text{ligand}}}{[R_T]} = \frac{[D]}{[D] + \frac{K_q+1}{2([D](K_b^2K_q+K_a^2)+(K_bK_q+K_a))}}. \quad (2.142)$$

Figure 2.26 shows a plot of the three different types of occupancy curves for the cooperative model. The parameters for the simulation are listed in the legend to the figure. The log concentration of agonist required for 50% receptor occupancy varies from a low value of −4.78 for the first method to a high value of −4.37 for the third.

These three occupancy curves exhibit Hill slopes greater than one (1.52–1.62), so they are inconsistent with the simple one-site model, and therefore, with the concept of an observed affinity constant, which specifies the shape of the binding curve. Hence, the concentration of ligand required for 50% receptor occupancy has been defined using a more empirical variable (K_{occ}) rather than the reciprocal of the standard K_{obs} notation (see Equation (2.131)). The Hill slope is an empirical measure of the steepness of the occupancy curve, and it is determined by fitting the following equation to the data using nonlinear regression analysis:

$$\text{occupancy} = \frac{[D]^n}{[D]^n + K_{\text{occ}}^n}. \quad (2.143)$$

When the value of the Hill slope (n) is one, Equation (2.143) is equivalent to the simple one-site model (Equation (2.12)). In this case, the empirical variable, K_{occ},

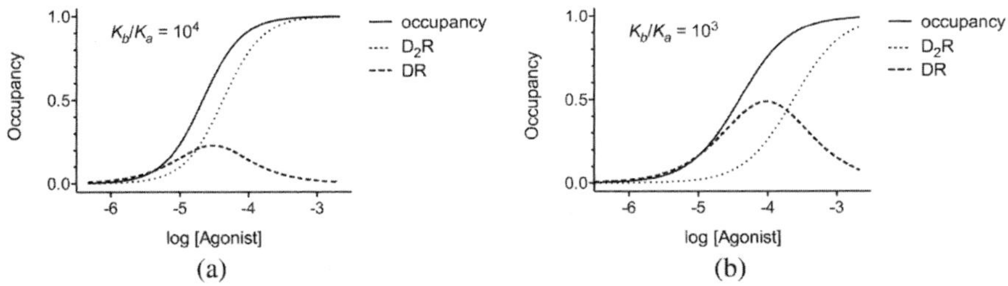

Figure 2.27: Relationships among receptor occupancy and the binary (DR_s) and ternary (D_2R) ligand-receptor complexes. The curves were generated using the appropriate equations in the text and the microscopic constants shown in the figure. The gating constant of the unliganded receptor (K_q) was 10^{-7}.

in the Hill equation represents the reciprocal of the observed affinity constant, and the change in the ligand concentration associated with an increase in occupancy from about 9% to 91% is 100-fold or two log units. When the occupancy curve exhibits cooperativity, the Hill slope is greater than one, and the change in the agonist concentration associated with an increase in occupancy from 9% to 91% is less than 100-fold. For highly efficacious agonists that elicit near complete activation of the receptor population, the Hill slope approaches two, whereas less efficacious agonists exhibit Hill slopes between 2 and 1.

This dependency of the Hill slope on efficacy is illustrated in Figure 2.27 for a receptor having a gating constant for the unliganded receptor (K_q) of 10^{-7}. Figure 2.27a shows the occupancy curve of a highly efficacious agonist having a K_b/K_a ratio of 10^4, which yields an empirical Hill coefficient of 1.52 when the occupancy curve (Equation (2.129)) is analyzed by Equation (2.143). Also shown are the binary and ternary complex components of the occupancy curve. Most of the bound ligand can be attributed to the ternary complex. Figure 2.27b shows the occupancy curve of a less efficacious agonist having one-tenth the K_b/K_a ratio (10^3), which results in a smaller Hill coefficient of 1.20. In this case, the binary complex contributes more to the occupancy curve.

In the remainder of this chapter, we will use the second definition of receptor occupancy as defined by Equation (2.129). This will facilitate estimation of the mean efficacy and the relative microscopic constants.

The meaning of observed efficacy

In the case of the one-site model, efficacy represents the fraction of the population of ligand-receptor complexes in the active state. It is calculated as the amount of active ligand-receptor complexes divided by the amount of receptors occupied. This fraction is constant for any level of receptor occupancy, although it is most often conceptualized as the relative amount of receptor complex in the active state when fractional occupancy is equivalent to 1.0 (i.e., 100% receptor occupancy). In the cooperative model, however, the efficacy of an agonist varies depending on the level of receptor occupancy.

To illustrate this relationship, ligand-induced receptor activation divided by receptor occupancy (i.e., efficacy) is expressed as a function of receptor occupancy. We have already developed equations for receptor occupancy (Equation (2.129)) and agonist-induced receptor activation (Equation (2.139)) as a function of the ligand concentration, but what we need is an equation that expresses activation as a function of occupancy. This can be derived by solving Equation (2.129) (occupancy as a function of D) for D, which yields a quadratic:

$$D^2a + Db + c = 0, \tag{2.144}$$

in which a, b and c represent expressions composed of occupancy and the microscopic constants, K_b, K_a and K_q. The appropriate solution to this quadratic is one of its roots:

$$D = \frac{-b + \sqrt{b^2 - 4ac}}{2a}, \tag{2.145}$$

in which

$$a = K_b^2 K_q + K_a^2, \tag{2.146}$$

$$b = 2(K_b K_q + K_a), \tag{2.147}$$

$$c = -\frac{\text{occupancy}\ (K_q + 1)}{1 - \text{occupancy}}. \tag{2.148}$$

Equation (2.145), which expresses the ligand concentration (D) as a function of occupancy, can be substituted for D in the ligand-induced receptor activation function (Equation (2.139)) to yield receptor activation as a function of occupancy (*activation* = f(*occupancy*)). Dividing this function by the independent variable, *occupancy*, yields *efficacy* as a function of occupancy:

$$\varepsilon = \frac{f(\text{occupancy})}{\text{occupancy}}. \tag{2.149}$$

Figure 2.28 shows a theoretical plot of efficacy against receptor occupancy for a group of agonists having a range of selectivities for the active state of the receptor (i.e., K_b/K_a) and a gating constant for the unliganded receptor (K_q) of 10^{-5}. In all

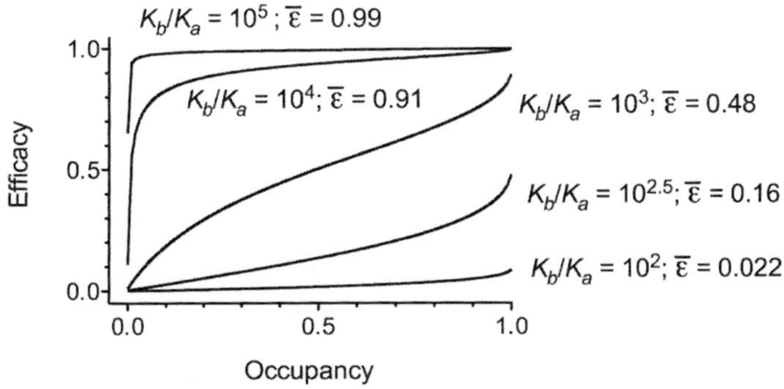

Figure 2.28: Relationship between efficacy and receptor occupancy for a group of agonists having varying selectivities for the active state of the receptor (K_b/K_a). The gating constant of the unliganded receptor (K_q) is 10^{-5}.

cases, efficacy increases with an increase in occupancy. For the most efficacious agonist, efficacy reaches a near maximal value at a low level of receptor occupancy and remains nearly constant at approximately 1.0 for all higher levels of receptor occupancy. For less efficacious agonists, efficacy increases throughout the occupancy interval. The mean efficacy value of the ligand represents the area under its efficacy versus occupancy plot, which can be estimated numerically by integrating the efficacy function (Equation (2.149)). This can be accomplished by dividing the area into a series of rectangles having a width equivalent to a small change in occupancy and a height equivalent to the value of efficacy at the start of the interval. The sum of the areas of these rectangles yields the estimate of mean efficacy, which is listed in the figure for each agonist.

Two agonist-binding sites are better than one

If the goal is to design a molecular switch that is "off" in the absence of endogenous ligand and "on" in its presence, then two ligand-binding sites per receptor are much better than one. This can be appreciated by considering how efficacy is determined by a ligand's ability to discriminate between active and inactive states of the receptor (K_b/K_a) at 100% receptor occupancy. For the one-site model, maximum receptor activation (activation$_{\text{max}}$) is given by Equation (2.19), which can be rearranged into the following form:

$$\text{activation}_{\text{max}} = \frac{\left(\frac{K_b}{K_a}\right) K_q}{\left(\frac{K_b}{K_a}\right) K_q + 1}. \tag{2.150}$$

In the case of the cooperative two-site model, receptor activation at 100% receptor occupancy can be determined by taking the limit of Equation (2.129) as D approaches infinity:

$$\text{activation}_{\max} = \frac{\left(\frac{K_b^2}{K_a}\right)^2 K_q}{\left(\frac{K_b^2}{K_a}\right)^2 K_q + 1}. \tag{2.151}$$

Thus, maximal receptor activation increases with an increase in the ratio K_b/K_a in the one-site model, but with the square of K_b/K_a in the cooperative two-site model.

From the perspective of designing an optimal switch, we need to place a reasonable limit on the maximum selectivity of a small neurotransmitter for the active state (e.g., $K_b/K_a = 10^4$) assuming a reasonable affinity for the ground state (e.g., $K_a = 10^4\ \text{M}^{-1}$). This yields K_b and K_a values (10^8 and $10^4\ \text{M}^{-1}$) not much different from those estimated by Anthony Auerbach and colleagues[23] for acetylcholine at the muscle-type nicotinic receptor (5.0×10^7 and $7.7 \times 10^3\ \text{M}^{-1}$, respectively). If we adjust K_q so that our ligand activates 95% of the receptor population at 100% receptor occupancy, what would be the basal noise in the "off" state when the receptor is unoccupied? From the perspective of a single switch, our goal amounts to a 95% probability of opening when the switch is occupied. For both models, receptor activation in absence of ligand ($K_q/(K_q + 1)$ is approximately equal to K_q, which is typically a very small number. Thus, given the design constraints just mentioned, the one-site and cooperative two-site models would exhibit constitutive activities of 0.003 and 0.00000003, respectively. The extra noise of the one-site model is costly, and hence, the evolution of two-sited ligand-gated ion channels probably represents an adaptation for minimizing wasteful basal activity while maximizing activity in the "on" state.[24]

Estimation of microscopic constants from the current response

Having an explicit function for receptor activation as a function of the agonist concentration (Equation (2.141)) enables one to determine how agonist activation depends on the microscopic affinity constants of the agonist. Thus, Equation (2.141) should predict the behavior of the concentration-response curve of an agonist for eliciting a whole-cell current response under voltage clamp conditions before desensitization occurs. If it were possible to fit Equation (2.141) to the whole-cell current data, we would be able to get information about the microscopic affinity constants of the agonist. These estimates would have a certain amount of error, however, depending on the degree of desensitization.

Before attempting this analysis, we need to modify Equation (2.141) because the dependent variable is fractional receptor activation, whereas the measured data is simply the whole cell current. There is no way to convert the raw data into fractional receptor activation unless the maximum current response of the receptor population is already known. The appropriate form of Equation (2.141) for the regression is:

$$i = \frac{i_{\text{max-sys}}}{1 + \frac{([D]K_a+1)^2}{K_q([D]K_b+1)^2}}, \quad (2.152)$$

in which i denotes the measured current and $i_{\text{max-sys}}$ the maximum current response if all of the ligand-gated ion channels were open (maximum response of the system).

But if you try to fit Equation (2.152) to the concentration-response curve shown in Figure 2.29a, for example, most nonlinear regression programs will not converge on a solution because there are an infinite number of combinations of parameter estimates (infinite solution set) that satisfy the least-squares criterion. This does not mean that *any* set of parameters will fit the data because different agonist concentration-response curves are described by different infinite solution sets.

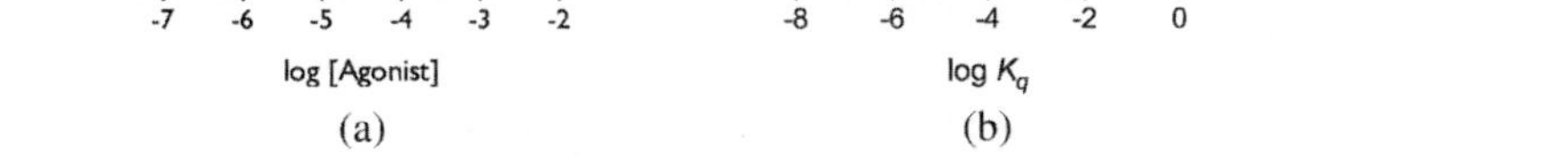

Figure 2.29: Relationship between the agonist-induced current response of a population of ligand-gated ion channels and the solution set of microscopic constants that provide the least-squares fit to the data. (a) Simulated agonist concentration-response curve for whole cell current. The curve was generated using Equation (2.141) with microscopic constants of K_a, 10^4 M^{-1}; K_b, 10^9 M^{-1} and K_q, 10^{-7}. A random error (±5%) was added. (b) Summary of nonlinear regression analysis of the data in panel (a) using Equation (2.152). A least-squares fit is obtained (minimum value of RSS) for the domain $K_q < 10^{-3}$. Over this domain, the estimate of $K_b^2K_q$ is constant. The estimate of K_a is also constant for most of the same domain (i.e., $K_q < 10^{-4}$).

There is something unique about the relationship among the parameters of a specific solution set, and the trick is to determine this relationship.

Consider the concentration-response curve of a highly efficacious agonist that elicits a maximum response nearly equivalent to that expected for activation of the entire receptor population (see Figure 2.29a). This theoretical receptor activation curve was generated using Equation (2.152) and the microscopic constants listed in the figure legend. A random error was added to the simulated data. In this case, there is no detectable constitutive activity, and hence, K_q must be very low. Recall that K_q represents the equilibrium constant describing the spontaneous conversion of the free receptor into the active state ($K_q = R_s^*/R_s$). But how low is K_q? It could be 10^{-3}, for example. This value represents about 0.1% receptor activation in the absence of agonist, which would be difficult to detect. Or it could be much lower because lower values of K_q would obviously yield undetectable levels of constitutive activity as well. To determine how the solution set depends on the value of K_q, Equation (2.152) is fitted to the data by nonlinear regression analysis with the parameter K_q constrained to a low value during the analysis. This process is repeated iteratively, each time constraining K_q to a new value so that a range of K_q values is explored.

A summary of this walk through parameter space is shown in Figure 2.29b. Each step of the walk represents one set of parameter estimates denoted by a set of plotted points forming a vertical line that intersects the abscissa at the value to which K_q was constrained during the regression. The plot shows how the parameter estimates depend on K_q. Also shown are the residual sum-of-squared deviations (RSS) of the data from the best fit of each repression step. RSS is at a constant minimum value whenever $K_q < 10^{-3}$. As the constant to which K_q is constrained decreases from 10^{-3}, the estimate of K_b increases in proportion to the square root of the reciprocal of K_q such that the estimate of the product of K_q and K_b^2 is constant. Over the domain $K_q < 10^{-4}$, the estimate of log K_a is also constant although it decreases somewhat between 10^{-4} and 10^{-3} and becomes inaccurate and essentially equivalent to zero for the domain $K_q > 10^{-3}$. Finally, over the domain of $K_q <$ 0.1, the estimate of $i_{\text{max-sys}}$ is constant. Thus, it is possible to estimate $K_b^2 K_q$ and $i_{\text{max-sys}}$ whenever RSS is at a minimum. If we have additional information showing that K_q is less than 10^{-4}, then it is also possible to estimate log K_a. Making this assumption yields the following parameter estimates: log K_a, 3.96 ± 1.26; log $K_b^2 K_q$, 11.01 ± 0.08 and $i_{\text{max-sys}}$, 1.00 ± 0.02. These values are essentially the same as those used to simulate the data (log K_a, 4.0; log $K_b^2 K_q$, 11 and $i_{\text{max-sys}}$, 1.00). When the i_{max} value of the agonist is close to the maximum possible current response ($i_{\text{max-sys}}$), there is substantial error in the estimate of log K_a as shown by the large standard error for log K_a in this example.

If this process were repeated with a partial agonist having a maximum current response (i_{max}) much less than that of the maximum response of the system, then it would be impossible to obtain reliable estimates of $i_{max\text{-}sys}$ and log $K_b^2K_q$, although the estimate of log K_a would be accurate. This is because there is insufficient information within the concentration-response curve of an agonist having a Hill slope close to 1.0 to give a reliable estimation of $i_{max\text{-}sys}$. The solution to this problem is to analyze the less efficacious agonist and the highly efficacious agonist simultaneously using global nonlinear regression analysis, sharing the estimate of $i_{max\text{-}sys}$ among the data and obtaining unique estimates of log $K_b^2K_q$ and log K_a for each agonist.

It is unnecessary to use the iterative procedure described in Figure 2.29b for this analysis, however. If there is sufficient reason to assume that the K_q value is very low, then regression analysis can be done with log K_q constrained to an arbitrarily low value (e.g., −6). Also, regression equation (2.152) can be modified so that the product, $K_b^2K_q/(1+K_q)$ represents a single parameter (P_b):

$$i = \frac{i_{max\text{-}sys}}{1 + \frac{([D]K_a+1)^2}{K_q\left([D]\sqrt{\frac{P_b(1+K_q)}{K_q}}+1\right)^2}}. \tag{2.153}$$

Global nonlinear regression analysis is done by sharing the estimate of $i_{max\text{-}sys}$ among the agonists and constraining K_q to an arbitrarily low value (e.g., 10^{-6}), provided that there is no constitutive activity. Figure 2.30 shows the results of this type of analysis when a series of simulated agonist concentration-response curves

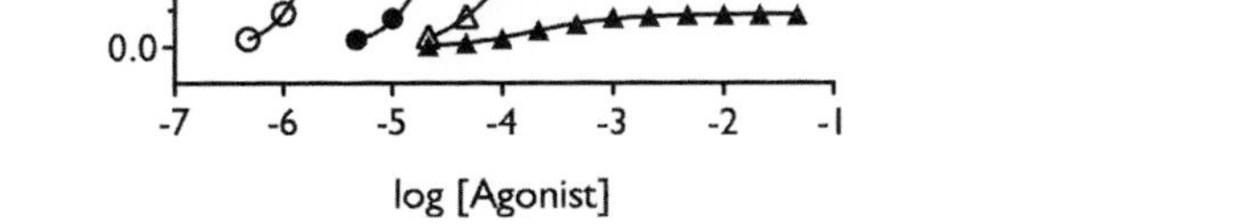

Figure 2.30: Simulated agonist concentration-response curves for eliciting whole cell current. The curves were generated using Equation (2.141) and parameter values of 10^{-7} (K_q) and 10^4 M^{-1} (K_a) for all of the agonists. The relative value of K_b for each agonist is listed in the figure. A random error (±5%) was added. The curves represent the least squares fit of Equation (2.141) to the data.

were analyzed simultaneously. Regression analysis yielded the following parameter estimates for the agonists: agonist A: log K_a, 3.93 ± 0.96; log $K_b^2 K_q/(1 + K_q)$, 11.01 ± 0.06; agonist B: log K_a, 3.98 ± 0.05; log $K_b^2 K_q/(1 + K_q)$, 8.97 ± 0.05; agonist C: log K_a, 4.00 ± 0.05; log $K_b^2 K_q/(1 + K_q)$, 7.94 ± 0.09 and agonist D: log K_a, 4.02 ± 0.25; log $K_b^2 K_q$, 7.03 ± 0.47. The estimate of $i_{\text{max-sys}}$ was 1.00 ± 0.012. These values are nearly the same as those used to simulate the data as can be determined from the microscopic parameters given in the legend to the figure.

Having a reliable estimate of $K_b^2 K_q/(1 + K_q)$ (P_b) enables one to estimate the microscopic affinity constants of an agonist relative to that of another agonist:

$$RA_i = \sqrt{\frac{P_{b-\text{B}}}{P_{b-\text{A}}}} = \frac{K_{b-\text{B}}}{K_{b-\text{A}}}. \qquad (2.154)$$

In this expression, the subscripts A and B designate the parameters of the standard agonist (agonist A) and the test agonist (agonist B). This relative estimate is known as *intrinsic relative activity* (RA_i). The corresponding estimate for the inactive state (intrinsic relative inactivity, RI_i) is calculated with the following equation:

$$RI_i = \frac{K_{a-\text{B}}}{K_{a-\text{A}}}. \qquad (2.155)$$

Again, subscripts (A and B) are used to designate the parameters of the test (agonists B) and standard (agonist A) agonists.

For the estimation of RA_i, the regression equation can be modified so that RA_i is a single parameter. This modification requires fitting the data for the standard agonist in Equation (2.153) with the parameters P_b and K_a of the standard agonist designated with the subscript A:

$$i = \frac{i_{\text{max-sys}}}{1 + \dfrac{([D]K_{a-\text{A}}+1)^2}{K_q\left([D]\sqrt{\frac{P_{b-\text{A}}(1+K_q)}{K_q}}+1\right)^2}}, \qquad (2.156)$$

while the data for the tests agonists are fitted are fitted to a different equation in which the test agonist parameters are denoted with a subscript B:

$$i = \frac{i_{\text{max-sys}}}{1 + \dfrac{([D]K_{a-\text{B}}+1)^2}{K_q\left([D]RA_{i-\text{B}}\sqrt{\frac{P_{b-A}(1+K_q)}{K_q}}+1\right)^2}}. \qquad (2.157)$$

Using global nonlinear regression analysis, the two equations are fitted simultaneously by 1) constraining K_q to an arbitrarily low value, 2) sharing the estimates of $i_{\text{max-sys}}$ and $P_{b-\text{A}}$ among the agonists, 3) obtaining a unique estimates of RA_i for each test agonist, and 4) obtaining a unique estimate of K_a for each of the agonists. When the data in Figure 2.30 were analyzed in this manner, the following values of log RA_i, expressed relative to agonist A, were estimated: agonist B, -1.02 ± 0.04; agonist C, -1.53 ± 0.05 and agonist D, -1.99 ± 0.24.

Although native ligand-gated ion channels usually lack constitutive activity, some mutant receptors of interest occasionally do exhibit constitutive activity, and hence, methods for analyzing these mutants are useful. If the receptor exhibits sufficient constitutive activity, the binding of one agonist molecule to the receptor may cause considerable receptor activation. Constitutive activity (ca) is defined as the fraction of the free receptor complex in the active state, which depends only on K_q:

$$ca = \frac{R^*}{R_s + R_s^*} = \frac{K_q}{1 + K_q}. \tag{2.158}$$

Constitutive activity, therefore, provides the necessary information for estimating K_q. It follows that it should be possible to estimate K_b and K_a from concentration-response curves of agonists measured at a ligand-gated ion channel that exhibits constitutive activity. The easiest way to do this is to fit Equation (2.152) to the data and obtain unique estimates of K_a, K_b, K_q and $i_{\max\text{-sys}}$. Again, it is difficult to estimate the K_a value of an agonist having an $i_{\max}$ equivalent to $i_{\max\text{-sys}}$, and this situation would be more common at a constitutively active receptor.

An example of an analysis of this type is shown for the theoretical data in Figure 2.31. The theoretical curves represent the least-squares fit of Equation (2.152) to the data. Global nonlinear regression analysis was done sharing the estimate of $i_{\max\text{-sys}}$ and K_q among the agonists and obtaining unique estimates of K_b and K_a for each agonist. The following K_b and K_a values were estimated from this analysis: agonist A, log K_b, 7.02 ± 0.02; agonist B, log K_b, 6.05 ± 0.03, log K_a, 4.55 ± 0.1; agonist C, log K_b, $4.72 \pm 0.0.04$, log K_a, 4.00 ± 0.03 and inverse agonist D, log K_a, 5.91 ± 0.53. Because of the high efficacy of agonist A, it was impossible to estimate its K_a value. It is also impossible to estimate the K_b value of an inverse agonist that nearly completely inhibits basal activity (e.g., Inverse Agonist D). The other parameter estimates, inclucing $i_{\max\text{-sys}}$ (1.01 ± 0.006) and log K_q (-1.32 ± 0.03), are similar to those used to simulate the data (see legend, Figure 2.31).

This approach of obtaining estimates of K_b and K_a for constitutively active receptors is particularly useful for the analysis of single channel recordings because the estimation of the open probability of the free receptor is more accurate than

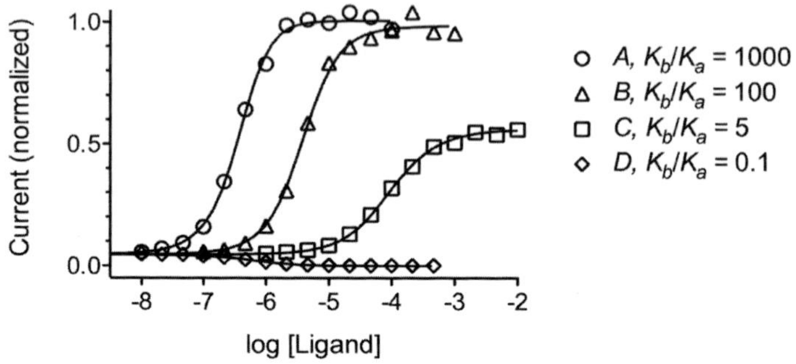

Figure 2.31: Simulated ligand concentration-response curves for eliciting whole cell current at a ligand-gated ion channel exhibiting constitutive activity. The curves were generated using Equation (2.141) and parameter values of 0.05 (K_q) and 10^4 M^{-1} (K_a) for all of the agonists except the inverse agonist ($K_a = 10^6$ M^{-1}). The relative value of K_b for each agonist is listed in the figure. A random error (±5%) was added. The curves represent the least squares fit of Equation (2.141) to the data.

the measurement of the constitutive current elicited by the receptor population. In addition, estimates of open probability can be corrected for desensitization.

Estimation of observed affinity and mean efficacy from the population response

The ultimate goal in the analysis of drug-receptor interactions is to determine the microscopic affinity constants of the ligand for the active and inactive states of the receptor. For these are the true independent variables that ultimately determine observed affinity and efficacy. These microscopic constants are most easily estimated from the population current response using the approach described in the prior section. Estimating observed affinity is more complicated because the measured current response is essentially proportional to the fraction of the receptor population in the ternary complex, whereas occupancy is proportional to the summation of the binary and ternary complexes. For agonists with low efficacy, the contribution of the binary complex can be substantial. Estimating mean efficacy is also complicated because efficacy depends on receptor occupancy. Despite these complications, I describe a method for estimating observed affinity and mean efficacy simply to show the relationship between these population parameters and the microscopic affinity constants (K_b and K_a).

The parameters estimated from the analysis of the current response can be used to estimate observed affinity ($1/K_{\text{occ}}$) and mean efficacy. When there is no measureable constitutive activity, it is possible to measure K_a and the product, $K_b^2 K_q/(1+K_q)$

(see prior section). From these products, provisional values of K_b' and K_q' are calculated. This is done by assigning K_q to an arbitrarily low value and calculating K_b by multiplying the estimate of $K_b^2 K_q/(1 + K_q)$ (P_{bq}) by the provisional value of $(1 + K_q')/K_q'$ and then taking the square root of the quotient. While these provisional estimates are unlikely to be the true estimates, they are, nonetheless, unique values of the infinite solution set that provides the best fit of the receptor activation function (Equation (2.139)) to the data. Hence, these provisional estimates can be used to estimate the empirical parameter, K_{occ}. This parameter is equivalent to the concentration of agonist yielding half-maximal receptor occupancy (i.e., 0.5), and it can be derived by solving the root to the quadratic given by Equation (2.145) when occupancy $= 0.5$. The reciprocal of K_{occ} is defined as observed affinity $(1/K_{\text{occ}})$. We will not use the variable, K_{obs}, for this purpose because of the empirical nature of $1/K_{\text{occ}}$. The estimates of the log $1/K_{\text{occ}}$ $(-\log K_{\text{occ}})$ values of agonists A–D in Figure 2.30 were 5.52, 4.63, 4.43 and 4.41, respectively.

The provisional estimates of K_b, K_a and K_q can also be used to estimate mean efficacy as described previously. The provisional microscopic constants are substituted into Equations (2.146)–(2.148) to solve Equation (2.145), which expresses the ligand concentration as a function of occupancy. Equation (2.145) is then substituted into the receptor activation function (Equation (2.139)) to express receptor activation as a function of occupancy. This function is divided by receptor occupancy to yield efficacy (see Equation (2.149)). The resulting function is integrated numerically over the interval of receptor occupancy from 0–1. Using this method, the mean efficacy values of agonists A–D in Figure 2.30 were 0.925, 0.487, 0.142 and 0.0217, respectively.

For the cooperative two-site receptor model, it can be shown the product of observed affinity and the square root of the mean efficacy of a test agonist expressed relative to that of a standard agonist is approximately equal to the corresponding ratio of microscopic affinity constants for the active state of the receptor:

$$\frac{K_{b-\text{B}}}{K_{b-\text{A}}} = RA_i \approx \frac{\sqrt{\bar{\varepsilon}_{\text{B}}}\,\frac{1}{K_{\text{occ}-\text{B}}}}{\sqrt{\bar{\varepsilon}_{\text{A}}}\,\frac{1}{K_{\text{occ}-\text{A}}}}. \tag{2.159}$$

In this equation, the symbol, ε with a horizontal line above it, denotes mean efficacy, and the subscripts A and B are used to denote the parameters of agonists A and B, respectively. Using Equation (2.159) and the estimates of observed affinity $(1/K_{\text{occ}})$ and mean efficacy described above, the following values for the log RA_i values of agonists B–D, expressed relative to that of agonist A, can be calculated for the data in Figure 2.30: Agonist B, −1.02; Agonist C, −1.50 and Agonist D, −1.93. These estimates are nearly the same as the theoretical values (−1.0, −1.5 and −2.0, respectively) and those estimated by regression analysis using Equations (2.156) and (2.157) as described above.

The corresponding relative ratio of microscopic affinity constants for the inactive state of the receptor (RI_i) is calculated by:

$$\frac{K_{a-\mathrm{B}}}{K_{a-\mathrm{A}}} = RI_i \frac{\sqrt{1-\bar{\varepsilon}_{\mathrm{B}}}\frac{1}{K_{\mathrm{occ-B}}}}{\sqrt{1-\bar{\varepsilon}_{\mathrm{A}}}\frac{1}{K_{\mathrm{occ-A}}}}. \quad (2.160)$$

When there is measureable constitutive activity, the estimates of observed affinity and mean efficacy can be calculated as just described, but using the actual estimates of K_b, K_a and K_q determined from regression analysis using Equation (2.158) rather than provisional estimates. While it may seem like an academic exercise to calculate K_b and K_a from the estimates of observed affinity and mean efficacy when these microscopic constants are already known from the initial analysis, it is, nonetheless, interesting to see how these observed population parameters are related to the microscopic constants. Thus, it can be shown that the microscopic affinity constant of the agonist for the active state of the receptor is approximately equal to:

$$K_b = \frac{\sqrt{\bar{\varepsilon}}\frac{1}{K_{\mathrm{occ}}}}{\sqrt{ca}}, \quad (2.161)$$

in which, ca denotes constitutive activity as defined by Equation (2.155). The microscopic affinity constant for the inactive state of the receptor is approximately equal to:

$$K_a \approx \frac{\sqrt{1-\bar{\varepsilon}}\frac{1}{K_{\mathrm{occ}}}}{\sqrt{1-ca}}. \quad (2.162)$$

By now the reader may have noticed that Equations (2.159)–(2.162) are similar to the corresponding equations described for the one-site model ((2.41), (2.44), (2.46) and (2.47)), except that the former equations for the cooperative model use the square root of the efficacy, inactivity (1–ε) and constitutive activity terms, whereas the latter use the first root or simply the untransformed efficacy and constitutive activity values. Thus, both sets of equations can be described by a single set in which all of the efficacy, inactivity (1–ε) and constitutive activity terms are raised to the nth root, where n denotes the number of linked binding sites.

Single Channel Analysis

If a model has been established for a given ligand-gated ion channel, a simple approach to estimate the relevant microscopic affinity constants is to analyze the relationship between open probability and the agonist concentration. The mathematical basis for this approach is analogous to that described under the prior section on population analysis. In this section I briefly summarize this method using simulated single-channel events. A quick summary of the simulation technique is described first.

Simulation of single channel events for the cooperative model

The method for simulating open and closed events using the cooperative two-site model is analogous to that already described for the one-site model. The difference lies in the larger set of conditional probabilities that define the more complicated cooperative two-site model. Before describing these, it will help if we simplify the cooperative two-site model.

The cooperative model shown in Figure 2.25b can be simplified because of the assumption that the ligand interacts with the two sites on the receptor in the same manner:

$$D + R_s \overset{K_a}{\longleftrightarrow} DR_s, \tag{2.163}$$

$$D + R_s \overset{K_a}{\longleftrightarrow} R_sD. \tag{2.164}$$

Thus, these two equilibria can be replaced with a single equilibrium in which the sum of the two types of receptor complexes (DR_s and R_sD) is denoted with a single complex in bold letters ($\boldsymbol{DR_s}$), which represents the summation of DR_s and R_sD. Because there are two ways that the agonist can form this composite binary complex from the unoccupied receptor, its rate of formation is twice that of each of its components:

$$D + R_s \xrightarrow{2k_{a+}} \boldsymbol{DR_s}. \tag{2.165}$$

On average, the amount of the composite complex ($\boldsymbol{DR_s}$) is about twice that of each of the two distinct binary complexes, and hence, the initial rate of decay of the binary complex ($k_{a-}\boldsymbol{DR_s}$) is twice that of each distinct binary complex ($k_{a-}DR_s$ or $k_{a-}R_sD$). The rate constant (k_{a-}) for dissociation of all of these is the same, however, because the rate constant is normalized relative to the amount of ligand bound. Hence, the rate of decay of $\boldsymbol{DR_s}$ is described as:

$$D + R_s \xleftarrow{k_{a-}} \boldsymbol{DR_s}. \tag{2.166}$$

The microscopic affinity constant for the composite, binary complex is, therefore:

$$D + R_s \overset{2K_a}{\longleftrightarrow} \boldsymbol{DR_s}. \tag{2.167}$$

From the perspective of the composite binary complex there is only one way that the ternary complex (D_2R_s) can form:

$$D + \boldsymbol{DR_s} \xrightarrow{k_{a+}} D_2R_s, \tag{2.168}$$

but there are two ways that it can dissociate:

$$D + DR_s \xleftarrow{k_{a-}} D_2R_s, \tag{2.169}$$

$$D + R_sD \xleftarrow{k_{a-}} D_2R_s. \tag{2.170}$$

Using the composite binary complex notation to describe both dissociation routes for the ternary complex yields:

$$D + \boldsymbol{DR_s} \xleftarrow{2k_{a-}} D_2R_s. \tag{2.171}$$

And therefore, the microscopic affinity constant for the formation of ternary complex is described by:

$$D + \boldsymbol{DR_s} \overset{K_a/2}{\longleftrightarrow} D_2R_s. \tag{2.172}$$

These same concepts apply to the active states of the binary and ternary complexes.

With these modifications, the cooperative two-site model can be simplified as shown in Figure 2.32a. The constraints imposed by this closed system of equilibria imply that the two conformational induction steps can be expressed in terms of three microscopic equilibrium constants as shown for the one-site model. The corresponding rate constants for the model are shown in Figure 2.32b. Equilibrium constraints dictate that the rate constants (k_{g-} and k_{g2}, respectively) for the conversion of the active binary and ternary complexes into their inactive states can be expressed in terms of the other rate constants ($k_{g-} = k_{g+}k_{q-}k_{a+}k_{b-}/k_{q+}k_{a-}k_{b+}$ and $k_{g2-} = k_{g+}k_{q-}k_{a+}^2k_{b-}^2/k_{q+}k_{a-}^2k_{b+}^2$).

The set of conditional probabilities that determine each transition of the model can be derived from the associated rate constants. These are listed in transition matrix (**T**) shown in Figure 2.33. Each conditional probability ($P_{i,j}$) is positioned in the array to identify the state of the receptor complex at the start (column j) and end (row i) of the transition. These are also indicated by the parenthetical expressions above and to the right of the matrix.

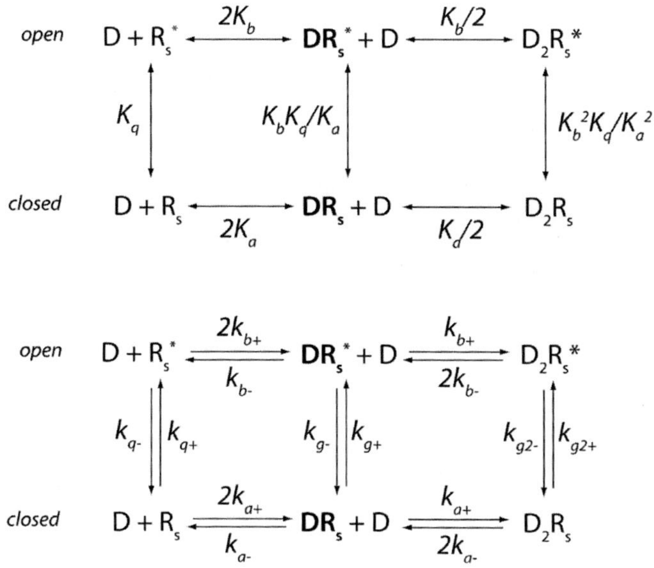

Figure 2.32: Two-site cooperative model for ligand-gated ion channels defined at the level receptor states and their associated microscopic equilibrium constants (a) and rate constants (b).

$$\mathbf{T} = \begin{array}{c} \\ \end{array}$$

	(R)	(R^*)	(DR)	(DR^*)	(D_2R)	(D_2R^*)
(R)	$1-\Delta t(k_{q+} + 2Dk_{a+})$	$k_{q-}\Delta t$	$k_{a-}\Delta t$	0	0	0
(R^*)	$k_{q+}\Delta t$	$1-\Delta t(k_{q-} + Dk_{b+})$	0	$k_{b-}\Delta t$	0	0
(DR)	$2Dk_{a+}\Delta t$	0	$1-\Delta t(k_{a-} + k_{g+} + DK_{a+})$	$k_{g-}\Delta t$	$2k_{a-}\Delta t$	0
(DR^*)	0	$2Dk_{b+}\Delta t$	$k_{g+}\Delta t$	$1-\Delta t(k_{b-} + k_{g-} + Dk_{b+})$	0	$2k_{b-}\Delta t$
(D_2R)	0	0	$Dk_{a+}\Delta t$	0	$1-\Delta t(2k_{a-} + k_{g2+})$	$k_{g2-}\Delta t$
(D_2R^*)	0	0	0	$Dk_{b+}\Delta t$	$k_{g2+}\Delta t$	$1-\Delta t(2k_{b-} + k_{g2-})$

Figure 2.33: Transition matrix ($\boldsymbol{T}$) for simulating the random single-receptor events of the cooperative two-site model (Figure 2.32) continuously in time.

The transition matrix is used to simulate single receptor events as described above for the one-site model. Essentially, the state of the system is determined numerically over a series of small time increments (Δt). A starting state is chosen (e.g., R_s), and the probability of the receptor staying in the state or transitioning to a new state is determined for each time increment using the conditional probabilities for the given starting state and a random variable. At some point, the receptor will transit to a new state, and the probability that the receptor will remain in the new state or transit to a different state is determined for each subsequent time increment using the set of conditional probabilities associated with the new state. This process is repeated iteratively.

Estimation of microscopic constants from measurements of open probability

Figure 2.34 shows an example of a simulation using a moderate concentration of agonist (0.63 mM) having a ratio of microscopic affinity constants for active and inactive states of 10^4. The open probability of the free receptor (i.e., $K_q/(1 + K_q)$) is only 10^{-7}, indicating very low constitutive activity. The plot in Figure 2.34b shows the transitions between the receptor states in time, and Figure 2.34a shows the closed and open states in time. Aside from a rare transition of the binary complex to the open state ($DR_s \rightarrow DR_s^*$; $K_g = 10^{-4}$), almost all channel openings occur in bursts associated with a series of oscillations between the D_2R and $D_2R_s^*$ states. The parameter values for this simulation are listed in the legend to Figure 2.34.

Using the same microscopic constants, a series of single channel traces were simulated for the condition of various concentrations of agonist (Figure 2.35). As expected, channel opening increases with an increase in the concentration of agonist. In Figure 2.36, open probability (time spent in the open state divided by the total time) is plotted against the agonist concentration (Agonist A) for the simulation

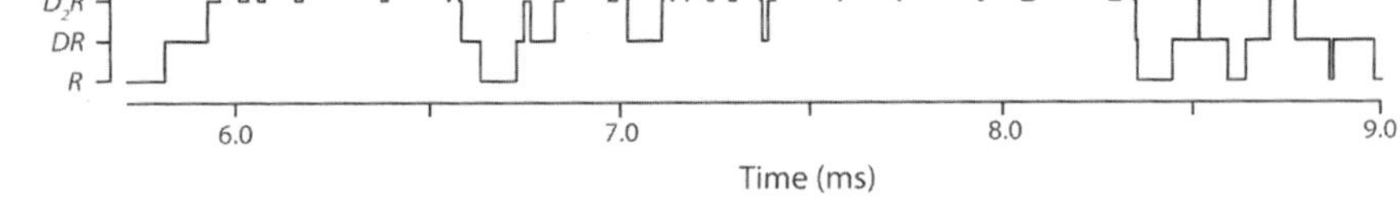

Figure 2.34: Simulation of single receptor events for the cooperative two-site model. The upper trace shows channel openings and closings, and the lower trace tracks the state of the receptor and number of ligands bound to it. The concentration of agonist is 0.63 mM. Parameters of the simulation: k_{a+}, 10^7 M^{-1} s^{-1}; k_{a-} 10^4 s^{-1}, k_{b+}, 10^7 M^{-1} s^{-1}; k_{b-}, 1.0 s^{-1}; k_{q+}, 1.0 s^{-1}; k_{q-}, 10^7 s^{-1}; k_{g+}, 2×10^3 s^{-1}; k_{g-}, 2×10^6 s^{-1}; k_{g2+}, 10^5 s^{-1} and k_{g2-}, 10^4 s^{-1}.

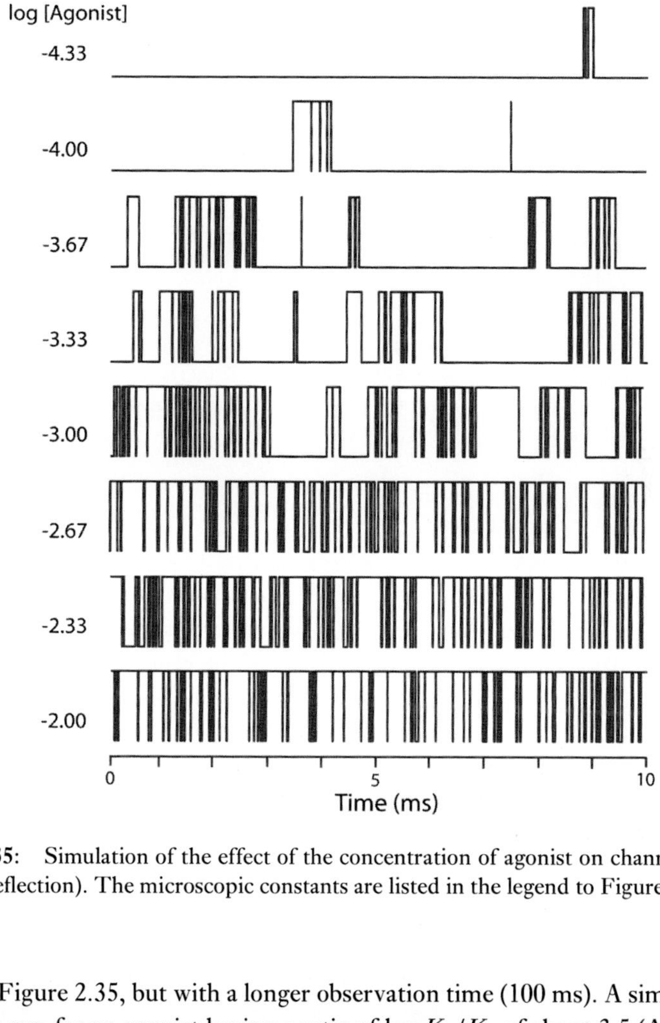

Figure 2.35: Simulation of the effect of the concentration of agonist on channel opening (upward deflection). The microscopic constants are listed in the legend to Figure 2.34.

shown in Figure 2.35, but with a longer observation time (100 ms). A similar simulation was run for an agonist having a ratio of log K_b/K_a of about 3.5 (Agonist B), and its open probability plot is also shown in Figure 2.36.

Each curve can be fitted by nonlinear regression analysis to an equation analogous to Equation (2.156), but with the equation modified to account for open probability (P_{open}):

$$P_{\text{open}} = \frac{1}{1 + \frac{([D]K_a+1)^2}{K_q\left([D]\sqrt{\frac{P_b(1+K_q)}{K_q}}+1\right)^2}}. \tag{2.173}$$

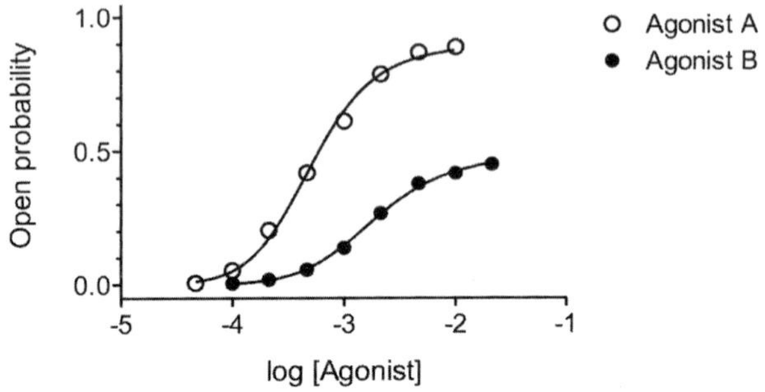

Figure 2.36: The relationship between open probability and the concentration of agonist. The theoretical data for Agonist A were calculated from the simulation shown in Figure 2.34, but for a longer time interval (100 ms). Those for agonist B were calculated similarly except for the following parameter changes: k_{b-}, 3.33 s^{-1}; k_{g-}, 6×10^6 s^{-1} and k_{g2-}, 1.1×10^5 s^{-1}.

The regression is done with K_q fixed to an arbitrarily low value (e.g., 10^{-6}). The analysis yields estimates of log P_b (log $K_b^2 K_q$) and log K_a for agonist A (log $P_b =$ 6.80 ± 0.051; log K_a, 2.93 ± 0.051) and agonist B (log P_b = 5.72 ± 0.037; log K_a, 2.87 ± 0.023). The log of the square root of the ratio of P_b values of the two agonists log $(P_{b-\mathrm{B}}/P_{b-\mathrm{A}})^{1/2}$ yields the log RA_i value of agonist B expressed relative to agonist A of −0.54.

All of these estimates are consistent with the microscopic constants used to simulate the data. For example, the gating constant for the free receptor (K_q) was 10^{-7} for this simulation. Subtracting log K_q from the estimates of the log P_b yields the log K_b^2 values (13.8 and 12.72) of Agonists A and B, respectively. Dividing the latter by two (i.e., taking the square root of their antilogs) yields log K_b values (6.90 and 6.36) that are not much different from those used in the simulation (7 and 6.48). The difference between the log K_b values of the two agonists (−0.52, log $K_{b-\mathrm{A}}/K_{b-\mathrm{B}}$) agrees very well with the estimate of log RA_i (−0.54). Also, the estimate of log K_a for each agonist is similar to that used in the simulation for both agonists (3).

The log RA_i value of Agonist B, expressed relative to that of Agonist A, can be estimated by fitting Equation (2.173) to the open probability plot of agonist A, while simultaneously fitting the following equation to the data for agonist B:

$$P_{\mathrm{open}} = \frac{1}{1 + \frac{([D]K_{a-\mathrm{B}}+1)^2}{K_q\left([D]RA_{i-\mathrm{B}}\sqrt{\frac{P_{b-\mathrm{A}}(1+K_q)}{K_q}}+1\right)^2}}. \tag{2.174}$$

This equation is equivalent to Equation (2.157), but modified for analysis of open probability. The modification simply involves dividing both sides of Equation (2.157) by $i_{\mathrm{max-sys}}$, which normalizes the right side of the equation and yields open probability (P_{open}) as the independent variable. The subscripts, A and B, are used to denote the parameters of standard (A) and test (B) agonists, respectively. This analysis yielded the same value of log RA_i (-0.54 ± 0.042) as estimated above.

The RI_i value is estimated in an analogous manner by fitting the following equation (analogous to Equation (2.174)) to the open probability measurements of the test agonist (B) while simultaneously fitting Equation (2.173) to the data for the standard agonist (A):

$$P_{\mathrm{open}} = \frac{1}{1 + \frac{([D]RI_{i-\mathrm{B}}K_{a-\mathrm{A}}+1)^2}{K_q\left([D]\sqrt{\frac{P_{b-\mathrm{B}}(1+K_q)}{K_q}}+1\right)^2}}. \tag{2.175}$$

This analysis yielded an estimate of log RI_i (-0.057 ± 0.06) not much different from the expected value of 0 (i.e., $K_{a-\mathrm{B}}/K_{a-\mathrm{A}} = 1$). The other parameter estimates were the same as those mentioned above.

Finally, if the receptor exhibits constitutive activity, then Equation (2.173) can be fitted to the open probability plot without constraining K_q. This approach enables the estimation of the gating constant of the free, constitutively active receptor (K_q) and the microscopic affinity constant of the active state of the receptor (K_b) in units of M^{-1}. One likely consequence is that it may be impossible to estimate K_a accurately. Recall that this difficulty arises when the open probability of the occupied receptor is close to 1 as in the case of a highly efficacious agonist. Because constitutive activity increases the open probability of both the occupied and unoccupied receptor, it makes it more difficult to estimate K_a using this approach.

Figure 2.37 shows a simulation of the effects of an agonist (0.21 mM; $K_b/K_a = 100$) on the single channel events of a receptor having constitutive activity ($K_q = 0.02$). The upper trace shows channel openings, and the lower trace tracts the state of the receptor. The high-frequency openings in the middle of the trace are rapid transitions between the unoccupied R_s and R_s^* states. In the absence of agonist, this constitutive activity only generates an open probability of about 0.02. The figure suggests a larger value, but this is because the thickness of the line used in the illustration spans a time interval on the x-axis much wider than that representing the mean open time of the R_s^* state ($0.1\,\mu$s). Agonist-induced openings of the channel have a much longer open time and occur at the beginning and end of the trace.

Normally, an agonist such as that used in the simulation would be unable to activate a wild type ligand-gated ion channel, like the nicotinic acetylcholine receptor of the neuromuscular junction. This is because the square of its K_b/K_a ratio is much smaller than the gating constant of the unliganded receptor. The latter has been estimated to be in the range of 7×10^{-7} for the muscle-type nicotinic acetylcholine receptor.[23] In this simulation, however, an agonist with a K_b/K_a ratio of 100 has enough selectivity for the open state so that channel openings are frequent for both the mono- and bi-liganded receptor complexes at moderate concentrations of agonist. A related phenomenon has been reported for mutant nicotinic acetylcholine receptors exhibiting constitutive activity and only one functional ligand-binding site. These receptors open when bound with only one molecule of acetylcholine.[25]

Additional simulations were run for different concentrations of the agonist, and open probability was estimated for each concentration. These estimates are plotted in Figure 2.38 (Agonist A) together with those of other simulations involving a less efficacious agonist (Agonist B) having a K_b/K_a ratio of only 10. While, the plot appears to exhibit no constitutive activity, the open probability approaches a plateau at of about 0.02 as the concentration of agonist approaches zero. The best fit of Equation (2.173) to the data yielded the following parameter estimates: Agonist A, log K_b, 5.09 ± 0.10 and log K_a, 3.09 ± 0.12 and Agonist B, log K_b, 4.11 ± 0.11 and log K_a, 2.93 ± 0.031. The estimate of log K_q was -2.05 ± 0.17. These are nearly identical to the corresponding ratios of rate constants used in the simulation: log K_q, -1.70; Agonist A, log K_b, 5 and log K_a, 3 and Agonist B, log K_b, 4 and log K_a, 3.

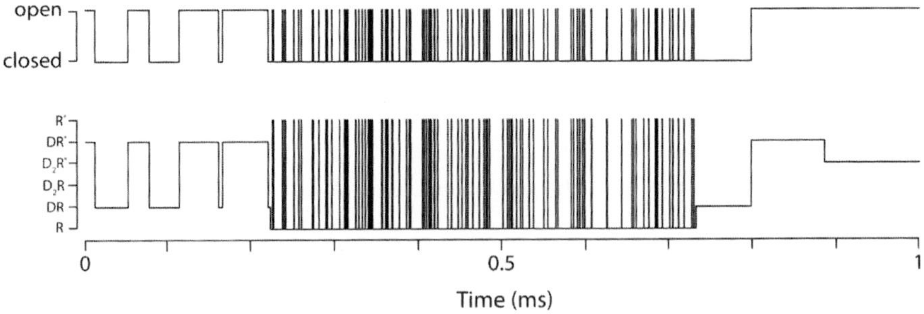

Figure 2.37: Simulation of single receptor events for the cooperative two-site model for the case of constitutive receptor activity. The upper trace shows channel openings and closings, and the lower trace tracks the state of the receptor and number of ligands bound to it. The concentration of agonist is 0.21 mM. Parameters of the simulation: k_{a+}, 10^7 M^{-1} s^{-1}; k_{a-} 10^4 s^{-1}, k_{b+}, 10^7 M^{-1} s^{-1}; k_{b-}, 10^2 s^{-1}; k_{q+}, 2×10^5 s^{-1}; k_{q-}, 10^7 s^{-1}; k_{g+}, 2×10^4 s^{-1}; k_{g-}, 10^4 s^{-1}; k_{g2+}, 10^5 s^{-1} and k_{g2-}, 5×10^2 s^{-1}.

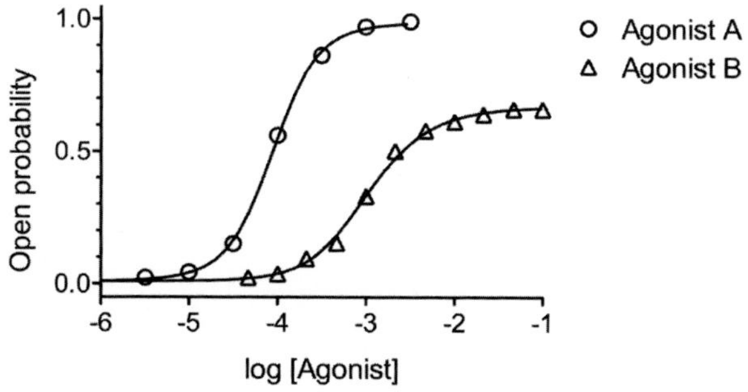

Figure 2.38: The relationship between open probability and the concentration of agonist for a receptor exhibiting constitutive activity. Open probability measurements were calculated from simulations of single receptor activity that were generated using the following microscopic constants: Agonist A: k_{a+}, 10^7 M^{-1} s^{-1}; k_{a-} 10^4 s^{-1}, k_{b+}, 10^7 M^{-1} s^{-1}; k_{b-}, 10^2 s^{-1}; k_{q+}, 2×10^5 s^{-1}; k_{q-}, 10^7 s^{-1}; k_{g+}, 2×10^3 s^{-1}; k_{g-}, 10^3 s^{-1}; k_{g2+}, 10^5 s^{-1} and k_{g2-}, 5×10^2 s^{-1}. Agonist B: k_{a+}, 10^7 M^{-1} s^{-1}; k_{a-} 10^4 s^{-1}, k_{b+}, 10^7 M^{-1} s^{-1}; k_{b-}, 10^3 s^{-1}; k_{q+}, 2×10^5 s^{-1}; k_{q-}, 10^7 s^{-1}; k_{g+}, 2×10^3 s^{-1}; k_{g-}, 10^4 s^{-1}; k_{g2+}, 10^5 s^{-1} and k_{g2-}, 5×10^4 s^{-1}. The theoretical curves represent the best fit of Equation (2.173) to the simulated open probability estimates.

Analysis of data from the literature

We can apply the methods just described to the analysis of real data. Figure 2.39 shows the results of a study by Akk and Auerbach[26] on the mouse muscle-type nicotinic acetylcholine receptor expressed in HEK 293 cells. These investigators measured the open probability of the nicotinic receptor in the presence of different concentrations of acetylcholine, carbachol and the selective muscarinic agonists, muscarine and oxotremorine-M. Open channel probability was measured from the analysis of single channel recordings, and the data were corrected for desensitization and missed events.

Global nonlinear regression analysis of the data using Equations (2.173) and (2.174) with log K_q constrained to an arbitrarily low value (10^{-6}) yielded the following estimates of log K_a: acetylcholine, 3.78 ± 0.10; carbachol, 2.89 ± 0.081; muscarine, 3.92 ± 0.64 and oxotremorine-M, 3.05 ± 0.21. The corresponding log RA_i values expressed relative to that of acetylcholine were 0.00, -1.22 ± 0.043, -1.18 ± 0.53 and -1.59 ± 0.13. The estimates of log K_a are similar to those reported by Akk and Auerbach: acetylcholine, 3.79; carbachol, 3.10; mucarine, 4.05 and oxotremorine-M, 3.20. When the K_b values reported by Akk and Auerbach are expressed relative to acetylcholine, the following log RA_i values are obtained: acetylcholine, 0.00; carbachol, -1.16, muscarinic, -0.98 and oxotremorine-M, -1.54.

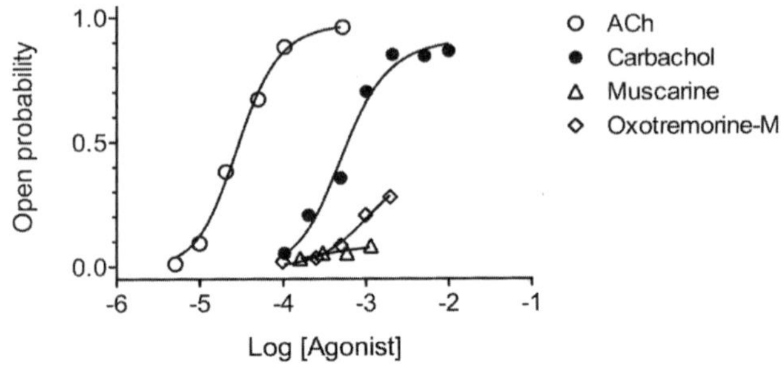

Figure 2.39: The effect of acetylcholine, carbachol, muscarine and oxotremorine-M the open probability of the mouse, muscle-type nicotinic acetylcholine receptor. The data have been redrawn from those previously published by Akk and Auerbach.[26] The theoretical curve represents the global, least-squares fit of Equations (2.173), (acetylcholine) and (2.174), (carbachol, muscarinic and oxotremorine-M) to the data with log K_q constrained to an arbitrarily low value of −6.

These values are similar to those estimated by regression analysis with Equations (2.173) and (2.174). Finally, if the data in Figure 2.39 are analyzed by Equation (2.173) with log K_q constrained to that estimated by Akk and Auerbach (−5.523), the following estimates of log K_b are obtained: acetylcholine, 7.36 ± 0.019; carbachol, 6.14 ± 0.014; muscarine, 6.18±0.15 and oxotremorine-M, 5.77 ± 0.057. These are nearly the same as the corresponding values reported by Akk and Auerbach[26]: 7.38, 6.22, 6.40 and 5.84.

Throwing Light on the Analysis of Agonist Action at *G* Protein Coupled Receptors

At the outset of this chapter, I described how the analysis of agonist action at ligand-gated ion channels and *G* protein-coupled receptors has progressed along different paths. The use of single channel and whole cell recording techniques by electrophysiologists provides a means of directly measuring the activation of single receptors and populations of receptors. These direct measures of receptor activity can be analyzed with various models to explain how agonists activate receptors based on the values of their microscopic affinity constants for the open and closed states of the channel.

Scientists working on *G* protein-coupled receptors, however, are usually restricted to measuring events downstream in the signaling pathway. Hence, the goal in this latter case has usually been to deduce the activation state of the receptor population by reverse engineering[27] or by applying a null method.[2,7] Thus, in a sense, the pharmacological analysis of ligand-gated ion channels begins where that of *G* protein-coupled receptors leaves off.

The analysis described in this chapter shows that it is possible to estimate microscopic constants from the behavior of a population of receptors. Specifically, K_a and a relative estimate of K_b (RA_i) can be determined from the population current. When there is constitutive activity, it is possible to estimate K_b and K_a in actual units of M^{-1} as well as the gating constant of the unliganded receptor (K_q, unitless parameter). The implications for the analysis of responses to *G* protein-coupled receptors are that it should also be possible to do the same after deducing the activation state of the receptor population (stimulus) (see Part II of this book). It is reassuring to know that the basis of this latter approach can be accurately applied to the analysis of the population and single receptor behavior of ligand-gated ion channels.

The next two chapters describe the analysis of competitive (Chapter 3) and allosteric (Chapter 4) interactions at ligand-gated ion channels.

References

1. Colquhoun, D. and Hawkes, A. G. On the stochastic properties of single ion channels. *Proc R Soc Lond B Biol Sci* **211**, 205–235 (1981).
2. Stephenson, R. P. A modification of receptor theory. *British J Pharmacol* **11**, 379–393 (1956).
3. Whorton, M. R., Bokoch, M. P., Rasmussen, S. G., Huang, B., Zare, R. N., Kobilka, B. and Sunahara, R. K. A monomeric *G* protein-coupled receptor isolated in a high-density lipoprotein particle efficiently activates its *G* protein. *Proc Natl Acad Sci USA* **104**, 7682–7687 (2007).
4. Colquhoun, D., Dowsland, K. A., Beato, M. and Plested, A. J. How to impose microscopic reversibility in complex reaction mechanisms. *Biophys J* **86**, 3510–3518 (2004).
5. Del Castillo, J. and Katz, B. Interaction at end-plate receptors between different choline derivatives. *Proc R Soc Lond B Biol Sci* **146**, 369–381 (1957).
6. Lape, R., Colquhoun, D. and Sivilotti, L. G. On the nature of partial agonism in the nicotinic receptor superfamily. *Nature* **454**, 722–727 (2008).
7. Furchgott, R. F. The use of β-haloalkylamines in the differentiation of receptors and in the determination of dissociation constants of receptor-agonist complexes. *Adv Drug Res* **3**, 21–55 (1966).
8. Colquhoun, D. Binding, gating, affinity and efficacy: The interpretation of structure-activity relationships for agonists and of the effects of mutating receptors. *British J Pharmacol* **125**, 924–947 (1998).
9. Tran, J. A., Chang, A., Matsui, M. and Ehlert, F. J. Estimation of relative microscopic affinity constants of agonists for the active state of the receptor in functional studies on M_2 and M_3 muscarinic receptors. *Mol Pharmacol* **75**, 381–396 (2009).

10. Ehlert, F. J., Griffin, M. T. and Suga, H. Analysis of functional responses at *G* protein coupled receptors: Estimation of relative affinity constants for the inactive receptor state. *J Pharmacol Exp Ther* (2011).
11. Hawkes, A. G., Jalali, A. and Colquhoun, D. Asymptotic distributions of apparent open times and shut times in a single channel record allowing for the omission of brief events. *Philos Trans R Soc Lond B Biol Sci* **337**, 383–404 (1992).
12. Katz, B. and Miledi, R. Further observations on acetylcholine noise. *Nat New Biol* **232**, 124–126 (1971).
13. Colquhoun, D. and Hawkes, A. G. Relaxation and fluctuations of membrane currents that flow through drug-operated channels. *Proc R Soc Lond Ser B* **199**, 231–262 (1977).
14. Colquhoun, D. and Hawkes, A. G. in *Single-Channel Recording*, 2nd Edn. (eds. B. Sakmann and E. Neher) (Plenum Press, 1995).
15. Colquhoun, D. and Hawkes, A. G. On the stochastic properties of bursts of single ion channel openings and of clusters of bursts. *Philos Trans R Soc Lond B Biol Sci* **300**, 1–59 (1982).
16. Colquhoun, D. and Hawkes, A. G. in *Microelectrode Techniques, The Plymouth Workshop Handbook* (eds. N. B. Standen, P. T. A. Gray and M. J. Whitaker) Ch. 7, 141–188 (Cambridge: Company of Biologists, 1987).
17. Qin, F., Auerbach, A. and Sachs, F. Maximum likelihood estimation of aggregated Markov processes. *Proc Biol Sci* **264**, 375–383 (1997).
18. Monod, J., Wyman, J. and Changeux, J. P. On the nature of allosteric transitions: A plausible model. *J Mol Biol* **12**, 88–118 (1965).
19. Culver, P., Fenical, W. and Taylor, P. Lophotoxin irreversibly inactivates the nicotinic acetylcholine receptor by preferential association at one of the two primary agonist sites. *J Biol Chem* **259**, 3763–3770 (1984).
20. Akk, G. and Auerbach, A. Inorganic, monovalent cations compete with agonists for the transmitter binding site of nicotinic acetylcholine receptors. *Biophys J* **70**, 2652–2658 (1996).
21. Colquhoun, D. and Ogden, D. C. Activation of ion channels in the frog end-plate by high concentrations of acetylcholine. *J Physiol* **395**, 131–159 (1988).
22. Salamone, F. N., Zhou, M. and Auerbach, A. A re-examination of adult mouse nicotinic acetylcholine receptor channel activation kinetics. *J Physiol* **516 (Pt 2)**, 315–330 (1999).
23. Auerbach, A. The gating isomerization of neuromuscular acetylcholine receptors. *J Physiol* **588**, 573–586 (2010).
24. Jackson, M. B. Perfection of a synaptic receptor: kinetics and energetics of the acetylcholine receptor. *Proc Natl Acad Sci USA* **86**, 2199–2203 (1989).
25. Jha, A. and Auerbach, A. Acetylcholine receptor channels activated by a single agonist molecule. *Biophys J* **98**, 1840–1846 (2010).
26. Akk, G. and Auerbach, A. Activation of muscle nicotinic acetylcholine receptor channels by nicotinic and muscarinic agonists. *Br J Pharmacol* **128**, 1467–1476 (1999).
27. Black, J. W. and Leff, P. Operational models of pharmacological agonism. *Proc R Soc Lond Ser B: Biol Sci* **220**, 141–162 (1983).
28. Chang, Y. and Weiss, D. S. Channel opening locks agonist onto the GABAC receptor. *Nat Neurosci* **2**, 219–225 (1999).

Competitive Interactions between Orthostheric Ligands at Ligand-Gated Ion Channels

3

If a drug binds to a ligand-gated ion channel and has no effect, it is still possible to estimate its affinity by measuring how it affects the ionic current elicited by an agonist that binds to the same site. This type of interaction is defined as competitive, and the method for its analysis is the same regardless of whether the two drugs interact at the primary orthosteric site or a secondary allosteric site. In this chapter, we will consider competitive interactions from the perspective of the orthosteric site.

Competitive interactions occur whenever two drugs bind reversibly to the same site on the receptor. This may seem like a straightforward definition of a mechanism, but complications arise with regard to what is meant by the same site. For example, acetylcholine binds to the muscle-type nicotinic acetylcholine receptor between the α–γ and α–δ subunit interfaces. It gains considerable binding affinity by interacting with the B and C loops on the counterclockwise face of the α subunit as described in Chapter 1. The nicotinic antagonist, tubocurarine, also binds at the same locus, but because of its larger size, tubocurarine interacts with additional loci surrounding the acetylcholine binding pocket including areas of the γ and δ subunits not seen by acetylcholine. In one sense, therefore, the binding pockets for acetylcholine and tubocurarine are different. Yet, for our purposes of defining competition, they are the same because the three-dimensional space occupied by each ligand when bound to the receptor intersects that of the other. Therefore, both ligands cannot bind to the receptor at the same time, and in this sense, the two ligands bind to the same site, and hence, their interaction is competitive.

The meaning of reversible within the context of competitive inhibition also generates confusion. If a drug readily dissociates from the receptor after it is removed from the surrounding medium then its binding mechanism exhibits reversibility.

Conversely, irreversible binding occurs if the effect of the drug outlasts its presence in the bathing medium. But how readily must a drug dissociate from a receptor to be classified as reversible, and how long must a drug remain bound to the receptor to be classified as irreversible? The answer depends somewhat on the frame of reference. Consider a scientist measuring the effect of an antagonist using different types of functional assays. An antagonist that dissociates from the receptor with a half time of two hours may appear to behave reversibly if its effects are studied *in vivo* over several hours. But if the same drug is investigated using an *in vitro* assay where the response to an agonist is typically measured within a few seconds, that same antagonist behaves essentially irreversibly over the brief time of the assay. This issue is critical at ligand-gated ion channels because the whole-cell current elicited by an agonist may desensitize rapidly, and the investigator may only have a few seconds to study its interaction with a competing ligand. Thus, regardless of what adjectives we choose to describe the action of a drug, we need to insure that our method of analysis is appropriate for how long it takes to equilibrate with the receptor and other competing drugs.

In this chapter, we will consider the condition where the two competing ligands reach equilibrium during the time that the response is measured. As in the prior chapter, the response under consideration is the current response of a single ligand-gated ion channel or a population of them. The reader may find the significance of this chapter limited because substantial desensitization occurs before many useful nicotinic antagonists reach equilibrium with the receptor and agonist. Nonetheless, one of the overarching goals of this chapter is to provide a theoretical understanding of how the activation state of a population of receptors in equilibrium with an agonist changes upon equilibration with a competitive antagonist. This insight will provide the basis for analysis of competitive interactions at *G* protein-coupled receptors as described in Part II of this book. Additional issues related to the effect of incomplete attainment of equilibrium are described in Chapter 10 within the context of *G* protein-coupled receptors.

When two drugs undergo competition at the orthosteric site of a ligand-gated ion channel, the interaction may exhibit cooperativity. This is because ligand-gated ion channels have multiple orthosteric sites that are allosterically linked (two in the case of the muscle type nicotinic acetylcholine receptor). If each ligand exhibits different affinities for the active and inactive states, then the binding of one ligand to one orthosteric site will affect the binding of the other ligand to the other orthosteric site. Thus, competitive interactions between different orthosteric ligands at a multisite ligand-gated ion channel can include a small allosteric component even though the binding of the interacting ligands is mutually exclusive. Thus, one of the goals of this chapter is to illustrate how to discriminate between competitive and allosteric interactions. The other is to illustrate how to estimate the microscopic and observed affinity constants of the competitive inhibitor.

I begin by deriving a model for competitive inhibition and the requisite equations for the analysis of the interaction of two orthosteric ligands with a population of ligand-gated ion channels. These equations will be applied in the analysis of simulated data. Finally, the equations will be modified and used in the analysis of open probability measurements estimated from single channel recordings.

Model for Competitive Inhibition

The model for the competitive interaction between two orthosteric ligands at a ligand-gated ion channel having two orthosteric binding sites is shown in Figure 3.1. Each of the two rectangular planes of equilibria that intersect at 90° at the back corner of the model represents the interaction of a single ligand with the receptor. Each plane is analogous to the model shown in Figure 2.32 of Chapter 2. The free ligands have been omitted to avoid crowding of symbols. The plane on the right illustrates the model for the primary agonist (D) and that on the left illustrates the model for the competitive inhibitor (B). The vertex projecting outward from the middle illustrates the receptor bound with both agonist and competitive ligand.

There are two different ways that a ligand can bind to the receptor to form the mono-liganded complex (e.g., *DR*), depending upon to which of the two sites

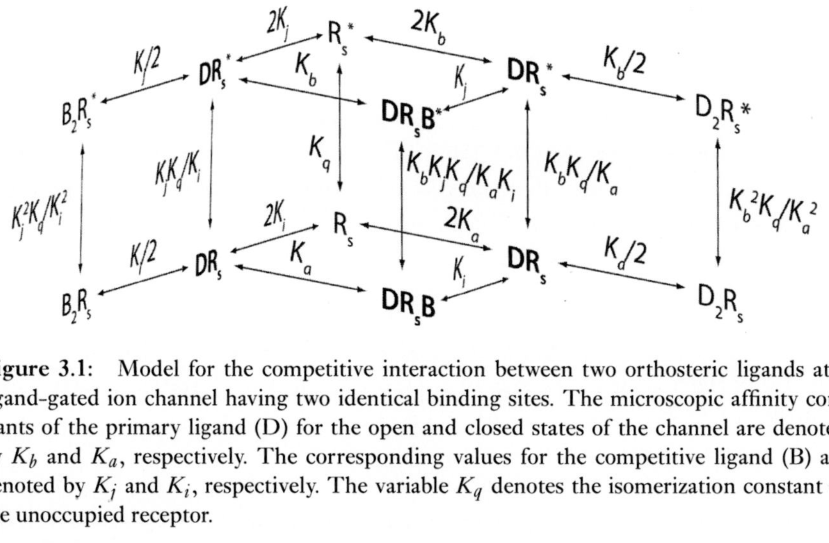

Figure 3.1: Model for the competitive interaction between two orthosteric ligands at a ligand-gated ion channel having two identical binding sites. The microscopic affinity constants of the primary ligand (D) for the open and closed states of the channel are denoted by K_b and K_a, respectively. The corresponding values for the competitive ligand (B) are denoted by K_j and K_i, respectively. The variable K_q denotes the isomerization constant of the unoccupied receptor.

the ligand binds. In this model, both of these binding sites exhibit the same affinity for the ligand. Therefore, both types of binary complexes are designated as a single binary complex in bold letters (e.g., $\boldsymbol{DR_s}$). Because there are two ways a ligand can form the binary complex, the microscopic affinity constant for this step is multiplied by two (e.g., $2K_a$). It follows that the binding of the second molecule of the same ligand is described by the same microscopic affinity constant but divided by 2 (e.g., $K_a/2$). The rationales for these manipulations are described in connection with the model shown in Figure 2.32 of Chapter 2.

The microscopic affinity constants of the two orthosteric ligands (D and B) are defined analogously to those defined for the model shown in Figure 2.32 of Chapter 2. Finally, the microscopic affinity constant for the binding of a given orthosteric ligand to a binary complex with the other orthosteric ligand is defined as follows for the four possible binding steps:

$$K_b = \frac{[DR_s^* B]}{[D][\boldsymbol{BR}^*]} \tag{3.1}$$

$$K_j = \frac{[BR_s^* D]}{[B][\boldsymbol{DR_s^*}]} \tag{3.2}$$

$$K_a = \frac{[DR_s B]}{[D][\boldsymbol{BR_s}]} \tag{3.3}$$

$$K_i = \frac{[BR_s D]}{[B][\boldsymbol{DR_s}]} \tag{3.4}$$

Derivation of Equilibrium Equations for Competitive Inhibition

To analyze the competitive inhibition of the current response of a population of ligand-gated ion channels all we need is an equation that expresses receptor activation as a function of the concentrations of the two competing ligands. This can be derived from the following equation, which denotes the summation of all of the active receptor species expressed as a fraction of the total amount of receptor:

$$\frac{active}{[R_T]} = \frac{[\boldsymbol{DR_s^*}] + [\boldsymbol{BR_s^*}] + [\boldsymbol{DR_s^*B}] + [D_2R_s^*] + [B_2R_s^*] + [R_s^*]}{\begin{array}{c}[\boldsymbol{DR_s}] + [\boldsymbol{BR_s}] + [\boldsymbol{DR_sB}] + [D_2R_s] + [B_2R] + [\boldsymbol{DR_s^*}] + [\boldsymbol{BR_s^*}] \\ +[\boldsymbol{DR_s^*B}] + [D_2R_s^*] + [B_2R_s^*] + [R_s^*] + [R_s]\end{array}}. \tag{3.5}$$

Using the strategy outlined in Chapter 2, each receptor complex can be substituted with an expression containing the concentration of the inactive state of the free receptor (R_s) and the appropriate microscopic constants and ligand concentrations. After factoring out $[R_s]$ and simplifying, the following equation is obtained:

$$\frac{active}{[R_T]} = \frac{1}{1 + \frac{([D]K_a+[B]K_i+1)^2}{K_q([D]K_b+[B]K_j+1)^2}}. \tag{3.6}$$

Multiplying both sides of the equation by a theoretical constant ($i_{\text{max-sys}}$) that represents the maximum current response of the cell, if all of the ligand-gated ion channel were activated at once, yields a form of the equation useful for our analysis:

$$i = \frac{i_{\max-sys}}{1 + \frac{([D]K_a+[B]K_i+1)^2}{K_q([D]K_b+[B]K_j+1)^2}}. \tag{3.7}$$

In this equation, i denotes the current response of a population of ligand-gated ion channels. We will use this equation to simulate and analyze the current response elicited by an agonist in the presence of a competitive orthosteric ligand.

Simulation and Analysis of Competitive Inhibition

In this section, theoretical concentration-response curves are simulated and then analyzed by nonlinear regression analysis using specific equations including (3.6) and (3.7) and additional ones described below. For regression analysis, the equations are transformed so that the affinity constants are expressed as logarithms. For example:

$$K_{\text{act}}^n = 10^{n \log K_{\text{act}}}. \tag{3.8}$$

The Hill slope (n) and parameter, K_{act}, are defined below. This analysis will enable us to estimate the affinity constants of the interacting ligands and determine if the nature of their interaction is consistent with competition.

Competition with a neutral antagonist

The simplest competitive interaction is that between an agonist and a neutral antagonist. Figure 3.2 shows a simulation of the effects of different concentrations of a neutral antagonist on the whole-cell current response elicited by an agonist at a population of ligand-gated ion channels. The curves were simulated using Equation (3.7) with no experimental error. Because a neutral antagonist has the same affinity for the active and inactive states of the receptor, K_i and K_j were set to the same value (i.e., $K_i = K_j = 10^8$ M^{-1}). The other parameters are listed in the

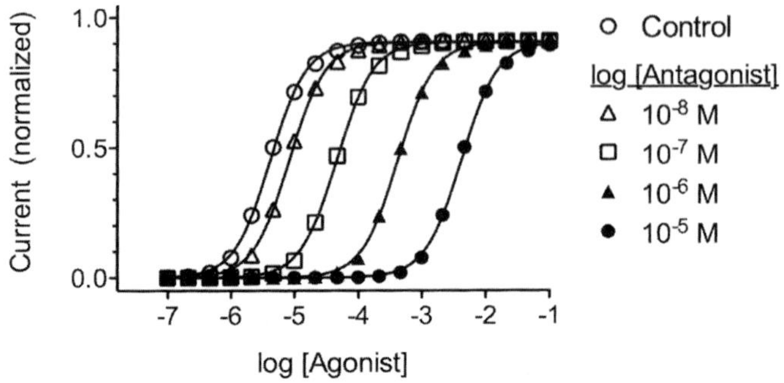

Figure 3.2: Effect of a neutral antagonist on the whole cell current response to an agonist. The theoretical plot shows the concentration-response curves of an agonist in the absence and presence of increasing concentrations of a neutral antagonist. The parameter values for this simulation are log K_a, 5; log K_b, 9; log K_i, 8; log K_j, 8 and log K_q, −7.

legend of the figure. The isomerization constant of the free receptor was set to a very low value ($K_q = 10^{-7}$) to illustrate the condition of a lack of constitutive activity.

The competitive antagonist shifts the agonist receptor activation curve to the right in a parallel fashion without affecting the slope of the curve. Equation (3.7) can be fitted to all of the data by global nonlinear regression analysis, although the method is complicated because the solution set is infinite. If the goal of the experiment is to estimate the affinity constant of the neutral antagonist for the receptor, then a much simpler approach is to fit the empirical Hill equation to the data. This equation was described in Chapter 2, and it is reproduced below in a slightly modified form:

$$\frac{activation}{[R_T]} = \frac{[D]^n}{[D]^n + K_{\text{act}}^n}. \tag{3.9}$$

Here, the dependent variable is fractional receptor activation, and K_{act} denotes the concentration of agonist eliciting half-maximal activation. Multiplying both sides of the equation by the maximum current response of the agonist (i_{max}) yields:

$$i = \frac{i_{\text{max}}[D]^n}{[D]^n + K_{\text{act}}^n}. \tag{3.10}$$

This equation was fitted to all of the data simultaneously by global nonlinear regression analysis sharing the estimate of i_{max} and n among the curves and obtaining a unique estimate of K_{act} for each curve. The estimates of log K_{act} are listed in Table 3.1. The estimates of i_{max} and n were 0.91 and 1.54, respectively. Note that the estimate, i_{max}, represents the maximum current response of the specific agonist and not the maximal response of the system, $i_{\text{max-sys}}$ (see Equation (3.7)).

Table 3.1: Effect of a neutral antagonist on the whole cell current response to an agonist. The parameter values were estimated from the theoretical data in Figure 3.2.

Antagonist (M)	LogK_{act} or K'^{a}_{act}	Log Shift (K'_{act}/K_{act})	LogK^{b}_{I}
0	−5.38	0	
10^{-8}	−5.08	0.30	8.00
10^{-7}	−4.34	1.04	8.00
10^{-6}	−3.37	2.04	8.00
10^{-5}	−2.38	3.00	8.00

[a]The concentration of agonist eliciting a half-maximal current response in the absence (K_{act}) and presence (K'_{act}) of the competitive antagonist.
[b]The K_I values of the antagonist was estimated using Equation (3.12).

When the competing ligand is a neutral antagonist, it can be shown that the shift (K'_{act}/K_{act}) in the agonist's concentration-response curve for eliciting the whole-cell current response is equivalent to:

$$\frac{K'_{act}}{K_{act}} = 1 + [B]K_I, \tag{3.11}$$

in which K'_{act} denotes the EC_{50} value of the agonist in the presence of a fixed concentration of the antagonist (B), and K_I denotes the observed affinity constant of the antagonist. Rearrangement of Equation (3.11) yields:

$$\frac{\frac{K'_{act}}{K_{act}} - 1}{B} = K_I. \tag{3.12}$$

Taking the reciprocal of both sides yields a more familiar form of the equation in which the reciprocal of the affinity constant ($1/K_I$) is replaced with the dissociation constant (K_B) in units of molarity:

$$\frac{[B]}{\frac{K'_{act}}{K_{act}} - 1} B = K_B. \tag{3.13}$$

Equation (3.13) was used to estimate K_B for each concentration of antagonist. These estimates are similar for each concentration of antagonist as shown in Table 3.1. A more rigorous way to analyze the data is to fit the following equation to the data by global nonlinear regression analysis sharing the estimates of i_{max}, n, K_I and K_{act}

among the curves:

$$i = \frac{i_{max}[D]^n}{[D]^n((1 + BK_I)K_{\text{act}})^n}. \tag{3.14}$$

When this was done the estimate of $\log K_I$ was 8. This value agrees with the predicted value because it is equivalent to the values of the microscopic affinity constants of both the ground and active states ($\log K_i = \log K_j = 8$), which are the same for a neutral antagonist.

From the foregoing analysis, it follows that if the neutral antagonist competitively inhibits the binding of the agonist, then the estimate of K_I should lack a significant difference for each concentration of competitive inhibitor used in the assay. Also, for any concentration of inhibitor exceeding the reciprocal of its affinity constant by a factor of 10 or so, a further increase in concentration will cause a nearly proportional shift to the right in the concentration-response curve of the agonist. Theoretically, if it were feasible to use an infinite concentration of antagonist with absolute selectivity for the orthosteric site, it would cause an infinite shift to the right in the concentration-response curve. In contrast, if the inhibitor reduced the effect of the orthosteric agonist through an allosteric mechanism, then there should be a limit to the maximal effect of the inhibitor as its concentration increases depending on the magnitude of its negatively cooperative effect. A rigorous analysis to discriminate between competitive inhibition and a negative allosteric mechanism is to fit all of the data simultaneously by global nonlinear regression analysis to determine if there is a significant improvement in residual error when an allosteric model is fitted to the data instead of a competitive model. The details of this analysis are described in Chapter 4.

Competition between an agonist and a less efficacious agonist

One can also study the competitive interaction between two agonists. This might be done, for example, if it is unclear whether both interact competitively at the orthosteric sites. A simulation of this type with random error ($\pm 5\%$) is shown in Figure 3.3a, which illustrates the concentration-response curves of a highly efficacious agonist ($K_b/K_a = 10^5$) in the absence (control) and presence of increasing concentrations of a less efficacious agonist ($K_j/K_i = 10^4$) that also binds to the orthosteric sites. The microscopic constants for this simulation are given in the legend to Figure 3.3.

In the presence of increasing amounts of a partial agonist, the concentration-response curve of the more efficacious agonist shifts to the right. There are some

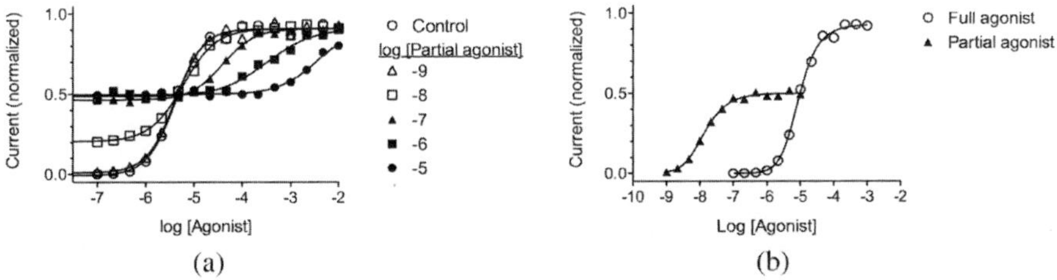

Figure 3.3: Effect of a partial agonist on the whole cell current response to a more efficacious agonist. (a) The theoretical plot shows the concentration-response curves of an agonist in the absence and presence of increasing concentrations of a partial agonist. The parameter values for this simulation are $\log K_a$, 5; $\log K_b$, 9; $\log K_i$, 8; $\log K_j$, 11.5 and $\log K_q$, −7. (b) The plot shows the independent concentration-response curves of the agonist and partial agonist from panel (a).

obvious differences between this interaction and that shown for the neutral antagonist in Figure 3.3a, however. When a partial agonist is present, the concentration-response curve of the full agonist begins at an elevated level corresponding to the response of the partial agonist. As the concentration of the partial agonist increases, there is a reduction in the span (range of current responses) of the concentration-response curve of the full agonist. At high concentrations of partial agonist, the concentration-response curve of the full agonist begins at the $i_{\max}$ value of the partial agonist and reaches a plateau at high concentrations corresponding to the $i_{\max}$ of the full agonist. There is also a reduction in the Hill slope of the concentration-response curve of the full agonist.

These empirical changes in the concentration-response curve of the full agonist can be measured by fitting the curves to the following logistic equation:

$$i = R_I + \frac{(i_{\max} - R_I)[D]^n}{[D]^n + K_{\text{act}}^n}, \tag{3.15}$$

in which, R_I denotes the response to the fixed concentration of the partial agonist, and K_{act} denotes the concentration of full agonist causing a half-maximal increase in the current response above the starting value (i.e., R_I). The other parameters are the same as those used in Equation (3.14). Table 3.2 lists the K_{act} values of the full agonist in the absence and presence of increasing concentrations of the partial agonist. At the two lowest concentrations of the partial agonist, the K_{act} value of the full agonist actually decreases below its control value, but a further increase in the amount of partial agonist causes a nearly proportional increase in the K_{act} value of the full agonist.

Table 3.2: Effect of a partial agonist on the whole cell current response to a more efficacious agonist. The parameter values were estimated from the theoretical data in Figure 3.3a.

Partial Agonist (M)	LogK_{act} or K'^{a}_{act}	Log Shift (K'_{act}/K_{act})	LogK^{b}_{I}
0	-5.39 ± 0.02	0	
10^{-9}	-5.40 ± 0.02	-0.006	
10^{-8}	-5.22 ± 0.03	0.17	7.67
10^{-7}	-4.40 ± 0.04	0.99	7.94
10^{-6}	-3.49 ± 0.06	1.90	7.90
10^{-5}	-2.46 ± 0.05	2.93	7.93

[a] The concentration of agonist eliciting a half-maximal current response in the absence (K_{act}) and presence ($K_{act}{'}$) of the partial agonist.
[b] The K_I values of the antagonist was estimated using Equation (3.12). This equation is only valid in the case of a competitive antagonist, however, although the equation does yield a good estimate of the K_I of an agonist whenever the concentration of the latter exceeds its corresponding value of K_{act} by a factor of 10.

If the simple competitive relationship for a neutral antagonist (Equation (3.12)) is used to analyze the data, a tentative K_I value can be estimated for the partial agonist based on its ability to antagonize the response to the full agonist. It is impossible to apply this calculation at the two lowest concentrations of partial agonist, however, because the measured K'_{act} values are actually lower than that measured under the control condition. This results in a negative estimate of K_I, which has no meaning. The estimates for the higher concentrations of the partial agonist are listed in Table 3.2. These values change with the concentration of partial agonist, and hence, they must not be reliable estimates of the observed affinity of the partial agonist for the ligand-gated ion channel. Recall than in the case of the neutral antagonist, the estimate of K_I was constant and independent of the agonist concentration (see Table 3.1). At the highest concentration of partial agonist, the $-\log K_I$ value approaches a limiting estimate of -7.93.

This limiting estimate is actually a good estimate of the log EC_{50} value (log K_{act}) of the partial agonist for eliciting a current response in the absence of the full agonist. This can be appreciated from the plot in Figure 3.3b, which shows the concentration-response curves of the partial agonist and the more efficacious agonist in isolation. Regression analysis of these data according to the simple Hill equation yields K_{act} values of -7.85 and -5.38 for the partial and more efficacious agonists, respectively. Thus, although the estimate of the apparent K_I value of a partial agonist for inhibiting the response to a full agonist changes with the concentration of partial agonist, its value does approach an accurate limiting estimate equivalent to its $1/K_{act}$ value.

A more rigorous way to analyze the interaction shown in Figure 3.3a is to fit a form of Equation (3.7) to the data by global nonlinear regression analysis. This equation describes the interaction in terms of the fundamental microscopic constants that determine drug-receptor interactions. As described in Chapter 2, it is impossible to estimate the K_b values of agonists and the K_q value of the receptor from the population current response unless the receptor exhibits constitutive activity. Nonetheless, it is possible to estimate the composite parameter of $K_b^2K_q/(1+K_q)$ of each agonist and the K_a values of all ligands except full agonists. To facilitate the regression, Equation (3.7) can be expressed in a form having the terms $K_b^2K_q/(1+K_q)$ and $K_j^2K_q/(1+K_q)$ denoted with the single parameters, P_b and P_j, respectively:

$$P_b = \frac{K_b^2 K_q}{1 - K_q}, \tag{3.16}$$

$$P_j = \frac{K_j^2 K_q}{1 - K_q}. \tag{3.17}$$

Rearranging these equations to solve for K_b and K_j yields:

$$K_b = \sqrt{\frac{P_b(1 - K_q)}{K_q}}, \tag{3.18}$$

$$K_j = \sqrt{\frac{P_j(1 - K_q)}{K_q}}. \tag{3.19}$$

Substituting Equations (3.18) and (3.19) for the variable K_b and K_j in Equation (3.7) yields:

$$i = \frac{i_{max-sys}}{1 + \frac{([D]K_a + [B]K_i + 1)^2}{K_q\left([D]\sqrt{\frac{P_b(1-K_q)}{K_q}} + [B]\sqrt{\frac{P_j(1-K_q)}{K_q}} + 1\right)^2}}. \tag{3.20}$$

Now, we have an equation that can be easily fitted to the data in Figure 3.3a to obtain estimates of the microscopic constants and to determine if the data are consistent with competitive inhibition. To accomplish this, Equation (3.20) is fitted to all of the curves simultaneously using global nonlinear regression analysis with K_q constrained to an arbitrarily low value (e.g., 10^{-8}) and the other microscopic constants shared among all of the data. When this was done, the following mean values ± SEM were obtained for the parameters $\log K_a(4.93 \pm 0.11)$, $\log K_i(7.99 \pm 0.02)$,

$\log P_b(10.95 \pm 0.04)$, $\log P_j(15.99 \pm 0.04)$ and $I_{\max}(0.98 \pm 0.033)$. The estimates of $\log K_a$ and $\log K_i$ are approximately the same as those used in the simulation (5 and 8, see legend to Figure 3.3). Similarly, the estimates of $\log P_b$ and $\log P_j$ are nearly the same as those predicted from Equations (3.16) and (3.17) and the values of K_b, K_j and K_q used to simulate the data (11 and 16). There is more error associated with the estimate of $\log K_a$ compared to that of $\log K_j$ because of the difficulty in obtaining a reliable estimate of the microscopic affinity constant for the inactive state of the receptor when the agonist is capable of eliciting a maximum response ($i_{\max}$) nearly equivalent to the maximum current response when all of the ligand gated ion channels are open ($i_{\max\text{-sys}}$) (see Chapter 2). The theoretical curves in Figure 3.3a represent the least squares fit of Equation (3.20) to the theoretical data.

From the definitions of P_b and P_j, it follows that:

$$\sqrt{\frac{P_j}{P_b}} = \frac{K_j}{K_b} = RA_i. \tag{3.21}$$

Substituting the estimates of P_b and P_j into Equation (3.21) yields a $\log K_j/K_b$ ratio of 2.52. This ratio is a relative estimate of the microscopic affinity constant of agonist B for the active state of the receptor (K_j) expressed relative to that of agonist $D(K_b)$ and is known as the RA_i value. This estimate of $\log RA_i$ is approximately the same as that calculated from the log ratio of K_j and K_b values used in the simulation (2.50).

The foregoing analysis illustrates that it is possible to obtain reliable estimates of K_a and RA_i from the competitive interaction between two agonists in an assay in which the current response of a population of ligand-gated ion channels is measured. If the goal were to simply estimate K_a and RA_i, then it would have been much simpler to analyze their individual concentration-response curves. When this was done for the theoretical data illustrated in Figure 3.3b, for example, similar estimates of $\log K_a$, $\log K_i$ and $\log RA_i$ were obtained ($4.77 \pm 0.13, 8.03 \pm 0.04$ and 2.83, respectively). The estimate of $\log K_a$ for the full agonist was less accurate, however, because of the problem of estimating this constant for a highly efficacious agonist. The larger amount of data for the analysis in Figure 3.3a enabled this parameter to be estimated more accurately. Regardless, the purpose of the latter experiment is to determine if the two agonists interact competitively. If there interaction were not competitive, then Equation (3.20) would not have described their interaction adequately or else the parameters estimated from the interaction (Figure 3.3a) would have differed significantly from those estimated from the individual behaviors of the agonists (Figure 3.3b).

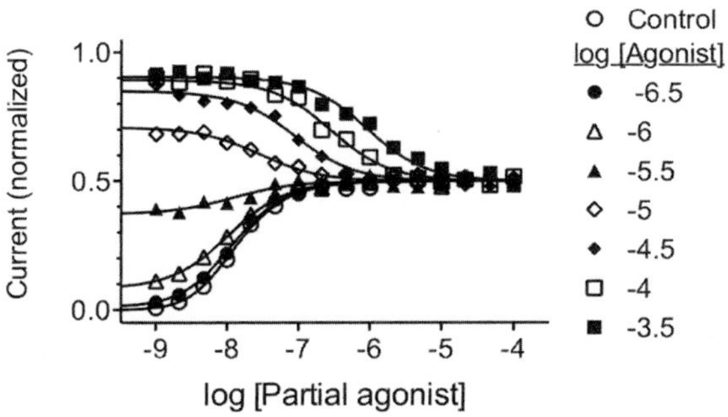

Figure 3.4: Effect of an efficacious agonist on the whole cell current response to a less efficacious agonist. (a) The plot shows the concentration-response curves of a partial agonist in the absence and presence of increasing concentrations of a more efficacious agonist. The parameter values for this simulation are the same as those of Figure 3.3a.

The competitive effect of a highly efficacious agonist on the current response of a less efficacious agonist

The functional responses for the competitive interaction between the two agonists shown in Figure 3.3a ($r_{i,j}$) could be organized into a two-dimensional matrix indexed by row (i) and column (j) to identify the concentration of the more efficacious agonist (i) and the less efficacious agonist (j). If each column of data is plotted as a function of the concentration of the more efficacious agonist, then the plot shown in Figure 3.3a is obtained. If each row of data is plotted, however, then a plot similar to that shown in Figure 3.4 is obtained. It should not matter how the data are plotted as far as parameter estimation is concerned. The purpose of this section is to illustrate this point.

The EC_{50} values or K_{act} values of the less efficacious agonist in the presence of increasing concentrations of full agonist can be estimated by nonlinear regression analysis using Equation (3.15). In this case, D denotes the concentration of less efficacious agonist. A nice feature of Equation (3.15) is that it can be fitted to increasing or decreasing concentration response curves. In the latter case, the value of the Hill slope is negative. The estimates of the log K_{act} values of the less efficacious agonist in the presence increasing concentrations of the full agonist are listed in Table 3.3.

From these estimates of the log K_{act} values of the less efficacious agonist, the tentative K_I values of the full agonist can be estimated from the curves measured in the presence of the four highest concentrations of the more efficacious agonist (see Table 3.3). The estimate of log K_I approaches a limiting value of 5.41

Table 3.3: Effect of an efficacious agonist on the whole cell current response to a less efficacious agonist. The parameter values were estimated from the theoretical data in Figure 3.4.

Agonist (M)	LogK_{act} or K'^{a}_{act}	Log Shift (K'_{act}/K_{act})	LogK_I^b
0	-7.88 ± 0.03	0	
3.16×10^{-7}	-7.93 ± 0.03	-0.043	
10^{-6}	-7.99 ± 0.04	-0.10	
3.16×10^{-6}	-7.76 ± 0.15	0.12	5.01
10^{-5}	-7.72 ± 0.11	0.16	4.65
3.16×10^{-5}	-7.00 ± 0.05	0.89	5.33
10^{-4}	-6.54 ± 0.06	1.35	5.33
3.16×10^{-4}	-5.97 ± 0.08	1.92	5.41

[a]The concentration of agonist eliciting a half-maximal current response in the absence (K_{act}) and presence (K'_{act}) of the agonist.
[b]The K_I values of the partial were estimated using Equation (3.12). This equation is only valid in the case of a competitive antagonist, however, although the equation does yield a good estimate of the K_I of an agonist whenever the concentration of the latter exceeds its corresponding value of K_{act} by a factor of 10.

at the highest concentration of the more efficacious agonist, and this estimate is approximately equivalent to its $-\log K_{act}$ value (5.38, see Figure 3.3b).

All of the data in Figure 3.4 can be analyzed by nonlinear regression analysis using Equation (3.20) in a manner analogous to that described above in connection with Figure 3.3a. In this case, however, the designation of the parameters and drug concentrations are reversed. That is, D and B now denote the concentrations of the less and more efficacious agonists, respectively, and the denotation of their respective parameters is also reversed (i.e., K_a for K_i, P_b for P_b and vice versa). Regression analysis yielded the following parameter estimates: $\log K_a$, 8.01 ± 0.02; P_b, 16.00 ± 0.03; $\log K_i$, 5.09 ± 0.09; $\log P_j$, 11.03 ± 0.03; and $\log RA_i$, -2.48. After taking into account the differences in notation, these agree with those estimated from the analysis shown in Figure 3.3a.

Single Channel Analysis

The most informative way to study the interaction of drugs with ligand-gated ion channels is through single channel analysis. The microscopic constants for any reaction mechanism can be determined from such measurements using the analysis technique described by David Colquhoun and his colleagues.[1–4] In this section,

I summarize a simple approach for analyzing competitive inhibition that is based on the assumption that the cooperative two-site model shown in Figure 3.1 has already been established. If so, then some of the microscopic constants describing competitive interactions can be gleaned from measurements of open probability.

The basis for this analysis rests on the assumption that the probability of channel opening is equivalent to the fraction of the receptor population in the open state as described in Chapter 2. Thus, one can apply population mathematics to study the competitive inhibition of single channel openings. An advantage of this approach is that it is possible to obtain a more reliable estimate of the theoretical fraction of the receptor population in the open state because the open probability is a direct measure of this fraction. In contrast, when measuring the population current, the estimate of the fraction of open channels is estimated as the current response divided by the theoretical current when all of the channels are open ($i_{\text{max-sys}}$). There can be significant error in the estimation of $i_{\text{max-sys}}$, particularly if there is desensitization or if only the responses of a partial agonist are measured. Thus, the analysis of open probability measurements leads to greater accuracy in the estimation of K_a and RA_i.

Open probability is equivalent to the time that the channel remains open divided by the total time of the observation. It can be estimated as such or by calculating open probability from the estimate of the total charge passed and the unitary conductance as described in Chapter 2. In the following analysis, it is assumed that the single channel record has been corrected for desensitization and missed events so that the estimates of open probability are unaffected by these factors.

Derivation of the equation for the analysis of competitive interactions at equilibrium

The equation describing the fraction of the receptor population in the active state in the presence of two competing ligands (Equation (3.6)) is appropriate for the analysis of open probability measurements. A minor modification will facilitate global nonlinear regression analysis. This involves introducing the parameters P_b and P_j and eliminating the affinity constants for the active state of the receptor. The modification is analogous to that used in the derivation of Equation (3.20):

$$P_{open} = \frac{1}{1 + \dfrac{([D]K_a + [B]K_i + 1)^2}{K_q\left([D]\sqrt{\dfrac{P_b(1-K_q)}{K_q}} + [B]\sqrt{\dfrac{P_j(1-K_q)}{K_q}} + 1\right)^2}}. \tag{3.22}$$

Competition between two agonists

The open probability measurements for the competitive interaction between two agonists yield concentration-response curves similar to those shown in Figure 3.3. The difference is that the ordinate always spans from 0 to 1.0, whereas the current responses span from 0 to whatever the maximum measured current is. The simulated data shown in Figure 3.3 has been normalized relative to $i_{\text{max-sys}}$, so that the plotted values of current are essentially the same as open probability. The difference in the regression analysis is that one need not extrapolate $i_{\text{max-sys}}$ for open probability measurements because the data are already expressed as fractions (probability measurements).

We can use the data in Figure 3.3a as an example of open probability measurements, because the current responses have been expressed relative to $I_{\text{max-sys}}$. The analysis of this competitive interaction involves fitting Equation (3.22) to the data by global nonlinear regression analysis. This is done by constraining K_q to an arbitrarily low value (e.g., 10^{-8}) and sharing the estimates of K_a, K_i, P_b and P_i among the curves. Regression analysis yielded the following parameter estimates: K_a, 4.98±0.02; K_i, 7.99±0.02; P_b, 10.96±0.02 and P_j, 15.98±0.04. These are very similar to those used to simulate the data in the first place. In this case, the estimate of the microscopic affinity constant of the most efficacious agonist for the inactive state of the receptor (K_a) has the less error compared with that estimated when the same data were as the population current or ensemble average. This illustrates the power of open probability measurements, not to mention the opportunity to correct these for missed events and desensitization.

Summary

Measuring the competitive inhibition of the current response to an agonist provide a means of estimating the affinity constant of an antagonist for the inactive state of the receptor. This approach can also be used to estimate the microscopic constants of two competing agonists. In the latter case, the experiment does not provide more information than what could be gleaned from measuring the effects of the agonists independently, but it does provide an opportunity to determine if the agonists compete for the same binding sites on the receptor. This chapter is written from the perspective of two linked orthosteric sites on a ligand-gated ion channel, but the analysis is also relevant to ligands competing at linked allosteric sites (e.g., different barbiturates acting on the $GABA_A$ receptor). In the next chapter, the interaction between allosteric and orthosteric ligands is described.

References

1. Colquhoun, D. and Hawkes, A. G. On the stochastic properties of bursts of single ion channel openings and of clusters of bursts. *Philos Trans R Soc Lond B Biol Sci* **300**, 1–59 (1982).
2. Colquhoun, D. and Sakmann, B. Fast events in single-channel currents activated by acetylcholine and its analogues at the frog muscle end-plate. *J Physiol* **369**, 501–557 (1985).
3. Colquhoun, D. and Hawkes, A. G. in *Microelectrode Techniques, The Plymouth Workshop Handbook* (eds. N. B. Standen, P. T. A. Gray and M. J. Whitaker) Ch. 7, 141–188 (Cambridge: Company of Biologists, 1987).
4. Colquhoun, D. and Hawkes, A. G. in *Single-Channel Recording*, 2nd edn., (eds. B. Sakmann and E. Neher) (Plenum Press, 1995).

Analysis of Allosteric Interactions at Ligand-Gated Ion Channels

4

Allosterism is a flexible mechanism of drug action that can be tailored to modify endogenous signaling with precision. This is accomplished through allosteric modulation of the binding of the endogenous ligand by the binding of a drug to another site on the receptor. This secondary drug site is defined as the allosteric site, and conformational changes in this site are associated with complementary changes in the conformation of the orthosteric site. Thus, both sites are allosterically linked, which provides the mechanism for the reciprocal modulation in ligand binding.

This mechanism is essentially the same as that described in Chapters 1 and 2 for the interaction of agonists with ligand-gated ion channels having two orthosteric sites. For example, it was shown that the binding of one molecule of acetylcholine to the nicotinic receptor enhances the binding of the second molecule, and that occupancy of both sites is required to induce the open state of the channel pore. In this chapter, the same mechanism is addressed, but for the condition where the interacting ligands and binding sites are different.

The word "allosteric" is often used in two contexts, and the meanings associated with these are succinctly described in the Oxford English Dictionary:

1. Pertaining to or involving a change in an enzyme's affinity for a substrate as a result of a change in its conformation by the attachment of a molecule not analogous to the substrate molecule at a site other than the substrate binding site.
2. Pertaining to any change in conformation caused by the attachment of a ligand.

My use of the term allosteric encompasses both of these meanings as illustrated in the following examples. We can say that acetylcholine allosterically enhances its own binding to the nicotinic acetylcholine receptor, and that this interaction allosterically opens the channel pore. Similarly, benzodiazepines allosterically enhance the binding of GABA to the $GABA_A$ receptor, which in turn, allosterically induces a chloride current.

Not surprisingly a large number of terms have appeared in our lexicon to describe the empirical behavior of allosteric ligands[1] This traditional approach rests on categorizing the details of drug behavior at superficial levels relating to the downstream response and the activation state of the receptor population. But at a deeper level, allosterism has fewer degrees of freedom and is determined by the selectivity of the allosteric ligand for the active and inactive states of the receptor. This selectivity is independent of the nature of the interacting ligands. Nonetheless, I will use nomenclature that is appropriate for the population level of analysis and requires specifying the interacting ligands.

A "positive allosteric modulator" is defined as an allosteric ligand that elicits little or no current response by itself but can enhance the affinity and efficacy of the orthosteric ligand. An "allosteric" agonist is defined as an allosteric ligand that has sufficient selectivity for the active state such that it causes activation by itself. It would also increase the affinity and efficacy of other agonists that signal through the same active state and inhibit the binding of inverse agonists. A "negative allosteric modulator" is defined as an allosteric ligand that can inhibit the affinity and efficacy of agonists. Finally, a neutral allosteric ligand is defined as a ligand that binds to the allosteric site and has no allosteric effect on the receptor but can inhibit the binding of other allosteric ligands competitively. These four terms will suffice to explain all of the interactions in this chapter.

So far I have considered allosteric effects as being mediated by conformational changes in the activation state of the receptor. Sometimes the effect of an allosteric ligand can be attributed, in part, to steric hindrance of receptor activation. For example, an allosteric ligand could bind to an allosteric site on a *G* protein-coupled receptor and sterically prevent the binding of a *G* protein. Alternatively, an allosteric ligand could bind selectively to the open state of a ligand-gated ion channel and plug the channel pore. This action could allosterically enhance the binding of an orthosteric agonist, but inhibit receptor activation by blocking the ion current.

Additional complexities can arise when there are multiple active states of the receptor, and these are described in Chapter 11.

As in Chapter 3, I focus my analysis of allosterism on the condition of equilibrium. I begin by briefly explaining allosterically induced changes in receptor conformation and the advantages this mechanism. I also address the question of why receptors have allosteric sites. The remainder of this chapter will address the analysis of allosteric interactions at ligand-gated ion channels that have two orthosteric sites and either one or two allosteric sites. The condition of one allosteric site linked to one orthosteric site is described in Chapter 11 within the context of

G protein-coupled receptors. Allosterism is addressed at the level of both single channels and the receptor population. This analysis enables one to understand why allosterism gives the illusion of causing a continuum of concentration-dependent changes in the shape and activity of the receptor complex even though these effects are caused by only two shapes of the receptor — active and inactive. Thus, despite the probe-dependent behavior of allosteric interactions at the level of the receptor population, a fixed set of microscopic constants can describe the variety of effects that a specific allosteric modulator has on a variety of different ligands.

Conformational Changes and Allosterism

One way to consider allosterism at a ligand-gated ion channel, like the muscle-type nicotinic acetylcholine receptor, is to imagine the receptor with two small fluorophores attached that are close enough for Forstner Resonance energy transfer when the channel is open but not when it is closed (see Figure 4.1). In the absence

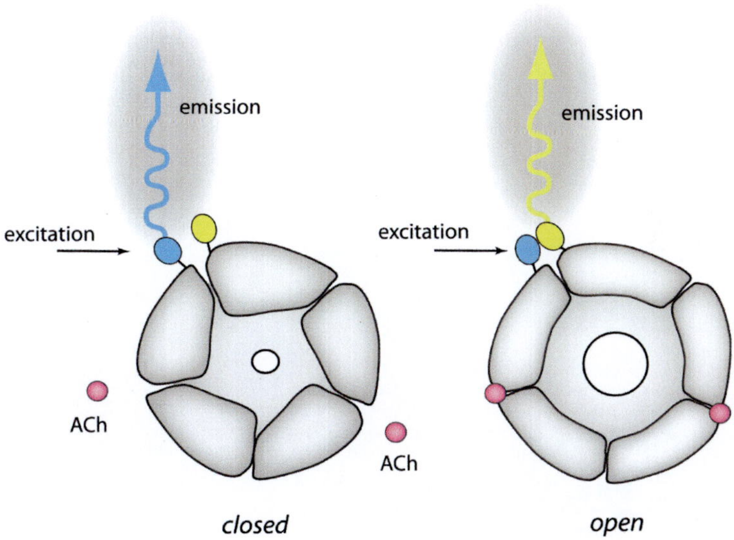

Figure 4.1: Hypothetical nicotinic acetylcholine receptor having two fluorophores that undergo Forstner resonance energy transfer depending upon whether the channel is open or closed. Light of the appropriate wave length excites the donor fluorophore and causes it to fluoresce blue whenever the channel is in the closed state. Channel opening results in a conformational change that brings the two fluorophore close enough together for a transfer of energy. This causes the acceptor fluorophore to fluoresce yellow whenever the blue fluorophore is excited by light that would otherwise cause it to fluoresce blue.

of acetylcholine, a cell expressing a population of this modified receptor would fluoresce blue when illuminated by light having the appropriate donor excitation wave length. This is because in the absence of ligands, the receptor is in the inactive state and the two fluorophores are sufficiently far from one another so that the donor fluorophore fluoresces at it characteristic emission wave length in the blue range. If acetylcholine is added to the bathing medium, it binds to the receptor and induces channel opening. This brings the donor fluorophore close to the acceptor fluorophore enabling a transfer of energy between the two and the emission of light at a wavelength characteristic of the acceptor fluorophore (yellow range). We can use this system to examine how orthosteric and allosteric drugs induce conformational changes in the receptor.

As illustrated in Figure 4.2, when a neutral antagonist is added to the bathing medium at a concentration that saturates the receptor population, the cell would still fluoresce blue because the antagonist does not induce the open state. When saturating concentrations of a partial agonist and a full agonist are added, however, the cell would fluoresce green and yellow respectively. One could imagine adding different agonists to the bathing medium and observing a continuum of fluorescence changes from blue to blue-green to green to greenish yellow to yellow. One could also imagine adding a positive allosteric modulator that has no effect by itself but enables a partial agonist to elicit yellow fluorescence instead of green. Still another positive

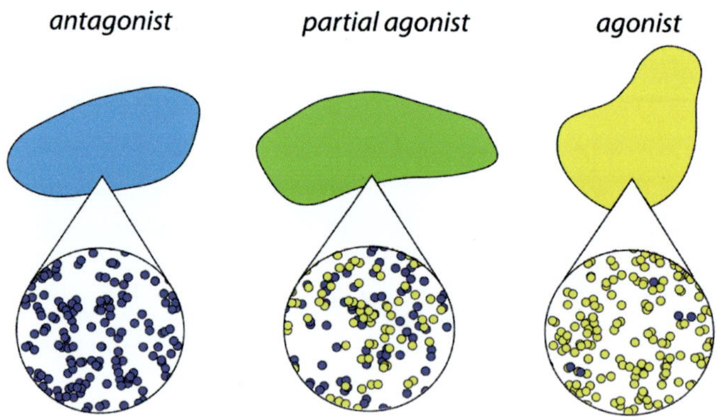

Figure 4.2: How ligands with different efficacies affect single receptors and their average population behavior. The figure shows three cells that express the theoretical receptor shown in Figure 4.1. Each cell is exposed to a receptor-saturating concentration of one of three different ligands — antagonist, partial agonist or agonist. At the level of the receptor population, the cells appear capable of fluorescing light on a continuum of wave lengths from blue to yellow. As shown by the magnification, however, individual receptors fluoresce only blue or yellow.

allosteric ligand could induce a fluorescence change by itself (e.g., blue-green) as well as increase the ability of a partial agonist to induce yellow fluorescence instead of green. From this perspective, it appears that conformational changes are on a continuum and that the magnitude of the shape change in the receptor depends on the efficacy of the ligand.

But this interpretation is an illusion based on a low-resolution population view. If we could turn up the zoom lens and visualize individual receptors we would see that the receptor behaves as a quantal switch that only shifts between on and off states as illustrated in Figure 4.2. In the presence of the antagonist, all of the individual receptors fluoresce blue, whereas in the presence of the full agonist, almost all of the receptors fluoresce yellow. In the presence of the partial agonist, however, approximately half of the receptors fluoresce yellow and half blue giving rise to green fluorescence from a population perspective. Thus, while the average behavior of the receptor population varies over a continuum of fluorescence changes, the behavior of individual receptors is quantal. Thus, drugs do not change the shape of the receptor. Rather, they select to varying degrees natural on–off states. Ultimately, the affinities of orthosteric and allosteric ligands for active and inactive states of the receptor determine the nature of their interaction. This chapter describes how to analyze allosteric interactions at both the population and receptor-state levels.

Benefits of Allosteric Modulation

In neural networks throughout the brain and periphery, information is encoded in both space and time, and it is the nature of this code that shapes our mind's experience of the world around us. Allosterism is an ideal mechanism of drug action for a condition that could be improved by tuning up or tuning down signaling through endogenous pathways without interfering with the timing of receptor activation.[2] Consider the case of a deficit of GABA neurotransmission, or alternatively, a condition that could be alleviated by enhancing GABA neurotransmission, like excessive anxiety. A positive allosteric modulator having little or no perceptible effect on $GABA_A$ receptors by itself would only enhance receptor activation when GABA is bound to the receptor, and thus, the information encoded in the temporal pattern of receptor activation would be preserved. In contrast, a directly acting agonist would activate receptors as long as its concentration near the receptor is sufficient to do so.

Another useful property of allosteric modulators is that the dynamic range of action that accompanies a large increase in their concentration can be much less than that of an orthosteric ligand. It follows that allosterism provides a means of protection against overdose.[3] The same could be said of a partial agonist, but the latter would also compete with the endogenous neurotransmitter and destroy the natural timing of signaling.

Of course there are situations when a directly acting agonist might be preferable. In the treatment of symptomatic asthma, for example, the activation of the β_2-adrenoceptor in the airways with an orthosteric agonist is desirable so that excessive bronchoconstriction is abated over long periods of time independently of whatever effect, if any at all, circulating levels of epinephrine might have.

Why do Ligand-Gated Ion Channels have Allosteric Sites?

In the Prologue, I presented the well-established idea that many drugs elicit their effects by acting on receptors for endogenous signaling molecules. With regard to the four families of physiological receptors, the locus of drug action is often the orthosteric site. It might seem odd, therefore, that an allosteric site can exist on a receptor that recognizes synthetic drugs (xenobiotics) but not an endogenous ligand.

We can throw some light on this question by reflecting on receptors that have evolved specifically to recognize xenobiotics. Some ligand-regulated transcription factors have binding sites for harmful xenobiotics ingested in the diet as well as a variety of drugs and endogenous compounds.[4] When activated, these receptors induce the transcription of enzymes that enhance the metabolism of xenobiotics, which usually reduces their activity and increases their renal excretion. Our special senses of vision, olfaction and taste provide additional examples of how G protein-coupled receptors detect xenobiotics in the form of photons and volatile and soluble compounds.[5–9] Some of the taste receptors in the tongue that mediate gustation are also expressed in the gastrointestinal tract where they relay afferent information to the brain to regulate metabolism.[10,11] In each of these examples, however, evolution has conscripted recognition sites cognate to the orthosteric site of other members of the receptor family. For example, odorant and taste molecules are thought to bind with odorant and taste receptors at the same locus where natural ligands bind to closely related members of the same *G* protein-coupled receptor family. In addition, rhodopsin detects light because it has a photo-activated ligand, 11-*cis*-retinal, covalently linked to a locus that is homologous to the orthosteric site of members of the amine class of *G* protein-coupled receptors.[12,13]

The picture that emerges from xenobiotic receptors is that sensing chemical information requires a special adaptation and that once nature solves this problem it uses a ligand-regulated allosteric protein to sense both endogenous and exogenous chemical messengers.

So how does this general concept relate to the presence of accessory allosteric sites on receptors? Before addressing this question, we need to consider one more question — what is the nature of the signal elicited by the site when occupied by an allosteric ligand? Does occupancy always disrupt the signaling of the receptor or does it induce or promote a unique conformational state that has high activity? With regard to the first mechanism, there are many ways that exogenous ligands can disrupt the function of various types of ion channels. Toxins can block the channel pore, disrupt the voltage sensor or interfere with desensitization and rectification.[14–16] Surely, these sites have not evolved for endogenous ligands, but rather, they are vulnerable parts of the channel machinery that are susceptible to disruption by foreign molecules. We might expect that molecular switches have evolved to be stealth and immune to such attacks, but at the same time, evolutionary forces are also at work to produce toxins that thwart receptor function. With regard to the second mechanism, it would seem that allosteric sites for ligands that elicit agonist-like effects are unique, and therefore, may represent sites for endogenous ligands, vestigial orthosteric sites from the evolutionary past or strategic loci capable of evolving into the orthosteric site of a molecular switch.

Something like this might account for the origin of the allosteric site for benzodiazepines on the $GABA_A$ receptor. Recall from Chapter 1 that orthosteric sites of the Cys-Loop receptor family are located between the amino terminal domains of adjacent receptor subunits (see Figures 1.7 and 1.12). Thus, acetylcholine binds to specific loops between β-strands on the counter clockwise surfaces of the two α subunits and also to specific residues on the opposing clockwise faces of the δ and γ subunits (see Chapter 1). With regard to $GABA_A$ receptors in brain, the orientation is reversed. That is, GABA binds to the clockwise faces of the two α subunits and the counter clockwise faces of the opposing β and γ subunits. If we turn up the zoom lens and focus in on the protein loops on the β and δ subunits that bind with GABA, these are homologous to those that interact with acetylcholine on the α subunits of the nicotinic receptor. The analogous relationship holds for the clockwise faces of the orthosteric sites for GABA and acetylcholine. It turns out that benzodiazepines interact with the $GABA_A$ receptor at a site that is homologous to the acetylcholine site on the nicotinic receptor. Perhaps this allosteric site for benzodiazepine is a vestigial orthosteric site from an earlier time in evolutionary history. Lowell Randall and coworkers at Hoffman Le Roche accidentally discovered the pharmacological consequences of activating this site through drug screening, although at the time, the mechanism for these allosteric effects was unknown.[17]

Thus, there appear to be strategic loci on oligomeric receptors that lie on the fault line between domains of the receptor that shift during activation. These hot spots may represent vestigial orthosteric sites that retain a structural function but are no longer activated by an endogenous ligand or have never been recruited for this purpose. Nonetheless, their capacity to function as an allosteric regulatory site may be discovered through drug screening. In the case of Cys-Loop receptors, opposite sides of the amino terminal domain of each receptor subunit may retain the capacity to mediate ligand-induced allosteric modulation even if the locus is not used for this purpose by an endogenous ligand. Thus, oligomeric receptors composed of different, but homologous, subunits would appear to be fruitful targets for allosteric modulation.

But how do we explain the allosteric sites for etomidate and other anesthetics on the transmembrane helices of the $GABA_A$ receptor? These ligands not only allosterically enhance the action of GABA, but they also elicit channel opening in the absence of GABA.[18] There is no evidence, however, that these allosteric agonists act on a site homologous to an orthosteric site, although recent evidence suggests that endocannabinoids act nearby.[19] Perhaps the anesthetic agents are the exception that proves the general rule about the relation between allosteric sites and vestigial orthosteric sites. Recall from Chapter 1 that the channel pore is formed by the pore-lining alpha helical M2 segments from each of the five receptor subunits. In the closed state, these are held in place through the formation of a hydrophobic plug in the middle of the M2 segments and also by the $\beta 1-\beta 2$ loop projecting downwardly from the amino terminal domain. Any allosteric ligand with the appropriate specificity for a binding pocket between the M2 helix and the surrounding M1, M3 and M4 helices might disrupt the highly ordered, closed structure of the M2 segments and cause the channel to open. Thus, a positive allosteric effect mediated by binding to the transmembrane alpha helices may result from a disruptive increase in entropy, whereas a similar allosteric action mediated through the amino terminus requires a more complex mechanism characteristic of an evolutionary adaptation.

Another way to look at this question is to recall the consequences of allosterism based on the two definitions of "allosteric" given at the outset of this chapter: 1) an interaction between binding sites and 2) an effect on the activity of an enzyme or receptor. In one sense, the effect of etomidate on the channel pore is hardly allosteric because it binds directly to the channel pore and disrupts it structure. It is only by default that etomidate allosterically modulates the binding of GABA because the orthosteric site is linked to the structure of the channel pore. Benzodiazepines, however, act at a distance to affect both the channel pore and the orthosteric site, and this interaction may represent a vestige of an evolutionary adaptation that is no longer used by current endogenous ligands.

Population Analysis

The main topic of this large section is the analysis of allosteric effects on the current response of a population of ligand-gated ion channels. I begin by describing the requisite mathematical models for a ligand-gated ion channel having two orthosteric sites and either one or two allosteric sites. These models will then be applied to the analysis of simulated data.

One of the well-known properties of allosteric interactions is that the each of the two interacting ligands has a reciprocal effect on the binding of the other ligand. That is, if allosteric ligand A enhances the binding of orthosteric ligand D, then ligand D should enhance the binding of ligand A to the same extent.[20] The application of this principle to the analysis of receptor function is not so obvious when the number of orthosteric and allosteric sites on the receptor are different, so I also develop a simpler model for allosteric binding interactions, which clarifies this issue.

Models for Allosterism at Ligand-Gated Ion Channels

Receptor activation models

The model describing allosteric interactions at a ligand-gated ion channel having two orthosteric sites and one allosteric site is shown in Figure 4.3. This model represents an extension of that developed in Chapter 2 for a ligand-gated ion channel having two orthosteric sites and is essentially the same as that described by Downing and coworkers.[21] An additional allosteric binding step was added to each receptor complex to yield the model in Figure 4.3. The result is useful for the analysis of the allosteric effects of benzodiazepines on $GABA_A$ receptor function, because there is one benzodiazepine-binding site per $GABA_A$ receptor (see Chapter 2).

The vertical panel on the back of the model in Figure 4.1 is essentially the same as the scheme in Figure 2.32. This part of the scheme describes how orthosteric ligand D binds to and activates the receptor. There are two ways that a single molecule of D can bind to form the mono-liganded receptor complex (e.g., DR_s), depending on which of the two orthosteric sites D binds. In our model, each of these has the same affinity for the ligand, so both are designated by a single complex in bold letters (e.g., $\boldsymbol{DR_s}$). Because there are two identical ways to form each DR complex, the microscopic affinity constant for these equilibria is multiplied by two (e.g., $2K_a$). It follows that the binding of the second molecule of the same ligand is

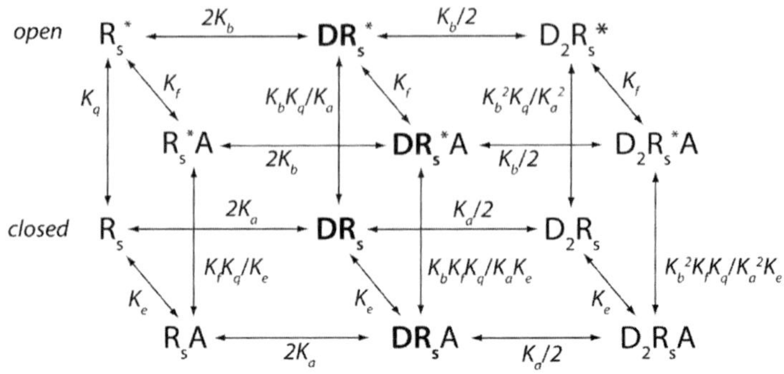

Figure 4.3: Equilibrium scheme for a ligand-gated ion channel with two orthosteric sites and one allosteric site. The drawing shows the different equilibrium binding steps for orthosteric (D) and allosteric (A) ligands. The receptor exists in two states, active (R_s^*) and inactive (R_s). The four receptor complexes in the middle of the scheme indicated in bold represent two species of receptor, each having a single orthosteric ligand bound to a different one of the two orthosteric sites. The parameters, K_a and K_e, denote the microscopic affinity constants of the orthosteric and allosteric ligands for the inactive state, respectively. The corresponding variables for the active state are K_b and K_f. The isomerization constant of the unoccupied receptor is denoted by K_q. The binding of one equivalent of orthosteric or allosteric ligand has the effect of altering the value of K_q by the scalar K_b/K_a or K_f/K_e, respectively.

described by the same microscopic affinity constant, but divided by 2 (e.g., $K_a/2$). A more complete description for this rationale is given in Chapter 2 in connection with the scheme shown in Figure 2.32.

Recall that when a drug-receptor interaction is defined at the level of receptor states, the microscopic affinity constant of a ligand for a receptor complex is determined solely by the state of the receptor and is unaffected by other ligands or proteins bound to the receptor. For example, the microscopic affinity constants of D for the R_s^* and R_s^*A complexes are the same (i.e., K_b). Similarly, the affinity of A for the R_s^* and DR_s^* complexes is also the same (i.e., K_f).

The front panel of the model in Figure 4.3 describes the binding of the allosteric ligand to the active and inactive states of the three different types of receptor complexes (R, DR and R_2R). The microscopic affinity constants of allosteric ligand A for the active (R_s^*) and inactive (R_s) states of the receptor are given by:

$$K_f = \frac{[R_s^*A]}{[A][R_s^*]}, \tag{4.1}$$

$$K_e = \frac{[R_sA]}{[A][R_s]}. \tag{4.2}$$

The other equilibria is defined as described in connection with the model in Figure 2.32.

The scheme in Figure 4.3 describes behavior at equilibrium, and consequently, each of the 11 squares of four equilibria is constrained such that there is no net cycling around the square at equilibrium (law of microscopic reversibility). Thus, all of the steps that describe the induction of the active state of the receptor are defined by constants that are present in the other three equilibria of the same square.

Let me describe the process of deriving the expressions for the isomerization steps with two examples. First, all of the ligand-binding steps on the top and bottom surfaces are assigned the appropriate microscopic constants (i.e., K_b and K_f on the top surface and K_a and K_e on the bottom surface). Next, the isomerization of the free receptor (R_s) into the active state (R_s^*) is assigned the constant K_q on the back, left-hand vertical edge of the model. This is the simplest equilibrium step, and R_s is the simplest species in this equilibrium. It is our starting point for defining the isomerization constants. From here we can proceed along two paths to reach the most complex isomerization step on the front, right-hand edge of the model. The first path proceeds to the right across the back side of the model and then forward to the quaternary complexes ($D_2R_s^*A$ and D_2R_sA), and the second proceeds in a forward direction and then to the right across the front of the model to the quaternary complexes. Let us begin with the second path. We start with the simplest species (R_s) and identify the two routes to the most complex species in the square of equilibria on the left-hand side of the model. These are:

$$R_s \overset{K_q}{\longleftrightarrow} R_s^* \overset{K_f}{\longleftrightarrow} R_s^*A, \tag{4.3}$$

$$R_s \overset{K_e}{\longleftrightarrow} R_sA \overset{?}{\longleftrightarrow} R_s^*A. \tag{4.4}$$

To satisfy the law of microscopic reversibility, the products of the two constants describing the two equilibria in the two different routes must be equal. The product for the upper scheme is K_qK_f, and therefore, the equilibrium expression for the process denoted by "?" in the second scheme must be K_qK_f/K_e.

Proceeding to the next square of equilibria on our path on the left-hand side of the front face, we see that the simplest receptor complex is R_sA. The two routes to the most complex species in this square of equilibria are:

$$R_sA \overset{K_qK_f/K_e}{\longleftrightarrow} R_s^*A \overset{2K_b}{\longleftrightarrow} DR_s^*A, \tag{4.5}$$

$$R_sA \overset{2K_a}{\longleftrightarrow} DR_sA \overset{?}{\longleftrightarrow} DR_s^*A. \tag{4.6}$$

Once again, to satisfy the law of microscopic reversibility, the products of the two expressions describing the two equilibria in each of the two routes must be equal. Hence, the expression denoted by "?" in scheme 4 must be $K_q K_f K_b / K_e K_a$. This process is repeated for the final square of equilibria on our path along the front side of the model. Then the entire process is repeated following the backside path from simplest species (R_s) to the most complex on the front right corner of the model.

And this is how the various equilibria of the model in Figure 4.3 are assigned their constants.

The next model describes a ligand-gated ion channel having both two orthosteric sites and two allosteric sites as in the case of the interaction of etomidate and other anesthetics with the $GABA_A$ receptor. This model is shown in Figure 4.4. It was derived by incorporating an additional allosteric binding step onto the front side of the model in Figure 4.1. The expressions of microscopic constants describing the various equilibria were solved using the strategy just described. The model is essentially the same as that described by Rusch and coworkers.[22]

An additional complexity relates to the actual number of species denoted by each receptor complex in the model. Only the complexes on the four outer corners of

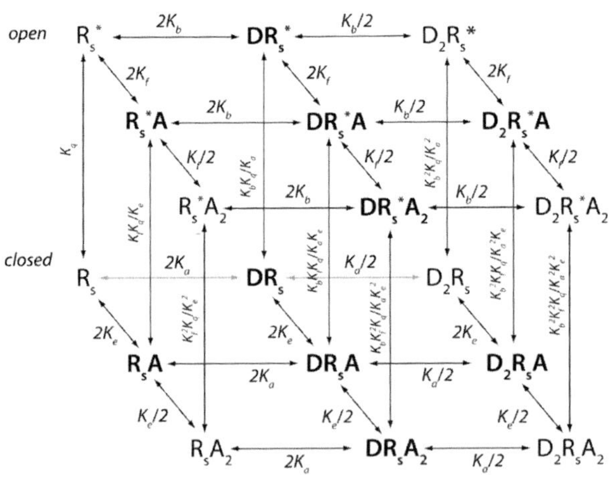

Figure 4.4: Equilibrium scheme for a ligand-gated ion channel with two orthosteric sites and two allosteric sites. The drawing shows the different equilibrium binding steps for the orthosteric (D) and allosteric (A) ligands to active (R_s^*) and inactive (R_s) states of the receptor. The microscopic constants are defined as in the legend to Figure 4.3. The receptor complexes indicated in bold represent two (middle of outer sides) or four (center) species, representing the different combinations of complexes, depending upon to which of the two orthosteric and allosteric sites each ligand is bound. Additional details are described in the text.

the model denote one type of receptor species (i.e., R_s, R_s^*, D_2R_s, $D_2R_s^*$, R_sA_2, $R_sA_2^*$, $D_2R_sA_2$ and $D_2R_s^*A_2$). Those in the center of the outer sides of the model denote two species each and are shown in bold (**DR_s**, **DR_s^***, **R_A**, **R_s^*A**, **D_2R_sA**, **$D_2R_s^*A$**, **DR_sA_2** and **$DR_s^*A_2$**). Finally, the two receptor complexes in the center of the model each denote four receptor species that arise from the four possible ways that one allosteric ligand and one orthosteric ligand can bind to the channel complex. These are also shown in bold (**DR_sA** and **DR_s^*A**). The rationale for assigning a value of four units to the latter complexes follows the same logic that was used to assign two units to the former complexes.

Ligand-binding model for cooperative interactions

We can use a simpler model to describe the binding of ligands to a receptor because the amount of ligand bound to the receptor includes both the active and inactive ligand-receptor complexes, and so, there is no need to discriminate between them. In a functional assay, of course, it is only the active state of the complex that generates a signal. Figure 4.5a shows the appropriate model for ligand binding in the case of two orthosteric sites and one allosteric site. Each receptor complex denotes the summation of the active and inactive states of a given receptor complex (i.e., D_2R

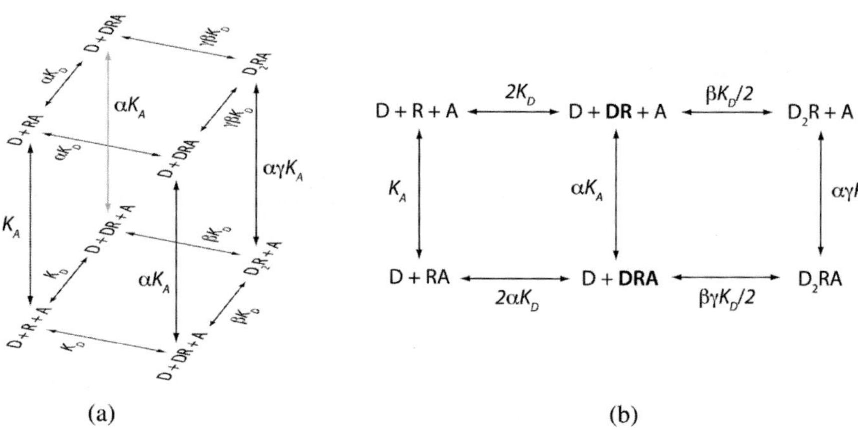

Figure 4.5: Model for the binding of ligands to a ligand-gated ion channel having two orthosteric sites and one allosteric site. The model is defined in terms of the observed affinity and cooperativity constants. (a) The binding of the orthosteric ligand to each of the two orthosteric sites is shown. (b) In this panel, the two different binding steps for a single equivalent of orthosteric ligand molecule are designated in a single step.

denotes the sum of D_2R_s and $D_2R_s^*$). At this population level of analysis, the observed affinity constant (K_D) describes how one equivalent of ligand binds to the receptor. It is defined as:

$$K_D = \frac{[DR]}{[D][R]}. \tag{4.7}$$

One way to think about the value of K_D is to imagine how the orthosteric ligand would bind if the second site on the receptor did not exist. In this thought experiment, the value of the reciprocal of K_D would be equivalent to the concentration of ligand required for half-maximal receptor occupancy. In reality, of course, there *is* a second site, and the binding of an agonist to the second site will enhance the affinity of the agonist for the first site. Hence, the actual concentration of agonist required for half-maximal receptor occupancy may be substantially less than the reciprocal of K_D. The value of K_D can be expressed in terms of the microscopic constants K_a, K_b and K_q as described below. In the case of partial agonists, the value of K_D may not much different from K_a, but in the case of a full agonists, the value is substantially greater than K_a. For antagonists and inverse agonists, K_D is equivalent to K_a as long as there is little or no constitutive receptor activity.

Because there are two orthosteric sites having the same observed affinity constant for D (K_D), the model in panel (a) of Figure 4.5 can be simplified as shown in panel (b). Here the summation of the two DR complexes in panel (a) are indicated as a single receptor complex in bold letters ($\boldsymbol{DR}$). The observed affinity constant for this complex is defined as:

$$2K_D = \frac{[\boldsymbol{DR}]}{[D][R]}. \tag{4.8}$$

Because the expression $[\boldsymbol{DR}]$ is equivalent to $2[DR]$, Equation (4.8) can be expressed as:

$$2K_D = \frac{2[DR]}{[D][R]}. \tag{4.9}$$

Factoring out the two yields Equation (4.7), which illustrates that there is no difference between Equations (4.7) and (4.8).

With regard to the allosteric ligand, the affinity constant K_A also denotes the overall observed affinity measured in the absence of the orthosteric ligand. But because there is only one allosteric site per receptor complex, we do not need to do a thought experiment to get an intuitive grasp of its value. Its value is equivalent to the reciprocal of the concentration of A required for half-maximal receptor occupancy in the absence of D. It is defined as:

$$K_A = \frac{[RA]}{[A][R]}. \tag{4.10}$$

One of the fundamental properties of allosterism is that the cooperative effect that one ligand has on the binding of another is reciprocal. This is shown by the cooperativity constants denoted by Greek letters in the model shown in Figure 4.5. For example, on the bottom side of the left-hand square of equilibria, the binding of D to either one of the two orthosteric sites on the receptor is defined by the affinity constant, K_D. If the allosteric site is occupied by A, however, then the affinity constant of D is altered by the scalar α as shown by the equilibrium on the upper side of the same square of equilibria. Similarly, in the absence of orthosteric ligand, the binding of the allosteric ligand is defined by the affinity constant K_A as shown on the left-hand side of the model. If one and only one of the orthosteric sites is occupied, however, then the affinity constant of A is also altered by the scalar α. Thus, α is the cooperativity constant that defines the allosteric interaction.

If the two interacting ligands increase the affinity of each other, then the allosteric effect is defined as positive heterotropic cooperativity, and the value of α is greater than one. For negative cooperativity, A and D reduce the affinity of each other, the interaction is defined as negative heterotropic cooperativity. For this type of interaction, the cooperativity is defined as heterotropic because it involves an interaction between different types of ligands binding to different types of sites. One convenient way to think about cooperativity is to consider log α. The value of this constant is zero for no cooperativity, positive for positive cooperativity and negative for negative cooperativity.

There are also homotropic interactions. The binding of the first equivalent of D to the receptor alters the binding of the second equivalent, and this cooperative effect is quantified by the scalar β as shown on the bottom side of the right-hand square of equilibria. If D is an agonist, the value of β is greater than 1, and the interaction is defined as positive homotropic cooperativity. If D is an inverse agonist, the value of β is nearly equal to 1 (very slightly greater than one) because an inverse agonist has selectivity for the inactive (closed) state of the receptor, and the receptor is usually already in the inactive state in the absence of ligands.

Now at this level of analysis, where we do not discriminate between active and inactive states of the receptor, the cooperativity between the allosteric ligand and the binding of the first (α) and second (γ) equivalents of agonist are different as explained below. The latter is designated by γ on the right and lower right equilibria of the model (Figure 4.5b). Thus, $\beta\gamma K_D$ denotes the affinity constant for the binding of the second equivalent of D to the DRA complex, and $\alpha\gamma K_A$ denotes the affinity constant of A for binding to the D_2R complex. The two affinity constants and cooperativities can be expressed in terms of the more fundamental microscopic constants shown in Figure 4.3 as shown in a later section.

The model illustrated in Figure 4.4 for the case of two allosteric sites can also be described in terms of cooperativity constants and observed affinity constants, but the effort in describing the additional complexity has no payoff in analyzing functional data so I will not present this model. Ligand binding equations are also described below based on the receptor state models shown in Figures 4.3 and 4.4.

Derivation of the equations for the models of allosterism

There are many equations that one can derive from the foregoing equilibrium schemes depending on which or how many components of ligand-receptor complex are of interest. With regard to the schemes shown in Figures 4.3 and 4.4, I derive equations for the binding of the orthosteric ligand, the binding of the allosteric ligand, and the summation of the receptor complexes in the active state. The latter equation is suitable for analyzing the effects of allosteric ligands on the current response of a ligand-gated ion channel because the current response is proportional to the fractional amount of receptor complex in the active state. For the ligand-binding scheme in Figure 4.5, I derive equations for the binding of the orthosteric and allosteric ligands. These equations are then used to illustrate the reciprocal allosteric effects on ligand binding. Other components of the models can be derived using a similar strategy. Those uninterested in these derivations may wish to skip to the next section, entitled "Simulation and analysis of allosteric binding interactions", on p. 205.

Receptor activation function for the case one allosteric site

In this section, an equation describing the fractional amount of receptor complex in the active state is described for the model shown in Figure 4.3. To begin, we need to define the microscopic affinity constants in the model. Those describing the binding of the orthosteric ligand on the back vertical panel of the model (i.e., K_b and K_a) have already been defined in Chapter 2. Once again, because affinity at the level of receptor states is determined by the state of the receptor, the microscopic affinity constants of the orthosteric ligand (K_b and K_a) also apply to receptor complexes involving the allosteric ligand:

$$2K_b = \frac{[\boldsymbol{DR_s^*A}]}{[D][R_s^*A]}, \tag{4.11}$$

$$\frac{K_b}{2} = \frac{[D_2R_s^*A]}{[D][\boldsymbol{DR_s^*A}]}, \tag{4.12}$$

$$2K_a = \frac{[\boldsymbol{DR_sA}]}{[D][R_sA]}, \tag{4.13}$$

$$\frac{K_a}{2} = \frac{[D_2R_sA]}{[D][\boldsymbol{DR_sA}]}. \tag{4.14}$$

The microscopic affinity constants of the allosteric ligand for the free forms of the active and inactive states of the receptor are defined earlier by Equations (4.1) and (4.2). It follows that these constants also apply to receptor complexes involving the orthosteric ligand:

$$K_f = \frac{[\boldsymbol{DR_s^*A}]}{[A][\boldsymbol{DR_s^*}]}, \tag{4.15}$$

$$K_f = \frac{[D_2R_s^*A]}{[A][D_2R_s^*]}, \tag{4.16}$$

$$K_e = \frac{[\boldsymbol{DR_sA}]}{[A][\boldsymbol{DR_s}]}, \tag{4.17}$$

$$K_e = \frac{[D_2R_sA]}{[A][D_2R_s]}. \tag{4.18}$$

Now that the microscopic constants are defined, we can derive the equation for the fraction of the agonist-receptor complex in the active state. We begin with the expression

$$\frac{[R_{\text{active}}]}{[R_T]} = \frac{[R_{\text{active}}]}{[R_{\text{active}}] + [R_{\text{inactive}}]}, \tag{4.19}$$

in which

$$[R_{\text{active}}] = [R_s^*] + [\boldsymbol{DR_s^*}] + [D_2R_s^*] + [\boldsymbol{DR_s^*A}] + [D_2R_s^*A] + [R_s^*A], \tag{4.20}$$

$$[R_{\text{inactive}}] = [R_s] + [\boldsymbol{DR_s}] + [D_2R_s] + [\boldsymbol{DR_sA}] + [D_2R_sA] + [R_sA]. \tag{4.21}$$

Using the strategy outlined in Chapters 2 and 3, each receptor complex can be substituted with an expression containing the concentration of the inactive state of the free receptor (R_s) and the appropriate microscopic constants and ligand concentrations. After factoring out $[R_s]$ and simplifying, the following equation is obtained:

$$\frac{[R_{\text{active}}]}{[R_T]} = \frac{1}{1 + \frac{(1+AK_e)(1+DK_a)^2}{K_q(1+AK_f)\left(1+D\sqrt{\frac{P_{b2}(1+K_q)}{K_q}}\right)^2}}, \quad (4.22)$$

in which

$$P_{b2} = \frac{K_b^2 K_q}{1+K_q}. \quad (4.23)$$

Equation (4.22) can be written in a slightly simpler form by substituting $K_b^2 K_q/(1+K_q)$ for P_{b2} and simplifying. When that form of Equation (4.22) is used for nonlinear regression analysis, it is possible to estimate $K_b^2 K_q/(1+K_q)$, but not necessarily the individual parameters, K_b and K_q. Hence, Equation (4.22) is in a useful form for regression analysis.

Open channel block

A special case of the allosteric mechanism just described is the open-channel block. In this case, the allosteric modulator has two effects — allosteric and steric. The allosteric effect is manifest as an increase in agonist affinity because the modulatory ligand (i.e., open channel blocker) binds selectively to the open state (active state) of the receptor and induces the active state, which has higher affinity for the agonist. When the modulator is bound to the active state, however, it plugs the channel and effectively eliminates the activity of the receptor. This mechanism is also known as uncompetitive within the context of enzyme activity.

To derive an equation for this mechanism, we begin with Equation (4.19), but define the active (R_{active}) and inactive (R_{inactive}) components differently from that described by Equations (4.20) and (4.21). Because the allosteric ligand plugs the channel when bound to it, the active states of the receptor bound with modulator (DR_s^*A, $D_2R_s^*A$ and R_s^*A) are functionally inactive. Hence, these are eliminated from the R_{active} component and added to R_{inactive}:

$$[R_{\text{active}}] = [R_s^*] + [\boldsymbol{DR_s^*}] + [D_2R_s^*], \quad (4.24)$$

$$[R_{\text{inactive}}] = [R_s] + [\boldsymbol{DR_s}] + [D_2R_s] + [\boldsymbol{DR_sA}] + [D_2R_sA] + [R_sA] + [\boldsymbol{DR_s^*A}] + [D_2R_s^*A] + [R_s^*A]. \quad (4.25)$$

Now, we proceed in the same manner as described in the prior section. Equations (4.24) and (4.25) are substituted into Equation (4.19), and then each receptor complex is substituted with an expression containing R_s and the requisite microscopic constants and ligand concentrations. These expressions are derived from the definitions of the microscopic constants. After simplification the following equation is obtained:

$$\frac{[R_{\text{active}}]}{[R_T]} = \frac{1}{1 + AK_f + \frac{(1+AK_e)(1+DK_a)^2}{K_q\left(1+D\sqrt{\frac{P_{b2}(1+K_q)}{K_q}}\right)^2}}. \tag{4.26}$$

Again, P_b is defined as in Equation (4.23), ($P_{b2} = K_b^2K_q/(1+K_q)$). This composite parameter is used to facilitate nonlinear regression analysis.

Receptor binding functions for the case of one allosteric site

We can also use the model shown in Figure 4.3 to derive equations for the binding of the orthosteric and allosteric ligands. We begin with the simple equation:

$$\frac{D_{\text{bound}}}{2[R_T]} = \frac{[D_{\text{bound}}]}{[D_{\text{bound}}] + [D_{\text{unbound}}]}, \tag{4.27}$$

in which,

$$\begin{aligned}[D_{\text{bound}}] = {} & [\boldsymbol{DR_s^*}] + 2[D_2R_s^*] + [\boldsymbol{DR_s^*A}] + 2[D_2R_s^*A] \\ & + [\boldsymbol{DR_s}] + 2[D_2R_s] + [\boldsymbol{DR_sA}] + 2[D_2R_sA],\end{aligned} \tag{4.28}$$

$$[D_{\text{unbound}}] = [R_s] + [R_sA] + [R_s^*] + [R_s^*A]. \tag{4.29}$$

Let us have a close look at this equation. The left-hand side shows that the amount of ligand bound to the receptor (D_{bound}) is divided by, or normalized to, twice the concentration of receptors ($2[R_T]$). This means that fractional binding is maximal (i.e., 1.0) or at 100% occupancy when each receptor complex is occupied by two orthosteric ligands. If we define receptor occupancy in this way, then each receptor complex bound with two molecules of D on the right-hand side of the equation needs to be multiplied by a factor of 2 to maintain the appropriate stoichiometry. This has been done as shown on the right-hand side of the equation. This equation is solved as described above by substituting each receptor complex with the appropriate expression composed of $[R_s]$, ligand concentrations and microscopic constants. After simplification the following equation is obtained:

$$\frac{D_{\text{bound}}}{2[R_T]} = \frac{[D]}{[D] + \frac{1+K_q+A(K_e+K_f K_q)}{2(K_a(1+DK_a)(1+AK_e)+K_b K_q(1+DK_b)(1+AK_f))}}. \tag{4.30}$$

The equation for the binding of A begins with:

$$\frac{A_{\text{bound}}}{[R_T]} = \frac{[A_{\text{bound}}]}{[A_{\text{bound}}] + [A_{\text{unbound}}]}, \tag{4.31}$$

in which:

$$[A_{\text{bound}}] = [R_s^* A] + [R_s A] + [\boldsymbol{DR_s A}] + [\boldsymbol{DR_s^* A}] + [D_2 R_s^* A] + [D_2 R_s A], \tag{4.32}$$

$$[A_{\text{unbound}}] = [R_s] + [R_s^*] + [\boldsymbol{DR_s}] + [\boldsymbol{DR_s^*}] + [D_2 R_s] + [D_2 R_s^*]. \tag{4.33}$$

In this case, binding is normalized relative to one unit of receptor because there is only one allosteric site per receptor. Also, there is no need to multiply the receptor complexes doubly bound with orthosteric ligand by two. After substituting each receptor complex with the appropriate expression and simplifying, the following equation is obtained:

$$\frac{A_{\text{bound}}}{[R_T]} = \frac{[A]}{[A] + \frac{(1+DK_a)^2+K_q(1+DK_b)^2}{K_e(1+DK_a)^2+K_f K_q(1+DK_b)^2}}. \tag{4.34}$$

A simpler way to envision a ligand-binding function is without consideration for the state of the receptor, which is by far the more common way that radioligand-binding data are analyzed. The relevant binding equations are equivalent to those described above, of course, but they are expressed in terms of the parameters of the model shown in Figure 4.5. These are the observed binding affinities of the allosteric (K_A) and orthosteric (K_D) ligands and the scalar cooperativity constants (α, β and γ) that describe their interaction. We have already defined K_D and K_A (see Equations (4.8) and (4.10), but we still need to define the other affinity constants shown in our model in Figure 4.5b:

$$\frac{1}{2}\beta K_D = \frac{[D_2 R]}{[D][\boldsymbol{DR}]}, \tag{4.35}$$

$$2\alpha K_D = \frac{[\boldsymbol{DRA}]}{[D][RA]}, \tag{4.36}$$

$$\frac{\beta\gamma K_D}{2} = \frac{[D_2RA]}{[D][\boldsymbol{DRA}]}, \tag{4.37}$$

$$\alpha K_A = \frac{[\boldsymbol{DRA}]}{[A][\boldsymbol{DR}]}, \tag{4.38}$$

$$\alpha\gamma K_A = \frac{[D_2RA]}{[A][\boldsymbol{DR}]}. \tag{4.39}$$

Now, we can derive the equation for the binding of orthosteric ligand. We begin with the following expression, which communicates the idea that fractional receptor occupancy by D is equivalent to the summation of the receptor complexes bound with D divided by the sum of all of the receptor complexes:

$$\frac{D_{\text{bound}}}{2[R_T]} = \frac{[\boldsymbol{DR}] + 2[D_2R] + [\boldsymbol{DRA}] + 2[D_2RA]}{[\boldsymbol{DR}] + 2[D_2R] + [\boldsymbol{DRA}] + 2[D_2RA] + [R] + [RA]}. \tag{4.40}$$

This equation normalizes fractional receptor occupancy with respect to two receptor equivalents because the receptor can bind two molecules of orthosteric ligand. Hence, the variable for bound ligand on the left-hand side of the equation is divided by $2[R_T]$, and each variable denoting a receptor complex bound with two molecules of orthosteric ligand on the right-hand side is multiplied by two as in the case of the derivation of Equation (4.27).

Now, we replace each receptor complex on the right side of the equation with an expression consisting of the free receptor concentration ($[R]$) and the appropriate constants and ligand concentrations. For example, using Equation (4.10), the complex $[RA]$ can be replaced with the expression $AK_A[R]$. Similarly, the complex $[\boldsymbol{DRA}]$ can be expressed as $2D\alpha K_D[RA]$ based on Equation (4.36). The $[RA]$ complex in this latter expression can be substituted with the expression given in the prior sentence to yield $2AD\alpha K_A K_D[R]$. After making similar replacements for the other receptor complexes, the following equation is obtained after simplification:

$$\frac{D_{\text{bound}}}{2[R_T]} = \frac{D}{D + \frac{1+AK_A}{2K_D((1+D\beta K_D)+A\alpha K_A(1+D\beta\gamma K_D))}}. \tag{4.41}$$

The equation for the binding of allosteric ligand A in the presence of orthosteric ligand D can be derived using an analogous strategy. We begin with the following expression:

$$\frac{A_{\text{bound}}}{[R_T]} = \frac{[RA] + [\boldsymbol{DRA}] + [D_2RA]}{[\boldsymbol{DR}] + [D_2R] + [\boldsymbol{DRA}] + [D_2RA] + [R] + [RA]}. \tag{4.42}$$

In this case, occupancy by the orthosteric ligand is normalized to one unit of receptor because there is only one allosteric binding site per preceptor. After substituting each receptor complex with the appropriate expression and simplifying, the following equation is obtained:

$$\frac{A_{\text{bound}}}{[R_T]} = \frac{[A]}{[A] + \frac{1+[D]K_D(2+[D]\beta K_D)}{K_A+[D]\alpha K_D K_A(2+[D]\beta\gamma K_D)}}. \qquad (4.43)$$

We shall use Equations (4.41) and (4.43) to explore the reciprocal binding interactions between the allosteric and orthosteric ligands.

It is possible to express the observed affinity (K_A and K_D) and cooperativity (α, β and γ) constants in terms of the more fundamental microscopic constants. For example, the equation for the definition of K_D (Equation (4.7)) is first rewritten by replacing each receptor complex in the equation with the summation of the active and inactive states of the receptor:

$$K_D = \frac{[DR_s + DR_s^*]}{[D][R_s + R_s^*]}. \qquad (4.44)$$

Then, the various receptor complexes are substituted with the appropriate equations expressed in terms of microscopic constants (e.g., $DR_s = DK_a[R_s]$), and the resulting expression is simplified:

$$K_D = \frac{K_a + K_b K_q}{1 + K_q}. \qquad (4.45)$$

Applying the same strategy yields the following equations for some of the other observed equilibrium constants:

$$K_A = \frac{K_e + K_f K_q}{1 + K_q}, \qquad (4.46)$$

$$\alpha K_D = \frac{K_a K_e + K_b K_f K_q}{K_e + K_f K_q}, \qquad (4.47)$$

$$\beta K_D = \frac{K_a^2 + K_b^2 K_q}{K_a + K_b K_q}, \qquad (4.48)$$

$$\beta\gamma K_D = \frac{K_a^2 K_e + K_b^2 K_f K_q}{K_a K_e + K_b K_f K_q}. \qquad (4.49)$$

The cooperativity constants can be expressed in terms of the microscopic constants by taking the ratio of some constants just described. For example, the cooperativity constant α is equivalent to $\alpha K_D/K_D$. Thus, substituting in Equations (4.47) and (4.45) for αK_D and K_D in the latter expression yields the following after simplification:

$$\alpha = \frac{(K_a K_e + K_b K_f K_q)(1 + K_q)}{(K_a + K_b K_q)(K_e + K_f K_q)}. \tag{4.50}$$

Using a similar approach for the derivation of β ($\beta = \beta K_D/K_D$) and γ ($\gamma = \gamma K_D/K_D$) yields:

$$\beta = \frac{(K_a^2 + K_b^2 K_q)(1 + K_q)}{K_a^2 + 2K_a K_b K_q + K_b^2 K_q^2}, \tag{4.51}$$

$$\gamma = \frac{(K_a^2 K_e + K_b^2 K_f K_q)(K_a + K_b K_q)}{(K_a^2 + K_b^2 K_q)(K_a K_e + K_b K_f K_q)}. \tag{4.52}$$

Receptor activation function for the case of two allosteric sites

The derivation of the receptor activation function for the case of two allosteric sites follows the same strategy as that described above for one allosteric site. We begin by defining the microscopic constants. Some of these have already been defined, and some that have been defined for the model in Figure 4.3 (Equation (4.22)) need to be rewritten in a different form for the model in Figure 4.4 to account for the two allosteric sites. These changes involve multiplying K_f and K_e by the scalar 2 to account for the two possible forms of some of the allosteric ligand-receptor complexes and identifying these receptor complexes in bold letters to account for their different forms.

Because there are two forms of the allosteric ligand-receptor complex, K_f and K_e can be defined in the following ways:

$$2K_f = \frac{[\mathbf{R_s^* A}]}{[A][R_s^*]}, \tag{4.53}$$

$$\frac{K_f}{2} = \frac{[R_s^* A_2]}{[A][\mathbf{R_s^* A}]}, \tag{4.54}$$

$$2K_f = \frac{[\mathbf{DR_s^* A}]}{[A]\left[\mathbf{DR_s^*}\right]}, \tag{4.55}$$

$$\frac{K_f}{2} = \frac{[\boldsymbol{DR_s^* A}_2]}{[A][\boldsymbol{DR_s^* A}]}, \tag{4.56}$$

$$2K_f = \frac{[\boldsymbol{D}_2\boldsymbol{R_s^* A}]}{[A][D_2R_s^*]}, \tag{4.57}$$

$$\frac{K_f}{2} = \frac{[D_2R_s^* A_2]}{[A][\boldsymbol{D}_2\boldsymbol{R_s^* A}]}, \tag{4.58}$$

$$2K_e = \frac{[\boldsymbol{R_s A}]}{[A][R_s]}, \tag{4.59}$$

$$\frac{K_e}{2} = \frac{[R_s A_2]}{[A][\boldsymbol{R_s A}]}, \tag{4.60}$$

$$2K_e = \frac{[\boldsymbol{DR_s A}]}{[A][\boldsymbol{DR_s}]}, \tag{4.61}$$

$$\frac{K_e}{2} = \frac{[\boldsymbol{DR_s A}_2]}{[A][\boldsymbol{DR_s A}]}, \tag{4.62}$$

$$2K_e = \frac{[\boldsymbol{D}_2\boldsymbol{R_s A}]}{[A][D_2R_s]}, \tag{4.63}$$

$$\frac{K_e}{2} = \frac{[D_2R_s A_2]}{[A][\boldsymbol{D}_2\boldsymbol{R_s A}]}. \tag{4.64}$$

Analogous changes are needed for K_b and K_a:

$$2K_b = \frac{[\boldsymbol{DR_s^* A}]}{[D][\boldsymbol{R_s^* A}]}, \tag{4.65}$$

$$\frac{K_b}{2} = \frac{[\boldsymbol{D}_2\boldsymbol{R_s^* A}]}{[D][\boldsymbol{DR_s^* A}]}, \tag{4.66}$$

$$2K_b = \frac{[\boldsymbol{DR_s^* A}_2]}{[D]pR_s^* A_2]}, \tag{4.67}$$

$$\frac{K_b}{2} = \frac{[D_2R_s^* A_2]}{[D][\boldsymbol{DR_s^* A}_2]}, \tag{4.68}$$

$$2K_a = \frac{[\boldsymbol{DR_s A}]}{[D][\boldsymbol{R_s A}]}, \tag{4.69}$$

$$\frac{K_a}{2} = \frac{[D_2R_sA]}{[D][DR_sA]}, \tag{4.70}$$

$$2K_a = \frac{[DR_sA_2]}{[D][R_sA_2]}, \tag{4.71}$$

$$\frac{K_a}{2} = \frac{[D_2R_sA_2]}{[D][DR_sA_2]}. \tag{4.72}$$

Having defined all of the microscopic constants, we can now derive the receptor activation function for a ligand-gated ion channel having two allosteric sites. Again, we begin with Equation (4.19) described above, but in this case the expressions for the summation of the active and inactive receptor complexes are defined by:

$$[R_{\text{active}}] = [R_s^*] + [DR_s^*] + [D_2R_s^*] + [DR_s^*A] + [D_2R_s^*A] \tag{4.73}$$
$$+ [R_s^*A] + [DR_s^*A_2] + [D_2R_s^*A_2] + [R_s^*A_2],$$

$$[R_{\text{inactive}}] = [R_s] + [DR_s] + [D_2R_s] + [DR_sA] + [D_2R_sA]$$
$$+ [R_sA] + [DR_sA_2] + [D_2R_sA_2] + [R_sA_2]. \tag{4.74}$$

Each receptor complex can be substituted with an expression containing the concentration of the inactive state of the free receptor (R_s) and the appropriate microscopic constants and ligand concentrations. After factoring out $[R_s]$ and simplifying, the following equation is obtained:

$$\frac{[R_{\text{active}}]}{[R_T]} = \frac{1}{1 + \frac{(1+DK_a)^2(1+AK_e)^2}{K_q\left(1+D\sqrt{\frac{P_{b2}(1+K_q)}{K_q}}\right)^2(1+AK_f)^2}}. \tag{4.75}$$

In this equation, P_{b2} is defined as in Equation (4.23). This variable replaces $K_b^2K_q/(1+K_q)$ to facilitate regression analysis.

Receptor binding functions for the case of two allosteric sites

We can also use the model shown in Figure 4.4 to derive equations for the binding of the orthosteric and allosteric ligands. The method follows the same strategy used above for the case of one allosteric site, and is quickly summarized below.

We begin with Equation (4.27), but with the those components of receptor complexes bound and unbound with the orthosteric ligand defined as:

$$\begin{aligned}[D_{\text{bound}}] = {} & \left[\boldsymbol{DR_s^*}\right] + [\boldsymbol{DR_s}] + 2\left[D_2R_s^*\right] + 2\left[D_2R_s\right] + \left[\boldsymbol{DR_s^*A}\right] \\ & + [\boldsymbol{DR_sA}] + 2\left[\boldsymbol{D_2R_s^*A}\right] + 2\left[\boldsymbol{D_2R_sA}\right] + \left[\boldsymbol{DR_s^*A_2}\right] \\ & + [\boldsymbol{DR_sA_2}] + 2\left[D_2R_s^*A_2\right] + 2\left[D_2R_sA_2\right], \end{aligned} \tag{4.76}$$

$$[D_{\text{unbound}}] = [R_s] + \left[R_s^*\right] + [\boldsymbol{R_sA}] + \left[\boldsymbol{R_s^*A}\right] + [R_sA_2] + \left[R_s^*A_2\right]. \tag{4.77}$$

Each receptor complex is replaced with an expression in terms of microscopic constants, and the equation is simplified to yield:

$$\frac{[D_{\text{bound}}]}{2\,[R_T]} = \frac{D}{D + \frac{1+K_q+A(K_e(2+AK_e)+K_fK_q(2+AK_f))}{2(K_a(1+DK_a)(1+AK_e)^2+K_bK_q(1+DK_b)(1+AK_f)^2)}}. \tag{4.78}$$

The derivation of the equation for the amount of bound allosteric ligand begins with the expression:

$$\frac{A_{\text{bound}}}{2\,[R_T]} = \frac{[A_{\text{bound}}]}{[A_{\text{bound}}] + [A_{\text{unbound}}]}, \tag{4.79}$$

in which

$$\begin{aligned}[A_{\text{bound}}] = {} & \left[\boldsymbol{R_s^*A}\right] + [\boldsymbol{R_sA}] + 2\left[R_s^*A_2\right] + 2\left[R_sA_2\right] + \left[\boldsymbol{DR_s^*A}\right] + [\boldsymbol{DR_sA}] \\ & + 2\left[\boldsymbol{DR_s^*A_2}\right] + 2\left[\boldsymbol{DR_sA_2}\right] + \left[\boldsymbol{D_2R_s^*A}\right] + [\boldsymbol{D_2R_sA}] \\ & + 2\left[D_2R_s^*A_2\right] + 2\left[D_2R_sA_2\right], \end{aligned} \tag{4.80}$$

$$[A_{\text{unbound}}] = [R_s] + \left[R_s^*\right] + [\boldsymbol{DR_s}] + \left[\boldsymbol{DR_s^*}\right] + [D_2R_s] + \left[D_2R_s^*\right]. \tag{4.81}$$

Each receptor complex is then replaced with an expression in terms of microscopic constants, and the equation is simplified to yield:

$$\frac{[A_{\text{bound}}]}{2\,[R_T]} = \frac{A}{A + \frac{1+K_q+D(K_a(2+DK_a)+K_bK_q(2+DK_b))}{2(K_e(1+AK_e)(1+DK_b)^2+K_fK_q(1+AK_f)(1+DK_b)^2)}}. \tag{4.82}$$

Now, let us use the algebra of this section to understand the analysis of the allosteric modulation of binding and function.

Simulation and Analysis of Allosteric Binding Interactions

Reciprocal modulation of observed binding affinity

One of the unique properties of allosterism mentioned above is that allosteric ligands have reciprocal effects on their respective binding affinities. This property follows from the law of microscopic reversibility also mentioned above in connection with the assignment of microscopic constants for the isomerization constants shown in Figure 4.3.

Another way to think about this property is illustrated by the simple scheme in Figure 4.6, which shows the interaction between allosteric (A) and orthosteric (D) ligands at a receptor having only one binding site for each type of ligand. The affinity constants for the orthosteric ligand for the free (R) and allosteric ligand-occupied (RA) receptor are denoted by K_1 and K_3. The analogous values for the allosteric ligand are K_2 and K_4. These constants are defined in the normal way by dividing the concentration of the ligand-receptor complex by the product of the two reactants (e.g., $K_1 = [DR]/([D][R])$. These constants are defined at the population level of analysis, in the same way that K_D and K_A are for the scheme in Figure 4.5.

One might think that these four constants could assume any value. A simple example shows that this is not the case. Suppose that the concentration of both ligands is 10^{-6} M and that affinity constants, K_1, K_2 and K_3, all have the same value of 10^6 M^{-1}, whereas K_4 has a value of 10^8 M^{-1}. Let us start at the simplest receptor species (R) in the left hand lower corner of the model and proceed in a clockwise manner and consider what happens at each equilibrium step. Because the concentration of D is 10^{-6} M and its affinity constant for R is 10^6 M^{-1}, then

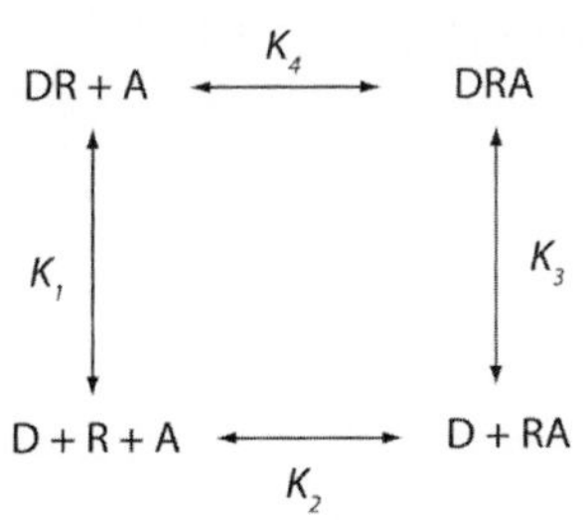

$$K_3 = K_1 K_4 / K_2$$

$$K_4 = K_2 K_3 / K_1$$

Figure 4.6: Simple allosteric model illustrating the property of reciprocal allosteric modulation in ligand affinity. The model is defined in terms of the observed affinity constants of the receptor population. Although there are four different binding steps, only three equilibrium constants describe the interaction.

at equilibrium, the ratio of $[R]$ to $[DR]$ would be 1. For the next step involving the binding of A to DR to yield the DRA complex, the ratio of $[DRA]$ to $[DR]$ at equilibrium is approximately 100, because the concentration of A is 10^{-6} and its affinity constant for the DR complex is 10^8 M^{-1}. The proportions of receptor complex in the three forms (DRA:DR:R) would now be 100:1:1. The next step involves the dissociation of the DRA complex into the RA complex. Because this step has an affinity constant of 10^6 M^{-1} and the concentration of D is 10^{-6} M, then the ratio of DRA to RA at equilibrium should be 1. So now the proportions of receptor complex in the four forms (RA:DRA:DR:R) would be 100:100:1:1. But look what happens at the next step. Because the concentration of A is 10^{-6} M and its affinity constant for the RA complex is 10^6 M^{-1}, then the ratio of RA to R at equilibrium should be 1. If we now have 100 units of RA, then we must also have 100 units of R. But as described four sentences earlier, we began with only one unit of R. This production of 100 units of R in one trip around the cycle would cause the system to cycle like a perpetual motion machine. Obviously this cannot happen for a system at equilibrium unless there is an input of energy.

From this example, we can see that the constants K_3 and K_4 cannot assume any value. Rather, they depend on K_1 and K_2 and the cooperative interaction between the two ligands. Let me describe the rule that governs these interactions and provide the proof later. For any starting and ending positions in the square of equilibria shown in Figure 4.6, there are always two routes to a given destination. Each route involves one to three equilibrium steps, depending on the origin and destination. Each equilibrium step is associated with an equilibrium term, and the product of the equilibrium terms for one route must equal the product of terms for the other route. If an equilibrium step on a route involves going in the direction of free to bound ligand, then the equilibrium term is simply equal to the observed affinity constant. But if the step is in the opposite direction (bound to free ligand), then the equilibrium term is equivalent to the reciprocal of the observed affinity constant.

For example, consider the transition from R to DR. This can occur in one step from reactants to product as determined by the constant K_1. The second route is more circuitous and proceeds from R to RA (free to bound A, K_2), then from RA to DRA (free to bound D, K_3), and finally, from DRA to DR (bound to free A, $1/K_4$). Based on the rule mentioned above, it follows that

$$K_3 = K_1 \frac{K_4}{K_2}. \tag{4.83}$$

This simple equation can be manipulated in various ways (e.g., $K_1K_4 = K_2K_3$) including the following:

$$\frac{K_4}{K_2} = \frac{K_3}{K_1} = \alpha. \tag{4.84}$$

This equation shows that the ratios of affinity constants on the two opposite sides of the square are equal. Therefore, the effect that allosteric ligand A has on the binding of D to RA is the same as the effect that D has on the binding of A to DR. This reciprocal allosteric effect is known as the cooperativity, and it is denoted by the variable α.

One way to prove this relationship is to substitute the equilibrium expressions for K_1 ($[DR]/([D][R])$), K_2 ($[RA]/([A][R])$), K_3 ($[DRA]/([A][\mathrm{D}R])$) and K_4 ($[\mathrm{DRA}]/([D](\mathrm{RA}])$) into Equation (4.84) while ignoring the α term:

$$\frac{[DRA][A][R]}{[A][DR][RA]} = \frac{[DRA][D][R]}{[D][RA][DR]}. \tag{4.85}$$

Canceling out the ligand concentrations on both sides yields the following equality, which proves the equation (Equation (4.84)) from which it was derived:

$$\frac{[DRA]}{[RA]} = \frac{[DRA]}{[RA]}. \tag{4.86}$$

If the rule mentioned above is valid, then we would expect that the effect of an allosteric modulator on the binding of an agonist to a ligand-gated ion channel should be equivalent to the effect of the agonist on the binding of the allosteric modulator. Figure 4.7 shows theoretical plots that address this question. In this case, the ligand-gated ion channel has a single allosteric site linked to two orthosteric sites. In Figure 4.7a, the binding curve of a highly efficacious agonist ($K_b/K_a = 10^4$) is plotted for the condition of varying concentrations of positive allosteric modulator including no modulator (control condition). Each binding curve consists of a series of theoretical binding measurements at different concentrations of orthosteric ligand and a fixed concentration of allosteric modulator. The theoretical values were calculated using Equation (4.30) with the addition of a random error. Figure 4.7b shows the analogous plot for the binding of the allosteric modulator in the presence of various concentrations of the orthosteric agonist as based on Equation (4.34). The legend of Figure 4.7 gives more details regarding the microscopic constants of the interacting ligands.

Two characteristic features of allosteric interactions are apparent from the plots in Figure 4.7. First, there is a limit to the effect of allosteric modulation. Increasing concentrations of allosteric modulator cause the binding curve of the orthosteric ligand in Figure 4.7a to shift to the left (increase in affinity), but there is a limit to the

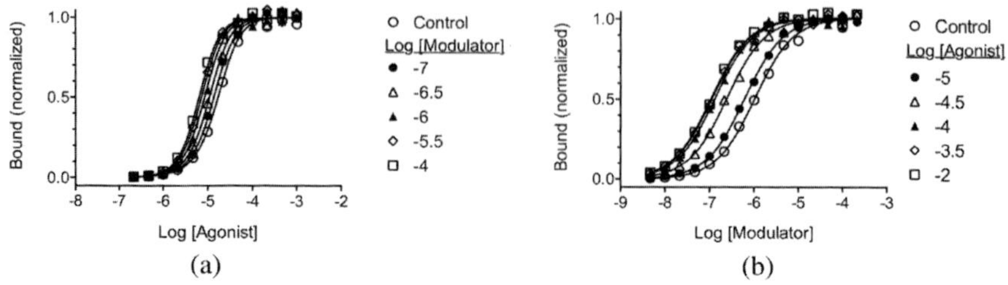

Figure 4.7: Reciprocal, positive allosteric interaction between the binding of orthosteric and allosteric ligands to a ligand-gated ion channel having two orthosteric sites and one allosteric site. (a) The binding of the agonist is plotted against its log concentration. Each binding curve was simulated for a different condition, corresponding to the absence (control) or presence of different concentrations of allosteric modulator as indicated in the legend. (b) The binding of allosteric ligand is plotted against its log concentration. Each curve was simulated for a different condition, corresponding to the absence (control) and presence of different concentrations of orthosteric agonist. The microscopic constants used in the simulation were: K_q, 10^{-7}; K_a, 10^4 M^{-1}; K_b, 10^8 M^{-1}; K_e, 10^6 M^{-1} and K_f, 10^7 M^{-1}.

shift for there are hardly any differences between the binding curves measured in the presence of modulator in the concentration range of $3.3 \times 10^{-6} - 10^{-4}$ M. The same can be said of the analogous data shown in Figure 4.7b except that the concentrations over which the effects of the orthosteric agonist reach a limit ($10^{-4} - 10^{-2}$ M) are greater because of its lower affinity. The second feature is that the increase in the observed affinity of the binding curve of the orthosteric agonist caused by the allosteric modulator (2.6-fold increase) is less than that caused by the agonist on the binding of the allosteric modulator (9.2-fold increase). While this may seem to conflict with the concept of reciprocal allosteric modulation, it does not because there is only one allosteric site linked to two orthosteric sites.

We can see the consequences of adding a second allosteric site to the ligand-gated ion channel in Figure 4.8. Figure 4.8a shows the effect of the allosteric modulator on the binding of agonist and Figure 4.8b shows the effect of the agonist on the binding of allosteric modulator. In both cases, the allosteric modulation is reciprocal, corresponding to a 10-fold increase in the observed affinity of one ligand by the other. Hence, in this case where the ratio of allosteric to orthosteric sites is 1:1, the allosteric modulation is clearly reciprocal.

These principles also apply to negative allosteric modulators as shown by the theoretical data in Figure 4.9 for a ligand-gated ion channel having both two orthosteric and two allosteric sites. The reciprocal reduction in observed affinity is readily apparent. While the allosteric modulation causes a parallel shift to the right in the occupancy curves, the mechanism is clearly different from competition because the inhibitory effect reaches a limit at high concentrations of modulator.

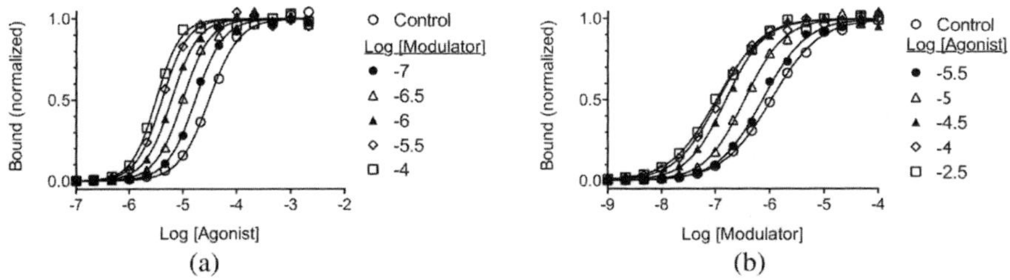

Figure 4.8: Reciprocal, positive allosteric interaction between the binding of orthosteric and allosteric ligands to a ligand-gated ion channel having two orthosteric sites and two allosteric sites. (a) The binding of the agonist is plotted against its log concentration. Each binding curve was simulated for a different condition, corresponding to the absence (control) or presence of different concentrations of allosteric modulator as indicated in the legend. (b) The binding of allosteric ligand is plotted against its log concentration. Each curve was simulated for a different condition, corresponding to the absence (control) and presence of different concentrations of orthosteric agonist. The microscopic constants are the same as those used in Figure 4.7.

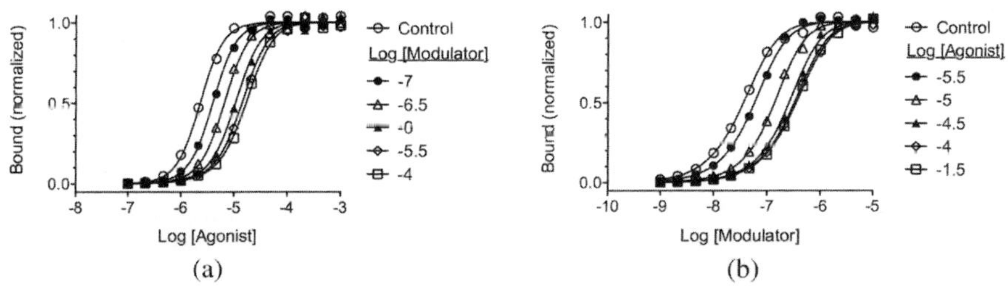

Figure 4.9: Reciprocal, negative allosteric interaction between the binding of orthosteric and allosteric ligands to a ligand-gated ion channel having two orthosteric sites and two allosteric sites. (a) The log concentration-binding curves of an orthosteric agonist are simulated under different conditions corresponding to the absence (control) and presence of different concentrations of negative allosteric modulator. (b) The log concentration-binding curves of a negative allosteric ligand are simulated under different conditions corresponding to the absence (control) and presence of different concentrations of the orthosteric agonist. The microscopic constants used in the simulation were: K_q, 10^{-7}; K_a, 10^4 M^{-1}; K_b, 10^9 M^{-1}; K_e, 10^7 M^{-1} and K_f, 10^6 M^{-1}.

Relationship between microscopic constants and observed cooperativities

Using Equations (4.50)–(4.52), it is possible to estimate the cooperativity constants from the values of the microscopic constants. For the simulated data in Figure 4.7, these estimates are 1.01, 11.0 and 9.10 for α, β and γ, respectively. The constant α represents the cooperativity between the binding of the first molecule of orthosteric ligand (D) and the modulator (A), whereas γ denotes the cooperativity between the binding of the second molecule of D and the modulator.

It may seem puzzling that these two constants differ because the two orthosteric sites are the same and have the same microscopic constants. These cooperativity constants reflect a more superficial level of analysis, however, and their values differ because of the very low level of constitutive receptor activation ($K_q = 10^{-7}$). When one orthosteric site is occupied, receptor activation increases from 0.0000001 to about 0.0001, or alternatively, the fraction of the receptor population in the inactive state decreases imperceptibly from 0.9999999 to 0.999. This is hardly a change, and consequently, the affinity of the allosteric ligand is nearly unaffected (i.e., $\alpha \approx 1$). Another way to look at this question is based on the premise that allosteric ligands have an effect indistinguishable from an alteration in the isomerization constant K_q as explained below in the section on single channel analysis. For the example in Figure 4.7, this amounts to increasing K_q from 10^{-7} to about 10^{-6}. This modified K_q value is still very low, and hence, the observed affinity of D (K_D) would hardly change (see Equation (4.45)).

But when the second molecule of orthosteric ligand binds to the receptor, there is a large increase in receptor activation, and hence, the observed binding affinity of the allosteric ligand is substantially increased (i.e., $\gamma = 9.10$). Similarly, because the allosteric modulator has the effect of increasing K_q about 10-fold, this would be manifest as a substantial increase in the observed binding affinity of the second molecule of orthosteric ligand, which in this case is a highly efficacious agonist (i.e., $K_b/K_a = 10^4$).

The estimate of the cooperativity constant γ for a given allosteric modulator varies depending upon the nature of the orthosteric ligand. Consider the theoretical data in Figure 4.10 for a ligand-gated ion channel having two orthosteric sites and one allosteric site. In these plots, the microscopic constants of the positive allosteric modulator are the same as those used in the simulation in Figure 4.7 ($K_f = 10^7$ M^{-1} and $K_e = 10^{-6}$ M^{-1}). It has a much smaller effect on the binding of the orthosteric ligand (Figure 4.7a) (1.2-fold increase in observed affinity), however, and the latter has a smaller effect on the binding of the allosteric modulator (1.8-fold increase in observed affinity, Figure 4.7b). The γ value for this interaction is only 1.8 as calculated by Equation (4.52). The reason for the decrease in the cooperativity is the lower selectivity of the orthosteric ligand for the active state of the receptor ($K_b/K_a = 10^3$).

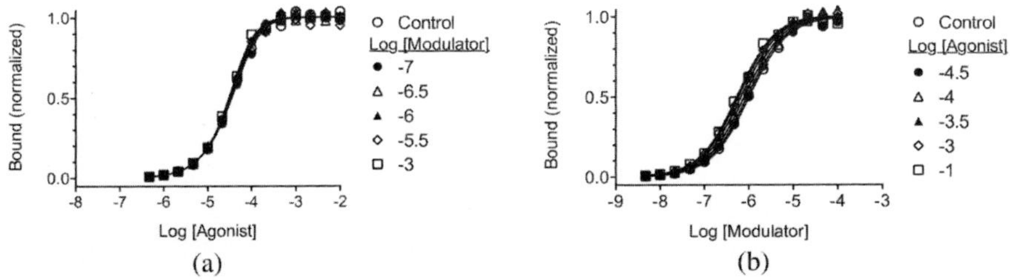

Figure 4.10: Reciprocal, positive allosteric interaction between the binding of an allosteric ligand and an orthosteric ligand with low efficacy to a ligand-gated ion channel having two orthosteric sites and one allosteric site. (a) The binding of orthosteric agonist is plotted against its log concentration. Each binding curve was simulated for a different condition, corresponding to the absence (control) or presence of different concentrations of allosteric modulator as indicated in the legend. (b) The binding of allosteric ligand is plotted against its log concentration. Each curve was simulated for a different condition, corresponding to the absence (control) and presence of different concentrations of orthosteric agonist. The microscopic constants used in the simulation were: K_q, 10^{-7}; K_a, 10^4 M^{-1}; K_b, 10^7 M^{-1}; K_e, 10^6 M^{-1} and K_f, 10^7 M^{-1}.

Estimation of observed affinity and cooperativity constants

Most investigators study allosteric interactions at ligand-gated ion channels by measuring the effect of the putative modulator on the current response to the endogenous ligand or another agonist. In contrast, it is often difficult to measure the binding of many agonists to ligand-gated ion channels because of their low affinity. Another problem is that the population parameters obtained from such an analysis (K_D, K_A, α, β and γ) may be perturbed because of desensitization. Nonetheless, it useful to understand what type of information can be gleaned from such data even if our experiment is only a thought experiment, like the simulations shown in some of the figures. Hence, in this section, I briefly describe how to analyze some of the simulated binding data illustrated in Figures 4.7–4.10. Additional details regarding the analysis of radioligand-binding data are described in Chapter 12, and the topic of the analysis of the current response is covered in the next section.

The analysis involves fitting the appropriate function to the ligand-binding data using nonlinear regression analysis. For the examples below, I focus on a ligand-gated ion channel having two orthosteric sites and one allosteric site. Hence, the appropriate functions for this analysis are Equations (4.41) and (4.43), which describe the binding of the orthosteric and allosteric ligands, respectively. These equations describe occupancy at the population level of analysis, and the parameters consist of the observed affinity (K_D and K_A) and cooperativity (α, β and γ) constants. It is impossible to measure occupancy directly, however. Usually we obtain a measure of binding based on the amount of radioactive ligand associated with the receptor. Ultimately, our ability to infer receptor occupancy rests on our ability to estimate the maximal amount of receptors. Thus, we need to modify Equations (4.41) and (4.43) by introducing the parameter, $B_{\max}$, which denotes the total amount of receptors. The equation for the binding of orthosteric ligand is:

$$\frac{D_{\text{bound}}}{2\,[R_T]} = \frac{DB_{\max}}{D + \frac{1+AK_A}{2K_D((1+D\beta K_D)+\alpha AK_A(1+D\beta\gamma K_D))}}. \tag{4.87}$$

In this equation, $B_{\max\text{-}D}$ denotes the total amount of orthosteric binding sites, which is equivalent to two times the amount of receptor molecules (i.e., $2[R_T]$). The equation used to analyze the binding of allosteric modulator is:

$$A_{\text{bound}} = \frac{[A]B_{\max-A}}{[A] + \frac{1+[D]K_D(2+[D]\beta K_D)}{K_A+[D]\alpha K_D K_A(2+[D]\beta\gamma K_D)}}. \tag{4.88}$$

In this equation, $B_{\max-A}$ denotes the amount of binding sites for the allosteric ligand, which is simply equal to the amount of receptor molecules ($[R_T]$).

Now let us use these equations to analyze the data in Figure 4.7. Again, the goal is to obtain the least-squares fit and as many parameter estimates, or relations among parameters, as possible. When Equation (4.87) is fitted to the theoretical data for the binding of the orthosteric ligand in Figure 4.7a, the results yield accurate estimates of all of the parameters. Global nonlinear regression analysis was done sharing the parameter estimates among all of the binding curves. The error in the estimation of the cooperativity constants is moderate, however. Because we know that the allosteric effect is moderate and that in many instances substantial receptor activation only occurs when two molecules of orthosteric ligand are bound to the receptor, it seems appropriate to assume that the cooperativity constant, α, is approximately equal to one. When regression analysis was repeated with α constrained to 1 and all of the other parameters shared, the following estimates were obtained: $\log K_D$, 4.00 ± 0.05; $\log K_A$, 6.12 ± 0.05; $\log \beta$, 1.02 ± 0.10; $\log \gamma$, 0.95 ± 0.03; and $B_{\max}$, 1.00 ± 0.004. These estimates are practically the same as the theoretical values calculated from the microscopic constants: $\log K_D$, 4.00; $\log K_A$, 6.00; $\log \alpha$, 0.0039; $\log \beta$, 0.96; $\log \gamma$, 0.96; and $B_{\max}$, 1.00.

It is possible to constrain α to different values over a modest range and still obtain a least-squares fit and the same values of K_D, K_A and β. The estimate of γ

significantly deviated from the true value, however, when α was constrained to values different from one. Nonetheless, the product, $\alpha\gamma$, was accurate ($\log \alpha\gamma = 0.95$). Thus, for the data in Figure 4.7a, it is possible, without any assumptions, to estimate K_D, K_A, β and $\alpha\gamma$, although the assumption that $\alpha = 1$ would appear valid in many instances.

Using the same approach it is possible to obtain estimates of the same parameters by analyzing the binding data for the allosteric ligand in Figure 4.7b using Equation (4.88). Global nonlinear regression analysis was done with α constrained to 1 ($\log \alpha = 0$) and all of the other parameters shared. The results yielded the following parameter estimates: $\log K_D$, 3.96 ± 0.24; $\log K_A$, 5.99 ± 0.018; $\log \beta$, 1.08 ± 0.41; $\log \gamma$, 0.99 ± 0.024; $\alpha\gamma = 0.99 \pm 0.024$ and $B_{\max}$, 1.01 ± 0.003. In this case, there was greater error in the estimation of $\log K_D$ and $\log \beta$, which was probably caused by the fewer number of agonist concentrations used in this simulation.

The limitation in this type of analysis is that it is impossible to get reliable estimates of K_D and β when the agonist is highly efficacious and causes maximal receptor activation. In this case only K_A and γ can be estimated accurately, assuming that $\alpha = 0$. Otherwise, only K_A and the product, $\alpha\gamma$, can be estimated.

Simulation and Analysis of Allosteric Modulation of the Agonist-Induced Current Response

In this section, the method for analyzing the effects of allosteric ligands on the population current response to an orthosteric agonist is described using simulated data. Again, we will assume that current is measured under equilibrium conditions and that desensitization is minimal. The analysis is done using the technique of global nonlinear regression analysis based on the equations derived for fractional receptor activation (Equations (4.22), (4.26) and (4.75). With real data, however, the measured variable is current (i), and there is no way to convert current into fractional receptor activation unless we have an estimate of the maximum current that occurs when all of the ligand gated ion channels open simultaneously ($i_{\text{max-sys}}$). Thus, we need to modify the equations just mentioned by multiplying both sides of them by $i_{\text{max-sys}}$:

$$i = \frac{i_{\text{max-sys}}}{1 + \frac{(1+AK_e)(1+DK_a)^2}{K_q(1+AK_f)\left(1+D\sqrt{\frac{p_{b2}(1+K_q)}{K_q}}\right)^2}}, \tag{4.89}$$

$$i = \frac{i_{\text{max-sys}}}{1 + AK_f + \frac{(1+AK_e)(1+DK_a)^2}{K_q\left(1+D\sqrt{\frac{P_{b2}(1+K_q)}{K_q}}\right)^2}}, \tag{4.90}$$

$$i = \frac{i_{\text{max-sys}}}{1 + \frac{(1+DK_a)^2(1+AK_e)^2}{K_q\left(1+D\sqrt{\frac{P_{b2}(1+K_q)}{K_q}}\right)^2(1+AK_f)^2}}. \tag{4.91}$$

The first two equations will be used to analyze allosteric modulation (Equation (4.89)) and open channel block (Equation (4.90)) at a ligand-gated ion channel having one allosteric site, and Equation (4.91) will be used to analyze allosterism at a ligand-gated ion channel having two allosteric sites.

It is impossible to obtain a reliable estimate of K_q because most ligand-gated ion channels have negligible constitutive receptor activity. Consequently, during regression analysis, K_q must be constrained to an arbitrarily low value (e.g., $K_q < 10^{-4}$) so that the predicted level of the basal current in the absence of agonist is immeasurable. In our case, we will arbitrarily constrain K_q to 10^{-8} in the examples below. Constraining K_q to different low values affects the estimate of K_b, but, as described in Chapter 2, the product $K_b^2K_q/(1+K_q)$ remains constant. This product is equivalent to the variable P_{b2} in Equations (4.89)–(4.91). Thus, in most instances, we should be able to estimate $i_{\text{max-sys}}$, K_a, P_{b2}, K_e and K_f.

Analysis of allosterism for the case of one allosteric site

For our first theoretical example, we will analyze the current response associated with the theoretical binding examples shown in Figure 4.7a. The theoretical data for the current response is shown in Figure 4.11a. These data were generated using Equation (4.22), with log K_a (4.0), log K_b (8.0), log K_e (6.0), log K_f (7.0) and log K_q (−7.0) assigned values the same as those used in Figure 4.7. A random error (±5%) was added to the data to demonstrate the feasibility of calculating parameters from actual data with experimental error. The plot shows that the positive allosteric modulator causes a concentration-dependent shift to the left in the agonist concentration-response curve, and that this effect is associated with a small increase in the maximum current response.

To analyze the data in Figure 4.11a, global nonlinear regression analysis is used to fit Equation (4.89) to all of the curves simultaneously, sharing the estimates of $i_{\text{max-sys}}$, K_a, P_{b2}, K_e and K_f among the curves with K_q constrained to a low value (i.e., log $K_q = -8$). It is more convenient to transform Equation (4.89) such

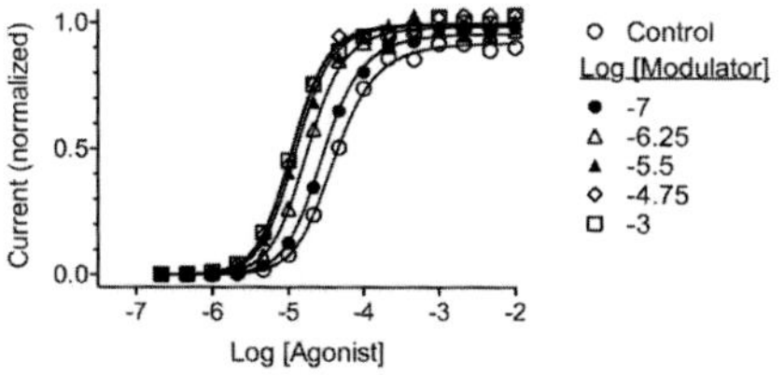

(a)

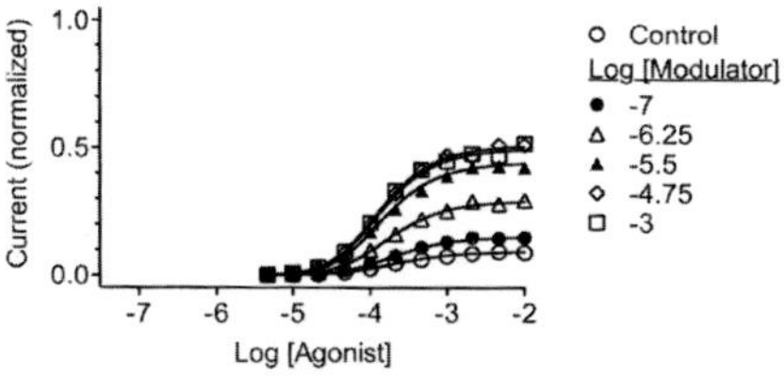

(b)

Figure 4.11: Positive allosteric modulation of the whole-cell current response of a highly efficacious agonist (a) and a less efficacious agonist (b) at a ligand-gated ion channel having two orthosteric sites and one allosteric site. The plots show simulations of the concentration-response curves of orthosteric agonists in the absence (control) and presence of different concentrations of modulator. The microscopic constants for the simulation were: K_q, 10^{-7}; K_e, $10^6\,\mathrm{M}^{-1}$ and K_f, $10^7\,\mathrm{M}^{-1}$; (a) K_a, $10^4\,\mathrm{M}^{-1}$; K_b, $10^8\,\mathrm{M}^{-1}$; (b) K_a, $10^4\,\mathrm{M}^{-1}$; K_b, $10^7\,\mathrm{M}^{-1}$.

that all of the parameters and independent variables are expressed as logarithms (e.g., $D = 10^{\log D}$, $DK_a = 10^{\log D + \log Ka}$ and $D^2 P_{b2} = 10^{2\log D + \log Pb}$). In this analysis, both D and A are the independent variables. Hence, A must be constrained to a different value for each curve.

Let me briefly describe how this can be done using the application Prism (GraphPad Software, Inc.). All of the data are entered into a Data Table. The log concentrations of agonist are entered into the far left column entitled "X". The current responses measured in the absence and presence of each concentration of allosteric modulator A are entered into different lettered columns (A–E). Replicate measurements at the same allosteric modulator concentration can be entered into subcolumns of the same lettered column. Each lettered column can be entitled with the log concentration of allosteric modulator. For the condition of no modulator (control), an arbitrarily large negative number (e.g., -20) can be entered into the column header because the log of zero is negative infinity. If measured, the basal response in the absence of orthosteric ligand can be entered into the appropriate column and row corresponding to a log orthosteric ligand concentration of -20. Thus, all of the current responses represent a two-dimensional matrix, with each entry ($y_{i,j}$) defined by ith row (log agonist concentration) and jth column (log modulator concentration).

Then the log form of Equation (4.89) is entered into Prism as a user defined equation. The equation is transformed as described two paragraphs above so that the parameters and independent variables are defined as logarithms. In Prism, the independent variable is defined as X. Therefore, in our equation, $\log D = \mathrm{X}$. In addition, Prism enables one to constrain any variable or parameter to a column title. This provides a simple means of constraining the log allosteric modulator concentration (log A) to the appropriate value for each curve. In other words, for the procedure described in the previous paragraph, the allosteric modulator concentration can be constrained to the column title. Nonlinear regression analysis is initiated with the parameters $i_{\text{max-sys}}$, log K_a, log P_{b2}, log K_e and log K_f shared among all data sets.

The regression analysis can be done, of course, without using Prism. To summarize, the essential steps of the regression are 1) analyze all of the data simultaneously using the log form of Equation (4.89), 2) share the parameter estimates of $i_{\text{max-sys}}$, log K_a, log P_{b2}, log K_e and log K_f among the data, 3) constrain log K_q to an arbitrarily low value, and 4) constrain the independent variable log A to the appropriate value for each concentration-response curve.

When this was done, the following parameter estimates were obtained for the data in Figure 4.11a: log K_a, 3.99 ± 0.035; log P_{b2}, 9.00 ± 0.033; log K_e, 6.01 ± 0.046; log K_f, 7.00 ± 0.05 and $i_{\text{max-sys}}$, 1.01 ± 0.005. These values are practically the same as those used to simulate the data.

Figure 4.11b shows the results of a simulation involving the same positive allosteric modulator (log K_e, 6; log K_f, 7), but the orthosteric agonist is less efficacious (log K_b, 7; log K_a, 4) than that used in Figure 4.11a. In this case, the maximum current response of the receptor population is not well-defined, and consequently, it is impossible to obtain reliable parameter estimates by regression analysis of these data only. One solution is to constrain $i_{\text{max-sys}}$ to the value obtained for the more efficacious agonist in Figure 4.11a. When this was done, the following estimates were obtained: log K_a, 4 ± 0.01; log P_{b2}, 7.01 ± 0.03; log K_e, 5.95 ± 0.015 and log K_f, 6.96 ± 0.025. Again, these values are nearly the same as those used to simulate the data.

The limitation in this type of analysis occurs when the agonist is highly efficacious and elicits a maximal response very close to $i_{\text{max-sys}}$. In this situation, it is difficult to obtain a reliable estimate of K_a, but it is possible to obtain accurate estimates of the other parameters.

As described in Chapter 2, although it is impossible to estimate the K_b value of the agonist from these data, it is possible to estimate the K_b value of a given agonist relative to that (K'_b) of a standard agonist. This ratio is known as the intrinsic relative activity (RA_i) and it can be calculated from the ratio of P_{b2} values as described by Equation (2.154) in Chapter 2. For convenience, this equation is repeated here:

$$\frac{K_b}{K'_b} = \sqrt{\frac{P_{b2}}{P'_{b2}}} = RA_i. \tag{4.92}$$

The simulations in Figure 4.11 illustrate one of the desirable properties of allosterism — it provides a mechanism for protection against overdose. As the concentration of modulator increases, the effect on the concentration-response curve of the agonist reaches a limit.

Imagine an allosteric modulator that only cased a three-fold increase in the observed affinity of the endogenous ligand. This is approximately equivalent to the effect of diazepam on the interaction of GABA with the $GABA_A$ receptor. Once the concentration of modulator in the extracellular fluid achieves a concentration about 10-fold higher than its observed affinity, it will cause about a three-fold increase in the observed affinity of the endogenous ligand. A huge increase in the concentration of modulator (e.g., 100-fold) would cause no further effect, assuming its action is confined to the allosteric site on the receptor. Thus, the dose administered to a patient could vary by two orders of magnitude without causing any difference in its pharmacological effect. Practically speaking, drugs often have off-target effects that limit the therapeutic dose range, but nonetheless, the allosteric mechanism has a built-in safety factor against overdose.[2,3]

Analysis of open channel block

The open channel block involves two mechanisms — the allosteric enhancement in the binding of the orthosteric agonist to the open state of the receptor and the plugging of the open state of the channel. The simulated data in Figure 4.12 show the effect of open-channel blockade on the current response to a highly efficacious agonist (log K_b, 8; log K_a, 4). The open channel blocker causes a large decrease in

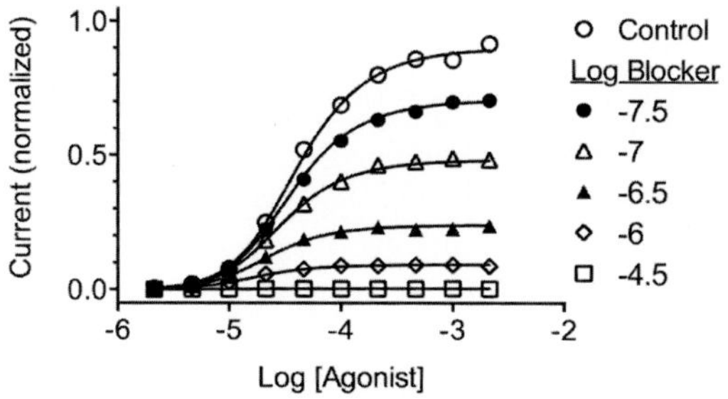

Figure 4.12: Effect of open channel block on the whole-cell current response of a highly efficacious agonist. The plots show simulations of the concentration-response curves of an orthosteric agonist in the absence (control) and presence of different concentrations of an open-channel blocker. The microscopic constants for the simulation were: K_q, 10^{-7}; K_a, 10^4 M^{-1}; K_b, 10^8 M^{-1}; K_e, 10^6 M^{-1} and K_f, 10^7 M^{-1}.

the maximal current response of the orthosteric agonist and a decrease in the EC_{50} value (increase in potency).

To estimate the relevant microscopic constants, Equation (4.90) is fitted to the data by global nonlinear regression analysis using a strategy analogous to that described in the prior section. That is, all of the data are analyzed simultaneously using the log form of Equation (4.90). During regression analysis, the parameter estimates of $i_{max-sys}$, $\log K_a$, $\log P_{b2}$, $\log K_e$ and $\log K_f$ are shared among the data, and $\log K_q$ is constrained to an arbitrarily low value (−8). Finally, the independent variable $\log A$ is constrained to the appropriate value for each concentration-response curve.

When this was done, the following parameter estimates were obtained: K_a, 4.22 ± 0.06; log P_{b2}, 9.06 ± 0.015; log K_e, 6.21 ± 0.10; log K_f, 7.02 ± 0.02 and $i_{max-sys}$, 1.12 ± 0.05. These are practically the same as those used to simulate the data.

The limitations in this analysis are the same as those described above for allosteric modulation. That is, it is difficult to estimate microscopic parameters accurately unless the estimate of $i_{max-sys}$ is accurate, which necessitates including a highly efficacious agonist in at least part of the analysis. For those analyses involving a highly efficacious agonist, it may be difficult to estimate $\log K_a$ accurately.

Analysis of allosterism for the case of two allosteric sites

In this section, I consider a few examples of allosterism involving a ligand-gated ion channel having two allosteric sites. Because the basal activity of many ligand-gated ion channels is so low ($K_q \approx 10^{-7}$), most small ligands cannot cause significant channel opening unless two molecules of agonist are bound (see Chapter 2). The same holds for positive allosteric modulators. Thus, in this section, I consider the case of an allosteric agonist, like the $GABA_A$ receptor ligand propofol, in addition to positive and negative allosteric modulators.

Figure 4.13 shows a simulation of the effects of positive (panel (a); log K_f, 7; log K_e, 6) and negative (panel (b); log K_f, 6; log K_e, 7) allosteric modulators on the current response of an efficacious agonist. The data in each panel can be analyzed by global nonlinear regression analysis with Equation (4.91) in a manner analogous to that described for the analysis of the simulated data in Figures 4.11 and 4.12. When this was done for the data in panel (a), the following parameter estimates were obtained: K_a, 3.99 ± 0.04; log P_{b2}, 9.00 ± 0.04; log K_e, 6.01 ± 0.03; log K_f, 7.00 ± 0.03 and $i_{max-sys}$, 1.01 ± 0.004. When the data in panel (b) were analyzed, the

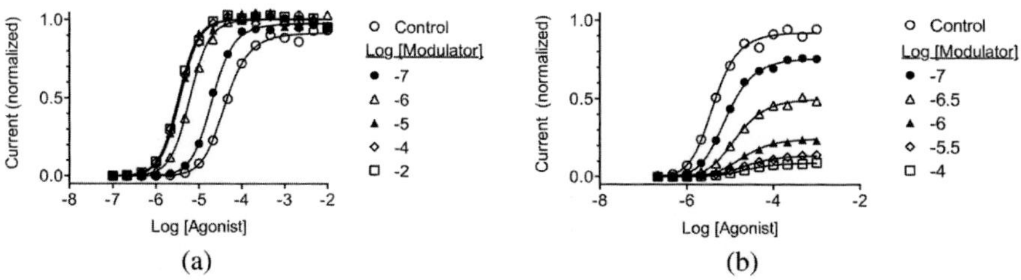

Figure 4.13: Positive (a) and negative (b) allosteric modulation of the whole-cell current response of highly efficacious agonists at a ligand-gated ion channel having two orthosteric sites and two allosteric sites. Each plot shows simulations of the concentration-response curves of an orthosteric agonist in the absence (control) and presence of different concentrations of a positive (a) and negative (b) modulator. The microscopic constants for the simulation were: (a) K_q, 10^{-7}; K_a, $10^4\ \mathrm{M}^{-1}$; K_b, $10^8\ \mathrm{M}^{-1}$; K_e, $10^6\ \mathrm{M}^{-1}$ and K_f, $10^7\ \mathrm{M}^{-1}$; (b) K_q, 10^{-7}; K_a, $10^5\ \mathrm{M}^{-1}$; K_b, $10^9\ \mathrm{M}^{-1}$; K_e, $10^7\ \mathrm{M}^{-1}$ and K_f, $10^6\ \mathrm{M}^{-1}$.

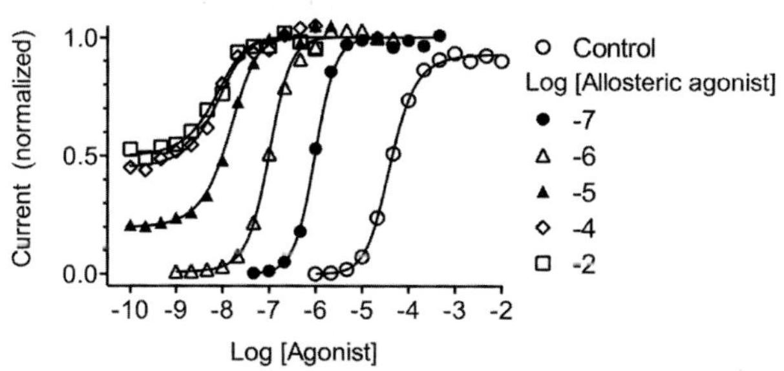

Figure 4.14: Effect of a partial allosteric agonist on the whole-cell current response of a highly efficacious agonist at a ligand-gated ion channel having two orthosteric sites and two allosteric sites. The plot shows simulations of the concentration-response curves of an orthosteric agonist in the absence (control) and presence of different concentrations of a partial allosteric agonist. The microscopic constants used in the simulation were: K_q, 10^{-7}; K_a, $10^4\ \mathrm{M}^{-1}$; K_b, $10^8\ \mathrm{M}^{-1}$; K_e, $10^5\ \mathrm{M}^{-1}$ and K_f, $3.2 \times 10^8\ \mathrm{M}^{-1}$.

following estimates were obtained: K_a, 5.01 ± 0.02; log P_{b2}, 10.99 ± 0.02; log K_e, 6.99 ± 0.02; log K_f, 6.01 ± 0.02 and $i_{\text{max-sys}}$, 1.02 ± 0.01. All of these estimates are practically the same as those used to simulate the data (see legend to Figure 4.13).

The simulated data in Figure 4.14 show the effects of a partial allosteric agonist ($K_f/K_e = 10^{3.5}$) on the current response to an efficacious agonist ($K_b/K_a = 10^4$). The allosteric agonist causes a huge increase in the potency of the orthosteric agonist because of its relatively high selectivity for the active state as compared to lower selectivity of the positive allosteric modulators in the prior simulations. It can also elicit a substantial current response in the absence of the orthosteric agonist. In this situation, it is possible to obtain estimates of all of the microscopic constants without constraining K_q to a constant during global nonlinear regression analysis. One can fit Equation (4.91) to the data, but a simpler approach is to substitute $K_b^2 K_q/(1+K_q)$ for P_b in Equation (4.91) and fit the data to the resulting equation after simplifying it:

$$i = \frac{i_{\text{max-sys}}}{1 + \frac{(1+DK_a)^2(1+AK_e)^2}{K_q(1+DK_b)^2(1+AK_f)^2}}. \quad (4.93)$$

When the data were analyzed using Equation (4.93), with all of the parameters shared, the following estimates were obtained: $\log K_q$, -7.04 ± 0.06; K_a, 3.93 ± 0.05; $\log K_b$, 8.01 ± 0.02; $\log K_e$, 5.00 ± 0.02; $\log K_f$, 8.51 ± 0.02 and $i_{\text{max-sys}}$, 1.00 ± 0.004. These are essentially the same as those used to simulate the data (see legend to Figure 4.14). If the data are analyzed with K_q constrained to lower (10^{-8}) or higher (10^{-6}) values, the residual sum of squares deviations of the data points from the regression equation increases substantially (3.8- and 4.7-fold, respectively). Thus, there appears to be sufficient information in this type of experiment to obtain an accurate estimate of the isomerization constant of the free receptor (K_q).

Part of the reason is based on the relationships among K_q, the microscopic constants of the allosteric ligand (K_f and K_e), and the maximum fractional response of the allosteric ligand ($i_{\text{max-}A}/i_{\text{max-sys}}$). This relationship is given by:

$$\frac{i_{max\text{-}A}}{i_{\text{max-sys}}} = \frac{K_q\left(\frac{K_f}{K_e}\right)^2}{K_q\left(\frac{K_f}{K_e}\right)^2 + 1}. \quad (4.94)$$

In this equation, $i_{\text{max-}A}$ denotes the maximum response of the allosteric ligand. Equation (4.94) is essentially equivalent to Equation (2.150) in Chapter 2 except that the microscopic constants of the orthosteric ligand (K_b and K_a) have been replaced with those of the allosteric ligand (K_e and K_f). Equation (4.94) can also be derived from Equation (4.93) by setting $D = 0$ and taking the limit as A approaches infinity.

The consequences of Equation (4.94) illustrate why it is impossible to estimate K_f and K_q when an allosteric agonist is used by itself and the receptor lacks constitutive activity. For the allosteric agonist to elicit a measureable response, the square of the ratio of microscopic affinity constants (K_f/K_e) must be in the same range as $1/K_q$. For example, when $(K_f/K_e)^2 = 1/K_q$, then the maximum fractional response of the allosteric agonist is 0.5. But any increase in K_f/K_e can be offset by a decrease in K_q. Suppose that the true values of K_q and K_f/K_e are 10^{-8} and 10^4, respectively. These values yield a maximum fractional response of 0.5. But the same maximal response could occur if K_q were 10^{-6} and K_f/K_e 10^3. So long as K_q is small enough to eliminate constitutive activity, there are an infinite number of combinations of K_f/K_e and K_q that yield a maximum fractional response of 0.5 so long as $K_q(K_f/K_e)^2 = 1$. This explains why it is only possible to estimate the expression $K_qK_b^2/(1 + K_q)$ for an orthosteric agonist and not the individual values of K_b and K_q. The same applies to the K_f and K_q values of an allosteric agonist.

But when the two types of ligands are used together to measure their combined effect on the whole-cell current response, it is always possible to estimate the ratio of microscopic affinity constants of the less efficacious ligand. In the examples shown in Figures 4.9–4.11, the less efficacious ligand was the allosteric modulator, and indeed, it was not only possible to estimate K_f/K_e, but the individual constants, K_e and K_f. Now suppose that the less efficacious ligand is an allosteric agonist with a measureable maximal response. This situation provides us with two of the three unknowns in Equation (4.94) (i.e., $i_{\text{max-}A}/i_{\text{max-sys}}I$ and K_f/K_e). Rearrangement of Equation (4.94) shows that K_q can be expressed as a function of the maximum fractional response ($i_{\text{max-}A}/i_{\text{max-sys}}$) and the ratio K_f/K_e:

$$\frac{\frac{i_{max-A}}{i_{\text{max-sys}}}}{1 - \frac{i_{max-A}}{i_{\text{max-sys}}}}\left(\frac{K_f}{K_e}\right)^2 = K_q. \tag{4.95}$$

Thus, it should be possible to estimate K_q by analyzing the interaction between an allosteric agonist and an orthosteric agonist in a functional assay. Knowing K_q, it should also be possible to estimate K_b.

I have described this process as if it requires a series of steps involving Equation (4.95), but it can be done in one step by global nonlinear regression analysis using Equation (4.93) as described for the analysis of the simulated data in Figure 4.14. The purpose of this digression with Equation (4.95) was to illustrate what information can be obtained from this type of experiment.

Later, we will use Equation (4.93) to estimate K_q of the $GABA_A$ receptor by analyzing the interaction between GABA and the anesthetic, etomidate, which behaves as an allosteric agonist.

Single Channel Analysis

As mentioned in Chapter 2, the most informative way to analyze drug-receptor interactions at a ligand-gated ion channel is by single channel analysis. The experimental record can be converted into a vector of idealized open and closed times, and the latter can be analyzed by Markov analysis using the maximum likelihood approach to obtain estimates of rate constants and the microscopic affinity constants as described by David Colquhoun, Anthony Auerbach and their respective colleagues. My intent here is not to summarize this powerful approach but to describe how measurements of open probability can be used to estimate microscopic affinity constants under equilibrium conditions. The rationale and basis of this approach is described in Chapters 2 and 3.

One of the benefits of this method is that the open probability of a single channel is proportional to the fraction of the receptor population in the active state. Thus, by measuring open probability, one can obtain estimates of fractional activation of the receptor population without having to estimate the maximal current associated with 100% channel opening ($i_{\text{max-sys}}$). This enables one to estimate the microscopic constants of partial agonists without having to obtain a separate measure $i_{\text{max-sys}}$ using a highly efficacious agonist. Open probability measurements can also be corrected for desensitization. Thus, the method of analysis is a substantial improvement of population analysis.

I begin this section by presenting a simulation of the allosteric modulation of single receptor events so that the reader can get an appreciation of how quantal events shape population behavior. Then, I briefly describe the approach for analyzing the allosteric modulation of ligand-gated ion channels using measurements of open probability.

Simulation of Single Channel Events

The effect of allosteric modulation on single channel events can be simulated using the technique described in Chapter 2. A brief summary of the process is described here for the case of a ligand-gated ion channel having one allosteric site and two orthosteric sites (see Figure 4.3). More details of the procedure are described in Chapter 2.

To begin, each equilibrium constant of the model in Figure 4.3 is replaced with a pair of rate constants describing the forward and reverse reactions rates (e.g., k_{a1} and k_{a2} replace affinity constant K_a). These rate constants are then used to calculate a set of conditional probabilities that determine the likelihood that the receptor will either remain in its current state or transition to a neighboring state during a small time increment. Thus, there are a total of four or five conditional probabilities for each of the two states of the various receptor complexes in the model. For example, a receptor in the R_s state could either remain in the R_s state or transition to R_s^*, DR_s or R_sA (four conditional probabilities), whereas a receptor in the DR_s state could remain in the DR_s state or transition to R_s, DR_s^*, D_2R_s or DR_sA (five conditional probabilities). All of the conditional probabilities ($P_{i,j}$) are assembled into a 12 by 12 transition matrix based on the state of the receptor at the start (column j) and end (row i) of the transition. For a given starting condition (e.g., R_s), a random variable between 0 and 1 is generated and used to determine whether the receptor remains in its current state or transitions to a new state. This process is repeated iteratively, and the state of the receptor is defined incrementally over a long time interval.

This procedure was used to simulate the effect of a positive allosteric modulator ($K_f/K_e = 10$) on the single channel events elicited by various concentrations of an orthosteric agonist with 3,000-fold selectivity for the active state ($K_b/K_a = 3,000$). Other details of the simulation are given in the legend to Figure 4.15. In the absence of modulation, increasing concentrations of agonist caused an increase in the time that the channel spends in the open state. The open probability was calculated at

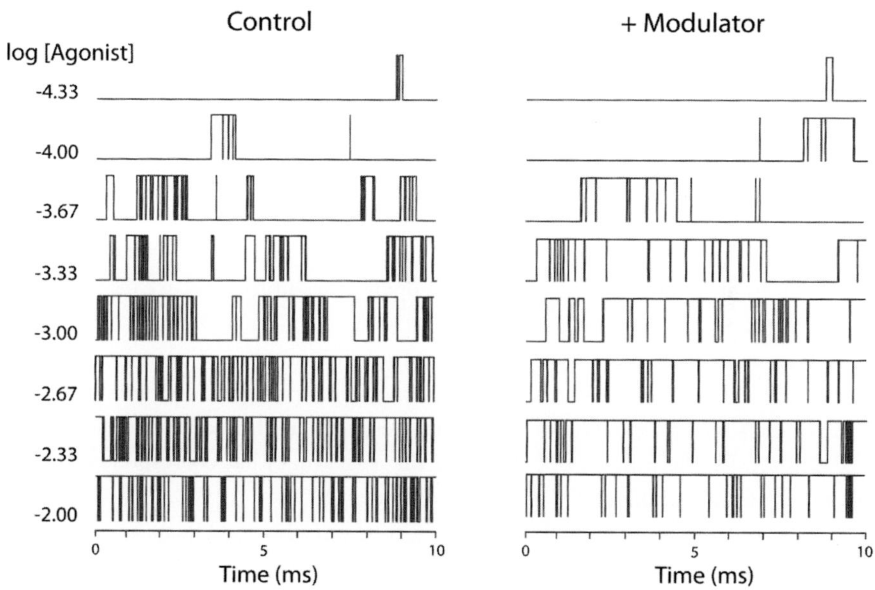

Figure 4.15: Simulation of the allosteric modulation of single channel events at a ligand-gated ion channel having one allosteric and two orthosteric sites. The plots show single channel events in the presence of increasing concentrations of agonist in the absence (Control) and presence (+ Modulator) of a receptor-saturating concentration of allosteric modulator. The transitions rate constants used in the simulation yield microscopic constants of K_q, 10^{-7}; K_a, 10^3 M^{-1}; K_b, 10^7 M^{-1}; K_e, 10^6 M^{-1} and K_f, 3×10^6 M^{-1}. The concentration of allosteric modulator was 1 mM.

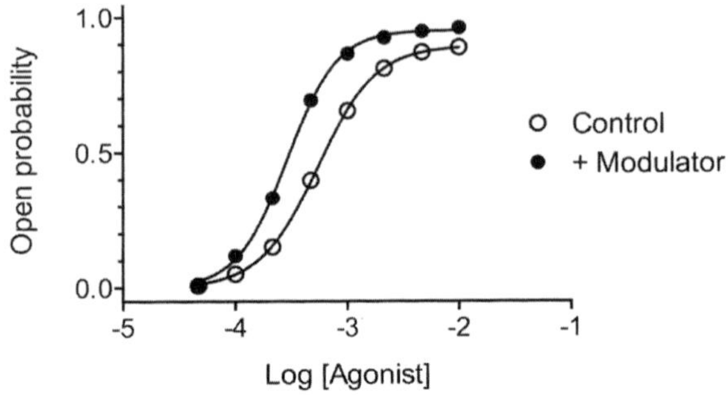

Figure 4.16: Positive allosteric modulation of the open probability of single ligand-gated ion channels having one allosteric site and two orthosteric sites. The estimates of open probability for the simulation in Figure 4.15 are plotted against the concentration of agonist in the absence (Control) and presence (+Modulator) of the positive allosteric modulator.

each agonist concentration over a long time interval, a representative fraction of which, is shown in the figure. These open probabilities are plotted in Figure 4.16. When simulated for the condition of a receptor-saturating concentration of allosteric modulator, open probability increased for each concentration of agonist, resulting in a shift to the left and increase in the maximum of the agonist's open-probability plot. These results are analogous to the effect of a positive allosteric modulator on the population current response (see Figure 4.7).

The primary effect of the allosteric modulator is to increase the mean open time as shown by the frequency distribution of open times in the absence (panel a) and presence (panel b) allosteric modulator (Figure 4.17). These exhibit an exponential distribution as described in Chapter 2. Plotting the number of events against log open time yields a more Gaussian distribution as shown in panels (c) and (d) for the control and allosteric modulation conditions, respectively. There is a small secondary distribution having a mean open time in the $1 \mu s$ range that corresponds to brief openings of the monoliganded receptor complex (DR_s^*).

One way to think about the data is to imagine that no allosteric modulator is present and ask the question, what change would have to occur in the microscopic constants of the model to explain the change caused by an allosteric ligand? From this perspective, the effect of the positive allosteric modulator is to cause an increase in the isomerization constant of the free receptor (K_q). If the allosteric modulator is present at a high concentration such that it is saturating the allosteric site, then the scalar change in K_q is equivalent to K_f/K_e. This relationship holds for both positive and negative allosteric modulators and allosteric agonists. If there

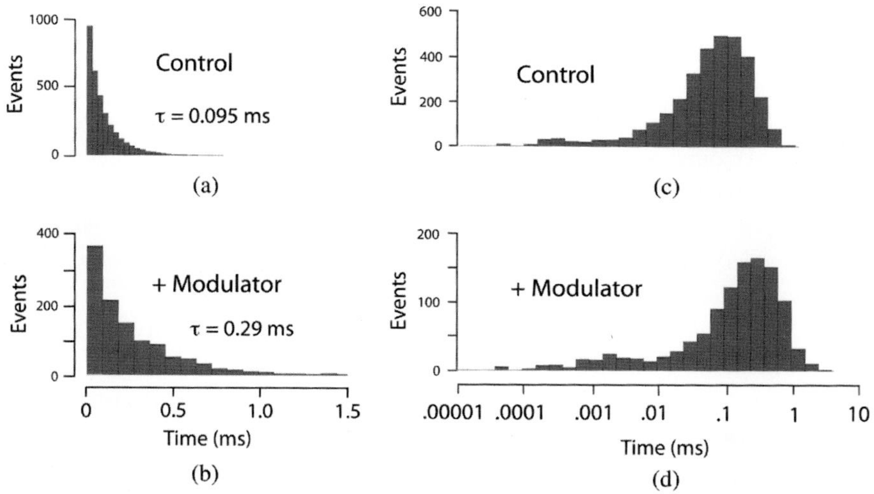

Figure 4.17: The effect of a positive allosteric modulator on the distribution of channel open times of a ligand-gated ion channel having one allosteric and two orthosteric sites. The number of channel events corresponding to an interval of open times is plotted against the bin interval on linear (a and b) and logarithmic (c and d) scales for the condition of no modulator (Control, a and c) and a receptor-saturating concentration of positive modulator (+ Modulator, b and d). The frequency distributions were generated from the simulated data in Figure 4.15 at log agonist concentration of −3.00 (control) and 3.33 (+ Modulator). Mean open times are indicated with the parameter τ in panels (a) and (b).

is no constitutive activity, and the allosteric modulator does not elicit a response by itself, one could liken the allosteric effect to a scalar change in K_b equivalent to the square root of the scalar changes mentioned above. For example, if a positive allosteric modulator was present at a saturating concentration and shifted the concentration-response curve to the left, this effect is indistinguishable for a scalar increase in K_q of K_f/K_e or a scalar increase in K_b of $(K_f/K_e)^{1/2}$. This relationship holds for the case of one allosteric site coupled to two orthosteric sites as simulated in the example above. For the case of two allosteric sites coupled to two orthosteric sites, the effect of a positive allosteric modulator is indistinguishable from scalar increase in K_q of K_f^2/K_e^2 or a scalar increase in K_b of K_f/K_e. Examples of these calculations are given below in the section on the analysis of published data.

Of course, the data can be analyzed with a more complete model that incorporates the allosteric ligand and eliminates the necessity of assuming changes in K_q or K_b. The method for doing this is described next.

Analysis of the Effect of Allosteric Modulation on Open Probability

As mentioned earlier in this chapter and in Chapters 2 and 3, the open probability of a single channel is also equivalent to the fraction of the receptor population in the open state. Therefore, we can use the equations that express the fraction of the receptor population in the active state as a function of the microscopic constants and the concentrations of agonist and allosteric ligand to analyze these data. The relevant equations are numbered (4.22), (4.26), (4.75) and (4.93). The first two are for the analysis of allosterism and open channel block at a ligand-gated ion channel having one allosteric site, and the third and fourth are for analyzing allosterism in the case of two allosteric sites. All of these equations are based on two orthosteric sites. For convenience, these equations have been rewritten below with the dependent variable open probability (P_{open}) replacing the fraction of the receptor population in the active state (R_{active}/R_T):

$$P_{\text{open}} = \frac{1}{1 + \dfrac{(1+AK_e)(1+DK_a)^2}{K_q(1+AK_f)\left(1+D\sqrt{\dfrac{P_{b2}(1+K_q)^2}{K_q}}\right)^2}}, \tag{4.96}$$

$$P_{\text{open}} = \frac{1}{1 + AK_f + \dfrac{(1+AK_e)(1+DK_a)^2}{K_q\left(1+D\sqrt{\dfrac{P_{b2}(1+K_q)^2}{K_q}}\right)^2}}, \tag{4.97}$$

$$P_{\text{open}} = \frac{1}{1 + \dfrac{(1+DK_a)^2(1+AK_e)^2}{K_q\left(1+D\sqrt{\dfrac{P_{b2}(1+K_q)^2}{K_q}}\right)^2(1+AK_f)^2}}, \tag{4.98}$$

$$P_{\text{open}} = \frac{1}{1 + \dfrac{(1+DK_a)^2(1+AK_e)^2}{K_q(1+DK_b)^2(1+AK_f)^2}}. \tag{4.99}$$

Because the open-probability plot is identical to a plot of the fraction of the receptor population in the active state, the simulated whole-cell current responses plotted in Figures 4.9–4.12 are also useful examples of open-probability plots. The latter have been normalized to the theoretical maximum current response caused by simultaneous activation of all of the channels ($i_{\text{max-sys}}$), so the range of the ordinate scale (0–1) is the same as that of an open-probability plot. Essentially, the open-probability measurement can only vary from 0–1.0, and hence, the need to estimate the maximum value of open probability (1) does not exist.

The data in Figure 4.11b serve as a useful example. Recall that when these data were treated as whole-cell current data and analyzed with Equation (4.89), it was impossible to obtain a reliable estimate of $i_{max-sys}$ because the maximum response of the partial agonist was much less. Without a reasonable estimate of $i_{max-sys}$, it is impossible to obtain accurate estimates of the other parameters. The solution was to constrain $i_{max-sys}$ to the value estimated with a highly efficacious agonist or to analyze the data simultaneously with those of the more efficacious agonist. In the case of open-probability measurements, however, one can fit Equation (4.96) to the data and obtain reliable parameter estimates for partial agonists. When this was done, the following parameter estimates were obtained: log K_a, 4.00 ± 0.01; log P_{b2}, 7.01 ± 0.03; log K_e, 5.95 ± 0.015 and log K_f, 6.96 ± 0.025.

To summarize, Equation (4.96) can also be used to analyze the effect of the positive allosteric modulator on the open probability plot of an efficacious agonist at a ligand-gated ion channel having one allosteric site. Equation (4.97) is used to analyze the effect of an open-channel blocker on open probability, and Equation (4.98) is used to analyze the effect of allosteric modulators on the open-probability plot of a ligand-gated ion channel having two allosteric sites. Finally, in the case of an allosteric agonist, Equation (4.99) can be used instead of Equation (4.98) for the case of two allosteric sites. In this latter situation, it is actually possible to estimate the isomerization constant of the free receptor complex. While both Equations (4.98) and (4.99) can be used in this analysis, Equation (4.97) is more convenient because one can obtain the estimate of K_b and its standard error directly.

Analysis of Allosteric Modulation of the GABA$_A$ Receptor

So how well do the models fit real data? In this section, I analyze some data previously published by other investigators on the allosteric modulation of GABA$_A$ receptor-mediated chloride currents.

The first example is from Erwin Sigel and Roland Baur who investigated the modulatory effects benzodiazepine ligands on GABA-induced whole cell-currents in *Xenopus* oocytes expressing chick brain mRNA (Figure 4.18). By itself, GABA elicits a robust, concentration-dependent current that is half-maximal at approximately 20 μM GABA ($pEC_{50} = 4.68$). In the presence of either diazepam (0.1 μM) or 6,7-dimethoxy-4-ethyl-β-carboline-3-carboxylic acid methyl ester (DMCM, 0.32 μM), the concentration-response curve shifts two-fold to the left or right, respectively. In both cases, there is little change in i_{max}. These results are consistent with the predicted effects of modest positive and negative allosteric modulation on the

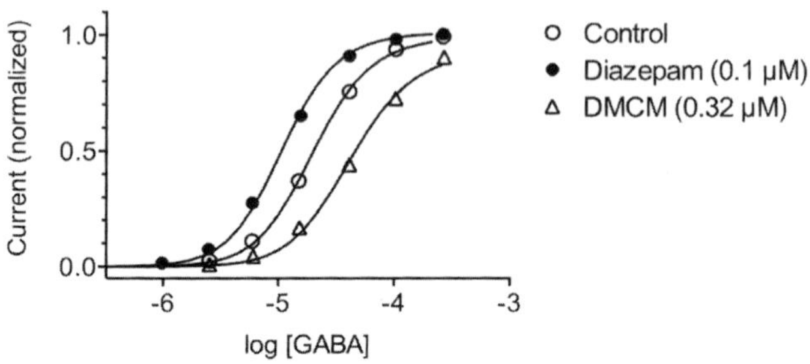

Figure 4.18: Effects of diazepam and DMCM (6,7-dimethoxy-4-ethyl-β-carboline-3-carboxylic acid methyl ester) on the whole cell current response to GABA at chick $GABA_A$ receptors expressed in *Xenopus* oocytes. The current values were estimated from a plot of the data published by Sigel and Baur.[26]

population current response of a highly efficacious agonist (see Figure 4.11a for positive modulation). That is, moderate allosteric modulation of the current response of a highly efficacious agonist is associated with shifts in EC_{50} with little effect on the $i_{\max}$.

Because only one concentration of each modulator was used, it is impossible to estimate their microscopic constants. But whenever a modulator is used at a high concentration relative to its observed affinity, there is sufficient information to estimate the ratio of affinity constants of the modulator for active and inactive states (K_f/K_e). When there is no constitutive activity and the modulator lacks activity by itself, the pair of concentration-response curves, measured in the absence and presence of modulator, can be analyzed according to Equation (2.153) as if the curves were generated by two different agonists in the absence of modulator. This equation is given below for convenience. It can also be derived from Equation (4.91) by setting $A = 0$ and simplifying:

$$i = \frac{i_{\max\text{-sys}}}{1 + \dfrac{(1+DK_a)^2}{K_q\left(1+D\sqrt{\frac{P_{b2}(1+K_q)}{K_q}}\right)^2}}. \tag{4.100}$$

Recall that P_{b2} is equivalent to $K_b^2K_q/(1+K_q)$. To estimate the K_f/K_e values of diazepam and DMCM, the three concentration-response curves in Figure 4.18 are analyzed simultaneously by global nonlinear regression analysis with the following provisions: 1) the estimates of $i_{\max\text{-sys}}$ and K_a are shared among the curves, 2) K_q is constrained to an arbitrarily low value (e.g., 10^{-8}), and 3) an estimate of P_{b2} is obtained for each curve. When this was done, the estimates of log P_{b2} were obtained for the control data (9.51 ± 0.03) and those measured in the presence of diazepam (10.00 ± 0.02) and DMCM (8.94 ± 0.04). Because the effect of an allosteric modulator is equivalent to that caused by a scalar change in either the value of K_q or K_b, and this scalar change is equivalent to K_f/K_e, then the latter ratio can be estimated from the corresponding ratio of P_{b2} values. For example, if the P_{b2}value of the curve measured in the presence of diazepam is divided by that of the control curve, then the ratio of K_f/K_e is obtained for diazepam. Taking these ratios yields log K_f/K_e estimates of 0.49 and -0.57 for diazepam and DMCM, respectively. Thus, diazepam exhibits about three-fold selectivity for the active state, whereas DMCM exhibits the opposite selectivity. The variance of the log K_f/K_e estimate is simply equal to the sum of the variances of the P_{b2} values from which it was calculated. Regression analysis also yields estimates of $i_{\text{max-sys}}$ (1.02 ± 0.01) and log K_a (3.92 ± 0.07).

The selectivities of diazepam and DMCM for the active and inactive states of the $GABA_A$ receptor are consistent with their pharmacological effects. Diazepam is a potent benzodiazepine having sedative, anticonvulsant and antianxiety effects that are mediated through an enhancement of GABA neurotransmission. In contrast, DMCM is pro-convulsant, and it acts by inhibiting GABA neurotransmission.

Another example for our analysis comes from Gabor Maksay, Sally Thompson and Keith Wafford who measured the effects of the tranquilizing benzodiazepine, chlordiazepoxide (30 μM), and the proconvulsant, DMCM (0.3 μM), on agonist-induced chloride currents in *Xenopus* oocytes expressing the α_1, β_3 and γ_2 $GABA_A$ receptor subunits (Figure 4.19). Again, moderate allosteric modulation mainly affects the EC_{50} value of a highly efficacious agonist, like GABA, and the i_{max} value of a less efficacious agonist, like P4S. This behavior is consistent with the

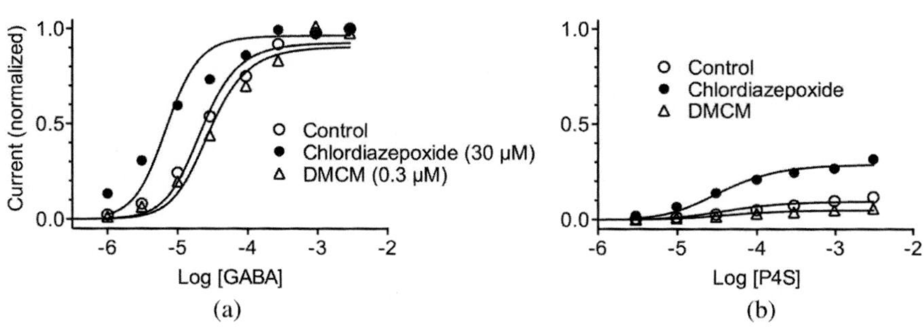

Figure 4.19: Effects of chlordiazepoxide and DMCM on the whole cell current response to GABA (a) and P4S (piperidine-4-sulfonic acid) (b) on chloride currents in *Xenopus* oocytes expressing α_1, β_3 and γ_2 $GABA_A$ receptor subunits. The current values have been estimated from a plot of the data published by Maksay and coworkers.[27]

theoretical data in Figure 4.11. Because the modulators were present at concentrations that nearly saturate the allosteric site for benzodiazepines, the data can be analyzed as just described for the data in Figure 4.18. Regression analysis of the data in panel (a) yielded estimates of $i_{max-sys}$ (0.97 ± 0.04) and the log K_a value of GABA (4.07 ± 0.24). The latter estimate has a large standard error because of the difficulty in estimating the K_a value of a highly efficacious agonist from the population current. The estimates of log P_{b2} for the data measured in the absence and presence of chlordiazepoxide and DMCM were 9.47 ± 0.06, 10.36 ± 0.10 and 9.29 ± 0.09, respectively. Taking the log ratio of P_{b2} for each modulator divided by the control P_{b2} value yields the estimate of log K_f/K_e for chlordiazepoxide (0.89) and DMCM (−0.19).

The same analysis can be done with the data for the less efficacious agonist piperidine-4-sulfonic acid (P4S) in panel (b) provided that regression analysis is done with $i_{max-sys}$ constrained to that estimated with GABA. Otherwise, it is unlikely that accurate parameter estimates will be obtained. (A better approach is to analyze the data in panels (a) and (b) simultaneously sharing the estimate of $i_{max-sys}$ across all of the data, but this more complicated approach will not be described here.) Regression analysis yields an estimate of 4.72 ± 0.06 for the log K_a of P4S. The estimates of log P_{b2} for the control, chlordiazepoxide and DMCM conditions were 8.45 ± 0.07, 9.04 ± 0.09 and 8.15 ± 0.03. Taking the appropriate log ratios of these values yield estimates of log K_f/K_e for chlordiazepoxide (0.59) and DMCM (−0.30). These latter estimates should agree with those estimated when GABA was used as the agonist. The small disagreement may be related to the large standard errors associated with the GABA concentration-response curve measured in the presence of chlordiazepoxide and the small effect of DMCM at the $\alpha_1\beta_3\gamma_2$ $GABA_A$ receptor.

The next example comes from Scott Downing, Yan Lee, David Farb and Terrell Gibbs who investigated the effects of the tranquilizing benzodiazepine, diazepam, on the concentration-response curves of GABA and kojic amine for eliciting chloride currents in chick neurons (Figure 4.20). Again, because the concentration of diazepam used in the experiment (1 μM) is likely to saturate the benzodiazepine site, the data can be analyzed as just described. Regression analysis yielded estimates of $i_{max-sys}$ (2.02 ± 0.11) and the log K_a values of GABA (3.08 ± 0.75) and kojic amine (2.35 ± 0.10). The log K_a value of GABA has a huge standard error because a small error in the estimate of $i_{max-sys}$ leads to large error in the estimate of the K_a of a highly efficacious agonist. The estimates of log P_{b2} for the control and diazepam conditions with GABA were 8.14 ± 0.12 and 8.77 ± 0.07, respectively. The corresponding values measured with kojic amine were 4.49 ± 0.08 and 4.97 ± 0.11. Taking the appropriate log ratios of these values yields estimates of log

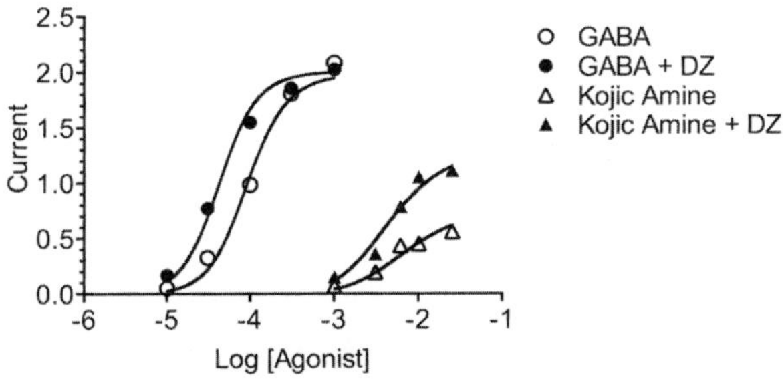

Figure 4.20: Effects of diazepam (DZ) on chloride currents elicited by GABA and kojic amine in HEK 293 cells expressing $\alpha_1\beta_3\gamma_2$ $GABA_A$ receptors. The current values have been estimated from a plot of the data published by Downing and coworkers.[21]

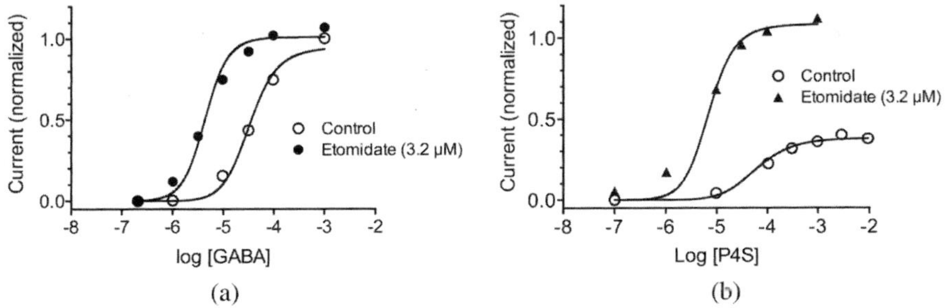

Figure 4.21: Effects of etomidate on the whole cell current response to GABA (a) and P4S (piperidine-4-sulfonic acid) (b) on chloride currents in *Xenopus* oocytes expressing α_1, β_3 and γ_2 $GABA_A$ receptor subunits. The current values have been estimated from a plot of the data published by Rusch and coworkers.[22]

K_f/K_e for diazepam when interacting with GABA (0.63) and kojic amine (0.48). These log ratios are expected to be the same, and indeed, they show no significant difference.

Our next two examples of data are from Dirk Rusch, Juijun Zhong and Stuart Forman who investigated the modulatory effect of the anesthetic, etomidate, on GABA agonist-mediated chloride currents in *Xenopus* oocytes expressing α_1, β_2, γ_{2L} $GABA_A$ receptors. Figure 4.21 shows the effect of etomidate (3.2 μM) on the concentration-response curves of GABA (panel (a)) and P4S (panel (b)). At this concentration, etomidate has no effect by itself, but nonetheless, causes large increases in the potency of GABA and both the potency and maximal effect of P4S. Regression analysis of the data with Equation (4.98) yields estimates of $i_{\text{max-sys}}$ (1.02 ± 0.04) and the log K_a values of GABA (3.99 ± 0.24) and P4S (4.46 ± 0.25).

The estimates of log P_{b2} for GABA under control conditions and in the presence of etomidate were 9.15 ± 0.09 and 10.75 ± 0.08, respectively. The corresponding estimates for P4S were 8.62 ± 0.11 and 10.44 ± 0.13. These estimates yield similar log P_{b2} ratios of 1.60 and 1.82 for the effects of etomidate on GABA- and P4S-mediated chloride currents, respectively. When there are two allosteric sites linked to two orthosteric sites, the change in P_{b2} is equivalent to K_f^2/K_e^2 if the concentration of allosteric modulator is high enough to saturate the allosteric site. But for the data in Figure 4.20, the concentration of etomidate is far from saturating, so we cannot assume that the latter ratios are an estimate of K_f^2/K_e^2. One way to think about the change in P_{b2} is that it is a measure of the allosteric change in receptor isomerization. In this case, isomerization is increased by the antilog of 1.7 (50-fold), approximately. Thus, the allosteric effect of etomidate is akin to an increase of 50-fold in the isomerization constant of the free receptor (K_q).

The same investigators performed a more complete analysis of the effect of etomidate, and the results are illustrated in Figure 4.22. Panel (b) shows the effects of various concentrations of etomidate (3.2–100 μM) on the concentration-response curve of GABA for eliciting chloride currents. By itself, etomidate activates $GABA_A$ receptors, although not to the same extent as GABA (Panel a). The combination of two agonists, each acting at a different site on the $GABA_A$ receptor, provides sufficient information to estimate all of the microscopic constants — including K_q — as long as the range of ligand concentrations is sufficient to cover the range of effects. The basis for this rationale is described above in connection with the analysis of the theoretical data shown in Figure 4.14. To estimate the microscopic constants, Equation (4.93) is fitted to the data by global nonlinear regression analysis sharing all of the parameter estimates among the curves. When this was done, the following parameter values were obtained: log K_q, -3.71 ± 0.18; K_a, 4.10 ± 0.12; log K_b, 6.54 ± 0.08; log K_e, 4.73 ± 0.09; log K_f, 6.42 ± 0.06 and $i_{\text{max-sys}}$, 1.03 ± 0.02.

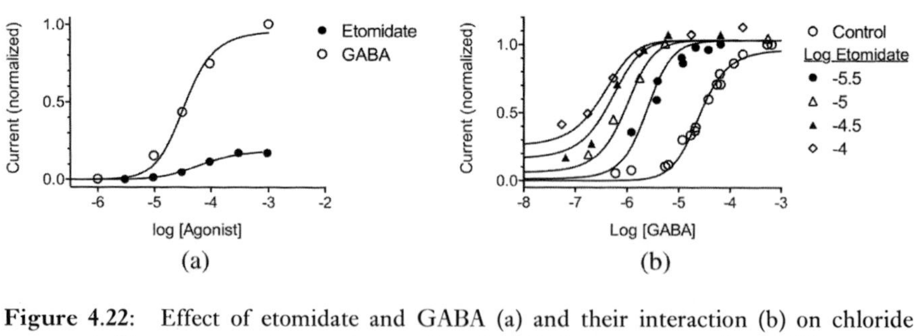

Figure 4.22: Effect of etomidate and GABA (a) and their interaction (b) on chloride currents in *Xenopus* oocytes expressing α_1, β_3 and γ_2 $GABA_A$ receptor subunits. The current values have been estimated from a plot of the data published by Rusch and coworkers.[22]

The results suggest that the isomerization constant of the $GABA_A$ receptor (K_q) is not that low ($\approx 10^{-4}$) and that the selectivity of GABA for the active state of the $GABA_A$ receptor is not that high ($K_b/K_a \approx 300$). This seems unlikely given that the analogous values for the muscle type nicotinic acetylcholine receptor indicate much lower constitutive activity ($K_q \approx 10^{-6}$) and a much greater selectivity of acetylcholine for the active state ($K_b/K_a \approx 10,000$). Would not a more efficient $GABA_A$ receptor with much less leak current in the absence of GABA have evolved through natural selection?

Before addressing possible explanations, let us consider how the behavior of allosteric modulation is related to the value of K_q. For the theoretical data in Figure 4.14, the positive modulator shifts the concentration-response curve to the left about 10^4-fold and increases constitutive activity to 0.5 or 50% of $i_{max-sys}$. The size of the shift is a measure of K_f/K_e. Again, thinking from the perspective of a simple model that does not include allosteric modulation, this effect would be indistinguishable from a scalar increase in K_q of 10^8 (i.e., K_f^2/K_e^2). Equation (2.151) in Chapter 2 shows that a constitutive activity value of 0.5 occurs when $K_q = 1.0$. Thus, the true value of K_q for the native receptor must be equal to the apparent value in the presence of modulator (1.0) divided by 10^8, which yields 10^{-8} — a value identical to that used in the simulation in Figure 4.12.

Let us apply the same logic to analyze the experimental data in Figure 4.22b. I estimate that the maximal shift in the GABA concentration-response curve caused by etomidate is a little over 1.5 log units (32-fold). This provides an estimate of K_f/K_e. The allosteric effect is indistinguishable from a 1000-fold increase in K_q (K_f^2/K_e^2) that causes the receptor to exhibit a level of constitutive activity of approximately 0.2. Equation (2.151) in Chapter 2 shows that this amount of constitutive activity is associated with a K_q value of 0.25. Dividing this apparent K_q value by the scalar effect of etomidate on K_q (1000) yields an estimate for the true value of K_q (0.00025; log K_q, −3.6) that is not much different from that (log K_q, −3.71) estimated by regression analysis.

Thus, if the two-state model with two orthosteric sites linked to two allosteric sites *does* account for the actions of etomidate, then the estimate of the K_q value of the $GABA_A$ receptor is approximately 10^{-4}. But are there other explanations? Perhaps a more complicated mechanism is at work. For example, etomidate may have a potent allosteric effect at low concentrations corresponding to a K_f/K_e ratio similar to that calculated for the data in Figure 4.19, but perhaps at higher concentrations, it acts at a different site not linked to the orthosteric site and disrupts the structure of the closed state causing an increase in conductance. This secondary action would cause over- and under-estimations of K_q and K_b, respectively. Another possibility is that perhaps etomidate selects for a different conformation of the open state having the same conductance as the native conformation, but lower affinity for GABA (i.e., log $K_b \approx 6.5$) compared to the native conformation (i.e., log $K_b \approx 8$). If so, then allosteric modulation by etomidate may differ depending on the orthosteric ligand. For if etomidate were to induce a novel active state, this different structure should exhibit a different pharmacological profile. Etomidate seems to affect GABA and P4S similarly, which provides no evidence for an alternative active state, but we would need to examine more agonists to address this question.

Of course, it is possible that the K_q value of the $GABA_A$ receptor *is* approximately 10^{-4} and that in the absence of GABA, the receptor helps to clamp the resting membrane potential to prevent excessive excitation in the same way that some constitutively active potassium channels do.

Probe-Dependence of Allosteric Interactions

The effects of an allosteric modulator depend on the orthosteric ligand with which it is interacting. For example, with regard to binding interactions, the positive cooperativity between an allosteric ligand and a partial agonist is less than that between the same allosteric ligand and a more efficacious agonist (compare simulations in Figure 4.7 and 4.10). Similarly, the effects of an allosteric ligand on the receptor activation functions of partial and highly efficacious agonists are different (Figure 4.11). The term, probe dependence, is used to refer to the differential effects of an allosteric modulator on the binding and functional responses of different orthosteric ligands, particularly within the context of *G* protein-coupled receptors.

This property of probe dependence represents a superficial level of analysis, however, relating to the average behavior of the receptor population. This is not to say that the property is unimportant, but rather, to emphasize that it is used to describe the ensemble average and not the properties of individual receptors. For at the level of receptor states, the estimates of the microscopic affinity constants of an allosteric modulator are the same regardless of whether it is modulating the action of a partial agonist or a highly efficacious agonist. This can be seen from the theoretical examples (Figures 4.7 and 4.10) and the published data with allosteric ligands of the $GABA_A$ receptor (Figures 4.19–4.21), where the estimates of the K_f and K_e values of the allosteric modulator are independent of the orthosteric ligand. This property is an inherent constraint of the simple two-state model and the analysis of receptor mechanisms at the level of receptor states.

The simple two-state model seems appropriate for the interaction of many endogenous ligands with their receptor, because most receptors evolved to be active when bound with endogenous ligand and inactive when unbound. Because high activity requires an optimum structure, multiple active states would reduce activity from the optimum and multiple inactive states with variable affinity would engender ligand induction of an inactive state over an active one. Thus, discrete active and inactive states with high and low affinity for the endogenous ligand, respectively, are likely properties of an efficient, ligand-dependent switch. That evolution selected efficient receptors can be seen in the uniform quantal currents of ligand-gated ion channels. Presumably, each time a channel opens or closes, there is a coincident quantal increase or decrease in the affinity of the endogenous ligand for the receptor (see Figure 4.2). Similarly, there would be quantal changes in the affinity of associated allosteric sites.

Within the global structure of the receptor, therefore, there is a quantal switch consisting of a linkage between the binding pocket for the endogenous ligand and the pore-lining M2 segments of the ligand-gated ion channel. Despite disordered regions of the active structure, continuous interconversion between states, and a multitude of evanescent transitions states, the muscle-type nicotinic acetylcholine receptor spends nearly 100% of its time in two states — active and inactive. The mathematical descriptions of the quantal switches for the case of one and two allosteric sites are shown in Figures 4.3 and 4.4. These schemes enable one to estimate the microscopic affinity constants of the quantal switch and were not intended, of course, to describe the global structure of the receptor.

The consequences of these two schemes are limited with regard to the probe dependence of allosteric ligands. With the exception of orthosteric ligands at the two ends of the efficacy spectrum, two-state models predict that allosteric modulation is always associated with changes in both affinity and efficacy. In the case of a highly efficacious agonist, the allosteric modulation consists mainly of affinity modulation, whereas in the case of an agonist with low efficacy, the modulation affects primarily efficacy. In the case of a neutral antagonist, allosteric modulation affects efficacy only. This principle is based on the idea that the efficacy of a neutral antagonist is equivalent to the constitutive activity of the receptor, however low that might be. Allosteric ligands modify constitutive activity, and therefore, the efficacy of neutral antagonists. The same is true for inverse agonists, although once again, the changes may be too small to measure. These issues are discussed in greater detail in Chapter 12 with regard to *G* protein-coupled receptors.

Another consequence of the two-state models is that they do not allow allosteric changes in the pharmacological specificity of a receptor, but rather, a shift in the direction of agonism or antagonism. That is, the two-state model predicts that a positive modulator of the endogenous ligand will enhance the affinity and efficacy of all agonists, whereas a negative modulator will have the opposite effect.

The two state-schemes described in Figures 4.3 and 4.4 seem to describe the interaction of endogenous ligands with their receptors accurately as well as the interaction of many small molecule allosteric ligands of the $GABA_A$ receptor. Nonetheless, these two-state schemes do not rule out the possibility that xenobiotics in the form of exogenous agonists and allosteric modulators may select for different conformations of the active and inactive states of a given receptor. For exogenous agonists often interact with residues different from those of the endogenous ligand, and hence, their interaction with the receptor has not been shaped by natural selection. Much evidence is presented in Chapter 9 that some agonists direct signaling though novel states of *G* protein-coupled receptor that exhibit selectivity for different *G* proteins. In most instances, however, an agonist of this type has much lower efficacy than the endogenous ligand, indicating that the novel active states are not as active as the native active state induced by the endogenous ligand. These principles also apply to allosteric ligands, and it was suggested in the prior section that perhaps etomidate selects for a unique active conformation of the $GABA_A$ receptor. Thus, some allosteric ligands might reduce the activity of one agonist and increase the activity of another through induction of a novel active state of the receptor.

Observed Affinity and Efficacy

It may seem odd that in a chapter in a book, entitled "Affinity and Efficacy", little effort has been given to discussing the allosteric modulation of the observed affinity and efficacy of an agonist in a quantitative way. Rather, most of the discussion is focused on microscopic constants. This is because the latter are both easier to estimate and have the advantage of being more fundamental constants that can be used to quantify probe dependence. In this section, however, I describe how allosteric modulation of the observed affinity and mean efficacy of the agonist-receptor complex is related to receptor occupancy by the allosteric ligand.

For a receptor having a single orthosteric site, the concepts of observed affinity and intrinsic efficacy are simple and straightforward as described in the first part of Chapter 2. The observed affinity constant (K_{obs}) defines receptor occupancy. It specifies that receptor occupancy is consistent with a simple one-site model and that half-maximal occupancy occurs when the orthosteric ligand concentration is equivalent to $1/K_{obs}$. Efficacy is a measure of the fraction of the population of agonist-receptor complexes in the active state. It is a simple scalar that ranges from 0 to 1, and its value is independent of the level of receptor occupancy. At 100% receptor occupancy, efficacy is equivalent to the fraction of the receptor population in the active state.

But when a receptor has more than one orthosteric site linked to receptor activation, the concepts of observed affinity and efficacy are more complicated. For example, if one considers a simple issue, like how to define occupancy, a few reasonable alternatives come to mind. If the receptor has two orthosteric sites, occupancy could be defined with respect to 1) the number of ligands bound to the receptor population, 2) the number of receptors bound with two equivalents of ligand or 3) the number of receptors bound with either one or two ligands. Each of these ways of measuring receptor occupancy yields a different value for the concentration of ligand required for half-maximal receptor occupancy. In addition, the shape of the occupancy curve is steeper than a one-site model for most agonists, but approaches a one-site model in the case of neutral antagonists, inverse agonists and agonists with low efficacy. In this section, I define occupancy by the third option as described in Chapter 2 and use the variable, K_{occ}, to denote the concentration of orthosteric ligand required for half-maximal receptor occupancy.

When there are two linked orthosteric sites on a receptor, the efficacy of an agonist can no longer be assigned a single value because efficacy changes with the level of receptor occupancy as described in Chapter 2. Integrating the efficacy function over the domain of receptor occupancy from 0 to 1 yields a useful parameter known as mean efficacy.

In this section, I analyze some of the prior theoretical examples to show how allosteric modulators affect the mean efficacy and the concentration of agonist required for half-maximal receptor occupancy at a ligand-gated ion channel having two orthosteric sites. The analysis shows that a given allosteric modulator exhibits a unique concentration-dependent modification in the product of observed affinity ($1/K_{\mathrm{obs}}$) and mean efficacy (RA) that is unique to the allosteric modulator and independent of the orthosteric ligand. The maximum change in RA is related to K_f/K_e.

Derivation of Equations for the Allosteric Modulation of Observed Affinity and Mean Efficacy

Allosteric modulators have an effect that is indistinguishable from an alteration in the observed affinity and mean efficacy of the agonist-receptor complex. In this section, equations for estimating the observed affinity and mean efficacy are developed so that we can determine how allosteric modulation affects these parameters.

This process will be accomplished in the following steps. 1) First, an equation is derived that expresses receptor occupancy as a function of the agonist concentration. In this case, occupancy is defined as the fraction of the population of ligand-receptor complexes that have either one or two ligands bound to them. 2) This function is used to derive an expression for the concentration of agonist required for half-maximal receptor occupancy (K_{occ}). 3) The occupancy function is inverted to express the agonist concentration as a function of receptor occupancy. 4) The next step involves deriving an equation that expresses the fraction of the population of agonist-receptor complexes in the active state as a function of the agonist concentration. 5) The equation from step 3 is substituted for the agonist concentration in the receptor activation function derived in step 4. This yields an equation for receptor activation expressed as a function of receptor occupancy. 6) This latter equation is divided by receptor occupancy to yield an equation for efficacy as a function of receptor occupancy. 7) Finally, the latter equation is integrated over the domain of receptor occupancy from 0 to 1 to determine mean efficacy.

This procedure is described below for ligand-gated ion channels having two orthosteric sites and either one or two allosteric sites.

Estimation of mean efficacy and observed affinity for the case of one allosteric site

The derivation of the receptor occupancy function is analogous to that described above for Equation (4.30). The only difference is that occupancy is expressed relative to the total receptor concentration ($[R_T]$ and not $2[R_T]$) because both the mono-liganded (e.g., DR) and bi-liganded (e.g., D_2R) receptor complexes contribute equivalently to receptor occupancy. Thus, there is no need to multiply the latter by two in tallying receptor occupancy. Thus, we begin with the following expression for receptor occupancy ($D_{\text{bound}}/[R_T]$):

$$\text{occupancy} = \frac{D_{\text{bound}}}{[R_T]} = \frac{[D_{\text{bound}}]}{[D_{\text{bound}}] + [D_{\text{unbound}}]}, \tag{4.101}$$

in which,

$$\begin{aligned}[D_{\text{bound}}] = {} & [DR_s^*] + [D_2R_s^*] + [DR_s^*A] + [D_2R_s^*A] + [DR_s] \\ & + [D_2R_s] + [DR_sA] + [D_2R_sA],\end{aligned} \tag{4.102}$$

$$[D_{\text{unbound}}] = [R_s] + [R_sA] + [R_s^*] + [R_s^*A]. \tag{4.103}$$

After making the appropriate substitutions for each receptor complex and simplifying, the following equation is obtained:

$$\frac{D_{\text{bound}}}{[R_T]} = \frac{[D]}{[D] + K_{\text{occ}}}, \tag{4.104}$$

in which,

$$K_{occ} = \frac{1 + K_q + A(K_e + K_f K_q)}{K_a(2 + DK_a)(1 + AK_e) + K_b K_q(2 + DK_b)(1 + AK_f)}. \qquad (4.105)$$

In this equation, K_{occ} denotes the concentration of orthosteric ligand required for half-maximal receptor occupancy in the presence of allosteric modulator.

Equation (4.104) can be rearranged into the form of a quadratic to express the orthosteric ligand concentration (D) as a function of receptor occupancy (occ):

$$D = \frac{-b + \sqrt{b^2 - 4ac}}{2a}, \qquad (4.106)$$

in which,

$$a = K_b^2 K_q(1 + AK_f) + K_a^2(1 + AK_e), \qquad (4.107)$$

$$b = 2(K_b K_q(1 + AK_f) + K_a(1 + AK_e)), \qquad (4.108)$$

$$c = \frac{\text{occ}}{1 - \text{occ}}(1 + K_q + A(K_e + K_f K_q)). \qquad (4.109)$$

We will use Equation (4.106) in the derivation of the equation for efficacy.

The first step in deriving an equation for agonist efficacy involves defining the relationship for the fraction of the population of agonist-receptor complexes in the active state. We begin with the following expression:

$$\frac{[DR_{\text{active}}]}{[R_T]} = \frac{[DR_{\text{active}}]}{[DR_{\text{active}}] + [nonDR_{\text{active}}]}, \qquad (4.110)$$

in which

$$[R_{\text{active}}] = [\mathbf{DR}_s^*] + [D_2R_s^*] + [\mathbf{DR}_s^*\mathbf{A}] + [D_2R_s^*A], \qquad (4.111)$$

$$[nonDR_{\text{active}}] = [R_s] + [R_s^*] + [\mathbf{DR}_s] + [D_2R_s] + [\mathbf{DR}_s\mathbf{A}] + [D_2R_sA] + [R_sA] + [R_s^*A]. \qquad (4.112)$$

The difference between the prior equation developed for receptor activation (Equation (4.22)) and Equation (4.110) is that the former describes total receptor activation, including the actives states of the free receptor (R_s^*) and the modulator-receptor complex (R_s^*A), as a function of the agonist concentration, whereas the latter focuses only on the active states of receptor complexes bound with orthosteric ligand. Substituting in the appropriate expressions for the various receptor complexes yields:

$$\frac{[R_{\text{active}}]}{[R_T]} = \frac{1}{1+\frac{1+K_q+DK_a(2+DK_a)(1+AK_e)+A(K_e+K_fK_q)}{DK_bK_q(2+DK_b)(1+AK_e)}}. \tag{4.113}$$

To derive an equation for agonist efficacy as a function of receptor occupancy, we substitute in Equation (4.106) for D into Equation (4.113) to derive a function ($f_1(\text{occ})$) that expresses fractional receptor activation as a function of receptor occupancy (occ). Dividing this function by occupancy yields efficacy as a function of receptor occupancy:

$$\text{efficacy} = \frac{f_1(\text{occ})}{\text{occ}}. \tag{4.114}$$

To estimate mean efficacy, Equation (4.14) is integrated over the domain of occ = 0 to occ = 1, which is done numerically as described in Chapter 2:

$$\text{efficacy}_{\text{mean}} = \int_0^1 \frac{f_1(\text{occ})}{\text{occ}}. \tag{4.115}$$

Estimation of mean efficacy and observed affinity for the case of two allosteric sites

The derivation of the receptor occupancy function for the case of two allosteric sites is analogous to that described above for Equation (4.78). The difference is that occupancy is expressed relative to the total receptor concentration ($[R_T]$ and not $2[R_T]$), and that both the mono-liganded (e.g., DR) and bi-liganded (e.g., D_2R) receptor complexes contribute equivalently to receptor occupancy. Thus, we begin with Equation (4.101) but with the $[D_{\text{bound}}]$ and $[D_{\text{unbound}}]$ components defined as:

$$\begin{aligned}[D_{\text{bound}}] = {} & [\boldsymbol{DR_s^*}] + [\boldsymbol{DR_s}] + 2[D_2R_s^*] + 2[D_2R_s] + [\boldsymbol{DR_s^*A}] \\ & + [\boldsymbol{DR_sA}] + 2[\boldsymbol{D_2R_s^*A}] + 2[\boldsymbol{D_2R_sA}] \\ & + [\boldsymbol{DR_s^*A_2}] + [\boldsymbol{DR_sA_2}] + 2[\boldsymbol{D_2R_s^*A_2}] + 2[\boldsymbol{D_2R_sA_2}],\end{aligned} \tag{4.116}$$

$$[D_{\text{unbound}}] = [R_s] + [R_s^*] + [\boldsymbol{R_sA}] + [\boldsymbol{R_s^*A}] + [R_sA_2] + [R_s^*A_2]. \tag{4.117}$$

After making the appropriate substitutions for each receptor complex and simplifying, Equation (4.104) is obtained, but with K_{occ} defined as:

$$K_{\text{occ}} = \frac{1+K_q+A(K_e(2+AK_e)+K_fK_q(2+AK_f))}{K_a(2+DK_a)(1+AK_e)^2+K_bK_q(2+DK_b)(1+AK_f)^2}. \tag{4.118}$$

In this equation, K_{occ} denotes the concentration of orthosteric ligand required for half-maximal receptor occupancy in the presence of an allosteric modulator.

The form of Equation (4.104) with K_{occ} defined by Equation (4.118) can be rearranged to express the orthosteric ligand concentration (D) as a function of receptor occupancy (occ) to yield a quadratic equation, like Equation (4.16), but with the following substitutions:

$$a = K_a^2 + K_b^2 K_q + A(K_a^2 K_e(2 + AK_e) + K_b^2 K_f K_q(2 + AK_f)), \quad (4.119)$$

$$b = 2\{K_a + K_b K_q + A(K_a K_e(2 + AK_e) + K_b K_f K_q(2 + AK_f))\}, \quad (4.120)$$

$$c = -\frac{\text{occ}}{1 - \text{occ}}\{1 + K_q + A(K_e(2 + AK_e) + K_f K_q(2 + AK_f))\}. \quad (4.121)$$

We will use this quadratic below in the derivation of the function for efficacy in the case of two allosteric sites.

To derive an equation for agonist efficacy, we begin with Equation (4.10) but with the $[DR_{\text{active}}]$ and $[non\text{-}DR_{\text{active}}]$ components defined as:

$$\begin{aligned}[DR_{\text{active}}] &= [\boldsymbol{DR_s^*}] + [D_2R_s^*] + [\boldsymbol{DR_s^*A}] + [\boldsymbol{D_2R_s^*A_2}] \\ &\quad + [\boldsymbol{DR_s^*A_2}] + [\boldsymbol{D_2R_s^*A}], \end{aligned} \quad (4.122)$$

$$\begin{aligned}[nonDR_{\text{active}}] &= [R_s] + [R_s^*] + [\boldsymbol{DR_s}] + [D_2R_s] + [\boldsymbol{DR_sA}] \\ &\quad + [D_2R_sA] + [\boldsymbol{DR_sA_2}] + [\boldsymbol{D_2R_sA_2}] \\ &\quad + [\boldsymbol{R_sA}] + [\boldsymbol{R_s^*A}] + [R_sA_2] + [R_s^*A_2]. \end{aligned} \quad (4.123)$$

Substituting in the appropriate expressions for the various receptor complexes yields:

$$\frac{[DR_{\text{active}}]}{[R_T]} = \frac{1}{1 + \frac{(1+DK_a)^2(1+AK_e(2+AK_e))+AK_f K_q(1+AK_f)}{DK_b K_q(2+DK_b)(1+AK_f)^2}}. \quad (4.124)$$

To express the fraction of the population of agonist-receptor complexes in the active state as a function of receptor occupancy ($f_2(\text{occ})$), we substitute in the quadratic described in relation to Equations (4.119)–(4.121) for D in Equation (4.124). This function is then divided by receptor occupancy to yield efficacy expressed as a function of receptor occupancy ($f_2(\text{occ})/\text{occ}$). Mean efficacy is calculated numerically by integrating the latter function over the domain of occupancy from 0–1 as described in Chapter 2.

Analysis of allosteric modulation of observed affinity and mean efficacy

The population parameters, observed affinity (K_{occ}) and mean efficacy (ε_{mean}), are determined by the more fundamental microscopic constants that define the ligand-receptor interactions. When the receptor has two cooperatively linked orthosteric sites, there does not appear to be a simple way to estimate K_{occ} and ε_{mean} from the empirical parameters of the current response of the receptor population. Nonetheless, K_{occ} and ε_{mean} can be calculated from the estimated microscopic constants. Recall that it is always possible to estimate K_f, K_e, K_a and the compound parameter, P_{b2} ($K_b^2K_q/(1+K_q)$), whenever data like those shown in the prior examples are analyzed. Knowing P_{b2}, one can assign an arbitrarily low value to K_q (e.g., 10^{-8}) and estimate a provisional value of K_b from P_{b2} (i.e., K_b = square root of $P_{b2}(1+K_q)/K_q$). These provisional values of K_b and K_q, together with the other microscopic constants, can then be used to calculate K_{occ} and ε_{mean} using the method described in the prior section. The estimates of K_{occ} and ε_{mean} are accurate even though the provisional estimates of K_b and K_q may be different from their true values. Because our goal in this section is to illustrate how allosteric modulators affect affinity and efficacy, I will use the theoretical parameters of some of the prior simulations to calculate observed affinity and mean efficacy.

Let us begin by reexamining the theoretical data in Figure 4.11. These plots show the whole cell current response (or open probability plot) of orthosteric agonists in the presence of various concentrations of a positive allosteric modulator having 10-fold higher affinity for the open state of the channel ($K_f = 10^7\ \text{M}^{-1}$) relative to the closed state ($K_e = 10^6\ \text{M}^{-1}$). Panel (a) shows the effects of a modulator on the current responses of a highly efficacious agonist ($K_b/K_a = 10^4$). Increasing concentrations of the modulator shift the concentration-response curve to the left without having much of an effect on the maximum current response. Figure 4.23a shows a summary of these effects on observed affinity ($1/K_{occ}$) and mean efficacy (ε_{mean}). The log changes in observed affinity (log ΔK_{occ}) and mean efficacy (log $\Delta\varepsilon_{mean}$) were calculated using the following equations:

$$\log \Delta K_{obs} = \log \frac{K'_{occ}}{K_{occ}}, \tag{4.125}$$

$$\log \Delta\varepsilon_{mean} = \log \frac{\varepsilon_{mean}}{\varepsilon'_{mean}}. \tag{4.126}$$

in which the parameters denoted with an apostrophe represent those estimated for the condition of no modulator. The parameter K_{occ} and ε_{mean} were calculated using Equations (4.105) and (4.115), respectively. Figure 4.23a shows that the modulator

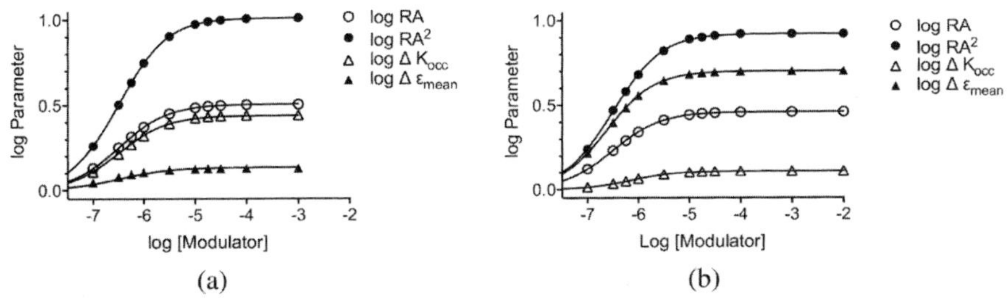

Figure 4.23: Summary of the effects of a positive allosteric modulator ($K_f/K_e = 10$) on the observed affinity ($1/K_{\text{occ}}$) and mean efficacy a highly efficacious agonist (a) and a less efficacious agonist (b) at a ligand-gated ion channel having one allosteric and two orthosteric sites. The plots show an analysis of the simulated data in Figure 4.11.

causes a substantial increase in observed affinity and a much smaller increase in mean efficacy. A useful measure of the combined effect of the modulator on observed affinity and mean efficacy is known as the RA value (relative activity). This parameter is a relative estimate of the product of the changes in observed affinity and the nth root of mean efficacy, where n denotes the number of orthosteric sites per receptor:

$$\log RA = \log\left(\Delta K_{\text{obs}}(\Delta\varepsilon_{\text{mean}})^{\frac{1}{n}}\right). \tag{4.127}$$

The plot in panel (a) shows that the modulator causes a concentration-dependence increase in log RA with the maximal effect being about 0.5 log units (3.3-fold). The plot also shows how the modulator affects RA^2. The observed affinity of the modulator in the absence of orthosteric ligand and its ratio of affinity constants for the open and closed state can be estimated by nonlinear regression analysis of the RA values according to the following equation:

$$\log RA = \log \frac{1 + A(RA_{\max} K_A)}{1 + AK_A}. \tag{4.128}$$

Regression analysis yields estimates of observed affinity ($\log K_A$) of the allosteric ligand in the absence of modulator of (5.99) and the maximal change in log RA (log $RA_{\max}$, 0.51). From Equation (4.46), it follows that K_A is approximately equal to K_e whenever $K_f/K_e \ll 1/K_q$, which is the case for an allosteric modulators. It can be shown that when there is one allosteric site for every two orthosteric sites, the relationship between $RA_{\max}$ and K_f/K_e is:

$$RA_{\max} \quad \sqrt{\frac{K_f}{K_e}}. \tag{4.129}$$

Thus, squaring the estimate of $RA_{\max}$ yields an estimate of K_f/K_e when there is one allosteric site per two orthosteric sites. This value is nearly the same as that used in the simulation ($K_f/K_e = 10$). It follows that regression analysis of the log RA^2 values according to the following equation:

$$\log RA^2 = \log\left(\frac{1 + A(RA_{\max})^2 K_A}{1 + AK_A}\right), \qquad (4.130)$$

should yield estimates of the observed affinity of the allosteric modulator (K_A) and $RA^2_{\max}$ (K_f/K_e). When this was done the following parameter estimates were obtained: log K_A, (5.99) and log RA^2 (1.01). The latter is a close estimate of the value of log K_f/K_e used in the simulation (1.0).

Because the modulator in this case has only modest selectivity for the open state of the channel relative to the value of K_q, then we can assume that K_A is approximately equal to K_e as just mentioned. Knowing K_e and K_f/K_e, it is possible to estimate log K_f (7). Thus, the foregoing analysis shows that it is possible to estimate K_e and K_f by analyzing the effects of the allosteric modulator on observed affinity and mean efficacy. Of course, this was done to illustrate how the allosteric modulation affects the observed affinity and mean efficacy of the orthosteric ligand. Obviously, it is much easier to estimate K_f and K_e from the analysis described earlier in this chapter.

Figure 4.23b shows the effect of the same allosteric modulator on the observed affinity and mean efficacy of a less efficacious agonist ($K_b/K_a = 10^3$). In this case, the effect of the modulator is due primarily to an increase in mean efficacy with only a small increase in observed affinity. The combined effects on affinity an efficacy are illustrated by the RA values. Regression analysis of the RA values according to Equation (4.128) yields an estimate of log K_A (5.90) and log $RA_{\max}$ (0.46). Squaring $RA_{\max}$ yields an approximate estimate of K_f/K_e (log $K_f/K_e = 0.92$). Hence, the estimates of log K_f (i.e., log K_A + log $RA^2_{\max}$, 6.82) and log K_e (i.e., log K_A, 5.90) determined from the interaction with the less efficacious agonist are almost the same as those determined from the interaction with the more efficacious agonist despite the large differences in affinity and efficacy modulation.

The same analysis also applies to the case of two allosteric sites, although a minor modification is required. The data in Figure 4.13 represent a suitable example for this analysis. A summary of the effects of a positive (panel a) and negative (panel b) modulator on the observed affinity and mean efficacy of an efficacious agonist for eliciting whole cell current responses are shown in Figure 4.24. The positive modulator mainly causes a concentration-dependent increase in observed affinity with little effect on mean efficacy. The maximal increase in RA is approximately 10. Regression

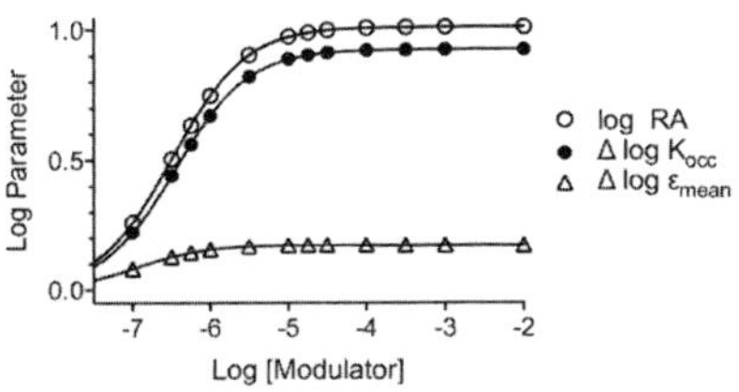

(a)

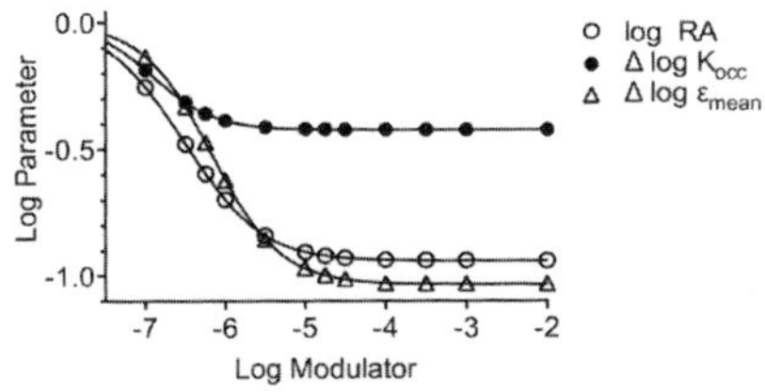

(b)

Figure 4.24: Summary of the effects of positive (a) and negative (b) allosteric modulation on the observed affinity ($1/K_{\mathrm{occ}}$) and mean efficacy of a highly efficacious agonist (K_b, 10^8 M^{-1} and K_a, 10^4 M^{-1}) at a ligand gated ion channel having two allosteric and two orthosteric sites. The microscopic constants of the allosteric ligand were: K_f, 10^7 and K_e, 10^6 M^{-1} (panel (a)) and K_f, 10^6 and K_e, 10^7 M^{-1} (panel (b)). The plots show an analysis of the simulated data in Figure 4.13.

analysis of the log RA values according to Equation (4.128) yielded estimates of log K_A (6.02) and RA_{max} (1.01). In this case, where the ratio of allosteric to orthosteric sites (2:2) is equal to 1, the value of RA_{max} is approximately equal to K_f/K_e. The estimates of log K_A and log RA_{max}, are approximately equivalent to the log K_e and log K_f/K_e values used in the simulation (6 and 1, respectively). The corresponding estimate of log K_f is 7.03, which is almost the same as that used in the simulation.

An analysis of the effects of the negative modulator is shown in Figure 4.24b. In this case, the modulator causes a concentration-dependent decrease in mean efficacy and a smaller reduction in observed affinity. Regression analysis of the log RA values using Equation (4.128) yields estimates of log K_A (6.8) and log RA_{max} (−0.93). These values are approximately equal to the values of log K_e (7) and log K_f/K_e (−1.0) used in the simulation. The corresponding estimate of log K_f is 5.87.

One final point is that when the ratio of allosteric sites to orthosterics site is equal to one, occupancy of the allosteric site by the modulator is equivalent to the normalized RA value, (RA_{norm}) which is defined as:

$$RA_{\mathrm{norm}} = \frac{RA - 1}{RA_{\mathrm{max}} - 1}. \tag{4.131}$$

Because the RA value in the absence of modulator is equal to one, this equation defines RA_{norm} as the change in RA caused by the modulator divided by the maximal change in RA. Figure 4.25 shows a plot of the normalized RA values from Figure 4.24 for the positive and negative modulator. Regression analysis of the normalized RA values according to a simple one-site model,

$$RA_{\mathrm{norm}} = \frac{A}{A + \frac{1}{K_A}}, \tag{4.132}$$

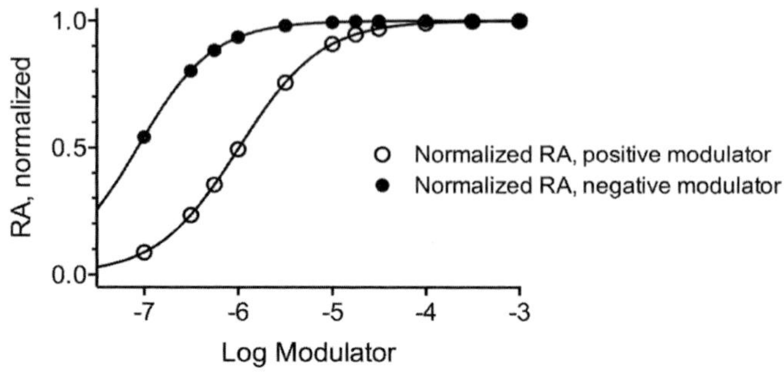

Figure 4.25: Plot of the normalized RA_i values for the positive and negative allosteric modulators shown in Figure 4.24. The RA_{norm} values were calculated from the RA_i values using equation (4.131).

yields estimates of the log K_A values for the positive (6) and negative (7.01) modulators. These are approximately equivalent to their corresponding log K_e values (6 and 7, respectively. In this case, where the ratio of $K_f/K_e \ll 1/K_q$, the Hill slope of the predicted occupancy cure is one.

There is an obvious parallel between the analysis of a series of different agonist concentration-response curves for eliciting whole-cell currents and the effect of an allosteric modulator on the current response of an agonist. In the latter case, the modulator causes changes in the observed affinity and efficacy of the agonist, which can be quantified by the RA value. With regard to a series of agonists, these can be quantified by their corresponding RA_i values as described in Chapter 2. Recall that the RA_i value represents the product of the observed affinity and the square root of mean efficacy of one agonist expressed relative to that of another. In the case of a series of agonists, the RA_i value is a relative measure of K_b, whereas in the case of modulation, the $RA_{\max}$ is a measure of K_f/K_e.

Relation to Endogenous Signaling and Protein-Protein Interactions

Because the effect of an allosteric modulator on a given ligand-gated ion channel depends on the agonist with which it is interacting, measuring the allosteric modulation of the endogenous ligand on the current response of a ligand-gated ion channel of interest is the most unequivocal way to determine how the modulator affects receptor function *in vivo*. A positive modulator primarily increases the observed affinity of a highly efficacious endogenous ligand. When the neurotransmitter is released, its peak concentration in the synapse is probably sufficient to saturate the receptor in the absence of modulation. Nonetheless, positive modulation would increase the mean open time and the length of channel bursts causing a faster rise and slower decay of population currents at the postsynaptic membrane. A more obvious role for affinity modulation is in extrasynaptic transmission where the degree of receptor activation is expected to be highly dependent on the concentration of neurotransmitter. Similarly, paracrine signaling and the release of neurotransmitters, like GABA, from microglia are prime examples of receptor signaling that is expected to be affected by allosteric affinity modulation.[23–25]

So far, we have considered allosteric modulation from the perspective of how a ligand affects receptor activation or how an allosteric modulator affects the ability of another ligand, like the endogenous ligand, to affect receptor activation. The principle of allosterism also applies to protein-protein interactions. For example, the binding of a neurotransmitter to a *G* protein-coupled receptor affects how the *G* protein interacts with the receptor. In addition, when bound to the α subunit of the *G* protein, GTP and GDP have opposite effects on how the $\beta\gamma$ subunits bind to the α subunits. These interactions ultimately determine whether signaling is turned on by GTP or turned off by GDP. Thus, the principles of allosterism that we have developed for ligand-gated ion channels will go a long way in explaining the multitude of protein–protein interactions that occur in signaling pathways elicited by *G* protein-linked receptors. These are explained in Part II of this book.

References

1. Schwartz, T. W. and Holst, B. Allosteric enhancers, allosteric agonists and ago-allosteric modulators: where do they bind and how do they act? *Trends in Pharmacol Sci* **28**, 366–373 (2007).
2. Ehlert, F. J., Roeske, W. R., Gee, K. W. and Yamamura, H. I. An allosteric model for benzodiazepine receptor function. *Biochem Pharmacol* **32**, 2375–2383 (1983).
3. Burgen, A. S. Conformational changes and drug action. *Federation Proceedings* **40**, 2723–2728 (1981).
4. Laudet, V. Evolution of the nuclear receptor superfamily: early diversification from an ancestral orphan receptor. *J Mol Endocrinol* **19**, 207–226 (1997).
5. Filipek, S., Stenkamp, R. E., Teller, D. C. and Palczewski, K. *G* protein-coupled receptor rhodopsin: a prospectus. *Annu Rev Physiol* **65**, 851–879 (2003).
6. Stryer, L. Cyclic GMP cascade of vision. *Annu Rev Neurosci* **9**, 87–119 (1986).
7. Sternini, C. Taste receptors in the gastrointestinal tract. IV. Functional implications of bitter taste receptors in gastrointestinal chemosensing. *Am J Physiol Gastrointest Liver Physiol* **292**, G457–G461 (2007).
8. Wu, S. V., Rozengurt, N., Yang, M., Young, S. H., Sinnett-Smith, J. and Rozengurt, E. Expression of bitter taste receptors of the T2R family in the gastrointestinal tract and enteroendocrine STC-1 cells. *Proc Natl Acad Sci USA* **99**, 2392–2397 (2002).

9. Fredriksson, R., Lagerstrom, M. C., Lundin, L. G. and Schioth, H. B. The *G*-protein-coupled receptors in the human genome form five main families. Phylogenetic analysis, paralogon groups, and fingerprints. *Mol Pharmacol* **63**, 1256–1272 (2003).
10. Rozengurt, E. Taste receptors in the gastrointestinal tract. I. Bitter taste receptors and alpha-gustducin in the mammalian gut. *Am J Physiol Gastrointest Liver Physiol* **291**, G171–G177 (2006).
11. Hao, S., Sternini, C. and Raybould, H. E. Role of CCK1 and Y2 receptors in activation of hindbrain neurons induced by intragastric administration of bitter taste receptor ligands. *Am J Physiol Regul Integr Comp Physiol* **294**, R33–R38 (2008).
12. McBee, J. K., Palczewski, K., Baehr, W. and Pepperberg, D. R. Confronting complexity: the interlink of phototransduction and retinoid metabolism in the vertebrate retina. *Prog Retin Eye Res* **20**, 469–529 (2001).
13. Palczewski, K. Chemistry and biology of vision. *J Biol Chem* **287**, 1612–1619 (2012).
14. Neher, E. and Steinbach, J. H. Local anaesthetics transiently block currents through single acetylcholine-receptor channels. *J Physiol* **277**, 153–176 (1978).
15. Leipold, E., Borges, A. and Heinemann, S. H. Scorpion beta-toxin interference with NaV channel voltage sensor gives rise to excitatory and depressant modes. *J Gen Physiol* **139**, 305–319 (2012).
16. Jeong, I., Choi, B. H. and Hahn, S. J. Rosiglitazone inhibits Kv4.3 potassium channels by open-channel block and acceleration of closed-state inactivation. *Br J Pharmacol* **163**, 510–520 (2011).
17. Randall, L. O. Chemistry and pharmacology of the benzodiazepines. *Nord Psykiatr Tidsskr* **25**, 9–11 (1971).
18. Olsen, R. W. and Li, G. D. GABA(A) receptors as molecular targets of general anesthetics: identification of binding sites provides clues to allosteric modulation. *Can J Anaes; J Can d'anesthesie* **58**, 206–215 (2011).
19. Baur, R., Gertsch, J. and Sigel, E. The cannabinoid CB1 receptor antagonists rimonabant (SR141716) and AM251 directly potentiate GABA(A) receptors. *Br J Pharmacol* **165**, 2479–2484 (2012).
20. Weber, G. Energetics of ligand binding to proteins. *Adv Protein Chem* **29**, 1–83 (1975).
21. Downing, S. S., Lee, Y. T., Farb, D. H. and Gibbs, T. T. Benzodiazepine modulation of partial agonist efficacy and spontaneously active GABA(A) receptors supports an allosteric model of modulation. *Br J Pharmacol* **145**, 894–906 (2005).
22. Rusch, D., Zhong, H. and Forman, S. A. Gating allosterism at a single class of etomidate sites on alpha1beta2gamma2L GABA A receptors accounts for both direct activation and agonist modulation. *J Biol Chem* **279**, 20982–20992 (2004).
23. Lee, M., McGeer, E. G. and McGeer, P. L. Mechanisms of GABA release from human astrocytes. *Glia* **59**, 1600–1611 (2011).
24. Velez-Fort, M., Audinat, E. and Angulo, M. C. Central role of GABA in neuron-glia interactions. *Neuroscientist* **18**, 237–250 (2012).
25. Lee, M., Schwab, C. and McGeer, P. L. Astrocytes are GABAergic cells that modulate microglial activity. *Glia* **59**, 152–165 (2011).

26. Sigel, E. and Baur, R. Allosteric modulation by benzodiazepine receptor ligands of the GABAA receptor channel expressed in Xenopus oocytes. *J Neurosci* **8**, 289–295 (1988).
27. Maksay, G., Thompson, S. A. and Wafford, K. A. Allosteric modulators affect the efficacy of partial agonists for recombinant GABA(A) receptors. *Br J Pharmacol* **129**, 1794–1800 (2000).

PART 2

G Protein-Coupled Receptors

G Protein-Coupled Receptors and *G* Protein and Arrestin Signaling

5

In his great book, *Origin of Species*, Darwin acknowledged that the evolution of an exquisite structure like the eye might seem too implausible so he addressed the issue in a section entitled, "Organs of extreme perfection and complication". Apparently it is not that implausible because contemporary biologists deduce that eyes must have evolved a minimum of 40 times to explain their distribution throughout the animal kingdom.[1] These organs differ vastly in their ingenious solution to the problem of catching photons and converting them into an electrical signal.[2,3] But at the heart of the eye in many diverse species is the remarkable light-detecting protein, rhodopsin. It is present in the eyes of vertebrates, arthropods and mollusks,[2–4] as well as in other light-detecting structures, like melanocytes,[5] pineal glands[6] and the eye spots of algae.[7] Rhodopsin triggers a different signaling pathway in vertebrates than in arthropods and mollusks,[3] but the first activation step and second transduction mechanism follow the same pattern in these groups. This general mechanism also accounts for signal transduction in a large class of proteins known as *G* protein-coupled receptors. In this chapter, I describe the elements of receptor-*G* protein signaling beginning with phototransduction in the mammalian retina.

Light Detection by Rhodopsin

The mammalian retina contains a two-dimensional array of about 125 million densely packed photocells lining the inner surface of the posterior chamber of each eye.[8,9] These cells are of two types: rods and cones. Rods express rhodopsin and are more sensitive to light, whereas cones express three related opsins, each maximally excited at a different wavelength of light to enable trichromatic vision. Much

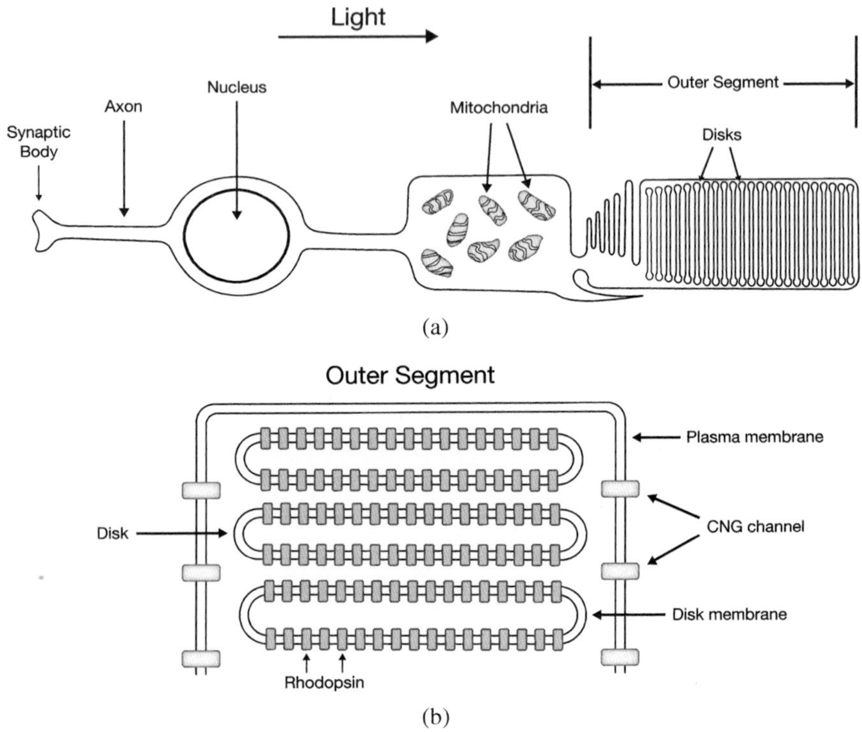

Figure 5.1: Structure of a photocell in the mammalian retina. (a) Longitudinal view of a rod cell. From this perspective, an incoming ray of light originates from the left as indicated in the figure. (b) An enlargement of the rod cell showing the disk membrane, the plasma membrane and the location of rhodopsin and CNG (cyclic nucleotide-gated) channels.

research has been done on rods because of their high sensitivity and great abundance and accessibility in the bovine retina. Figure 5.1 shows a rod cell positioned so that its long axis is normal to the surface of the retina. At the end of the cell farther away from the source of light is a stack of about 2000 disks enclosed within the plasma membrane. This section forms the rod outer segment. Each disk develops as an invagination in the cell membrane, and ultimately forms a flattened reservoir (disk) bounded by membrane separate from the plasma membrane.[10,11] Rhodopsin is expressed in high density on the two membrane surfaces of each disk. An incoming photon, therefore, has approximately 4000 chances to collide with a rhodopsin molecule. If the eye has a reflecting layer or tapetum behind the retina as in the eyes of many nocturnal mammals, then an incoming photon can be reflected back and have another 4000 chances to hit rhodopsin if it is unsuccessful on its first inward attempt. This anatomy partially explains why the threshold for light detection in humans only amounts to 5–14 photons.[12] Because this number is so small, each photon presumably interacts with a different photocell under the threshold condition. Thus, the single-photon response of only a few rods is sufficient for humans to detect light.

Rhodopsin is a conjugate consisting of the protein, opsin and the light-absorbing molecule, retinal.[13,14] Retinal is covalently attached to a specific lysine residue in opsin to form a protonated Schiff base. The chromophore or light-absorbing moiety of retinal consists of a series of conjugated double-bond carbon atoms. When rhodopsin is in the inactive state, its prosthetic retinal (11-*cis*-retinal) has all but one of its conjugated double bond carbons in the *trans* form. Activation by light causes 11-*cis*-retinal to isomerize into all-*trans*-retinal (Figure 5.2). This structural change induces rhodopsin to undergo a short series of conformational changes into its active state (metarhodopsin II). When activated, rhodopsin interacts with a nearby protein, transducin, which is a member of a large group of proteins known as the *G* protein family.[9] Transducin is expressed in the retina and is composed of three subunits, $G\alpha_t$, $G\beta$ and $G\gamma$. The $G\alpha_t$ subunit binds GTP and GDP, and the $G\beta$ and $G\gamma$ subunits bind tightly to each other to form a functional $G\beta\gamma$ dimer. In its inactive state, transducin is tightly bound with GDP in the form of a GDP–$G\alpha_t\beta\gamma$ aggregate or holoprotein. It is only in this trimeric state that transducin can interact with rhodopsin.

Why transducin is saturated with GDP in the resting state is not obvious. The cell, after all, expends energy converting GDP into GTP such that GTP accounts for the major fraction of the total pool of guanine nucleotides in the cytosol. One

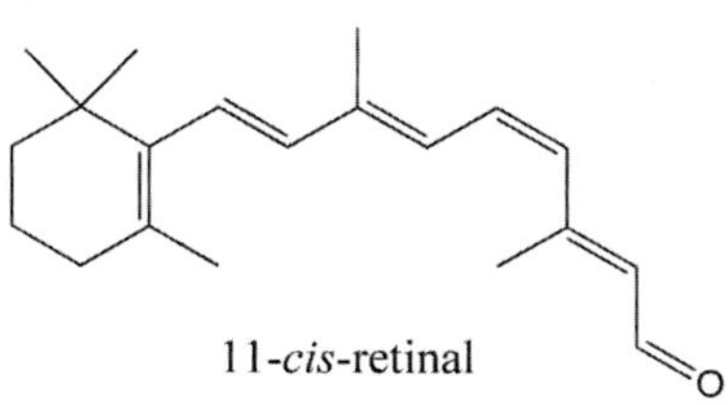

11-*cis*-retinal

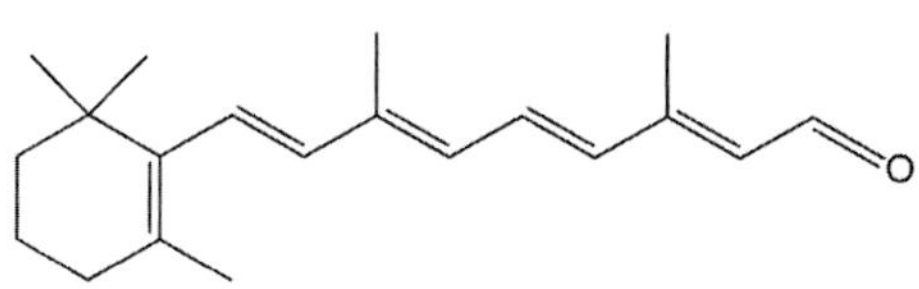

all *trans*-retinal

Figure 5.2: Structure of 11-*cis*-retinal and all *trans*-retinal.

might expect, therefore, that most G proteins would have GTP bound to them in the resting state, particularly because its concentration in most cells is approximately 30 μM, and the dissociation constant of the GTP-G protein complex is in the range of 1 nM. Two factors push the G protein complex into a steady state dominated by the GDP-bound form. First, any GTP bound to the α_t subunit is rapidly hydrolyzed to GDP because of the intrinsic GTPase activity of the $G\alpha_t$ subunit, which is enhanced by a retinal RGS (regulators of G protein signaling) protein.[15] Second, the resulting GDP-transducin complex is nearly irreversible. Thus, the rapid hydrolysis of GTP and the negligible rate of dissociation of GDP establish a steady state in which G proteins are saturated with GDP. The hydrolysis of GTP provides the energy for the maintenance of this steady state.

When activated by light, rhodopsin interacts with the GDP-transducin complex, causing it to undergo a conformational change so that it loosens its grip on GDP.[16] This enables GTP to displace GDP. When GTP binds to transducin, it causes the protein to split apart into its $G\beta\gamma$ and GTP bound $G\alpha_t$ subunits (GTP–$G\alpha_t$). A single photo-excited rhodopsin receptor activates about 500 transducin proteins per 0.5 s. This process represents the activation of transducin (Figure 5.3).

In the dark state, the rod cell is depolarized because the concentration of cGMP in the cytosol is high and cGMP binds to a cGMP-gated channel in the plasma membrane causing it to open and conduct a net inward current of cations (Na^+ and Ca^{2+}).[9] Ca^{2+} levels remain low (0.5 μM) in the rod cell, however, because of an effective Ca^{2+} pump driven by Na^+ influx and K^+ efflux. When transducin is activated in the light state, GTP–$G\alpha_t$ subunits accumulate and bind to and activate a cyclic GMP phosphodiesterase, causing this enzyme to hydrolyze cGMP into GMP (Figure 5.3). A single photo-excited rhodopsin causes the hydrolysis of about 10^6 cGMP molecules. This action reduces the concentration of cGMP in the cell, resulting in a closing of the cGMP-gated cation channels, a decrease in Na^+ and Ca^{2+} influx, and a hyperpolarization of the rod cell. The hyperpolarization is graded, with the magnitude depending upon the intensity of the light stimulus. A bright flash of light can completely suppress the dark current, whereas a dim flash may halve it. The hyperpolarization causes a decrease in the release of glutamate from the rod cell at its synapse with bipolar and horizontal cells. This information is processed by the neuronal network of the retina before being relayed to the visual system of the brain through the optic nerves.

The turn-off mechanism for the rod cell is complicated because all-*trans*-retinal does not isomerize back into 11-*cis*-retinal in the dark, and hence, rhodopsin remains in the active conformation after a brief light stimulus.[17] Within 100 ms or so of the formation of all-*trans* retinal, however, an enzyme that recognizes the active state of

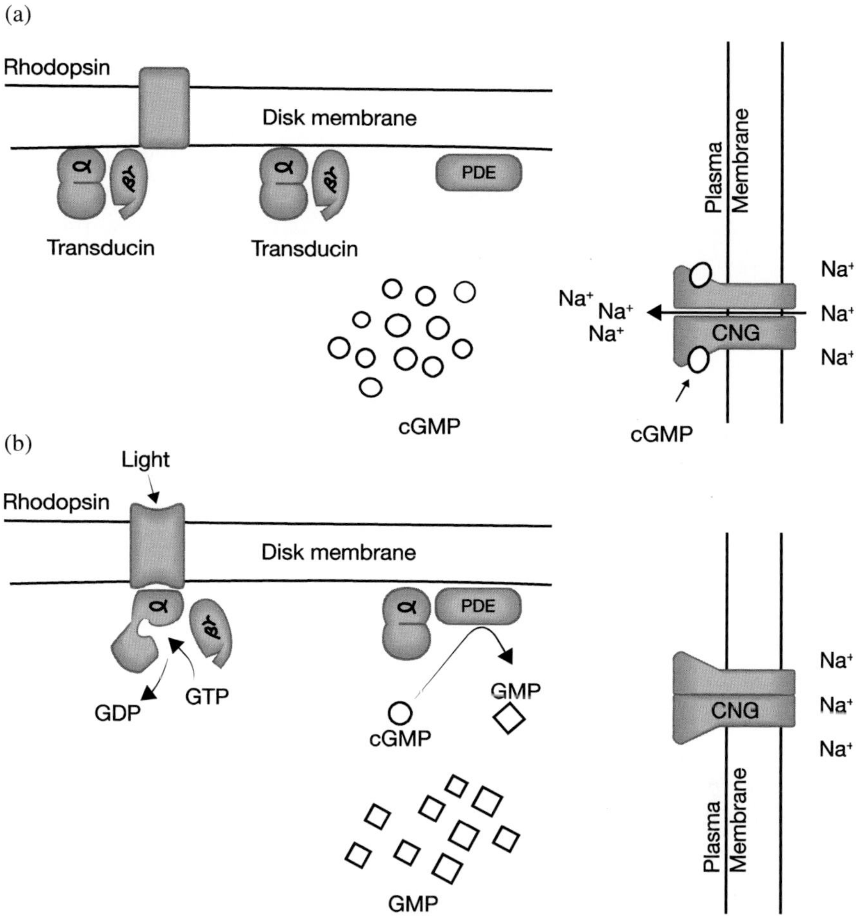

Figure 5.3: Light-induced rhodopsin signaling in a rod cell. (a) Resting state in the dark. (b) Light-induced rhodopsin signaling. Abbreviations: PDE, retinal phosphodiesterase; cGMP, cyclic GMP; and CNG, cyclic nucleotide-gated channel.

rhodopsin (rhodopsin kinase) begins to catalyze the transfer of the terminal phosphate of ATP onto specific serine and threonine residues in the C-terminal region of rhodopsin. These phosphorylated residues act as docking sites for another protein, visual arrestin (arrestin 1), which quickly binds to phosphorylated rhodopsin. Both the initial phosphorylation and the subsequent binding of arrestin 1 prevent the interaction of rhodopsin with transducin, and hence, the rate of formation of new GTP–$G\alpha_t$ subunits nearly stops. All existing GTP–$G\alpha_t$ subunits rapidly decay into GDP-$G\alpha_t$ because of the GTPase activity of the $G\alpha_t$ subunit and its enhancement by the retinal RGS protein.[15] The $G\beta\gamma$ subunits then coalesce with the GDP–$G\alpha_t$ subunit to form the inactive holoprotein, GDP-transducin.

With the halt in the formation of GTP-$G\alpha_t$ subunits, the activity of the cGMP phosphodiesterase stops as well, and cGMP levels rise due to the conversion of GTP into cGMP by the enzyme guanylate cyclase. The prior light-induced lowering in cytosolic Ca^{2+} increases the activity of guanylate cyclase through a mechanism involving the Ca^{2+} binding protein, recoverin.[18] As described above, the Ca^{2+} concentration falls during light activation because of the closing of cGMP-gated channels and the net efflux of Ca^{2+} through the Ca^{2+} pump. With a fall in its concentration, Ca^{2+} dissociates from recoverin, and free recoverin binds with high affinity to the regulatory subunits of guanylate cyclase causing a marked increase in cyclase activity. This leads to a rapid restoration in cGMP levels and a depolarization of the rod cell through the binding of cGMP to the cGMP-gated, cation channel.

The active metarhodopsin II conformation relaxes into the inactive state, metarhodopsin III, which lacks affinity for arrestin 1 regardless of whether it is still phosphorylated. The phosphorylated serine and threonine residues in the carboxyl terminus of rhodopsin are cleaved by phosphatases, and the Schiff base linkage of all-*trans*-retinal with opsin hydrolyzes in minutes. All-*trans*-retinal is converted back to 11-*cis*-retinal through a series of enzymatic reactions in nearby retinal epithelial cells. From there, 11-*cis*-retinal is transported back to rods where it is conjugated with opsin. The entire recovery process for rhodopsin takes tens of minutes in rods.[13,19]

By now you may be wondering how it is possible to see continuously because rhodopsin kinase rapidly blocks the interaction of light-activated rhodopsin with transducin, and each activated rhodopsin is replenished with new 11-*cis*-retinal only after minutes. In other words, once a rhodopsin receptor generates a light signal for about a half a second, it cannot signal again until much later, so how is it possible to see longer than a second? The answer lies in the massive redundancy of the rod outer segment with its many rhodopsin-studded disks. During continuous illumination at moderate intensity, only a tiny fraction of the total number of rhodopsin receptors is activated in a single rod cell at any given instant of time. As time progresses, more of the large depot of rhodopsin receptors is used, but the rate is low relative to the total amount of rhodopsin. This enables the restoration of rhodopsin to keep pace with photobleaching. To work efficiently, it would seem ideal for the signal elicited by each rhodopsin receptor to be highly amplified so that the rate of utilization of rhodopsin receptors can be minimal. This would reduce the total amount of rhodopsin required for signaling during constant illumination. Indeed, the rhodopsin signaling pathway does exhibit marked amplification as indicated by the ability of a rod cell to register a response to a single photon. This remarkable sensitivity is often interpreted as an adaptation to low light intensity. But high sensitivity is also useful for a visual transduction mechanism based on the irreversible photoactivation of a chromophore that can only be restored on a time-scale of minutes.[a] Thus, the high sensitivity of rod cells may be, in part, an adaptation for continuous vision under normal illumination, rather than only an adaptation to dim light.

G Protein-Coupled Receptors

Although rhodopsin is unique in its ability to respond to light, it belongs to a superfamily of membrane-bound proteins that signal through *G* proteins and have seven hydrophobic stretches of 25 to 35 consecutive amino acids, each forming a transmembrane α helix (TM).[20,21] Other members of this superfamily detect taste or odorant molecules or mediate responses to endogenous neurotransmitters, eicosanoids, neuropeptides and peptide hormones.

Their common signaling mechanism is shown in Figure 5.4. The process usually begins with the binding of an agonist ligand to the receptor, which induces the receptor to assume its active state. The active state exhibits high affinity for specific *G* proteins that are either free of guanine nucleotide or bound with GDP.

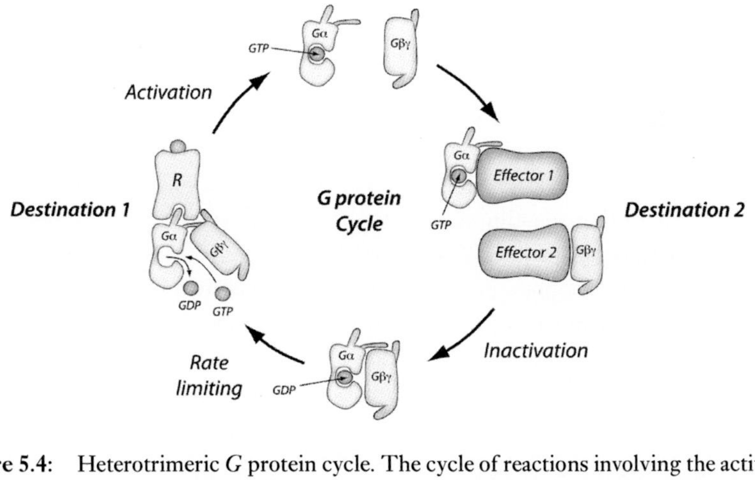

Figure 5.4: Heterotrimeric *G* protein cycle. The cycle of reactions involving the activation and inactivation of heterotrimeric *G* proteins is shown. The site of receptor-mediated guanine nucleotide exchange and effector activation are arbitrarily named as Destinations 1 and 2, respectively.

---

[a] Cones recover from photobleaching much more quickly.

The resulting interaction causes a *G* protein bound with GDP to open its guanine–nucleotide-binding pocket and enable the displacement of GDP with GTP. Once GTP is bound to the *G* protein, it splits the *G* protein into its GTP–$G\alpha$ and $G\beta\gamma$ subunits, which diffuse away from the receptor because of their low affinity for it. They bind with high affinity to various proteins known generically as effectors and regulate their activity. These include ion channels, enzymes that synthesize second messengers and numerous other signaling proteins. The activation by a single GTP–$G\alpha$ subunit is transient because of its GTPase activity, which converts it back into the GDP–$G\alpha$ form. This species exhibits high affinity for the $G\beta\gamma$ subunits so the two coalesce into GDP–$G\alpha\beta\gamma$ holoprotein. If the receptor is still bound with agonist, the GDP-bound form of the *G* protein can undergo another round of activation. Otherwise the cycle stops when the agonist is removed from the microenvironment of the receptor. This causes receptor-bound agonist to dissociate from the receptor and enables the receptor to relax back into its inactive state. In isolation, many *G* proteins have low intrinsic GTPase activity, but this is enhanced, in some instances, by the specific effector or very often by a member of a class of RGS (regulators of *G* protein signaling) proteins.

Like rhodopsin, *G* protein-coupled receptors elicit a rapid activation of *G* proteins, but unlike rhodopsin, this is inhibited on a much slower time-scale by receptor phosphorylation and the subsequent binding of arrestin.[22] There is usually no need for a rapid, receptor-kinase inactivation mechanism because endogenous ligands typically bind reversibly, and the activation state of the receptor is regulated by the controlled release of neurotransmitter or chemical messenger. That is, activation coincides with neurotransmitter release, and inactivation occurs when release stops. Recruitment of arrestin does occur, nonetheless, on a much longer time-scale after continuous receptor activation. This mechanism causes homologous desensitization of receptor-*G* protein signaling. The recruitment of arrestin not only leads to an uncoupling with heterotrimeric *G* proteins, but also couples the receptor with proteins involved in endocytosis of the receptor within clathrin-coated pits. For some receptors, this mechanism enables the activation of mitogen-activated protein kinase (MAPK) cascade.[23] Thus, not all of the actions of *G* protein coupled-receptors are mediated by *G* proteins, and not all members of the superfamily have actually been shown to interact with *G* proteins. Because of this ambiguity some investigators have suggested that heptahelical or seven-transmembrane is a more appropriate name for a *G* protein-coupled receptor.

What defines a *G* protein-coupled receptor is its predicted structure, which consists of seven TM domains exhibiting varying degrees of homology, if only remote, and usually a putative disulfide bridge linking cysteines located on extracellular loops (EL): EL1 and EL2. In humans, these proteins have been divided into five major families — rhodopsin (class A), secretin (class B), adhesion, frizzled/taste2 and glutamate (class C) — some of which correspond with the classification scheme of Attwood and Findlay[24] and Kolakwoski[25] as indicated in parentheses. A brief summary of these families follows.

Rhodopsin Family

The rhodopsin family is the largest group of *G* protein-coupled receptors (about 701).[20] The majority (about 460) represents a homologous group of olfactory receptors that are expressed in the nose and presumably detect volatile compounds. The remainder (241) includes receptors for amine neurotransmitters, eicosanoids, neuropeptides and some large peptide hormones. Some members have an additional amphipathic helix in the carboxyl terminus that runs parallel to the inner surface of the membrane. It may be tethered to the membrane through palmitoylation of a conserved cysteine residue near the C-terminal end of the helix. Almost all members of the rhodopsin family have a set of a few highly conserved residues that probably maintain interactions between TM helixes in the resting state, and their disruption is involved in receptor activation (see Figure 5.5). At the cytosolic end of TM3 is a highly conserved arginine residue flanked by an acidic aspartate or glutamate residue on the amino terminal side and by a tyrosine residue on the other side to form the so-called "D/ERY" motif. In the inactive state of rhodopsin, the acidic glutamate forms an ion pair with the adjacent basic arginine. When rhodopsin is activated the glutamate is protonated, causing a disruption in the ionic bond with arginine. Another highly conserved set amino acids is the $X_1NPX_2X_3Y$ motif at the cytosolic end of TM6. It consists of variable residues (X), the first of which (X_1) is often acidic and the third (X_3) often polar (threonine). In some instances an acidic X_1 and a polar X_3 may form ionic and hydrogen bonds, respectively, with the arginine of the D/ERY motif on TM3, thereby stabilizing the inactive state of the receptor. Members of the rhodopsin-like family often contain a highly conserved asparagine in TM1 and an aspartate in TM2 that are thought to be part of a hydrogen bond network. There is also a highly conserved proline residue in each of TM5 through TM7 that is thought to cause a bend in the TM helix and impart essential structural properties.

The lengths of EL1 and EL3 as well as intracellular loops, (IL) IL1 and IL2 are highly conserved and relatively short. The length of EL2 is also highly conserved, but it is somewhat longer. Near the middle it contains a highly conserved cysteine residue that is thought to form a disulfide bond with another highly conserved

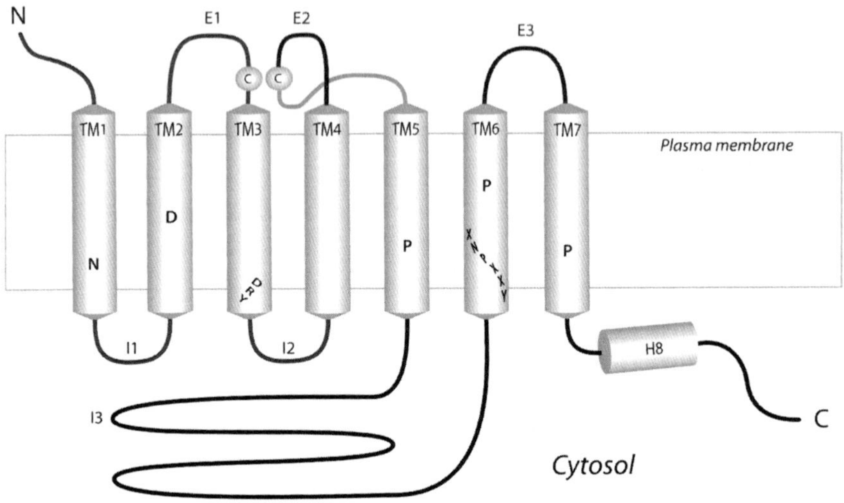

Figure 5.5: Schematic drawing of conserved structural features within the rhodopsin family of *G* protein-coupled receptors. The approximate location of conserved residues is indicated using single letter notation with "X" denoting any amino acid. Abbreviations: N, amino terminus; C, carboxyl terminus; E1–E3, EL1–EL3; I1–3, IL1–IL3; TM1–TM7, transmembrane α helices 1–7; and H8, amphipathic helix 8.

cysteine near the extracellular end of TM3. This disulfide bond essentially converts EL2 into two shorter loops that tether TM3 and TM4 together as well as TM3 and TM5. The result is that TM1 through TM5 are held together by a system of relatively short, intra- and EL. In contrast, the lengths of the N-terminus, IL3 and the C-terminus are more variable.

For those receptors for which the question has been investigated, the location of the endogenous-ligand binding site correlates with the size of the ligand. Monoamines and small peptides bind to their respective receptors at a location analogous to the site where retinal interacts with rhodopsin. For most receptors, this conclusion is based primarily on the results of numerous studies employing the technique of site-directed mutagenesis to identify residues important for agonist binding. In the case of a few receptors, like the β_2-adrenergic[26] and M_2 muscarinic[27] receptors, the crystal structure has been solved with a small orthosteric ligand bound to the receptor at the predicted location within the binding pocket (see Figure 5.6). For monoamine receptors, the binding pocket is encircled by TM3, TM5, TM6 and TM7 at a level within the span of the plasma membrane, but close to the extracellular surface. This location enables the endogenous ligand to interact simultaneously with the different TM domains surrounding the binding pocket and

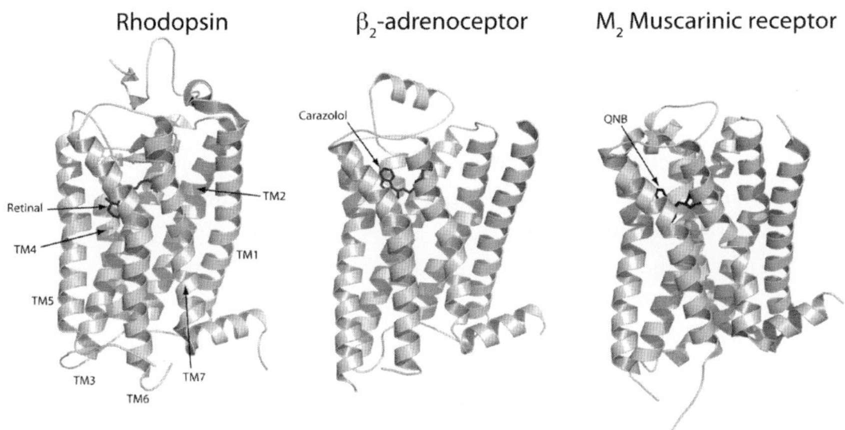

Figure 5.6: Crystal structures of the inactive states of rhodopsin and the β_2—adrenergic and M_2 muscarinic receptors.

cause them to move relative to one another and generate the active conformation. Intermediate and small sized neuropeptides interact with the EL, N-terminus and extracellular ends of the TM domains. Large glycoprotein hormones, like TSH and LH, interact almost exclusively with the N-terminus. It is possible, nonetheless, that part of the hormone interacts with the EL of the receptor to cause activation. In these receptors, the N-terminus is relatively large and contains several cysteine residues that form disulfide bonds with each other to create a globular structure. Protease-activated receptors contain a constitutive ligand as part of the sequence of the N-terminus. The native form of the receptor is inactive, but after part of the N-terminus is cleaved by a protease, like thrombin, the remaining truncated segment binds intramolecularly to the EL of the receptor to cause activation. Inactivation involves the recruitment of arrestin and the subsequent internalization and degradation of the receptor in lysosomes.

Members of the rhodopsin family interact with several *G* proteins, and many receptors have a preference for one of three major subfamilies, G_i, G_s or G_q. In a given cellular context, a productive receptor-*G* protein interaction may depend on whether the contact surfaces between the receptor and *G* protein are complementary or simply whether the two proteins reside near each other. To identify the parts of the receptor that interact with a *G* protein, scientists have altered the amino acid sequence of the receptor and measured the functional consequences. This usually involves mutagenesis, deletion or the construction of a chimera from two related receptors having different functions. For example, a chimera of the M_1 muscarinic receptor with its IL3 swapped with that of the β-adrenoceptor of the turkey erythrocyte turns on adenylate cyclase when activated by the muscarinic agonist, carbachol.[28] Normally, activation of the M_1 receptor elicits phosphoinositide hydrolysis, and not stimulation of adenylate cyclase. These results show that the seven TM domains determine ligand binding and that IL3 has a large influence on *G* protein coupling. Studies such as these on many receptors have shown that the membrane-proximal ends of the IL3, the C-terminus, and parts of the IL2 are important for *G* protein coupling.

One part of the *G* protein that has a large influence on receptor coupling is the short sequence of about seven amino acids at the C-terminus.[29,30] Crystals of rhodopsin have been grown in the presence of the corresponding C-terminal fragment of transducin, and analysis of the presumed active structure shows some striking differences from that obtained with the protein locked in its ground state with 11-*cis*-retinal.[31,32] One difference is that the ends of TM5 and TM6 project farther into the cytoplasm in the active state and interact with the C-terminal fragment of transducin. This can partially explain how the active state engages transducin, and presumably, how other members of the rhodopsin family engage their *G* proteins. More strikingly, the crystal structure of the β_2-adrenoceptor in a complex with G_s shows that the C-terminus of the *G* protein α subunit interacts with TM5 and TM6.[33]

Secretin Receptor Family

The secretin family of *G* protein-coupled receptors shows no homology with the rhodopsin family and consists of about 15 receptors for neuropeptides and hormones, some of which are involved in glucose or Ca^{2+} metabolism (see Figure 5.7). The secretin group is so named because the secretin receptor was the first member to be cloned. Members of this group have a large N-terminus that probably forms a globular structure maintained by disulfide linkages among six conserved cysteine residues. This domain contains the binding site for the endogenous ligand. A conserved cysteine residue is present in both EL1 and EL2, which together are thought to form a disulfide linkage similar to that found in the rhodopsin family. Conserved prolines are present in TM4, TM5 and TM6, but these are not homologous to those of the rhodopsin family, which are located in TM5, TM6 and TM7. Other prototypically conserved residues of the rhodopsin family are lacking in the secretin group. Once activated by their endogenous ligands, the secretin group of receptors elicits responses through G_s.

How endogenous ligands turn on receptors of the secretin group depends on the particular receptor. For the secretin receptor, a three-amino-acid sequence

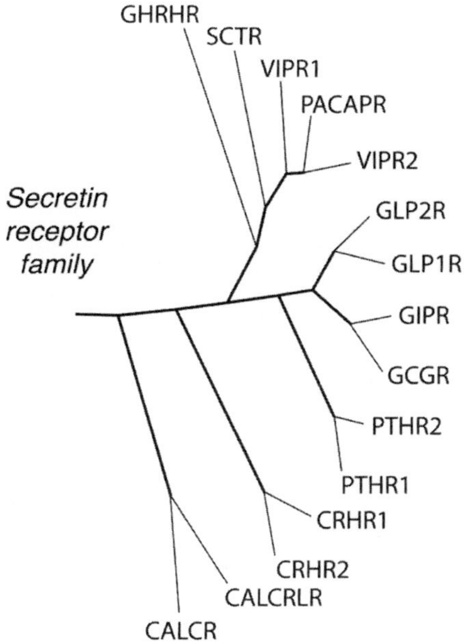

Figure 5.7: Evolutionary relationship among members of the secretin family of *G* protein-coupled receptors. Abbreviations: GHRHR, Growth-hormone-releasing-hormone receptor; SCTR, secretin receptor; VIPR1 and VIPR2, vasoactive intestinal peptide receptors 1 and 2; PACAPR, pituitary adenylate cyclase activating peptide receptor; GLP1R and GLP2R, glucagon-like peptide receptors 1 and 2; GIPR, gastric inhibitory polypeptide receptor; GCGR, glucagon receptor; PTHR1 and PTHR2, parathyroid hormone receptors 1 and 2; CRHR1 and CRHR2, corticotropin-releasing hormone receptors 1 and 2; CALR, calcitonin receptor; and CALRLR, calcitonin receptor-like receptor.

within the N-terminus may act as a tethered agonist. When secretin binds to the N-terminus, it is thought to expose the hidden agonist sequence, enabling it to bind with and activate the seven-transmembrane domain.[34] Isolated peptides containing the tripeptide sequence act as agonists for the secretin receptor. Thus, the secretin receptor resembles protease-activated receptors of the rhodopsin family in the part where its N-terminus acts as the endogenous ligand. This activation mechanism may apply to other members of the secretin group, but not all. One example where it does not is the PTH type 1 receptor, whose endogenous ligand binds to the N-terminus as well as the seven-transmembrane domains. The latter interaction is thought to turn on the receptor, whereas the former increases binding affinity.[35,36] Peptide ligands still activate a PTH type 1 receptor lacking its N-terminus, but their potency for doing so is greatly reduced.

The calcitonin receptor and calcitonin receptor-like receptor (CRLR) dimerize with chaperones known as receptor-activity-modifying proteins (RAMPs).[37,38] RAMPs have an influence on the ligand specificity of these receptors and are required for plasma membrane expression of the CRLR.[39] By itself, the calcitonin receptor is functional and is activated by calcitonin. When it dimerizes with RAMP1, RAMP2 or RAMP3, however, it forms the amilyn 1–3 receptors, respectively. When the CRLR dimerizes with RAMP1, it forms the calcitonin gene-related peptide 1 (CGRP1) receptor, which is selectively activated by CGRP1. When the CRLR dimerizes with RAMP2 or RAMP3, however, it forms the adrenomedullin 1 and 2 receptors.

Adhesion Receptor Family

Members of the adhesion receptor family total to 33 and are unique in having extremely long N-termini of 200–2800 amino acids.[20] The N-terminus often contains a cysteine-box motif adjacent to TM1 and a nearby region rich in serines and threonines (mucin stalk) that are likely to be O-glycosylated. Otherwise, there is great variation in the N-terminus. It is frequently composed of one or more protein modules from the following list that often occur in repeats: cadherin, epidermal growth factor-like, galactose-binding lectin, immunoglobulin, laminin A, olfactomedin and thrombospondin. This modular structure suggests that the N-terminus arose from exon shuffling of adhesion family receptor genes with those of the parental genes for the protein modules. In other contexts, the protein modules of the N-terminus are involved in cell adhesion and migration, suggesting that the adhesion receptor family communicates information through cell–cell interactions and the extracellular matrix. Some members of the adhesion receptor family are expressed in leukocytes and macrophages, and many are expressed in brain and peripheral tissues.

The endogenous ligands that have been identified for these receptors, so far, are consistent with an adhesion mechanism.[40] For example, one member of the family, the CD97 receptor, is expressed on the surface of leukocytes, particularly during complement activation, and it has been shown to interact specifically with CD55. The latter protein is attached to the plasma membrane of many blood cells through a glycophosphatidylinositol anchor. The activation of leukocytes leads to their adhesion with blood cells through the CD97–CD55 interaction. Other adhesion receptor family members (EMR2, EMR3, EMR4 and GPR56) have been shown to interact with molecules expressed in the extracellular matrix or on the surface of cells, including macrophages and activated neutrophils.

Almost all adhesion receptor family members contain a *G*-protein-coupled-receptor proteolytic (GPS) motif located proximal to TM1 on the N-terminus.[21] Adhesion family receptors are cleaved at the PPS motif after exit from the endoplasmic reticulum, but the two products are reattached prior to insertion in the plasma membrane through interactions that can be disrupted under denaturing conditions.

How these receptors trigger cellular responses is unclear. One member of the family, GPR56, forms a complex with G_q and G_{11}, suggesting signaling through these *G* proteins.[41] But currently there is little evidence that other members of the family can actually signal through *G* proteins. The cytoplasmic *C*-terminus of many of these receptors is quite long and contains consensus sequences for motifs involved in protein–protein interactions, suggesting a possible interaction with other signaling proteins. One family member (LEC1) is also a high affinity receptor for the black widow spider toxin, α-latrotoxin.[42] This toxin is a potent stimulator of neurotransmitter and hormone release from neurons and neuroendocrine cells. *G* proteins are unlikely to mediate the response, however, because a truncated form of the receptor, lacking all sequence C-terminal to TM1, is just as effective as the wild type receptor in mediating the response.

Frizzled/Taste2 Family

This family of receptors is composed of two groups, frizzled and taste2. These unlikely cousins share certain consensus sequences in their TM domains that place them at nearby branching points on the phylogenetic tree of *G* protein-coupled receptors despite their dissimilar functions.[20]

In humans, the frizzled group consists of 10 frizzled receptors (FZD1–FZD10) plus one related smoothened receptor. These receptors are required for embryogenesis, and they also maintain tissue homeostasis in the adult.[43] This is accomplished through the communication of information regarding cell fate and polarity, which ultimately determines how cells are positioned and oriented within tissues. In brain, frizzled receptors are involved in axonal guidance, dendrite morphology and synapse formation. The N-terminus of frizzled receptors is 200–320 amino acids long, and it contains a conserved cysteine-rich region that acts as the binding site for a group of secreted glycoproteins known as Wingless-Int (Wnt) as well as other soluble proteins. The smoothened receptor has a homologous N-terminal domain, but it forms a complex with the patched sonic hedgehog protein.[44]

Wnt proteins turn on frizzled receptors and elicit cellular responses through different mechanisms immediately downstream from receptor activation.[43] The process is thought to involve the recruitment of signaling proteins to the C-terminus.

Frizzled receptors contain consensus sequences in their C-termini that bind PDZ domains, and several proteins containing PDZ domains have been shown to interact with the C-terminus. One such protein is known as disheveled, and it becomes phosphorylated upon receptor activation by Wnts. There is also evidence that *G* proteins are involved in the signaling of both frizzled and smoothened receptors.

Taste2 receptors comprise a group of 13 receptors in humans.[20] They are expressed in the papillae of the tongue and are sensors for bitter tasting compounds, which bind to the EL of the receptor.[45] There is great sequence diversity in EL1 and EL2, which explains how taste2 receptors sense a diverse range of compounds. Several taste2 receptors are expressed on single sensory cells in taste buds showing that there is convergence in sensory input. Taste2 receptors are also expressed in the gastrointestinal tract where they are thought to regulate nutrient metabolism and signal the ingestion of potentially harmful xenobiotics.[46,47]

Both the N- and C-termini are very short, and taste2 receptors are sometimes difficult to express in heterologous cells unless additional N-terminal sequence is spliced onto the coding sequence.[48] Unlike other members of the *G* protein-coupled receptor superfamily, taste2 receptors lack the disulfide-bond-forming cysteines in EL1 and EL2. Taste2 receptors signal though the olfactory *G* protein gustducin (G_{olf}) as well as transducin (G_t) and G_i, and these *G* proteins are co-expressed with receptors in the tongue and gastrointestinal tract.[49]

Glutamate Receptor Family

The glutamate receptor family consists of 22 receptors including eight metabotropic glutamate receptors (mGluRs1–mGluRs8), three taste1 receptor proteins (TAS1R1–TAS1R 3), a Ca^{2+} sensing receptor (CASR) and two $GABA_B$ receptor subunits that combine to form a single receptor.[21] Members of this family have a unique, N-terminal, ligand-binding domain of approximately 280–580 amino acids that is structurally related to a bacterial amino acid transporter (Figure 5.8). Its upper and lower lobes (lobes 1 and 2, respectively) form a so-called Venus flytrap domain that engulfs the endogenous ligand when it binds. The lower lobe is closer to the plasma membrane. Most family members also have a cysteine-rich domain containing nine highly conserved cysteines residues between the ligand binding domain and the first TM domain. Eight of the cysteines form disulfide bonds within the cysteine-rich domain, and the ninth forms a disulfide bond with a cysteine located on the lower lobe of the Venus flytrap domain. The $GABA_B$ receptor proteins lack the cysteine-rich domain.

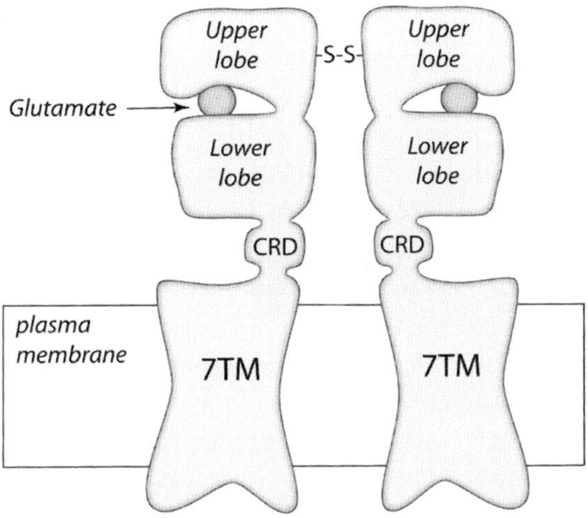

Figure 5.8: Schematic drawing of a metabotropic glutamate receptor. The figure shows the upper and lower lobes of the Venus flytrap domain that constitutes the ligand-binding site. The dimeric receptor complex is linked by a disulfide bond between the upper lobes as indicated (-S-S-). Abbreviations: CRD, cysteine-rich domain; 7TM, seven transmembrane spanning domain.

Receptors of the glutamate receptor family function as dimers. Usually the two protomers are linked through a disulfide bond between conserved cysteines in the upper lobe of the N-terminus.[21] In the case of the two $GABA_B$ receptor proteins ($GABA_{B1}$ and $GABA_{B2}$), the heterodimer lacks the disulfide bond and is held together through non-covalent interactions. The $GABA_{B1}$ receptor protein has an endoplasmic-reticulum-retention signal in its C-terminus, and its expression on the plasma membrane depends on its dimerization with the $GABA_{B2}$ receptor protein. Neither protein forms a functional receptor on the plasma membrane when expressed by itself in heterologous cells. The three taste receptor proteins ($T1R_1$, $T1R_2$ and $T1R_3$) form sweet- and bitter-molecule-sensing heterodimeric receptors ($T1R_2/T1R_3$ and $T1R_1/T1R_3$, respectively), and the eight metabotropic glutamate receptors form homodimers when expressed in heterologous cells.

How ligands bind with metabotropic glutamate receptors is known in detail because truncated receptors consisting of only the N-terminal domain bind glutamate, and their crystal structures have been determined in both the absence and presence of agonists and antagonists.[50–52] In the case of the mGluR3, the crystal structure of a complete N-terminal domain including both the ligand-binding and the cysteine-rich domains has been solved, whereas that of only the ligand-binding domain of mGluR1 and mGluR7 have been solved. These structures reveal two Venus flytrap domains connected by a disulfide bond between the upper lobes of each protomer. In the resting state, the dimer defines an angle formed by the two protomers with the vertex representing the disulfide bond between them. Two major conformational changes occur in the dimer when the agonist binds. The first is an intramolecular closing of the two lobes of the Venus flytrap domain around the agonist, and the second is an intermolecular change in the orientation of the two protomers. These rotate relative to each other on an axis perpendicular to the interfacial plane between the two upper lobes. This rotation partially closes the angle between the two protomers so that the two lower lobes of the ligand-binding domain are much closer when bound with agonist. This change should bring the two seven-transmembrane-domain bundles of each receptor protomer closer together as well, which is thought to cause activation of G protein signaling. The crystal structure of the mGluR1 receptors suggests that activation can occur with only one of the Venus flytrap domains closed. This active structure may be stabilized further by the closing of both Venus flytrap domains, provided that a metal cation, like Gd^{3+} or Ca^{2+}, is present to coordinate negatively charged aspartic acid residues on the surface of the two opposing lower lobes of the Venus flytrap domains.

The crystal structure of the Venus flytrap domain suggests that receptor activation occurs when the two lobes of the Venus flytrap domain close around the agonist. This hypothesis is consistent with how site-directed mutagenesis affects the activity of two competitive antagonists of the mGluR8 – ACPT-II (1-aminocyclopentane-1,3,4-tricarboxylic acid) and MAP4 (2-amino-2-methyl-4-phosphonobutanoic acid).[53] When bound, these antagonists act like a wedge to prop open the Venus flytrap domain by making contact with aspartate 309 (ACPT-II) and tyrosine 227 (MAP4) in the lower lobe. When the side chains of these residues are eliminated by mutation to alanine, ACPT-IIa and MAP4 act as agonists. Presumably, the two lobes of the receptor are able to close around ACPT-II or MAP4 when the bulky side chains of aspartate 309 or tyrosine 227 have been removed, respectively. Also, the introduction of cysteine residues at matching locations in the upper and lower lobes of the Venus flytrap domain of the mGluR1 receptor causes it to become active in the absence of agonists. The two cysteines are likely to form a disulfide linkage that forces the two lobes shut resulting in constitutive receptor activation. Thus, these results are consistent with the idea that closure of the Venus flytrap domains is necessary for receptor activation.

How the TM domains of mGluRs are activated and whether activation requires a closing of one or both Venus flytrap domains in the dimer has been addressed through the investigation of the function of receptors composed of modified protomers.[54–56] These include chimerical receptors with TM and N-terminal domains from different receptors, receptors with mutations in the ligand binding domain to prevent activation, receptors with mutations in the TM domain enabling allosteric regulation by noncompetitive antagonists, and receptors with a mutation in IL3 to prevent *G* protein coupling. By expressing dimers of unique combinations of these protomers and examining their properties, Jean-Philippe Pin and his colleagues have addressed how receptor activation occurs. The results show that receptor activation can occur with the closing of only one ligand-binding domain of the dimer, although in some receptors maximal activation occurs with the closing of both ligand-binding domains. Depending on the receptor, activation of the binding-domain of one protomer can trans-activate the TM domain of its partner, whereas in some dimers only *cis*-activation occurs. There is also circumstantial evidence suggesting that at any given instant in time, only one TM domain can be activated.

Trans-activation is the obligate mechanism for the $GABA_B$ receptor.[57] Only the $GABA_{B1}$ receptor subunit has a functional ligand-binding domain, whereas the TM domains of the $GABA_{B2}$ subunit are required for *G* protein activation. The ligand-binding domain of the $GABA_{B2}$ subunit lacks certain residues conserved in the $GABA_{B1}$ subunit that are required for GABA binding. The ligand-binding domain of the $GABA_{B2}$ subunit is required nonetheless for the formation of a functional dimer between the two $GABA_B$ receptor subunits.

Once activated, mGluR1 and five receptors interact with G_q to stimulate phosphoinositide hydrolysis, whereas the mGluR2, mGluR 3, mGluR 4, mGluR 6, mGluR 7 and mGluR 8 subtypes signal through G_i and G_o.[58] Taste1 receptors are thought to interact with G_q and G_i,[59,60] whereas the $GABAB_B$ receptor interacts with G_i and G_o.[61,62]

Heterotrimeric *G* Proteins

The *G* proteins that interact with *G* protein-coupled receptors are termed heterotrimeric because they are composed of three different subunits, $G\alpha$, $G\beta$ and $G\gamma$.[63,64] The $G\alpha$ subunit binds guanine nucleotides (GTP, GDP and their analogs), has GTPase activity and has a large influence on determining the receptors with which the trimeric complex interacts. Part of the $G\alpha$ subunit is homologous to the small *G* protein Ras mentioned in the Prologue in connection with receptor tyrosine kinases. Ras belongs to a large family of small *G* proteins (monomeric GTPases) that mediate a variety of functions within the cell.[65] Simply stated, *G* proteins mediate effects by traveling back and forth between two destinations.

In the active state, the *G* protein is bound with GTP and exhibits high affinity for an effector protein located at destination 2. The active state is short lived because of the GTPase activity of the *G* protein, which hydrolyzes bound GTP to GDP. The resulting GDP–*G* protein complex lacks high affinity for the effector so it diffuses away. In the GDP bound form, however, the *G* protein exhibits high affinity for an activated guanine nucleotide exchange protein located at destination 1. The exchange protein catalyzes the replacement of GDP with GTP, rendering the *G* protein into the active state again and enabling it to associate with its effector again at destination 2. The cycle continues so long as the guanine nucleotide exchange protein is active. This general mechanism is utilized by a large group of small *G* proteins to activate MEK kinase (Ras), to guide vesicular trafficking (Rab), to assemble the coats of transport vesicles at ER and Golgi membranes (Sar and Arf), to improve the recognition of aminoacyltransfer RNA by ribosomes (EF-Tu) and to accomplish many other functions in the cell. The role of heterotrimeric *G* proteins in conveying information from a family of guanine nucleotide-exchange proteins known as *G*-protein-coupled receptors represents a specialization of the same general mechanism.

Structure of G protein subunits

The crystal structures of several *G* proteins in various configurations ($G\alpha$ bound with GDP[66,67] or a nonhydrolyzable GTP analog[68,69] or the entire holoprotein, $G\alpha\beta\gamma$ bound with GDP[70,71]) have been determined (Figure 5.9). The $G\alpha$ subunit has about 350–400 residues and is composed of three domains, an N-terminal α helix, a helical domain and a Ras-like domain. $G\alpha$ is tethered to the plasma membrane by a fatty acid anchor (myristic or palmitic acid) esterified to cysteine residues in the N-terminal domain. The binding pocket for GTP is formed by the Ras-like domain, which is adjacent to a helical domain composed of five alpha helixes. The Ras-like domain contains residues critical for the hydrolysis of GTP. In the $G\alpha_{i1}$ sequence, these include glutamine 204, which coordinates the nucleophilic attack of the terminal phosphoester bond in GTP by an incoming water molecule, arginine 178, which stabilizes the leaving γ-phosphate of GTP, and threonine 181, which coordinates a Mg^{2+} that also stabilizes the leaving γ-phosphate (Figure 5.10). When bound with GDP, the helical domain closes on the Ras-like domain to trap GDP.

The $G\beta\gamma$ subunits form an obligate dimer (i.e. $G\gamma$ is required for the proper folding of $G\beta$) that separates only under denaturing conditions.[63,64] One exception is $G\beta 5$, which can be easily dissociated from $G\gamma$. The $G\beta$ subunits have about 350 residues and are composed of an extended, alpha-helical, N-terminal

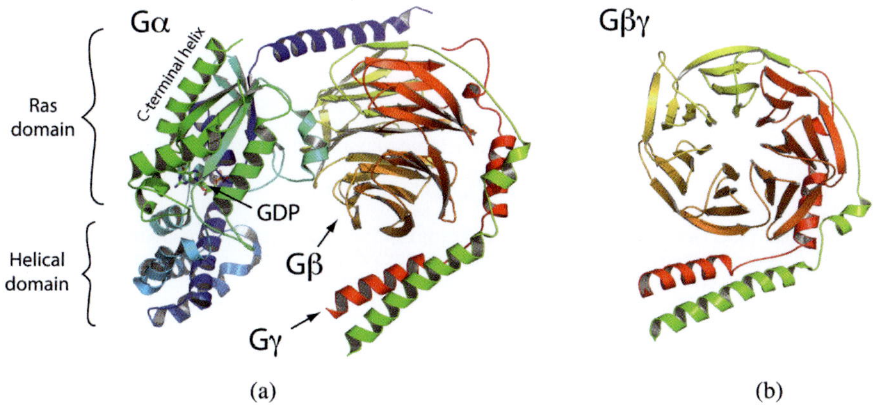

Figure 5.9: Crystal structure of the *G* protein heterotrimer $G\alpha_{i1}$, $G\beta_1$ and $G\gamma_2$ bound with GDP. (a) Structure of the trimeric complex. (b) Structure of the $G\beta\gamma$ dimer. The structure was determined by Wall *et al.*[71] (pdb 1gp2).

domain and a major β-propeller domain consisting of seven four-stranded, antiparallel β-sheets (Figure 5.9). $G\gamma$ has about 70 residues and consists of an extended helical domain, the N-terminus of which interacts tightly with the corresponding N-terminal domain of $G\beta$. These two entwined domains may form a lid that extends to the $G\alpha$ subunit and obstructs the release of GDP from its binding pocket. The C-terminal end of $G\gamma$ is tethered to the membrane via a fatty acid anchor esterified to a terminal cysteine residue.

In the inactive, GDP-bound state, the $G\alpha$ subunit is in a conformation that favors its interaction with the $G\beta\gamma$ subunits at two main sites of attachment.[63,64] A strong interaction occurs between the Ras-like domain and a complementary surface on the β-propeller that is also involved in the binding of $G\beta\gamma$ to effectors. Another interaction occurs between the N-terminal helix of $G\alpha$ and the β-propeller domain. These interactions compete with the interactions of $G\alpha$ and $G\beta\gamma$ with their respective effectors and ultimately cause a termination of effector activation when $G\alpha$ is bound with GDP. It is only in the trimeric state that the *G* protein interacts with the receptor.

Receptor-catalyzed guanine nucleotide exchange

When activated by an agonist, *G* protein-coupled receptors transmit a signal through contact surfaces with the trimeric *G* protein complex to cause GDP release.[63,64,72] These points of interaction on $G\alpha$ include its C-terminal helix, extended N-terminus and a loop ($\alpha 4/\beta 6$) within the Ras domain. During activation, TM3 and TM5 of

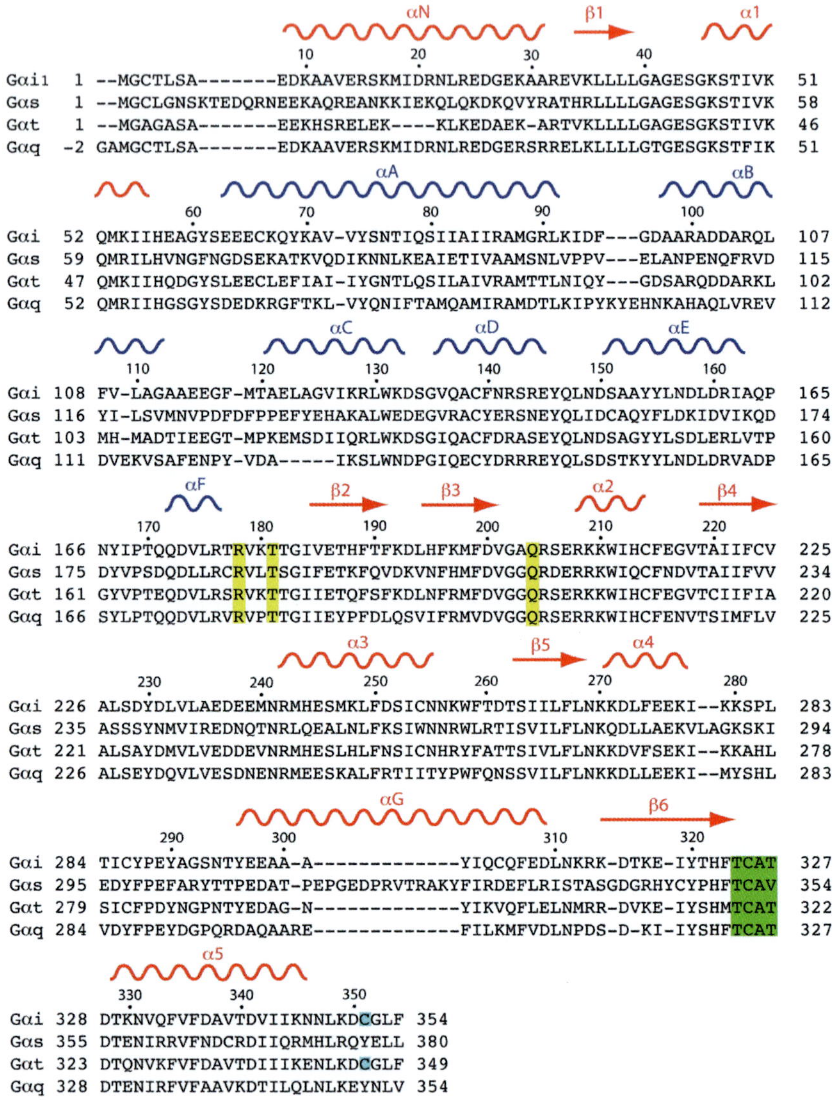

Figure 5.10: Sequence alignment of the αsubunits of the heterotrimeric G proteins, G_{i1}, G_s, transducin ($G\alpha_{t1}$) and G_q. The numbering above the sequence corresponds to that of $G\alpha_{i1}$. Secondary structure (α helix and β strand) is indicated above the sequence of $G\alpha_i$ for the Ras (red) and helical (blue) domains, respectively. Highly conserved residues in the guanine nucleotide binding pocket are indicated with yellow background, and the location of the guanine nucleotide binding motif (TCAT) is indicated with green background. The cysteine residues in the C terminal region of $G\alpha_i$ and $G\alpha_t$ that are ADP ribosylated by pertussis toxin are indicated with cyan background.

the receptor engage the C-terminal tail of $G\alpha$ and transmit a conformational change to the adjoining helix 5 and $\beta 6/\alpha 5$ loop of the Ras domain. This loop contains a guanine-ring binding motif (TCAT) that interacts with GDP, and the conformational change is thought to induce the release of GDP (Figure 5.10). The $G\beta\gamma$ subunit makes contacts with IL3 and C-terminal domain of the receptor, and these interactions are thought to act as a lever to transmit conformational changes to the *G* protein during activation. These include the tilting of the $G\beta\gamma$ subunit away from $G\alpha$ and a reorientation of another loop ($\beta 3/\alpha 2$) within the Ras domain to open up the GDP binding pocket. In the resting state, this $\beta 3/\alpha 2$ loop prevents the dissociation of GDP.

The crystal structure of β_2-adrenoceptor-G_s complex shows that, in the active state, TM6 and TM5 of the β_2-adrenocptor are extended and provide an opening for C-terminal helix of $G\alpha_s$ to engage the receptor through an interaction with the highly conserved DRY (TM3) and $X_1NPX_2X_3Y$ (TM6) motifs (Figure 5.11).[33] There are also interactions between the C-terminal helix of G_s and TM3 and TM5 of the receptor. Additional interactions involve IL2, TM5 and TM6. The net effect

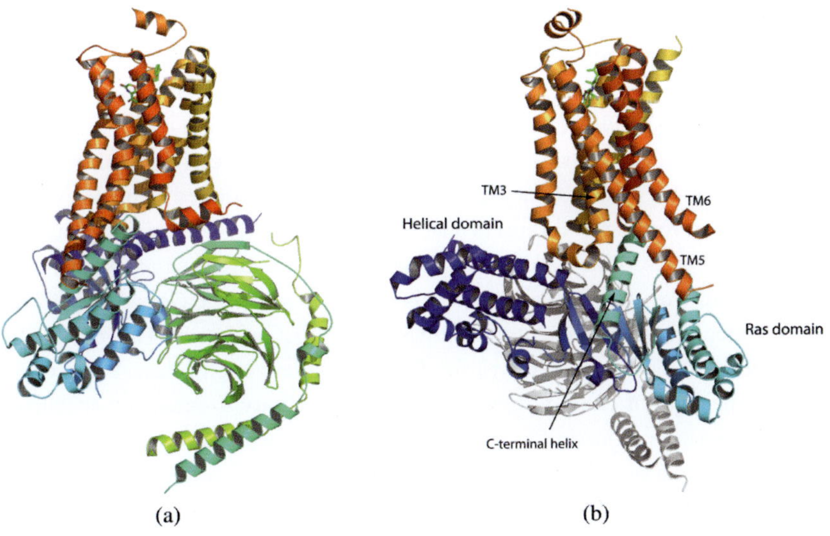

Figure 5.11: (Color online) Crystal structure of the active state of the β_2-adrenoceptor in a complex with G_s. (a) In this view, $G\alpha$ (left) and $G\beta\gamma$ are located below the β_2-adrenoceptor on the left and right, respectively. The helical domain is hidden behind the Ras domain. (b) In this view, the receptor-*G* protein complex has been rotated about 90° on a vertical axis. The helical (dark blue) and Ras (cyan) domains are located on the left and right, respectively. Behind is $G\beta\gamma$ colored in light gray.

is that the $G\alpha - G\beta\gamma$ interface opens and the guanine nucleotide-binding pocket on G_s opens through a separation of the Ras and helical domains, thus enabling a rapid exchange of GTP for GDP.

The binding of GTP induces a conformational change in $G\alpha$ that disrupts its interaction $G\beta\gamma$ and causes the dissociation of these subunits.[63,64] This change is based on the production of a favorable interaction of $G\alpha$ with the γ-phosphate of GTP, which ultimately causes an elimination of the $G\beta\gamma$ binding surface. The resulting conformation of the GTP–$G\alpha$ complex favors its interaction with downstream effectors, and specific examples are described below in this chapter.

G protein inactivation and hydrolysis of GTP

The GTP–$G\alpha$ complex is short lived because of the intrinsic GTPase activity of $G\alpha$. This fast inactivation is a fundamental requirement of any signaling pathway capable of high temporal resolution. It is thought that when the agonist is removed, the rate of inactivation of a response depends on speed at which GTP–$G\alpha$ is converted to GDP–$G\alpha$. The latter coalesces with $G\beta\gamma$ to terminate the response. GTPase activity can be quantified as the rate constant for hydrolysis of GTP, and this constant has been estimated for various $G\alpha$ subunits in their purified state.[64] The values for $G\alpha_s$, $G\alpha_i$ and Ga_q are 3.5, 2 and 0.8 min^{-1}, which translate into half times of about 12, 20 and 50 s, respectively. Assuming that hydrolysis follows first-order kinetics for a population of $G\alpha$ subunits, the process is expected to be nearly complete in about four half times or 50–200 s for the $G\alpha$ proteins just mentioned. This is extremely slow compared to the rapid rates of inactivation measured *in situ* for various intact signaling pathways. There are two explanations for the discrepancy. In the case of $G\alpha_q$, $G\alpha_s$ and $G\alpha_t$, the GTPase activity is enhanced when the $G\alpha$ subunit is bound to some effectors. These are phospholipase C–β (PLC–β1 and PLC–β2) for $G\alpha_q$, retinal phosphodiesterase (PDEγ) for $G\alpha_t$, and adenylate cyclase V for $G\alpha_s$. In the great majority of cases, however, the GTPase activity is enhanced or enhanced further by the interaction of $G\alpha$ with RGS proteins, of which there are many.[73] RGS proteins contain a so-called RGS motif that binds to the Ras domain and stabilizes a water molecule near the γ-phosphate of GTP. This configuration represents the transition state of $G\alpha$, and its stabilization by RGS proteins accelerates the hydrolysis of GTP. Proteins that increase the GTPase activity of a $G\alpha$ subunit are known generically as GTPase activating proteins (GAP) and their capacity to do so is often referred to as GAP activity.

RGS proteins also contain modules, like PDZ and PTB domains, that bind to other proteins that participate in the signaling cascade.[74] In so doing, RGS proteins help to anchor the signaling complex (receptosome) and increase its efficiency.

Types of G protein

The number of different types of G protein is far less than that of G protein-coupled receptors. In mammals there are 16 genes for $G\alpha$ subunits, five genes for $G\beta$ subunits, and 12 genes for $G\gamma$ subunits.[64] It follows that many receptors use the same G protein to initiate signaling.

G proteins are designated by the $G\alpha$ subunit of the trimeric complex. They can be divided into four families based on sequence homology, and these are named after the prototypical member of each family — G_s, G_i, G_q and G_{12}.[75] What follows is a brief description of their properties, tissue distribution and the proximate effectors that they stimulate or inhibit through the direct binding of their $G\alpha$ or $G\beta\gamma$ subunits.

The G_s family is composed of two members, $G\alpha_s$ and $G\alpha_{\text{olf}}$, that stimulate adenylate cyclase. $G\alpha_s$ has a widespread distribution and occurs in four splice variants.[76] The $G\alpha_s$ subunit stimulates all nine isoforms of adenylate cyclase. This stimulation was the criterion for the initial functional definition of this G protein, hence the subscript "s" in G_s. Ga_{olf} is limited to the olfactory epithelium where it activates adenylate cyclase III following olfactory receptor activation by odorants. Members of the G_s family are ADP-ribosylated by cholera toxin on arginine 187 in the $G\alpha_s$ sequence (Figure 5.10). This residue is critical for GTP hydrolysis, and the introduction of the bulky ADP-ribose group inhibits the GTPase activity of $G\alpha_s$ and causes a persistent activation of G_s signaling. Mutations in arginine 187 cause a loss in GTPase activity, and these have been found in $G\alpha_s$ from human pituitary tumors.[77]

The G_i family has eight members, $G\alpha_{i1-3}$, $G\alpha_o$, $G\alpha_z$, $G\alpha_{t1,2}$ and $G\alpha_{\text{gust}}$.[76] The three $G\alpha_i$ subunits are widely expressed at relatively high levels and mediate an inhibition of type V and VI adenylate cyclase. This inhibition was the functional criterion for the definition of G_i, which accounts for the subscript "i" in G_i. Because of their high expression, G_{i1-3} can provide a substantial source of $G\beta\gamma$ subunits, which have numerous effects depending upon the subunit composition of the dimer. These subunits inhibit adenylate cyclase I and enhance the stimulatory effect of $G\alpha_s$ on the II, IV and VII isoforms of adenylate cyclase. They also activate inwardly rectified potassium channels and isoforms 2 and 3 of phospholipase C–β. G_o has two splice variants and is expressed abundantly in brain. When activated, its $G\beta\gamma$ subunits mediate a voltage-dependent inhibition of neuronal Ca^{2+} channels, and they interact with SNARE proteins to inhibit exocytosis. These mechanisms account for presynaptic receptor-mediated inhibition of neurotransmitter release. The two transducin $G\alpha$ subunits ($G\alpha_{t1}$ and $G\alpha_{t2}$) are expressed in rods and cones, respectively, and activate a retinal phosphodiesterase (PDEγ). $G\alpha_{\text{gust}}$ is expressed mainly in taste cells in the tongue and also in chemosensitive cells in the gastrointestinal tract where it mediates responses by taste receptors that sense bitter-tasting chemicals. The mechanism involves $G\alpha_{\text{gust}}$ and $G\beta\gamma$ stimulation of phosphodiesterase and phospholipase C–β respectively. G_z is expressed in brain, retina, adrenal medulla, pituitary and pancreatic islets and platelets, and it inhibits adenylate cyclase.

$G\beta\gamma$ subunits are known to bind to other proteins and effectors.[76] Because some members of the G_i family are highly expressed, it seems likely that they are involved in signaling through these other $G\beta\gamma$ targets. For example, $G\beta\gamma$ recruits β-adrenergic receptor kinase to the membrane where it phosphorylates activated β-adrenoceptors. $G\beta\gamma$ subunits activate several kinases including Raf1 protein kinase, a phosphatidylinositol-3-kinase isoform, and the Bruton and Tsk tyrosine kinase. $G\beta\gamma$ also binds to monomeric GTPases including Rho, Rac and Arf.

All members of the G_i family except G_z are ADP-ribosylated by pertussis toxin on a cysteine located four residues in from the C-terminus (see Figure 5.12).[76] This region is critical for coupling to the receptor, and the introduction of a bulky ADP-ribose group prevents receptor-G protein interactions. Thus, pertussis toxin is a useful tool for determining if a G protein-coupled receptor signals through members of the G_i family. G_t and G_{gust} are also ADP-ribosylated by cholera toxin on an arginine cognate to the one in G_s that undergoes the same reaction (Figure 5.10). The result is an inhibition of GTPase activity and a persistent activation of G_t

$G_{s1\text{-}4}$	RMHLRQYELL
$G_{olf1,2}$	RMHLKQYELL
$G_{i1,2}$	KNNLKDCGLF
G_{i3}	KNNLKECGLF
G_{o1}	ANNLRGCGLY
G_{o2}	AKNLRGCGLY
G_t	KENLKDCGLF
G_z	QNNLKYIGLC
$G_{q/11}$	QLNLKEYNLV
G_{12}	QENLKDIMLQ
G_{13}	HDNLKQLMLQ
G_{14}	QLNLREFNLV
G_{15}	ARYLDEINLL

Figure 5.12: Sequence of the last ten residues of the C-terminus of heterotrimeric G protein α subunits. The cysteine residues that are ADP ribosylated by pertussis toxin are enclosed within a gray box.

and G_{gust} just as in the case of G_s. All *G* proteins have this arginine residue, but steric factors prevent its ADP ribosylation by cholera toxin in G_i, G_o and G_z and in members of the G_q and G_{12} families.

The G_q family includes four members in humans, G_q, G_{11}, G_{14} and G_{16}. $G\alpha_q$ and $G\alpha_{11}$ are widely expressed and mediate an activation of the 1, 3 and 4 isoforms of phospholipase C–β through the binding of their $G\alpha$ subunits to the enzyme.[76] The remaining members of the G_q family have a restricted distribution. $G\alpha_{14}$ is limited to epithelial cells, and $G\alpha_{16}$ (rat homolog is $G\alpha_{15}$) are limited to cells of hematopoietic origin. These *G* proteins also activate phospholipase C–β through their $G\alpha$ subunits.

The G_{12} family has two members, $G\alpha_{12}$ and $G\alpha_{13}$, that are widely expressed.[76] $G\alpha_{12}$ and $G\alpha_{13}$ activate adenylate cyclase VII and a guanine nucleotide exchange factor for the monomeric GTPase, Rho.

Specificity of the Receptor-*G* Protein Interaction

Many *G*-protein-coupled receptors exhibit a preference for one the three families of *G* protein (G_s, $G_{i/o}$ or $G_{q/11}$). For example, activation of β-adrenoceptors increases heart rate (β_1) and ventricular contractility (β_1) and relaxes the smooth muscle (β_{1-3}) of the airways, gastrointestinal tract, urinary tract and some blood vessels. Each of these tissues expresses G_s, $G_{i/o}$ and $G_{q/11}$, yet the response to β-adrenoceptor activation is typically G_s-like.

In the heart G_s stimulates adenylate cyclase, which synthesizes cAMP from ATP.[78] The cAMP directly activates a voltage-sensitive, cAMP-gated cation channel (HCN channel) that depolarizes pacemaker cells and increases heart rate. cAMP also activates protein kinase A, which phosphorylates voltage-sensitive Ca^{2+} channels, causing them to conduct a greater Ca^{2+} flux and increase the force of contraction of the myocardium. The heart also expresses G_i. Its $G\alpha$ subunit inhibits adenylate cyclase in the myocardium and its $G\beta\gamma$ subunits activate inwardly rectified K^+ channels in pacemaker cells. Activation of G_i, therefore, reduces the force of contraction and slows heart rate.

In smooth muscle, G_s also activates adenylate cyclase, but the resulting increase in cAMP causes relaxation.[79] G_s also activates Ca^{2+}-activated K^+ channels in smooth muscle, and these cause relaxation as well. In contrast, $G_{q/11}$ mediates contraction in smooth muscle and $G_{i/o}$ inhibits Ca^{2+}-activated K^+ channels[80] and mediates other effects that promote contraction indirectly.[80,81]

Thus, β-adrenoceptors in heart and smooth muscle are coexpressed with other *G* proteins that often mediate opposite effects, yet the *G* protein-mediated responses to β-adrenoceptor activation are mainly mediated through G_s. Other G-protein-coupled receptors also selectively activate G_s.

Another group of receptors interact preferentially with $G_{i/o}$, such as α_2-adrenoceptros and M_2 and M_4 muscarinic receptors. These often mediate presynaptic inhibition of neurotransmitter release through activation of G_o, which causes an inhibition of exocytosis and neuronal Ca^{2+} channels.[76] In the heart, M_2 muscarinic receptors interact with G_i to inhibit heart rate and decrease the force of contraction. In smooth muscle, M_2 receptors interact with G_i to inhibit adenylate cyclase and Ca^{2+}-activated K^+ channels and to stimulate a nonselective cation channel. These actions have little effect by themselves but oppose the action of receptors linked to G_s and promote contraction elicited by other G_q-linked receptors.

Finally, other receptors prefer to interact with $G_{q/11}$ to mediate phosphoinositide hydrolysis. Often these receptors have excitatory actions through their mobilization of Ca^{2+}. They frequently excite neurons and stimulate exocrine secretion and contraction of smooth muscle.[76]

A few examples of receptors that preferentially signal through one family of G proteins are listed in Table 5.1. In each case, the receptor is often expressed in native tissues with various G proteins, yet it prefers to interact with only one family. This preference is often maintained when the receptor is expressed in cell lines, although overexpression can lead to some interaction with additional G proteins. G_{15} and G_{16} are promiscuous members of the $G_{q/11}$ family that do not follow the pattern.[82] Some of the receptors shown in Table 5.1 that interact preferentially with G_s or $G_{i/o}$ also stimulate phosphoinositide hydrolysis in cell lines expressing G_{15} or G_{16}, although the magnitude of this stimulation is usually not as great as that seen with receptors that preferentially interact with $G_{q/11}$.

One part of the $G\alpha$ subunit that has a large effect on receptor coupling is the sequence of the last four residues of the C-terminus ($G\alpha$-C-tail).[29,30] Recall that

Table 5.1: Examples of receptor-G protein preferences.

Gprotein	Receptor
G_s	A_2 and A_3 adenosine, $\beta_1 - \beta_3$-adrenergic, D_1 and D_5 dopamine, H_2 histamine, MC_{1-5} melanocortin
$G_{i/o}$	A_1 adenosine, α_2-adrenergic, M_2 and M_4 muscarinic, CB_1 and CB_2 cannabinoid, D_2–D_3 dopamine, $GABA_B$, H_3 and H_4 histamine, $MGlu_{2-4,6-8}$, opioid, P2Y, AT_2 angiotensin
$G_{q/11}$	α_1-Adrenergic, M_1, M_3 and M_5 muscarinic, H_1 histamine, $MGlu_{1,5}$, P2Y, prokineticin, AT_1 angiotensin

when the receptor is activated, it engages the *G* protein and transmits a signal from the C-tail α-helix of $G\alpha$ to the $\beta 6/\alpha 5$ loop of the Ras domain, which opens the guanine-nucleotide-binding pocket and enables the release of GDP. Short peptides incorporating the tetrapeptide sequence of the $G\alpha$-C-tail of transducin selectively inhibit the transducin–rhodopsin interaction, illustrating the specificity of the receptor-$G\alpha$-C-tail interaction.[83] If the $G\alpha$-C-tail of one $G\alpha$ is switched with that of another, the chimerical *G* protein retains its effector selectivity, but often switches its receptor preference to that of its $G\alpha$-C-tail. For example, the α_2-adrenoceptor normally interacts with the G_i family of *G* proteins to inhibit adenylate cyclase and neuronal Ca^{2+} receptors and activate inwardly rectified K^+ channels. When expressed with a chimeric *G* protein consisting of $G\alpha_q$ with the last four residues of the C-terminus replaced with those of $G\alpha_{i2}$, however, the α_2-adrenoceptor mediates a stimulation of phosphoinositide hydrolysis. Table 5.2 lists some additional receptor-*G* protein chimeras pairs that elicit productive responses even though the receptors in their native context only interact with the wild type *G* protein from which the $G\alpha$-C-tail of the chimera was derived. Table 5.2 also lists some $G\alpha/G\alpha-$C-tail chimeras that fail to engage receptors that normally prefer the *G* protein from which the $G\alpha$-C-tail was derived. This is not unexpected because the $G\alpha$-C-tail is not the only part of the *G* protein that interacts with the receptor.

Since a Ca^{2+} signal can be measured rapidly, $G\alpha/G\alpha_q$-C-tail chimeras have been used in high-throughput screening to measure activation of receptors that normally signal through G_i or G_s. For example, some metabotropic glutamate receptors

Table 5.2: Examples of productive and nonproductive signaling through carboxyl-terminal *G* protein chimeras.

Receptor	Preferred *G* protein	Chimera	Productive signaling?
A_1 adenosine	$G_{i/o}$	G^a_{q-i5}	PLCβ, yes
α_2 adrenergic	$G_{i/o}$	G_{q-i5}	PLCβ, yes
D_2 dopamine	$G_{i/o}$	G_{q-i5}	PLCβ, yes
M_2 muscarinic	$G_{i/o}$	G_{q-i5}	PLCβ, yes
Somatostatin	$G_{i/o}$	G_{q-i5}	PLCβ, no
β_2-adrenergic	G_s	G^b_{q-s5}	PLCβ, no
V_2 vasopressin	G_s	G_{q-s5}	PLCβ, yes
V_{1a} vasopressin	$G_{q/11}$	G^c_{s-q5}	Adenylate cyclase, yes
Oxytocin	$G_{q/11}$	G_{s-q5}	Adenylate cyclase, no

[a] $G\alpha_q$ with the five carboxyl-terminal amino acids replaced with those of $G\alpha_i$.
[b] $G\alpha_q$ with the five carboxyl-terminal amino acids replaced with those of $G\alpha_s$.
[c] $G\alpha_s$ with the five carboxyl-terminal amino acids replaced with those of $G\alpha_q$.

(mGluR2– mGluR4, mGluR6–mGluR8) signal through $G_{i/o}$ to cause many effects in brain including an inhibition of neurotransmitter release (G_o), an activation of inwardly rectified K^+ channels (G_i), and an inhibition of adenylate cyclase (G_i). In cell lines stably transfected the mGluR4, it is possible to measure agonist-mediated inhibition of cAMP accumulation in the intact cell or a stimulation in [^{35}S]GTPγS binding in a cellular homogenate, but these measurements take time and require several steps involving invasive manipulation of cells or fractions thereof. If the receptor is coexpressed with $G_i/G\alpha q$-C-tail, however, receptor activation causes phosphoinositide hydrolysis and a subsequent release of Ca^{2+} from intracellular stores.[84] This rise in Ca^{2+} can be measured at a distance as an increase in fluorescence or luminescence using a fluorescent dye or expressed luminescent proteins that convert the rise in Ca^{2+} into a light signal. The assay can be automated, and the activity of hundreds or even thousands of compounds at a receptor can be measured in a single day based on their ability to generate a light signal.

In some respects, *G*-protein-coupled receptors behave as on-off switches that relay information through the receptor-*G*-protein interface. If there is only one on-state of the receptor, then responses mediated by the same receptor, but through different *G* proteins, should exhibit the same rank order of agonist potency. There may be absolute differences in EC_{50} and $E_{\max}$ due to a difference in the sensitivities of different signaling cascades, but the overall activity profile of a group of agonists should be the same. Some agonists, however, exhibit a preference and cause the receptor to deviate from its normal *G* protein preferences. An example is shown in Figure 5.13, where the concentration-response curves of agonists for mediating responses through the M_2 muscarinic receptor are shown. This receptor normally prefers to signal through the G_i family to mediate, among other things, an inhibition of adenylate cyclase (Figure 5.13a), but the receptor can also couple to G_{15} to mediate phosphoinositide hydrolysis (Figure 5.13b). Collectively, the muscarinic agonists oxotremorine-M, carbachol and R- and S-aceclidine generate the same profile for the two responses, but the compound McN-A-343 falls out of the pattern. Relative to carbachol, its potency is one-tenth for signaling through G_{15}, but only one-hundredth for signaling through G_i.[85] This implies that the interface between an activated M_2 receptor and the *G* protein can change depending upon the agonist occupying the receptor. In other words, there must be more than one active conformation of the M_2 receptor, and McN-A-343 appears to select a conformation that favors G_{15}. This phenomenon is known as agonist bias or ligand-directed signaling, and methods to quantify it are described in Chapter 9. Ligand-directed signaling is not limited to receptor-*G* protein responses. Some striking examples involve the interaction of the active receptor state with a *G*-protein-coupled-receptor

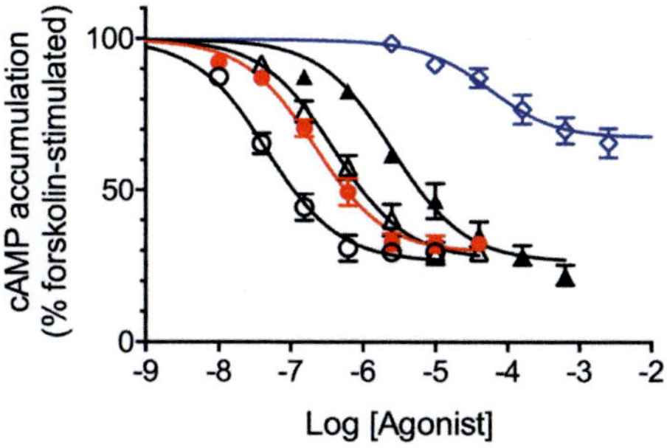

(a)

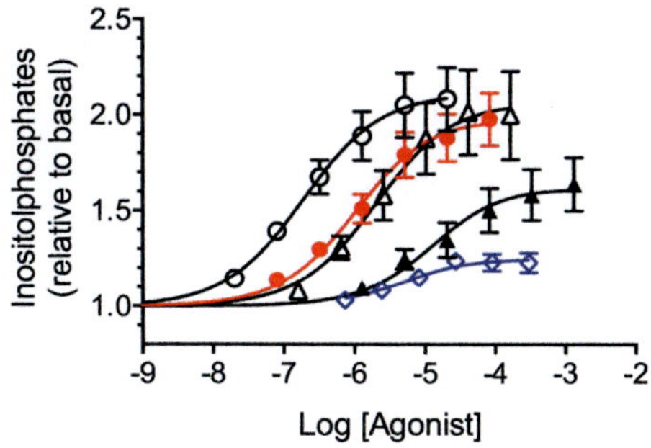

(b)

Figure 5.13: The effects of muscarinic agonists on phosphoinositide hydrolysis and cAMP accumulation in HEK 293 cells expressing the human M_2 muscarinic receptor and$G\alpha_{15}$. (a) cAMP accumulation was measured in the presence of muscarinic agonists and forskolin (10 μ M). *b*, [^{3}H]Inositolphosphate accumulation was measured in the presence of LiCl (1 mM). The data represent previously unpublished exploratory experiments associated with a prior study by Griffin and coworkers.[113]

kinase (GRK) yielding a unique profile for ligand-mediated internalization of the receptor.

The specificity of receptor-*G* protein interactions in native tissues often exceeds that observed in cell lines and artificial membranes in which receptors and *G* proteins have been expressed or reconstituted. For example, α_2-adrenoceptors couple to $G_{i/o}$, causing stimulation of inwardly rectified K^+ channels and inhibition of adenylate cyclase and Ca^{2+} channels. In sympathetic neurons, the α_2-adrenoceptor effect on inwardly rectified K^+ channels is mediated through G_i and that on Ca^{2+} currents is through G_o.[86] Both of these effects are prevented by pertussis toxin-treatment, but each can be rescued by expression of any one of the pertussis toxin-resistant forms of $G\alpha_o$ and $G\alpha_{i1-3}$. These recombinant *G* proteins have been engineered with a point mutation replacing the cysteine in the C-tail that is normally ADP-ribosylated by pertussis toxin in the wild type sequence. Thus, the assembly of receptors, *G* proteins and effectors in signaling complexes determines with which member of the G_i family that α_2 receptor interacts to modulate different channels.

$G\beta\gamma$ subunits also have a role in governing receptor-*G* protein interactions, particularly in their native context. One example occurs in the pituitary gland, which releases hormones and peptides depending upon the balance of positive and negative inputs. Negative regulation is mediated through *G* protein-coupled receptors that couple to G_o to inhibit Ca^{2+} influx and ultimately the exocytotic release of hormones. Inhibition of the Ca^{2+} current can also be measured in cell lines derived from the pituitary gland. In GH3 pituitary tumor cells, the M_4 muscarinic signals through $G\alpha_o$ $\beta_1\gamma_3$ to inhibit Ca^{2+} currents, whereas the somatostatin receptor signals through $G_o\beta_3\gamma_4$ to mediate the same effect.[87]

The Receptosome

The speed and specificity with which an activated *G* protein-coupled receptor elicits a response depends on its proximity to *G* proteins and downstream effectors. These do not simply diffuse freely in the plasma membrane during signaling, but rather, they are held close together in a signaling complex known as a receptosome. This enables the receptor to elicit a response quickly through a specific *G* protein and effector that are anchored nearby. In pacemaker cells of the heart, for example, stimulation of the M_2 muscarinic receptor causes an activation of inward rectified K^+ channels through the $G\beta\gamma$ subunits of G_i with only a 50–150 ms delay.[88] Presumably, it is the close association of G_i with the M_2 muscarinic receptor and the K^+ channel in a receptosome that enables this fast response and explains why the response is mediated through G_i as opposed to G_o. Receptosomes are maintained by binding interactions among scaffolding proteins, downstream effectors and the receptor, usually through its C-terminus. Part of the receptosome is assembled in the ER, and it ultimately inserts in the plasma membrane and associates with other anchoring proteins that may attach to the cytoskeleton. Receptosomes vary depending upon the cell type, receptor and signaling pathway involved.

One example is the postsynaptic density (PSD) of excitatory synapses in the dendritic spines of central neurons.[89] It contains an orderly array of many signaling and scaffolding proteins. Lying at a level deep in the postsynaptic density are a group of Shank proteins that act as higher order scaffolding proteins. These bind with each other as well as with other lower order scaffolding proteins that ultimately bind with receptors and other signaling proteins on or near the plasma membrane. On the internal side, Shank proteins bind with actin-binding proteins that ultimately

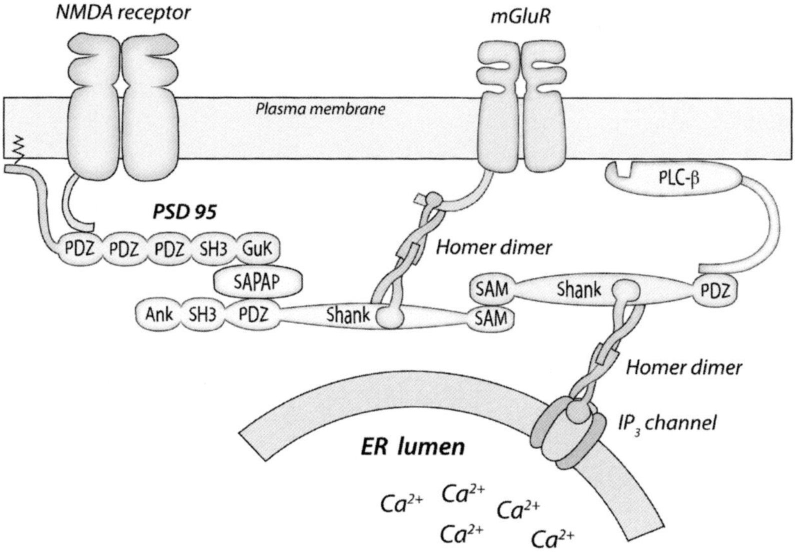

Figure 5.14: Role of the scaffolding proteins, PSD-95, Shank, homer and the guanlylate kinase associated protein (SAPAP), in the maintenance of a signaling platform for ionotropic and metabotropic glutamate receptors.

tether the postsynaptic density to the F-actin-based cytoskeleton of the dendritic spine.

Shank proteins contain a number of protein-protein interaction motifs including ankyrin, SH3- and PDZ-domains, a long proline-rich domain and a C-terminal SAM (sterile alpha motif) domain. These enable Shank to bind with itself as well as with other lower order scaffolding proteins. Self-association occurs primarily through SAM, but also through other motifs (PDZ, ankyrin and SH3).

Figure 5.14 illustrates an example of how excitatory ionotropic and metabotropic glutamate receptors are anchored in the postsynaptic density ultimately through interactions with different Shank proteins. The C-terminus of the NMDAR2 subunit of the NMDA receptor binds with the PDZ domain in the PSD-95 scaffolding protein. AMPA receptors also associate with PSD-95, but indirectly through an accessory protein known as stargazin. PSD-95 contains a guanylate kinase-like domain (GuK) that lacks enzymatic activity, but functions as a binding domain for a guanylate kinase associated proteins known as SAPAP. This protein interacts with a PDZ domain in Shank. Thus, although Shank does not directly bind with glutamate receptors, it maintains their location in the PSD indirectly through interactions with other scaffolding proteins.

G protein-coupled receptors for glutamate (mGluRs) also interact with Shank through Homer (Figure 5.14). This protein dimerizes, and the corresponding ends of the dimer bind to pentapeptide consensus sequences (PPXXF) in the C-terminus of mGluRs and in the proline-rich sequence of Shank, thereby cross-linking mGluRs to Shank.[89] Homer also cross-links Shank to Ca^{2+} channels gated by inositol trisphosphate (IP_3) in cisternae of the endoplasmic reticulum. Finally, an isoform of phospholipase C–β (PLC-β3) interacts with a PDZ domain in Shank. Thus, with the exception of *G* proteins, most of the downstream elements (PLC-β3, IP_3 gated Ca^{2+} channels and ER cisternae) for the signaling of mGluRs to Ca^{2+} release from the endoplasmic reticulum are held in close proximity to each other by Shank in the PSD.

The cardiac myocyte provides another example of how scaffolding proteins determine the actions of *G* protein-coupled receptors.[90] Both β_1- and β_2-adrenoceptors are expressed in the heart, but they signal differently despite the ability of both to couple effectively to G_s in other tissues and cell lines. The β_1-adrenoceptor signals through G_s to increase cAMP levels, heart rate and the force of contraction, whereas activation of β_2-adrenoceptors has little effect on cAMP levels and can actually inhibit heart rate and contractility and promote myocyte survival by inhibiting apoptosis. Many types of scaffolding proteins have been shown to interact with the C-termini of β_1- and β_2-adrenoceptors, and presumably, the different signaling pathways of these receptors can be attributed to differences in scaffolding.

Two examples of proximate scaffolds for β-adrenoceptors in the heart are shown in Figure 5.15. The C-terminus of the β_1-adrenoceptor interacts with a PDZ domain in SAP97, which acts as a scaffold for A kinase anchor protein 79 (AKAP79) and protein kinase A (PKA). Protein kinase A is activated by cAMP,

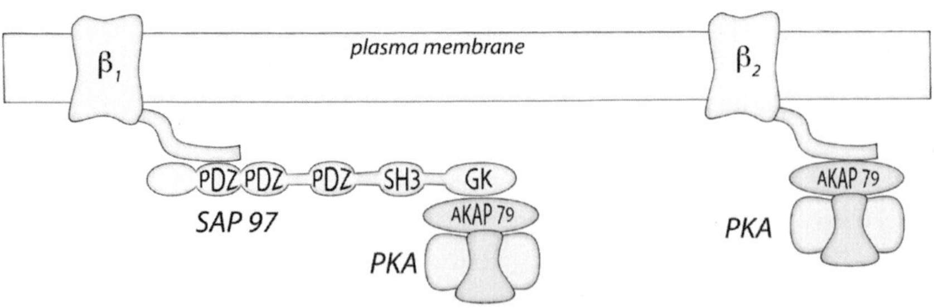

Figure 5.15: Role of synapse-associated protein (SAP97) and A kinase anchoring protein (AKAP) in the maintenance of a signaling platform for β_1- and β_2-adrenodceptors in the heart.

and the close proximity of the β_1-adrenoceptor with this kinase facilitates signaling through the cAMP-protein kinase A pathway. The β_2-adrenoceptor interacts directly with AKAP79/150 to bring the receptor in close proximity to protein kinase A as well. The β_2 adrenoceptor is phosphorylated by protein kinase A, however, and this phosphorylation is thought to enable the receptor to signal through G_i and activate mitogen activated protein kinase. Several other differences in the signaling pathways of β_1- and β_2-adrenoceptors have been described in the heart including their differential localization in caveolae and lipid rafts.

Examples of the Proximate Signaling Pathways of *G* Protein-Coupled Receptors

G protein-coupled receptors elicit a variety of signaling pathways throughout the body. The more proximate aspects of these pathways are somewhat less diverse, and in this section, I describe a handful of prototypical examples that are discussed in various contexts throughout the remainder of this book.

β_1-*Adrenoceptor signaling in the heart*

As just mentioned, the β_1-adrenoceptor is expressed in the heart. It is the most abundant β-adrenoceptor in the heart, and its activation leads to well-known excitatory effects on heart rate and contractility that are elicited by activation of sympathetic nerves.[76]

In pacemaker cells, activation of the β_1-adrenoceptor by norepinephrine causes activation of G_s and the liberation of GTP-bound $G\alpha_s$ subunits (Figure 5.16). These activate adenylate cyclase, which catalyzes the conversion of ATP into cAMP. The most abundant isozymes of adenylate cyclase in the heart are Types V and VI[91], and $G\alpha_s$ has been shown to bind directly with the catalytic subunit of the type V enzyme.[92,93] The resulting increase in cAMP causes an activation of hyperpolarization-activated, nucleotide-gated (HCN) channels.[94,95] These tetrameric channels contain a cyclic nucleotide-binding domain on the cytosolic C-terminal region that is homologous to the nucleotide-binding site on CNG channels and protein kinases A and G. The latter are activated by cAMP and cGMP, respectively. The binding of cAMP to the HCN channel causes it to open and conduct an inward current of Na^+. The resulting depolarization of pacemaker cells increases their firing rate, and ultimately, heart rate.

In the ventricular myocardium (Figure 5.16), the β_1-adrenoceptor also causes an increase in cAMP, but in this case, an important functional consequence is the

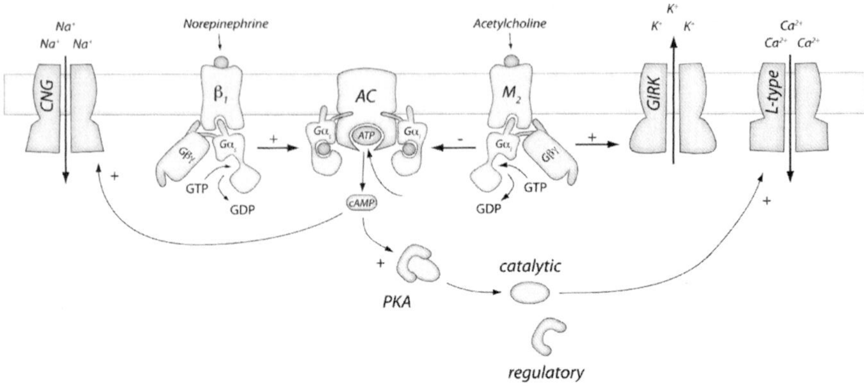

Figure 5.16: Signaling pathways of β_1−adrenergic and M_2 muscarinic receptors in the heart. Abbreviations: AC, adenylate cyclase; CNG, cyclic nucleotide-gated ion channel; GIRK, G protein-coupled receptor inwardly rectified potassium channel; L-type, L-type Ca^{2+} channel and PKA, protein kinase A.

activation of protein kinase A (PKA).[76] In the inactive state, PKA consists of a complex of catalytic and regulatory subunits. The regulatory subunit has two binding sites for cAMP, and in the inactive state, these are wrapped around the catalytic subunit.[96] The binding of cAMP to the regulatory subunit allosterically reduces its affinity for the catalytic subunit causing the two to dissociate. The free catalytic subunit is active and catalyzes the phosphorylation of downstream signaling proteins on serine and threonine residues. In the myocardium, an important target is the voltage sensitive Ca^{2+} channel.[76] Its phosphorylation causes an increase in the inward Ca^{2+} current during systole and increases the force of contraction of the heart.

M_2 muscarinic receptor signaling in the heart

Activation of vagal nerves to the heart causes the release of acetylcholine, which acts on the M_2 muscarinic receptor to inhibit heart rate and contractility. The M_2 muscarinic receptor is the main muscarinic subtype expressed in the heart, and it signals through the $G_{i/o}$ family of G proteins to oppose the effect of β_1-adrenoceptor stimulation.[76]

In pacemaker cells, M_2 receptor stimulation causes an activation of G_i resulting in the liberation of $G\beta\gamma$ and GTP-bound $G\alpha_i$ subunits (Figure 5.16). The $G\beta\gamma$ subunits bind to the cytosolic domain of a tetrameric, inwardly rectified K^+ (GIRK) channel[97,98] causing an efflux of K^+ and driving the potential difference across the plasma membrane closer to the K^+ equilibrium potential (about −80 mv).[88] Under physiological conditions, these channels only conduct a K^+ current at voltages near the resting membrane potential. This causes a decrease in the rate of depolarization and a consequent slowing in heart rate. But once the pacemaker cell reaches threshold and depolarizes, there is little K^+ efflux through the GIRK channel because of its inward rectification. This protects the cell from an unnecessary loss of K^+ that would otherwise occur with each heartbeat during periods of vagal stimulation if there were no rectification. Rectification is probably caused by the binding of intracellular polyamines to negatively charged residues lining the channel on the intracellular side. Electrophysiologists often study GIRK channels at hyperpolarizing potentials (e.g., −130 mV) because of the much larger inward current at these potentials.

The release of GTP-bound $G\alpha_i$ subunits that accompanies M_2 receptor activation in pacemaker cells inhibits adenylate cyclase activity causing a reduction in cAMP and an inhibition of the stimulatory effect of the β_1-adrenoceptor on the HCN channel.[76] In the myocardium, M_2 receptor activation also inhibits the stimulatory effect of the β_1-adrenoceptor on adenylate cyclase, and thus, opposes the inward current of Ca^{2+} and causes a decrease in contractility (Figure 5.16). The $G\alpha_i$ subunit directly inhibits the catalytic domain of adenylate cyclase V [99,100], and presumably, this mechanism accounts for the inhibition of adenylate cyclase by $G\alpha_{i/o}$ following M_2 receptor activation in the heart.

M_3 muscarinic receptor mediated phosphoinositide hydrolysis in smooth muscle

A large group of G protein-coupled receptors trigger cellular responses by signaling through the $G_{q/11}$ to mediate phosphoinositide hydrolysis.[101] One example is the M_3 muscarinic receptor.[102] It elicits contraction of smooth muscle and secretion from exocrine glands through this mechanism (Figure 5.17).

Phosphatidylinositol occurs on the inner leaflet of the plasma membrane. It is sequentially phosphorylated by kinases in two steps to yield phosphatidylinositol-4,5-*bis*phosphate (PIP_2).[100] In smooth muscle, the M_3 muscarinic receptor signals through G_q to activate phospholipase C-β, which hydrolyzes PIP_2.[102] The hydrolysis products, inositol-1,4,5-*tris*phosphate (IP_3) and diacylglycerol (DAG), act as second messengers. IP_3 binds to an IP_3-gated Ca^{2+} channel on the ER to conduct an efflux of Ca^{2+} from the ER [103], whereas DAG activates protein kinase C.[104] Both pathways are involved in mediating a variety of cellular responses.

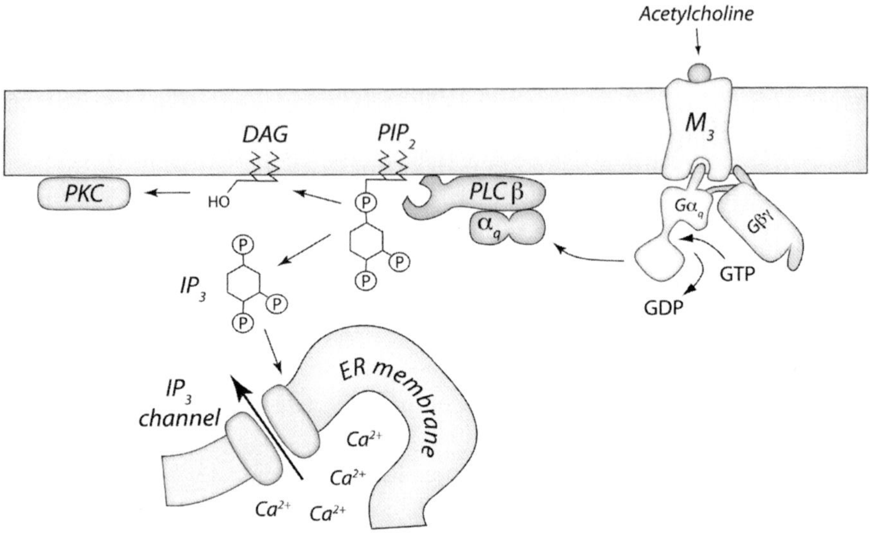

Figure 5.17: Signaling pathway of M_3 muscarinic receptor in smooth muscle. Abbreviations: DAG, dacylglycerol; IP_3, inositol-1,4,5-*tris* phosphate; PIP_2, phosphatidylinositol-4,5-*bis*phosphate and PLCβ, phospholipase-Cβ.

There are four isozymes of PLC-β (PLC-β1–4).[105–107] $G\alpha_q$ activates all of these, and $G\beta\gamma$ activates PLC-β2 and 3. The structure of PLC-β consists of a central catalytic domain composed of eight α helices alternating with eight β strands that form a TIM barrel with a central linker connecting its two halves.[108] In the inactive state, the linker occludes the TIM barrel and prevents its catalytic activity. The binding of GTP-$G\alpha_q$ to PLC-β somehow causes a displacement of the linker, which enables the catalytic domain to cleave PIP_2. Motifs on the N-terminal (EF hands) and C-terminal (C2 domain) flanks of the catalytic domain stimulate the GTPase activity of $G\alpha_q$ and rapidly terminate its action.

In smooth muscle, the formation of IP_3 and the resulting Ca^{2+} release elicited by M_3 receptor activation probably elicits the initial phasic contraction. The more prominent tonic phase of contraction depends on an influx of Ca^{2+} through voltage-dependent Ca^{2+} channels, however.[109] The initial depolarization that activates the voltage-dependent Ca^{2+} influx is thought to be caused by a nonselective cation conductance, and likely candidate channels include members of the transient receptor potential (Trp) family.[110,111] Some of these are inhibited by PIP_2, and consequently, M_3 receptor-mediated phosphoinositide hydrolysis is expected to activate such a channel. In intestinal smooth muscle, the cation conductance also depends on activation of G_i by the M_2 muscarinic receptor.

Recruitment of arrestin

When stimulated with an agonist for several minutes, most *G* protein-coupled receptors undergo dynamic movements on the plasma membrane. One way to observe this behavior is to express a gene for a recombinant fusion protein consisting green fluorescent protein (GFP) spliced to the N-terminal sequence of the receptor of interest. Using total internal reflection fluorescence microscopy (TIRF), only those receptors on the surface of the plasma membrane are illuminated and visualized. Cells examined in this way appear to have low fluorescence uniformly distributed over the plasma membrane. Following addition of agonist, however, the fluorescence gradually concentrates in tiny randomly distributed foci. Over time, these aggregates increase in size and each suddenly disappears randomly. At other sites, aggregates burst onto the membrane, and their fluorescence dissipates as the receptors diffuse away from the initial focus. This sequence is explained by agonist-induced clustering of receptors followed by their endocytosis and recycling to the plasma membrane.

The process begins with the phosphorylation of the agonist-bound receptor by a *G* protein receptor kinase (GRK)[22] (Figure 5.18). There are seven isoforms of this kinase (GRK1–7), including two that are expressed in the retina (GRK1, rhodopsin kinase, and GIRK7, cone opsin kinase) and four (GRK2, GRK3, GRK5 and GRK6) that are widely distributed throughout the body.[22] These kinases phosphorylate the active state of a variety of *G* protein-coupled receptors on serine and threonine residues located on IL3 and the C-terminal domain. The phosphoserine and phosphothreonine residues act as docking sites for arrestin. Additional proteins bind to arrestin including clathrin and the AP2 adaptor complex, which binds clathrin and dynamin and recruits the receptor to clathrin-coated pits. These invaginate as more receptors are recruited, and ultimately, the pit pinches off through the action of dynamin to yield an endocytic vesicle. This process causes an interruption in *G* protein signaling and a loss of receptors from the plasma membrane. Once inside the cell, the vesicles can be recycled back to the plasma membrane or ultimately be absorbed by lysozymes resulting in receptor degradation.

Another possibility is that the receptor-arrestin complex functions as a platform for recruiting additional signaling proteins (Figure 5.19). Some examples include Raf-1, MEK and MAPK.[22,112] Thus, in these latter cases, the recruitment of arrestin switches receptor signaling from *G* proteins to intracellular kinases.

What is the Measure of Receptor Activation at *G* Protein Coupled Receptors?

Now that we know all about *G* protein-coupled receptors and their signaling pathways, what constitutes an accurate measure of receptor activation? This issue needs

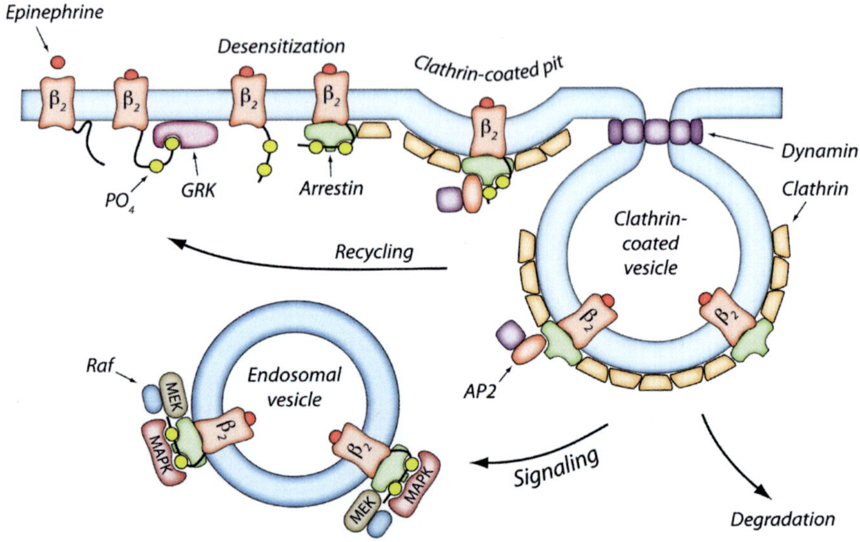

Figure 5.18: Agonist-induced internalization of the β_2-adrenoceptor. Abbreviations: β_{arr}, β-arrestin; C, clathrin; D, dynamin and MAPK, mitogen activated protein kinase.

to be clarified if we are to have a measure of the observed affinity and efficacy of ligands.

In the case of ligand-gated ion channels, the answer to this question is a simple one — the open state of the receptor protein. All that is required for an agonist to induce a current through a ligand-gated ion channel on a cell is the ligand-gated ion channel itself. But *G* protein-coupled receptors do not function in isolation. They need a coupling protein to transduce receptor activation into a signal, and the *G* protein usually fulfills this function.

In next chapter, I argue that the fraction of the agonist-receptor population associated with the *G* protein bound with GDP is a useful measure of receptor activation. This complex is the immediate precursor to the formation of free $G\beta\gamma$ and GTP-bound Gα subunits that trigger the response to the agonist. Hence, the rate of production of $G\beta\gamma$ and GTP-$G\alpha$ subunits is proportional to the fraction of the receptor population in a quaternary complex consisting of GDP, *G* protein and the active state of the receptor bound with agonist. The next chapter shows that the amount of this complex is a simple increasing function of receptor occupancy by the agonist.

References

1. Salvini-Plawen, L. and Mayr, E. *On the Evolution of Photoreceptors and Eyes* (Plenum Press, 1977).
2. Gehring, W. J. Historical perspective on the development and evolution of eyes and photoreceptors. *Int J Dev Biol* **48**, 707–717 (2004).
3. Arendt, D. and Wittbrodt, J. Reconstructing the eyes of Urbilateria. *Philos Trans R Soc Lond B Biol Sci* **356**, 1545–1563 (2001).
4. Lamb, T. D., Collin, S. P. and Pugh, E. N., Jr. Evolution of the vertebrate eye: opsins, photoreceptors, retina and eye cup. *Nat Rev Neurosci* **8**, 960–976 (2007).
5. Provencio, I., Jiang, G., De Grip, W. J., Hayes, W. P. and Rollag, M. D. Melanopsin: an opsin in melanophores, brain, and eye. *Proc Natl Acad Sci USA* **95**, 340–345 (1998).
6. Schomerus, C., Ruth, P. and Korf, H. W. Photoreceptor-specific proteins in the mammalian pineal organ: immunocytochemical data and functional considerations. *Acta Neurobiol Exp (Wars)* **54 Suppl**, 9–17 (1994).
7. Kateriya, S., Nagel, G., Bamberg, E. and Hegemann, P. "Vision" in single-celled algae. *News Physiol Sci* **19**, 133–137 (2004).
8. Filipek, S., Stenkamp, R. E., Teller, D. C. and Palczewski, K. G protein-coupled receptor rhodopsin: a prospectus. *Annu Rev Physiol* **65**, 851–879 (2003).
9. Stryer, L. Cyclic GMP cascade of vision. *Annu Rev Neurosci* **9**, 87–119 (1986).
10. Steinberg, R. H., Fisher, S. K. and Anderson, D. H. Disc morphogenesis in vertebrate photoreceptors. *J Comp Neurol* **190**, 501–508 (1980).
11. Anderson, D. H., Fisher, S. K. and Steinberg, R. H. Mammalian cones: disc shedding, phagocytosis, and renewal. *Invest Ophthalmol Vis Sci* **17**, 117–133 (1978).
12. Hecht, S., Shlaer, S. and Pirenne, M. H. Energy, Quanta, and Vision. *J Gen Physiol* **25**, 819–840 (1942).
13. McBee, J. K., Palczewski, K., Baehr, W. and Pepperberg, D. R. Confronting complexity: the interlink of phototransduction and retinoid metabolism in the vertebrate retina. *Prog Retin Eye Res* **20**, 469–529 (2001).
14. Palczewski, K. Chemistry and biology of vision. *J Biol Chem* **287**, 1612–1619 (2012).
15. Chen, C. K., Wieland, T. and Simon, M. I. RGS-r, a retinal specific RGS protein, binds an intermediate conformation of transducin and enhances recycling. *Proc Natl Acad Sci USA* **93**, 12885–12889 (1996).
16. Fung, B. K., Hurley, J. B. and Stryer, L. Flow of information in the light-triggered cyclic nucleotide cascade of vision. *Proc Natl Acad Sci USA* **78**, 152–156 (1981).
17. Fu, Y. and Yau, K. W. Phototransduction in mouse rods and cones. *Pflugers Archiv: Eur J Physiol* **454**, 805–819 (2007).
18. Stryer, L. Visual excitation and recovery. *J Biol Chem* **266**, 10711–10714. (1991).
19. Leibrock, C. S., Reuter, T. and Lamb, T. D. Molecular basis of dark adaptation in rod photoreceptors. *Eye (Lond)* **12 (Pt 3b)**, 511–520 (1998).
20. Fredriksson, R., Lagerstrom, M. C., Lundin, L. G. and Schioth, H. B. The *G*-protein-coupled receptors in the human genome form five main families. Phylogenetic analysis, paralogon groups, and fingerprints. *Mol Pharmacol* **63**, 1256–1272 (2003).

21. Lagerstrom, M. C. and Schioth, H. B. Structural diversity of *G* protein-coupled receptors and significance for drug discovery. *Nat Rev Drug Discov* **7**, 339–357 (2008).
22. Luttrell, L. M. and Lefkowitz, R. J. The role of beta-arrestins in the termination and transduction of *G*-protein-coupled receptor signals. *J Cell Sci* **115**, 455–465 (2002).
23. Shenoy, S. K., Drake, M. T., Nelson, C. D., Houtz, D. A., Xiao, K., Madabushi, S., Reiter, E., Premont, R. T., Lichtarge, O. and Lefkowitz, R. J. beta-arrestin-dependent, *G* protein-independent ERK1/2 activation by the beta2 adrenergic receptor. *J Biol Chem* **281**, 1261–1273 (2006).
24. Attwood, T. K. and Findlay, J. B. Fingerprinting *G*-protein-coupled receptors. *Protein Eng* **7**, 195–203 (1994).
25. Kolakowski, L. F., Jr. GCRDb: a *G*-protein-coupled receptor database. *Receptors Channels* **2**, 1–7 (1994).
26. Rasmussen, S. G., Choi, H. J., Rosenbaum, D. M., Kobilka, T. S., Thian, F. S., Edwards, P. C., Burghammer, M., Ratnala, V. R., Sanishvili, R., Fischetti, R. F., Schertler, G. F., Weis, W. I. and Kobilka, B. K. Crystal structure of the human beta2 adrenergic*G*-protein-coupled receptor. *Nature* **450**, 383–387 (2007).
27. Haga, K., Kruse, A. C., Asada, H., Yurugi-Kobayashi, T., Shiroishi, M., Zhang, C., Weis, W. I., Okada, T., Kobilka, B. K., Haga, T. and Kobayashi, T. Structure of the human M2 muscarinic acetylcholine receptor bound to an antagonist. *Nature* (2012).
28. Wong, S. K., Parker, E. M. and Ross, E. M. Chimeric muscarinic cholinergic: beta-adrenergic receptors that activate *G*s in response to muscarinic agonists. *J Biol Chem* **265**, 6219–6224 (1990).
29. Conklin, B. R., Farfel, Z., Lustig, K. D., Julius, D. and Bourne, H. R. Substitution of three amino acids switches receptor specificity of Gq alpha to that of Gi alpha. *Nature* **363**, 274–276 (1993).
30. Conklin, B. R., Herzmark, P., Ishida, S., Voyno-Yasenetskaya, T. A., Sun, Y., Farfel, Z. and Bourne, H. R. Carboxyl-terminal mutations of G_q alpha and G_s alpha that alter the fidelity of receptor activation. *Mol Pharmacol* **50**, 885–890 (1996).
31. Scheerer, P., Park, J. H., Hildebrand, P. W., Kim, Y. J., Krauss, N., Choe, H. W., Hofmann, K. P. and Ernst, O. P. Crystal structure of opsin in its *G*-protein-interacting conformation. *Nature* **455**, 497–502 (2008).
32. Park, J. H., Scheerer, P., Hofmann, K. P., Choe, H. W. and Ernst, O. P. Crystal structure of the ligand-free *G*-protein-coupled receptor opsin. *Nature* **454**, 183–187 (2008).
33. Rasmussen, S. G., DeVree, B. T., Zou, Y., Kruse, A. C., Chung, K. Y., Kobilka, T. S., Thian, F. S., Chae, P. S., Pardon, E., Calinski, D., Mathiesen, J. M., Shah, S. T., Lyons, J. A., Caffrey, M., Gellman, S. H., Steyaert, J., Skiniotis, G., Weis, W. I., Sunahara, R. K. and Kobilka, B. K. Crystal structure of the beta2 adrenergic receptor-$G_s$protein complex. *Nature* **477**, 549–555 (2011).
34. Dong, M., Pinon, D. I., Asmann, Y. W. and Miller, L. J. Possible endogenous agonist mechanism for the activation of secretin family *G* protein-coupled receptors. *Mol Pharmacol* **70**, 206–213 (2006).
35. Luck, M. D., Carter, P. H. and Gardella, T. J. The (1-14) fragment of parathyroid hormone (PTH) activates intact and amino-terminally truncated PTH-1 receptors. *Mol Endocrinol* **13**, 670–680 (1999).
36. Gensure, R. C., Shimizu, N., Tsang, J. and Gardella, T. J. Identification of a contact site for residue 19 of parathyroid hormone (PTH) and PTH-related protein analogs in

transmembrane domain two of the type 1 PTH receptor. *Mol Endocrinol* **17**, 2647–2658 (2003).

37. Armour, S. L., Foord, S., Kenakin, T. and Chen, W. J. Pharmacological characterization of receptor-activity-modifying proteins (RAMPs) and the human calcitonin receptor. *J Pharmacol Toxicol Methods* **42**, 217–224 (1999).
38. McLatchie, L. M., Fraser, N. J., Main, M. J., Wise, A., Brown, J., Thompson, N., Solari, R., Lee, M. G. and Foord, S. M. RAMPs regulate the transport and ligand specificity of the calcitonin-receptor-like receptor. *Nature* **393**, 333–339 (1998).
39. Parameswaran, N. and Spielman, W. S. RAMPs: the past, present and future. *Trends Biochem Sci* **31**, 631–638 (2006).
40. Stacey, M., Lin, H. H., Gordon, S. and McKnight, A. J. LNB-TM7, a group of seven-transmembrane proteins related to family-B *G*-protein-coupled receptors. *Trends Biochem Sci* **25**, 284–289 (2000).
41. Little, K. D., Hemler, M. E. and Stipp, C. S. Dynamic regulation of a GPCR-tetraspanin-*G* protein complex on intact cells: central role of CD81 in facilitating GPR56-Galpha q/11 association. *Mol Biol Cell* **15**, 2375–2387 (2004).
42. Lelianova, V. G., Davletov, B. A., Sterling, A., Rahman, M. A., Grishin, E. V., Totty, N. F. and Ushkaryov, Y. A. Alpha-latrotoxin receptor, latrophilin, is a novel member of the secretin family of G protein-coupled receptors. *J Biol Chem* **272**, 21504–21508 (1997).
43. Schulte, G. and Bryja, V. The Frizzled family of unconventional *G*-protein-coupled receptors. *Trends in Pharmacol Sci* **28**, 518–525 (2007).
44. Ruiz-Gomez, A., Molnar, C., Holguin, H., Mayor, F., Jr. and de Celis, J. F. The cell biology of Smo signalling and its relationships with GPCRs. *Biochim Biophys Acta* **1768**, 901–912 (2007).
45. Adler, E., Hoon, M. A., Mueller, K. L., Chandrashekar, J., Ryba, N. J. and Zuker, C. S. A novel family of mammalian taste receptors. *Cell* **100**, 693–702 (2000).
46. Wu, S. V., Rozengurt, N., Yang, M., Young, S. H., Sinnett-Smith, J. and Rozengurt, E. Expression of bitter taste receptors of the T2R family in the gastrointestinal tract and enteroendocrine STC-1 cells. *Proc Natl Acad Sci USA* **99**, 2392–2397 (2002).
47. Sternini, C. Taste receptors in the gastrointestinal tract. IV. Functional implications of bitter taste receptors in gastrointestinal chemosensing. *Am J Physiol Gastrointest Liver Physiol* **292**, G457–G461 (2007).
48. Chandrashekar, J., Mueller, K. L., Hoon, M. A., Adler, E., Feng, L., Guo, W., Zuker, C. S. and Ryba, N. J. T2Rs function as bitter taste receptors. *Cell* **100**, 703–711 (2000).
49. Rozengurt, E. and Sternini, C. Taste receptor signaling in the mammalian gut. *Curr Opin Pharmacol* **7**, 557–562 (2007).
50. Muto, T., Tsuchiya, D., Morikawa, K. and Jingami, H. Structures of the extracellular regions of the group II/III metabotropic glutamate receptors. *Proc Natl Acad Sci USA* **104**, 3759–3764 (2007).
51. Kunishima, N., Shimada, Y., Tsuji, Y., Sato, T., Yamamoto, M., Kumasaka, T., Nakanishi, S., Jingami, H. and Morikawa, K. Structural basis of glutamate recognition by a dimeric metabotropic glutamate receptor. *Nature* **407**, 971–977 (2000).
52. Tsuchiya, D., Kunishima, N., Kamiya, N., Jingami, H. and Morikawa, K. Structural views of the ligand-binding cores of a metabotropic glutamate receptor complexed with

an antagonist and both glutamate and Gd3+. *Proc Natl Acad Sci USA* **99**, 2660–2665 (2002).

53. Bessis, A. S., Rondard, P., Gaven, F., Brabet, I., Triballeau, N., Prezeau, L., Acher, F. and Pin, J. P. Closure of the Venus flytrap module of mGlu8 receptor and the activation process: insights from mutations converting antagonists into agonists. *Proc Natl Acad Sci USA* **99**, 11097–11102 (2002).
54. Kniazeff, J., Bessis, A. S., Maurel, D., Ansanay, H., Prezeau, L. and Pin, J. P. Closed state of both binding domains of homodimeric mGlu receptors is required for full activity. *Nat Struct Mol Biol* **11**, 706–713 (2004).
55. Pin, J. P., Kniazeff, J., Goudet, C., Bessis, A. S., Liu, J., Galvez, T., Acher, F., Rondard, P. and Prezeau, L. The activation mechanism of class-C G-protein coupled receptors. *Biol Cell* **96**, 335–342 (2004).
56. Hlavackova, V., Goudet, C., Kniazeff, J., Zikova, A., Maurel, D., Vol, C., Trojanova, J., Prezeau, L., Pin, J. P. and Blahos, J. Evidence for a single heptahelical domain being turned on upon activation of a dimeric GPCR. *Embo J* **24**, 499–509 (2005).
57. Pin, J. P., Kniazeff, J., Binet, V., Liu, J., Maurel, D., Galvez, T., Duthey, B., Havlickova, M., Blahos, J., Prezeau, L. and Rondard, P. Activation mechanism of the heterodimeric GABA(B) receptor. *Biochem Pharmacol* **68**, 1565–1572 (2004).
58. Conn, P. J. and Pin, J. P. Pharmacology and functions of metabotropic glutamate receptors. *Annu Rev Pharmacol Toxicol* **37**, 205–237 (1997).
59. Li, X., Staszewski, L., Xu, H., Durick, K., Zoller, M. and Adler, E. Human receptors for sweet and umami taste. *Proc Natl Acad Sci USA* **99**, 4692–4696 (2002).
60. Margolskee, R. F. Molecular mechanisms of bitter and sweet taste transduction. *J Biol Chem* **277**, 1–4 (2002).
61. Jones, K. A., Borowsky, B., Tamm, J. A., Craig, D. A., Durkin, M. M., Dai, M., Yao, W. J., Johnson, M., Gunwaldsen, C., Huang, L. Y., Tang, C., Shen, Q., Salon, J. A., Morse, K., Laz, T., Smith, K. E., Nagarathnam, D., Noble, S. A., Branchek, T. A. and Gerald, C. GABA(B) receptors function as a heteromeric assembly of the subunits GABA(B)R1 and GABA(B)R2. *Nature* **396**, 674–679 (1998).
62. Xu, J. and Wojcik, W. J. Gamma aminobutyric acid B receptor-mediated inhibition of adenylate cyclase in cultured cerebellar granule cells: blockade by islet-activating protein. *J Pharmacol Exp Ther* **239**, 568–573 (1986).
63. Johnston, C. A. and Siderovski, D. P. Receptor-mediated activation of heterotrimeric G-proteins: current structural insights. *Mol Pharmacol* **72**, 219–230 (2007).
64. Oldham, W. M. and Hamm, H. E. Structural basis of function in heterotrimeric G proteins. *Q Rev Biophys* **39**, 117–166 (2006).
65. Takai, Y., Sasaki, T. and Matozaki, T. Small GTP-binding proteins. *Physiol Rev* **81**, 153–208 (2001).
66. Lambright, D. G., Noel, J. P., Hamm, H. E. and Sigler, P. B. Structural determinants for activation of the alpha-subunit of a heterotrimeric G protein. *Nature* **369**, 621–628 (1994).
67. Mixon, M. B., Lee, E., Coleman, D. E., Berghuis, A. M., Gilman, A. G. and Sprang, S. R. Tertiary and quaternary structural changes in G_i alpha 1 induced by GTP hydrolysis. *Science* **270**, 954–960 (1995).

68. Coleman, D. E., Berghuis, A. M., Lee, E., Linder, M. E., Gilman, A. G. and Sprang, S. R. Structures of active conformations of G_i alpha 1 and the mechanism of GTP hydrolysis. *Science* **265**, 1405–1412 (1994).
69. Noel, J. P., Hamm, H. E. and Sigler, P. B. The 2.2 A crystal structure of transducin-alpha complexed with GTP gamma S. *Nature* **366**, 654–663 (1993).
70. Lambright, D. G., Sondek, J., Bohm, A., Skiba, N. P., Hamm, H. E. and Sigler, P. B. The 2.0 A crystal structure of a heterotrimeric *G* protein. *Nature* **379**, 311–319 (1996).
71. Wall, M. A., Coleman, D. E., Lee, E., Iniguez-Lluhi, J. A., Posner, B. A., Gilman, A. G. and Sprang, S. R. The structure of the *G* protein heterotrimer G_i alpha 1 beta 1 gamma 2. *Cell* **83**, 1047–1058 (1995).
72. Oldham, W. M. and Hamm, H. E. Heterotrimeric *G* protein activation by *G*-protein-coupled receptors. *Nat Rev Mol Cell Biol* **9**, 60–71 (2008).
73. Berman, D. M. and Gilman, A. G. Mammalian RGS proteins: barbarians at the gate. *J Biol Chem* **273**, 1269–1272 (1998).
74. Hepler, J. R. Emerging roles for RGS proteins in cell signalling. *Trends Pharmacol Sci* **20**, 376–382 (1999).
75. Simon, M. I., Strathmann, M. P. and Gautam, N. Diversity of *G* proteins in signal transduction. *Science* **252**, 802–808 (1991).
76. Wettschureck, N. and Offermanns, S. Mammalian *G* proteins and their cell type specific functions. *Physiol Rev* **85**, 1159–1204 (2005).
77. Landis, C. A., Masters, S. B., Spada, A., Pace, A. M., Bourne, H. R. and Vallar, L. GTPase inhibiting mutations activate the alpha chain of G_s and stimulate adenylyl cyclase in human pituitary tumours. *Nature* **340**, 692–696 (1989).
78. Bers, D. M. Cardiac excitation-contraction coupling. *Nature* **415**, 198–205 (2002).
79. Andersson, R. and Nilsson, K. Cyclic AMP and calcium in relaxation in intestinal smooth muscle. *Nat New Biol* **238**, 119–120 (1972).
80. Kotlikoff, M. I., Kume, H. and Tomasic, M. Muscarinic regulation of membrane ion channels in airway smooth muscle cells. *Biochem Pharmacol* **43**, 5–10 (1992).
81. Ehlert, F. J. Pharmacological analysis of the contractile role of M_2 and M_3 muscarinic receptor in smooth muscle. *Receptors and Channels* **9**, 261–277 (2003).
82. Offermanns, S. and Simon, M. I. $G\alpha 15$ and $G\alpha 16$ couple a wide variety of receptors to phospholipase C. *J Biol Chem* **270**, 15175–15180 (1995).
83. Hamm, H. E., Deretic, D., Arendt, A., Hargrave, P. A., Koenig, B. and Hofmann, K. P. Site of G protein binding to rhodopsin mapped with synthetic peptides from the alpha subunit. *Science* **241**, 832–835 (1988).
84. Blahos, J., 2nd, Mary, S., Perroy, J., de Colle, C., Brabet, I., Bockaert, J. and Pin, J. P. Extreme C terminus of G protein alpha-subunits contains a site that discriminates between Gi-coupled metabotropic glutamate receptors. *J Biol Chem* **273**, 25765–25769 (1998).
85. Griffin, M. T., Figueroa, K. W., Liller, S. and Ehlert, F. J. Estimation of agonist activity at *G* protein-coupled receptors: analysis of M2 muscarinic receptor signaling through $G_{i/o}$,G_s and G_{15}. *J Pharmacol Exp Ther* **321**, 1193–1207 (2007).
86. Fernandez-Fernandez, J. M., Abogadie, F. C., Milligan, G., Delmas, P. and Brown, D. A. Multiple pertussis toxin-sensitive *G*-proteins can couple receptors to

GIRK channels in rat sympathetic neurons when expressed heterologously, but only native G(i)-proteins do so in situ. *Eur J Neurosci* **14**, 283–292 (2001).
87. Schneider, T., Igelmund, P. and Hescheler, J. *G* protein interaction with K+ and Ca2+ channels. *Trends Pharmacol Sci* **18**, 8–11 (1997).
88. Brown, A. M. and Birnbaumer, L. Ionic channels and their regulation by *G* protein subunits. *Annu Rev Physiol* **52**, 197–213 (1990).
89. Kreienkamp, H. J. Scaffolding proteins at the postsynaptic density: shank as the architectural framework. *Handb Exp Pharmacol*, 365–380 (2008).
90. Chudasama, N. L., Marx, S. O. and Steinberg, S. F. Scaffolding proteins in cardiac myocytes. *Handb Exp Pharmacol* 301–325 (2008).
91. Sunahara, R. K., Dessauer, C. W. and Gilman, A. G. Complexity and diversity of mammalian adenylyl cyclases. *Annu Rev Pharmacol Toxicol* **36**, 461–480 (1996).
92. Sunahara, R. K., Dessauer, C. W., Whisnant, R. E., Kleuss, C. and Gilman, A. G. Interaction of Gsalpha with the cytosolic domains of mammalian adenylyl cyclase. *J Biol Chem* **272**, 22265–22271 (1997).
93. Mou, T. C., Masada, N., Cooper, D. M. and Sprang, S. R. Structural basis for inhibition of mammalian adenylyl cyclase by calcium. *Biochemistry* **48**, 3387–3397 (2009).
94. Kaupp, U. B. and Seifert, R. Molecular diversity of pacemaker ion channels. *Annu Rev Physiol* **63**, 235–257 (2001).
95. Kaupp, U. B. and Seifert, R. Cyclic nucleotide-gated ion channels. *Physiol Rev* **82**, 769–824 (2002).
96. Kim, C., Cheng, C. Y., Saldanha, S. A. and Taylor, S. S. PKA-I holoenzyme structure reveals a mechanism for cAMP-dependent activation. *Cell* **130**, 1032–1043 (2007).
97. Nishida, M. and MacKinnon, R. Structural basis of inward rectification: cytoplasmic pore of the G protein-gated inward rectifier GIRK1 at 1.8 A resolution. *Cell* **111**, 957–965 (2002).
98. Whorton, M. R. and MacKinnon, R. Crystal structure of the mammalian GIRK2 K+ channel and gating regulation by G proteins, PIP2, and sodium. *Cell* **147**, 199–208 (2011).
99. Taussig, R., Tang, W. J., Hepler, J. R. and Gilman, A. G. Distinct patterns of bidirectional regulation of mammalian adenylyl cyclases. *J Biol Chem* **269**, 6093–6100, (1994).
100. Taussig, R., Iniguez-Lluhi, J. A. and Gilman, A. G. Inhibition of adenylyl cyclase by G_i alpha. *Science* **261**, 218–221, (1993).
101. Berridge, M. J. Inositol trisphosphate and diacylglycerol as second messengers. *Biochem J* **220**, 345–360, (1984).
102. Candell, L. M., Yun, S. H., Tran, L. L. and Ehlert, F. J. Differential coupling of subtypes of the muscarinic receptor to adenylate cyclase and phosphoinositide hydrolysis in the longitudinal muscle of the rat ileum. *Mol Pharmacol* **38**, 689–697 (1990).
103. Mikoshiba, K., Furuichi, T., Miyawaki, A., Yoshikawa, S., Nakagawa, T., Yamada, N., Hamanaka, Y., Fujino, I., Michikawa, T., Ryo, Y. *et al.* Inositol trisphosphate receptor and Ca2+ signalling. *Philos Trans R Soc Lond B Biol Sci* **340**, 345–349 (1993).
104. Nishizuka, Y. The role of protein kinase C in cell surface signal transduction and tumour promotion. *Nature* **308**, 693–698, (1984).
105. Smrcka, A. V., Hepler, J. R., Brown, K. O. and Sternweis, P. C. Regulation of polyphosphoinositide-specific phospholipase C activity by purified G_q. *Science* **251**, 804–807 (1991).

106. Taylor, S. J., Chae, H. Z., Rhee, S. G. and Exton, J. H. Activation of the beta 1 isozyme of phospholipase C by alpha subunits of the G_q class of G proteins. *Nature* **350**, 516–518 (1991).
107. Waldo, G. L., Boyer, J. L., Morris, A. J. and Harden, T. K. Purification of an AlF4- and G-protein beta gamma-subunit-regulated phospholipase C-activating protein. *J Biol Chem* **266**, 14217–14225 (1991).
108. Harden, T. K., Waldo, G. L., Hicks, S. N. and Sondek, J. Mechanism of activation and inactivation of Gq/phospholipase C-beta signaling nodes. *Chem Rev* **111**, 6120–6129 (2011).
109. Bolger, G. T., Gengo, P., Klockowski, R., Luchowski, E., Siegel, H., Janis, R. A., Triggle, A. M. and Triggle, D. J. Characterization of binding of the Ca^{++} channel antagonist, [3H]nitrendipine, to guinea-pig ileal smooth muscle. *J Pharmacol Exp Ther* **225**, 291–309 (1983).
110. Hardie, R. C. TRP channels and lipids: from Drosophila to mammalian physiology. *J Physiol* **578**, 9–24 (2007).
111. Hilgemann, D. W., Feng, S. and Nasuhoglu, C. The complex and intriguing lives of PIP2 with ion channels and transporters. *Sci STKE* **2001**, re19 (2001).
112. Shenoy, S. K. and Lefkowitz, R. J. beta-Arrestin-mediated receptor trafficking and signal transduction. *Trends in Pharmacol Sci* **32**, 521–533 (2011).
113. Griffin, M. T., Figueroa, K. W., Liller, S. and Ehlert, F. J. Estimation of agonist activity at G protein-coupled receptors: analysis of M2 muscarinic receptor signaling through $G_{i/o}$, G_s and G_{15}. *J Pharmacol Exp Ther* **321**, 1193–1207 (2007).

Affinity and Efficacy of the Agonist-Receptor-*G*-Protein Interaction

6

When a drug binds to and activates a GPCR, the potential outcomes are numerous. For, depending on the location of its receptor, the drug could cause diuresis, hypotension, pain relief, motivation, alertness, appetite suppression, a reduction in airway resistance in the lung, and many other effects. Each of these is the result of a train of events beginning with the initial receptor activation and ending with the final response. One could also measure the different steps that lead up to the final response, which may include things like, adenylate cyclase activity, phospholipase $C\beta$ activity, Ca^{2+} mobilization, activation of ion channels, recruitment of arrestin, phosphorylation of proteins, activation of contractile filaments, and modulation of the firing rate of a neuron in a neuronal network. Any of these responses, whether proximal or distal in the signaling pathway, could be analyzed to glean information about the initial agonist-receptor interaction and how this interaction can be quantified in terms of the two components, affinity and efficacy.

But how could all of the steps from receptor activation to the final response be modeled accurately so that one could deduce the former through measurement of the latter? There are two simple solutions to this problem, and both depend on having an accurate model for the receptor-*G* protein-interaction function (stimulus). One approach involves reverse engineering.[1] That is, if we know the input function to a system and can measure its output, we can empirically describe the mathematical function that converts the input into the output (transducer function). The other approach (response-clamp analysis) involves determining the relationship between concentrations of agonist that generate equivalent responses under different conditions.[2] Equivalent responses imply that equal stimuli generated them, regardless of the nature of the transducer function. The latter is irrelevant to the analysis.

Both of these methods are described in the Chapter 7, and they depend on having an accurate equation that incorporates the components of affinity and efficacy into the initial receptor-activation function. In this chapter, therefore, I explain how to relate agonist affinity and efficacy to the first step in the signaling pathway — receptor-*G* protein activation. Ultimately, I show that the agonist activation, expressed as a function (f) of the agonist concentration (D), is simply:

$$f(x) = \frac{\varepsilon D R_T}{D + K_D}, \tag{6.1}$$

in which ε represents the efficacy of the agonist, R_T the concentration of functional receptors and K_D, the observed dissociation constant of the agonist.

I begin with a quick summary of receptor-*G* protein interactions, which were described in more detail in Chapter 5.

The Receptor-*G* Protein Activation Cycle

Figure 6.1 shows a hypothetical receptosome consisting of receptor, *G* protein, and adenylate cyclase. The receptor is tethered by its C-terminus to a scaffolding protein, which in turn is bound to lower order scaffolds. Adenylate cyclase is also attached to the same scaffolding protein as the receptor, while the *G* protein is anchored to the

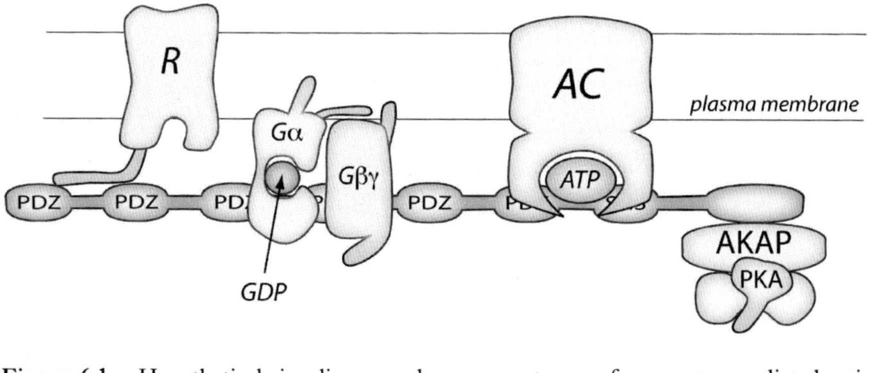

Figure 6.1: Hypothetical signaling complex or receptosome for receptor-mediated activation of protein kinase A. The figure illustrates a GPCR tethered via its C terminus to a scaffolding protein containing many PDZ domains (see Chapter 5). Upon activation, the receptor interacts with a heterotrimeric *G* protein consisting of Gα and $G\beta\gamma$ subunits. The latter activate downstream effectors including adenylate cyclase (AC), which generates cAMP from ATP. The cAMP binds to the regulatory subunit of protein kinase A (PKA), causing it to dissociate from the catalytic subunit. This releases the catalytic subunit from inhibition and enables it to phosphorylate various substrates on serine and threonine residues.

plasma membrane. The close proximity of the proteins ensures the rapid transfer of information from the receptor to the adenylate cyclase through the intermediary *G* protein.

In the resting state, the concentration of GTP in the cytosol can be high (0.01–0.1 mM) relative to its dissociation constant for the *G* protein (submicromolar).[3] This situation leads to full occupancy of the guanine nucleotide-binding site on the *G* protein. It might seem that the guanine nucleotide in question is GTP, but the *G* protein acts like an enzyme and quickly converts any bound GTP to GDP.[4] When GDP is bound, it is engulfed between the ras and helical domains of the Gα subunit, and its pathway for release from the *G* protein is blocked by the associated $G\beta\gamma$ subunits (see Chapter 5, Figure 5.9). In the resting state, therefore, practically all *G* proteins are inactive with GDP bound to them nearly irreversibly. GTP is nearby in abundance ready to jump into the guanine nucleotide-binding pocket of the *G* protein as soon as it becomes vacant.

The binding of an agonist molecule causes the receptor to associate with the *G* protein. It may be that the *G* protein is already touching the receptor, but the agonist-induced change in the shape of the receptor generates a more productive binding interface between the C-terminus of the $G\alpha$ subunit and the cytoplasmic extensions of helix 3 and 5 of the receptor. This interaction causes the *G* protein to open and release tightly bound GDP. A molecule of GTP rapidly replaces GDP, and this causes the *G* protein to split into its GTP-$G\alpha$ and $G\beta\gamma$ subunits, which are free to bind with and activate an effector, like adenylate cyclase (see Chapter 5, Figure 5.4). A given GTP-$G\alpha$ subunit is short-lived, however, because of the GTPase activity of the $G\alpha$ subunit, which converts GTP-$G\alpha$ to GDP-$G\alpha$. The GDP-$G\alpha$ subunit lacks high affinity for adenylate cyclase, so it dissociates from it and coalesces with a free $G\beta\gamma$ subunit to form the holoprotein complex. In this form the *G* protein is inactive.

What I have described so far represents one cycle of activation for the Gα subunit. The cycle began with the $G\alpha$ subunit in the resting state bound with GDP and $G\beta\gamma$ and ended in the same condition, but with new GDP and perhaps new $G\beta\gamma$ as well. This *G* protein cycle represents one quantum of activation in time. As long as the agonist is occupying a significant fraction of the population of receptors, the cycle continues to cycle at each agonist-occupied receptor, and there is a steady production of new GTP-$G\alpha$ subunits at a rate that increases as a function of receptor occupancy. There is also a corresponding rate of decay of the pool of GTP-Gα subunits to GDP-$G\alpha$. The net result is a steady-state concentration of GTP-$G\alpha$ as long as receptor activation is constant and desensitization is minimal. Upon removal of the agonist, the formation of GTP-$G\alpha$ ceases and the activity

Step 1 $D + R + G\text{-}GDP \rightleftharpoons DR + G\text{-}GDP \rightleftharpoons DRG\text{-}GDP$

Step 2 $D + RG\text{-}GDP \rightleftharpoons DRG\text{-}GDP \rightleftharpoons DRG + GDP$

Step 3 $D + RG\text{-}GTP \rightleftharpoons DRG\text{-}GTP \rightleftharpoons DR + G\text{-}GTP$

Step 4 $GTP + G\alpha G\beta\gamma \rightleftharpoons GTP\text{-}G\alpha G\beta\gamma \rightleftharpoons GTP\text{-}G\alpha + G\beta\gamma$

Step 5 $GTP + G\alpha + AC \rightleftharpoons GTP\text{-}G\alpha + AC \rightleftharpoons GTP\text{-}G\alpha AC$

Step 6 $GDP\text{-}G\alpha AC \rightleftharpoons GDP\text{-}G\alpha + AC \rightleftharpoons GDP + G\alpha + AC$

Step 7 $GDP + G\alpha + G\beta\gamma \rightleftharpoons GDP\text{-}G\alpha + G\beta\gamma \rightleftharpoons GDP\text{-}G\alpha G\beta\gamma$

Figure 6.2: The seven allosteric interactions that enable an agonist (*D*) to activate a GPCR (R), its *G* protein (G), and ultimately a downstream signaling protein, like adenylate cyclase (AC).

of adenylate cyclase decays coincidently with the GTPase-mediated conversion of GTP-$G\alpha$ into GDP-$G\alpha$.

This sequence is illustrated in Figure 6.2. Not shown are the important scaffolding proteins, but the figure does identify the essential steps in receptor-*G* protein-effector interactions that we need to explain if we are to understand how GPCRs work and what the molecular determinants of affinity and efficacy are. These steps are:

1. The binding of agonist to the receptor causes the receptor to associate with the GDP-bound *G* protein.
2. The binding of agonist to the orthosteric site on the receptor reduces the affinity of GDP for its binding site on the *G* protein. This enables the competitive displacement of GDP by GTP.
3. GTP reduces the affinity of the agonist-receptor complex for the *G* protein and enables the latter to dissociate from the receptor.
4. The binding of GTP to the $G\alpha$ subunit causes the $G\beta\gamma$ subunits to dissociate from each other.
5. The binding of GTP to $G\alpha$ causes it to bind with an effector.
6. Following the enzymatic conversion of the GTP-$G\alpha$ to GDP-$G\alpha$ by the GTPase activity of $G\alpha$, GDP-$G\alpha$ dissociates from the effector.
7. GDP-$G\alpha$ associates with $G\beta\gamma$ to form the GDP-$G\alpha G\beta\gamma$ holoprotein complex.

Each step of this cycle represents an allosteric interaction between proteins and ligands. Recall from Chapter 4 that an allosteric interaction is a reciprocal modulation in the binding of one ligand to a receptor by another ligand. Small molecules (e.g., agonist, GTP, and GDP) as well as proteins behave as ligands in these allosteric interactions. For example, step 7 involves an allosteric interaction between the binding of GDP (ligand A) and $G\beta\gamma$ (ligand B) to different allosteric sites on $G\alpha$ (allosteric receptor protein *C*). This interaction alters the interface through which $G\beta\gamma$ and $G\alpha$ interact and influences whether they remain bound to each other with high affinity or come apart because of their mutual low affinity for each other. The problem of understanding the *G* protein cycle involves determining what types of allosteric interactions in steps 1–7 will enable the agonist to catalyze the generation of free GTP-$G\alpha$ and $G\beta\gamma$ subunits that bind to and activate various effectors, like adenylate cyclase. Thus, one can formulate a question about the nature of the allosterism for each step in the *G* protein cycle. These are listed below, with the components that behave as ligands denoted by "*L*" and those that behave as allosteric receptors denoted by "*R*".

The seven questions are, "What is the nature of the allosteric interaction between

1. the binding of agonist (*L*) and the GDP-*G* protein complex (*L*) to the receptor (*R*)?"
2. the binding of agonist (*L*) and GDP (*L*) to the receptor-*G* protein complex (*R*)?"
3. the binding of agonist (*L*) and the GTP-*G* protein complex (*L*) to the receptor (*R*)?"
4. the binding of GTP (*L*) and $G\beta\gamma$ (*L*) to $G\alpha$ (*R*)?"
5. the binding of GTP (*L*) and the effector (*L*) to $G\alpha$ (*R*)?"
6. the binding of GDP (*L*) and the effector (*L*) to $G\alpha$ (*R*)?"
7. the binding of GDP (*L*) and $G\beta\gamma$ (*L*) to $G\alpha$ (*R*)?"

If we can answer these questions, we should be able to understand why the agonist-receptor complex catalyzes the production of free GTP-$G\alpha$ and $G\beta\gamma$ subunits and why these subunits bind to and turn on their effectors.

It might seem that these questions are of the type that can only be answered by obtaining experimental evidence. But given the set of conditions outlined in Figure 6.2, there is usually only one way that most of steps 1–7 could possibly work if information is to flow through the cycle. All that is required to explain these steps is a little armchair philosophy. In the sections that follow, each of these steps is explained, and some experimental evidence illustrating the nature of these allosteric interactions is presented.

Analysis of the Receptor-*G* Protein-Effector Interaction as a Series of Linked Allosteric Steps

Let me first give a brief review of some of the principles of allosterism that were previously presented in Chapter 4. We will need to adapt our analysis for the situation where one of the two interacting ligands is a protein that binds with its receptor in or near the plasma membrane.

At the population level of analysis, the observed dissociation constant is equivalent to the concentration of ligand required for half-maximal occupancy of the receptor population. The reciprocal of the observed dissociation constant is equivalent to the observed affinity constant, and the affinity of a ligand for its site on a receptor is proportional to its affinity constant.

As described in Chapter 4, allosteric interactions can occur when two ligands bind to different sites on the same receptor protein. If the binding of one ligand (A) alters the binding of the other (B), then the two ligands interact allosterically. These interactions are reciprocal. That is, if the binding of A to its site increases the affinity of B for its site, then the binding of B will enhance the affinity of A for its site to the same extent. This interaction is defined as positive cooperativity. Negative cooperativity refers to the situation where the ligands allosterically inhibit the binding of each other.

Allosteric ligands have a limit to their effects. Consider a hypothetical example in which ligand A allosterically enhances the binding of ligand B. At the population level of analysis, this effect is manifest as a concentration-dependent increase in the observed affinity constant of B as the concentration of A increases. When A has completely saturated its allosteric site, it will have a maximal effect on the observed affinity constant of B. Let us say that this effect represents a ten-fold increase in the affinity of B. At lower concentrations of A, the effect on the affinity of B will be less. Thus, it appears that increasing concentrations of A gradually morph the receptor into a shape that exhibits higher affinity for B.

This behavior is really an illusion because we are monitoring a population of receptors, and we have defined affinity in terms of the observed affinity constant. Each receptor can only exist in discrete conformations (at least two) that have unique microscopic affinity constants. At the single receptor level, the binding of the allosteric ligand A increases the probability that the receptor will assume a conformation that exhibits higher affinity for B. Each receptor is randomly undergoing quantal changes in affinity and not gradual changes. At the population level, the summation of these random events yields a gradual change in the observed affinity of B in the presence of increasing concentrations of A (see Chapter 4, Figure 4.2).

This creates a problem in explaining population behavior with an illustration of a single receptor undergoing conformational changes (usually two). The picture would seem to imply only quantal events, but the mathematics of population behavior predicts a gradual, concentration-dependent change in the affinity of ligand B by ligand A. The reader should be aware of this dichotomy inherent to the upcoming illustrations (e.g., Figure 6.4) describing allosteric interactions for steps 1–7 of the *G* protein cycle. In a subsequent section of this chapter, the *G* protein cycle will be explained at the level of conformational states of the receptor and *G* protein.

Another issue relates to how the allosteric interactions of the *G* protein cycle are described mathematically. In the context of drug-receptor interactions, allosterism involves the binding of two relatively small ligands to a larger receptor with two binding sites, one for each ligand. The ligands are usually dissolved in solution and are present in great excess relative to the receptor, so that the binding of the ligand to the receptor does not alter its free concentration in solution to a significant extent. With regard to the seven steps of the *G* protein cycle, only step 2 behaves in this manner. In the other six steps, one of the ligands is a protein comparable in size to its operational receptor. It is usually not free in solution, but tethered to the membrane or a scaffolding protein. When it binds to its receptor, its "free" concentration in the local environment is greatly reduced. Thus, the allosteric interaction alters the availability of the protein ligand in the membrane.

The final issue relates to the use of affinity constants to describe the allosteric interactions of the *G* protein cycle. For drug-receptor interactions, the observed affinity constant (K_{obs}) is defined as, $K_{\text{obs}} = [DR]/[D][R]$, in which $[D]$ represents the concentration of the drug in solution, $[R]$ the concentration of receptors, and $[DR]$ the concentration of the drug-receptor complex. The observed dissociaton constant (K_D) is simply equal to the reciprocal of K_{obs}. For receptor-*G* protein interactions, the binding of the *G* protein to the receptor occurs within or near the plasma membrane, and it is useful to normalize the concentration of the *G* protein relative to that of the receptor. Thus, for the interaction, $G + R \leftrightarrow GR$, in which G and R are present in equal amounts, an affinity constant of 10^{-2} means that at any given instant, about one-percent (0.99%) of the receptor population is bound with *G* protein. At the single receptor level, the *G* protein is bound to the receptor 1% of the time on average.

Now let us return to our discussion of the allosteric nature of the individual steps of the *G* protein activation cycle.

***Step 1: Formation of the agonist-receptor-G protein-GDP quaternary complex*:** In the resting state, *G* proteins are bound with GDP nearly irreversibly. There is plenty of GTP in the cytosol, which could bind to the *G* protein if it were

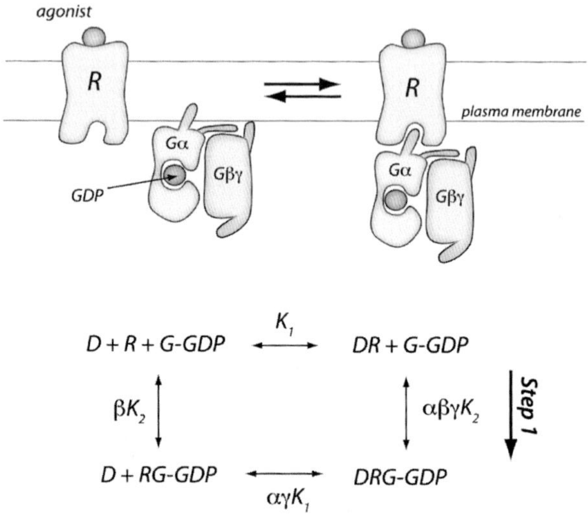

Figure 6.3: Step 1, the allosteric interaction between the binding of agonist and GDP-bound *G* protein (G-GDP) with a GPCR (G). The parameters K_1 and βK_2 denote the microscopic affinity constants for the binding of agonist and GDP-bound *G* protein (*GDP-G*) with the receptor, respectively. The cooperativity constant for this interaction is denoted by the product, $\alpha\gamma$.

not occupied by GDP. In fact, under equilibrium conditions, GTP could easily out-compete GDP and cause a near complete activation of the pool of *G* proteins without any help from the agonist-receptor complex. GDP dissociates very slowly from the *G* protein, however. Any GTP that does bind after a rare GDP-dissociation event is rapidly hydrolyzed to GDP by the GTPase activity of $G\alpha$. Thus, there is little formation of the GTP-$G\alpha$ complex under resting conditions because the dissociation of GDP is extremely slow relative to the rapid rate of hydrolysis of GTP. The first step in the process of displacing GDP from the *G* protein is the agonist-induced association of the receptor with the GDP-*G* protein complex as shown in Figure 6.3.

When the agonist binds to the receptor, it induces a conformational change that enables the receptor to bind more effectively with the *G* protein. Only the trimeric form of the *G* protein ($G\alpha G\beta\gamma$) can associate with the receptor — free $G\alpha$ and $G\beta\gamma$ cannot.

The scheme beneath the illustration tracks the allosteric changes in ligand affinity. K_1 and βK_2 denote the affinity constants of the agonist (*D*) and the GDP-*G* protein complex (*GDP-G*) for the receptor, respectively. The value of βK_2 is much less than one because there is little pre-coupling of receptor to the *G* protein in the absence of agonist. We could have simply used the constant K_2 to describe this equilibrium, but K_2 is used later for the interaction between R and the free G protein (G). Other constants also appear more complicated for an analogous reason. The binding of the agonist to the receptor increases the affinity of *GDP-G* for the receptor by the factor $\alpha\gamma$, which is a measure of the positive cooperativity. Analogously, the binding of *GDP-G* to R increases the affinity of D for the receptor by the same cooperativity factor ($\alpha\gamma$) because allosteric interactions are reciprocal. Thus, the affinity constant for the binding of *GDP-G* to *DR* is denoted by $\alpha\beta\gamma K_2$, and the affinity constant of the agonist for the RG-GDP complex is denoted by $\alpha\gamma K_1$. The net result is that the addition of agonist causes more of the receptor to associate with G-GDP. Alternatively, the G protein may already be associated with the unoccupied receptor in the absence of agonist and the binding of agonist causes this interaction to strengthen.

Step 2: Negative cooperativity between the binding of agonist and GDP to the receptor-G protein complex: Having persuaded the receptor to associate with the *G*-GDP complex, the agonist now needs to remove GDP. This is accomplished through negative cooperativity as shown in Figure 6.4. When the agonist is bound, it induces a conformational change in the receptor-*G* protein complex that reduces the affinity of GDP for its binding site. This increases the rate of dissociation of GDP, which gives GTP more opportunities to bind onto the *G* protein whenever it

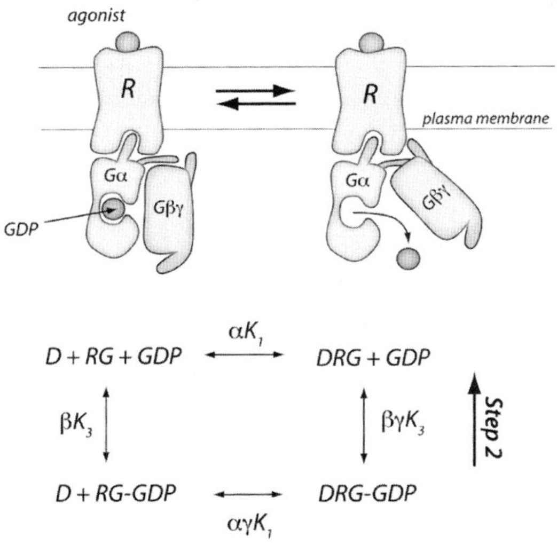

Figure 6.4: Step 2, The allosteric interaction between the binding of agonist (D) and GDP with the receptor-G protein complex (RG). The parameters αK_1 and βK_3 denote the microscopic affinity constants for the binding of agonist (D) and GDP to the receptor, respectively. The cooperativity constant for this interaction is denoted by the scalar, γ.

is vacant. The agonist causes the rate constant for dissociation of GDP to be in the same range or greater than that of the GTPase activity of the *G* protein. This pushes the system away from the steady state dominated by the GDP-*G* protein complex closer to an equilibrium where GTP is better able to occupy the *G* protein. At a true equilibrium (i.e., when there is no GTPase activity), *G* proteins would be nearly saturated with GTP because its concentration in the cytosol is high.

One might think that if the agonist ultimately reduces the affinity of GDP and enables GTP to bind, then the agonist must increase the affinity of GTP for the *G* protein. This is certainly not the case because if the agonist increased the affinity of GTP, this would stabilize the agonist-receptor-*G* protein-GTP quaternary complex and prevent the *G* protein from stimulating the effector. In other words, the agonist would inhibit signaling. It turns out that the agonist reduces the affinity of both GDP and GTP for the *G* protein. We know this because the reciprocal interaction (i.e., GDP- and GTP-induced reduction in agonist affinity) has been demonstrated for many receptors.[5–8] The agonist-induced reduction in the affinity of GTP does not lead to a reduction in the equilibrium binding of GTP because its K_D for the *G* protein is in the submicromolar range[9], and its concentration in the cytosol is near 10^{-4} M[3]. Even if the agonist reduced the affinity of GTP to only 0.1%, GTP could still nearly saturate the *G* protein under equilibrium conditions.

We do not have to assume that the concentration of GTP is nearly saturating. In the absence of agonist, the *G* protein is in a steady-state condition with GDP locked in its binding site. The amount of GTP bound to the *G* protein under this condition is essentially zero. The agonist pushes the system closer to an equilibrium where any reasonable concentration of GTP will equilibrate with the *G* protein. The resulting increase in the binding of GTP would cause an increase in stimulation over the basal condition.

The scheme beneath the drawing in Figure 6.4 quantifies the observed affinities and negative cooperativity between the binding of the agonist and GDP to the receptor-*G* protein complex. The affinity constant of the agonist for the receptor-*G* protein complex is denoted by αK_1 and that for GDP is denoted by βK_3. The binding of the agonist to the receptor reduces the affinity of GDP for its site on the *G* protein by the factor γ, which is a measure of the negative cooperativity. Since the cooperativity is negative, then $0 < \gamma < 1$. GDP also reduces the affinity of the agonist for the receptor by the same factor.

The ability of the agonist to reduce the affinity of GDP for the *G* protein and thereby increase its dissociation kinetics is a measure of agonist efficacy. The efficacy of the agonist, therefore, is inversely proportional to γ.

There is much evidence for agonist-induced reduction in the affinity of both GDP and GTP for the *G* protein. A simple experiment to detect this effect would be to measure the binding of labeled GDP to the *G* protein in the presence and absence of the agonist to determine if the agonist allosterically modifies the binding affinity of GDP. This approach does not work very well in many tissues because there are many different types of *G* proteins that bind GDP, and the agonist-mediated modulation of a small subset of these may be difficult to measure.

Even if we could use a selective GDP analog that only bound to the same type *G* protein that interacted with the receptor in question, the approach still might not work. A given GPCR interacts with many *G* proteins and the ratio of *G* protein to receptor is usually quite large.[10] As we saw in Chapter 4, the magnitude of the allosteric effect of a ligand is greatly reduced if the ratio of the number of sites that it modulates relative to that to which it binds is large. For example, the allosteric effect of a benzodiazepine on the binding of GABA is expected to be only half that of the effect of GABA on the binding of benzodiazepine because the ratio of benzodiazepine : GABA binding sites on the $GABA_A$ receptor is 1:2 (e.g., see Chapter 4, Figure 4.7). There could also be spatial differences in the distribution of the total pool of *G* protein compared to the pool of receptor that interacts with the *G* protein. Other types of receptors might interact with different pools of the same *G* protein at different loci in the cell.

Nonetheless, Jakubik and coworkers[8] have clearly shown that muscarinic agonists allosterically inhibit the binding of $[^3H]$GDP to membranes prepared from Chinese hamster ovary (CHO) cells expressing the M_2 muscarinic receptor. The negative cooperativity between the binding of GDP and muscarinic agonists is correlated with the efficacy of the agonists at the M_2 receptor. Muscarinic agonists also increase the dissociation rate of $[^3H]$GDP from $G_{i/o}$ in a manner that is proportional to efficacy.

Another approach for studying allosteric interactions between receptor agonists and the binding of guanine nucleotides to *G* proteins is based on the reciprocity of allosteric interactions — if the agonist reduces the affinity of GDP, then GDP should reduce the affinity of the agonist. Often there is a selective way to measure the binding of an agonist to a specific receptor in the absence and presence of GDP. It does not matter that the GDP in this type of experiment will be interacting with many *G* proteins other than the specific pool that interacts with the receptor. Only cooperative interactions at the specific receptor will be measured because of the specificity of the radioligand used to label the receptor. This type of experiment has been done with numerous *G* protein-coupled receptors, although a much more common related experiment has been to measure the effect of GTP analogs on agonist binding.

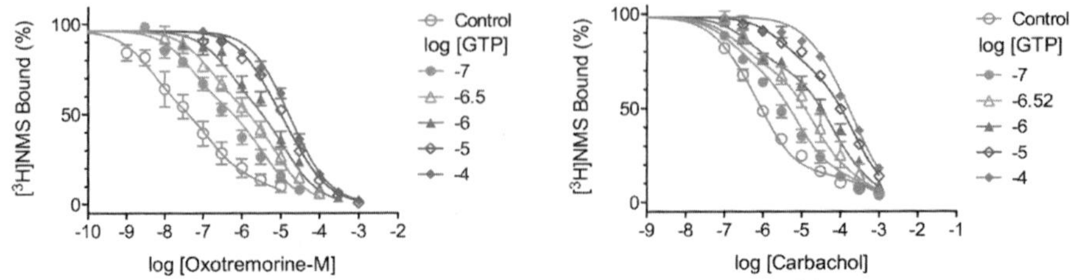

(a) (b)

Figure 6.5: The competitive inhibition of the binding of [^{3}H]NMS to the rat myocardium by the agonists, oxotremorine-M (a) and carbachol (b), in the presence of a nucleotide regenerating system and various concentrations of GTP. The data are from Ehlert and Rathbun[11] with permission. The theoretical curves represent the least squares fit to a three-site model. Two of the sites are allosterically regulated by GTP and undergo large reductions in observed affinity in the presence of GTP. The third site is unaffected by GTP. The parameter values are described in the text under the section entitled, "How well does the model predict the behavior of *G* protein-coupled Receptors."

An example of the effect of GTP on the binding of agonists to the M_2 muscarinic receptor is shown in Figure 6.5. In this experiment, the antagonist, [^{3}H]N-methylscopolamine ([^{3}H]NMS), was used to label muscarinic receptors in the rat myocardium.[11] This tissue expresses a nearly homogeneous population of the M_2 receptor. The specific binding of [^{3}H]NMS was measured at a fixed concentration (0.5 nM) in the presence of increasing concentrations of the muscarinic agonist oxotremorine-M (panel *a*) and carbachol (panel *b*). These agonists compete with [^{3}H]NMS for occupancy of the M_2 receptor, causing a competitive inhibition of the binding. The measurements are expressed as a percentage of control-specific binding measured in the absence of agonist. Thus, one can equate percent displacement of [^{3}H]NMS binding with percent receptor occupancy after correcting for the competitive effect of [^{3}H]NMS. The experiment was repeated in the presence of various concentrations of GTP. The figure shows that the competition curves shift to the right about 140- (oxotremorine-M) and 46-fold (carbachol) in the presence of a maximally effective concentration of GTP (0.1 mM).

Additional experiments showed that under the conditions of the assay, there was little effect of GTP on the binding of [^{3}H]NMS. Thus, the effect of GTP on agonist binding cannot be attributed to direct competition at the orthosteric binding site; otherwise the guanine nucleotide would have inhibited the binding of both the agonists and [^{3}H]NMS. Rather, the effect of GTP is mediated allosterically through another binding site, presumably the guanine nucleotide-binding pocket of G_i. In addition, the large dextral shifts in the agonist/[^{3}H]/NMS competition curves can be attributed entirely to a reduction in the affinity of the agonists, and not to an increase in the affinity of [^{3}H]NMS. When the experiment was repeated using GDP instead of GTP, a similar result was obtained by other investigators using the agonist, carbachol.[5,8,12] Thus, when GTP or GDP binds to the *G* protein, it reduces the affinity of agonists for the receptor, but not that of the antagonist, NMS. It follows from the reciprocity of allosteric interactions that when carbachol and oxotremorine-M bind to the M_2 muscarinic receptor, they allosterically inhibit the binding of GTP and GDP to the *G* protein.

This type of experiment has been repeated with a group of muscarinic agonists having varying degrees of intrinsic efficacy as determined from their ability to mediate an inhibition of adenylate cyclase activity in the myocardium.[13] It was found that the inhibitory effect of GTP (and presumably that of GDP) on agonist binding affinity is proportional to the efficacy of the agonist (see Figure 6.6). These results are consistent with the idea that the agonist-receptor complex elicits a signal by reducing the affinity of GDP for the *G* protein and thereby enabling GTP to bind and activate the system.

Step 3: Dissociation of the GTP-G protein complex from the receptor: Once GTP has replaced GDP on the guanine nucleotide binding site of the *G* protein, the resulting GTP-bound form of the *G* protein needs to dissociate from the agonist-receptor complex if it is to be free to activate the effector. Figure 6.7 shows this dissociation step. It essentially represents the reverse of step 1 but with GTP bound to the *G* protein. The affinity of the agonist-receptor complex (*DR*) for the *G* protein-GTP complex (*G-GDP*) is defined as $\alpha\beta^*\gamma^*K_2$. In the absence of agonist, the affinity of the *G-GDP* for *R* is represented by β^*K_2, and the negative

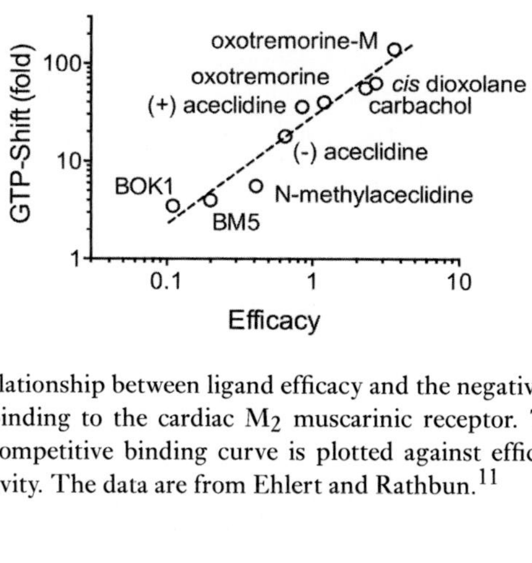

Figure 6.6: The relationship between ligand efficacy and the negatively cooperative effect of GTP on ligand binding to the cardiac M_2 muscarinic receptor. The fold-shift in the agonist/[^{3}H]NMS competitive binding curve is plotted against efficacy for inhibition of adenylate cyclase activity. The data are from Ehlert and Rathbun.[11]

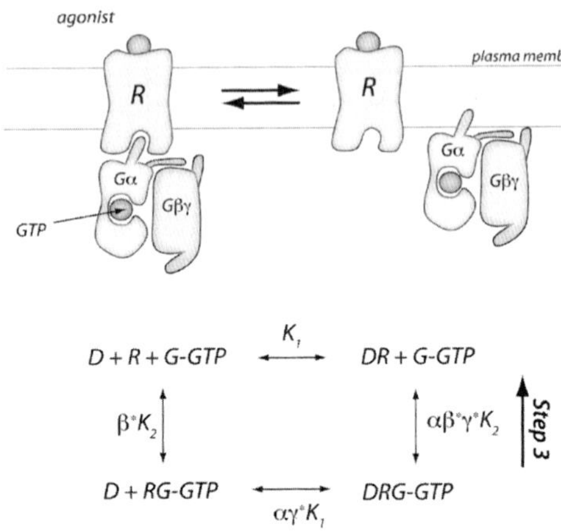

Figure 6.7: Step 3, The allosteric interaction between the binding of agonist (D) and the GTP-bound *G* protein (G-GTP) with the receptor (*R*). The parameters K_1 and $\beta^* K_2$ denote the microscopic affinity constants for the binding of agonist (*D*) and the GTP-bound *G* protein (*G*-GTP) to the receptor, respectively. The cooperativity constant for this allosteric interaction is denoted by the scalar, $\alpha\gamma^*$.

cooperativity between the binding of *G-GDP* and *D* to *R* is, therefore, $\alpha\gamma^*$. While this scalar may not be much less than one, it is clear that the DRG-GTP complex is very unstable because the holoprotein form of *G-GTP* is very unstable as shown by the following step 4.

Step 4: Negative cooperativity between the binding of GTP and $G\beta\gamma$ to $G\alpha$: After the *GTP-G* protein complex dissociates from the receptor, it splits apart into its GTP-$G\alpha$ and $G\beta\gamma$ subunits. This is accomplished through negative cooperativity between the binding of GTP and $G\beta\gamma$ to their respective binding sites on $G\alpha$ as shown in Figure 6.8. When GTP is bound, it induces a conformational change in the $G\alpha$ subunit, which reduces the affinity of $G\beta\gamma$ for $G\alpha$. This allosteric effect causes the holoprotein form of the *G* protein to dissociate into its two subunits.[4,14,15]

This allosteric interaction is represented by the scheme beneath the illustration of the *G* protein. K_4^* and K_5 denote the affinity constants of GTP and $G\beta\gamma$ for their binding sites on $G\alpha$, respectively. The binding of GTP to $G\alpha$ reduces the affinity of $G\beta\gamma$ for $G\alpha$ by the factor, π^*, which is a measure of the negative cooperativity. Similarly, the binding of $G\beta\gamma$ to $G\alpha$ reduces the affinity of GTP by the same factor. Thus, the affinity constant for the binding of GTP to $G\alpha G\beta\gamma$ is $\pi^* K_4^*$, and the affinity constant of $G\beta\gamma$ for GTP-$G\alpha$ is $\pi^* K_5^*$. The net result is that the binding of GTP causes more of the *G* protein to split apart into GTP-Gα and $G\beta\gamma$. The free subunits of the *G* protein are now able to interact with their effectors.

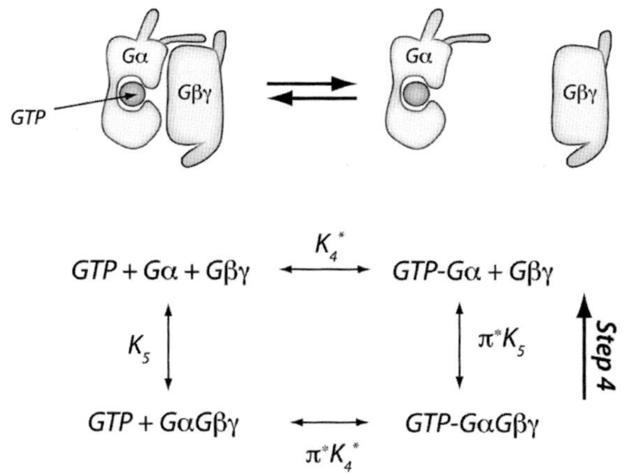

Figure 6.8: Step 4, The allosteric interaction between the binding of GTP and $G\beta\gamma$ with $G\alpha$. The parameters K_4^* andK_5 denote the microscopic affinity constants for the binding of GTP and $G\beta\gamma$ to the Gα subunit of the G protein, respectively. The cooperativity constant for this interaction is denoted by the scalar, π^*.

$G\beta\gamma$ probably alters the activity of its effector by inducing a conformational change in it. In the case of GTP-$G\alpha$, it also does the same, but its interaction involves allosterism between GTP and the effector as described in step 5.

***Step 5: Positive cooperativity between the binding of GTP and the effector to Gα*:** The interaction of $G\alpha$ with the effector is also influenced by allosteric interactions between guanine nucleotides and the effector. When GTP is bound, $G\alpha$ has relatively high affinity for its effector as shown in Figure 6.9 for the case of GTP-$G\alpha_s$ and the effector, adenylate cyclase. Thus, there must be positive cooperativity between the binding of GTP and adenylate cyclase with $G\alpha_s$. This is evident in the crystal structure of the catalytic domains of adenylate cyclase in a complex with $G\alpha_s$ bound with GTPγS.[16]

The scheme of equilibria beneath the illustration in Figure 6.9 describes these interactions. The constant, K_4^*, denotes the affinity of GTP for free $G\alpha$, K_6 denotes the affinity of the effector for free $G\alpha$. When GTP is bound, it increases the affinity of $G\alpha$ for the effector by a factor of δ. Similarly, when the effector is bound to $G\alpha$, it increases the affinity of GTP for $G\alpha$ by the same amount. This factor δ is a measure of the cooperativity, which is positive in this case (i.e., $\delta > 1$). The binding of GTP-$G\alpha$ to adenylate cyclase alters its activity. In the case of the interaction of GTP-$G\alpha_s$ with adenylate cyclase, the result is a stimulation of the enzyme. Other types of GTP-$G\alpha$ subunits (e.g., GTP-$G\alpha_i$) inhibit adenylate cyclase.

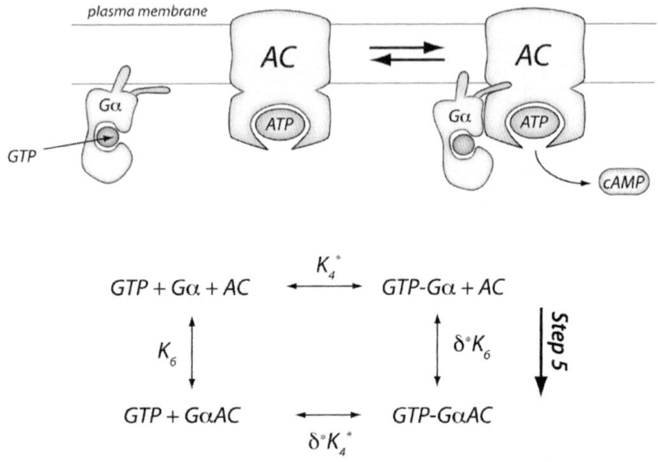

Figure 6.9: Step 5, The allosteric interaction between the binding of GTP and adenylate cyclase with Gα. The parameters K_4^* and K_6 denote the microscopic affinity constants for the binding of GTP and adenylate cyclase to the Gα subunit of the G protein, respectively. The cooperativity constant for this interaction is denoted by the scalar, π^*.

Step 6: Negative cooperativity between the binding of GDP and the effector to Gα: The activation (or inhibition) of the effector by GTP-$G\alpha$ is short-lived because of the GTPase activity of $G\alpha$. In some instances, the hydrolysis of GTP is enhanced by the effector, whereas in others, an RGS protein increases the activity (see Chapter 5). This GTPase activity converts GTP-$G\alpha$ to GDP-$G\alpha$, and changes the allosterism at the effector as shown in Figure 6.10 for the case of GTP-$G\alpha_s$ and the effector, adenylate cyclase. When GDP is bound, $G\alpha$ exhibits low affinity for the effector. Thus, as described for step 5, $G\alpha$ behaves as a receptor with two allosterically linked sites, one for the effector and the other for GDP. In this case, there is negative cooperativity between the binding of GDP and the effector. This causes GDP-$G\alpha$ to dissociate from the effector and terminates its action.

This allosteric interaction is described by the scheme beneath the illustration in Figure 6.10. This scheme is identical to that shown in Figure 6.9, except that the cooperativity is negative ($0 < \delta < 1$).

Step 7: Positive cooperativity between the binding of GDP and G$\beta\gamma$ to Gα: Following the GTPase-mediated inactivation of the G protein, the GDP-$G\alpha$ subunit combines with $G\beta\gamma$ to form the holoprotein (GDP-$G\alpha G\beta\gamma$). This process is facilitated by a positive allosteric interaction between the binding of GDP and $G\beta\gamma$ to $G\alpha$. The binding of GDP causes $G\beta\gamma$ to bind more tightly to $G\alpha$ as shown in Figure 6.11.[17]

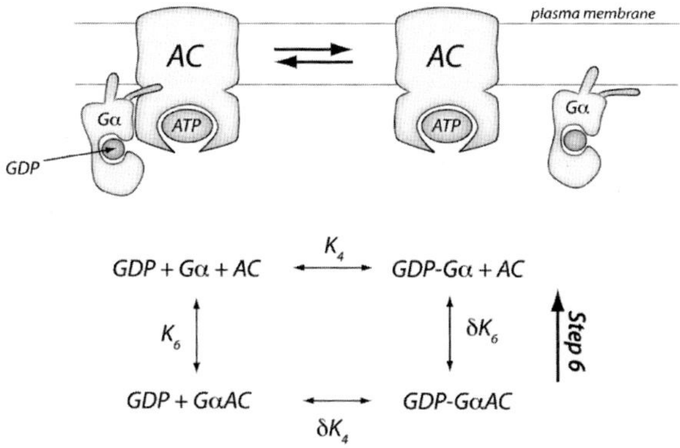

Figure 6.10: Step 6, The allosteric interaction between the binding of GDP and adenylate cyclase with $G\alpha$. The parameters K_4 andK_6 denote the microscopic affinity constants for the binding of GDP and adenylate cyclase to the Gα subunit of the G protein, respectively. The cooperativity constant for this interaction is denoted by the scalar, δ.

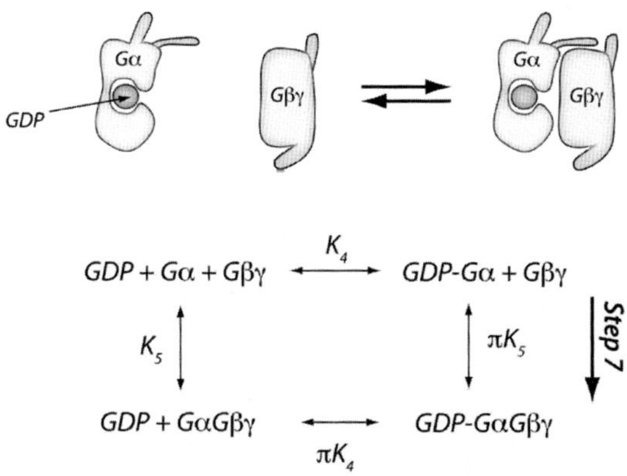

Figure 6.11: Step 7, The allosteric interaction between the binding of GDP and$G\beta\gamma$ with $G\alpha$. The parameters K_4 andK_5 denote the microscopic affinity constants for the binding of GDP and $G\beta\gamma$ to the Gα subunit of the G protein, respectively. The cooperativity constant for this interaction is denoted by the scalar, π.

The scheme beneath the illustration defines the nature of the allosteric interaction. It is the same as that shown in step 5, except in this instance, the cooperativity is positive ($\pi > 1$). The formation of the holoprotein prevents the free $G\alpha$ subunit from interacting with the effector and readies the G protein for another round of activation should the receptor be activated by an agonist. It is only the holoprotein form of the G protein, which can interact with the receptor.

Intuitive Explanation of the Receptor-*G* Protein Cycle

We can get an intuitive understanding of the overall process of how *G*-protein-coupled receptors activate their effectors by integrating the behavior of the seven allosteric interactions described above. These steps were described at equilibrium, so we will have to consider how the steady-state condition in the cell alters these steps.

Consider again the resting cell. The concentration of GTP is usually high, and kinase activity continually phosphorylates any GDP to form GTP in the cytosol. Thus, GTP is the predominant guanine nucleotide. It will bind to any unoccupied *G* protein readily. GTPase activity rapidly converts bound GTP to GDP, and the structure of the GDP-*G* protein complex is such that GDP cannot readily dissociate from it (see Chapter 5, Figure 5.9). This leads to a steady state where almost all of the *G* proteins are bound with GDP as illustrated in the histogram in Figure 6.12 by the bar labeled, *steady state*, *resting condition*.

Imagine what would happen if we could turn off the GTPase activity of the *G* protein. During the rare instances in which GDP spontaneously dissociates from the *G* protein, it would be rapidly replaced with GTP. The binding of GTP would cause a dissociation of the *G* protein into GTP-$G\alpha$ and $G\beta\gamma$. These would be stable because of the lack of GTPase activity. Consequently, their concentration would slowly rise as GDP slowly dissociated from the *G* protein. At equilibrium, nearly all of the *G* proteins would be bound with GTP, and they would exist in the dissociated state of GTP-$G\alpha$ and $G\beta\gamma$. This would lead to a large activation of effectors, like adenylate cyclase. A small proportion of the *G* proteins would exist in the form of GDP-$G\alpha G\beta\gamma$ because of the presence of a low concentration of GDP in the

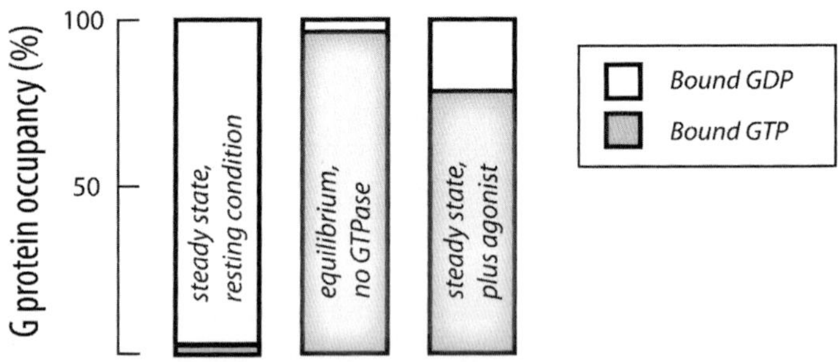

Figure 6.12: Theoretical occupancy of *G* proteins by GTP and GDP in the presence of a *G* protein-saturating concentration of GTP and under three conditions — resting steady state, equilibrium with no GTPase activity, and steady state with agonist. The rationale is described in the text under the section, entitled "Intuitive explanation of the receptor-*G* protein cycle."

cell. This equilibrium condition is illustrated in the histogram in Figure 6.12 by the bar labeled, *equilibrium, no GTPase*. The addition of agonist would not alter the system at equilibrium. It would simply increase the rate at which equilibrium is achieved by catalyzing the exchange of GTP for GDP at the *G* protein. Also removing the agonist would not alter the equilibrium level of activation. That is, the system would remain permanently activated upon removal of the agonist, because there is no GTPase activity to convert GTP-$G\alpha$ to GDP-$G\alpha$.

The absence of GTPase activity can be simulated experimentally in broken cell preparations by replacing GTP with a nonhydrolyzable analog of it, like Gpp(NH)p or GTPγS. These compounds cause a slow activation of *G* proteins, depending upon how much GDP is occupying the *G* protein.[4] The more GDP, the slower the rate of activation. The activation step is enhanced in the presence of an agonist.

Another way to eliminate GTPase activity at some *G* proteins (i.e., G_s and G_t) is through the action of cholera toxin, which ADP-ribosylates a cysteine residue near the guanine nucleotide binding pocket causing an inhibition of GTPase activity.[18,19] Cholera toxin treatment causes a slow, permanent activation of G_s in the absence of agonist. The addition of agonist accelerates the rate of activation, but the maintenance of the activation does not require agonist.

Thus, ignoring the energy used to convert GDP into GTP in the cytosol, one could say that the natural tendency of the system is to be fully activated and that it is the hydrolysis of GTP to GDP that provides the energy to keep the system inactive.

Enter the agonist-receptor complex. Its purpose is to engage the GDP-*G* protein complex and reduce the affinity of GDP for it as described in step 2. This is accomplished through negative cooperativity between the binding of agonist and GDP to the receptor-*G* protein complex. Thus, under normal conditions with GTPase activity, the agonist-receptor complex will speed up the dissociation kinetics of GDP from the *G* protein and enable GTP to compete more effectively for occupancy of the *G* protein. This drives the system closer to the equilibrium condition because the dissociation rate of GDP from the *G* protein is now in the same range or greater than GTPase activity. Occupancy of *G* proteins by GTP is now increased as shown in the histogram in Figure 6.12 by the bar labeled, *steady state, plus agonist*. The system is still in a steady state because of the GTPase activity, but now the level of GTP-$G\alpha$ is much greater because of the agonist-induced guanine nucleotide exchange is fast relative to GTPase activity. The addition of the agonist increases GTPase activity, but this is an indirect effect, secondary to the increased occupancy of the *G* protein by GTP.

Having a knowledge of how the *G* protein cycle works enables one to understand how labeled guanine nucleotides can be used to measure agonist activation of the cycle. This topic is briefly summarized next.

Measuring the *G* Protein Cycle with Labeled Guanine Nucleotides

One of the simplest ways to assess agonist activation of the *G* protein cycle is to measure GTPase activity in homogenates of cells expressing the receptor of interest. If the receptor-*G* protein interaction accounts for a substantial fraction of the nucleotidease activity in the membranes, then agonist-stimulated GTPase activity can be easily measured.

As mentioned above, the rate-limiting step in the *G* protein cycle is the dissociation of GDP from the *G* protein following the hydrolysis of GTP. In the absence of receptor activation, GTP is a poor substrate for the *G* protein because the hydrolysis product, GDP, binds tightly to the substrate-binding site and prevents the binding of new GTP. Thus, the holoprotein form of the *G* protein hydrolyzes GTP very slowly after the first molecule of GTP has been hydrolyzed. With the addition of agonist, the receptor complex engages the *G* protein and causes the dissociation of GDP. Ultimately, this greatly increases GTPase activity by removing GDP from the *G* protein and allowing new substrate to bind.

This reaction can be measured by incubating homogenates of cells with [γ-^{32}P]GTP and measuring the release of [^{32}P]-labeled inorganic phosphate in the supernatant after precipitating the parent labeled nucleotide with charcoal.[20] Agonist-mediated receptor activation causes an increase in the appearance of [^{32}P]-labeled phosphate in the supernatant.

Cassel and Selinger[19] were the first to demonstrate that the mechanism for receptor-mediated activation of adenylate cyclase involved an opening of the guanine-nucleotide binding site thereby enabling GDP to dissociate and GTP to bind. These investigators incubated membranes from turkey erythrocytes with [^{3}H]GTP in the presence of the β-adrenergic agonist, isoproterenol, to increase the labeling of G_s with guanine nucleotide. Following incubation with unlabeled GTP in the absence of agonist to reduce background labeling, the membranes were washed once and then incubated with agonist and the nonhydrolyzable GTP analog, Gpp(NH)p. The difference between [^{3}H]GDP released in the presence of isoproterenol and Gpp(NH)p and that released with Gpp(NH)p only was defined as agonist-induced release of [^{3}H]GDP.

A common method of assessing the G protein cycles is the measurement of agonist-stimulated $[^{35}S]GTP\gamma S$ binding. If a purified GPCR is reconstituted with the relevant G protein in phospholipid vesicles or HDL particles, agonist stimulation is likely to cause an increase in the rate of $[^{35}S]GTP\gamma S$ binding to the membranes.[21–23] Usually, low nanomolar concentrations of $[^{35}S]GTP\gamma S$ are used in the assay. Presumably, agonist-receptor activation leads to an opening of the guanine nucleotide binding pocket on the G protein, which facilitates the binding of $[^{35}S]GTP\gamma S$. In contrast, the binding of $[^{35}S]GTP\gamma S$ to the free holoprotein is expected to proceed at a slower rate.

If the same method is used to study a receptor expressed in a native tissue or cell line, it is often difficult to detect an agonist signal over background labeling. Presumably, $[^{35}S]GTP\gamma S$ would readily bind to the large population of available G proteins in the tissue, and agonist occupancy of a given population of GPCRs in the tissue would have little effect on this outcome. If a relatively high concentration of GDP ($1.0-100\ \mu M$) is added to the assay, however, the rate of $[^{35}S]GTP\gamma S$ binding to all of the G proteins in the membranes is greatly suppressed. Under this condition, agonist-mediated receptor activation would provide a means for $[^{35}S]GTP\gamma S$ to occupy those G proteins that communicate with the agonist-receptor complex.[24–27] Although the most common event catalyzed by the agonist would be an exchange of GDP for GDP, occasionally $[^{35}S]GTP\gamma S$ would have the opportunity to bind to the G protein. Once bound, it would cause the G protein to dissociate into its $[^{35}S]GTP\gamma S$-$G\alpha$ and $G\beta\gamma$ subunits. With continued receptor activation, $[^{35}S]GTP\gamma S$-$G\alpha$ subunits would accumulate because $[^{35}S]GTP\gamma S$ is nonhydrolyzeable and it binds nearly irreversibly to $G\alpha$. The specific activity of the accumulated $[^{35}S]GTP\gamma S$-$G\alpha$ subunits is not diluted by agonist-mediated exchange of GDP for $[^{35}S]GTP\gamma S$ because free $G\alpha$ subunits do not interact with GPCRs.

Definition of Observed Affinity and Intrinsic Efficacy at GPCRs at the Population Level of Analysis

Knowing how an agonist triggers a signal at GPCRs enables one to define observed affinity and efficacy at the population level of analysis. Observed affinity and efficacy are parameters that describe the receptor-activation function, which is also known as the stimulus. The concentration of agonist that causes half-maximal receptor activation is defined as the observed dissociation constant (reciprocal of the observed affinity constant), and the maximum of the stimulus function is proportional to observed efficacy. But what represents receptor activation at a GPCR? Recall that the purpose of the agonist-receptor complex is to engage the GDP-*G* protein complex and reduce the affinity of GDP or increase its rate of dissociation. This effect enables GTP to compete GDP off the *G* protein rapidly. The magnitude of this agonist effect is proportional to the amount of agonist-receptor-*G* protein-GDP complex (quaternary complex). The actual degree of activation is equivalent to the proportion of the quaternary complex with the receptor in the active state as described below under "Conformation analysis of the ternary complex model with guanine nucleotide." This receptor activation function is known as the stimulus. It represents the independent variable — active state of the receptor in a quaternary complex with GDP-bound *G* protein — expressed as a function of the agonist concentration. The stimulus of a GPCR is analogous to the whole cell current response at a ligand-gated ion channel. The EC_{50} value of the agonist for generating the quaternary complex represents its observed dissociation constant and the maximal amount of quaternary complex formed at high concentrations of agonist is proportional to observed intrinsic efficacy.

One might ask, what if a ligand were very effective in persuading the receptor to form the quaternary complex, but much less effective in reducing the affinity of GDP for it? Such a ligand would be defined as highly efficacious even though it is ineffective in catalyzing the exchange of GTP for GDP on the *G* protein. When the agonist engages the *G* protein, it selects a conformation that has low affinity for GDP and GTP. This is why it is much easier for the agonist to generate a high affinity, stable ternary complex (agonist-receptor-*G* protein) in the absence of guanine nucleotides. The latter complex is a biological adaptation, which evolved specifically because of its high affinity for the endogenous ligand and low affinity for guanine nucleotides. Otherwise, it would not relay information efficiently. Thus, the formation of the quaternary complex is inextricably linked to a reduction in the affinity of GDP for the *G* protein. In a later section of this chapter on conformational analysis, the relationship between affinity and efficacy and the different states of the receptor and *G* protein is described.

Modeling the Agonist-Receptor-*G* Protein Interaction

Let me begin by explaining what I do not mean by modeling. The model developed below is not a representation of all of the proteins required for receptor-*G* protein signaling and how they work. For example, no scaffolding or RGS proteins are described. Rather, the model is a useful device for measuring the activity of the receptor in its various complexes with ligands and *G* protein subunits. As such, it represents a type of accounting device.

There are two reasons why developing a model for the receptor-*G* protein interaction is important. First, having a model will increase our understanding of how the system works and help us design experiments, interpret data, and answer a host of questions like, "What is the likelihood of obtaining a crystal structure of the active state of a GPCR in the absence of *G* protein?" Second, a model is required for the goal of estimating observed affinity and efficacy through the analysis of a downstream response to an agonist. This goal can be reformulated in terms of our definition of observed affinity and efficacy. That is, we need a model for the receptor-*G* protein interaction so that we can estimate the maximum (observed efficacy) and EC_{50} value (observed dissociation constant) of the agonist for generating the quaternary complex simply through the analysis of a downstream response to an agonist (e.g., contraction of smooth muscle or increase in fluorescence of a Ca^{2+}-sensitive probe).

Our model should enable us to predict how changing the concentration of agonist and guanine nucleotide alters the formation of the quaternary complex. This can be accomplished with an equation that expresses the amount of quaternary complex formed as a function of the agonist concentration and other parameters, like the concentration of GTP and *G* protein relative to receptor. The first step involves drawing a scheme that includes all possible complexes or combinations of receptor, *G* protein, and small ligands. There are three rules for generating this scheme, (1) each equilibrium should involve only one process (a ligand-protein or protein-protein interaction), (2) each equilibrium should be defined by an equilibrium constant, and (3) any closed circuit of equilibria must not allow a net flux in one direction around the cycle. For a square of equilibria, like that shown for step 1 of the *G* protein cycle, Rule 3 prescribes that only three constants are required to describe the four unique equilibria. For example, in step 1 (see Figure 6.3), K_1, βK_2 and $\alpha\gamma$ behave as single constants (i.e., C_1, C_2 and C_3, respectively). The other two equilibrium constants, $\alpha\gamma K_1$ and $\alpha\beta\gamma K_3$, are equivalent to C_1C_3 and C_2C_3, respectively.

To begin, consider the interactions among the agonist (*D*), receptor (*R*), and *G* protein in the form of $G\alpha G\beta\gamma$ (*G*) (Figure 6.13, *middle* scheme). This portion of the model is known as the Ternary Complex Model of DeLean, Stadel, and Lefkowitz[28], and it is a modification of the "Mobile Receptor Hypothesis" of Jacobs and Cuatrecasas[29] The parameter K_1 denotes the affinity constant of the agonist for the receptor, and K_2 denotes the constant describing the binding of the *G* protein to the receptor. The binding of the agonist to the receptor allosterically enhances the binding of the *G* protein to the receptor. Similarly, the binding of the *G* protein to the receptor allosterically enhances the binding of the agonist to the receptor. This

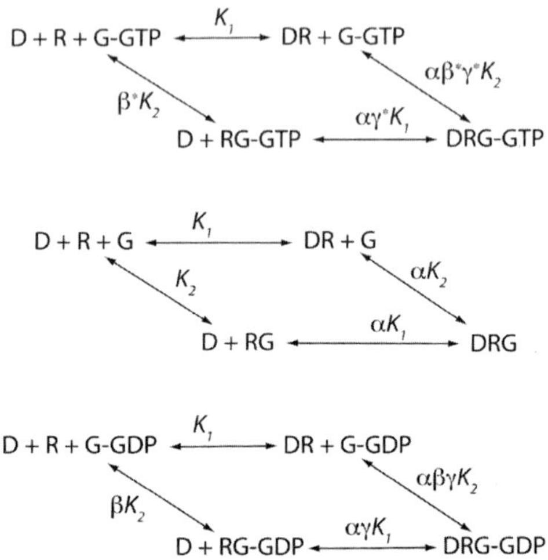

Figure 6.13: Three representations of the Ternary Complex Model. The Ternary Complex Model is illustrated for the conditions of ambient levels of GTP (top), no guanine nucleotides (middle) and ambient levels of GDP (bottom).

mutual enhancement of binding represents positive cooperativity and is denoted by the symbol "α" in the middle scheme of Figure 6.13. Thus, the affinity constants for the four equilibria are K_1, K_2, αK_1 and αK_2.

It is also necessary to account for the same square of equilibria, but with GDP bound to the *G* protein as shown in Figure 6.13, lower scheme. In this scheme, K_1 again denotes the affinity constant of the agonist for the receptor, and βK_2 denotes that for the binding of the GDP-bound form of the *G* protein (*G-GDP*) with the receptor. The cooperativity between the binding of agonist and the *G-GDP* complex with the receptor is denoted by $\alpha\gamma$. Thus, the binding of agonist to *RG-GDP* is described by the affinity constant, $\alpha\gamma K_1$, and the binding of *G-GDP* to the DR complex is described by the affinity constant, $\alpha\beta\gamma K_2$. This scheme represents the allosteric interaction in Step 1 of the *G* protein cycle as previously described (Figure 6.3).

The upper scheme in Figure 6.13 accounts for the binding of GTP to the *G* protein. It is analogous to that shown in lower scheme except that the parameters corresponding to β and γ have been replaced by β^* and γ^* to denote potential differences in the way GTP influences the interaction of the *G* protein with the receptor. This allosteric interaction represents Step 3 in the *G* protein cycle (Figure 6.7).

The three schemes in Figure 6.13 are not isolated, but connected through additional equilibria shown in Figure 6.14. The result resembles two cubes with

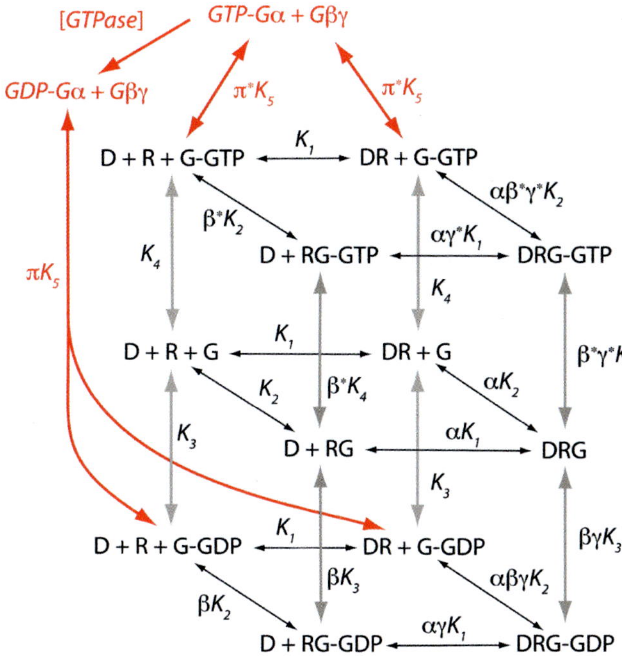

Figure 6.14: Complete description of the binding of a ligand (D) to a receptor (R) interacting with a *G* protein (G) in the presence of GTP and GDP. The sequence indicated in red denotes the dissociation of the GTP-bound holoprotein form of the *G* protein into its GTP-$G\alpha$ and $G\beta\gamma$ subunits and the conversion of GTP-$G\alpha$ to GDP-$G\alpha$ by the GTPase activity of $G\alpha$.

their horizontal surfaces corresponding to the equilibria shown in Figure 6.13. In connecting the schemes together, cooperative interactions need to be addressed. For example, consider the left-hand face of the lower cube. K_3 represents the affinity constant for the binding of GDP to the *G* protein, and K_2 represents the affinity constant for the binding of the receptor to the *G* protein. The receptor has an allosteric effect on the binding of GDP to the *G* protein, and GDP has a reciprocal effect on the binding of the receptor to the *G* protein. This cooperativity is denoted as β. Hence, the binding of the receptor to the *G-GDP* complex is described by the affinity constant βK_2, and the binding of GDP to the *RG* complex is described by βK_3. The same type of rationale applies to the other faces of the bi-cubic structure. Another example is the front face of the lower cube. This represents the allosteric interaction between the binding of agonist and GDP to the receptor-*G* protein complex in Step 2 of the *G* protein cycle, which was described previously. The only part of the scheme lacking allosteric interactions is the back side, which represents the binding of small ligands (agonist or guanine nucleotide) to their respective receptors (*R* and *G*) before the latter have a chance to interact.

So far only interactions of the holoprotein form of the *G* protein have been described in Figure 6.14. The G-GTP complex readily splits into free $G\alpha$-GTP and $G\beta\gamma$ subunits because of the negative cooperativity between the binding of GTP and $G\beta\gamma$ to $G\alpha$ (step 4 of the *G* protein cycle). The resulting free subunits then activate or inhibit their respective effectors. The $G\alpha$-GTP complex is short-lived, however, because of the GTPase activity of $G\alpha$, which converts the $G\alpha$-GTP complex into $G\alpha$-GDP. The latter coalesces with $G\beta\gamma$ to form the holoprotein, *G*-GDP, because of the positive cooperativity between the binding of GDP and G$\beta\gamma$ with $G\alpha$ (step 7 of the*G* protein cycle). These additional steps have been indicated in red in Figure 6.14.

In the presence of agonist, the *G* protein is under a steady-state condition. The *G-GDP* form of the *G* protein ultimately engages the agonist-receptor complex to form the quaternary complex (*DRG-GDP*). The negative cooperativity between the binding of agonist and *GDP* causes dissociation of the latter to yield the ternary complex (*DRG*). This guanine nucleotide-free complex is very transient because of the high concentration of GTP in the cytosol. GTP rapidly binds to the ternary complex to form the *DRG-GTP* complex. The dashed red arrow tracks this sequence in Figure 6.14. When the *G-GTP* complex dissociates from the receptor, it readily splits into the free $G\alpha$-GTP and $G\beta\gamma$ subunits. These activate their effectors, and the GTPase activity of $G\alpha$ terminates this stimulation and promotes the formation of the holoprotein *G-GDP* as indicated by the solid red arrow in Figure 6.14. Thus, *G* proteins are constantly cycling through the system driven by GTPase activity as indicated by the red arrows. In the absence of agonist, the *G* protein is also in a steady state, although the flux through the pathway is reduced to a trickle.

The receptor may not be far from equilibrium, however, because in many instances, GTP and GDP have similar effects on the binding of agonists to GPCRs. The engine that drives the steady state (GTPase activity) converts G-GTP into G-GDP, yet both of these may have the same or nearly the same effect on the binding of the agonist to the receptor.[5–7] Thus, with regard to our goal of formulating an equation expressing the quaternary complex as a function of the agonist concentration, we can ignore the part of the scheme indicated in solid red arrows. We can also replace GTP and GDP with a single nucleotide symbol (*X*), because both guanine nucleotides have similar effects on the binding of agonists. The revised model is shown in Figure 6.15. In practice, *X* primarily represents GDP because it promotes the formation of the holoprotein form of the *G* protein and only the latter interacts with the receptor. This model is referred to in the text as the "Quaternary Complex

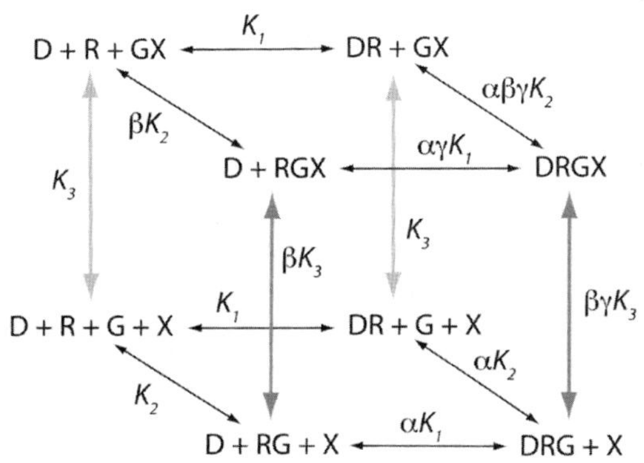

Figure 6.15: Quaternary Complex Model. The scheme illustrates the equilibrium binding of ligand (D) to a receptor (R) interacting with a G protein (G) that binds guanine nucleotide (X).

Model". The formulation of equations that describe the ternary and quaternary complexes as well as other receptor complexes in Figure 6.15 are described in the following sections. Those uninterested in the mathematical derivation of the model may want to skip ahead to the section, entitled "Behavior of the Quaternary Complex Model, Population Analysis".

Derivation of Equations for the Quaternary Complex Model

Equations can be derived for different components of the Quaternary complex model. I begin by deriving an equation for receptor-bound agonist ($DR + DRG + DRGX$) as a function of the agonist concentration. Readers uninterested in the details of the mathematics may want to skip to the next section, entitled "Behavior Quaternary Complex Model: Population Analysis."

The first step is to set up a simple equation for total receptor-bound agonist (D_B) normalized relative to the total concentration of receptors (R_T):

$$\frac{D_B}{R_T} = \frac{DR + DRG + DRGX}{DR + DRG + DRGX + R + RG + RGX}. \tag{6.2}$$

The numerator on the right side of the equation lists the summation of all of the agonist-bound receptor complexes in the model, and the denominator lists all possible receptor complexes including the free receptor (R).

Before continuing with step 2, let me outline the overall strategy. Each receptor complex (e.g., DRG) in the numerator and denominator is replaced with an algebraic expression in terms of the free receptor (R) and one or more affinity constants (e.g., αK_1) and perhaps one or more ligands (D and X) and the total receptor concentration (R_T). The resulting expression is then simplified into a more manageable form usually resembling a fractional occupancy equation (i.e., occupancy $= DK_{obs}/(DK_{obs} + 1)$). The steps to achieve this goal follow.

The next step (step 2) involves generating an equation for the affinity constant of each receptor complex shown in Figure 6.15. The affinity constant is simply derived by taking the ratio of the bound species divided by the product of the reactants. The microscopic affinity constants for the binding of D to the various agonist-receptor complexes are:

$$K_1 = \frac{[DR]}{[D][R]}, \tag{6.3}$$

$$\alpha K_1 = \frac{[DRG]}{[D][RG]}, \tag{6.4}$$

$$\alpha\gamma K_1 = \frac{[DRGX]}{[D][RGX]}. \tag{6.5}$$

Those for the binding of guanine nucleotide (X) to the various complexes of G protein are:

$$K_3 = \frac{[GX]}{[X][G]}, \tag{6.6}$$

$$\beta K_3 = \frac{[RGX]}{[X][RG]}, \tag{6.7}$$

$$\beta\gamma K_3 = \frac{[DRGX]}{[X][DRG]}. \tag{6.8}$$

For the equilibria involving the binding of the receptor with the G protein, the concentration of G protein is normalized relative to the total receptor concentration (e.g., G/R_T). This is done because their interactions occur within or near the plasma membrane, and their local concentrations are unrelated to their total concentrations in solution. The calculations involving the free receptor (R) and G protein (G) are:

$$K_2 = \frac{[RG]}{[R][G/R_T]} = \frac{[RG][R_T]}{[R][G]}. \tag{6.9}$$

The three other equilibria between the receptor and G protein complexes are calculated analogously:

$$\alpha K_2 = \frac{[DRG][R_T]}{[DR][G]}, \tag{6.10}$$

$$\beta K_2 = \frac{[RGX][R_T]}{[R][GX]}, \tag{6.11}$$

$$\alpha\beta\gamma K_2 = \frac{[DRGX][R_T]}{[DR][GX]}. \tag{6.12}$$

In the third step, each receptor complex on the right hand side of Equation (6.2) is substituted with an expression consisting of affinity constants, ligands, and the free receptor concentration. Each expression is derived from the equation for the appropriate affinity constant. For example, it is possible to rearrange Equation (6.3) to yield an expression for the drug-receptor complex ($[DR]$):

$$[DR] = DK_1[R]. \tag{6.13}$$

In most cases, more than one affinity constant is required. For example, rearranging Equation (6.4) yields an expression for the ternary complex ($[DRG]$):

$$[DRG] = D\alpha K_1[RG]. \tag{6.14}$$

But this equation contains the $[RG]$ complex, which can be replaced with an expression derived from Equation (6.9):

$$[RG] = K_2\frac{[G]}{[R_T]}[R]. \tag{6.15}$$

Substituting Equation (6.15) into Equation (6.14) yields an expression for the ternary complex:

$$[DRG] = D\alpha K_1 K_2\frac{[G]}{[R_T]}[R]. \tag{6.16}$$

The equations for the other receptor complexes are derived using the same strategy and are listed below:

$$[DRGX] = DX\alpha\beta\gamma K_1 K_2 K_3\frac{[G]}{[R_T]}[R], \tag{6.17}$$

$$[RGX] = X\beta K_2 K_3\frac{[G]}{[R_T]}[R]. \tag{6.18}$$

Each receptor complex in Equation (6.2) can now be substituted with the appropriate equation (6.13, 6.15, 6.16, 6.17 or 6.18) to yield:

$$\frac{D_B}{R_T} = \frac{DK_1[R] + D\alpha K_1K_2\frac{[G]}{[R_T]}[R] + DX\alpha\beta\gamma K_1K_2K_3\frac{[G]}{[R_T]}[R]}{DK_1[R] + D\alpha K_1K_2\frac{[G]}{[R_T]}[R] + DX\alpha\beta\gamma K_1K_2K_3\frac{[G]}{[R_T]}[R] + [R] + K_2\frac{[G]}{[R_T]}[R] + X\beta K_2K_3\frac{[G]}{[R_T]}[R]}. \quad (6.19)$$

Dividing the numerator and denominator on the right-hand side of the equation by $[R]$ and rearranging yields:

$$\frac{D_B}{R_T} = \frac{D\left(K_1 + \alpha K_1K_2\frac{[G]}{[R_T]} + X\alpha\beta\gamma K_1K_2K_3\frac{[G]}{[R_T]}\right)}{D\left(K_1 + \alpha K_1K_2\frac{[G]}{[R_T]} + X\alpha\beta\gamma K_1K_2K_3\frac{[G]}{[R_T]}\right) + 1 + K_2\frac{[G]}{[R_T]} + X\beta K_2K_3\frac{[G]}{[R_T]}}. \quad (6.20)$$

Further rearrangement yields:

$$\frac{D_B}{R_T} = \frac{D}{D + \frac{1+K_2\frac{[G]}{[R_T]}(1+X\beta K_3)}{K_1+\alpha K_1K_2\frac{[G]}{[R_T]}(1+X\beta\gamma K_3)}}. \quad (6.21)$$

This equation is in the form of a simple occupancy equation:

$$\frac{D_B}{R_T} = \frac{D}{D + \frac{1}{K_{obs}}}, \quad (6.22)$$

in which K_{obs} is defined as:

$$K_{obs} = \frac{K_1 + \alpha K_1K_2\frac{[G]}{[R_T]}(1 + X\beta\gamma K_3)}{1 + K_2\frac{[G]}{[R_T]}(1 + X\beta K_3)}. \quad (6.23)$$

K_{obs} represents the observed affinity constant, and it is usually expressed in inverse concentration units, like M^{-1}. The reciprocal of the observed affinity constant represents the observed dissociation constant (K_D), which has concentration units.

At a fixed concentration of guanine nucleotide (X), the value of K_{obs} will remain nearly constant when the agonist concentration changes provide that the ratio of free G protein to the total receptor concentration ($[G]/[R_T]$) is high or if the concentration of guanine nucleotide is high. Under such conditions, the binding of agonist will be consistent with a simple occupancy model (Equation (6.22)) with K_{obs} described by Equation (6.23). Each component of binding would also resemble a simple occupancy function having the same K_{obs} value. For example, the concentration of agonist required for half-maximal formation of the quaternary complex (*DRGX*) would also be equivalent to the K_{obs} defined in Equation (6.23).

We need a more robust equation for the situation when both the concentration of G protein relative to receptor is not large and the concentration of guanine nucleotide is low or equal to zero. Under this condition, the value of K_{obs} changes because the concentration of G protein ($[G]$ and $[GX]$) available for interacting with the receptor decreases as agonist occupancy increases. This occurs because agonist occupancy drives the formation of receptor-G protein complexes (DRG and $DRGX$), causing depletion in free $[G]$ and $[GX]$ in the local environment of the receptor. The depletion is minimal relative to the total pool of G protein when the ratio of G protein to receptor is high or when the concentration of guanine nucleotide is high because the latter limits the total amount of $DRGX$ formed. What is needed is an equation that expresses K_{obs} as a function of the total concentration of G protein ($[G_T]$) and not $[G]$. One approach is to develop a separate, substitution equation that expresses $[G]$ as a function of $[G_T]$ and other constants in the model. The variable $[G]$ in Equation (6.23) can then be replaced with this substitution equation (Equation (6.41)).

The derivation of the substitution equation for [G] is based on the conservation of mass relationship:

$$[G_T] = [G_B] + [GX] + [G]. \tag{6.24}$$

This is another way of expressing the idea that the total amount of G protein (G_T) is equal to the sum of the free (G), receptor-bound (G_B) and guanine nucleotide-bound (GX) forms of the G protein. The goal is to solve this equation for $[G]$ after replacing $[G_B]$ and $[GX]$ with functions in terms of the parameters of the model. Rearranging Equation (6.6) yields an expression for $[GX]$:

$$[GX] = \frac{XK_3[G]}{XK_3[G]+1}. \tag{6.25}$$

The derivation of an equation for the amount of receptor-bound G protein ($[G_B]$) expressed relative to the total receptor concentration begins with the following equation:

$$\frac{[G_B]}{[R_T]} = \frac{[RG]+[RGX]+[DRG]+[DRGX]}{[RG]+[RGX]+[DRG]+[DRGX]+[R]+[DR]}. \tag{6.26}$$

The numerator on the right-hand side of Equation (6.26) lists all possible receptor-G protein complexes and the denominator lists all receptor complexes including the free receptor. Each receptor complex on the right hand side of Equation (6.26) can be replaced with an equation derived above (i.e., Equations (6.14), (6.15), (6.16),

(6.17) and (6.18)):

$$\frac{G_B}{R_T} = \frac{\begin{array}{c} K_2\frac{[G]}{[R_T]}[R] + X\beta K_2K_3\frac{[G]}{[R_T]}[R] + D\alpha K_1K_2\frac{[G]}{[R_T]}[R] \\ + DX\alpha\beta\gamma K_1K_2K_3\frac{[G]}{[R_T]}[R] \end{array}}{\begin{array}{c} K_2\frac{[G]}{[R_T]}[R] + X\beta K_2K_3\frac{[G]}{[R_T]}[R] + D\alpha K_1K_2\frac{[G]}{[R_T]}[R] \\ + DX\alpha\beta\gamma K_1K_2K_3\frac{[G]}{[R_T]}[R] + [R] + DK_1[R] \end{array}}. \quad (6.27)$$

Dividing the numerator and the denominator on the right-hand side of the equation by $[R]$ and simplifying yields:

$$\frac{G_B}{R_T} = \frac{\frac{[G]}{[R_T]}(K_2 + X\beta K_2K_3 + D\alpha K_1K_2 + DX\alpha\beta\gamma K_1K_2K_3)}{\frac{[G]}{[R_T]}(K_2 + X\beta K_2K_3 + D\alpha K_1K_2 + DX\alpha\beta\gamma K_1K_2K_3) + 1 + DK_1}. \quad (6.28)$$

Further simplification yields:

$$[G_B] = \frac{[G][R_T]}{[G] + [R_T]\frac{1+DK_1}{K_2+X\beta K_2K_3+D\alpha K_1K_2+DX\alpha\beta\gamma K_1K_2K_3}}. \quad (6.29)$$

This equation can be rewritten in the form:

$$[G_B] = \frac{[G][R_T]}{[G] + [R_T]K_G}, \quad (6.30)$$

in which

$$K_G = \frac{1 + DK_1}{K_2 + X\beta K_2K_3 + D\alpha K_1K_2 + DX\alpha\beta\gamma K_1K_2K_3}. \quad (6.31)$$

Now it is possible to substitute Equations (6.25) and (6.29) for $[GX]$ and $[G_B]$, respectively, in conservation-of-mass Equation (6.24). Making these substitutions yields:

$$[G_T] = \frac{[G][R_T]}{[G] + [R_T]K_G} + \frac{XK_3[G]}{XK_3[G] + 1} + [G]. \quad (6.32)$$

This equation can be rearranged into the form of a quadratic. First, $[G_T]$ is subtracted from both sides of the equation:

$$0 = \frac{[G][R_T]}{[G] + [R_T]K_G} + \frac{XK_3[G]}{XK_3[G] + 1} + [G] - [G_T]. \quad (6.33)$$

Multiplying both sides of the equation by $[G] + [R_T]K_G$ yields:

$$0 = \left(\frac{1}{1 + XK_3)}\right) - 2[G]^2 + K_g R_T[G]\left(\frac{1}{(1 + XK_3)}\right) - 2 - [R_T][G] + [G_T][G] + [R_T]K_G[G_T]. \quad (6.34)$$

Further simplification yields a quadratic in $[G]$:

$$0 = \left(\frac{1}{\frac{(1+XK_3)}{[R_T]}} - 2\right)[G]^2 + \left(1 - \frac{[G_T]}{[R_T]} + K_G\left(\frac{1}{\frac{(1+XK_3)}{[R_T]}} - 2\right)\right)[G] + K_G[G_T]. \tag{6.35}$$

A quadratic equation in X has the form:

$$0 = AX^2 + BX + C. \tag{6.36}$$

The general solution is:

$$X = \frac{-B \pm \sqrt{B^2 - 4AC}}{2A}. \tag{6.37}$$

Therefore the solution to Equation (6.35) is:

$$[G] = \frac{-\left(1 - \frac{[G_T]}{[R_T]} + K_G(1+XK_3)\right) + \sqrt{\left(1 - \frac{[G_T]}{[R_T]} + K_G(1+XK_3)\right)^2 - \left(\frac{1}{\frac{(1+XK_3)}{[R_T]}} - 2\right)K_G[G_T]}}{2\left(\frac{1}{\frac{(1+XK_3)}{[R_T]}} - 2\right)}. \tag{6.38}$$

Equation (6.38) can be substituted into Equation (6.23) for $[G]$. Recall that Equation (6.23) defines the parameter K_{obs} for Equation (6.22), which represents the agonist occupancy function for binding to a receptor that conforms to the model described in Figure 6.15. Making these substitutions yields:

$$[D_B] = \frac{D[R_T]}{D + \frac{1}{K_{\text{obs}}}}, \tag{6.39}$$

in which

$$K_{\text{obs}} = \frac{K_1 + \alpha K_1 K_2 [G](1 + X\beta\gamma K_3)}{1 + K_2[G](1 + X\beta K_3)}, \tag{6.40}$$

$$[G] = \frac{B + \sqrt{B^2 + 4AC}}{2A}, \tag{6.41}$$

$$A = \left(\frac{1}{\frac{(1+XK_3)}{[R_T]}} - 2\right), \tag{6.42}$$

$$B = R - 1 - K_G A, \tag{6.43}$$

$$C = [G_T]K_G, \tag{6.44}$$

$$K_G = \frac{1 + DK_1}{K_2 + X\beta K_2K_3 + D\alpha K_1K_2 + DX\alpha\beta\gamma K_1K_2K_3}, \tag{6.45}$$

$$R = \frac{[G_T]}{[R_T]}. \tag{6.46}$$

The equation describing the amount of drug bound in the form of quaternary complex (*DRGX*) is derived using an analogous approach, which yields:

$$\frac{[DRGX]}{R_T} = \frac{[D]}{[D] + \frac{1}{K_{DRGX}}}, \tag{6.47}$$

in which

$$K_{DRGX} = \frac{[X][G]\alpha\beta\gamma K_1K_2K_3}{1 + [D]K_1 + K_2[G] + [D]\alpha K_1K_2[G] + [X][G]\beta K_2K_3}. \tag{6.48}$$

In this equation, $[G]$ is defined by Equation (6.41).

This completes the derivation of the equations that describe drug binding to a receptor that conforms to the model shown in Figure 6.15.

It is also possible to derive equations for the individual components of drug binding (*DRG*, *DR*) and of the receptor-*G* protein complex (*RG* and *RGX*). Using conformational analysis, it is possible to divide each of these receptor complexes into their active and inactive components. The derivation of these components has been described previously[30–32] and the method is analogous to that described for total receptor occupancy.

Behavior of the Quaternary Complex Model, Population Analysis

Having a model for drug-receptor interactions at GPCRs, it is possible to simulate data and determine how agonist affinity and efficacy are manifest by the parameters of the model.

One might think that a sensible approach would be to assign random values to the parameters of the model and determine how these affect the ability of the agonist to bind with the receptor and generate the quaternary complex. For example, parameter values that increase the amount of quaternary complex (increase in α) and the negative cooperativity with guanine nucleotide (decrease in γ) should cause the ligand to exhibit higher efficacy. The increase in α enhances the ability of the agonist to engage the *G* protein, and the reduction in γ increases the negative cooperativity between the agonist and GDP. This latter effect enables the agonist to increase the exchange of GTP for GDP on the *G* protein. But an increase in α opposes the effect of a reduction in γ, so what is the combination of α and γ, which resembles a real receptor? And why should some ligands not be able to generate the quaternary complex (large α) and increase the affinity of GDP for the *G* protein (large γ)? Such drugs would behave as antagonists because of their stabilization of the G-GDP bound form of the *G* protein and their inhibition of the exchange of GTP for GDP.

The answer lies in how the receptor works as mentioned above from the perspective of a biological adaptation. For the receptor to function, its active state must have higher affinity for the free *G* protein relative to the GDP-bound form. Otherwise the active state of the receptor would neither reduce the affinity of GDP for the *G* protein nor catalyze GTP/GDP exchange. This is a fundamental property of the receptor-*G* protein interaction, just as conduction of an ionic current is a property of the open state of a ligand-gated ion channel. For an agonist to trigger an effect at a GPCR, it must have much higher affinity for the active state of the receptor relative to the inactive state. This implies that the agonist will also exhibit negative cooperativity with GDP based on the properties of the receptor and *G* protein. Thus, for any agonist, the values of α and γ will be inversely proportional to each other, just as they are for the endogenous ligand. The true independent parameters of the model are the microscopic constants of the resting and active states of the receptor. These determine the values of the dependent parameters shown in Figure 6.15, like α and γ. Thus, the model shown in Figure 6.15 can be analyzed at a deeper level that addresses the resting and active states of the receptor as described under "Conformational analysis of the Quaternary Complex Model".

In simulating the behavior of the model shown in Figure 6.15, only parameters allowed by conformational analysis are used.

Behavior of the model with excess G protein: Figure 6.16 shows the behavior of the model when there is excess *G* protein relative to receptor ($G_T/R_T = 10$) and little constitutive activity ($K_2 << 1$). The low K_2 value means that there is little spontaneous formation of either the *RG* or *GRX* complexes. The theoretical curves show total receptor occupancy (*a*), occupancy of the quaternary complex (*b*) and occupancy of the ternary complex (*c*) by a highly efficacious agonist in the presence of increasing concentrations of guanine nucleotide. All of the occupancy curves are consistent with a simple one-site model. The concentration of agonist required for half-maximal occupancy ($1/K_{obs}$) is the same for the different types of receptor complexes (*DR*, *DRG* and *DRGX*) at a given concentration of guanine nucleotide. As the concentration of guanine nucleotide increases, the observed affinity decreases, the maximum of the quaternary complex function (observed efficacy) increases, and that of the ternary complex decreases. The bar graphs in Figure 6.17 summarize the values of the observed dissociation constants (panel *a*) and maximums (panel *b*) of the components of the theoretical curves shown in Figure 6.16.

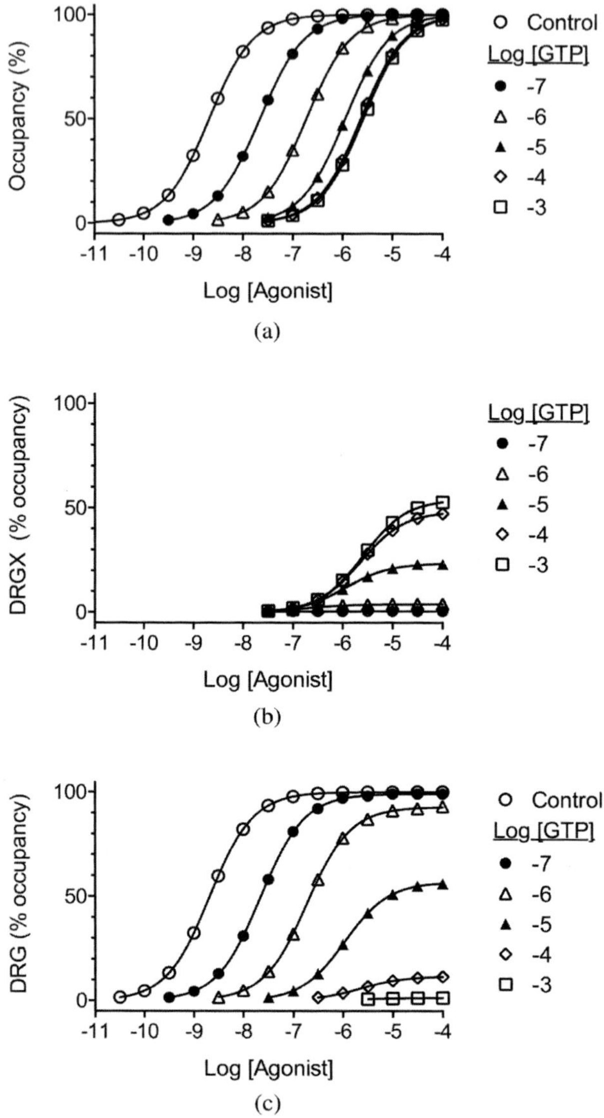

Figure 6.16: Simulation of agonist occupancy curves in the presence of various concentrations of GTP using the Quaternary Complex Model for the condition of a ten-fold excess of *G* protein relative to receptor. (a) Agonist-receptor occupancy. (b) Fraction of the agonist-receptor complex in the form of quaternary complex (*DRGX*). (c) Fraction of the agonist-receptor complex in the form of ternary complex (*DRG*). The parameter values for this simulation were: K_1, 1.8×10^5 M^{-1}; K_2, 1.8×10^{-3}; K_3, 10^8 M^{-1}; α, 2.47×10^4; β, 5.56×10^{-1} and γ, 7.42×10^{-4}.

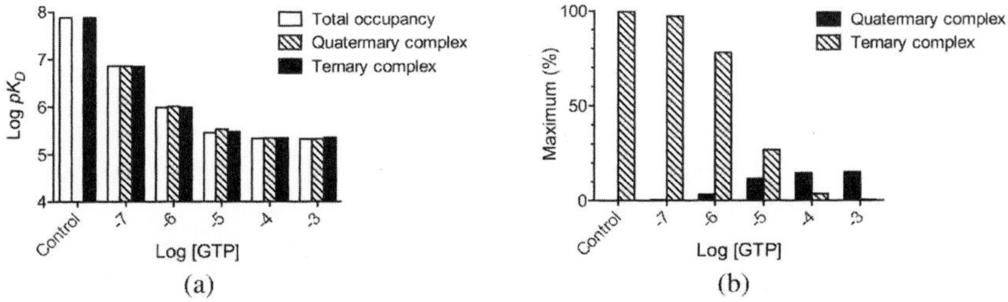

Figure 6.17: Empirical parameters of the simulated occupancy curves in Figure 6.16. (a) The value of the negative log of the observed dissociation (pK_D) of the agonist for different components of receptor occupancy — the total pool of receptors (open bars), the quaternary complex (stippled bars), and the ternary complex (closed bars) is plotted against the value of the concentration of GTP used in the simulation. (b) The maximum fractional amount of quaternary and ternary complexes formed by the agonist is plotted against the value for the concentration of GTP used in the simulation.

A comparison of the behavior of a highly efficacious agonist (*a*), a less efficacious agonist (*b*), a neutral antagonist (*c*), and an inverse agonist (*d*) is shown in Figure 6.18 for a receptor with little constitutive activity ($K_2 = 0.0126$). Behavior in the absence and presence of a high concentration of guanine nucleotide was simulated. The figure shows plots of total receptor occupancy and occupancy of the quaternary complex. In the absence of guanine nucleotide, no quaternary complex forms and the two agonists exhibit relatively high affinity. In the presence of a *G* protein-saturating concentration of guanine nucleotide, the total occupancy curves of the agonists shift to the right and a component of occupancy includes the quaternary complex. Both the quaternary complex and total occupancy curves for a given agonist exhibit one-site behavior and the same K_{obs} value. Both the maximum of the quaternary complex function and the guanine nucleotide-induced shift in agonist affinity are proportional to efficacy. In contrast, the neutral antagonist and inverse agonist do not generate significant amounts of the *DRGX* complex nor are their occupancy curves affected by guanine nucleotide. The histograms in Figure 6.19 summarize the observed parameters for the theoretical curves shown in Figure 6.18.

In the prior two examples, there is little or no spontaneous formation of the *RG* and *RGX* complexes. Any receptor that does associate with the *G* protein is almost always occupied by the agonist. When there is significant constitutive activity, however, a fraction of the unoccupied receptor population spontaneously couples with the *G* protein, and *G* protein activation includes agonist-occupied and -unoccupied receptor components (*DRGX* and *RGX*). Figure 6.20 illustrates this condition for a highly efficacious agonist (*a*), a less efficacious agonist (*b*), a neutral antagonist (*c*) and an inverse agonist (*d*). The plots show total receptor occupancy ($DR + DRGX$),

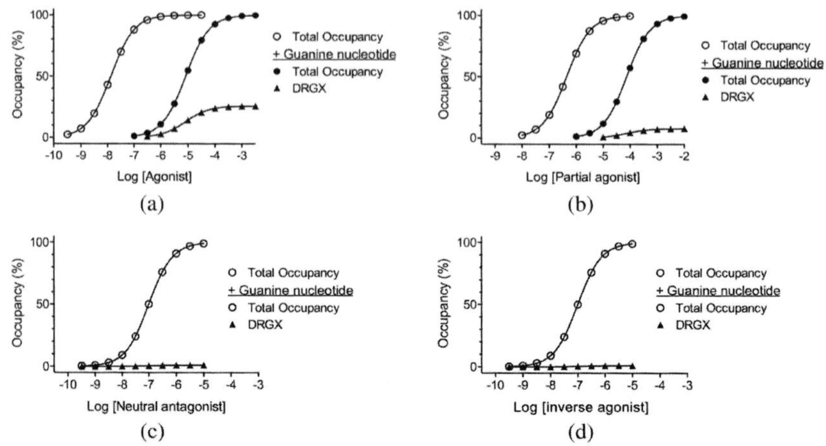

Figure 6.18: Simulation of ligand-occupancy curves in the absence and presence of a maximally effective concentration of GTP using the Quaternary Complex Model for the condition of a ten-fold excess of *G* protein relative to receptor. The plots show the sum of all ligand-receptor complexes (total receptor occupancy) and the fraction of the ligand-receptor complex in the form of quaternary complex (*DRGX*) as a function of the ligand concentration. The simulations were run for an agonist (a), a less efficacious agonist (b), a neutral antagonist (c) and an inverse agonist (d). The microscopic affinity constants of the ligands for the active (K_b) and inactive (K_a) states of the receptor are: agonist, K_b, 10^{10} M^{-1}; K_a, 10^4 M^{-1}; partial agonist, K_b, 3×10^8 M^{-1}; K_a, 10^4 M^{-1}; neutral antagonist, K_b, 10^7 M^{-1}, K_a, 10^7 M^{-1} and inverse agonist, K_b, 10^5 M^{-1}; K_a, 10^7 M^{-1}. These microscopic constants yielded the following level 2b parameters for the ligands: agonist, K_1, 9×10^4 M^{-1}; α, 4.94×10^4 and γ, 7.22×10^{-4}; partial agonist, K_1, 1.24×10^4 M^{-1}; α, 1.08×10^4 and γ, 7.95×10^{-4}, neutral antagonist, K_1, 1.0×10^7 M^{-1}; α, 1.0 and γ, 1.0, and inverse agonist, K_1, 1.0×10^7 M^{-1}; α, 5.6×10^{-1} and γ, 1.79. The other values of the parameters for the quaternary complex model were: K_2, 1.8×10^{-3} and K_3, 1.0×10^8 M^{-1}.

the component of the receptor population in the quaternary complex (*DRGX*) and the summation of the latter and the unoccupied receptor-*G* protein bound with guanine nucleotide (*DRGX* + *RGX*). Because of the complexity of the figure, only the condition of a G-protein-saturating concentration of guanine nucleotide is shown. The parameter values of the model are identical to those used in Figure 6.17 except that the value of K_2 is larger ($K_2 = 45.5$), which simulates constitutive activity. Also the efficacies of the two agonists are less. The increase in constitutive activity increases the observed affinity of the agonists, has no effect on the affinity of the neutral antagonist, and reduces the affinity of the inverse agonist. In all cases, the occupancy curves of the ligands are consistent with a one-site model, and the components of receptor occupancy (*DR*, *DRGX*) have the same observed affinity constant (K_{obs}).

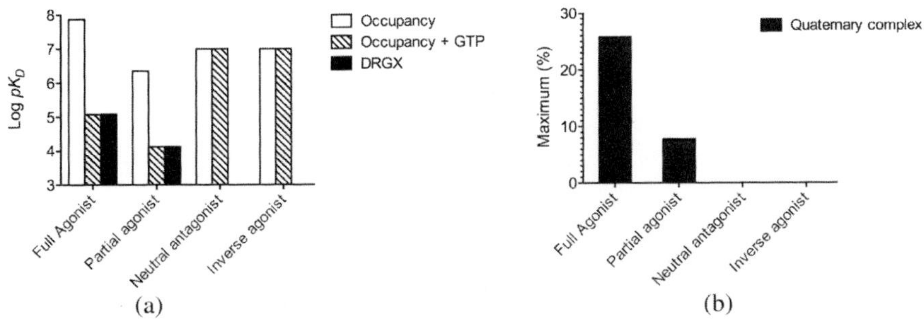

Figure 6.19: Empirical parameters of the simulated occupancy curves in Figure 6.18. (a) The value of the negative log of the observed dissociation (pK_D) of the agonist for different components of receptor occupancy — the total pool of receptors (open bars), in the presence of GTP (stippled bars), and the quaternary complex (closed bars) is plotted against the type of ligand used in the simulation. (b) The maximum fractional amount of quaternary complex formed by the different ligands is shown.

For the example in Figure 6.20, the total stimulus function includes both constitutive and ligand-mediated components. The stimulus at receptor-saturating concentrations of ligand is attributed entirely to the ligand and is, therefore, proportional to the observed efficacy. The concentration of the ligand causing a half-maximal change in the summation of $DRGX$ and RGX is equivalent to the observed dissociation constant, K_D. Observed efficacy is greatest for the highly efficacious agonist, less for the less efficacious agonist, less still for the neutral antagonist, and least for the inverse agonist. It follows that when there is constitutive activity, the stimulus of a neutral antagonist is equivalent to the constitutive stimulus, whereas those of an agonist and an inverse agonist are greater and less than the constitutive stimulus, respectively.

The effect of a G protein-saturating concentration of guanine nucleotide on total receptor occupancy for the condition shown in Figure 6.20 is illustrated in the next figure (Figure 6.21). The guanine nucleotide-induced shift is proportional to the efficacy of the ligand. That is, the larger rightward shift is observed with the more efficacious agonist, a smaller rightward shift with the less efficacious agonist and no shift with the neutral antagonist. When there is constitutive activity, the guanine nucleotide causes an increase in the observed affinity of the inverse agonist, resulting in a leftward shift of its occupancy curve.

***Behavior of the model with limiting G protein*:** Figure 6.22 shows the behavior of the model when there is limiting G protein relative to receptor ($G_T/R_T = 0.8$) and little constitutive activity ($K_2 << 1$). The theoretical curves show total receptor occupancy (*a*) and occupancy of the quaternary complex (*b*), ternary complex (*c*),

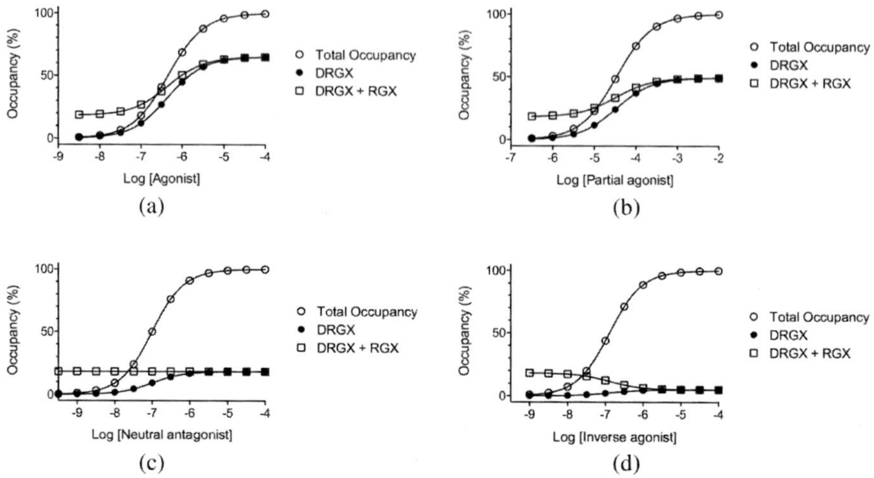

Figure 6.20: Simulation of ligand-occupancy curves in the presence of a maximally effective concentration of GTP using the Quaternary Complex Model for the condition of constitutive activity and a ten-fold excess of *G* protein relative to receptor. The plots show the sum of all ligand-receptor complexes (total receptor occupancy), the fraction of the ligand-receptor complex in the form of quaternary complex (*DRGX*), and the summation of occupied and unoccupied receptor associated with the guanine nucleotide-bound *G* protein (*RGX* and *DRGX*) as a function of the ligand concentration. The simulations were run for an agonist (a), a less efficacious agonist (b), a neutral antagonist (c) and an inverse agonist (d). The microscopic affinity constants of the ligands for the active (K_b) and inactive (K_a) states of the receptor are: agonist, K_b, 10^7 M^{-1}; K_a, 10^4 M^{-1}; partial agonist, K_b, 10^5 M^{-1}; K_a, 10^4 M^{-1}; neutral antagonist, K_b, 10^7 M^{-1}, K_a, 10^7 M^{-1} and inverse agonist, K_b, 10^5 M^{-1}; K_a, 10^7 M^{-1}. These microscopic constants yielded the following level 2b parameters for the ligands: agonist, K_1, 9.18×10^5 M^{-1}; α, 10.9 and γ, 0.8; partial agonist, K_1, 1.82×10^4 M^{-1}; α, 5.5 and γ, 0.82, neutral antagonist, K_1, 1.0×10^7 M^{-1}; α, 1.0 and γ, 1.0, and inverse agonist, K_1, 9.1×10^6 M^{-1}; α, 1.11×10^{-2} and γ, 20.6. The other values of the parameters for the quaternary complex model were: K_2, 45.5 and K_3, 1.0×10^8 M^{-1}.

and binary complex (*d*) by a highly efficacious agonist in the presence of increasing concentrations of guanine nucleotide. In the absence of guanine nucleotide, the total occupancy curve of the agonist has high- and low-affinity components, corresponding to the ternary (*DRG*) and binary (*DR*) complexes. In the presence of increasing concentrations of guanine nucleotide, there is a progressive shift in the total occupancy curve to the right until it resembles a one-site model when the guanine nucleotide saturates the *G* protein. This occurs because of a guanine nucleotide-induced reduction in the affinity and maximum of the high affinity component of receptor occupancy. This component is composed of both the quaternary

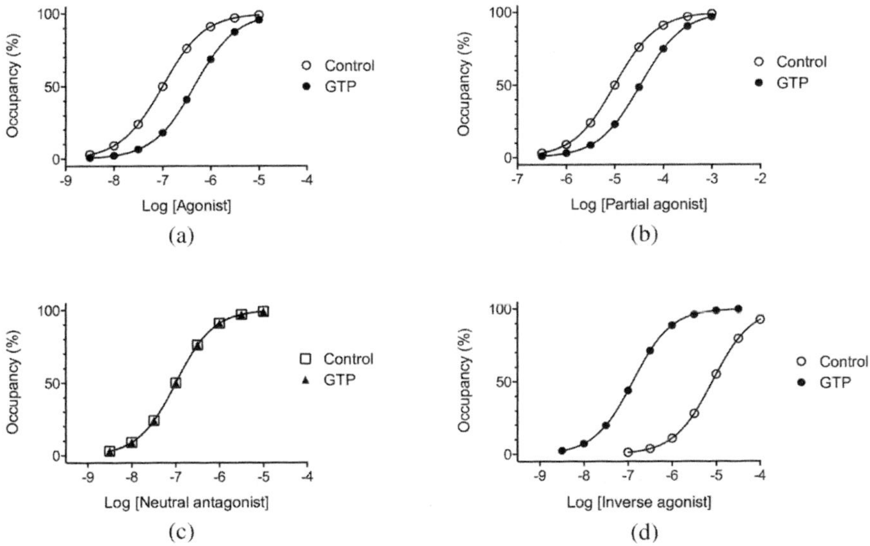

Figure 6.21: Simulation of ligand-occupancy curves in the absence and presence of a maximally effective concentration of GTP using the Quaternary Complex Model for the condition of constitutive activity and a ten-fold excess of G protein relative to receptor. The plots show receptor occupancy by the various ligands in the absence and presence of guanine nucleotide. The simulations were run for an agonist (a), a less efficacious agonist (b), a neutral antagonist (c) and an inverse agonist (d). The microscopic affinity constants of the ligands for the active (K_b) and inactive (K_a) states of the receptor are: agonist, K_b, 10^7 M^{-1}; K_a, 10^4 M^{-1}; partial agonist, K_b, 10^5 M^{-1}; K_a, 10^4 M^{-1}; neutral antagonist, K_b, 10^7 M^{-1}, K_a, 10^7 M^{-1} and inverse agonist, K_b, 10^5 M^{-1}; K_a, 10^7 M^{-1}. These microscopic constants yielded the following level 2b parameters for the ligands: agonist, K_1, 9.18×10^5 M^{-1}; α, 10.9 and γ, 0.8; partial agonist, K_1, 1.82×10^4 M^{-1}; α, 5.5 and γ, 0.82, neutral antagonist, K_1, 1.0×10^7 M^{-1}; α, 1.0 and γ, 1.0, and inverse agonist, K_1, 9.1×10^6 M^{-1}; α, 1.11×10^{-2} and γ, 20.6. The other values of the parameters for the quaternary complex model were: K_2, 45.5 and K_3, 1.0×10^8 M^{-1}.

and ternary complexes that share the same K_D value. The increase in the guanine nucleotide concentration reduces the observed affinity of both complexes and causes the maximum of the quaternary complex to increase and that of the ternary complex to decrease. Figure 6.23 summarizes the values of the observed dissociation constants (panel a) and maximums (panel b) of the components of the theoretical curves shown in Figure 6.22.

A comparison of the occupancy curves of a highly efficacious agonist (a), a less efficacious agonist (b), a neutral antagonist (c), and an inverse agonist (d) is shown in Figure 6.24 for a receptor-G protein complex like that shown in the

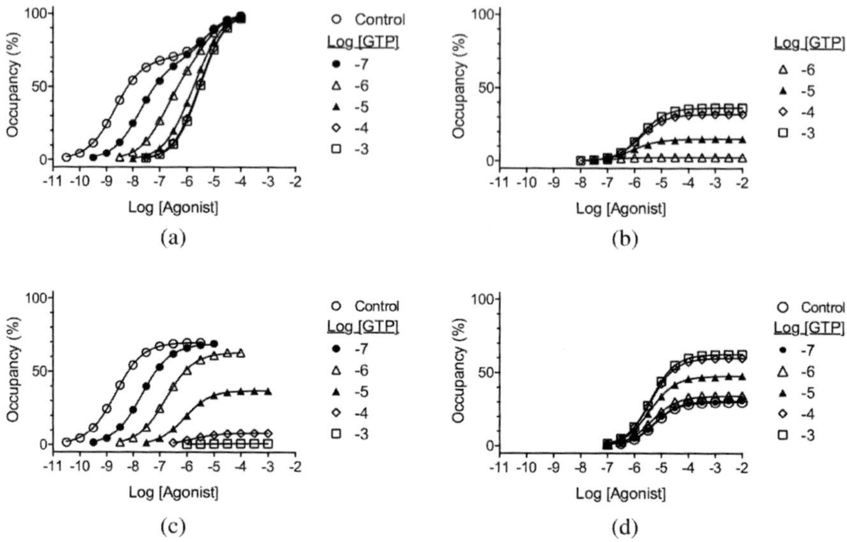

Figure 6.22: Simulation of agonist occupancy curves in the presence of various concentrations of GTP using the Quaternary Complex Model for the condition of limiting G protein relative to receptor ($G_T/R_T = 0.7$). (a) Agonist-receptor occupancy. (b) Fraction of the agonist-receptor complex in the form of binary complex (DR). (c) Fraction of the agonist-receptor complex in the form of ternary complex (DRG). (d) Fraction of the agonist-receptor complex in the form of quaternary complex ($DRGX$). The parameter values for this simulation were: K_1, 1.8×10^5 M^{-1}; K_2, 9×10^{-2}; K_3, 10^8 M^{-1}; α, 4.49×10^4; β, 1.11×10^{-1} and γ, 3.6×10^{-3}.

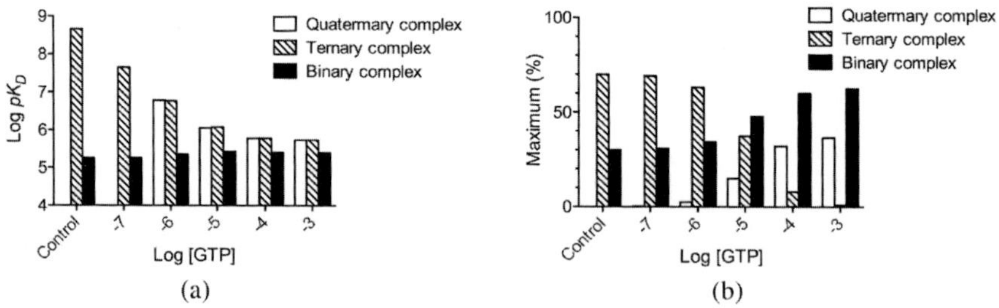

Figure 6.23: Empirical parameters of the simulated occupancy curves in Figure 6.22. (a) The value of the negative log of the observed dissociation (pK_D) of the agonist for different components of receptor occupancy — the quaternary complex (open bars), the ternary complex (stippled bars), and the binary complex (closed bars) is plotted against the value of the concentration of GTP used in the simulation. (b) The maximum fractional amount of quaternary, ternary and binary complexes formed by the agonist is plotted against the value for the concentration of GTP used in the simulation.

prior figure (Figure 6.22). Occupancy curves were simulated in the absence and presence of a high concentration of guanine nucleotide. The figure shows plots of total receptor occupancy and occupancy of the quaternary complex. In the absence of guanine nucleotide, no quaternary complex forms and the two agonists exhibit relatively high affinity. In the presence of a *G* protein-saturating concentration of guanine nucleotide, the total occupancy curves of the agonists shift to the right and a component of agonist binding includes the quaternary complex. The quaternary complex and total occupancy curves for a given agonist exhibit one-site behavior and a similar K_{obs} value when the concentration of guanine nucleotide is high. Both the maximum of the quaternary complex function and the guanine nucleotide-induced shift in agonist affinity are proportional to efficacy. In contrast, the neutral antagonist and inverse agonist do not generate significant amounts of the *DRGX* complex nor are their occupancy curves affected by guanine nucleotide.

When there is both constitutive activity and limiting *G* protein, the model behaves similarly to that shown in the last figure (Figure 6.24) except that there is high basal activity. Also, the total occupancy curve for the inverse agonist exhibits high- and low-affinity components, corresponding to the *DR* and *DRG* receptor complexes, respectively. Addition of guanine nucleotide increases the affinity of the inverse agonist by increasing the amount of the *DR* component of binding and reducing that of *DRG*.

Summary of the simulations: All of the simulations show that, in the presence of guanine nucleotide, the agonist occupancy of the quaternary complex resembles a simple one-site model. The concentration of agonist required for half-maximal formation of the quaternary complex represents the reciprocal of the observed affinity constant. Also, the maximum of the quaternary complex function at high agonist concentrations is proportional to the observed efficacy of the agonist and to the negatively cooperative effect of the guanine nucleotide on agonist binding. It follows that agonist efficacy is also proportional to the ability of the agonist to reduce the affinity of guanine nucleotides because allosteric interactions are reciprocal.

How Well Does the Model Predict the Behavior of *G* Protein-Coupled Receptors?

Many *G* protein-coupled receptors that interact with G_i or G_s exhibit ligand-binding properties that, on the surface, resemble those shown in the simulations in Figures 6.16–6.24. The interaction of ligands with the M_2 muscarinic receptor in the myocardium serves as an example. As mentioned above, increasing concentrations

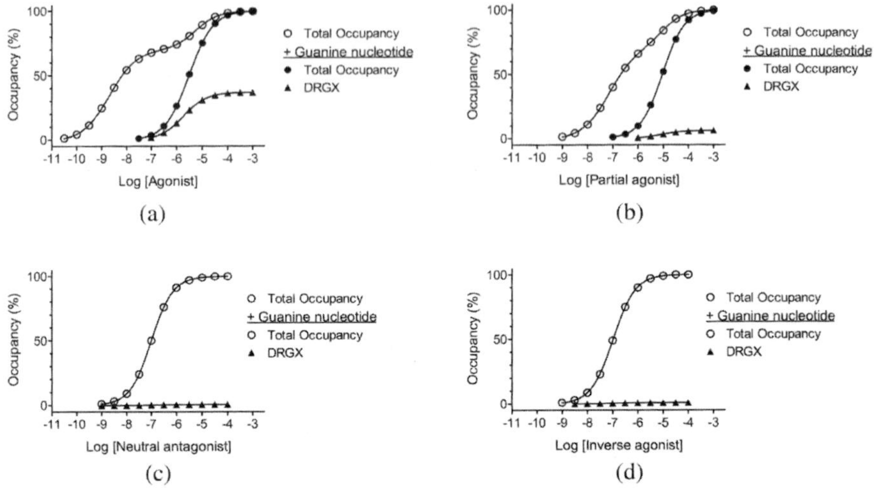

Figure 6.24: Simulation of ligand-occupancy curves in the absence and presence of a maximally effective concentration of GTP using the Quaternary Complex Model for the condition of limiting G protein relative to receptor ($G_T/R_T = 0.7$). The plots show the sum of all ligand-receptor complexes (total receptor occupancy) and the fraction of the ligand-receptor complex in the form of quaternary complex (*DRGX*) as a function of the ligand concentration. The simulations were run for an agonist (a), a less efficacious agonist (b), a neutral antagonist (c) and an inverse agonist (d). The microscopic affinity constants of the ligands for the active (K_b) and inactive (K_a) states of the receptor are: agonist, K_b, 10^{10} M^{-1}; K_a, 10^5 M^{-1}; partial agonist, K_b, 3×10^8 M^{-1}; K_a, 10^5 M^{-1}; neutral antagonist, K_b, 10^7 M^{-1}, K_a, 10^7 M^{-1} and inverse agonist, K_b, 10^5 M^{-1}; K_a, 10^7 M^{-1}. These microscopic constants yielded the following level 2b parameters for the ligands: agonist, K_1, 1.8×10^5 M^{-1}; α, 4.94×10^4 and γ, 3.6×10^{-3}; partial agonist, K_1, 1.02×10^5 M^{-1}; α, 2.6×10^3 and γ, 4.0×10^{-3}, neutral antagonist, K_1, 1.0×10^7 M^{-1}; α, 1.0 and γ, 1.0, and inverse agonist, K_1, 1.0×10^7 M^{-1}; α, 1.2×10^{-1} and γ, 8.3. The other values of the parameters for the quaternary complex model were: K_2, 9.0×10^{-2} and K_3, 1.0×10^8 M^{-1}.

of GTP reduce the affinity of highly efficacious agonists for the M_2 muscarinic receptor in the heart (Figure 6.5), and the magnitude of this negative cooperativity is proportional to the efficacy of the ligand (Figure 6.6).

Figure 6.25 illustrates another example of the effect of a high concentration of GTP on the binding of ligands to the M_2 muscarinic receptor in the heart.[13] The competitive displacement of the specific binding of [^{3}H]NMS was measured in the presence of different concentrations of three ligands. [^{3}H]NMS binding is expressed as a percentage of that measured in the absence of the competitors. Thus, the monotonically decreasing competitive inhibition curves also define increasing

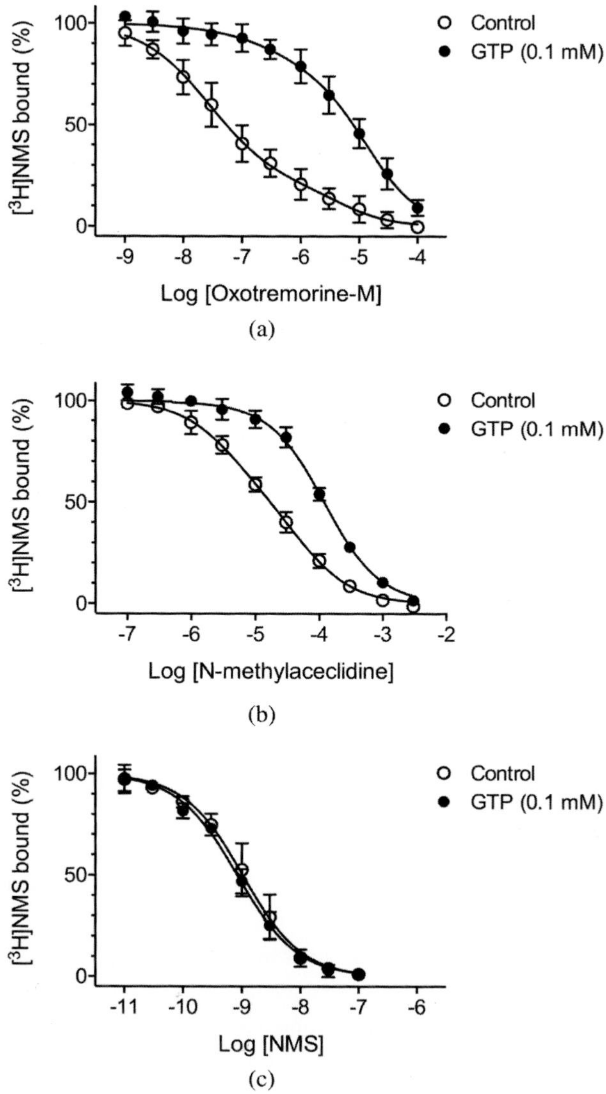

Figure 6.25: The competitive inhibition of the binding of [^{3}H]NMS to the rat myocardium by oxotremorine-M (a), N-methylaceclidine (b) and nonlabeled NMS in the absence and presence of a near maximally effective concentration of GTP. The data are from Ehlert[13] with permission. The theoretical curves for the control condition represent the least squares fit to ternary complex model.

receptor occupancy curves for the nonlabeled ligands, with percent receptor occupancy equivalent to percent displacement of [^{3}H]NMS binding. GTP (0.1 mM) caused 140- and 5.6-fold rightward shifts in the competition curves of the agonists, oxotremorine-M and S-aceclidine, while having no effect on the binding of NMS. Given that oxotremorine-M is more efficacious than S-aceclidine and that NMS is an antagonist, the results are consistent with the postulate that the negative allosteric effects of GTP are proportional to the efficacy of the orthosteric ligand.

After transforming the competition curves in Figure 6.25 using the rule that percent displacement of radioligand binding is equivalent to percent receptor occupancy by the competitor, the resulting curves are not much unlike those shown in Figure 6.24a–c. The competition curves of the agonists, oxotremorine-M and N-methylaceclidine, are clearly biphasic, consisting of high and low affinity components. GTP has the effect of causing the occupancy curve to nearly resemble a simple one-site curve with an observed affinity similar to that of the low affinity component. In contrast, the competition curve for NMS resembles a simple one-site occupancy curve that is unaffected by GTP.

If the sodium in the assay buffer is greatly reduced, then the M_2 muscarinic receptor displays constitutive activity in the absence of agonist, which can be measured as a stimulation in the binding of [^{35}S]GTPγS.[33] This activity is potently inhibited by NMS revealing its inverse agonist character. When binding is measured in sodium-free buffer, [^{3}H]NMS exhibits low affinity for the M_2 receptor, and GTP causes an increase in its observed affinity.[34,35] These binding results are similar to that illustrated for the inverse agonists in Figure 6.21d.

Using nonlinear regression analysis, it is possible to obtain a close fit of the model to the binding data measured in the absence of GTP. But this analysis only yields an estimate of about 80% for the amount of *G* protein relative to receptor. Is the concentration of *G* protein in the heart really only 80% that of the M_2 receptor? It is also possible to obtain a reasonable fit of the model to all of the data, but there are significant discrepancies at the higher concentrations of GTP. The model predicts one-site behavior, but the data predicts more complicated binding properties. What is the significance of this finding? Other questions come to mind. Do single receptors interact with single *G* proteins in a one-to-one manner, and what role do scaffolding proteins have in the receptor-*G* protein interaction?

Concerning the ratio of *G* protein to receptor, it is well known that the *G* proteins that interact with M_2 receptor in the heart (members of the G_i family) exceed the number of M_2 receptors by at least two orders of magnitude.[10] If the parameter for the G_T/R_T ratio were set to 100, the quaternary complex model would generate one-site occupancy curves under all conditions and not two-component curves like those seen with real data in Figures 6.5 and 6.25. Native tissue is heterogeneous, however, and there is probably a small population of receptors (about 20%), which does not have access to G proteins. Applying the model to the population that interacts with the G protein would yield a fit closer to the expected one-site model.

But even if the model is applied to the subset of receptors, which interacts with the G protein, the parameter for the ratio of G protein to receptor (R; see Equation 36) is ambiguous because of its status as a paremeter in a standard equilibrium expression. This type of expression applies to the binding of a ligand to a receptor in solution where the ligand can collide with and bind to the receptor from any trajectory in three-dimensional space. Increasing the concentration of the ligand in solution causes a proportional increase in the number of hits with the receptor. In contrast, a G protein can only interact with nearby receptors within the two-dimensional constraints of the membrane. The situation is analogous to standing in a large crowd of people — your interactions with those in your immediate surroundings are unaffected by the size of the crowd over wide range. If some receptors are close together and compete for the same nearby G proteins, the local concentration of available G proteins decreases leading to a low R value for the model even though the actual ratio of G protein to receptor may be high.

Heterogeneity in receptosomes may be an explanation for some of the discrepancies between the predictions of the quaternary complex model and the binding properties of the cardiac M_2 receptor. For example, the binding data in Figure 6.5*a* and *b* are consistent with the presence of three types of sites having high, medium, and low affinity for agonist in the absence of GTP. An increase in the concentration of GTP causes a reduction in the observed affinity constant of oxotremorine-M for the high affinity site (log K_H, 8.68) such that its observed affinity in the presence of a maximally effective concentration of GTP is indistinguishable from that of the low affinity site (log K_L, 5.31). This effect corresponds to a log negative cooperativity value (log γ_1) of -3.4, which represents a reduction in affinity of 1/2300. A maximally effective concentration of GTP also reduces the log affinity constant of the medium affinity site for oxotremorine-M (log K_M, 7.09) to a value lower than that of the low affinity site (4.60). This effect corresponds to a log negative cooperativity value (log γ_2) of -2.49, which represents a reduction in affinity of 1/310.

To summarize, the data are consistent with two populations of receptors that conform to the quaternary complex model with an excess of G protein relative to receptor (see Figure 6.16a). These sites undergo a reduction in agonist affinity in the presence of increasing concentrations of GTP without a reduction in relative abundance. The low affinity site in the absence of GTP may represent a population of receptors that does not communicate with G protein.

It may seem that the quaternary complex model is incomplete because it does not address the role of scaffolding proteins in maintaining the speed and specificity of receptor-G protein signaling. The model was designed specifically to enable the measurement of allosteric interactions, however, and not to describe all of the molecular components. Thus, the quaternary complex model is not so much a model of the receptosome so much as an accounting scheme for measuring observed affinity and cooperativity. Ultimately, we want to use the scheme to estimate the affinity and efficacy of agonists, antagonists, and inverse agonists through measurement of a downstream response, regardless of how the signaling pathway is constructed.

One way to appreciate the significance of the model is to consider a system that contains no proteins other than the receptor and G protein. Tatsuya Haga and colleagues have done this by reconstituting purified muscarinic receptors in artificial membranes in the absence and presence of purified G proteins.[36–38] Figure 6.26 illustrates the results of some of their experiments on the binding of ligands to cerebral muscarinic receptor reconstituted in this manner. In artificial membranes containing only the receptor, acetylcholine binds to the receptor with relatively low affinity. In membranes reconstituted *with* both the receptor and G_i, however, acetylcholine exhibits a progressive increase in affinity with an increase in the concentration of G_i. In membranes reconstituted with receptor and a high concentration of G_i, the addition of a stable GTP analog causes a large reduction in agonist affinity. In membranes containing only the receptor, the guanine nucleotide has no effect. These results are consistent with the postulate that the agonist induces a conformation of the receptor that exhibits high affinity for the free G protein and that there is

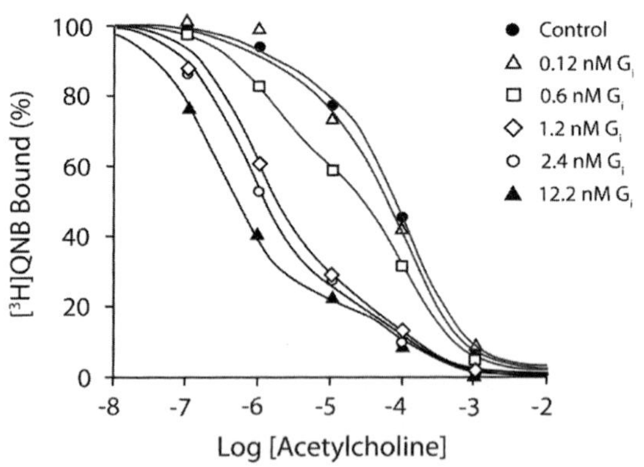

Figure 6.26: The competitive inhibition by acetylcholine of the binding of [^{3}H]NMS to purified muscarinic receptor from rat forebrain reconstituted in phospholipid vesicles with various amounts of G_i. The data are from Haga and coworkers[37] with permission.

negative cooperativity between the binding of guanine nucleotide and agonist with the receptor-*G* protein complex. These are the essential properties of the ternary complex model with guanine nucleotide, which are required for guanine nucleotide exchange and, ultimately, the transduction of a signal by the agonist. The same behavior is apparent in the binding studies on native membranes from the heart and other tissues.

Another approach to assess the cooperative interactions between receptor and *G* protein involves the use of discoidal high-density lipoprotein (HDL) particles.[39–41] These are composed of a phospholipid bilayer of about 160 phospholipids circumscribed by a ring of two apolipoprotein A-I molecules. The inner diameter of this small disk is only about 84 angstroms so that only one receptor and *G* protein can fit in it simultaneously. Roger Sunahara, Brian Kobilka, and their colleagues have reconstituted the β_2-adrenoceptor and G_s in HDL particles and examined their binding properties.[23] When reconstituted with both the β_2-adrenoceptor and G_s, the HDL particles bind the efficacious agonist, isoproterenol, with high affinity, and the observed affinity is greatly reduced by GTP. When the particles are reconstituted with the receptor only, however, the affinity of isoproterenol is low and unaffected by GTP. These results indicate that a single receptor and *G* protein exhibit the allosteric interactions of the ternary complex model with guanine nucleotide. Analogous results have been obtained with the μ opiate receptor reconstituted with G_{i2} in HDL particles.[22]

The experiments with HDL particles, of course, are not inconsistent with the oligomerization of the β_2-adrenoceptor (probably tetramers) in its native state in the cell.[42] This oligomeric clustering of receptors surely increases the efficiency of signaling through the receptosome by reducing the interaction distances between a group of clustered receptors, *G* proteins, and their effectors. Does oligomerization have any effect on the ability of the agonist to generate the quaternary complex (aside from increasing the availability of G-GDP), and if so, how does this affect agonist-receptor-*G* protein interactions and the usefulness of our model? If oligomerization has a constitutive effect that enhances or inhibits receptor function, then this would not invalidate the model as a tool for measuring receptor activity. But if agonist occupancy of one receptor in the oligomer influences the occupancy of another, then this would alter the slope of the quaternary complex function, and we would have to revise the model. Such a modification might be required for some metabotropic glutamate where negative homotropic cooperative binding has been described.[43]

James Wells and his colleagues[44] have investigated the binding properties of purified M_2 muscarinic receptors in solution and reconstituted into artificial membranes and have observed a number of additional complexities not predicted by the quaternary complex model. Nonetheless, these investigators do observe the negatively cooperative effect of guanine nucelotides on agonist binding.

This section began with the question of how well the Quaternary Complex Model describes the behavior of real receptors. Although some important components have not been addressed (receptor oligomerization, RGS proteins, and receptor-*G* protein stoichiometry), the purpose of our model is not to describe the biology of the signaling complex, but rather, to be a useful tool for measuring its allosteric transitions. From the latter perspective, our model should be simple so that we can use it to measure a variety of receptor structures, just as a pair of calipers can be used to measure the size of a variety of objects. From what has been reviewed, it seems likely that the receptor activation function for GPCRs represents the agonist-receptor-*G* protein-GDP complex through which negatively cooperative effects of the agonist and GDP is communicated. It also seems likely that this function is well approximated by a one-site occupancy function in many, but perhaps not all, instances. Ultimately, we will use this function in analyzing downstream responses as described in the next chapter. But first, we will need to understand how the agonist selects for the active state of the receptor.

Conformational Analysis of the Ternary Complex Model with Guanine Nucleotide

In the prior simulations of the model (Figures 6.16–6.24), the values of the various affinity constants and cooperativity factors were not allowed to assume any value, but were constrained to a subset that are consistent with those allowed by ligand-induced conformational changes in the receptor and *G* protein. In the three subsections that follow, I explain what this means. Before doing so I explain the different types of affinity constants used in modeling GPCRs. Readers already familiar with the differences between observed and microscopic affinity constant may wish to skip to the next subsection entitled, "Model for conformational changes of GPCRs."

Three types of affinity constants: Regarding G-protein-coupled receptors, there are three types of affinity constants that describe the binding of orthosteric and allosteric ligands. The most superficial is the observed affinity constant, which describes occupancy of a given receptor complex, like the quaternary complex. The second type describes the microscopic affinity constant of an agonist for a particular type of receptor complex in isolation. This type of constant was used in the models shown in Figures 6.13–6.15 and in all of the equations used under "Derivation of equations for the Quaternary Complex Model." For example, if it were somehow possible to attach the receptor to the G protein irreversibly, the concentration of agonist required for half-maximal occupancy of this complex in the absence of guanine nucleotide would be equivalent to the reciprocal of αK_2, the microscopic affinity constant of the agonist for the receptor-G-protein complex. But this complex is not static. It changes conformations between resting and active states. If the receptor were locked into the active state, the concentration of agonist required for half-maximal occupancy would be equivalent to the reciprocal of the microscopic affinity constant for the active state of the receptor. The parameters at this deepest level of analysis are the true independent parameters of the model, and they determine the more superficial parameters as described in the following three subsections.

Mathematically speaking, an affinity constant is defined as the concentration of the ligand-receptor complex at equilibrium divided by product of the equilibrium concentrations of ligand and receptor. Simple enough, you might say, but this definition leads to two seemingly harmless corollaries that have generated much confusion:

(1) The reciprocal of the affinity constant determines the location on the ligand concentration scale where half-maximal receptor occupancy occurs.
(2) The affinity constant is a measure of the complementariness of the structure of the ligand with that of its binding site on the receptor (e.g., a large affinity constant implies a tight fit of the former into the latter).

When the ligand-receptor complex is static, these two aspects of the affinity constant represent real properties of the ligand-receptor interaction. But when a conformational change in the receptor occurs, at least one of them is imaginary.

For example, consider the binding of an agonist to a GPCR under the condition of a large pool of available G protein and no GTP or GDP (see control curve of Figure 6.16a). The open bar labeled control on the histogram shown in Figure 6.17a represents the negative log dissociation constant of this curve. For the purposes of discussion, this curve has been illustrated again in Figure 6.27. The value of $-\log K_D$ (pK_D) is also equivalent to the log affinity constant ($\log K_{obs}$) because K_{obs} is equivalent to the reciprocal of K_D. The value of K_D or the reciprocal of K_{obs} equals the concentration of agonist required for half-maximal receptor occupancy. This relationship is illustrated in Figure 6.27. Thus, there is a real occupancy curve that satisfies the first corollary defined above. But there is no receptor structure that actually exhibits this affinity constant, because the receptor is in equilibrium between at least two different shapes, and K_{obs} represents the weighted average of the

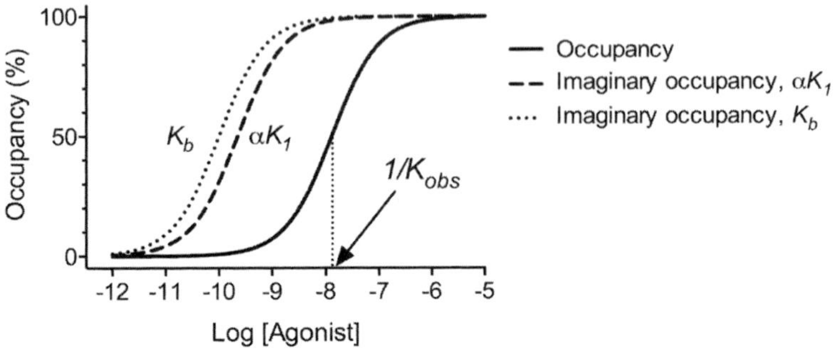

Figure 6.27: Relationship between the observed affinity constant (K_{obs}) and the microscopic affinity constants for the active state of the receptor (K_b) and the ternary complex (αK_1).

affinities of the different shapes. According to the second corollary, the conformation of the binding site with an affinity constant of K_{obs} is imaginary.

With regard to the active state of the receptor, it has an affinity constant known as the microscopic affinity constant of the active state of the receptor (K_b). This is a true measure of the affinity of the agonist for the active state of the receptor. But there is no real occupancy curve that exhibits this affinity constant. The dashed line in Figure 6.27 labeled with K_b shows the imaginary occupancy curve for the receptor if it were fixed in the active conformation.

With regard to the *DRG* complex, it has a microscopic affinity constant known as αK_1. If the *G* protein were affixed to the receptor, αK_1 would define the occupancy curve for DRG, which is illustrated by the dashed line labeled with αK_1 in Figure 6.27. There is no conformation of the receptor that exhibits this affinity constant, nor is there any real occupancy curve that exhibits this affinity. Thus, aK_1 represents an imaginary occupancy curve for an imaginary conformation of the binding site. It would represent the observed affinity constant of the ligand for the *RG* complex if the latter were stable.

There is no shortage of confusion in the literature because many authors fail to appreciate whether receptor occupancy and the shape of the binding pocket represent real or imaginary properties for a given type of affinity constant. For example, some authors have stated that the observed affinity constant should not be used because it does not describe the affinity of the inactive state of the receptor. But is not this constant very useful for determining receptor occupancy? Another puzzling locution is the common statement that "agonists bind to the active state of the receptor." But does not the agonist also bind to the inactive state of the receptor? How else could it occupy the receptor if it did not first bind to the receptor in the

Hierarchical Levels of Analysis of Drug-Receptor Interactions

Level 1: Empirical parameters of the concentration-response curve

EC_{50}, E_{max}, *Hill slope*

Level 2a: Observed parameters of the receptor population

K_{obs} and *efficacy* (ε)

Level 2b: Microscopic affinity constants and cooperativity values of receptor complexes

K_1 (*D* and *R*); K_2 (*R* and *G*); K_3 (*G* and *X*)
α (*D* and *G on R*); β (*R* and *X on G*); γ (*D* and *X on RG*)

Level 3: Microscopic affinity constants of receptor states and the isomerization constant of the unoccupied receptor

K_a (R_s); K_b (R_s^*); K_q (R_s^*/R_s)

Figure 6.28: Parameters for characterizing drug-receptor interactions at different hierarchical levels of analysis. Drug-receptor interactions can be categorized on four hierarchical levels, and parameters for characterizing drug action at each level are indicated in the figure.

inactive state? A more correct description is, "An agonist binds to a receptor and exhibits the same observed affinity for the inactive and active states, but more of the receptor is in the active state after the agonist binds."

The different types of affinity constants used for describing GPCRs are summarized in Figure 6.28. These not only include constants for ligand-receptor and receptor-G protein interactions, but also those for the spontaneous interconversion of the receptor between ground and active states. The parameters are divided into three hierarchical categories. To maintain consistency throughout this book, these levels are designated as 2a, 2b and 3 (Level 1 parameters relate to the EC_{50} and $E_{\max}$ values of the functional response). The level 2a parameters are the most superficial and include the observed affinity constant for the ligand-receptor complex. The level 2b parameters describe the microscopic constants for the various receptor and G protein complexes described for the ternary complex model with guanine nucleotide as shown in Figure 6.15. Finally, the level 3 parameters define the properties of the resting and active states of the receptor.

A given parameter at one hierarchical level of analysis is described by a combination of deeper level parameters. Our goal in the following subsections is to define the level 2b parameters in terms of level 3 parameters. Also the value of a given parameter (e.g., level 2a parameter $K_{\text{obs}} = 10^6\ \text{M}^{-1}$) at one level could be generated mathematically by an infinite set of parameter values at the next deeper level of analysis. Thus, it is usually impossible to estimate most of the deeper level parameters from the analysis of functional or binding data. However, it is possible to estimate all of these from functional data involving allosteric interactions under the condition of partial receptor inactivation (see Chapter 11). Another classification is based on whether the parameter defines properties of the ligand (ligand-dependent) or properties of the receptor and *G* protein (ligand-independent). Examples of the latter include the spontaneous interaction between the receptor and *G* protein and the spontaneous interconversion between resting and active states of the receptor. For a given set of ligand-independent parameters at level 3, there is only a finite range of behavior at more superficial levels, even if an infinite set of ligand-dependent parameters is considered. Thus, to define the subset of real outcomes of the ternary complex model with guanine nucleotide, we need to define the level 2b parameters in terms of level 3 parameters.

Model for conformational changes of GPCRs: In this section, a two-state model is developed to explain the ligand-induced activation of the receptor. The model represents a receptor with a single conformation of both the active and inactive states of the receptor. Many authors have stated that two-state models fail to explain the behavior of GPCRs because there is substantial evidence for more than one active receptor conformation. My reason for describing a two-state model is not because I think that there are only two conformations of GPCRs that represent the active and inactive states, but rather, to explain how conformational changes occur and how they determine intrinsic efficacy. In other words, explaining how one component of a multi-component system works helps our understanding of how the entire system functions. Thus, the common statement about the failure of two-state models needs to be turned on its head. Not only are two-state models useful for understanding multiple-receptor conformations, but we need to apply the principles of the two-state model as many times as there are distinct active conformations of the receptor.

If each receptor species in Figure 6.15 is replaced with an equilibrium describing the interconversion between active and inactive states of the receptor, we will have our model for conformational changes in GPCRs. Adding a conformational change to the model increases its dimensions from three to four, which is difficult to illustrate with a two-dimensional drawing. A simple solution is to flatten the current cubic model (Figure 6.15) into two concentric squares (Figure 6.29a). The outer square represents the top of the cube, the inner square the bottom, and the diagonals connecting the vertices of the two squares represent the four vertical edges of the cube. Two such flattened cubes are used to designate the two states of

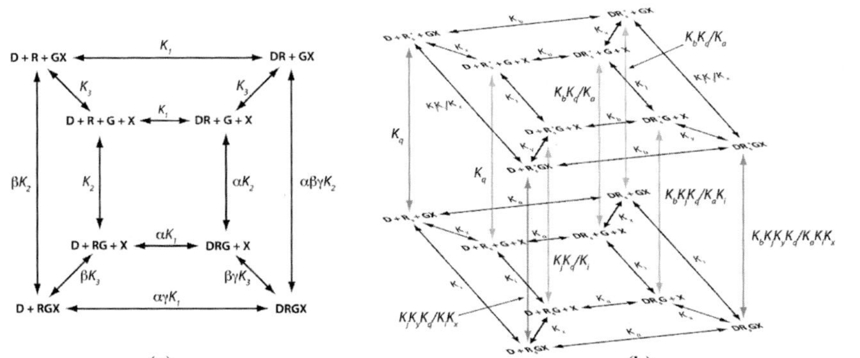

(a) (b)

Figure 6.29: Conformational analysis of the quaternary complex model. (a) The cubic quaternary complex model shown in Figure 6.15 has been compressed onto a single plane. (b) The active (upper plane) and inactive (lower plane) states of the quaternary complex model shown in panel *a* are illustrated with their associated microscopic constants.

the receptor (active or inactive) for each receptor complex in the model. The active and inactive states correspond to the top and bottom surfaces of the complicated model shown in Figure 6.29b and the vertical lines interconnecting them represent quantal transitions between the inactive and active states.

To solve our model, an affinity constant or transition constant is assigned to each equilibrium step. These parameters represent the microscopic constants of the receptor states (level 3 parameters). Then equations that define level 2b parameters in terms of level 3 parameters are derived. Finally, equations that express the amount of a given receptor state as a function of the orthosteric ligand and guanine nucleotide concentrations are derived. I will summarize how to express some level 2 parameters in terms of level 3 parameters and how to develop an equation for the active state of the quaternary complex. In the next section, the behavior of the model and the ultimate determinants of affinity and efficacy are explained.

The entire model is complicated, so I will describe how to assign microscopic constants to the inner cube, which represents the binding of drug to the receptor in the absence of guanine nucleotide. This cube is derived from the simple ternary complex model shown by the middle scheme in Figure 6.13. This simple scheme has been reproduced in Figure 6.30a. Each of the four equilibria (E_1–E_4) has been expanded in Figure 6.30b to identify the inactive (R_s) and active (R_s^*) states of the receptor. These are assembled into a cube as shown in Figure 6.30c to yield the central cube shown in Figure 6.29b.

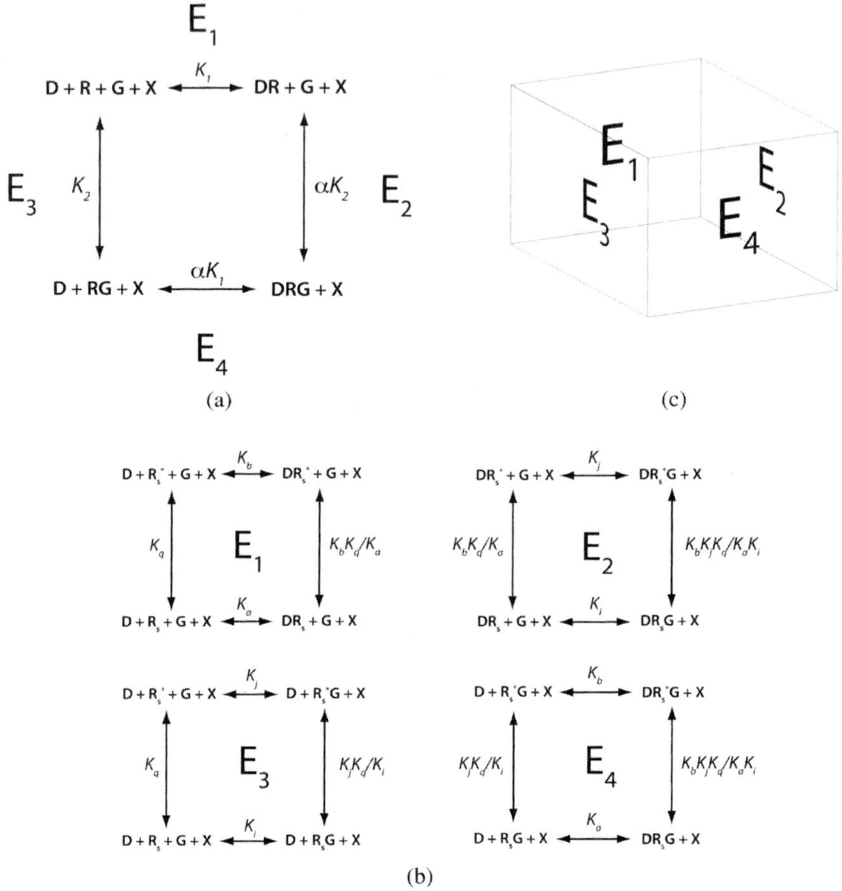

Figure 6.30: Derivation of the states of the receptor in the Ternary Complex Model and their associated microscopic constants. (a) The ternary complex model and its associated level 2b parameters are shown. Each of the four equilibrium steps are designated by the capital letter "E" and the corresponding subscript. (b) The receptor complexes for each equilibrium shown in panel *a* have been expanded into their active and inactive states, and their associated microscopic constants are shown. (c) The orientation of each expanded equilibrium in panel *b* is illustrated for the complete ternary complex model defined at the level of receptor states. The cube represents the central cube of equilibrium shown in Figure 6.29b.

At the level of receptor states, the simple binding of the drug to the receptor ($D + R \leftrightarrow DR$) can involve four steps, only three of which represent fundamental properties of the receptor or the ligand-receptor complex. These are the spontaneous interconversion of the receptor between inactive and active states (step 1) and the binding of the drug to the inactive (step 2) and active (step 3) states of the receptor. The three microscopic constants that define these processes are K_q, K_a and K_b, respectively:

$$K_q = \frac{[R_s*]}{[R_s]}, \tag{6.49}$$

$$K_a = \frac{[DR_s]}{[D][R_s]}, \tag{6.50}$$

$$K_b = \frac{[DR_s^*]}{[D][R_s^*]}. \tag{6.51}$$

These constants have been applied to the left-most square of equilibria in Figure 6.30b. The fourth process is the drug-induced conversion of the receptor into the active state, which is determined by the three microscopic constants just mentioned:

$$\frac{K_b K_q}{K_a} = \frac{[DR_s^*]}{[DR_s]}. \tag{6.52}$$

As described for the model in Figure 6.15, there is no net flux around the circuit of equilibrium, and consequently, only three constants are required to describe the four equilibria. Before continuing with the development of the model, I digress on two important aspects of conformational analysis.

First, the reader may have noticed that in developing the prior model in Figure 6.15, two affinity constants and a cooperativity factor were used to describe each square of equilibria. The cooperativity factor represents the maximum amount by which one ligand changes the observed affinity of another within the context of a population of receptors. At the single receptor level, however, there are only quantal changes in affinity corresponding to the active and inactive conformations. These have defined affinities (K_b and K_a, respectively), and the free receptor has a defined tendency to interconvert between the two states (K_q). The ability of the drug to induce the active conformation is determined by these fundamental properties as described in Equation (6.52). That is, conformational induction is proportional to the drug's affinity for the active state relative to the inactive state (K_b/K_a) and is inhibited by the inherent tendency of the receptor to prefer the inactive state ($0 < K_q < 1$). Thus, each affinity constant of the drug for each state of the receptor is assigned a unique constant.

The second issue relates to the two pathways that a drug can take to activate the receptor. One is by binding selectively to the active state after the receptor spontaneously assumes this conformation (conformational selection):

$$D + R_s \overset{K_q}{\longleftrightarrow} D + R_s^* \overset{K_b}{\longleftrightarrow} DR_s^*. \tag{6.53}$$

The other is to bind to the receptor in the inactive state and then induce the active conformation (conformational induction):

$$D + R_s \overset{K_a}{\longleftrightarrow} DR_s \overset{\frac{K_b K_q}{K_a}}{\longleftrightarrow} DR_s^*. \tag{6.54}$$

At equilibrium, both processes yield the same result, so there is no disadvantage to including both even if only one of them actually occurs. The spontaneous conversion of the receptor into the active state in the absence of agonist is usually a rare event for any single receptor. If a drug had to wait for the receptor to isomerize into the active state before it bound, then the rate of drug action would be very slow. Another possibility is that spontaneous receptor activation may occur frequently. The mean time that the unoccupied receptor spends in the active is usually very short, however, so that the probability that the drug will bind to the unoccupied active state is still very low. It is known that activation of many GPCRs occurs immediately upon addition of agonist. This implies that conformational induction is the main route for receptor activation. If so, why complicate the model with conformational selection?

This issue has been addressed previously in Chapter 2 in regard to ligand-gated ion channels. For convenience, I revisit the question here. One reason to consider conformational selection is that this step defines the affinity constant of the ligand for the active state of the receptor, and we need this estimate for quantifying conformational induction as shown by Equations (6.52) and (6.54) even if conformational selection is impossible.

For example, suppose that when the agonist induces the active state, it is engulfed by the receptor and cannot dissociate. Would this imply that the agonist has infinitely high affinity for the active state? After all, the affinity constant is defined as the ligand-receptor complex divided by the product of the concentrations of ligand and receptor at equilibrium. If the ligand-receptor complex cannot dissociate, then the affinity constant would be infinite. This would be the case if each agonist-receptor complex remained permanently activated even after the agonist was washed away. But the odds are great that the agonist-receptor complex would revert back to the inactive state, and the agonist would either dissociate or undergo one or more rounds of receptor activation before ultimately dissociating from the receptor. If so then the agonist would not have infinitely high affinity for the active state. Its microscopic affinity constant is a measure of the strength of the bond that it has with the binding pocket of the active state. It is equivalent to the value of K_b that would accurately predict the degree of conformational induction by Equation (6.49) given that K_a and K_q are known. Thus, having a conformational selection step in the model is useful for specifying the theoretical value of K_b even if the agonist cannot directly bind to or dissociate from the active state.

Now let us return to the development of the conformational model. Microscopic affinity constants are added to the other squares of equilibria as shown in Figure 6.30b. The affinity of the agonist for the receptor is determined by the state of the receptor and not whether the latter is associated with a G protein. For example the affinity of D for R_s is the same as its affinity for R_sG (K_a). Similarly, the affinity of D for R_s^* is the same as its affinity for R_s^*G (K_b). Conformational changes in the G protein are described in a later section. Here it is assumed that the affinities of G for R_s and R_s^* are K_i and K_j, respectively. The two isomerization steps ($DR_s \leftrightarrow DR_s^*$ and $DR_sG \leftrightarrow DR_s^*G$) are described in terms of other constants so that there is no net flow around the equilibrium circuit. The four squares of equilibrium are assembled into a cube as shown in Figure 6.30*c*. The top surface of the cube represents receptor complexes in the active state and the bottom side the inactive receptor complexes. The process is repeated following the same procedure for all of the equilibrium in the model shown in Figure 6.29*a* to yield the complete model in Figure 6.29*b*.

Defining level 2b parameters in terms of level 3 parameters: A simple way to define the parameters of the model in Figure 6.15 (level 2*b* parameters) in terms of the microscopic constants for receptor states (level 3 parameters) is to replace each receptor complex in the equations that define level 2b parameters with the summation of the two receptor-states that comprise the complex. For example, the microscopic affinity constant of the agonist for the free receptor complex is defined by Equation (6.3) at the beginning of this chapter. Replacing each receptor complex in this equation with the summation of its two states yields

$$K_1 = \frac{[DR_s] + [DR_s^*]}{[D]\left([R_s] + [R_s^*]\right)}. \tag{6.55}$$

Each receptor-state complex in Equation (6.55) can be expressed as a function of the concentration of ligand and G protein and the appropriate microscopic constants for the receptor states. For example, rearrangement of Equations (6.49)–(6.51) yields

$$K_q[R_s] = [R_s^*] \tag{6.56}$$

$$[DR_s] = DK_a[R_s] \tag{6.57}$$

$$[DR_s^*] = DK_bK_q[R_s]. \tag{6.58}$$

Substituting these expressions for the appropriate variables in Equation (6.55) yields

$$K_1 = \frac{DK_a[R_s] + DK_bK_q[R_s]}{[D]\left([R_s] + K_q[R_s]\right)}. \tag{6.59}$$

Simplifying yields

$$K_1 = \frac{K_a + K_bK_q}{1 + K_q}. \tag{6.60}$$

Using an analogous approach for the solution for K_2 yields

$$K_2 = \frac{\left(K_i + K_jK_q\right)[R_T]}{1 + K_q}. \tag{6.61}$$

To solve the equation for α, one can first determine the solution for αK_1, and then divide the equation for latter by that for K_1 to estimate α. The parameter αK_1 is solved in a manner similar to that described for K_1 and K_2:

$$\alpha K_1 = \frac{K_aK_i + K_bK_jK_q}{K_i + K_jK_q}. \tag{6.62}$$

Solving for α yields

$$\alpha = \frac{\alpha K_1}{K_1} = \frac{\frac{K_aK_i+K_bK_jK_q}{K_i+K_jK_q}}{\frac{K_a+K_bK_q}{1+K_q}} = \frac{\left(1 + K_q\right)\left(K_aK_i + K_bK_jK_q\right)}{\left(K_a + K_bK_q\right)\left(K_i + K_jK_q\right)}. \tag{6.63}$$

The other cooperativity factors can be solved in an analogous manner:

$$\beta = \frac{\left(K_iK_x + K_jK_qK_y\right)}{K_x\left(K_i + K_jK_q\right)}. \tag{6.64}$$

$$\gamma = \frac{\left(K_aK_iK_x + K_bK_jK_qK_y\right)\left(K_i + K_jK_q\right)}{\left(K_iK_x + K_jK_qK_y\right)\left(K_aK_i + K_bK_jK_q\right)}. \tag{6.65}$$

To avoid complexity, the different states of the G protein have not been included in the model, so it is unnecessary to define K_3 in terms of the microscopic affinity constants of GTP and GDP for the different states of the G protein. It follows that K_3 is equivalent to K_x.

This concludes the derivation of the equations for the level 2b parameters in the model in Figure 6.15. All of the simulations in Figures 6.16–6.24 were generated by first choosing values for the level 3 parameters (K_a, K_b, K_e, K_f, and K_q) and then using these to determine the values of the level 2b parameters (K_1, K_2, α, β, and γ). Then these parameters were used in Equations (6.39) and (6.47) to simulate the behavior of the ternary complex model with guanine nucleotide (Figure 6.15).

Determination of occupancy of receptor states: It is also possible to develop state equations that define agonist occupancy of the active and inactive states of the various receptor complexes in terms of the level 3 parameters. Given that the quaternary complex component of the model has already been described in Figures 6.16–6.24, it is possible to divide each component into its active and inactive state using a simple approach. The equilibrium between the active (DR_s^*GX) and inactive (DR_sGX) forms of the quaternary complex, for example, is defined by the following level three parameters:

$$\frac{K_bK_jK_yK_q}{K_aK_iK_x} = \frac{[DR_s^*GX]}{[DR_sGX]}. \qquad (6.66)$$

The fraction of the quaternary complex in the active state is given by

$$\frac{[DR_s^*GX]}{[DR_s^*GX]+[DR_sGX]} = \frac{\frac{[DR_s^*GX]}{[DR_sGX]}}{\frac{[DR_s^*GX]}{[DR_sGX]}+1}. \qquad (6.67)$$

Substituting Equation (6.66) for the ratio $[DR_s^*GX]/[DR_sGX]$ and simplifying yields

$$\frac{[DR_s^*GX]}{[DR_s^*GX]+[DR_sGX]} = \frac{1}{\frac{K_aK_iK_x}{K_bK_jK_yK_q}+1}. \qquad (6.68)$$

Multiplying this equation by the fractional amount of receptor occupancy yields an equation for the fractional amount of the total receptor population in the active state of the quaternary complex:

$$\frac{[DRG_s^*X]}{R_T} = \frac{1}{\frac{K_aK_iK_x}{K_bK_jK_yK_q}+1} \times \frac{[D]}{[D]+\frac{1}{K_{DRGX}}}, \qquad (6.69)$$

in which K_{DRGX} is defined by Equation (6.48).

The various receptor complexes of the Quaternary Complex Model can be divided into active and inactive states using an analogous approach. The only disadvantage of Equation (6.70) is that it combines both level 2b and 3 parameters. But given that we have already defined the amount of $DRGX$ complex using level 2b parameters, Equation (6.70) is convenient for calculating the active and inactive components of this complex.

Figure 31 shows a simulation of the fractional amount of agonist-receptor complex in the active and inactive states for the condition of a lack of guanine

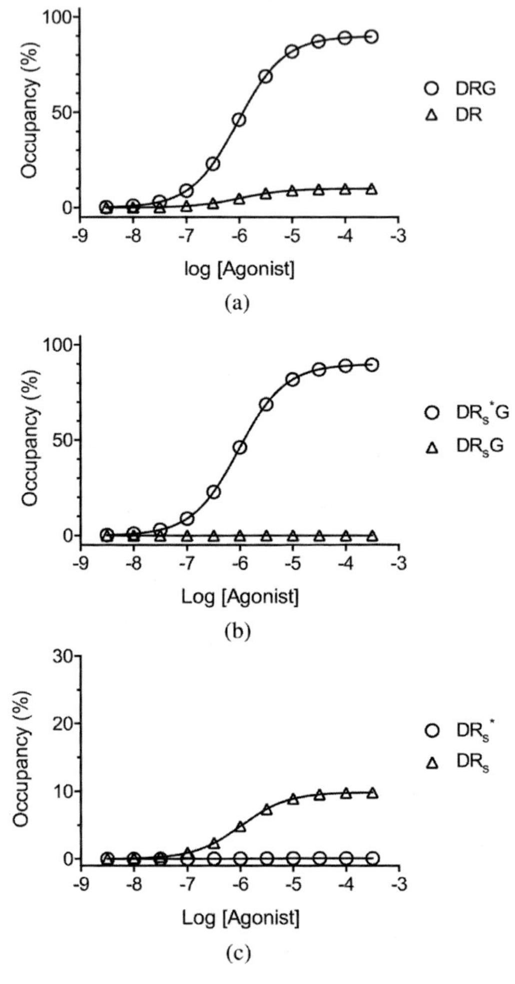

Figure 6.31: Conformational analysis of the Quaternary Complex Model for the condition of no guanine nucleotide and a ten-fold excess of G protein relative to receptor. The plots represent simulations of the different components of receptor occupancy for an agonist that binds to a receptor conforming to the consequences of the Quaternary Complex Model. (a) The fractional receptor occupancies of the ternary (DRG) and binary (DR) complexes are plotted against the concentration of agonist. (b) The fractional amounts of ternary complex in the active (DR_s^*G) and inactive (DR_sG) states are plotted against the concentration of agonist. (c) The fractional amounts of binary complex in the active (DR_s^*) and inactive (DR_s) states are plotted against the concentration of agonist. The microscopic constants of the receptor states for this simulation were: K_b, 10^9 M^{-1}; K_a, 10^5 M^{-1}; K_i, 10^{-3}; K_j, 10^2; and K_q, 10^{-6}. These microscopic constants yield the following level 2b parameters: K_1, 1.01×10^5 M^{-1}; K_2, 1.1×10^{-3}; and α, 9.01×10^2.

nucleotide and an abundance of G protein relative to receptor. The simulations are for an efficacious agonist ($K_b/K_a = 10^4$). Panel *a* shows the components of agonist occupancy in the form of the binary (DR) and ternary (DRG) complexes. Because the agonist is highly efficacious, most of the drug-receptor complex is in the form of DRG. Panel *b* divides the ternary complex into its active (DR_s^*G) and inactive (DR_sG) states, and panel *c* does the same for the active (DR_s^*) and inactive (DR_s) states of the binary complex. The scale of the ordinate in panel *c* has been enlarged so that the low amounts of the binary complex can be seen more easily. Almost the entire population of the ternary complex is in the active state, whereas the binary complex is mainly in the inactive state. This simulation shows that it is primarily the active state of the receptor that engages the G protein.

Figure 32 describes the analogous theoretical curves for the condition of a G protein-saturating amount of guanine nucleotide. Panel *a* shows the components of receptor occupancy in the form of binary and quaternary complex. Even though the selectivity of the agonist for the active state ($K_b/K_a = 10^5$) is ten-fold greater than that of the agonist in Figure 31, the maximum of the quaternary complex function corresponds to only 10% receptor occupancy because of the negatively cooperative effects of the guanine nucleotide. Panel *b* shows the components of the population of quaternary complex in the active (DR_s^*GX) and inactive states (DR_sGX), and panel *c* shows the corresponding amounts for the binary complex (DR_s^* and DR_s). Whereas most of the quaternary complex is in the active state, nearly equal amounts of the binary complex are in the active and inactive states.

The determinants of agonist efficacy: The observed efficacy of a ligand is proportional to how much of the receptor population it converts into the active state bound with G protein and guanine nucleotide (DR_s^*GX). How this ability is related to the selectivity of the ligand for the active state of the receptor is described in Figure 33a. The maximal amount DR_s^*GX formed at a receptor-saturating concentration of ligand ($DR_s^*GX_{\mathrm{max}}$) is plotted against the ratio of the microscopic affinity constant of the ligand for the active state of the receptor divided by that of the inactive state (K_b/K_a). The simulation represents a system with little constitutive activity under the condition of a G-protein-saturating amount of guanine nucleotide. As the selectivity of the ligand for the active state increases above values greater than the reciprocal of K_q, $DR_s^*GX_{\mathrm{max}}$ increases until reaching a maximal plateau that depends on the properties of the system. Once the ratio of K_b/K_a reaches a certain magnitude, the amount of the receptor population in the form of $DR_s^*GX_{\mathrm{max}}$ is at the maximum possible level, and a further increase in K_b/K_a leads to no further increase in $DR_s^*GX_{\mathrm{max}}$. Thus, all agonists with K_b/K_a values exceeding that required for maximal formation of DR_s^*GX would exhibit the same

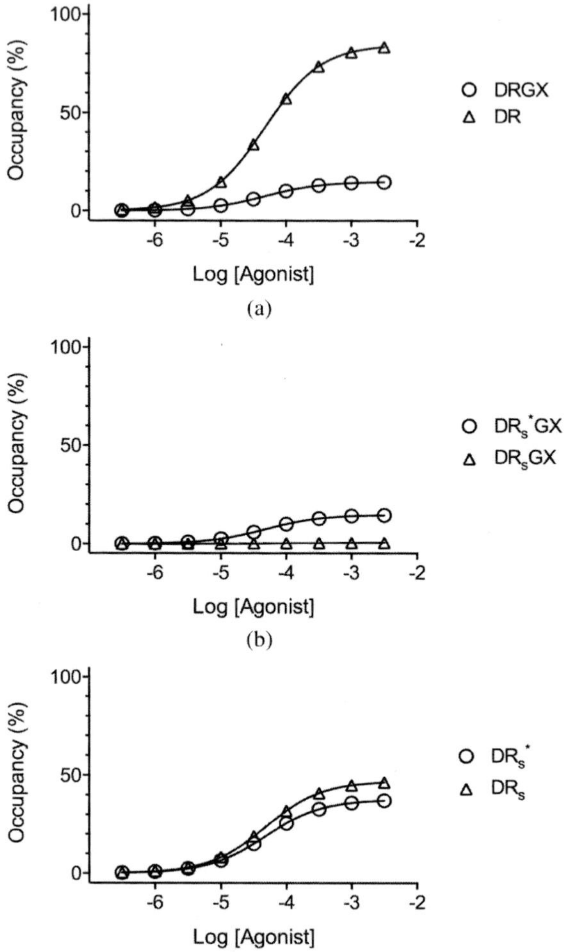

Figure 6.32: Conformational analysis of the Quaternary Complex Model for the condition of a maximally effective concentration of guanine nucleotide and a ten-fold excess of *G* protein relative to receptor. The plots represent simulations of the different components of receptor occupancy for an agonist that binds to a receptor conforming to the consequences of the Quaternary Complex Model. (a) The fractional receptor occupancies of the quaternary (*DRGX*) and binary (*DR*) complexes are plotted against the concentration of agonist. (b) The fractional amounts of quaternary complex in the active (DR_s^*G) and inactive (DR_sG) states are plotted against the concentration of agonist. (c) The fractional amounts of binary complex in the active (DR_s^*) and inactive (DR_s) states are plotted against the concentration of agonist. The microscopic constants of the receptor states for this simulation were: K_b, 10^9 M^{-1}; K_a, 10^4 M^{-1}; K_i, 10^{-3}; K_j, 10^2; K_q, 8×10^{-6}; K_x, 10^8 M^{-1} and K_y, 10^4 M^{-1}. These microscopic constants yield the following level 2b parameters: K_1, 1.8×10^4 M^{-1}; K_2, 1.8×10^{-3}; K_3, 10^8 M^{-1}; α, 2.47×10^4; β, 5.56×10^{-1} and γ, 7.42×10^{-4}.

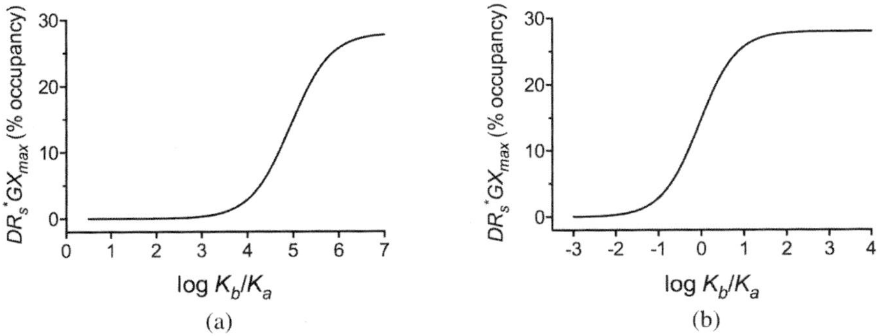

Figure 6.33: Relationship between the selectivity of the ligand for the active state of the receptor (K_b/K_a) and the maximal amount of the active state of the quaternary complex ($DR_s^*GX_{\max}$) generated by the ligand. The simulations are for the condition of a maximally effective concentration of guanine nucleotide and receptors exhibiting little or no (a) and substantial (b) constitutive activity. Plots show the maximal amount of quaternary complex associated with the active state of the receptor expressed as a function of the selectivity of the agonist for the active state. The microscopic constants for this simulation are, panel *a*: K_i, 7×10^{-3}; K_j, 7×10^2; K_q, 8×10^{-6}; K_x, 10^8 M^{-1} and K_y, 4×10^4 M^{-1} and panel *b*: K_i, 10^{-3}; K_j, 10^2; K_q, 0.8; K_x, 10^8 M^{-1} and K_y, 4×10^4 M^{-1}.

observed efficacy even though their selectivities for the active state (K_b/K_a) may be different.

Figure 6.33b shows the same relationship when there is constitutive activity. In the absence of ligands, the amount of receptor in the R_s^*GX state represents 15% receptor occupancy for this example. When the K_b/K_a ratio is less than one, the ligand (i.e., inverse agonist) selects for the inactive receptor state and exhibits a $DR_s{}^*GX_{\max}$ value less than the constitutive stimulus (R_s^*GX) in the absence of ligand. When the ligand has a K_b/K_a ratio of one (i.e, neutral antagonist), it has no influence on receptor activation and the amount of $DR_s'GX_{\max}$ is equivalent to the constitutive stimulus. As the K_b/K_a value of the ligand (i.e., agonist) increases, there is a gradual increase in $DR_s{}^*GX_{\max}$ until the maximal level is reached.

Two other parameters that determine observed efficacy are the negative cooperativity between the binding of agonist and GDP to the RG complex (γ) and positive cooperativity between the binding of agonist and G protein to the receptor (α). The latter parameter is proportional to the ability of the agonist to persuade the receptor to associate with the G protein (αK_2):

$$DR + G \overset{\alpha K_2}{\longleftrightarrow} DRG$$

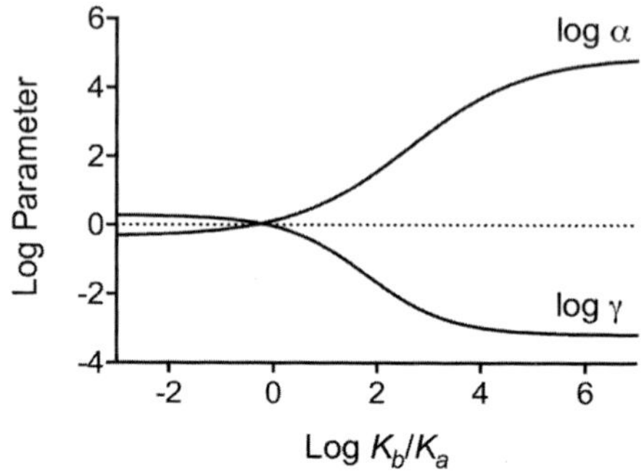

Figure 6.34: Relationships between the selectivity of the ligand for the active state of the receptor (K_b/K_a) and the positive cooperativity between the binding of the ligand and the G protein with the receptor (α) and the negative cooperativity between the binding of ligand and guanine nucleotide the receptor-G protein complex (γ). The simulations are for the condition of a maximally effective concentration of guanine nucleotide and a receptor exhibiting little or no constitutive activity. Plots show the values of the log cooperativity constants plotted against the selectivity of the ligand for the active state of the receptor. The microscopic constants for this simulation are, K_i, 7×10^{-3}; K_j, 7×10^2; K_q, 8×10^{-6}; K_x, 10^8 M^{-1} and K_y, 4×10^4 M^{-1}.

The values of both log α and log γ are plotted against the K_b/K_a ratio of the ligand in Figure 6.34 for a system with little constitutive activity. As the K_b/K_a value of the ligand increases above one, the value of α increases and that of γ decreases until they asymptotically approach their limiting values. The inverse relationship between these parameters shows that as the selectivity of the agonist for the active state increases, the corresponding increase in efficacy is associated with an increase in the ability of the agonist to persuade the receptor to associate with the G protein (increase in α) and cause the latter to loosen its grip on GDP (decrease in γ).

Multiple conformations of active and inactive receptor states: The explanation of receptor function rests on the assumption that the receptor can exist in a small number of discrete states. One of the most striking pieces of evidence for this idea is the behavior of single, ligand-gated ion channels, which switch between open and closed states rapidly with no evidence of intermediate states (see Prologue, Figure 7). This type of mechanism enables a high degree of fidelity in the control of receptor function. In the absence of ligand, the receptor is off, and it switches on to a fixed level of conductance when occupied by the endogenous ligand. The summation of the digital output from hundreds and thousands of such channels yields a graded output proportional to the magnitude of receptor activation.

This fine control arises from the bifunctional structure of the receptor. That is, the structure of the channel gate is linked to the structure of the ligand-binding site. When the conformation of the ligand-binding site has high affinity for the endogenous ligand, the structure of the receptor is such that the channel gate is open. An abundance of receptor states would imply a large number of gate conformations with different single-channel conductances, each linked to a different conformation of the ligand-binding site. It might be argued that a single conformation of the channel gate is associated with a number of different conformations of the ligand-binding site. But this is a recipe for a non-functional receptor because major conformational changes in the ligand-binding site would not be linked to changes in conductance.

Imagine the case of a large number of discrete receptor states for a GPCR, each having a different capacity to engage the *G* protein and exchange GDP for GTP. In the absence of endogenous ligand, the receptor would be more likely to assume different conformations. Some may have activity, and thus, there would be more spontaneous basal activity, resulting in misfiring of the receptor in the absence of ligand. When the endogenous ligand bound, the probability that the resulting ligand-receptor complex would exhibit high activity would decrease in proportion to the total number of receptor conformations. For example, if there were several conformations exhibiting high affinity for the ligand, then only a minor subset would have high activity because of the unique structural requirements for guanine nucleotide exchange. The time that the agonist-receptor complex spent in less active states would reduce the rate of *G* protein activation. If an infinitely large number of states is proposed, then the endogenous ligand ceases to have any control over receptor activity. Conformational selection or induction would be impossible because the agonist would be unable to discriminate between inactive and active states of the receptor. Clearly, a small number of discrete receptor states is an adaptation that enables an endogenous ligand to cause a receptor to switch from a nearly completely inactive state to a fully active one. The summation of many such quantal events gives rise to a graded control of function.

The idea of discrete states, however, does not imply that the conformations of the active and inactive states are fixed. There is certainly Brownian and thermal motion of the receptor protein around a preferred conformation. This implies that the microscopic affinity constant of the agonist for a discrete receptor state represents a weighted average of a potentially large number of similar sub-conformations depending upon how a sub-conformation is defined.

If so, why not use cooperativity factors to illustrate that the discrete states of the receptor can assume a range of conformations with varying affinities for a given ligand? Weiss and colleagues[45] have described a model for receptor-*G* protein interactions using this approach. Their model is useful for illustrating a range of receptor conformations of the active and inactive states, but it does not enable one to predict why the receptor behaves the way it does. The cooperativity factor α, for example, is defined in terms of a number of other cooperativity factors, but the origin of these sub-factors is not defined in terms of receptor states. Consequently, it is not obvious what represents an appropriate set of parameter values for the model. For example, without defining the level 2b cooperativity factors α and γ in terms of the microscopic affinity constants (level 3 parameters), one would never know that the solution set of real parameter values only includes those showing a reciprocal relationship between the two (see Figure 34). Thus, to define and explain the behavior of any receptor model, each discrete conformation of a receptor state should be defined with a unique microscopic affinity constant for the ligand even though there may be range of conformational motion associated with the state.

There is certainly evidence that there are multiple discrete conformations of the active state of GPCRs with unique selectivity for signaling proteins including *G* proteins and *G* protein receptor kinases.[46] This situation leads to the possibility that some ligands may bias signaling through one pathway relative to another. Numerous cases of agonist-biased signaling have been reported. It is possible to expand the Quaternary Complex model with additional conformations of the active state of the receptor to account for this mechanism. The functional consequences of this model are described in Chapter 9.

Conformational Changes of the *G* Protein

In the foregoing conformational analysis of the Quaternary Complex model, the different states of the *G* protein were not addressed. These were omitted to reduce the complexity of the presentation of the model. In this section, the consequences of the different states of the *G* protein are described.

To explain the interaction of heterotrimeric *G* proteins with receptor and guanine nucleotides, no less than three states of the G_α subunit must be considered. In the absence of guanine nucleotides, highly efficacious agonists induce the formation of a stable, ternary complex between the receptor and the holoprotein ($G_\alpha G_{\beta\gamma}$) implying that there is a state of G_α (G_α^*) that exhibits high affinity for both $G_{\beta\gamma}$ and the active state of the receptor (DR_s^*). Because GDP stabilizes the holoprotein form of the *G* protein and reduces the affinity of agonists for the receptor in a manner proportional to their intrinsic efficacy, there must be another state of G_α (G_α^{**}) that exhibits high affinity for GDP and $G_{\beta\gamma}$ and low affinity for DR_s^*. Finally, GTP is known to destabilize the holoprotein and reduce the affinity of agonists for the

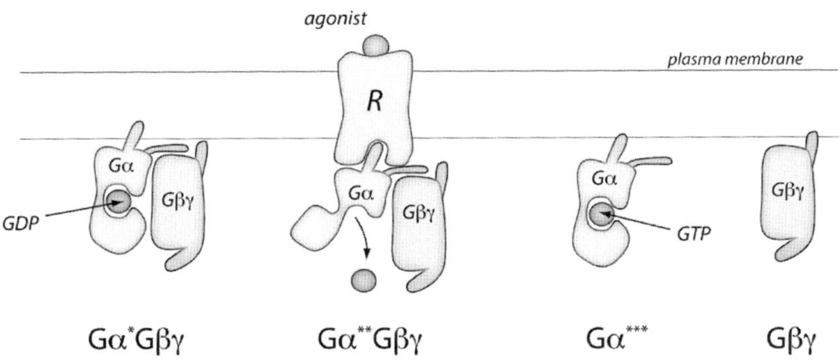

Figure 6.35: The three states of the $G\alpha$ subunit of a heterotrimeric G protein and their association with guanine nucleotides and the active state of the receptor. The $G\alpha^*$ state exhibits high affinity for GDP and $G\beta\gamma$ and relatively low affinity for GTP. It might also exhibit some affinity for the inactive state of the receptor. The $G\alpha^{**}$ state exhibits high affinity for the active state of the receptor and relatively low affinity for both GTP and GDP. The $G\alpha^{***}$ state exhibits high affinity for GTP and low affinity for $G\beta\gamma$ and GDP.

receptor. Thus, there must be a state of the G_α (G_α^{***}) that exhibits high affinity for GTP and low affinity for $G_{\beta\gamma}$ and the active state of the receptor. The numerous crystal structures of the different forms of G proteins are consistent with these conclusions.[47–52]

A schematic drawing of the three states of G_α is shown in Figure 6.35. Their interactions with DR_s^* and guanine nucleotides can explain the cooperative mechanisms described near the beginning of this chapter under "Analysis of the G protein cycle as a series of linked allosteric interactions." The high affinity of GDP and $G_{\beta\gamma}$ for G_α^* accounts for the positive cooperativity between the binding of GDP and $G_{\beta\gamma}$ to G_α^* and explains why GDP stabilizes the holoprotein form of the G protein. The high affinity of $G_\alpha^{**}G_{\beta\gamma}$ for DR_s^* explains why occupancy of the receptor by a highly efficacious agonist causes the receptor to interact with $G_\alpha G_{\beta\gamma}$ (Figure 6.4). The opposite selectivities of DR_s^* and GDP for $G_\alpha^*$and G_α^{**} account for the negative cooperativity between the binding of the agonist and GDP to the receptor-$G_\alpha G_{\beta\gamma}$ (Figure 6.4). Finally, the high affinity of GTP for G_α^{***} and the low affinity of the latter for $G_{\beta\gamma}$ and DR_s^* explain why GTP causes the holoprotein to dissociate into GTP$-G_\alpha$ and $G_{\beta\gamma}$(Figure 6.8) and why there is negative cooperativity between the binding of agonist and GTP with the receptor-$G_\alpha G_{\beta\gamma}$ complex (Figure 6.7). It is possible to incorporate these states into a complete model for receptor-G protein interactions. The results of this more complicated model are consistent with the behavior of the Quaternary Complex Model described above in that both models predict that receptor activation of the G protein cycle is proportional to the amount of the quaternary complex ($DR_s^* G^* X$).

Non-*G* Protein Signaling

Not all of the effects of GPCRs are mediated through G proteins. As described in Chapter 5, some members of the adhesion and frizzled families of GPCRs interact with signaling proteins (e.g., disheveled) through protein-binding motifs in their C terminus. Also, many members of the rhodopsin family of GPCRs recruit G protein receptor kinases (GRKs) when activated by agonist. The kinase phosphorylates the receptor in several places enabling other proteins (e.g., arrestin) to interact with it. Generally speaking, these responses are triggered by the formation of a ternary complex consisting of an agonist-receptor-effector protein complex. Thus, our ternary complex model in the absence of guanine nucleotide can adequately model this system. In other words, we would expect that agonist efficacy is related to how much of the ternary complex is formed at 100% receptor occupancy and that affinity is related to the concentration of agonist required for half-maximal formation of the ternary complex.

A Simple Function for Agonist Activation of GPCRs

At the beginning of this chapter, I said that it was possible to estimate the observed affinity and intrinsic efficacy of an agonist through analysis of a response downstream from receptor activation, provided that we have an accurate function for receptor activation. We now have that function (Equation (6.69)), and it will be used in the next three chapters to estimate the observed affinity and intrinsic efficacy of agonists (Chapter 7), the affinity of antagonists (Chapter 10), and the affinity and cooperative effects of allosteric ligands (Chapter 11). In most instances, it is unnecessary to use the complete function with all of its associated microscopic constants, because the actual shape of the function can be approximated by the simple one-site equation that I described at the beginning of this chapter, which is shown below for convenience:

$$f(x) = \frac{\varepsilon D R_T}{D + K_D}. \tag{6.1}$$

One can think of this equation as a function for the amount of active agonist-receptor complex in the form of quaternary complex. As shown in Figure 6.16, the quaternary complex function is consistent with a simple one-site model. This will enable us to replace Equation (6.69) with the simple function (Equation (6.1)).

The maximum of this function is proportional to the scalar ε, observed efficacy, and the concentration of agonist generating half-maximal formation of the quaternary complex is essentially equal to the dissociation constant, K_D, or the reciprocal of the observed affinity constant ($1/K_{obs}$).

This function was first described by Stephenson[53] and Furchgott[54] in the 1950s and 60s at a time when G proteins had yet to be discovered. Why where they able to come up with a valid function even though I have gone through more than a dozen pages of complicated algebra to account for the receptor, G protein, and guanine nucleotide as well as the different states of the receptor? The answer has two parts. First, the goal of these investigators was to estimate population parameters; that is, the observed affinity and intrinsic efficacy and not microscopic constants for receptor states. This makes the problem simpler, but nonetheless, still complicated as shown by the model in Figure 6.14 and its solution in Equation (6.39). Second, the authors of the equation assumed that the law of mass-action governed ligand-receptor interactions. While they did not know that there was another ligand that interacted with the receptor (i.e., the G protein) that had two ligands of its own (i.e., GDP and GTP), all of these ligand-protein interactions follow the law of mass-action. Consequently, with regard to the pool of receptor in equilibrium with G protein, all of the components of agonist binding should have the same observed affinity and resemble a simple one-site model. It follows that receptor activation can be described as the product of a scalar, observed intrinsic efficacy, and receptor occupancy.

Our efforts in generating the complete model have not been in vain, however, for it is reassuring to prove that a simple model can describe the complex system. Also, we will use the complex model to show how it is possible to obtain a relative estimate of the microscopic affinity constants of agonists and allosteric modulators for the active state of the receptor, and how this constant is proportional to the product of the observed affinity and efficacy. All of this can be done through the analysis of responses that are downstream from receptor activation as explained in the next three chapters.

References

1. Leff, P. Analysis of agonist action using the operational model. *Trends Pharmacol Sci* **9**, 395–398 (1988).
2. Furchgott, R. F. and Bursztyn, P. Comparison of dissociation constants and of relative efficacies of selected agonists acting on parasympathetic receptors. *Annals of the New York Academy of Sciences* **144**, 882–899 (1967).

3. Traut, T. W. Physiological concentrations of purines and pyrimidines. *Mol Cell Biochem* **140**, 1–22 (1994).
4. Gilman, A. G. *G* proteins: transducers of receptor-generated signals. *Annu Rev Biochem* **56**, 615–649 (1987).
5. Berrie, C. P., Birdsall, N. J., Burgen, A. S. and Hulme, E. C. Guanine nucleotides modulate muscarinic receptor binding in the heart. *Biochemical and Biophysical Research Communications* **87**, 1000–1005 (1979).
6. Childers, S. R. and Snyder, S. H. Guanine nucleotides differentiate agonist and antagonist interactions with opiate receptors. *Life Sci* **23**, 759–762 (1978).
7. Freedman, S. B., Poat, J. A. and Woodruff, G. N. Effect of guanine nucleotides on dopaminergic agonist and antagonist affinity for [3H]sulpiride binding sites in rat striatal membrane preparations. *J Neurochem* **37**, 608–612 (1981).
8. Jakubik, J., Janickova, H., El-Fakahany, E. E. and Dolezal, V. Negative cooperativity in binding of muscarinic receptor agonists and GDP as a measure of agonist efficacy. *Br J Pharmacol* **162**, 1029–1044 (2011).
9. Damonte, G., Morelli, A., Piu, M., Longone, P. and De Flora, A. "In situ" characterization of guanine nucleotide-binding properties of erythrocyte membranes. *Biochemical and Biophysical Research Communications* **159**, 41–47 (1989).
10. Sternweis, P. C. and Robishaw, J. D. Isolation of two proteins with high affinity for guanine nucleotides from membranes of bovine brain. *Journal of Biological Chemistry* **259**, 13806–13813 (1984).
11. Ehlert, F. J. and Rathbun, B. E. Signaling through the muscarinic receptor-adenylate cyclase system of the heart is buffered against GTP over a range of concentrations. *Mol Pharmacol* **38**, 148–158 (1990).
12. Jakubik, J., Bacakova, L., el-Fakahany, E. E. and Tucek, S. Constitutive activity of the M1-M4 subtypes of muscarinic receptors in transfected CHO cells and of muscarinic receptors in the heart cells revealed by negative antagonists. *FEBS Lett* **377**, 275–279 (1995).
13. Ehlert, F. J. The relationship between muscarinic receptor occupancy and adenylate cyclase inhibition in the rabbit myocardium. *Mol Pharmacol* **28**, 410–421 (1985).
14. Hanski, E., Sternweis, P. C., Northup, J. K., Dromerick, A. W. and Gilman, A. G. The regulatory component of adenylate cyclase. Purification and properties of the turkey erythrocyte protein. *The Journal of Biological Chemistry* **256**, 12911–12919 (1981).
15. Sternweis, P. C., Northup, J. K., Hanski, E., Schleifer, L. S., Smigel, M. D. and Gilman, A. G. Purification and properties of the regulatory component (G/F) of adenylate cyclase. *Adv Cyclic Nucleotide Res* **14**, 23–36 (1981).
16. Tesmer, J. J., Sunahara, R. K., Gilman, A. G. and Sprang, S. R. Crystal structure of the catalytic domains of adenylyl cyclase in a complex with Gsalpha.GTPgammaS. *Science* **278**, 1907–1916 (1997).
17. Brandt, D. R. and Ross, E. M. GTPase activity of the stimulatory GTP-binding regulatory protein of adenylate cyclase, Gs. Accumulation and turnover of enzyme-nucleotide intermediates. *The Journal of Biological Chemistry* **260**, 266–272 (1985).
18. Abood, M. E., Hurley, J. B., Pappone, M. C., Bourne, H. R. and Stryer, L. Functional homology between signal-coupling proteins. Cholera toxin inactivates the GTPase activity of transducin. *The Journal of Biological Chemistry* **257**, 10540–10543 (1982).

19. Cassel, D. and Selinger, Z. Mechanism of adenylate cyclase activation by cholera toxin: inhibition of GTP hydrolysis at the regulatory site. *Proc Natl Acad Sci USA* **74**, 3307–3311 (1977).
20. Cassel, D. and Selinger, Z. Catecholamine-stimulated GTPase activity in turkey erythrocyte membranes. *Biochim Biophys Acta* **452**, 538–551 (1976).
21. Senogles, S. E., Benovic, J. L., Amlaiky, N., Unson, C., Milligan, G., Vinitsky, R., Spiegel, A. M. and Caron, M. G. The D2-dopamine receptor of anterior pituitary is functionally associated with a pertussis toxin-sensitive guanine nucleotide binding protein. *The Journal of Biological Chemistry* **262**, 4860–4867 (1987).
22. Kuszak, A. J., Pitchiaya, S., Anand, J. P., Mosberg, H. I., Walter, N. G. and Sunahara, R. K. Purification and functional reconstitution of monomeric mu-opioid receptors: allosteric modulation of agonist binding by Gi2. *The Journal of Biological Chemistry* **284**, 26732–26741 (2009).
23. Whorton, M. R., Bokoch, M. P., Rasmussen, S. G., Huang, B., Zare, R. N., Kobilka, B. and Sunahara, R. K. A monomeric *G* protein-coupled receptor isolated in a high-density lipoprotein particle efficiently activates its *G* protein. *Proc Natl Acad Sci U S A* **104**, 7682–7687 (2007).
24. Hilf, G., Gierschik, P. and Jakobs, K. H. Muscarinic acetylcholine receptor-stimulated binding of guanosine 5'-O-(3-thiotriphosphate) to guanine-nucleotide-binding proteins in cardiac membranes. *Eur J Biochem* **186**, 725–731 (1989).
25. Clark, E. A. and Hill, S. J. Sensitivity of histamine H3 receptor agonist-stimulated [35S]GTP gamma[S] binding to pertussis toxin. *European Journal of Pharmacology* **296**, 223–225 (1996).
26. Sim, L. J., Selley, D. E., Xiao, R. and Childers, S. R. Differences in G-protein activation by mu- and delta-opioid, and cannabinoid, receptors in rat striatum. *European Journal of Pharmacology* **307**, 97–105 (1996).
27. Lazareno, S. Measurement of agonist-stimulated [35S]GTP gamma S binding to cell membranes. *Methods in Molecular Biology* **106**, 231–245 (1999).
28. De Lean, A., Stadel, J. M. and Lefkowitz, R. J. A ternary complex model explains the agonist-specific binding properties of the adenylate cyclase-coupled beta-adrenergic receptor. *Journal of Biological Chemistry* **255**, 7108–7117 (1980).
29. Jacobs, S. and Cuatrecasas, P. The mobile receptor hypothesis and "cooperativity" of hormone binding. Application to insulin. *Biochim Biophys Acta* **433**, 482–495 (1976).
30. Ehlert, F. J. in *Biomedical Applications of Computer Modeling* (ed. A. Christopoulos) 21–85 (CRC Press, 2000).
31. Ehlert, F. J. On the analysis of ligand directed signaling at *G* protein coupled receptors. *Naunyn Schmiedebergs Archive Pharmacology* **377**, 549–577 (2008).
32. Tran, J. A., Chang, A., Matsui, M. and Ehlert, F. J. Estimation of relative microscopic affinity constants of agonists for the active state of the receptor in functional studies on M_2 and M_3 muscarinic receptors. *Mol Pharmacol* **75**, 381–396 (2009).
33. Hilf, G. and Jakobs, K. H. Agonist-independent inhibition of *G* protein activation by muscarinic acetylcholine receptor antagonists in cardiac membranes. *European Journal of Pharmacology* **225**, 245–252 (1992).
34. Hulme, E. C., Berrie, C. P., Birdsall, N. J. and Burgen, A. S. Two populations of binding sites for muscarinic antagonists in the rat heart. *European Journal of Pharmacology* **73**, 137–142 (1981).

35. Gerstin, E. H., Jr., Luong, T. and Ehlert, F. J. Heparin, dextran and trypan blue allosterically modulate M2 muscarinic receptor binding properties and interfere with receptor-mediated inhibition of adenylate cyclase. *J Pharmacol Exp Ther* **263**, 910–917 (1992).
36. Haga, K., Haga, T. and Ichiyama, A. Reconstitution of the muscarinic acetylcholine receptor. Guanine nucleotide-sensitive high affinity binding of agonists to purified muscarinic receptors reconstituted with GTP-binding proteins (Gi and Go). *Journal of Biological Chemistry* **261**, 10133–10140 (1986).
37. Haga, K., Uchiyama, H., Haga, T., Ichiyama, A., Kangawa, K. and Matsuo, H. Cerebral muscarinic acetylcholine receptors interact with three kinds of GTP-binding proteins in a reconstitution system of purified components. *Mol Pharmacol* **35**, 286–294 (1989).
38. Shiozaki, K. and Haga, T. Effects of magnesium ion on the interaction of atrial muscarinic acetylcholine receptors and GTP-binding regulatory proteins. *Biochemistry* **31**, 10634–10642 (1992).
39. Bayburt, T. H. and Sligar, S. G. Self-assembly of single integral membrane proteins into soluble nanoscale phospholipid bilayers. *Protein Sci* **12**, 2476–2481 (2003).
40. Baas, B. J., Denisov, I. G. and Sligar, S. G. Homotropic cooperativity of monomeric cytochrome P450 3A4 in a nanoscale native bilayer environment. *Arch Biochem Biophys* **430**, 218–228 (2004).
41. Leitz, A. J., Bayburt, T. H., Barnakov, A. N., Springer, B. A. and Sligar, S. G. Functional reconstitution of Beta2-adrenergic receptors utilizing self-assembling Nanodisc technology. *Biotechniques* **40**, 601–602, 604, 606, passim (2006).
42. Fung, J. J., Deupi, X., Pardo, L., Yao, X. J., Velez-Ruiz, G. A., Devree, B. T., Sunahara, R. K. and Kobilka, B. K. Ligand-regulated oligomerization of beta(2)-adrenoceptors in a model lipid bilayer. *Embo J* **28**, 3315–3328 (2009).
43. Suzuki, Y., Moriyoshi, E., Tsuchiya, D. and Jingami, H. Negative cooperativity of glutamate binding in the dimeric metabotropic glutamate receptor subtype 1. *The Journal of Biological Chemistry* **279**, 35526–35534 (2004).
44. Shivnaraine, R. V., Huang, X. P., Seidenberg, M., Ellis, J. and Wells, J. W. Heterotropic cooperativity within and between protomers of an oligomeric M(2) muscarinic receptor. *Biochemistry* **51**, 4518–4540 (2012).
45. Weiss, J. M., Morgan, P. H., Lutz, M. W. and Kenakin, T. P. The Cubic Ternary Complex Receptor-Occupancy Model .1. Model Description. *Journal of Theoretical Biology* **178**, 151–167 (1996).
46. Kenakin, T. Functional selectivity and biased receptor signaling. *J Pharmacol Exp Ther* **336**, 296–302 (2011).
47. Coleman, D. E., Berghuis, A. M., Lee, E., Linder, M. E., Gilman, A. G. and Sprang, S. R. Structures of active conformations of Gi alpha 1 and the mechanism of GTP hydrolysis. *Science* **265**, 1405–1412 (1994).
48. Lambright, D. G., Noel, J. P., Hamm, H. E. and Sigler, P. B. Structural determinants for activation of the alpha-subunit of a heterotrimeric *G* protein. *Nature* **369**, 621–628 (1994).
49. Lambright, D. G., Sondek, J., Bohm, A., Skiba, N. P., Hamm, H. E. and Sigler, P. B. The 2.0 A crystal structure of a heterotrimeric *G* protein. *Nature* **379**, 311–319 (1996).
50. Mixon, M. B., Lee, E., Coleman, D. E., Berghuis, A. M., Gilman, A. G. and Sprang, S. R. Tertiary and quaternary structural changes in Gi alpha 1 induced by GTP hydrolysis. *Science* **270**, 954–960 (1995).

51. Noel, J. P., Hamm, H. E. and Sigler, P. B. The 2.2 A crystal structure of transducin-alpha complexed with GTP gamma S. *Nature* **366**, 654–663 (1993).
52. Wall, M. A., Coleman, D. E., Lee, E., Iniguez-Lluhi, J. A., Posner, B. A., Gilman, A. G. and Sprang, S. R. The structure of the *G* protein heterotrimer Gi alpha 1 beta 1 gamma 2. *Cell* **83**, 1047–1058 (1995).
53. Stephenson, R. P. A modification of receptor theory. *Br J Pharmacol* **11**, 379–393 (1956).
54. Furchgott, R. F. The use of β-haloalkylamines in the differentiation of receptors and in the determination of dissociation constants of receptor-agonist complexes. *Advances in Drug Research* **3**, 21–55 (1966).

Estimating Affinity and Efficacy by Reverse-Engineering and Response-Clamp Analysis

7

G protein-coupled receptors number about a thousand, and the functions they mediate throughout the body are limitless. At any one receptor, agonist activation elicits a train of biochemical events that culminate in a physiological response. Each event represents a potentially measurable response. For example, when β_2 adrenoceptors are activated in tracheal smooth muscle, one could measure changes in adenylate cyclase activity, Ca^{2+}-activated potassium channel currents, Ca^{2+} influx, smooth-muscle contraction, airway resistance and many others. The β_2-adrenoceptor could also be expressed in a cell line, and an endogenous signaling pathway can be measured (e.g., cAMP accumulation). Alternatively, a recombinant receptor and reporter system (CREB/luciferase) could be introduced into the cell and its signal can be monitored. Thus, the output from receptor activation is diverse, but the input for a given agonist-receptor pair consists of only two components, affinity and efficacy. The goal of this chapter is to explain how to estimate these components from the output.

This may seem like an academic exercise of reducing a remarkable train of events into two mathematical quantities, affinity and efficacy, but knowing how to measure affinity and efficacy enables one to recognize the signature of an agonist for eliciting a response through a specific receptor because the agonist-receptor complex has a unique combination of affinity and efficacy. Thus, a sufficiently selective agonist could be used as a probe to identify the receptor mediating a physiological response. The right combination of affinity and efficacy is essential in designing drugs with selectivity and minimal side effects. An agonist may have a different signature, depending upon the signaling pathway that it elicits (see Chapter 9), but this only increases the need for measuring its affinity and efficacy at the different signaling pathways.

Reverse Engineering and Receptor Theory

Determining how a complex response is generated from receptor activation is engineering in reverse. This is what computational biologists do when they try to figure out how the retina works or how spiral Ca^{2+} waves are generated in an oocyte or in a syncytium of cardiac myocytes. The problem of estimating affinity and efficacy from a downstream response is much simpler because it is unnecessary to identify each step in the pathway and to model its output based on its input, and ultimately, to link all of these together accurately so that one can predict the final output from the initial input. All that is necessary is a mathematical function that reliably predicts how the initial input (affinity and efficacy) is converted into the final output (response). This transducer function has come to be known as the operational model, and I cannot think of a better explanation of its basis than that given by its originator, Paul Leff[1]:

> "The operational model was derived using the simple piece of logic that if one knows about the beginning and end of a process one can deduce what happens in the middle."

With regard to the input function, recall that in the prior chapter, an equation was developed to define how receptor activation depends on the concentration of the agonist (independent variable, D) and the observed affinity and efficacy of the agonist-receptor complex. This equation represents the input function ($f(D)$), known in pharmacology as the stimulus (S). It describes receptor activation as a function of the agonist concentration (D). At the population level of analysis the relevant parameters in the equation are observed affinity (K_{obs}) and efficacy (ε). The equation is numbered 6.1 in chapter 6 and is shown below for convenience in a slightly modified form. The modification simply involves showing that the function is equivalent to the stimulus (S) and that the latter is equivalent to the produce of ε and occupancy:

$$f(D) = S = \varepsilon \times \text{occupancy} = \frac{\varepsilon D R_T}{D + \frac{1}{K_{\text{obs}}}}. \tag{7.1}$$

It is also possible to describe $f(D)$ at a deeper level of analysis, taking into account the microscopic affinity constants of the agonist for the active and inactive states of the receptor. This will be done in upcoming sections. For now, we will use Equation (7.1) to describe the input to the response.

With regard to the output function, a near universal description of concentration-response curves is that these resemble nearly symmetrical S-shaped curves when the response is plotted against the log concentration of the agonist. That is, if the concentration of cAMP or the percent relaxation of smooth muscle elicited by the β_2-adrenoceptor agonist, isoproterenol, in tracheal smooth muscle is plotted against the log concentration of isoproterenol, the data points tend to describe a smooth curve ($r(D)$):

$$r(D) = \text{response} = \frac{E_{\max}}{1 + 10^{n(\log EC_{50} - \log D)}}, \tag{7.2}$$

in which $E_{\max}$ denotes the maximum response elicited by the agonist at high concentrations, EC_{50} the concentration of agonist causing a half-maximal response, and n (slope factor or Hill slope), a parameter related to the slope of the function at its inflection point (i.e., when $D = EC_{50}$). I find it easier to understand the behavior of the equation if the logarithms are eliminated. This can be done by replacing $10_{50}^{\text{nlog}EC}$ and $10^{\text{nlog}x}$ with EC_{50}^{n} and D^n, respectively. With rearrangement, Equation (7.2) becomes:

$$r(D) = \text{response} = \frac{D^n E_{\max}}{D^n + EC_{50}^{n}}. \tag{7.3}$$

A plot of the initial input function ($f(D)$, stimulus) and the final response ($r(D)$) is shown in Figure 7.1. The ordinate has been normalized for each function. The maximum possible stimulus ($f(D) = 1$) is defined as that generated by an agonist that is able to activate the receptor completely (100% receptor activation) at 100%

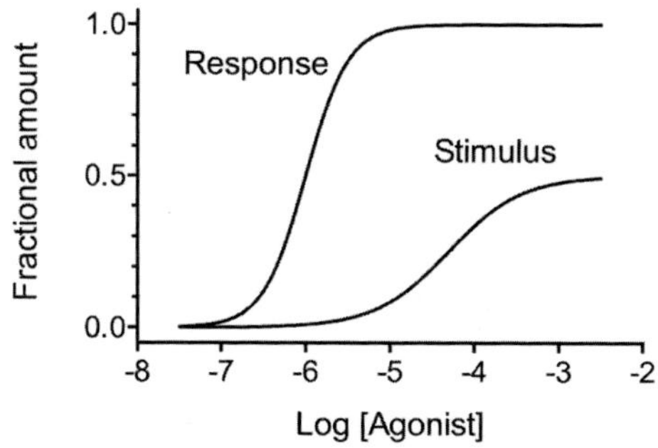

Figure 7.1: The relationship between stimulus and response for a highly efficacious agonist acting on a receptor coupled to a sensitive signaling pathway. The stimulus represents the product of receptor occupancy and efficacy. The theoretical concentration-response curve was generated using the operational model with the following parameters: efficacy, 0.5; log K_{obs}, 4.301; M_{sys}, 1, log K_E, −2.0 and m, 1.8.

receptor occupancy, and the maximal response of the system ($r(D) = 1$) corresponds to the maximum possible response that the system could generate by an agonist with infinite selectivity for the active state of the receptor. Often, this maximal level of response is equivalent to that caused by a highly efficacious agonist, but it could be higher and unachievable by any known agonist at 100% receptor occupancy. Figure 7.1 shows that the potency of the agonist for eliciting the response can be much greater than that for activating the receptor population. This behavior occurs because the signal transduction pathway for the response typically amplifies the signal elicited by the receptor population.

Our reverse-engineering problem consists of determining how the stimulus, $f(D)$, is converted into the final response. The problem can be restated in mathematical terms as finding the function, $g(S)$, that converts $f(D)$ into the response function, $r(D)$:

$$\text{response} = r(D) = g(f(D)). \tag{7.4}$$

Black and Leff[2] showed that if the input function is a simple one-site model of the agonist concentration (Equation (7.1)) and the output a logistic function of the agonist concentration, then the transducer function ($g(S)$) must also be logistic:

$$\text{response} = g(S) = \frac{S^m M_{\text{sys}}}{S^m + K_E^m}. \tag{7.5}$$

In this equation, M_{sys} denotes the maximum possible response of the system (i.e., that mediated by an agonist with infinite efficacy), m the transducer-slope factor and K_E a constant inversely proportional to the sensitivity of the signaling pathway. Substituting Equation (7.1) for S in Equation (7.5) yields the response as a function of the agonist concentration:

$$\text{response} = \frac{\left(\frac{\varepsilon D R_T}{D + \frac{1}{K_{\text{obs}}}}\right)^m M_{\text{sys}}}{\left(\frac{\varepsilon D R_T}{D + \frac{1}{K_{\text{obs}}}}\right)^m + K_E^m}. \tag{7.6}$$

The method for simplifying this equation is more apparent when it is written in the following form:

$$\text{response} = \frac{\frac{(\varepsilon D R_T)^m}{\left(D + \frac{1}{K_{\text{obs}}}\right)^m} M_{\text{sys}}}{\frac{(\varepsilon D R_T)^m}{\left(D + \frac{1}{K_{\text{obs}}}\right)^m} + K_E^m}. \tag{7.7}$$

Multiplying the numerator and denominator by $(D + K_{obs})^n/(\varepsilon R_T)^n$ yields:

$$\text{response} = \frac{D^m M_{sys}}{D^m + \frac{\left(D+\frac{1}{K_{obs}}\right)^m}{(\varepsilon R_T)^m} K_E^m} \tag{7.8}$$

Minor rearrangement yields:

$$\text{response} = \frac{D^m M_{sys}}{D^m + \frac{\left(D+\frac{1}{K_{obs}}\right)^m}{\left(\frac{\varepsilon R_T}{K_E}\right)^m}}. \tag{7.9}$$

Because ε, R_T and K_E are all constants, the ratio, $\varepsilon R_T/K_E$ can be replaced with a single constant, τ:

$$\text{response} = \frac{D^m M_{sys}}{D^m + \frac{\left(D+\frac{1}{K_{obs}}\right)^m}{\tau^m}}, \tag{7.10}$$

in which τ is defined as:

$$\tau = \frac{\varepsilon R_T}{K_E}. \tag{7.11}$$

Thus, Equation (7.10) represents our reverse-engineering goal. It defines the response as a function of the agonist concentration and other parameters including the observed affinity and efficacy of the agonist. A form of this equation was initially developed by Black and Leff.[2] In their analysis, intrinsic efficacy was considered as a component of K_E so that the parameter τ was defined simply as the ratio of R_T/K_E. In my use of the operational model, ε is defined as a separate variable, so that it will be possible to attribute changes in observed efficacy and the sensitivity of the signaling pathway to different variables (ε and K_E, respectively). In the next section, the method of using the operational model to estimate the observed affinity and intrinsic efficacy is described.

Another way to tackle the reverse-engineering problem is by using a so-called "null method". This approach is applied in other areas of science including, for example, physics (e.g., Wheatstone bridge) and electrophysiology (e.g., voltage clamp). The method can be used to measure an unknown amount of a particular trait whenever that trait can be generated independently in known amounts and compared with the unknown amount. In the voltage clamp technique described earlier in the Prologue and Chapter 2, a current is applied to the cell to keep the voltage constant during a stimulus. Because the applied current keeps the voltage constant, then it must be equal, but opposite in sign, to that generated by endogenous ion channels activated by the stimulus.

In the case of agonist concentration-response curves, one can assume that a specific level of receptor activation causes a specific level of response. Therefore, one can assume further that equivalent responses to two different agonists imply equal levels of receptor activation. Alternatively, one can assume that equivalent responses to the same agonist, before and after the size of the receptor population has been reduced, also imply equivalent levels of receptor activation. In both cases, one can write a mathematical equality describing that the two levels of receptor activation are equal. The unknown transducer function cancels out of the relationship. One is left with an equation consisting of the parameters affinity and efficacy. This "null equation" can then be used to extract some or all of the affinity and efficacy parameters from pairs of equi-active agonist concentrations. Even though the data are usually not collected under "response-clamp" conditions, the calculations involve a response-clamp analysis.

In the following sections, the details of these two approaches (i.e., operational model and null method) for estimating agonist affinity and efficacy are described.

Operational Model

Before describing how to use the operational model, let me first address some of its basic properties, many of which have been described previously by Black and Leff.[2,3]

Properties

With regard to the units of the parameters in the operational model, the input function ($f(D)$, stimulus) describes the number of active receptors. Therefore, the units of R_T are of number or density (e.g., fmol or fmol/mg protein), and intrinsic efficacy (ε) is a unitless scalar that denotes the fraction of the occupied receptors that are in the active state. The affinity constant has units of inverse concentration (e.g., M^{-1}), with the concentration units being the same as those of the agonist concentration (D).

The output of the transducer function ($g(S)$) has units equivalent to those of the response. Therefore, M_{sys} has the same units as the response (e.g., force of contraction (mN), concentration of cAMP (pmol mg^{-1} tissue) or cytosolic concentration of Ca^{2+} (nM)), and K_E has units equivalent to those of the input function (number or density of receptors). Finally, the parameter τ and the exponent m (transducer slope factor) are unitless.

The parameters of the operational model include those that are unique to the agonist-receptor complex and those that are properties of the signaling pathway or transducer function. The parameters, ε and K_{obs}, are unique properties of the agonist receptor complex and potentially other proteins that physically interact with the receptor (e.g., G proteins), whereas K_E, M_{sys} and m are properties of the transducer function. The amount of receptors R_T is an obvious receptor property, but different signaling pathways often have different sizes of an active receptor pool. Thus, R_T can also be considered as a system property. While the parameter τ incorporates both agonist-receptor- and system-dependent parameters (see Equation (7.11)), the ratio of the τ value of one agonist (τ) relative to that of a standard agonist (τ') is an agonist-receptor-dependent property known as relative efficacy (ε/ε'):

$$\frac{\tau}{\tau'} = \frac{\frac{\varepsilon R_T}{K_E}}{\frac{\varepsilon' R_T}{K_E}} = \frac{\varepsilon}{\varepsilon'}. \tag{7.12}$$

This equation for relative efficacy was derived by substituting Equation (7.11) for the value of τ and τ' and simplifying.

Because the operational model resembles an S-shaped function of the log concentration of agonist, it can be described not only by the parameters of the operational model, but also by the empirical parameters of E_{max}, EC_{50} and a slope factor (n) (see Equation (7.3)). To determine E_{max} in terms of the parameters of the operational model, the limit of the operational model as the agonist concentration approaches infinity is taken:

$$E_{max} = \lim_{D\to\infty} \frac{D^m M_{sys}}{D^m + \frac{\left(D+\frac{1}{K_{obs}}\right)^m}{\tau^m}}. \tag{7.13}$$

Simplifying yields:

$$E_{max} = \frac{M_{sys}\tau^m}{1+\tau^m}. \tag{7.14}$$

Whenever $\tau^m \gg 1$, the right-hand side of the equation is approximately equal to M_{sys}. Thus, agonists with vary large τ values elicit maximal responses that are approximately equal to the value of M_{sys}. Because τ is proportional to the intrinsic efficacy of the agonist, we can rephrase this statement in the following form. Highly efficacious agonists elicit a maximal response approximately equivalent to the value of M_{sys}.

This is an appropriate point to introduce some empirical definitions often used in pharmacology. A *full agonist* has an E_{max} value equivalent to M_{sys} (the maximal response of the system), whereas a *partial agonist* elicits an $E_{\max}$ value less than that of the system. These terms are superficial and depend on the type of response being measured. For example, the same agonist could behave as a full agonist in a highly sensitive system ($1 \gg K_E > 0$), as a partial agonist in a less sensitive system ($K_E \approx 1$) and as an antagonist in a very insensitive system ($K_E > 1$). These terms relate to how the agonist behaves at the most superficial level (level 1) of the hierarchical analysis system introduced in the Prologue.

The EC_{50} value of the agonist can also be defined in terms of the parameters of the operational model. Solving for the value of D that yields a half-maximal response yields an equation for EC_{50}. The first step involves making appropriate substitutions into the operational model (Equation (7.10)):

$$0.5\frac{M_{\text{sys}}\tau^m}{1+\tau^m} = \frac{EC_{50}^m M_{\text{sys}}}{EC_{50}^m + \frac{\left(EC_{50}+\frac{1}{K_{\text{obs}}}\right)^m}{\tau^m}}. \tag{7.15}$$

The left-hand side of this equation is equal to the $E_{\max}$ value multiplied by 0.5 (half-maximal response), with $E_{\max}$ defined in Equation (7.14). The right-hand side of the equation is equivalent to the operational model (Equation (7.10)), but with the independent variable D replaced with EC_{50}. Solving for EC_{50} yields:

$$EC_{50} = \frac{1}{K_{\text{obs}}((\tau^m + 2)^{1/m} - 1)}. \tag{7.16}$$

This equation shows that the EC_{50} value of a highly efficacious agonist ($EC_{50\text{-full}}$; $\tau \gg 1$) is approximately equal to:

$$EC_{50\text{-full}} \approx \frac{1}{\tau K_{\text{obs}}}. \tag{7.17}$$

That of a partial agonist ($EC_{50\text{-partial}}$; $\tau \approx 1$) is approximately equal to:

$$EC_{50\text{-partial}} \approx \frac{1}{K_{\text{obs}}}. \tag{7.18}$$

The EC_{50} values of partial agonists can be greater or less than the reciprocal of the affinity constant ($1/K_{\text{obs}}$). This reciprocal is commonly known as the dissociation constant (K_D). When there is little threshold for the response ($m \approx 1$), the EC_{50} value of a partial agonist with a low $E_{\max}$ value is often two- to three-fold greater than the K_D. When there is a substantial threshold ($m > 1$), then the EC_{50} value for the partial agonist can be less than K_D. For example, contraction of smooth muscle often requires a depolarization of the smooth muscle cell and an activation of voltage sensitive Ca^{2+} channels. A low amount of receptor activation, however, does not cause sufficient depolarization of the sarcolemma to activate the voltage sensitive Ca^{2+} channels. A very weak partial agonist may have to occupy 90% of the receptor population before the membrane potential has been depolarized to the threshold level required to activate voltage sensitive Ca^{2+} channels. Therefore, as receptor occupancy increases from 0 to 90%, no response is generated, but as occupancy increases from 90–100%, the entire range of response values of the concentration-response curve is generated. The EC_{50} concentration of the partial agonist in this case will be substantially greater than the observed dissociation constant or the reciprocal of the observed affinity constant.

The transducer slope factor (m) is a system-dependent parameter, and it is related to how the stimulus is converted into the final response. The presence of cooperativity or a threshold in the transduction mechanism causes an increase in the value of m above 1. Because m is a system-dependent parameter, its value will be the same regardless of the agonist. Nonetheless, the observed slope factor (n) of the logistic equation describing the concentration-response curve (Equation (7.3)) can change, depending upon the intrinsic efficacy of the agonist and the value of m. In situations where the transducer-slope factor of the system is greater than one (e.g., $m = 1.8$), the slope factor (n) of the concentration-response curve will be approximately equal to m (i.e., $m \approx n$ for highly efficacious full agonist with $EC_{50} \ll K_D$). In contrast, less efficacious agonists and partial agonist will exhibit concentration-response curves with slope factors that are less than the transducer-slope factor of the operational model (i.e., $m > n > 1$ for less efficacious and partial agonists). By now it should be obvious that the value of the slope factor of the concentration-response curve (n in Equation (7.3)) is not equivalent to the transducer-slope factor of the operational model (m in Equation (7.10)) unless the agonist is highly efficacious and the EC_{50} value is much less than the observed dissociation constant K_D.

This positive correlation between the slope factor of the concentration-response curve and the efficacy of the agonist has led some to conclude erroneously that intrinsic efficacy is proportional to the amount of positive cooperativity in the binding of the agonist to the receptor. But the magnitude of a downstream response in a signaling pathway is not proportional to receptor occupancy. In fact, the input to the operational model is a simple one-site model ($f(D)$, Equation (7.1)). It is the nature of the downstream signaling events that introduces a steep slope ($n > 1$) into the concentration-response curve and not the binding of the agonist to the receptor. Only when receptor activation is measured directly, as in the case of the whole-cell current response of a ligand-gated ion channel, does the cooperative concentration-response curve imply cooperative binding.

One final point is that when the transducer slope factor in the operational model (m) is equal to one, then all agonist concentration-response curves for that system also exhibit slope factors (n) equivalent to one.

What can you get from a single concentration-response curve?

Having a function (Equation (7.10)) that expresses the response in terms of the independent variable (agonist concentration) and other parameters, including affinity and efficacy, enables one to analyze concentration-response curves and estimate affinity and efficacy by nonlinear regression analysis. But if you try to fit the operational model to the concentration-response data of a single agonist, most nonlinear-regression applications will not converge on a solution set, or else the parameters of the solution set will have enormous confidence intervals. By solution set I mean the best estimates of the parameters in the operational model (Equation (7.10)). There is simply insufficient information in the concentration-response curve of a single agonist to estimate affinity and efficacy.

Nonetheless, there is a solution to this nonlinear regression problem; it is just that the solution set is infinite. For example, consider the concentration-response curve of the muscarinic agonist, pilocarpine, for eliciting contraction of the isolated guinea pig ileum (Figure 7.2a). The smooth curve through the data points illustrates the least-squares fit of the operational model (Equation (7.10)). If the solution set is

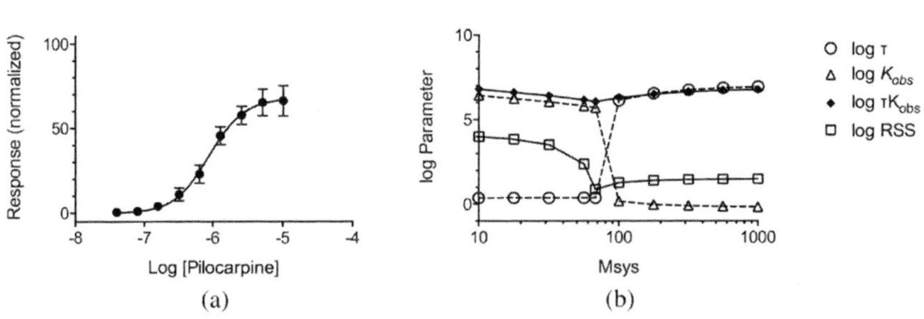

Figure 7.2: Analysis of a single concentration-response curve using the operational model. (a) The concentration-response curve of the muscarinic agonist, pilocarpine, for eliciting contraction of the guinea pig ileum is shown. The data are from Ehlert *et al.*[7]. (b) Nonlinear regression analysis of the data in panel (a) using the operational model (Equation (7.10)). The operational model was fitted to the data repetitively, each time constraining the parameter M_{sys} to a constant and determining the values of the other parameters that minimized the residual sum of squared deviations of the data points from theoretical curves (RSS). The diagram shows the values of the parameter estimates for each regression plotted against the value to which M_{sys} was constrained.

infinite, one might ask, of what value is it because so many different combinations of parameter values (i.e., values of M_{sys}, K_{obs} and τ) will yield a good fit to the data? If the goal is to determine unique estimates of the parameters, then the least-squares fit is of no value. But the solution set does throw light on what can be extracted from the data and what additional data we need to obtain unique estimates of M_{sys}, K_{obs} and τ.

The concept of an infinite solution set does not imply that *any* set of parameter values can fit the data. Rather, it implies that there is a unique relationship among the parameter values of any set that *does* fit the data. Solving the nonlinear regression problem involves finding this relationship. There are some simple properties of the solution set that are intuitively obvious. All of the parameter values must be greater than one, for negative values have no meaning in a real receptor-signaling system with regard to agonists that elicit a response above that which is measured in the absence of agonist. In addition, the value of M_{sys} must be equal to or greater than the E_{max} value of the concentration-response curve, because M_{sys} represents the maximum possible response of the system.

With these constraints, it is possible to define the solution set to the data in Figure 7.2a by first setting M_{sys} to an arbitrary constant value within its domain ($M_{sys} \geq E_{max}$) and then using nonlinear regression analysis to estimate the other parameters (K_{obs}, τ and m) that yield a least-squares fit to the data. This process is repeated over and over again, setting M_{sys} to a new value within its domain each time until we have covered the entire range of M_{sys} values. Theoretically, this process should take forever because M_{sys} could be any value over the range $E_{max} \to \infty$. But we know that a guinea pig ileum is incapable of eliciting an infinite amount of tension, so we can explore values of M_{sys} over a reasonable range. The process could still be very time-consuming, depending upon the size of the steps taken on the M_{sys} scale. Usually, moderate sized steps can be taken because of the smooth predictable relationship among the best fitting parameter values.

The results of this walk through parameter space are illustrated in Figure 7.2b. Each step represents one possible solution and is indicated by a set of plotted points (best estimates of log K_{obs} and log τ) corresponding to the constrained value of M_{sys} on the abscissa. To avoid clutter, the estimates of the transducer slope factor (m) have been omitted from the plot. The dashed and solid lines were simply drawn through the points. Any vertical line drawn through the graph over this domain intersects the curves and abscissa at a provisional solution (unique estimates of M_{sys}, log K_{obs} and log τ), depending on how well the best fitting curve fits the observed data. Essentially, the plot shows the estimates of log K_{obs} and log τ as the value of M_{sys} is set as a constant over the range 10–1000%. The plot also shows the log residual sum of the squared deviations (log *RSS*) that correspond to each provisional solution.

Log RSS is at a minimum when the M_{sys} value is constrained to values close to the E_{max} of pilocarpine. Log RSS increases abruptly as M_{sys} is constrained to values less than the E_{max} of pilocarpine. When M_{sys} is constrained to values in the range where RSS is lowest (i.e., 67–100%), the estimates of log τ and log K_{obs} change abruptly, indicating that it is impossible to obtain reliable estimates of these parameters. Over the entire range of reasonable M_{sys} values (i.e., $M_{sys} > 60\%$), the log of the product of τ and K_{obs} (log τK_{obs}) is nearly constant. Thus, the answer to the question posed in the title of this section is, for any concentration-response curve, it is usually possible to obtain a reasonable estimate the product of K_{obs} and τ.

While this may seem unsatisfying, the product of K_{obs} and τ for agonist A, expressed relative to that of another agonist B is equivalent to the ratio of the products of affinity and efficacy of the two agonists ($\tau_A K_{obs\text{-}A}/\tau_B K_{obs\text{-}B}$). This product is known as the intrinsic relative activity (RA_i). It has many useful properties as described in later sections of this and other chapters. It is also equal to the ratio of microscopic affinity constants of the two agonists for the active state of the receptor. Thus, through comparison of the concentration-response curves of two agonists, it is possible to obtain a relative estimate of the affinity of an agonist for the active state of the receptor. The basis for this estimation is described later in this chapter.

Estimating the affinity of a partial agonist

The results of our walk through parameter space showed that if the maximum response of the system (M_{sys}) is known, then it is possible to estimate K_{obs} and τ from the concentration-response data of a partial agonist (i.e., an agonist with an E_{max} less than M_{sys}) using nonlinear regression analysis with the operational model. This conclusion suggests that it should be possible to estimate the τ and K_{obs} values of a partial agonist by analyzing its concentration-response curve simultaneously with that of a full agonist, because the latter has an E_{max} equivalent to M_{sys}.

Stephenson[4] was one among the first to realize that there is enough information in the concentration-response curves of a full and a partial agonist to estimate the affinity constant of the partial agonist. His method involved a comparison of equi-active concentrations of a full and partial agonist (null method). In the process, he rewrote receptor theory by disposing the well-accepted, but clearly implausible, idea that the relative magnitude of a downstream response was proportional to receptor occupancy. Rather, he introduced the idea that the magnitude of the response is an unknown function of receptor activation and that the latter was equivalent to the product of efficacy and receptor occupancy (stimulus).

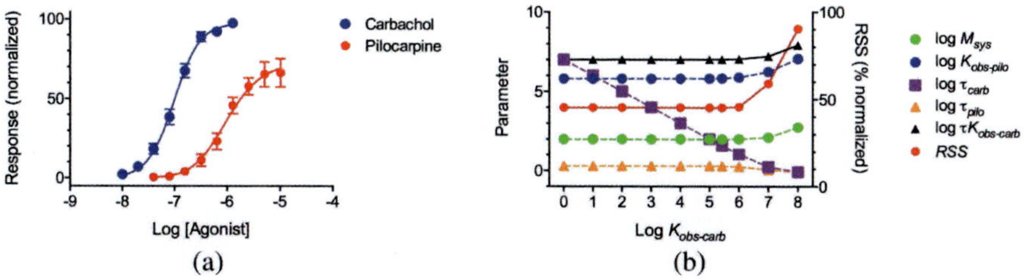

Figure 7.3: Simultaneous analysis of the concentration-response curves of a full agonist (carbachol) and a partial agonist (pilocarpine) using the operational model. (a) The concentration-response curves of carbachol and pilocarpine, for eliciting contraction of the guinea pig ileum are shown. The data are from Ehlert *et al.*[7]. (b) Global nonlinear regression analysis of the data in panel (a) using the operational model (Equation (7.10)). The operational model was fitted to the data repetitively, each time constraining the K_{obs} value of carbachol to a constant and determining the values of the other parameters that minimized the residual sum of squared deviations of the data points from theoretical curves (*RSS*). The estimates of M_{sys} and m (not shown) were shared between the two sets of data during regression analysis. The diagram shows the values of the parameter estimates for each regression plotted against the value to which $K_{obs\text{-carb}}$ was constrained.

A more rigorous way to achieve Stephenson's goal is to fit the operational model to the concentration-response curves of the full and partial agonists simultaneously by global nonlinear regression analysis, sharing the estimates of the system-dependent parameters (M_{sys} and m) between the two sets of data and obtaining unique estimates of the agonist-receptor dependent parameters (K_{obs} and τ) for each agonist. Figure 7.3a shows the results of such an analysis for the concentration-response curves of pilocarpine and the full agonist, carbachol, for eliciting contraction in the guinea pig ileum. The data for pilocarpine are the same as those shown in Figure 7.2a. The theoretical curves represent the least-squares fit of the operation model to the data points. It is possible that a given nonlinear regression application will not converge on a specific solution when the data are analyzed simultaneously because the E_{max} of carbachol is similar to M_{sys}. As described above, the solution set is usually infinite in this situation.

Using the Prsim application (GraphPad Software, Inc.) and the detailed procedures described under Example 1 in the appendix, I was able to obtain a unique solution set for these data. Global nonlinear regression analysis yielded best estimates of the system parameters (M_{sys}, $98.3 \pm 5.7\%$ and m, 1.78 ± 0.25) and the K_{obs} and τ values of pilocarpine (log K_{obs}, $5.83 \pm 0.17\,\text{M}^{-1}$; log τ, 0.27 ± 0.10) and carbachol (log K_{obs}, $5.43 \pm 3.33\,\text{M}^{-1}$; and log τ, 1.59 ± 3.27) as indicated in parenthesis. Although Excel converged on unique parameters for carbachol, these estimates are highly unreliable as shown by the colossal SEM values (over three log units) and the analysis of parameter space in Figure 6.3b. This plot shows the results of a walk through parameter space analogous to that shown in Figure 6.2b, but in this case, steps were taken along the K_{obs} scale for carbachol ($K_{obs\text{-}carb}$). That is, global nonlinear regression analysis was done by constraining the $K_{obs\text{-}carb}$ value to a constant and determining the estimates of the other parameters that yielded a least-squares fit to the data. This process was repeated several times, each time constraining $K_{obs\text{-}carb}$ to a different value and determining the best-fitting parameter estimates for the other parameters. The results of each regression step are indicated by the plotted points. The estimates of the transducer slope factor (m) are not shown to improve the clarity of the other estimates. Lines were drawn to connect adjacent estimates of a given parameter over the range of $K_{obs\text{-}carb}$ values investigated.

As shown in Figure 6.3b, when the value of $K_{obs\text{-}carb}$ was constrained to any constant over the range $\log K_{obs\text{-}carb} < 6$, a least-squares fit was obtained as indicated by the constant minimum value of RSS. It is impossible, therefore, to estimate $K_{obs\text{-}carb}$ accurately because the same least-squares fit is obtained over a wide range of $K_{obs\text{-}carb}$ values. As the value of $\log K_{obs\text{-}carb}$ increases above 6, however, RSS increases, indicating a poor fit to the data. Over the domain of $K_{obs\text{-}carb}$ that yields a least-squares fit (i.e., $K_{obs\text{-}carb} < 6$), the estimates of the system parameters are constant (M_{sys} and m) as well as the observed affinity constant ($K_{obs\text{-}pilo}$) and τ value of pilocarpine. Thus, the best estimates of these parameters (M_{sys}, 98 ± 5.7; m, 1.78 ± 0.25; $\log K_{obs\text{-}pilo}$, 5.83 ± 0.17; $\log \tau_{pilo}$, 0.27 ± 0.10) can be obtained from the data. Over the domain of $\log K_{obs\text{-}carb}$ values that yield a least squares fit, the best estimate of the $\log \tau$ value of carbachol ($\log \tau_{carb}$) increases proportionately such that the product ($\log \tau_{carb} K_{obs\text{-}carb}$) is constant. Thus, although it is impossible to obtain individual estimates of $\log \tau_{carb}$ and $\log K_{obs\text{-}carb}$, it is possible to estimate their product. One might ask why the domain for $\log K_{obs\text{-}carb}$ was not explored below a value of 0. There is a mountain of data showing that carbachol has much higher affinity than $\log K_{obs} = 0$. Regardless, the same least-squares fit is obtained when $\log K_{obs\text{-}carb} < 0$.

From these parameter estimates, the product of the affinity and efficacy of pilocarpine, expressed relative to that of carbachol can be estimated. This estimate is known as the RA_i value of pilocarpine, and it can be calculated as:

$$RA_{i-\text{pilo}} = \frac{\tau_{\text{pilo}} K_{\text{obs-pilo}}}{\tau_{\text{carb}} K_{\text{obs-carb}}}. \tag{7.19}$$

Substituting in Equation (7.11) for τ yields:

$$RA_{i-\text{pilo}} = \frac{\frac{\varepsilon_{\text{pilo}} R_T}{K_E} K_{\text{obs-pilo}}}{\frac{\varepsilon_{\text{carb}} R_T}{K_E} K_{\text{obs-carb}}}. \tag{7.20}$$

Simplifying yields:

$$RA_{i-\text{pilo}} = \frac{\varepsilon_{\text{pilo}} K_{\text{obs-pilo}}}{\varepsilon_{\text{carb}} K_{\text{obs-carb}}}. \tag{7.21}$$

Thus, the product of the τ and K_{obs} value of one agonist expressed relative to that of another is equivalent to the corresponding products of affinity and efficacy. This estimate can always be obtained from the concentration-response curves of two agonists. For the data in Figure 7.3, the log RA_i value of pilocarpine, espressed relative to that of carbachol, is -0.92.

Estimating affinity and relative efficacy

So far we have seen that there is insufficient information in a single concentration-response curve to estimate the affinity and efficacy of an agonist, but if we analyze the concentration-response curves of a full and a partial agonist simultaneously, it is possible to estimate the affinity constant of the partial agonist and its RA_i value relative to that of the full agonist. One way to estimate the affinity constant of a full agonist is to analyze the effect of partial inactivation of the receptor population on the concentration-response curve of the agonist. This approach can be used to estimate the affinity constant of any agonist, regardless of whether it is a full or partial agonist. Partial receptor inactivation is usually accomplished by treating the receptor preparation with an irreversible antagonist that blocks a portion of the receptor population, but in principle, it could also be achieved by interrupting part of the signaling pathway.

The late Robert Furchgott, winner of the Nobel prize in Physiology and Medicine together with Louis Ignarro and Ferid Murad for their work on nitric oxide, was one of the first to realize that there is enough information in pairs of equi-active-agonist concentrations, measured before and after a fraction of the receptor population has been inactivated, to estimate the affinity constant of the agonist.[5,6] He also showed that by knowing the affinity constants of more than one agonist, it is possible to plot their responses implicitly in terms of receptor occupancy and compute the relative efficacies of the agonists from the location of the occupancy-response relationship (transducer functions) on the occupancy scale. His approach, like that of Stephenson, is based on a null method, which is described in an upcoming section.

A rigorous way to achieve Furchgott's goal is to analyze the concentration-response curves measured before and after partial receptor inactivation with the operational model. This way the model is fitted directly to the measured data, rather than to estimates of agonist concentrations interpolated from the control concentration-response curve (see section below on Null Method). Using the operational model is appropriate provided that the concentration-response curve is a logistic function of the agonist concentration (i.e., a symmetric S-shaped curve when response is plotted against the log agonist concentration). This condition is so common that it hardly represents a limitation, but it does highlight an advantage of the null method approach in that the latter does not depend on a logistic-transducer function; it can be any continuously increasing function of the agonist concentration.

The plots in Figure 7.4a show an example of how the operational model can be used to estimate the affinity constant of an agonist. The data represent measurements of contraction of the isolated ileum from the M_2 KO mouse in response to various concentrations of the muscarinic agonist, oxotremorine-M. The curve on the left represents the control concentration-response curve while that on the right was measured after treating the ileum with the irreversible muscarinic antagonist, 4-DAMP mustard, followed by washing. The theoretical curves represent the least-squares fit of the operational model to the data. Global nonlinear regression analysis was done sharing the estimates of the affinity constant of carbachol (K_{obs}) and the system parameters (M_{sys} and m) between the two curves and estimating a unique value of τ for each curve. Using the procedure described in the appendix, it was

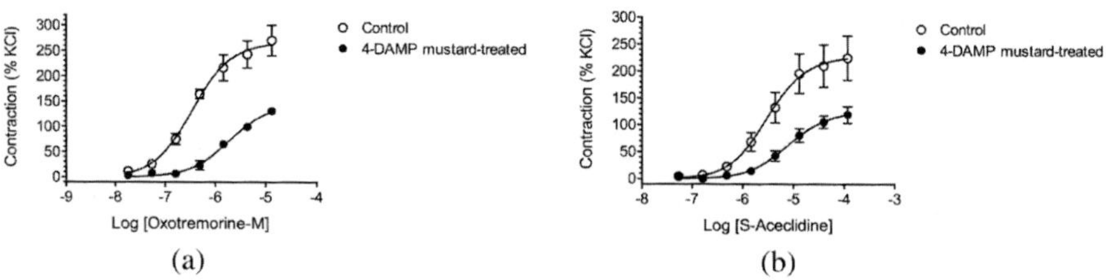

Figure 7.4: Method of partial receptor inactivation for the estimation of the observed affinities and relative efficacies of oxotremorine-M (a) and S-aceclidine (b) for eliciting contraction of the mouse ileum. The isolated ileum was obtained from M_2 muscarinic receptor knockout mice. Receptor inactivation was achieved by incubating the tissue with the irreversible muscarinic antagonist, 4-DAMP mustard, for a period of time and then washing the tissue. For each panel, the theoretical curves represent the global fit of the operational model (Equation (7.10)) to the control and 4-DAMP mustard treated curves. The data are from Tran *et al.*[13]

possible to obtain unique estimates of the shared parameters of the operational model: M_{sys}, 288 ± 23; m, 1.17 ± 0.22; and log K_{obs}, 5.49 ± 0.30. The estimates of log τ for the control curve and that measured after 4-DAMP mustard treatment were 0.95 ± 0.30 and 0.020 ± 0.14, respectively. The fraction of the receptor population not inactivated by 4-DAMP mustard is equivalent to the ratio of the τ value measured after receptor inactivation divided by that measured for the control curve. This value is equivalent to the antilog of the difference (0.02–0.95) in the two log τ values, 0.12 ($10^{-0.93}$).

The reason why it was possible to obtain a unique solution for the operational model in this case is that a single estimate of K_{obs} was estimated from both concentration-response curves, and the E_{max} value of the curve measured after partial receptor inactivation was substantially lower than that of the control curve. If 4-DAMP mustard had only shifted the concentration-response curve of carbachol to the right without depressing its E_{max}, then the solution set to the operational model would have been infinite. In practice, it is useful to treat with the irreversible antagonist so that the E_{max} value is depressed within the range of 80–20% of control, approximately. A more rigorous approach is to treat with the irreversible antagonist for different periods of time or different concentrations so that two or three different levels of receptor inactivation are achieved. The operational model can then be fitted to all of the data simultaneously sharing the estimate of K_{obs} among the curves and obtaining unique values of τ for each curve.

It is possible to estimate the fraction of the receptor population not inactivated by 4-DAMP mustard by taking the ratio of τ values. Recall from Equation (6.11) that τ is equivalent to the product of ε and R_T divided by K_E. Therefore, the ratio of the τ value measured after receptor inactivation (τ'), expressed relative to that measured before (τ'), is equivalent to the fraction of residual receptors not inactivated:

$$\frac{\tau'}{\tau} = \frac{\frac{\varepsilon R'_T}{K_E}}{\frac{\varepsilon R_T}{K_E}} = \frac{R'_T}{R_T}. \tag{7.22}$$

In this equation, R_T and R'_T denote the number of receptors before and after partial receptor inactivation, respectively.

But how do we estimate efficacy? This parameter seems inextricably linked with R_T and K_E because the three parameters define τ. It is not a simple matter to determine the actual value of the unitless scalar, ε, but it is possible to estimate the efficacy of one agonist relative to that of another. If the method of partial receptor inactivation is used to estimate the τ value of two agonists, then relative efficacy is simply equivalent to the ratio of their τ values. For this calculation, the τ value of the agonist estimated from the concentration-response curve measured before receptor inactivation is used. The following equation, which is analogous to Equation (6.22), illustrates that the τ value of agonist A (τ_A) expressed relative to that of agonist B (τ_B) is equal to the corresponding ratio of intrinsic efficacies:

$$\frac{\tau_A}{\tau_B} = \frac{\frac{\varepsilon_A R_T}{K_E}}{\frac{\varepsilon_B R_T}{K_E}} = \frac{\varepsilon_A}{\varepsilon_B}. \tag{7.23}$$

The data in Figures 7.4a and 7.4b serve as an example of how to estimate relative efficacy. The data in Figure 7.4b are analogous to those in Figure 7.4a except that contractions to the less efficacious agonist, S-aceclidine, were measured in the ileum from the M_2 KO mouse. Concentration-response curves were measured under controlled conditions and after a portion of the receptor population were inactivated. The theoretical curves represent the least-squares of the operational model to the data. The parameter estimates are for M_{sys}, m and log K_{obs} are 282 ± 75, 1.24 ± 0.47 and 4.98 ± 0.46, respectively. The estimates of log τ before and after partial receptor inactivation are 0.50 ± 0.44 and -0.076 ± 0.24, respectively. The estimate of the intrinsic efficacy of S-aceclidine (ε_{acc}) relative to that of carbachol (ε_{carb}) is equal to the ratio of the corresponding τ values of their control concentration-response curves ($\log \varepsilon_{ace}/\varepsilon_{carb} = \log \tau_{ace}/\tau_{carb} = 0.50 - 0.95 = -0.45$). Therefore, the efficacy of S-aceclidine relative to oxotremorine-M is about one-third (0.35).

The reader may notice that the SEM values for the log τ values of both oxotremorine-M and S-aceclidine are large (0.30–0.46 log units). This variation is due in part to the variability in the data and to the lack of a clear plateau value for the near maximal responses measured after partial receptor inactivation. If higher concentrations of agonist were used, the estimates of log K_{obs} and log τ would have been more reliable.

Null Methods

The logic behind a null method for estimating agonist affinity and efficacy rests on the assumption that the downstream response elicited by agonist-receptor activation is a function ($g(S)$) of the stimulus function ($f(D)$) (see Equation (7.4)). Provided that the nature of the stimulus function is known, one can estimate its relevant parameters (affinity and relative efficacy) even if the transducer function is unknown. The trick is to set up an equality describing equivalent responses to different agonists (Stephenson's idea[4]) or to the same agonist under different conditions involving a predictable modification of the stimulus function (Furchgott's idea[5,6]). Because the transducer function is the same for different agonists or for the same agonist under different conditions, it cancels out of the equality.

One advantage of the null method approach is that it is valid under conditions where the operational model is invalid, that is, when the transducer function is not logistic. All that is required for use of the null method is that the transducer function is a continuous increasing function of the stimulus, and that the stimulus can be described by Equation (7.1). In the following two sections, methods for using Furchgott's and Stephenson's ideas to estimate the affinities and efficacies of agonists are described.

Estimating affinity and efficacy

As mentioned above Furchgott realized that it was possible to estimate the affinity constant of an agonist by analyzing its control concentration-response curve and that measured after a portion of the receptor population had been inactivated. He reasoned, as Stephenson did, that the response to the agonist could be expressed as a function of the stimulus, with the latter defined by Equation (7.1). If a fraction of the receptor population were inactivated, then this condition could also be described by the same function but with the receptor density term multiplied by a scalar representing the fraction of the residual receptor population which is still active. This logic is represented by the following equality, which describes the relationship between equi-active-agonist concentrations before and after partial receptor inactivation:

$$g\left(\frac{\varepsilon D R_T}{D + \frac{1}{K_{\mathrm{obs}}}}\right) = g\left(\frac{\varepsilon D' q R_T}{D' + \frac{1}{K_{\mathrm{obs}}}}\right). \tag{7.24}$$

In this equation, g denotes the unknown transducer function, q, the fraction of residual receptors after inactivation, and D and D', equi-active agonist concentrations before and after receptor inactivation, respectively. R_T represents receptor density and ε and K_{obs} the efficacy and observed affinity constant of the agonist-receptor complex as defined previously. The function on the left side of the equality defines the response before receptor inactivation and that on the right defines the response after receptor inactivation.

A pictorial representation of this equality is shown in Figure 7.5a. The control concentration-response-curve is shown on the left and that measured after partial receptor inactivation is on the right. The data are from Figure 7.4a and represent the concentration-response curves of oxotremorine-M for eliciting contraction of the ileum, before and after partial receptor inactivation. The D' values denote the

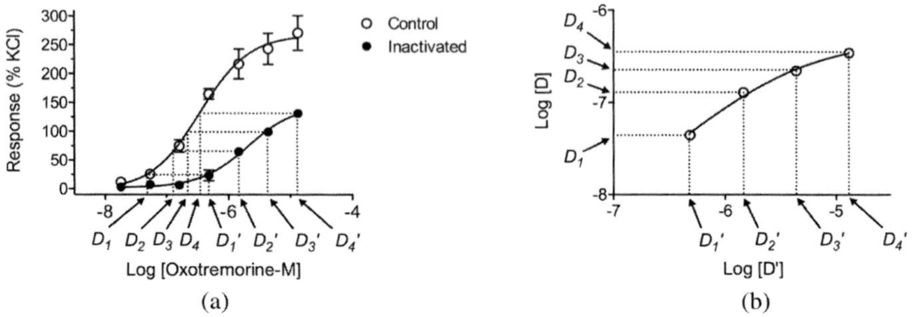

Figure 7.5: Null method for the estimation of the observed affinity of carbachol for eliciting contraction of the mouse ileum. The method is based on the analysis of equi-active agonist concentrations before and after inactivation of a fraction of the receptor population. The source of the ileum data and method of receptor inactivation is the same as described in Figure 7.4. (a) The concentration-response curve of oxotremorine-M for eliciting contraction of the mouse ileum, before (control) and after (inactivated) partial inactivation of the receptor population is shown. Equi-active agonist concentrations before (D) and after (D') partial receptor inactivation are indicated on the abscissa. (b) Plot of equi-active agonist concentrations before (ordinate) and after (abscissa) partial receptor inactivation. The theoretical curve represents the least-squares fit of Equation (7.32) to the data. The data are from Tran *et al.*[13]

concentrations of oxotremorine-M used to measure the responses after inactivation of a proportion of the receptor population. As mentioned previously, receptor inactivation is usually accomplished by treating the preparation with an irreversible antagonist for a period of time and then washing it away. It is possible to interpolate response values on the control concentration-response curve that are equivalent to those measured after receptor inactivation. The oxotremorine-M concentrations corresponding to these interpolated responses on the control curve are denoted by D. In this way, pairs of equi-active oxotremorine-M concentrations before and after receptor inactivation (D and D') are determined. These correspond to the variables D and D' in Equation (7.24).

Now let us manipulate Equation (7.24) into a more useful form. Furchgott manipulated the equation into a linear form so that it would be possible to estimate K_{obs} and q by linear regression analysis. This was a sensible approach in the days when computer algorithms for nonlinear regression analysis were not readily available, but a better approach is to express log D as a function of D' and the other parameters so that we can apply nonlinear regression analysis to the nontransformed measurement, log D. The following sequence tracks the algebraic conversion of Equation (7.24) into a form suitable for nonlinear regression analysis.

Because the same transducer function $g(S)$ operates on the entire contents of both sides of Equation (7.24), it can be eliminated:

$$\frac{\varepsilon D R_T}{D + \frac{1}{K_{obs}}} = \frac{\varepsilon D' q R_T}{D' + \frac{1}{K_{obs}}}. \tag{7.25}$$

Dividing both sides of the equation by εR_T yields:

$$\frac{D}{D + \frac{1}{K_{obs}}} = \frac{D' q}{D' + \frac{1}{K_{obs}}}. \tag{7.26}$$

Cross multiplying yields:

$$DD' + \frac{D}{K_{obs}} = qDD' + \frac{qD'}{K_{obs}}. \tag{7.27}$$

Further rearrangement yields Equations (7.28)–(7.32):

$$DD' + \frac{D}{K_{obs}} - qDD' = \frac{qD'}{K_{obs}}, \tag{7.28}$$

$$D\left(D' + \frac{1}{K_{obs}} - qD'\right) = \frac{qD'}{K_{obs}}, \tag{7.29}$$

$$D = \frac{\frac{qD'}{K_{obs}}}{\left(D' + \frac{1}{K_{obs}} - qD'\right)}, \tag{7.30}$$

$$D = \frac{qD'}{D'K_{obs}(1-q) + 1}, \tag{7.31}$$

$$\log D = \log\left(\frac{qD'}{D'K_{obs}(1-q) + 1}\right). \tag{7.32}$$

Equation (7.32) is our goal. It can be fitted to the estimates of pairs of equi-active agonist concentrations shown in Figure 7.5a. These pairs are plotted as points in Figure 7.5b with the x and y coordinates of each point representing a pair of log D' and log D values. The smooth curve through the points represents the least-squares fit of Equation (7.32) to the data. The parameter estimates from regression analysis are log K_{obs}, 5.43 $\pm$ 0.08 and q, 0.11 $\pm$ 0.0087. These values are similar to those estimated using the operational model, 5.49 $\pm$ 0.30 and 0.12, respectively.

The intrinsic efficacy of an agonist (e.g., S-aceclidine) can be estimated relative to that of another agonist (e.g., oxotremorine-M) by examining the occupancy-response relationships of the two agonists. For a given agonist, occupancy values are calculated for every agonist concentration on the control concentration-response curve. Regarding the control data for oxotremorine-M in Figure 7.4a, the log molar

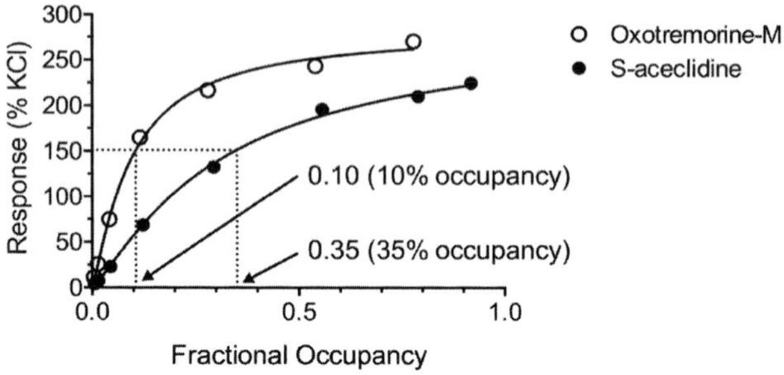

Figure 7.6: Occupancy-response plot for estimation of relative efficacy. Knowing the observed affinity constants of oxotremorine-M and S-aceclidine, it is possible to plot their control responses in Figure 7.5 agoinst receptor occupancy instead of the agonist concentration. The resulting occupancy-response relationships can be used to determine the levels of receptor occupancy that yield equivalent responses by the two agonists. These occupancy values are inversely related to efficacy as described in the text.

concentrations of this agonist are −7.75, −6.79, −6.31, −5.84, −5.36 and −4.88. The antilogs of these values are 1.80×10^{-8}, 5.40×10^{-8}, 1.62×10^{-7}, 4.86×10^{-7}, 1.46×10^{-6}, 4.38×10^{-6} and 1.31×10^{-5} molar. Fractional receptor occupancy at each concentration is calculated from the equation, $D/(D+1/K_{\text{obs}})$. Thus, the corresponding occupancy values of the control concentration response curve, expressed as a fraction are: 0.0048, 0.014, 0.042, 0.115, 0.281, 0.540 and 0.779. A plot of the responses of the control concentration-response curve against these occupancy values is shown in Figure 7.6. The curve on the left represents the data from Figure 7.4a for oxotremorine-M and that on the right represents a similar analysis for the S-aceclidine data in Figure 7.4b.

To estimate the efficacy of S-aceclidine relative to that of oxotremorine-M, consider the level of receptor occupancies of the two agonists required to produce a response equivalent to 150% that is elicited by KCl, for example. The exact level of response is unimportant, so long as it is possible to obtain a reliable estimate of occupancy for the designated level of response. It takes about 35% receptor occupancy by S-aceclidine to produce a response 150% that of KCl, whereas it only requires 10% occupancy by carbachol to elicit the same response. Because it requires 3.5-fold greater occupancy by S-aceclidine to produce a response equivalent to that of oxotremorine, then the efficacy of S-aceclidine must be only 1/3.5 that of carbachol (i.e., 0.29). Thus, the efficacy of agonist A (ε_A) expressed relative to that of agonist B (ε_B) is equivalent to the reciprocal of the corresponding ratios of receptor occupancies yielding equivalent responses:

$$\frac{\varepsilon_A}{\varepsilon_B} = \frac{BR}{AR}. \tag{7.33}$$

BR and AR denote levels of receptor occupancy by B and A that elicit equivalent responses.

The plots in Figure 6.6 actually represent a form of the transducer function, $g(S)$. The plotted function differs in a minor way from $g(S)$ because the response is plotted against receptor occupancy instead of the stimulus. Regardless, when $g(S)$ is logistic, then these curves are described by the function:

$$\text{response} = \frac{O^m M_{\text{sys}}}{O^m + \frac{1}{\tau^m}}, \tag{7.34}$$

in which O represents receptor occupancy and m and τ are the parameters of the operational model described previously. Nonlinear regression analysis of the data in Figure 7.6 according to Equation (7.34) yielded estimates of the shared parameters, M_{sys} and m of 282% ± 7.9 and 1.21 ± 0.06, respectively, and values of log τ for carbachol and pilocarpine of 1.03 ± 0.035 and 0.51 ± 0.029, respectively. These values are essentially the same as those estimated from the data in Figure 7.4.

Estimating the affinity of a partial agonist by analysis of its concentration-response curve and that of a full agonist

As described above, Stephenson realized that there was enough information in the concentration-response curves of full and partial agonists to estimate the dissociation constant of the partial agonist. The basis for his method rests on the assumption that the response is a function of the known stimulus function and that equivalent responses to different agonists imply equal stimuli. Using the logic applied in the prior section, it is possible to derive an equality describing equi-active responses to the full (A) and partial (B) agonists:

$$g\left(\frac{\varepsilon_{\text{A}} A R_T}{A + \frac{1}{K_{\text{obs-A}}}}\right) = g\left(\frac{\varepsilon_{\text{B}} B R_T}{B + \frac{1}{K_{\text{obs-B}}}}\right). \tag{7.35}$$

In this equation, g denotes the unknown transducer function, $K_{\text{obs-A}}$ and $K_{\text{obs-B}}$, the observed affinity constants of A and B, and ε_{A} and ε_{B}, the corresponding efficacies. The function on the left side of the equality defines the response to A and that on the right defines the response to B.

Figure 7.7a shows the relationship between equi-active concentrations of the full and partial agonist. In this case, the full agonist is more potent, but the method also applies when it is less potent. The data represent the concentration-response

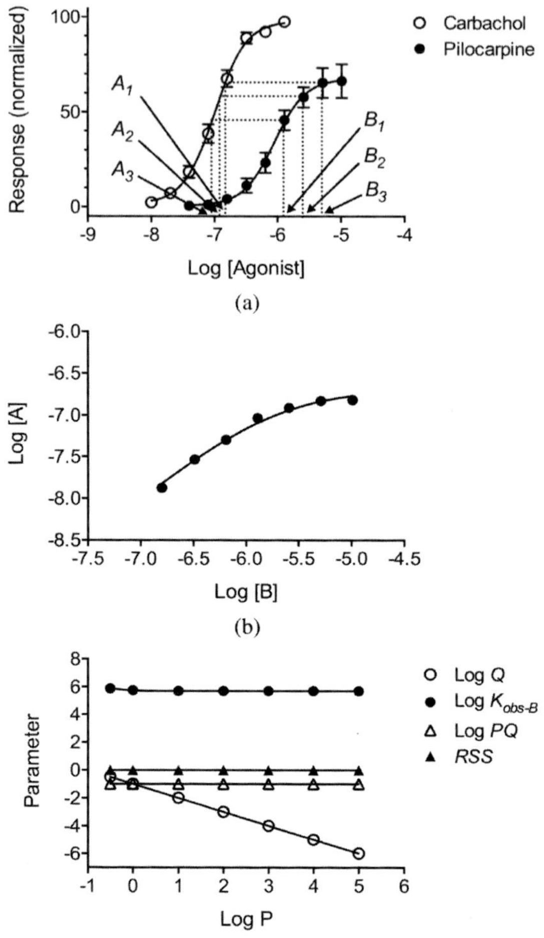

Figure 7.7: Null method for the estimation of the observed affinity constant of a partial agonist by analysis of its concentration-response curve with that of a full agonist. (a) Concentration-response curves of carbachol and pilocarpine for eliciting contraction of the guinea pig ileum. The data are the same as those shown in Figure 7.3a. The plot illustrates the relationship between equi-active concentrations of carbachol (D) and pilocarpine (D'). (b) The data points represent equi-active concentrations of carbachol (ordinate) and pilocarpine (abscissa). The theoretical curve represents the least-squares fit of Equation (7.41) to the data. (c) Summary of nonlinear regression analysis of the data in panel (c). Equation (7.41) was fitted to the data points in panel (c) repetitively, each time constraining the value of log P to a constant and determining the values of the other parameters that that minimized the residual sum of squared deviations of the data points from the regression equation (RSS). The diagram shows the values of the parameter estimates for each regression plotted against the value to which M_{sys} was constrained.

curves of carbachol and pilocarpine for eliciting contraction in the guinea pig ileum and are the same as those shown in Figure 7.3.

The B values denote the concentrations of pilocarpine used to measure its concentration-response curve. It is possible to interpolate response values on the carbachol concentration-response curve that are equivalent to those measured for pilocarpine. The carbachol concentrations corresponding to these interpolated responses are denoted by A. In this way, pairs of equi-active concentrations of carbachol and pilocarpine are determined, which correspond to the variables A and B in Equation (7.35). Only a handful pairs of equi-active concentrations of agonist are shown on the plot to avoid overcrowding.

Now, we need to manipulate Equation (7.24) into form useful for nonlinear regression analysis. This involves expressing log A as a function of B and other parameters. Because the transducer function acts on the entire contents of each side of the equation, it can be eliminated:

$$\frac{\varepsilon_A A R_T}{A + \frac{1}{K_{\text{obs-A}}}} = \frac{\varepsilon_B B R_T}{B + \frac{1}{K_{\text{obs-B}}}}. \tag{7.36}$$

Eliminating R_T from both sides of the equation and cross-multiplying yields:

$$\varepsilon_A AB + \frac{\varepsilon_A A}{K_{\text{obs-B}}} = \varepsilon_B AB + \frac{\varepsilon_B B}{K_{\text{obs-A}}}. \tag{7.37}$$

Further rearrangement yields Equations (7.38)–(7.40):

$$A\left(\varepsilon_A B + \frac{\varepsilon_A}{K_{\text{obs-B}}} - \varepsilon_B B\right) = \frac{\varepsilon_B B}{K_{\text{obs-A}}}, \tag{7.38}$$

$$A = \frac{\varepsilon_B B}{K_{\text{obs-A}}\left(\varepsilon_A B + \frac{\varepsilon_A}{K_{\text{obs-B}}} - \varepsilon_B B\right)}, \tag{7.39}$$

$$A = \frac{\frac{\varepsilon_B}{\varepsilon_A} B}{BK_{\text{obs-A}}\left(1 - \frac{\varepsilon_B}{\varepsilon_A}\right) + \frac{K_{\text{obs-A}}}{K_{\text{obs-B}}}}. \tag{7.40}$$

This equation can be rewritten in the form:

$$\log(A) = \log\left(\frac{BPQ}{BK_{\text{obs-B}}(1-Q)+1}\right), \tag{7.41}$$

in which:

$$Q = \frac{\varepsilon_B}{\varepsilon_A}, \tag{7.42}$$

$$P = \frac{K_{\text{obs-B}}}{K_{\text{obs-A}}}. \tag{7.43}$$

Equation (7.41) is in the appropriate form to analyze equi-active concentrations of the full and partial agonists. These pairs of agonist concentrations are plotted as points in Figure 7.7b with the x and y coordinates of each point representing a pair of A and B values. The smooth curve through the points represents the least-squares fit of Equation (7.41) to the data. In this case, the solution set is infinite, and we need to explore parameter space using an approach analogous to those used in connection with Figure 7.3.

A summary of the nonlinear regression analysis is shown in Figure 7.7c. Regression analysis was done repeatedly with the value of P constrained to a constant each time. The values of $K_{obs\text{-}B}$ and Q that yielded a least-squares fit to the plot in Figure 7.7b were determined and these are plotted as points corresponding to the constrained value of log P on the abscissa in Figure 7.7c. The plot shows that a least-squares fit could be obtained over the domain of log P values on the abscissa, because the sum of the squared deviations of the data points from the theoretical curve (RSS) was at a constant minimum. It was impossible to obtain a reliable estimate of P because the same least squares fit was obtained for the range of P values shown on the abscissa. As the constrained value of P increased, the estimate of Q decreased such that the product of the two parameters remained constant. Thus, it was possible to estimate log $K_{obs\text{-}B}$ (5.88 ± 0.074) and the product PQ (log PQ, -0.987). The estimate of the K_{obs} value for pilocarpine is essentially identical to that determined previously using the operational model to analyze the data (5.835 ± 0.17).

The parameters P and Q represent the affinity and efficacy of pilocarpine, respectively, expressed relative those of carbachol (see Equations (6.42) and (6.43)). Thus, the product of these two constants is equivalent to the RA_i value of pilocarpine expressed relative to carbachol. The value estimated using the null method as described in Figure 7.7 (log RA_i, -0.99) is essentially equivalent to that estimated using the operational model in Figure 7.3 (-0.92).

Philosophical significance of the null method approach

From the forgoing analyses illustrated in Figures 7.5–7.7, it is clear that the null methods for estimating the K_{obs} value of an agonist (Figure 7.5) or that of a partial agonist (Figure 7.7) are more tedious than the corresponding methods employing the operation model as illustrated in Figures 7.3 and 7.4. My intention in presenting these analyses is not to encourage their use but to illustrate the simple logic of their basis and how they illustrate what information can be gleaned from concentration-response curves.

Recall that these methods are based on the assumption that agonist-receptor occupancy at a GPCR is consistent with a one-site model and that equivalent tissue responses imply equivalent stimuli (product of receptor occupancy and efficacy). If the stimuli are equivalent, one can derive a null equation describing the relationship between equi-active agonist concentrations of a full and partial agonist (Equation (7.41)) or between equi-active concentrations of the same agonist before and after partial receptor inactivation (Equation (7.32)). This approach effectively eliminates the unknown relationship between occupancy and response from the analysis and defines the functional relationship between equi-active agonist concentrations. There is sufficient information in the pairs of equi-active agonist concentrations to estimate some of the parameters in the function (observed affinity, relative efficacy and RA_i) as summarized above. In those rare instances when the concentration-response curves of agonists are inconsistent with a logistic equation, then the null approach provides a useful method of estimating affinity and efficacy.

Estimation of Agonist RA_i Values

While reading a book about how the effects of drugs can be deduced from the affinity and efficacy components of the initial drug-receptor complex, you would probably assume that the methods described above for estimating these parameters are widely used. If you did, you would be wrong, for it would not be unlikely that a search of dozens of current pharmacology publications would yield not a single article describing the use of these methods. If so, why devote so much effort to describing them?

One answer is obvious. The methods should probably be used more. But there are two other reasons, based on the widespread measurement of agonist concentration-response curves in pharmacological research. First, an agonist that acts on a single type of receptor often elicits different responses in different systems, and the behavior of its concentration-response curve can vary substantially. This variation can be attributed to differences in the transducer function as well as potential differences in the affinity and efficacy of the agonist-receptor complex, based on different active conformations of the receptor. Knowing how to measure affinity and efficacy imparts an intuitive understanding of the transducer function and how it can vary in different systems. In a sense, one can consider a less sensitive system as one having a degree of receptor inactivation. If all agonists exhibit lower potency or maximal response or both in this system, then the transducer function is less sensitive. If not, then different active conformations of the receptor might be involved in eliciting the different responses (ligand-directed signaling, see

Chapter 9). Having an intuitive knowledge of how to discriminate ligand-directed signaling from a difference in the sensitivity of the transduction pathway is an essential skill in pharmacology, and it is acquired from knowledge of the estimation of affinity and efficacy.

The other reason for describing the estimation of affinity and efficacy is that one of the methods simply involves an analysis of the concentration-response curves of full and partial agonists. No receptor inactivation is required for this approach. This method can be modified to estimate the RA_i values of any set of two or more agonists provided that their concentration-response curves are known. Because concentration-response curves are frequently measured in pharmacological research, the requisite information for measuring RA_i is already being widely measured. The RA_i value is simply an estimate of the product of affinity and efficacy of one agonist expressed relative to that of another. At a deeper level of analysis, it is equivalent to the ratio of microscopic affinity constant of the agonists for the active state of the receptor as described in a later section of this chapter. In this section the methods and properties of the RA_i value are described in greater detail.

Estimating RA$_i$ values from a series of agonist concentration-response curves

An example of a common analysis in pharmacology is the measurement of the concentration-response curves of a series of agonists for eliciting a particular response, like those shown in Figure 7.8. Usually, the data are analyzed by measuring the E_{max}, EC_{50} and perhaps the Hill slope or slope factor of the concentration response curves (see Equation (7.3)). From this analysis, potency ratios (ratio of EC_{50} values) can be estimated, and an intuitive sense of the activity of these agonists at the receptor mediating the response can be gleaned. But what about receptor-dependent parameters, and how do we compare agonists that differ in E_{max} values? The potency ratio does not address changes in E_{max}. In this situation, estimation of the agonist RA_i values is very useful. The parameter is a purely receptor-dependent (i.e., a relative estimate of the microscopic affinity constant of the active state of the receptor), and it is a convenient way to convert the two disparate parameters of EC_{50} and E_{max} into a single measure of agonist activity.

Because RA_i is a relative measure, the concentration-response curves of at least two agonists are required for its determination. In the analysis below, the agonist to which all other agonists are compared is denoted as the "standard agonist", whereas other agonists are referred to as "test agonists". It is better if the concentration-response curves of the standard agonist and the test agonists are measured in the

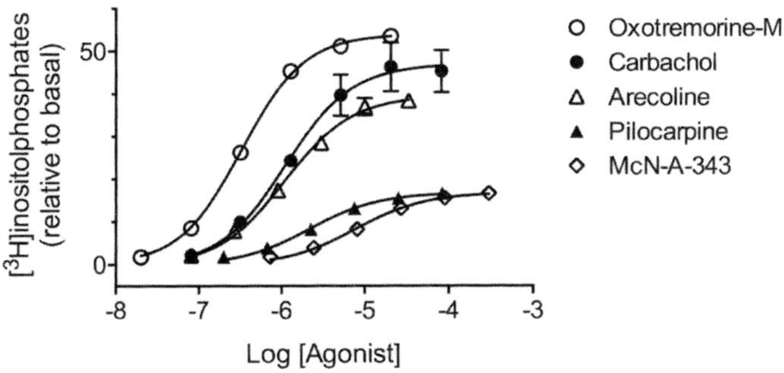

Figure 7.8: Global nonlinear regression analysis of the concentration-response curves of muscarinic agonists for eliciting phosphoinositide hydrolysis in Chinese hamster ovary cells stably transfected with the M_3 muscarinic receptor. The theoretical curves represent the global fits of Equations (7.47) (all agonists except oxotremorine-M) and (7.49) (oxotremorine-M) to the data. The estimates of M_{sys}, m and R where shared among the curves and unique estimates of $K_{obs}$and RA_i were estimated for all of the agonists except oxotremorine-M.

same experiment. For a large group of test agonists, a standard agonist can be run each time a new subset of test agonists is investigated.

Before continuing with the method of RA_i analysis, let me describ the overall logic. As described above, it is possible to fit the operational model to the concentration-response curves of a full agonist and a partial agonist and obtain an estimate of the RA_i value of the partial agonist relative to that of the full agonist. Also, for any one concentration-response curve, it is possible to estimate the product of τK_{obs}. It follows that it should be possible to estimate the product of τK_{obs} for two concentration-response curves having the same E_{max} value. Thus, it should be possible to estimate the RA_i value of one agonist relative to the other when their E_{max} values are the same because the RA_i value is equivalent to the ratio of τK_{obs} values of the two agonists. In conclusion, therefore, it should be possible to estimate the RA_i values of a group of agonists with various EC_{50} and E_{max} values by fitting the operational model to the data and estimating RA_i from the various estimates of τ and K_{obs} or their product.

To expedite the process, all of the concentration-response curves can be analyzed simultaneously by global nonlinear regression analysis, sharing the system-dependent parameters of the operational model (M_{sys} and m) among the curves, and obtaining unique estimates of the receptor-dependent parameters of each agonist. Also, the equation for the operational model can be modified so that the RA_i value of each test agonist is a parameter. Because the RA_i value of the standard agonist expressed relative to itself is one, a slightly modified form of the operational model is used for the standard agonist. In the following sequence of paragraphs, I first describe the modifications in the operational model for the standard and test agonists and then the results of global nonlinear regression analysis of the data in Figure 7.8 are described using the modified equations. Further details of the method are described in several publications from my laboratory[7–13] and also in the appendix to this chapter, Example 3.

To modify the operational model for each test agonist, we begin by writing the operational model (Equation (7.10)) with the subscript "t" added to the receptor dependent parameters, K_{obs} and τ, to designate that these are "test agonist" parameters:

$$\text{response} = \frac{D^m M_{\mathrm{sys}}}{D^m + \frac{\left(D + \frac{1}{K_{\mathrm{obs}-t}}\right)^m}{\tau_t^m}}. \tag{7.44}$$

With slight modification, Equation (7.44) becomes:

$$\text{response} = \frac{D^m M_{\mathrm{sys}}}{D^m + \frac{(DK_{\mathrm{obs}-t}+1)^m}{(\tau_t K_{\mathrm{obs}-t})^m}}. \tag{7.45}$$

To introduce the parameter RA_i into this equation, we first turn to Equation (7.19), which shows that the RA_i value of a test agonist, expressed relative to that of a standard agonist, is equivalent to the ratio of the product of τK_{obs} of the test agonist divided by that of the standard agonist. Using the subscript "s" for the parameters of the standard agonist, Equation (7.19) can be rewritten in the following form:

$$RA_i = \frac{\tau_t K_{\mathrm{obs}-t}}{\tau_s K_{\mathrm{obs}-s}}. \tag{7.46}$$

We can now substitute the product of RA_i and $\tau_s K_{\mathrm{obs}-s}$ into Equation (7.45) for the product $\tau_t K_{\mathrm{obs}-t}$:

$$\text{response} = \frac{D^m M_{\mathrm{sys}}}{D^m + \frac{(DK_{\mathrm{obs}-t}+1)^m}{(R^* RA_i)^m}}, \tag{7.47}$$

in which

$$R = \tau_s K_{\mathrm{obs}-s}. \tag{7.48}$$

Equation (7.47) is the form of the operational model that will be fitted to the concentration-response curves of each test agonist. Because the RA_i value of the standard agonist expressed relative to itself is one, the equation for the standard agonist can be obtained by substituting in the value of one for RA_i in Equation (7.47) to yield:

$$\text{response} = \frac{D^m M_{sys}}{D^m + \frac{(DK_{obs-s}+1)^m}{R^m}}. \tag{7.49}$$

We are now ready to analyze the data in Figure 7.8 to estimate the RA_i values of the agonists. To do this, global nonlinear regression analysis is done fitting Equation (7.49) to the data for the standard agonist (carbachol) and Equation (7.47) to the other agonists. It is convenient to define the standard agonist as the agonist with the largest E_{max} value. The system parameters (M_{sys} and m) are shared across the data, and unique estimates of RA_i and R are obtained for the test agonists and the standard agonist, respectively. It will also be possible to estimate the K_{obs} of any partial agonist. If the standard agonist and any test agonist are full agonists, then it will be impossible to obtain reliable estimates of their K_{obs} values as described earlier in this chapter. In this situation, it may be necessary to constrain their K_{obs} values to arbitrarily low values (e.g., 1) to get the nonlinear regression algorithm to converge on a solution set. The best fit of the operational model to the data in Figure 7.8 is shown by the theoretical curves. The corresponding parameter estimates are indicated in the following list: oxotremorine-M, log R, 6.48 ± 0.045; carbachol, log K_{obs}, 5.19 ± 0.14; log RA_i, −0.56 ± 0.064; arecoline, log K_{obs}, 5.49 ± 0.12, log RA_i, −0.61 ± 0.075; pilocarpine, log K_{obs}, 5.58 ± 0.16, log RA_i, −1.19 ± 0.15 and McN-A-343, log K_{obs}, 5.02 ± 0.16, log RA_i, −1.76 ± 0.15. The estimates of the system parameters are: M_{sys}, 53.9 ± 1.3 and m, 1.15 ± 0.09. Further details of this analysis are described in Ehlert[8] and Ehlert *et al.*[10]

Relationship of the RA$_i$ value to the EC$_{50}$ and E$_{max}$ values of agonist

When the E_{max} value of the test agonist (E_{max-t}) is equivalent to that of the standard agonist ($E_{max-s}a$), then the ratio of the EC_{50} value of the standard agonist (EC_{50-s}) divided by that of the test agonist (EC_{50-t}) is equivalent to the RA_i value of the test agonist. That is, when $E_{max-t} = E_{max-s}$, then

$$RA_i = \frac{EC_{50-s}}{EC_{50-t}}. \tag{7.50}$$

The proof of this corollary can be grasped by examining the equation for the EC_{50} value of an agonist in terms of the parameters of the operational model (Equation (7.16)). Partial agonists with equivalent E_{max} values must have the same τ values (see Equation (7.14)). It follows from Equation (7.16), that the ratio of EC_{50-s}/EC_{50-t}

for partial agonists with equivalent E_{max} values is equivalent to $\tau_t K_{obs-t}/\tau_s K_{obs-s}$. This latter ratio is equivalent to the RA_i value of the test agonist. For full agonists, EC_{50} is a good approximation of τK_{obs}; hence, the ratio of EC_{50-s}/EC_{50-t} for full agonists is also equivalent to the RA_i value of the test agonist. This is a simple way to estimate the RA_i value of a full test agonist relative to a full standard agonist. There is no need to do the complicated regression analysis shown in Figure 7.8 in this instance.

In situations where the slope factors of the concentration-response curves are equal to one, the RA_i can be estimated from the EC_{50} and E_{max} values of the agonists as shown in the next series of Equations (7.51)–(7.58). Recall that when the slope factor of the concentration-response curve (n in Equation (7.3)) is equivalent to one, the transducer slope factor in the operational model (m in Equation (7.6)) is also equal to one. Setting m in Equation (7.6) equal to one yields:

$$\text{response} = \frac{\frac{\varepsilon D R_T}{D+\frac{1}{K_{obs}}} M_{sys}}{\frac{\varepsilon D R_T}{D+\frac{1}{K_{obs}}} + K_E}. \tag{7.51}$$

Algebraic manipulation of this equation yields:

$$\text{response} = \frac{D\frac{M_{sys}}{1+\frac{K_E}{\varepsilon R_T}}}{D + \frac{1}{K_{obs}(\varepsilon R_T + K_E)}}. \tag{7.52}$$

This equation is equivalent to a simple logistic equation (Equation (7.3)) with $n = 1$:

$$\text{response} = \frac{DE_{max}}{D + EC_{50}}, \tag{7.53}$$

with the EC_{50} and E_{max} equivalent to:

$$E_{max} = \frac{M_{sys}}{1 + \frac{K_E}{\varepsilon R_T}}, \tag{7.54}$$

$$EC_{50} = \frac{1}{K_{obs}\left(\frac{\varepsilon R_T}{K_E} + 1\right)}. \tag{7.55}$$

Note that these equations for E_{max} and EC_{50} can also be derived from those described for the general case when m is not constrained to one (i.e., Equations (7.14) and (7.15)). Setting $m = 1$ and substituting in $\varepsilon R_T/K_E$ for τ in Equations (7.14) and (7.16) yields Equations (7.54) and (7.55). When the slope factor of the concentration-response curve is equal to one, the RA_i value of the test agonist expressed relative to the standard agonist is equivalent to:

$$RA_i = \frac{E_{\text{max-}t} EC_{50\text{-}s}}{E_{\text{max-}s} EC_{50\text{-}t}}. \tag{7.56}$$

To prove that this equation yields a valid estimate of the product of affinity and efficacy of the test agonist expressed relative to that of the standard, the E_{max} and EC_{50} values defined in Equations (7.54) and (7.55) are substituted into it:

$$RA_i = \frac{\frac{M_{\text{sys}}}{1+\frac{K_E}{\varepsilon_t R_T}} \times \frac{1}{K_{\text{obs-s}}(\varepsilon_s R_T + K_E)}}{\frac{M_{\text{sys}}}{1+\frac{K_E}{\varepsilon_s R_T}} \times \frac{1}{K_{\text{obs-s}}(\varepsilon_t R_T + K_E)}}. \tag{7.57}$$

Simplification yields:

$$RA_i = \frac{\varepsilon_t K_{\text{obs-}t}}{\varepsilon_s K_{\text{obs-}s}}. \tag{7.58}$$

This equation is essentially the same as Equation (7.21). It proves that Equation (7.56) is a valid method for estimating the product of affinity and efficacy of an agonist, expressed relative to that of a standard, when the slope factor of the concentration response curve is equal to one.

Estimating the relative affinities and efficacies of partial agonists is easy when the slope factor is equivalent to one

It is possible to estimate the relative affinities of partial agonists when the slope factors of their concentrations response curves are equal to one. Knowing relative affinity, it is also possible to estimate their relative efficacies by factoring out the affinity component from the RA_i value (Equation (7.58)).

The method involves the derivation of a simple equation for a relative estimate of the affinity of a partial agonist ($K_{\text{obs-}t}$) expressed relative to that of a standard partial agonist ($K_{\text{obs-}s}$). I proceed in reverse order by showing the final equation first and proving subsequently that it is equivalent to a relative estimate of affinity:

$$P = \frac{\left(M_{\text{sys}} - E_{\text{max-}t}\right) EC_{50\text{-}s}}{\left(M_{\text{sys}} - E_{\text{max-}s}\right) EC_{50\text{-}t}}. \tag{7.59}$$

In this equation, P denotes the affinity constant of the test agonist expressed relative to that of the standard agonist ($K_{\text{obs-}t}/K_{\text{obs-}s}$). Because M_{sys} is approximately equal to the E_{max} of a full agonist ($E_{\text{max-full}}$), the latter can be substituted into

Equation (7.59) for M_{sys} to estimate the relative affinities of partial agonists provided that their respective EC_{50} and E_{max} values are known.

$$P = \frac{(E_{max\text{-}full} - E_{max-t})\,EC_{50-s}}{(E_{max\text{-}full} - E_{max-s})\,EC_{50-t}}. \tag{7.60}$$

It is possible to prove Equation (7.60) by substituting in the appropriate equations for E_{max} and EC_{50}. When the slope factor (m) in the operational model is equivalent to one, the equations for E_{max} (7.14) and EC_{50} (7.16) can be simplified:

$$E_{max} = \frac{M_{sys}\tau}{1+\tau} \tag{7.61}$$

$$EC_{50} = \frac{1}{K_{obs}(\tau+1)}. \tag{7.62}$$

Substituting these into Equation (7.60) yields:

$$P = \frac{\left(M_{sys} - \left(\frac{M_{sys}\tau_t}{1+\tau_t}\right)\right)\frac{1}{K_{obs-s}(\tau+1)}}{\left(M_{sys} - \left(\frac{M_{sys}\tau_s}{1+\tau_s}\right)\right)\frac{1}{K_{obs-t}(\tau+1)}}. \tag{7.63}$$

This equation is simplified in the following steps:

$$P = \frac{\left(1 - \left(\frac{\tau_t}{1+\tau_t}\right)\right)K_{obs-t}(\tau_t+1)}{\left(1 - \left(\frac{\tau_s}{1+\tau_s}\right)\right)K_{obs-s}(\tau_s+1)}, \tag{7.64}$$

$$P = \frac{\left(\frac{1}{1+\tau_t}\right)K_{obs-t}(\tau_t+1)}{\left(\frac{1}{1+\tau_s}\right)K_{obs-s}(\tau_s+1)}, \tag{7.65}$$

$$P = \frac{K_{obs-t}}{K_{obs-s}}. \tag{7.66}$$

Thus, Equation (7.59) or (7.60) can be used to estimate the affinity constant of a partial agonist, expressed relative to that of another partial agonist provided that their EC_{50} and E_{max} values are known and the slope factors of their concentration-response curves are equivalent to one.

Because the RA_i value is a relative estimate of the product of affinity and efficacy and the P value is an estimate of relative affinity, it is possible to divide the former by the latter to estimate relative efficacy:

$$\frac{\frac{\varepsilon_t K_{obs-t}}{\varepsilon_s K_{obs-s}}}{\frac{K_{obs-t}}{K_{obs-s}}} = \frac{RA_i}{P} = \frac{\varepsilon_t}{\varepsilon_s}. \tag{7.67}$$

This equation was derived from Equations (7.58) and (7.66) and is a mathematical expression for the idea that the relative estimate of the product of affinity and efficacy divided by the relative estimate of affinity is equal to the RA_i value divided by the P value. Substituting in Equations (7.56) and (7.59) for RA_i and P, respectively, yields:

$$\frac{\frac{\varepsilon_t K_{obs-t}}{\varepsilon_s K_{obs-s}}}{\frac{K_{obs-t}}{K_{obs-s}}} = \frac{\frac{E_{max-t} EC_{50-s}}{E_{max-s} EC_{50-t}}}{\frac{(M_{sys} - E_{max-t}) EC_{50-s}}{(M_{sys} - E_{max-s}) EC_{50-t}}}. \tag{7.68}$$

Simplifying both sides of the equations yields:

$$\frac{\varepsilon_t}{\varepsilon_s} = \frac{E_{max-t}\left(M_{sys} - E_{max-s}\right)}{E_{max-s}\left(M_{sys} - E_{max-t}\right)}. \tag{7.69}$$

In principle, this equation applies to the concentration-response curves of any agonists provided that their slope factors are equivalent to one. But because there is often no significant difference between the E_{max} values of different full agonists, the equation has practical application only in the analysis of partial agonists. Equation (7.69) can be written in the following form because the E_{max} value of a full agonist is approximately equal to M_{sys}:

$$\frac{\varepsilon_t}{\varepsilon_s} = \frac{E_{max-t}\left(M_{max-full} - E_{max-s}\right)}{E_{max-s}\left(M_{max-full} - E_{max-t}\right)}. \tag{7.70}$$

In summary, it is possible to use Equations (7.60) and (7.70) to estimate the relative affinities and efficacies of partial agonists whenever their concentration-response curves have slope-factors equivalent to one.

The relationship of RA_i value to the initial slope of the agonist concentration-response curve

When the response is a logistic function of the stimulus, there is a simple relationship between the initial slope of the agonist concentration-response curve and the RA_i value. This is by far the most common situation because agonist concentration-response curves usually resemble a logistic function of the agonist concentration (i.e., Equation (7.3)). To understand how the RA_i value relates to the initial slope of the agonist concentration-response curve, consider how the operational model behaves at very low agonist concentrations. This behavior can be defined by taking the limit of the operational model as the agonist concentration approaches zero:

$$\lim_{D \to 0} \frac{D^m M_{\text{sys}}}{D^m + \frac{\left(D + \frac{1}{K_{\text{obs}}}\right)^m}{\tau^m}} = D^m(\tau K_{\text{obs}})^m. \tag{7.71}$$

The left-hand side of the equation shows that we want to express the operational model (Equation (7.10)) in a simpler form that accurately describes its behavior when the agonist concentration is almost zero. The right-hand side of the equation is the solution; that is, it is a simplified version of the operational model that applies when the value of D is practically zero. It is derived by setting the two occurrences of the variable D in the denominator of the left-hand side of the equation equal to zero and simplifying.

The slope of a function is defined as the incremental change in the value of the function (Δy) that occurs when there is an incremental change in x (Δx) divided by the incremental change in x (i.e., slope $= \Delta y/\Delta x$). To determine the initial slope of a concentration-response curve, we can evaluate the operation model as the agonist concentration increase from zero ($x = y = 0$) to some small value of x (Δx). Because the Δx increment begins at 0, the value of the operational model at Δx is equivalent to Δy. Thus, the initial slope (slope_{i-t}) of the operational model for a test agonist is simply equal to:

$$\text{slope}_{i-t} = \frac{(\Delta D)^m(\tau_t K_{\text{obs}-t})^m}{\Delta D}. \tag{7.72}$$

Therefore, the initial slope of the test agonist, expressed relative to that of the standard agonist (slope_{i-s}) is described by:

$$\frac{\text{slope}_{i-t}}{\text{slope}_{i-s}} = \frac{\frac{(\Delta D)^m(\tau_t K_{\text{obs}-t})^m}{\Delta D}}{\frac{(\Delta D)^m(\tau_s K_{\text{obs}-s})^m}{\Delta D}}. \tag{7.73}$$

Simplifying yields:

$$\frac{\text{slope}_{i-t}}{\text{slope}_{i-s}} = \frac{\tau_t K_{\text{obs}-t}}{\tau_s K_{\text{obs}-s}}. \tag{7.74}$$

Because the product of the τ and K_{obs} values of an agonist, expressed relative to that of another agonist, is equaivalent to the RA_i value (Equation (7.19)),

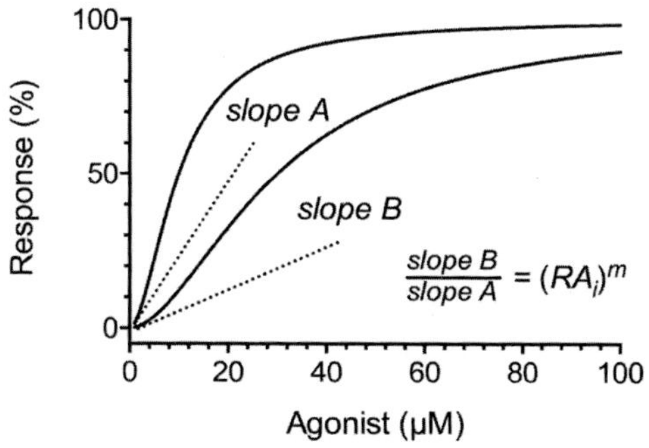

Figure 7.9: Relationship between the initial slopes of agonist concentration-response curves and the RA_i value.

Equation (7.74) reduces to:

$$\frac{\text{slope}_{i-t}}{\text{slope}_{i-s}} = (RA_i)^m. \tag{7.75}$$

Thus, the initial slope of the agonist's concentration-response curve divided by that of the standard agonist is equivalent to the RA_i value raised to the power m. Figure 7.9 illustrates the relationship between the initial slopes of the concentration response curves of two agonists and the RA_i value. Because the slope is defined relative to a linear change in the agonist concentration, the concentration-response curves have been plotted against D instead of log D.

When seen from this perspective, the RA_i value is a relative measure of the sensitivity of the signal generated by the agonist. Agonists with high RA_i values will generate a signal at low agonist concentrations. Whether an agonist is capable of generating a signal over the entire dynamic range of the response (M_{sys}) depends on the efficacy of the agonist (ε) and the sensitivity of the system (K_E). In an imaginary system with high sensitivity such that even partial agonists have E_{max} values equivalent to M_{sys}, all agonist would behave as full agonists and the ratio of their EC_{50} values relative to that of a standard agonist (Equation (7.50)) would be equivalent to their RA_i values.

Agonist Activity in Different Assay Systems

Now that we have learned about the estimation of affinity and efficacy, it is time to use this knowledge to interpret the behavior of an agonist for eliciting responses in different tissues and assay systems. We will have a look at just a tiny sample of the limitless number of responses that could be measured. These not only vary in the nature of the final output, but also in the nature of the step in the signaling pathway that is measured and whether the response is natural or artificial. Our goal is to understand how the affinity and efficacy components of the initial receptor interaction influence the agonist's behavior in different systems. Given the sheer abundance of different types of responses that could be measured, it is not surprising that confusion often exists with regard to whether an agonist is selective for one signaling pathway relative to another, or whether an agonist exhibits selectivity for one of two closely related receptor subtypes. One common mistake is to confuse the differential activity of a partial agonist in sensitive and insensitive systems as evidence that the agonist exhibits ligand bias for the more sensitive system or selectivity for the receptor subtype mediating the more sensitive response. We will see that a given agonist-receptor-G protein complex leaves a characteristic signature across different assay systems in terms of how its EC_{50} and E_{max} change with the sensitivity of the system, provided that the same active conformation of the receptor is involved in mediating the different responses. Being able to identify this signature is an essential skill in drug research.

The reverse engineering approach is appropriate for a variety of assay procedures

The reverse engineering methods described in the prior sections are robust and can be applied to different types of measurements. My example of "how robust these methods are" works better when couched in terms of the null method. This approach is based on the idea that the output function is an unknown function of the stimulus. This "black box" as some refer to it, not only accounts for the transduction pathway inside the cell, but also completely arbitrary features of the assay related to what technique is used to measure the response, how the response is defined, whether the output is linear in time, or even whether the response is measured at the same time that the stimulus is generated. All of these modifications can have colossal effects on EC_{50} and E_{max}, but because they are part of the unknown stimulus-response function that is conveniently eliminated from the analysis, they do not perturb the estimation of affinity and efficacy.

In the simplest case, the response to the agonist is constant over a reasonable period of time. One example is the contraction of an isolated strip of smooth muscle suspended in a tissue bath and bathed in a physiological salt solution. When an agonist is added to the bath that elicits contraction through activation of receptors on the muscle, the increase in tension can be measured with a transducer and recording device. The tension increases in time and reaches a stable plateau within seconds or minutes, depending upon how long it takes the agonist to diffuse into the tissue and equilibrate with the receptors. Although high concentrations of the agonist sometimes elicit a biphasic response, a stable level of contraction usually follows the initial phasic response. When the agonist is removed from the bath, the response decays to resting tension. This behavior implies that the stable level of contraction in the presence of agonist represent a steady state.

In this situation, the steady-state level of the response can be measured at different concentrations of the agonist to determine the concentration-response curve. If the response of the tissue equilibrates quickly, then it is often easier to measure a cumulative concentration-response curve by adding increasing concentrations of agonist to the tissue bath at a constant interval of time without removing the prior addition of agonist. This method works well provided there is little desensitization during the measurements. A plot of the level of contraction against the cumulative concentration of agonist represents the concentration-response curve. An example of this type of measurement is shown in Figure 7.10. The recording of tension in the mouse urinary bladder during the cumulative addition of the muscarinic agonist oxtremorine-M is shown in Figure 7.10a, and these tension measurements are plotted against the cumulative concentration of agonist in Figure 7.10b after subtracting the basal tension measured in the absence of agonist from the recordings. The cumulative approach saves time in experiments on isolated tissues because the

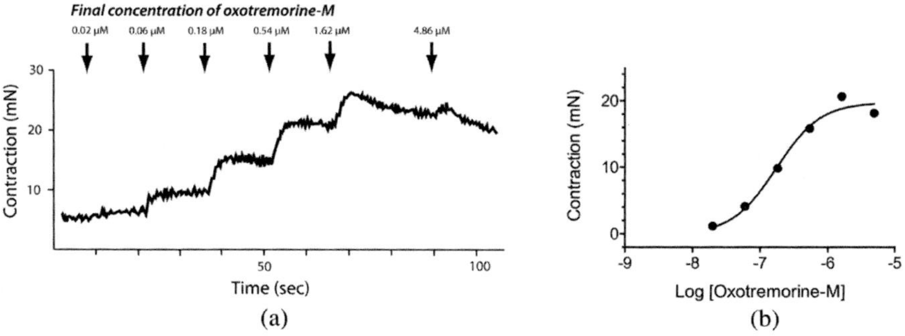

Figure 7.10: The cumulative technique for measurement of concentration-response curves. (a) The plot shows a real time trace of the contraction of the mouse ileum during periodic additions of the muscarinic agonist, oxotremorine-M, to the bathing medium. (b) Plot of oxotremorine-M-induced tension (total tension minus basal tension in the absence of agonist) plotted against the cumulative concentration of oxotremorine-M for the data in panel (a).

preparation does not have to be washed and equilibrated after each agonist addition. Measuring steady-state responses in a homogenous population of isolated cells, however, is usually easier if the cells are divided into different samples and a single concentration of agonist tested in each sample.

The change in the accumulation of a second messenger or other product is often measured as a response to agonist-receptor activation. Usually there is a mechanism for the metabolic breakdown or transformation of signaling molecules, particularly second messengers, so that in the face of constant receptor activation, second messenger levels should reach a steady-state level as shown in Figure 7.11a. At steady state, the rate of formation of the product is equivalent to its rate of breakdown.

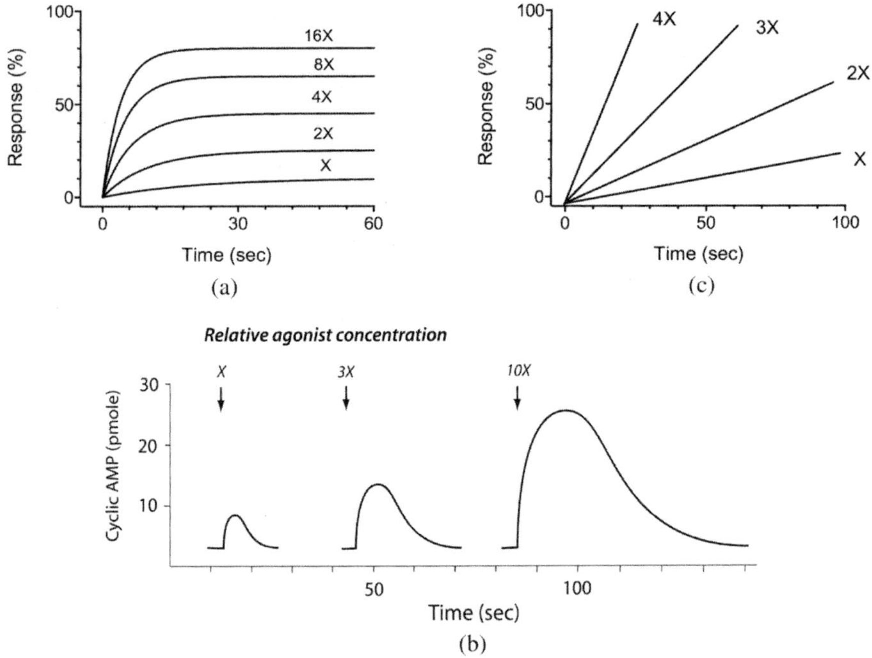

Figure 7.11: Different methods for collecting data for agonist concentration-response curve. The plots show responses to different concentrations of agonist plotted against time. The concentration of agonist is indicated by the product of a scalar and X. (a) Plot of response against time for a response that achieves a steady state in time. (b) Example of a real time plot of agonist-induced production of a second messenger. In this case the response desensitizes and the second messenger is degraded. (c) Real time plot of agonist-induced production of a second messenger. In this case, there is no desensitization, and degradation of the second messenger is inhibited.

But with some receptors, particularly those that stimulate cAMP levels by activating adenylate cyclase through G_s, significant desensitization occurs within minutes. An example of agonist activation of cAMP levels is shown in Figure 7.11b. The level of cAMP reaches a peak in seconds or minutes and then slowly decays. The decay implies a decrease in the rate of formation of cAMP (e.g., receptor desensitization), or possibly, an enhanced rate of degradation. If the peak responses to a range of agonist concentrations are measured, the data can be plotted in the form of a concentration-response curve, and analyzed by the methods described earlier in this chapter.

A useful way of increasing the amount of second messenger (e.g., cAMP) is to add an inhibitor of the enzyme that breaks it down (phosophodiesterase inhibitor). By preventing breakdown, the levels of cAMP continually increase with receptor activation until complete desensitization occurs. Figure 7.11c shows the effect of adding an inhibitor to the steady-state example shown in Figure 7.11a. In the presence of the inhibitor, the amount of product increases linearly in time because of the constant rate of synthesis in the absence of degradation. In the case of receptor desensitization (Figure 7.11b), addition of the inhibitor would cause cAMP levels to rise to a plateau, at which point, no further synthesis of cAMP occurs. The level of cAMP at the plateau corresponds to the total amount of cAMP produced. When an inhibitor is added to the assay shown in Figure 7.11b, the resulting data would look like that shown in Figure 7.11a.

If the goal of the experiment is to analyze the concentration-response curves of agonists and estimate their affinity and efficacy, then any of the methods shown in Figure 6.11 will suffice, provided that the same procedure is used for all of the agonists under consideration. But if the goal is to measure the actual rate of synthesis of cAMP caused by receptor mediated-stimulation of adenylate cyclase, then the initial rate of cAMP accumulation needs to be measured. The method used to collect the data in Figure 7.11c is appropriate for this purpose, although with the appropriate exponential curve fitting procedure, the same information could be gleaned from the experimental paradigms illustrated in Figures 7.11a and 7.11b.

The addition of inhibitor to prevent the breakdown of a measured product is a useful trick for enhancing the signal of an insensitive response. If the response undergoes desensitization for whatever reason, then the EC_{50} value of the agonist can change substantially, depending upon the length of the incubation with agonist. Nonetheless, this does not alter the estimates of affinity and efficacy obtained using the methods described earlier in this chapter.

A useful example that illustrates this point is agonist-stimulated phosphoinositide hydrolysis. Many GPCRs, like the M_3 muscarinic receptor, couple with G_q to stimulate phosphoinositide hydrolysis. The mechanism involves receptor-mediated activation of $G\alpha_q$, the activation of phospholipase-Cβ by $G\alpha_q$, and the subsequent hydrolysis of phosphatidylinositol-4,5-bisphosphate (PIP_2) into diacylglyceraol and inositol-1,4,5-trisphosphate (IP_3) by phospholipase-Cβ. One of the products of the reaction, IP_3, is easy to measure, but it is rapidly converted into inositol-1-phosphate (IP_1), and ultimately to inositol, by phosphatases in the cell. Lithium inhibits the final dephosphorylation of IP_1 by myoinositol-1-phosphatase. In the presence of lithium, activation of the M_3 receptor causes a large accumulation in IP_1. Normally, IP_1 is broken down into inositol, and the inositol is recycled to form new PIP_2, which is needed to maintain adequate levels of PIP_2 for receptor signaling. In the presence of lithium, the levels of inositol and PIP_2 gradually decrease, which ultimately causes a cessation in M_3 receptor-mediated phosphoinositide hydrolysis because there is little PIP_2 to hydrolyze.

Figure 7.12 shows how this phenomenon can affect the concentration-response curve of an agonist for stimulating phosphoinositide hydrolysis through the M_3 muscarinic receptor depending upon when the data are measured after addition of the agonist. Figure 7.12a shows the theoretical accumulation of IP_1 as a function of the time of incubation with muscarinic agonist. In the presence of a low concentration of agonist, IP_1 levels slowly accumulate in time in a near-linear fashion. In the presence of a high concentration of agonist, IP_1 levels increase rapidly and achieve a plateau in about 5 min, further which minimal increase in IP_1 occurs.

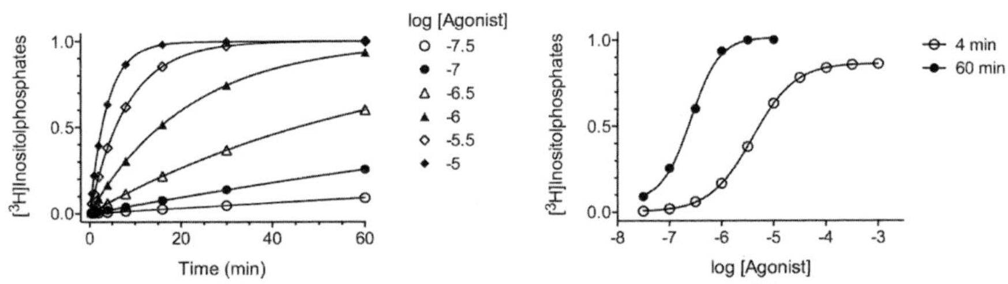

Figure 7.12: Effect of incubation time on agonist-mediated [^{3}H]phosphoinositide hydrolysis in the presence of lithium in cells prelabeled with [^{3}H]inositol. (a) The plots show the theoretical time course for the accumulation of [^{3}H]inositolphosphates in the presence of different concentrations of agonist. (b) Theoretical concentration-response curves for agonist-mediated phosphoinositide hydrolysis following a brief (4 min) and lengthy (60 min) incubation with agonist.

If IP_1 levels are measured after a short period of time and the data are displayed as a concentration-response curve, then the low potency plot shown in Figure 7.12b is obtained. When the data are measured after a much longer period of incubation, the IP_1 levels are much higher and the EC_{50} value of the concentration-response curve is much lower (higher potency) (see Figure 7.12b). Despite these substantial changes in the concentration-response curve, there is no difference in the estimates of the affinity and relative efficacy values of agonist when the data from the two different assay conditions are analyzed by the methods described earlier in this chapter.

My final example is a thought experiment that illustrates how changing the definition of the response alters the concentration-response curve, but not the estimates of affinity and efficacy. Consider a hypothetical experiment where an agonist is administered *in vivo* to activate a receptor in the brain that elicits an increase in locomotion in mice. The goal is to estimate the observed affinities and efficacies of a series of agonists for eliciting this response. In the real world, the methods described earlier in this chapter are clearly inappropriate for the analysis of this response. The concentration of the agonist in microenvironment of the receptor is unknown so how could a concentration-response relationship be determined? Also, how does the concentration of the agonist in the local environment of the receptor change in time, and is this relationship different for different agonists? Even if the agonist acted on only one type of GPCR, this receptor could have different effects on behavior in different regions of the brain, giving rise to a complex concentration response curve (e.g., biphasic or bell shaped). The methods described earlier in this chapter only apply if the response is a continuously increasing function of the agonist concentration. The agonist might also have effects outside the brain that influence the response.

But for the sake of argument, let us assume that we could estimate the steady-state concentration of the agonist in the microenvironment of the receptor and that none of the other problems just mentioned were present. If so, the response (motor behavior) could be measured as a function of the agonist concentration in the brain, and the resulting concentration-response curves could be analyzed to estimate affinity and efficacy. But how should the response in the mice be measured or defined? One could measure the number of times the mouse interrupted a light beam, the amount of time spent ambulating, the total distance traveled or many other measures of motor activity. Each of these measurements could yield a different concentration-response curve. Nonetheless, the different measures should not influence the estimation of affinity and efficacy because the relationship between the stimulus and the final response is irrelevant to the analysis.

Signaling pathways amplify the response and increase the potency of the agonist

A common property of many signaling pathways is that the information traveling through them is converted into a new form and usually amplified at each step of the pathway. One spectacular example is light detection by photocells in the mammalian retina (see the introduction to chapter 5). Another is the regulation of the force of contraction of the heart by the β_1-adrenoceptor (see Chapter 5, Figure 5.16). The initial form of the information is the molecule isoproterenol, which has a structure exhibiting complementariness for both the resting and active states of the β_1-adrenodeptor. Its ability to bind to and activate the β_1-adrenoceptor is determined by its affinity for the ground state and its relative selectivity for the active state, respectively. Once activated, the β_1-adrenoceptor associates with G_s, causing it to loosen its grip on GDP, enabling GTP in the cytosol to bind to G_s. The negative cooperativity between the binding of GTP and $G\beta\gamma$ to $G\alpha_s$ causes the holoprotein to split apart, and the free GTP–$G\alpha_s$ subunit then binds with and activates a nearby adenylate cyclase enzyme. The activated enzyme converts ATP in the cytosol into cAMP, and the latter binds to the regulatory subunit of cAMP-dependent protein kinase. Because of the negative cooperativity between the binding of cAMP and the catalytic unit to the regulatory subunit, the subunits dissociate, and the catalytic subunit is activated and phosphorylates Ca^{2+} channels in the cardiac myocyte. This causes a greater influx of Ca^{2+} during the depolarization phase of the cardiac action potential, and ultimately, an increase in the force of contraction of the heart.

This process is caused by a tiny amount of isoproterenol acting on a still smaller number of unseen receptors, yet the final response is easily perceived by touching the chest or the skin above an artery, or by visualizing the heart using ultrasound or observing it with the unaided eye during thoracic surgery, for example. Somewhere between the binding of isoproterenol to the β_1-adrenocpetor and the increase in the force of contraction of the heart, there must have been a huge amplification, otherwise it would be impossible to perceive the response so easily.

The series of steps leading to the inotropic response of the heart serves as an example of how information flows through a signaling pathway. The general principle for gain-of-function responses is that as the signal travels down the pathway, it is transformed into increasingly larger responses that are elicited by lower and lower concentrations of the agonist. The net result is that the EC_{50} value of the agonist for eliciting the final response of the pathway can be much lower (e.g., 100) than the concentration required for occupying half of the receptors. A schematic illustration of this principle is shown in Figure 7.13, in which receptor occupancy, cAMP

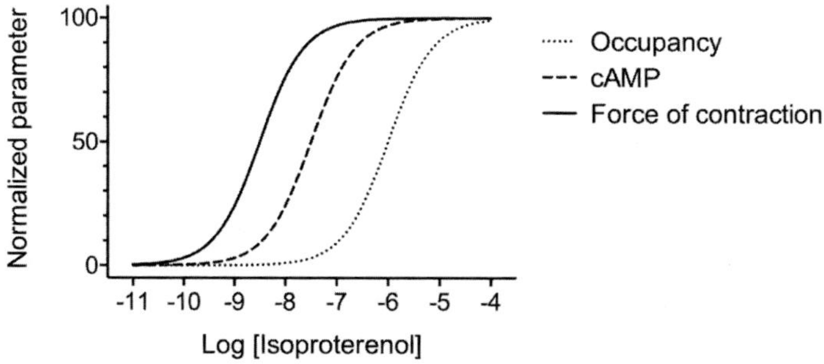

Figure 7.13: Theoretical relationships among β_1-adrenoceptor occupancy, cAMP accumulation and inotropy in isolated cardiac cells stimulated with isoproterenol.

and force of contraction are plotted against the concentration of the β-adrenergic agonist, isoproterenol. Half-maximal occupancy of the β_1-adrenoceptor in an intact cardiac myocyte containing 30 μM GTP occurs when the concentration of isoproterenol is about 1 μM ($K_D = 1\,\mu$M, $K_{\text{obs}} = 10^6$ M^{-1}), whereas the EC_{50} value for increasing cAMP levels is about 0.1 μM and that for increasing the force of contraction of the heart is only about 0.01 μM.[14]

If each step only involved amplification, then the EC_{50} value of the final response should be the same as the dissociation constant ($1/K_{\text{obs}}$) for receptor occupancy. What is it about the signaling pathway that causes both an amplification of the response and an increase in its sensitivity to isoproterenol? Signaling pathways involve a series of ligand–protein and protein–protein interactions. A ligand binds in a saturable manner to its proximate signaling protein. A signaling protein can act in two ways, as ligand or as a catalyst. When acting as a catalyst, it produces a scalar increase in the amount of the next ligand or signaling protein in the cascade. The scalar increase is usually many-fold greater than the amount of the catalytic protein. Thus, catalytic signaling proteins amplify the signal without increasing the potency of the agonist for eliciting the signal. Whatever their form, ligands bind in a saturable manner to their proximate signaling protein. If there is an excess of functional ligand (e.g., second messenger or G protein subunit), it will cause half-maximal activation of its proximal signaling protein at a concentration that is much less than half of its total concentration. Ultimately, this type of saturable interaction increases the potency of the agonist for eliciting the final response. Thus, the elements of the signaling pathway usually have one of two roles: ligand or catalyst. An excess of ligand ultimately increase the potency of the agonist for eliciting the response, and a catalyst amplifies the response.

Let me illustrate these points within the context of the inotropic pathway relating to Figure 7.13. Although the β_1-adrenoceptor is not an enzyme and G_s has enzymatic activity, the former acts catalytically and the latter acts as a ligand with short life. For when the β_1-adrenoceptor is activated by isoproterenol, it produces GTP–$G\alpha_s$ subunits at a steady rate. If the mean life of the GTP–$G\alpha_s$ complex is divided by the time it takes the receptor to convert GDP–$G\alpha_s G\beta\gamma$ into GTP-$G\alpha_s$, a measure of how many G proteins activated per activated receptor is obtained. This represents signal amplification. Each GTP–$G\alpha_s$ subunit acts as a ligand that activates adenylate cyclase. Because there is an excess of G_s relative to adenylate cyclase, this step increases the potency of isoproterenol for activating adenylate cyclase. Adenylate cyclase catalyzes the formation of a huge excess of cAMP, which acts as a ligand to activate cAMP-dependent protein kinase. The production of cAMP amplifies the signal, and its activation of the kinase increases the potency of isoproterenol for eliciting this response. The protein kinase acts catalytically to activate Ca^{2+} channels, and the latter act catalytically to allow an influx of Ca^{2+}. Ca^{2+} acts as a saturable ligand to cause depolarization, and ultimately, to cause contraction. Thus, through a series of scalar amplifications and saturable (logistic) increases in sensitivity, the final response to isoproterenol is highly amplified and exhibits high potency.

The entire sequence can be described mathematically as a series of nested logistic functions, with some multiplied by a scalar conversion factor (c_j). Each logistic function is of the form:

$$g_j(x) = \frac{xM_j}{x + K_j}. \tag{7.76}$$

In this equation, the index j simply keeps track of the different logistic functions (g), sensitivity constants (K) and maximal conversion factors (M). The variable x represents the amount or concentration of the input signaling ligand or protein, K is a sensitivity constant ($0 < K < 1$) having units equivalent to x, and M is a conversion factor representing the maximum output of the function as the input (x) approaches infinity. The lower the value of K, the more sensitivity is the system. This output simply represents the amount or concentration of the next signaling ligand or protein. The function is overly simplified because it does not take into account potential cooperativity in the transduction process. The multiplication of a logistic function by a scalar can be represented in Equation (7.76) by multiplying M_j by a constant (c_j) to yield a new conversion factor representing the conversion of the output from the logistic function into the ouput of the scalar function. The two steps, therefore, can be described by a single logistic with a larger M_j value (c_jM_j). An analogous manipulation can convert the sequence of a scalar (c_1) times a scalar

(c_2) times a logistic function into a single logistic with a maximal conversion factor equal to $c_1 c_2 M_j$. Thus, the entire signaling pathway can be modeled by a series of nested logistic functions, with the initial input function being the stimulus:

$$\text{response} = r(D) = g_4(g_3(g_2(g_1(f_\beta(D))))). \quad (7.77)$$

In this equation, $f_\beta(D_1)$ represents the stimulus function of the β_1-adrenoceptor when occupied by isoproterenol (D_1). Any series of nested logistic functions can be expressed as a single logistic. For example,

$$g_r(x) = g_4(g_3(g_2(x))), \quad (7.78)$$

in which $g_r(x)$ represents the response expressed as a function of the concentration of cAMP(x). The sensitivity constant (K_r) of g_r (x) is approximately equivalent to:

$$K_r = K_2 K_3 K_4. \quad (7.79)$$

The constants, K_2–K_4 represent the corresponding sensitivity constants for logistic functions g_2–g_4. Thus, the entire signaling pathway can be described as:

$$\text{response} = r(D) = g_r(g_1(f_\beta(D_1))), \quad (7.80)$$

in which the function g_1 defines the output of cAMP as a function of the stimulus ($f_\beta(D)$). The relationships among the final response (increase in the force of contraction), the cAMP concentration, and receptor occupancy are illustrated in Figure 7.13 as described above. In this figure, the output of each function has been normalized to its maximum. If the stimulus, $f_\beta(D)$, is normalized to its maximum, the functions in Figure 7.13 can be described by Equation (7.80) with both values of K_r and K_1 being equivalent to 0.1 and 0.01, respectively, and the observed affinity constant (K_{obs}) of isoproterenol equivalent to 10^6 M^{-1}. Ultimately, this relationship shows that the entire signaling process, which consists of a series of nested logistic equations multiplied by scalars, can be described a single operational model (i.e., a logistic function of the stimulus). Note that because occupancy and the stimulus have been normalized to their respective maximums in Figure 7.13, both of these functions are the equivalent.

The relationships shown in Figure 7.13 apply to a pathway undergoing a series of gains of function. But what happens when a receptor mediates an inhibition of the cAMP accumulation elicited by isoproterenol? The answer depends on how much the inhibitory receptor is able to oppose the effect of isoproterenol over its dynamic range. A robust inhibitory receptor would be one that could cause at least a 90% inhibition of the maximal effect of isoproterenol. This is a powerful inhibition because the relationship between cAMP and the force of contraction is such that it only requires a 10% level of cAMP to cause a half-maximal increase in the force of contraction. When cAMP levels are maximally stimulated to 100%, the inhibitory receptor would have to reduce cAMP levels to 10% (90% inhibition) and less than 1% (> 99% inhibition) to cause 50% and 90% inhibition of the inotropic response to isoproterenol, respectively.

The inhibitory effect of the M_2 muscarinic receptor is mediated by G_i inhibition of adenylate cyclase activity. The inhibition represents a logistic function of the stimulus:

$$\text{inhibition} = g_m(f_m(D_2)), \tag{7.81}$$

in which, $fm(D_2)$ represents the stimulus of the M_2 receptor when occupied by acetylcholine, and g_m represents the function that converts the M_2 stimulus into an inhibition of andenylate cyclase. It is analogous to g_1 of the β_1-adrenoceptor mechanism except that g_m inhibits activity, whereas $_{g1}$ stimulates activity. Because g_m is inhibitory, its normalized form is:

$$g_m(x) = 1 - \frac{x}{x + K_m}, \tag{7.82}$$

in which x denotes the stimulus, $f_m(D_2)$, and K_m denotes the sensitivity constant. To model the inhibitory effect of acetylcholine on isoproterenol-stimulated cAMP accumulation, the function g_1 is multiplied by g_m:

$$\text{cAMP} = g_m(f_m(D_2)) * g_1(f_\beta(D_1)). \tag{7.83}$$

This equation describes the level of cAMP as a function of the concentrations of isoproterenol and acetylcholine. The force of contraction is determined by the function $g_r(x)$. Inserting Equation (7.83) for x in the $g_r(x)$ function yields the equation describing the force of contraction (response) as a function of the concentrations of isoproterenol and acetylcholine:

$$\text{response} = g_r(g_m(f_m(D_2)) * g_1(f_\beta(D_1))). \tag{7.84}$$

This equation was used to model the concentration-response curve of the muscarinic agonist, oxotremorine-M, for inhibiting the inotropic response to a maximal concentration of isoproterenol (10 μM) (see Figure 7.14). The stimulus of oxotremorine-M was normalized to a maximum value of 1 (100%), like that of isoproterenol. The K_{obs} value of oxotremorine-M was assigned a value of 10^5 M^{-1}. The values of the other parameters were the same as those used for the simulation in Figure 7.13. For oxotremorine-M to cause at least 90% inhibition of the inotropic response to isoproterenol, it was necessary to set K_m to 0.01 or lower. The result of this simulation is illustrated in Figure 7.14, which shows a plot of the concentration-response

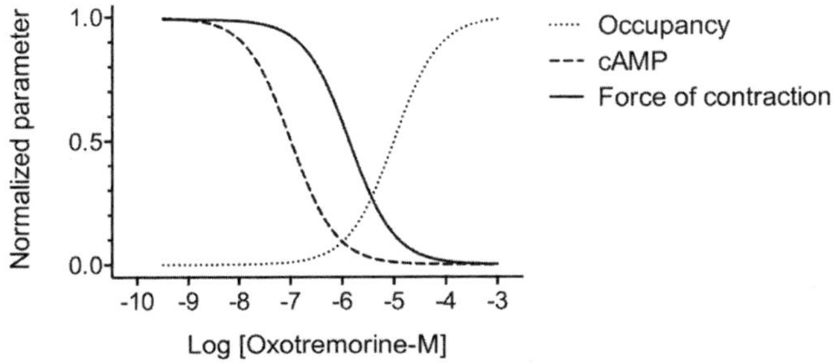

Figure 7.14: Theoretical relationships among M_2 muscarinic receptor occupancy by oxotremorine-M, cAMP accumulation elicited by a maximally effective concentration of isoproterenol and inotropy in isolated cardiac cells.

curve of oxotremorine-M for inhibiting cAMP levels and the force of contraction of the heart. Oxotremorine-M is more potent at inhibiting cAMP levels than at inhibiting the response. This behavior is opposite to that shown for isoproterenol in Figure 7.13. The inverted relationship between cAMP levels and response is an inevitable consequence of an inhibitory mechanism acting at an early point in a signaling pathway that exhibits a gain in sensitivity downstream from the point of inhibition. The proximal M_2 receptor signaling mechanism has to be very effective to oppose the downstream increase in sensitivity.

From an engineering perspective, an efficient inhibitory system should interrupt the stimulatory signal early before it has a chance to be amplified. Given that evolution bequeathed organisms with a modular M_2 receptor signaling system, based on a neurotransmitter-receptor-G protein transduction mechanism, inhibition of adenylate cyclase is the earliest point in the pathway that the M_2 receptor can oppose the β_1-adrenoceptor signal. To cause a 90% inhibition of the final response, the M_2 receptor transduction mechanism $(g_m(f(D_2)))$ must be at least ten-fold more sensitive than the β_1-adrenoceptor system. Perhaps this explains why G proteins with inhibitory functions (e.g., G_i and G_o) are usually expressed in such high abundances relative to G_s and G_q.

Effect of the sensitivity of the assay system on agonist activity

For any given assay, the EC_{50} and $E_{\max}$ values of an agonist depend on the affinity and efficacy of the agonist and the sensitivity of the assay. This principle applies to different experimental preparations and to the different steps within the signaling pathway of a single system.

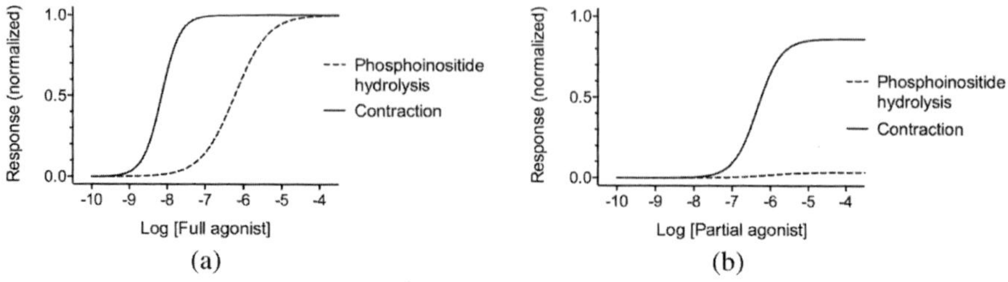

Figure 7.15: Simulation of the effects of full (a) and partial (b) agonists on contraction and phosphoinositide hydrolysis in smooth muscle. The theoretical curves were generated using the operational model (Equation (7.10)). The values of the maximum response of the system (M_{sys}, 1) and of the observed affinity constant (K_{obs}, 10^6 M^{-1}) were the same for all of the curves. The efficacy values of the full and partial agonists were 0.7 and 0.014, respectively. The parameter K_E in the operational model was assigned values of 0.001 and 1 for the contractile and phosphoinositide responses, respectively. The corresponding values for the transducer slope factor (m) were 1.8 and 1.

Another example of this latter phenomenon is illustrated in Figure 7.15. This diagram shows the concentration-response curves of two agonists for eliciting phosphoinositide hydrolysis and contraction of smooth muscle. Figures 7.15a and 7.15b show the behaviors of a full and a partial agonist, respectively. The curves were generated using the operational model with the observed affinity constants of the two agonists set to the same value ($K_{obs} = 10^5$ M^{-1}) and the observed efficacy of the agonist in Figure 7.15a 50-fold greater than that in Figure 7.15b. In smooth muscle, the potency of agonists for eliciting contraction is usually far greater than that for eliciting phosphoinositide hydrolysis, so the K_E parameter in the operational model was set to 0.005 and 1 for these two responses, respectively. The responses of the less efficacious agonist were normalized to the E_{max} values of the more efficacious agonist shown in Figure 7.15a. For a given agonist, observed affinity and efficacy are the same for both responses. The greater activity in the contractile assay is caused by the greater amplification of the signal for this response. This is simulated mathematically by a lower K_E value, which represents an increase in the sensitivity of the signaling pathway.

One might conclude that because the less efficacious agonist does not elicit a significant phosphoinositide response, then this pathway must not be involved in the contractile response. This conclusion does not survive close scrutiny, however. In the case of the more efficacious agonist, the EC_{50} value for contraction is only about 1/100th the value of the observed dissociation constant (K_D, i.e., $1/K_{obs}$). This means that the more efficacious agonists elicits a half-maximal response when receptor occupancy is only about 0.5%. Thus, whatever the initial signaling step, we know that a tiny amount of it is sufficient to cause a substantial contraction, and it would not be unreasonable to assume that it is immeasurable at low agonists concentrations that elicit measurable contractions. In the case of the more efficacious agonist, the phosphoinositide response starts to become detectable when the agonist concentration increases to about 30 nM. Much lower concentrations of the agonist elicit robust contractile responses, however. Thus, it is not unreasonable that a partial agonist requiring full receptor occupancy to elicit a maximal contraction would be unable to elicit a detectable phosphoinositide response.

The same type of behavior often occurs when different signaling responses are measured in the same cell, like cAMP accumulation, extracellular-receptor-activated kinase (ERK) activation and recruitment of arrestin. cAMP accumulation is a highly amplified response so many agonists are able to elicit this response, even if they have very low efficacy. Recruitment of arrestin involves the phosphorylation of the active state of the agonist-receptor complex followed by the binding of arrestin to the receptor. The response is the accumulation of the arrestin-receptor complex; hence, it is proportional to receptor occupancy and does not exhibit amplification. Partial agonists that elicit a robust cAMP response would not be expected to elicit significant recruitment of arrestin.

Analogous behavior occurs when a group of agonists are tested for their ability to elicit the same response in different tissues. Often, the signaling pathway may undergo different levels of amplification in different tissues. Any number of factors could be responsible for this including the density of receptors, *G* proteins, adenylate cyclase, phospholipase-Cβ or voltage-dependent ion channels or even the shape of the cell. Figure 7.16 shows the theoretical concentration-response curves of three agonists, A, B and C, for eliciting a response in three different tissues. The affinity constants and relative efficacies of the agonists are 10^7, 10^5 and 10^5 M^{-1} and 0.1, 1 and 0.01, respectively. In the highly sensitive tissue (Figure 7.16a) all three agonists behave as full agonists and elicit the maximal response of the system. The ratio of EC_{50} values of agonists A, B and C expressed relative to that of B are 0.1, 1 and 100. The corresponding RA_i values represent the reciprocal of these potency ratios (10, 1 and 0.01). In the less-sensitive tissue (Figure 7.16b), A and B behave as full agonists, and C behaves as a partial agonist. In the least-sensitive tissue (Figure 7.16c), both A and B behave as a partial agonists and C as an antagonist. This variation in behavior occurs even though the affinity and efficacy of each agonist is the same across the different systems.

This is a convenient time to introduce the term "spare receptors". The term does not refer to a type of receptor, but rather, to a condition so a better locution is "receptor reserve". The receptor reserve for a response depends on the agonist, the

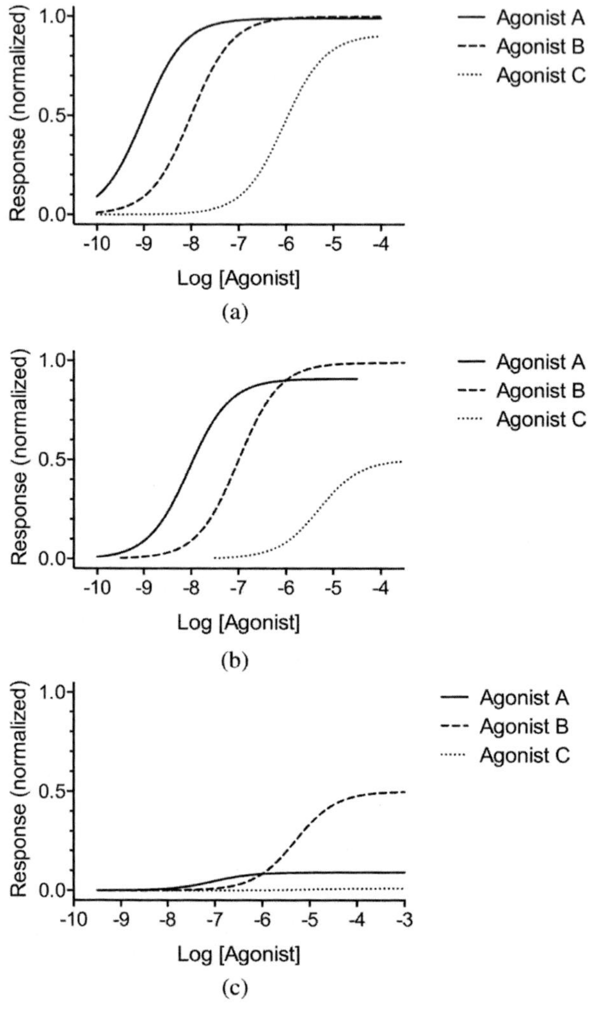

Figure 7.16: Simulation of the effects of three agonists (A, B and C) on a given response in three different tissues exhibiting high (a), medium (b) and low (c) sensitivity, respectively. The theoretical curves were generated using the operational model (Equation (7.10)). The values of K_{obs} for agonists A, B and C were 10^7, 10^5 and 10^5 M^{-1}, respectively. The corresponding efficacy values were 0.1, 1.0 and 0.01. These values were constant for the three different tissues. The values of the maximum response of the system (M_{sys}, 1) and the transducer slope factor (m, 1) were also constant for the three tissues. The sensitivity constant (K_E) was assigned values of 0.001 (a), 0.01 (b) and 1.0 (c).

response and the level of response being measured. A receptor reserve exists whenever the fractional level of response is greater than the fractional level of receptor occupancy that elicited it. For example, if agonist B elicits a 50% response at only 0.5% receptor occupancy, then there is a receptor reserve for B for this level of response. If agonist A elicits the same level of response at 5% receptor occupancy, there is also receptor reserve for agonist A, but it is not as great as that for agonist B. Also there is no receptor reserve for the E_{max} response of a partial agonist, whereas there usually is for a full agonist.

Is the inactivity of a partial agonist caused by the insensitivity of the system?

Given that a group of agonists can exhibit a range of behaviors when their concentration-response curves are measured across different assay systems for the same receptor, how do we determine if their receptor-dependent properties (observed affinity and efficacy) differ across the different systems? It is conceivable that in one assay system, the agonist may select a unique conformation of the active state of the receptor that elicits the response. To address this question, the observed affinities and efficacies can be estimated using the procedures described earlier in this chapter. A simpler approach, however, is to estimate the RA_i values of the agonists from their respective concentration-response curves. The RA_i value (relative estimate of the product of affinity and efficacy) is very likely to be the same across different systems if the agonist affinity and efficacy are the same. The only exception would be if the reduction in affinity in one assay were associated with a corresponding increase in efficacy, such that the product of the two remains the same.

If a test agonist elicits a response in one assay (system 1) but not in another (system 2), then it may be that the loss of function is caused by a decrease in the sensitivity of the response. This can be determined by measuring the activity of a more efficacious agonist (standard) in both systems. If the activity of the standard decreases in system 2 to an extent that can account for the inactivity of the test agonist, then there is no evidence that either agonist exhibits a change in efficacy across the two systems.

An example of this type analysis is shown in Figure 7.17, which illustrates the effects of carbachol and McN-A-343 on M_2 muscarinic receptor-mediated inhibition (Figure 7.17a) and stimulation (Figure 7.17b) of cAMP accumulation. The data in Figure 7.17b were measured in pertussis toxin-treated cells in the presence of forskolin.[12] Pertussis toxin uncouples the M_2 receptor from G_i to prevent inhibition of adenylate cyclase, and forskolin amplifies a weak G_s-mediated stimulation

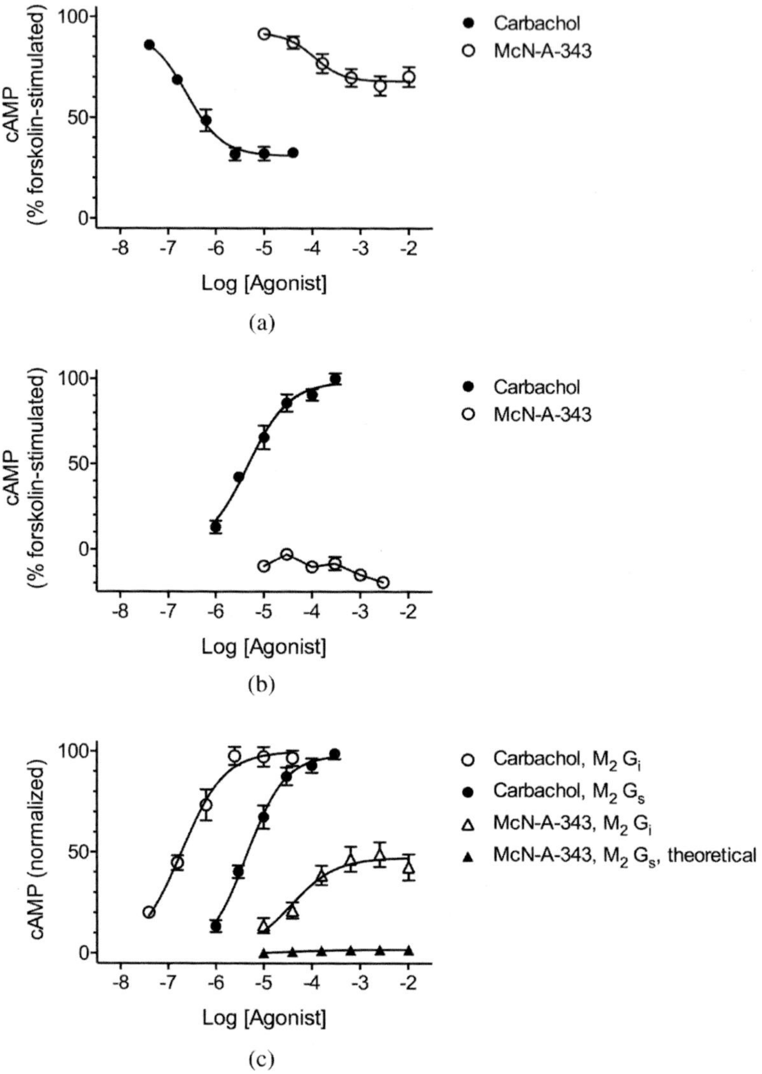

Figure 7.17: Effects of carbachol and McN-A-343 on forskolin (10 μM) stimulated cAMP accumulation in control (a) and pertussis toxin-treated (b) Chinese hamster ovary cells stably expressing the human M_2 muscarinic receptor. The data are from Griffin *et al.*[12]. (c) Analysis of the effects of carbachol and McN-A-343 on cAMP accumulation using the operational model. The τ values of carbachol were estimated in control ($\tau_{control}$) and pertussis toxin-treated tissue (τ_{PTX}). The theoretical τ value for McN-A343 in pertussis toxin treated tissue was calculated by multiplying its τ value in control cells by the ratio ($\tau_{PTX}/\tau_{control}$) estimated for carbachol. This value was then used to calculate its theoretical concentration-response curve.

of adenylate cyclase by the M_2 receptor. Carbachol elicits both responses, whereas McN-A-343 only elicits the G_i response.

The data for each response were analyzed simultaneously with other agonists to estimate the maximum response of the system (M_{sys}) for each assay using the operational model and the procedures described earlier in this chapter. The responses to carbachol and McN-A-343 were normalized relative to M_{sys} in each system and plotted in Figure 7.17c. For the G_i response, the data are expressed as percent inhibition. Clearly, carbachol is less potent at stimulating cAMP accumulation. To estimate the difference in the sensitivities of the two assays, the operational model is fitted to the concentration-response curves of carbachol in the two systems, sharing the estimate of K_{obs} and the system parameters (K_E and m) and obtaining a unique estimate of τ for each curve. Depending upon the nature of the two assays, it may be necessary to obtain a unique estimate of m for each assay. The ratio of the τ value in the G_s assay expressed relative to that of the G_i assay is equivalent to the loss sensitivity, assuming that the affinity and efficacy of carbachol are unchanged between the two assays. This loss in sensitivity is manifest in the operational model as a corresponding increase in the K_E parameter. An increase in K_E causes a reduction in τ (see Equation (7.11)). Because carbachol behaves as a full agonist in both assays, this ratio can be estimated easily from the ratio of the EC_{50} value for inhibition of cAMP accumulation divided by that for stimulation, instead of estimating the actual τ values. If there were no change in the affinity and efficacy of McN-A-343 between the two assays, then its predicted concentration-response curve in the stimulatory assay would be equivalent to the theoretical curve shown in Figure7.17c, which shows no response. This curve was generated using the operational model with the parameters m and K_{obs} the same as those for inhibition of cAMP accumulation but with the value of τ reduced by a factor of equivalent to the difference in the sensitivities of the two assays. Because the operational model predicts no response for McN-A-343 in the stimulatory assay, there is no evidence that McN-A-343 exhibits greater efficacy for inhibiting cAMP accumulation.

Influence of cytosolic GTP on agonist responses

An important cofactor for a GPCR is GTP, for without it there is no G protein-mediated response. As the concentration of GTP increases, the observed intrinsic efficacy of the agonist-receptor complex increases, whereas observed affinity decreases. For a given change in the concentration of GTP, these two effects oppose each other so that there is little change in receptor-activation at low agonist-receptor occupancy. Because an increase in sensitivity often occurs as the receptor-activation signal travels down the signaling pathway, agonists elicit many downstream responses at low levels of receptor occupancy. It follows that an alteration in the concentration of GTP may have little effect on the concentration-response curve of an agonist.

Recall that the amount of the active state of the agonist-receptor complex associated with the GDP-bound form of the *G* protein (quaternary complex, DR_s^*GX) is a measure of receptor activation at a GPCR as far as *G* protein-mediated responses are concerned (see Chapter 6). In the absence of GTP, no guanine nucleotides occupy *G* proteins. In the presence of GTP, a steady state develops in which most *G* proteins are bound with guanine nucleotide in the form of GDP, because GTP hydrolysis is much faster than the rate of dissociation of GDP from the *G* protein. Agonist-receptor activation is proportional to how well the agonist persuades the receptor to associate with the *G*-GDP complex to form the quaternary complex consisting of agonist-receptor-*G* protein-GDP. This is the intermediate for the exchange of GTP for GDP. As shown in Figure 7.18a, there is little or no quaternary complex when the concentration of GTP is very low, but as the concentration of GTP increases the maximum of the quaternary-complex function increases and its potency decreases.

To simulate how this affect of GTP alters the response to an agonist, the quaternary complex function can be substituted for the stimulus in the operational model. That is, we can substitute Equation (6.69) from Chapter 6, for *S* in Equation (7.5) of this chapter. The result is a function that expresses the response in terms of the concentrations of agonist and GTP (independent variables) as well as the system parameters (K_E and m) and the various microscopic constants of the ternary complex model with guanine nucleotide. The results of this simulation are shown in Figure 6.18b. The most obvious result is that a huge change in the concentration of GTP has little or no effect on the concentration-response curve. It is only when the concentration of GTP falls to very low levels that the maxim of the response declines. Thus, agonist-receptor signaling through GPCRs is buffered against the wide variation in the intracellular concentration of GTP.

Figure 6.18c shows how the variation in the concentration of GTP affects receptor occupancy by the agonist. As the concentration of GTP increases from 3.32 μM (log molar, 5.5) to 10 mM (log molar, 2), the occupancy curve shifts to the right 100-fold. The maximal shift in the curve caused by the highest concentration of GTP (10 mM) is 500-fold relative to control. Thus, the concentration-response curve of a highly efficacious agonist for eliciting a highly sensitive response is unaffected by a variation in the intracellular concentration of GTP over a wide range.[15]

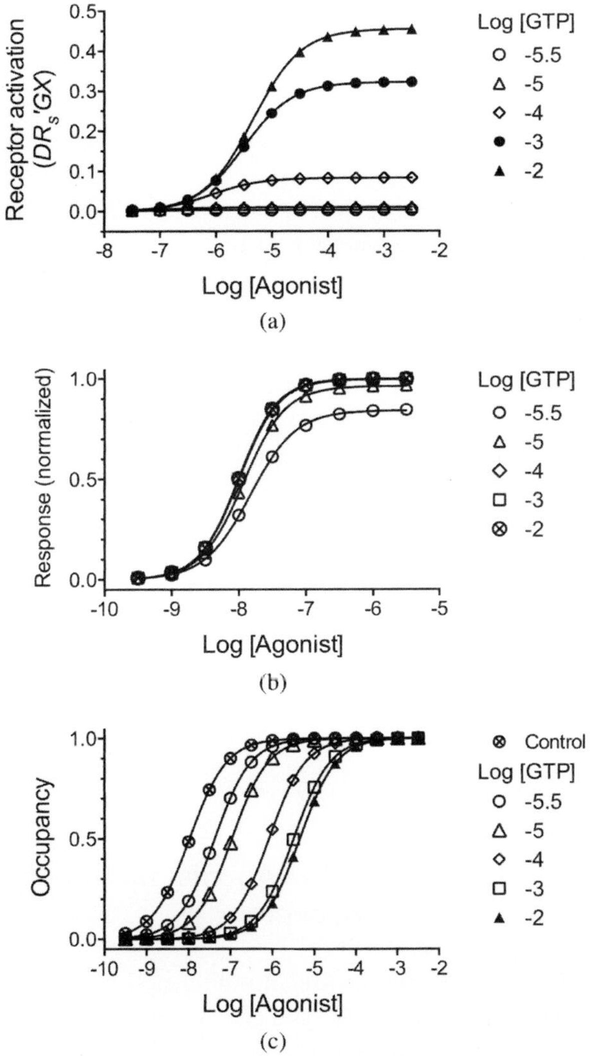

Figure 7.18: Simulation of the influence of various concentrations of GTP on the ability of a highly efficacious agonist to activate a receptor population (a), to elicit a highly sensitive downstream response (b) and to occupy the receptor population (c). The quaternary complex model described in Chapter 6 was used to simulate receptor occupancy (c) and the fraction of the quaternary complex in the active state (a). The activation function was then substituted into the operational mode to generate the theoretical concentration-response curves shown in (b).

Recapitulation of the determinants of agonist selectivity

As illustrated earlier in this section the selectivity of an agonist for eliciting different responses through a given GPCR depends on the receptor and its downstream signaling pathway.

When the response to the agonist is measured at different points in a given signaling pathway, there is a tendency for the potency of the agonist to increase the more distal in the signaling pathway the response is measured. It may be difficult to detect a response to an agonist with very low efficacy if an initial signaling step is measured. Yet the same agonist may elicit a substantial response downstream. These same principles also apply to the same response in different cells and tissues. Thus, low efficacy in the face of tissue variation in signal amplification and sensitivity provides a mechanism for the selectivity of agonists with low efficacy. This mechanism is entirely dependent on the signaling pathway.

If the latter were the sole mechanism for selectivity at a single receptor, then a given agonist would exhibit the same relative estimates of affinity and efficacy at all responses mediated by the same receptor. There is abundant data, however, that many agonists do not follow this simple prediction. The likely explanation is that there are different conformations of the active state of the receptor that exhibit selectivity for different different signaling proteins. This mechanism of selectivity is termed biased agonism or ligand-directed signaling, and it will be covered in detail in Chapter 9.

Measuring the Selectivity of the Agonist for the Active and Inactive States of the Receptor

The ability of an agonist to activate a GPCR is ultimately determined by its microscopic affinity constants for active (K_b) and inactive (K_a) states of the receptor (see Chapter 6). While these individual constants have been estimated for ligand-gated ion channels through analysis of single-channel recordings, it is difficult to glean the same information from a downstream response of a GPCR unless there is constitutive activity or allosteric modulation. Remarkably, it is possible to estimate a relative measure of K_b, K_a and the ratio, K_b/K_a, from the concentration-response curves of agonists for eliciting responses through GPCRs.

One way to explain how this is done is through simulation based on the receptor activation function developed in Chapter 6 (Equation (6.69)). If this function is substituted for the stimulus in the operational model (S in Equation (7.5)), the resulting equation describes the response as a function of the agonist concentration (independent variable) and the microscopic constants for the active and inactive states of the receptor (parameters). The theoretical curves generated with the equation can then be analyzed to illustrate how the RA_i value and the observed affinity and efficacy of the agonist are related to the affinity constants of the receptor states.

This approach is used in the following sections to explain how observed intrinsic efficacy is related to the ratio K_b/K_a and how RA_i is related to K_b. I will also describe how to estimate a relative measure of K_a. For the simulations, the amount of the active state of the receptor in the form of quaternary complex ($DR_s{}^*GX$) was simulated using Equation (6.69) in Chapter 6. This equation requires the further substitution of equation 6.48 for K_{DRGX}. The latter equation is defined in terms of the level 2*b* parameters of the ternary complex model in the presence of guanine nucleotide, and these must be defined by the microscopic constants of receptor states as described in Equations (6.55), (6.60)–(6.66). After all of these substitutions are made, it is possible to express $DR_s{}^*GX$ as a function of the agonist's microscopic constants. The section entitled "Conformational analysis of the ternary complex model with guanine nucleotide" in Chapter 6 explains these calculations in detail.

Efficacy and selectivity for the active state

As proposed in Chapter 6, the efficacy of an agonist can be defined in two ways depending on the level of analysis. Observed efficacy (ε) refers to the maximum of the receptor activation function. At a deeper level of analysis, efficacy is a measure of the selectivity of the agonist for the active state of the receptor. This latter measure of efficacy is defined as the ratio of the microscopic affinity constant of the active state divided by that of the inactive (K_b/K_a). Using the simulation technique just described, it is possible to observe how the ratio K_b/K_a affects the observed affinity (K_{obs}) and intrinsic efficacy (ε) of the concentration-response curve.

An example of such a simulation is shown in Figure 7.19a. This figure illustrates how a change in the K_b/K_a ratio affects the concentration-response curve of the agonist. The affinity of the ground state of the receptor (K_a) was set to a constant value of 10^4 M^{-1}. As the K_b/K_a ratio increases from a low value of about 300 (log K_b/K_a, 2.5), the maximal response and potency increase until the maximum response is equivalent to the maximal response of the system ($E_{max} = M_{sys}$). A further increase in K_b/K_a causes an increase in potency only. The values of the other parameters for this simulation are given in the figure legend.

To determine how observed affinity and intrinsic efficacy are related to microscopic constants, each curve in panel (a) can be simulated after a fraction of the

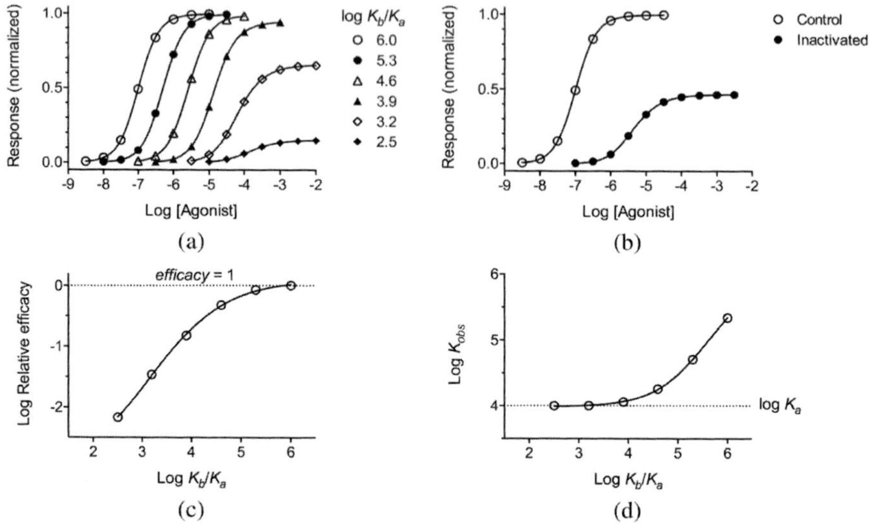

Figure 7.19: Influence of the selectivity of an agonist for the active state of the receptor on the concentration-response curve for a downstream response (a), efficacy (c) and the observed affinity constant (d). (a) Simulation of a series of concentration-response curves for a group of agonists having the indicated ratio of microscopic affinity constants for active and inactive receptor states (K_b/K_a). The microscopic affinity constant for the inactive state was assigned a log value of 4 for all of the agonists. Using these microscopic constants, the concentration response curves were generated as described in the legend to Figure 7.18. The parameters of the operational model were M_{sys}, 1; log K_E, -2.0 and m, 1.5. (b) The concentration response curve of the most efficacious agonist was simulated for control conditions as in panel (a) and also after inactivation of 98% of the receptor population. (c) The log of the efficacy of each agonist is expressed relative to that of the most efficacious agonist is plotted against the selectivity for the active state (i.e., log K_b/K_a). (d) The observed affinity constant of each agonist is plotted against log K_b/K_a. The log affinity constant of each agonist for the inactive state of the receptor (log K_a) was assigned the same value (4), which is indicated by the horizontal dotted line.

receptor population has been inactivated. This is shown in panel (b) for the single condition of log $K_b/K_a = 6$. The consequence of reducing the active receptor population to only 1% is about a 40-fold increase in EC_{50} and a 50% depression in the $E_{\max}$ for this particular simulation. Analysis of these data using the operational model and the method described earlier in this chapter yields estimates of 5.34 and 1.66, for log K_{obs} and log τ, respectively. This τ value represents that of the control curve (i.e., that simulated without receptor inactivation). This process was repeated for each curve in panel (a), each time simulating a degree of receptor inactivation causing about a 50% depression in $E_{\max}$ and analyzing the simulated "control" and "inactivated" conditions simultaneously with the operational model to estimate log K_{obs} and log τ values of the control curve. The τ values of all the control curves were normalized relative to the one with the highest K_b/K_a ratio (i.e., log $K_b/K_a = 6$). The normalized τ value is equivalent to relative efficacy. A plot of relative efficacy and of K_{obs} against the K_b/K_a ratio is illustrated in panels (c) and (d) respectively.

The plot in panel (c) shows that relative efficacy increases with K_b/K_a over a range of values and then reaches a plateau corresponding to 100% receptor activation as K_b/K_a increases further. This relationship shows that relative efficacy is proportional to K_b/K_a at the low end of the K_b/K_a scale shown in Figure 6.19c. In thinking about how GPCRs work, it seems that efficacy may often be proportional to K_b/K_a for highly efficacious agonists because the concentration of GTP in the cytosol is probably high relative to its affinity for the G protein in many types of cells. This means that the level of formation of the quaternary complex ($DR_s{}^*GX$) is probably much less than submaximal, even for the most highly efficacious agonists. If so, the relative efficacy would be proportional to K_b/K_a for most agonists. Of course, this may not apply in cells with low intracellular concentrations of GTP or in experimental assays on broken cells where the concentration of GTP can be manipulated to lower levels (e.g., adenylate cyclase measurements, [^{35}S]GTPγS binding).

Panel (d) shows K_{obs} plotted against K_b/K_a. Over the range where K_b/K_a is proportional to efficacy, K_{obs} is approximately equal to K_a or a little greater. As K_b/K_a increase further, K_{obs} increases and ultimately exhibits a proportional relationship with K_b/K_a.

Thus, the simulation in Figure 6.19 shows that when the concentration of GTP in the cytosol is high, the observed intrinsic efficacy of most agonists is proportional to the selectivity of the agonist for the active state of the receptor (K_b/K_a) and K_{obs} is approximately equal to K_a.

Although the efficacy of an agonist relative to another can be estimated, the absolute quantity of the scalar, intrinsic efficacy (ε), is difficult to measure unless there is allosteric modulation. The simulations in Figure 6.19c suggest how it might be estimated for a hypothetical group of agonists with unusually high efficacy, particularly in a system having a low concentration of GTP. The reason why the τ value reaches a plateau as K_b/K_a increases is because the receptor becomes maximally activated when the agonist K_b/K_a ratio exceeds some high level. It follows that the value of ε must be 1 in this case, because ε represents the fractional amount of the agonist-receptor complex in the active state. Another way to conceptualize ε is that it equals the ratio of activated receptors occupied by the agonist divided by the sum of active and inactive receptors occupied by the agonist. This ratio is equivalent to one when every receptor occupied by the agonist is activated. Panel (c) shows that the only condition where agonists with varying K_b/K_a ratios can exhibit the same relative efficacy is when they all have a sufficiently large K_b/K_a ratios to activate the receptor nearly maximal. It seems unlikely that a structurally diverse group of agonists would exhibit the same K_a or K_b values, and hence, the same K_b/K_a ratios. Thus, if a group of highly efficacious agonists with diverse structural variation all exhibit the same relative efficacy, it would seem likely that they all maximally activate the receptor, and if so, their intrinsic efficacy values would be equivalent to 1. Normalizing the efficacy of a test agonist relative to that of an agonist with an intrinsic efficacy of 1 yields the intrinsic efficacy of the test agonist. Thus, in this hypothetical situation, the efficacy values of agonists could be measured.

RA$_i$ and the microscopic affinity constant of the active state

A common way of characterizing the activity of an agonist at a receptor is to measure its concentration-response curve for eliciting a downstream response. When this is done for a series of agonists, data similar to those shown in Figure 7.20a are obtained. These were simulated using the technique described above which involves substituting the receptor-activation function (Equation (6.69)) for the stimulus (S)

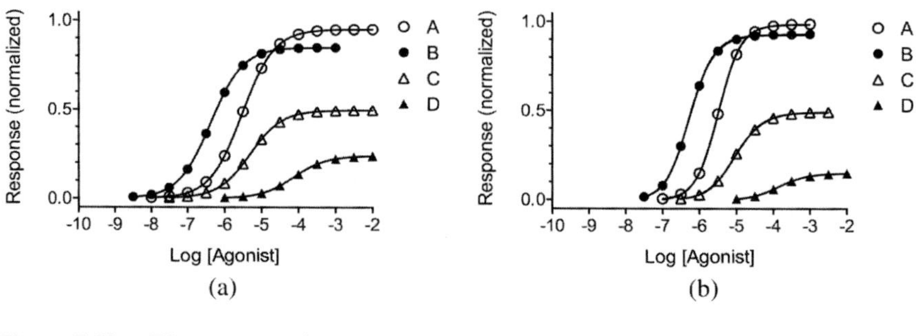

Figure 7.20: The concentration-response curves for eliciting a response downstream in the signaling pathway of a GPCR exhibiting a transducer slope factor (m) of 1 (a) and 1.5 (b). The data were simulated to illustrate the estimation of agonist RA_i values. The simulations were run using the quaternary complex model described in Chapter 6. The following parameter values were used in the simulations. The log affinity constants of agonists A–D for the active state of the receptor (log K_b) were 8.5, 9.3, 8 and 6.5. The corresponding values for the inactive state of the receptor (log K_a) were 4, 5.5, 5 and 4. The log value of the isomerization constant of the unoccupied receptor (log K_q) was -5.0. The values of M_{sys} and K_E were 1 and 0.01, respectively. The results of the regression analysis are summarized in Tables 7.1 and 7.2 for panels (a) and (b), respectively.

Table 7.1: The parameters of the agonist concentration-response curves illustrated in Figure 7.20.

Agonist	Efficacy	logK_{obs}	Hill Slope	logRA_i	$\log \frac{E_{max}EC_{50-A}}{E_{max-A}EC_{50}}$	$\log \frac{E_b}{K_{b-A}}$
A	0.19	4.22	1	0.0	0.0	0.0
B	0.056	5.55	1	0.80	0.80	0.80
C	0.0098	5.01	1	−0.50	−0.50	−0.50
D	0.0031	4.01	1	−2.00	−2.00	−2.00

Table 7.2: The parameters of the agonist concentration-response curves illustrated in Figure 7.21.

Agonist	Efficacy	log K_{obs}	Hill Slope	log RA_i	$\log \frac{E_{max}EC_{50-A}}{E_{max-A}EC_{50}}$	$\log \frac{K_b}{K_{b-A}}$
A	0.19	4.22	1.42	0.0	0.0	0.0
B	0.056	5.55	1.32	0.80	0.75	0.80
C	0.0098	5.01	1.16	−0.50	−0.75	−0.50
D	0.0031	4.01	1.06	−2.00	−2.44	−2.00

in the operational model (Equation (7.5)). The parameter values for the simulations are given in the legend to Figure 7.20.

For each agonist, observed efficacy (ε) was calculated as the maximum of the receptor activation function and K_{obs} as the reciprocal of the EC_{50} for causing half-maximal receptor activation. These estimates are listed in Table 7.1 for each agonist. The transducer-slope factor in the operational model (m) was assigned a value of 1 for this simulation. Consequently, the slope factors (Hill slopes, n) of the concentration-response curves are all equivalent to one, and these estimates are also listed in Table 7.1. The product of the observed affinity and intrinsic efficacy of each agonist expressed relative to that of agonist A is also listed in the table. This latter estimate is known as the relative intrinsic activity (RA_i), and it can be estimated by analyzing all of the curves simultaneously with the operational model as described earlier in this chapter. These RA_i estimates are also listed in Table 7.1. When the slope factors of the concentration-response curves are equivalent to one, it is possible to estimate RA_i from the E_{max} and EC_{50} values of each agonist using Equation (7.56). This estimate is also shown in Table 7.1. Finally, the ratio of the microscopic affinity constant of each agonist for the active state of the receptor expressed relative to that of agonist A is given in Table 7.1. Inspection of the tabular values shows that this ratio is equivalent to the RA_i value. Thus, RA_i is a relative estimate of the microscopic affinity constant of the agonist for the active state of the receptor.

This principle still holds when the slope factors of the concentration-response curves differ from one. An example of this situation is shown in Figure 6.20b. These simulations were made with the same parameters as those used in Figure 6.20a except that the transducer slope factor (m) in the operational model was set to 1.5. This caused changes in the $E_{\max}$ and EC_{50} values of the concentration-response curves, but not of the estimates of observed affinity, efficacy and the RA_i value because the latter are receptor-dependent parameters (see Table 7.2). In this situation, however, there is a modest discrepancy between the estimation of the quantity, $E_{\max-t}EC_{50-s}/E_{\max-s}EC_{50-t}$, and the RA_i value for partial agonists. Nonetheless, there is agreement between RA_i and the relative estimate of K_b. Thus, the RA_i value is equivalent to:

$$RA_i = \frac{K_{b-t}}{K_{b-s}}. \tag{7.85}$$

In this equation, K_{b-t} and K_{b-s} denote the microscopic affinity constants of the test and standard agonists for the inactive state of the receptor, respectively.

If an agonist elicits a downstream response through a particular signaling pathway, then the RA_i values of the agonist for eliciting responses at different points in the signaling pathway should be the same. An example of this relationship is shown in Figure 7.21. In panel (a), the concentration-response curves of selected agonists for eliciting contraction of the guinea pig ileum are shown. It is known that the M_3 muscarinic receptor elicits this response through a G_q mechanism. In panel (b), the concentration-response curves of the agonists for eliciting phosphoinositide hydrolysis in CHO cells expressing the M_3 muscarinic receptor are illustrated. The RA_i values of the agonists were calculated for each response relative to carbachol, and these estimates are illustrated in panel (c). The agreement between the two sets of data indicate that the agonists exhibits the same affinity for the active state of the M_3 receptor that elicits both responses, which is not unexpected because both are mediated through G_q.

The RA_i value is entirely receptor-dependent because it is a relative measure of the affinity constant of the agonist for the active state of the receptor. This is the conformation of the receptor that engages the G protein or other signaling protein, and hence, the RA_i value is useful in the characterization of agonist bias or ligand-directed signaling as described in Chapter 9. The RA_i values of some agonists change depending on the signaling pathway, which might appear like evidence that the RA_i value is at least partially depending on the system. But this argument can be turned on its head because different active states of the receptor are required for the

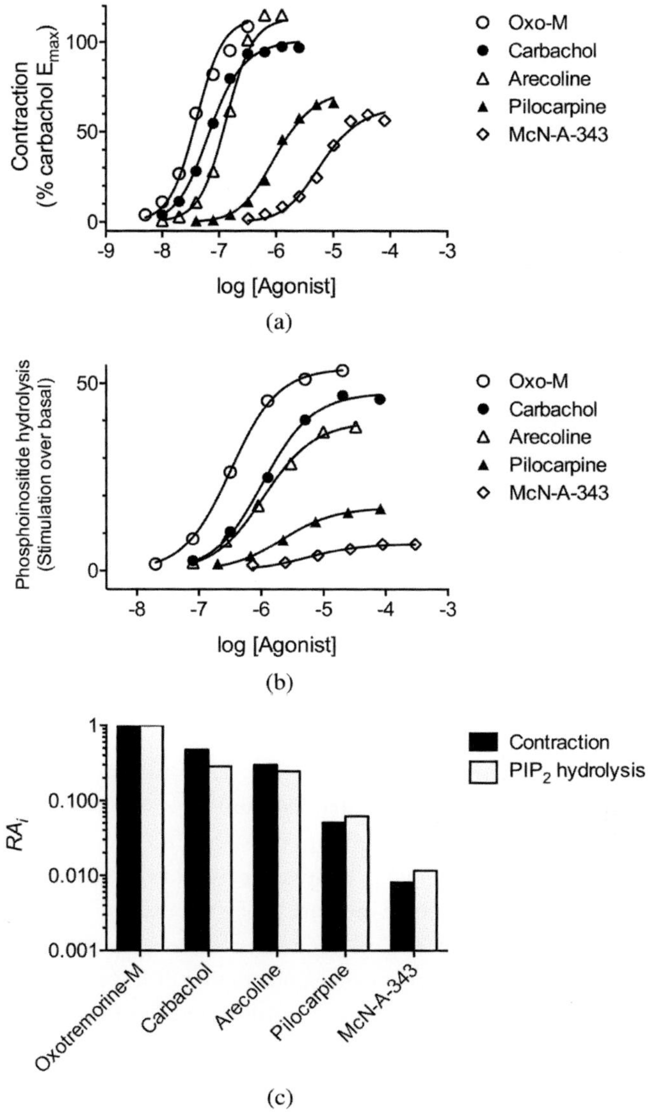

Figure 7.21: Comparison of the RA_i values of a group of muscarinic agonists estimated from their contractile activity on the guinea pig ileum (a) and for eliciting phosphoinositide hydrolysis in Chinese hamster ovary cells stably expressing the M_3 muscarinic receptor (b). The estimates of the RA_i values are illustrated in panel (c). The data are from Ehlert *et al.*[7]

phenomenon of ligand-directed signaling. Thus, different G proteins act as sensors for detecting the selectivity of an agonist for different active states of the receptor. A practical way of stating this point within the context of drug discovery is that different cellular signaling assays can be engineered into high throughput systems to detect different active states of the receptor.

RI_i and the microscopic affinity constant of the inactive state

Conventional receptor theory focuses on observed affinity (K_{obs}) and the fraction of the ligand-receptor complex in the active state (observed intrinsic efficacy). This enables the estimation the RA_i value and the observed affinity and intrinsic efficacy of the agonist-receptor complex through analysis of a downstream response. To fully account for the activity of an orthosteric agonist, however, it is necessary to estimate the microscopic affinity constant of the inactive state of the receptor.

This might seem unnecessary because, in the intact cell, GPCRs typically function in the presence of a high concentration of GTP, which reduces the observed affinity constant of the agonist to a value that is probably not much different from that of the microscopic affinity constant of the inactive state of the receptor (K_a). Nonetheless, it is useful to obtain a more precise estimate of K_a if only to prove that K_{obs} is a good approximation of K_a.

When there is constitutive activity, however, having a relative estimate of K_a is essential for analysis of the actions of agonists and inverse agonists, and it is even possible to estimate the microscopic affinity constants, K_a and K_b, in units of M^{-1} under these conditions and not just in relative terms. For these reasons, I describe a method for the estimation a relative measure of the microscopic affinity constant of the inactive state of the receptor. This parameter is known as the intrinsic relative inactivity (RI_i), and it has properties analogous to that of RA_i.

The relationship between receptor occupancy and the fraction of the ligand-receptor complex in the active and inactive states is shown in Figure 7.22 for an agonist having an observed affinity constant of 10^5 M^{-1}. The graph shows the latter parameters plotted against the log concentration of the agonist. The amount of the two types of agonist-receptor complexes (active and inactive) are expressed as a fraction of total amount of receptors on a normalized occupancy scale. Observed intrinsic efficacy (ε) represents the fraction of the population of agonist-receptor complexes in the active state (DR_s^*), which is also equivalent to the fractional amount of the agonist in the active state at 100% receptor occupancy. The observed inactivity (ε_i) represents the fraction of the population of agonist-receptor complexes in the

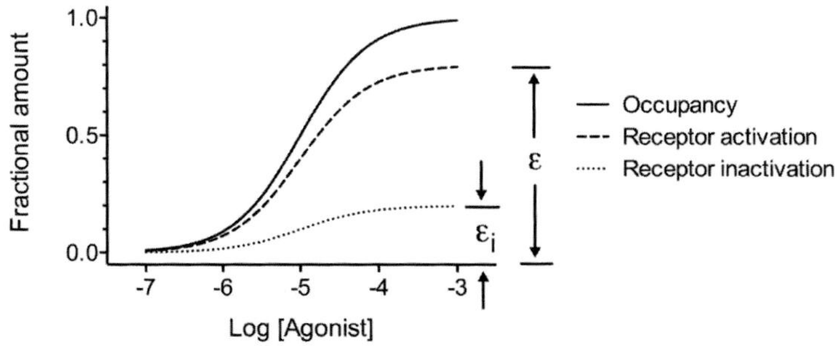

Figure 7.22: Relationships among receptor occupancy and the fraction of the occupied receptor population in the active (ε) and inactive (ε_i) states.

inactive state (DR_s), which is also equivalent to the fraction of the agonist-receptor complex in the inactive state at 100% receptor occupancy. Thus, the relationship between ε and ε_i is:

$$\varepsilon_i = 1 - \varepsilon. \tag{7.86}$$

There are two other useful parameters (τ_i and RI_i) that are related to ε_i in a manner analogous to how τ and RA_i are related to ε. The parameter τ_i is defined as:

$$\tau_i = \frac{\varepsilon_i R_T}{K_E}. \tag{7.87}$$

The parameter RI_i is defined as the product of observed intrinsic inactivity (ε_i) and observed affinity, expressed relative to that of a standard agonist:

$$RI_i = \frac{\varepsilon_{i-t} K_{\text{obs}-t}}{\varepsilon_{i-s} K_{\text{obs}-s}}. \tag{7.88}$$

In this equation, the subscripts t and s denote the parameters of the test agonist and the standard agonist. In the following paragraphs, it is shown through simulation that RI_i is a relative estimate of the microscopic affinity constant of the agonist for the inactive state of the receptor:

$$RI_i = \frac{K_{a-t}}{K_{a-s}}. \tag{7.89}$$

To estimate RI_i from agonist concentration-response curves, it is necessary to derive a form of the operational model that incorporates the observed intrinsic inactivity of the agonist (ε_i). This can be done by defining the τ value of the operational model in terms of ε_i and other parameters. From Equation (7.11) and the definition of ε_i

described above (Equation (7.86)), it is possible to define τ as:

$$\tau = \frac{(1 - \varepsilon_i)R_T}{K_E}. \tag{7.90}$$

To simplify the analysis, it is useful to normalize K_E relative to the receptor concentration (R_T):

$$K_{E\text{-obs}} = \frac{K_E}{R_T}, \tag{7.91}$$

in which $K_{E-\text{obs}}$ denotes the observed value of K_E. Solving this latter equation for K_E yields $K_E = R_T K_{E\text{-obs}}$. Substituting $R_T K_{E\text{-obs}}$ for K_E in Equation (7.90) followed by rearrangement yields:

$$\tau = \frac{1 - \varepsilon_i}{K_{E\text{-obs}}}. \tag{7.92}$$

It follows that τ and τ_i can also be defined as:

$$\tau = \frac{\varepsilon}{K_{E\text{-obs}}}, \tag{7.93}$$

$$\tau_i = \frac{\varepsilon_i}{K_{E\text{-obs}}}. \tag{7.94}$$

Substituting Equation (7.92) for τ in the operational model (Equation (7.10)) yields:

$$\text{response} = \frac{D^m M_{\text{sys}}}{D^m + \frac{D + \left(\frac{1}{K_{\text{obs}}}\right)^m}{\left(\frac{1-\varepsilon_i}{K_{E\text{-obs}}}\right)^m}}. \tag{7.95}$$

Equation (7.95) is our goal, and it can be used in the analysis of agonist concentration-response curves to estimate the RI_i values of agonists.

Unlike the standard form of the operational model, however, Equation (6.94) cannot be simplified further to eliminate both ε_i and K_E and replace them with a single parameter, like τ_i. Consequently, an infinite number of combinations of inversely correlated ε_i and $K_{E\text{-obs}}$ can yield a least squares fit to a concentration-response curve. When ε_i is constrained to its minimum value, zero, a least squares fit can be obtained that yields the maximum possible value of $K_{E\text{-obs}}$. With these constraints, the τ_i values $(1 - \varepsilon_i)/K_{\text{obs}}$ of all but the most efficacious agonist in a series are accurate provided that their τ values are less than half that of the most efficacious agonist.

Thus, one approach is to use global nonlinear regression analysis sharing the estimate of $K_{E\text{-obs}}$ among the data and constraining ε_i to zero for the most efficacious agonist. This process amounts to fitting the following equation to the concentration-response curve of the most efficacious agonist:

$$\text{response} = \frac{D^m M_{\text{sys}}}{D^m + \frac{D + \left(\frac{1}{K_{\text{obs}-s}}\right)^m}{\left(\frac{1-\varepsilon_i}{K_{E-\text{obs}}}\right)^m}}, \tag{7.96}$$

and fitting Equation (7.95) to the curves of the less efficacious agonists. Equation (7.96) was derived from Equation (7.95) by setting ε_i to zero in the latter equation. The parameter, $K_{\text{obs}-s}$, in Equation (7.96) denotes the observed affinity constant of the most efficacious agonist. This regression analysis requires estimates of K_{obs} of all of the agonists. This does not require additional data in the case of partial agonists because their K_{obs} values can be estimated through analysis of their concentration-response curves in combination with that of a full agonist as described earlier in this chapter. In the case of full agonists, however, it is necessary to use the method of partial receptor inactivation to estimate K_{obs}.

When this approach was used to analyze the simulated data in Figure 6.20a, a value of -1.28 was estimated for log $K_{E-\text{obs}}$. Again, this estimate represents the maximum possible value of $K_{E-\text{obs}}$. The true value is most likely smaller. Regression analysis also yielded estimates of log ε_i for agonists B (-0.151), C (-0.023) and D (-0.0072). The corresponding value for agonist A is negative infinity (log of zero). With these ε_i values, it is possible to estimate τ_i values using Equation (7.94). The log τ_i estimates for agonists B, C and D were 1.13, 1.26 and 1.27, respectively. Multiplying the τ_i value of each agonist by its corresponding K_{obs} value yields $\tau_i K_{\text{obs}}$, and expressing these relative to that of the least efficacious agonist yields the log RI_i values of agonists B, C and D (1.4, 0.99 and 0.00, respectively). These values are approximately equivalent to the log ratio of the K_a values of the agonists divided by that for agonist D (1.5, 1.0 and 0, respectively). Note that the log K_a values of the agonists B–D were 5.5, 5.0 and 4.0, respectively (see legend to Figure 7.20a).

With regard to GPCRs, it may be that even the most efficacious agonists are unable to generate more than 50% receptor activation because of the large negatively cooperative effects of GTP on agonist binding in the intact cell. If so, then the method for estimation of RI_i should have broad application. An obvious exception would be the case of measuring responses in broken cell preparations (e.g., adenylate cyclase assay, [^{35}S]GTPγS binding) when the concentration of GTP or both GTP and GDP is low. Regardless, it should be possible to determine whether the RI_i estimate is accurate by examining a series of agonists and expressing the RI_i value relative to an agonist with sufficiently low efficacy. If the RI_i estimates are constant

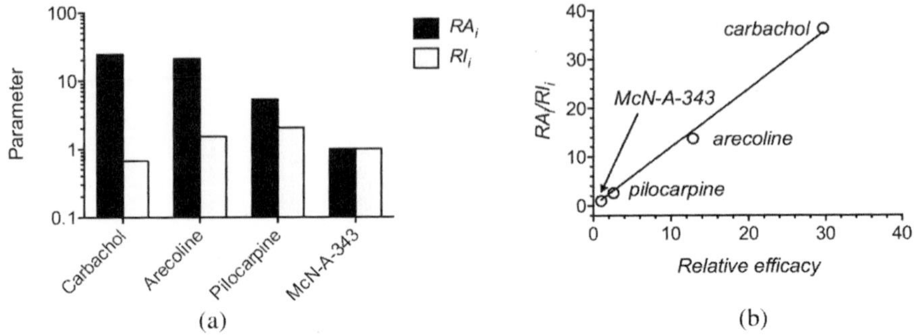

Figure 7.23: Relationships among the RA_i, RI_i and relative efficacy values of a group of agonists. (a) The log RA_i and log RI_i values were estimated from the concentration-response curves for M_3 muscarinic receptor-mediated phosphoinositide hydrolysis shown in Figure 7.21b. (b) The ratio, RA_i/RI_i is plotted against relative efficacy.

over the domain of $K_{E\text{-obs}}$ values used in regression analysis, then the values are accurate.

An example of the estimation of RI_i values from the concentration-response curves of agonists for eliciting phosphoinositide hydrolysis through activation of the M_3 muscarinic receptor is shown in Figure 6.23a together with the corresponding RA_i analysis. The actual data from which these estimates were made are shown in Figure 6.21b. Both the RA_i and RI_i values were estimated using the techniques described above (global nonlinear regression analysis using Equations (7.47) and (7.49), RA_i; Equations (7.95) and (7.96), RI_i, respectively) and expressed relative to the values for McN-A-343.

It is possible to estimate the observed affinities (K_{obs}) and relative efficacies of the agonists using the analysis described in Figure 7.3. That is, all of the data can be analyzed by global nonlinear regression analysis, sharing the estimate of the system parameters (M_{sys} and m) among the curves and obtaining unique estimates of τ and K_{obs} for each of the test agonists. For the standard agonist (oxotremorine-M), the method of partial receptor inactivation is used to estimate τ and K_{obs} as shown in Figure 7.4. Knowing the τ values, the relative efficacy of each agonist is estimated by expressing its τ value relative to that of the standard agonist (see Equation (7.23)). For consistency, efficacy was estimated relative to McN-A-343. Taking the ratio of RA_i/RI_i yields a relative measure of K_b/K_a. Relative efficacy is plotted against the ratio of RA_i/RI_i in Figure 7.23b. There is a linear relationship between the two estimates, and there is no indication that efficacy reaches a plateau as the ratio of the microscopic affinity constant of the active state, divided by that of the inactive state, increases (i.e., RA_i/RI_i, a relative measure of K_b/K_a). These data suggest that even the most efficacious muscarinic agonists do not come close to activating the M3 muscarinic receptor completely in the intact CHO cell. This is not surprising because GPCRs work in the presence of intracellular concentrations of GTP that inhibit the generation of the active state of the receptor.

The K_{obs} value of any partial agonist is likely to be indistinguishable from its K_a value. Thus, one can estimate the K_a values of a series of agonists from their respective RI_i values using the following relationship:

$$K_{a-t} = \frac{RI_{i-t}}{K_{obs-s}}, \tag{7.97}$$

in which, K_{a-t} and RI_{i-t} denotes the K_a and RI_ivalues of a given test agonist and K_{obs-s} denotes the K_{obs} value to which the RI_{i-t} value is normalized. Note that in the initial estimation of RA_i and RI_i values, RA_i is usually normalized relative to the most efficacious agonist, whereas RI_i is usually normalized relative to an agonist of low efficacy. Thus, in the equations described above, the parameters indicated with a subscript "*s*" denote those of the most efficacious agonist in the RA_ianalsyis and those of a less efficacious partial agonist in the RI_i analysis.

Using Equation (7.97), the log K_a values of the agonists shown in Figures 7.21b and 7.23 can be calculated, and these are illustrated in the bar graph in Figure 7.24. There is general agreement between K_{obs} and K_a for all of the agonists, but it can be seen that there is an increasing systematic deviation between the estimates of the two affinity constants as the efficacy of the agonist increases, with the value of K_{obs} being larger than that of K_a.

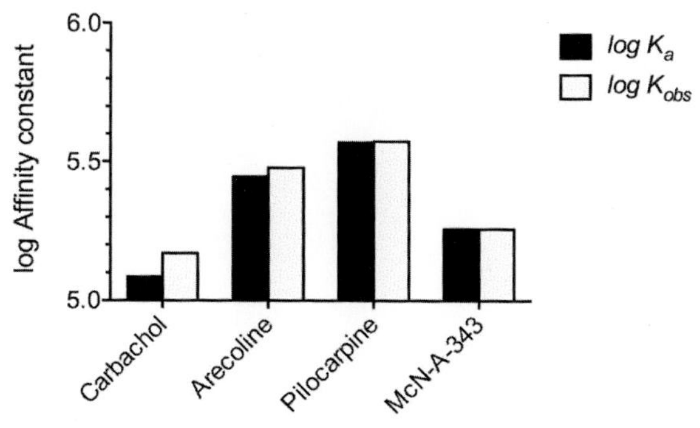

Figure 7.24: Relationship between the K_a and K_{obs} values of a group of agonists. The values of log K_a and log K_{obs} are shown. These were estimated from the concentration-response curves for M3 muscarinic receptor-mediated phosphoinositide hydrolysis shown in Figure 7.21b.

Proof of the relationships among microscopic affinity constants and RA_i and RI_i

The validity of the estimates of RA_i and RI_i can also be verified analytically by first deriving equations for K_{obs}, ε and ε_i in terms of the microscopic affinity constants of the agonist and then substituting these equations into the two equations that define RA_i and RI_i in terms of K_{obs}, ε and ε_i. To begin, the K_{obs} and ε values could be defined as the EC_{50} and E_{max} values of the agonist for generating the active state of the quaternary complex (DR_s^*GX, see Chapter 6). These two parameters could then be expressed in terms of the microscopic affinity constants of the two states of the receptor. A much simpler, yet equally valid, way to start the proof involves the model shown in Figure 2.1 in Chapter 2. The DR_s^* component of this model behaves just like the quaternary complex function even though the G protein and guanine nucleotide are not part of the model. These latter components have effects attributable to a change in the K_q value of the simpler model.

The sequence of the proof is as follows. First the microscopic affinity constants are defined, and these are used to derive equations for the various types of receptor states and complexes (i.e., R_s, R_s^*, DR_s and DR_s^*). These equations are subsequently used to develop functions describing receptor occupancy and the fractional amount of the agonist-receptor complex in the active and inactive states, with the agonist concentration being the independent variable. Next, these functions are used to generate equations for K_{obs}, ε and ε_i, respectively, in terms of the microscopic constants illustrated in Figure 2.1. Finally these state equations for the latter parameters are substituted for K_{obs}, ε and ε_i in the equations that define RA_i and RI_i (Equations (7.21) and (7.88)). The resulting equations are then simplified to yield expressions for RA_i and RI_i in terms of K_b and K_a, respectively.

Now back to the proof. The equilibrium between the free forms of the active (R_s^*) and inactive (R_s) states of the receptor is defined by K_q:

$$K_q = \frac{[R_s^*]}{[R_s]}. \tag{7.98}$$

The microscopic affinity constants of the agonist for the active (K_b) and inactive (K_a) states of the receptor are defined as:

$$K_a = \frac{[DR]}{[D][R_s]}, \tag{7.99}$$

$$K_b = \frac{[DR_s^*]}{[D][R_s^*]}. \tag{7.100}$$

Equations (7.98) and (7.99) can be solved for $[R_s^*]$ and $[DR_s]$:

$$[R_s^*] = K_q[R_s], \tag{7.101}$$

$$[DR_s] = DK_a[R_s]. \tag{7.102}$$

Equation (7.100) can be solved for $[DR_s^*]$ and then Equation (7.101) can be substituted for $[R_s^*]$ to yield the equation for $[DR_s^*]$:

$$[DR_s^*] = DK_bK_q[R_s]. \tag{7.103}$$

The expressions for fractional receptor occupancy and the fractional amount of the agonist receptor complex in the active and inactive states are:

$$\frac{[DR_s] + [DR_s^*]}{R_T} = \frac{[DR_s] + [DR_s^*]}{[DR_s] + [DR_s^*] + [R_s] + [R_s^*]}, \tag{7.104}$$

$$\frac{[DR_s^*]}{R_T} = \frac{[DR_s^*]}{[DR_s] + [DR_s^*] + [R_s] + [R_s^*]}, \tag{7.105}$$

$$\frac{[DR_s]}{R_T} = \frac{[DR_s]}{[DR_s] + [DR_s^*] + [R_s] + [R_s^*]}. \tag{7.106}$$

Substituting in Equations (7.101)–(7.103) for the various receptor species in these equations yields:

$$\frac{[DR_s] + [DR_s^*]}{R_T} = \frac{DK_a[R_s] + DK_bK_q[R_s]}{DK_a[R_s] + DK_bK_q[R_s] + [R_s] + K_q[R_s]}, \tag{7.107}$$

$$\frac{[DR_s^*]}{R_T} = \frac{DK_bK_q[R_s]}{DK_a[R_s] + DK_bK_q[R_s] + [R_s] + K_q[R_s]}, \tag{7.108}$$

$$\frac{[DR_s]}{R_T} = \frac{DK_a[R_s]}{DK_a[R_s] + DK_bK_q[R_s] + [R_s] + K_q[R_s]}. \tag{7.109}$$

Simplifying Equation (7.107) yields receptor occupancy:

$$\frac{[DR_s] + [DR_s^*]}{R_T} = \frac{D}{D + \frac{1}{K_{\text{obs}}}}. \tag{7.110}$$

In this equation, K_{obs} is defined as:

$$K_{\text{obs}} = \frac{K_a + K_bK_q}{1 + K_q}. \tag{7.111}$$

Simplifying Equations (7.108) and (7.109) yields:

$$\frac{[DR_s^*]}{R_T} = \frac{1}{1 + \frac{1+K_q+DK_a}{DK_bK_q}}, \tag{7.112}$$

$$\frac{[DR_s]}{R_T} = \frac{1}{1 + \frac{1+K_q+DK_bK_q}{DK_a}}. \tag{7.113}$$

The maximum values of these two functions represent the maximal fraction amounts of the active and inactive forms of the agonist-receptor complex. These values are equivalent to ε and ε_i and can be derived by taking the limit of these functions as D approaches infinity:

$$\varepsilon = \lim_{D\to\infty} \frac{1}{1 + \frac{1+K_q+DK_a}{DK_bK_q}}, \tag{7.114}$$

$$\varepsilon_i = \lim_{D\to\infty} \frac{1}{1 + \frac{1+K_q+DK_bK_q}{DK_a}}. \tag{7.115}$$

Solving these limits yields:

$$\varepsilon = \frac{1}{1 + \frac{K_a}{K_bK_q}}, \tag{7.116}$$

$$\varepsilon_i = \frac{1}{1 + \frac{K_bK_q}{K_a}}. \tag{7.117}$$

The equations for K_{obs}, ε and ε_i can be substituted into the equations define RA_i and RI_i to yield equations for the latter in terms of microscopic affinity constants. Substituting Equations (7.111) and (7.116) for K_{obs} and ε into Equation (7.58) yields the RA_i value expressed in terms of microscopic constants:

$$RA_i = \frac{\frac{K_{a-t}+K_{b-t}K_q}{1+K_q} \times \frac{1}{1+\frac{K_{a-t}}{K_{b-t}K_q}}}{\frac{K_{a-s}+K_{b-s}K_q}{1+K_q} \times \frac{1}{1+\frac{K_{a-s}}{K_{b-s}K_q}}}. \tag{7.118}$$

The subscripts "t" and "s" denote the microscopic constants of the test and standard agonists, respectively. Simplifying this equation yields:

$$RA_i = \frac{K_{b-t}}{K_{b-s}}. \tag{7.119}$$

Thus, the RA_i value is equivalent to the microscopic affinity constant of the test agonist for the active state of the receptor divided by that of the standard agonist.

Substituting Equations (7.111) and (7.117) for K_{obs} and ε_i in Equation (7.88) yields the RI_i value expressed in terms of microscopic constants:

$$RI_i = \frac{\frac{K_{a-t}+K_{b-t}K_q}{1+K_q} \times \frac{1}{1+\frac{K_{b-t}K_q}{K_{a-t}}}}{\frac{K_{a-s}+K_{b-s}K_q}{1+K_q} \times \frac{1}{1+\frac{K_{b-s}K_q}{K_{a-s}}}}. \tag{7.120}$$

This equation can be simplified to:

$$RI_i = \frac{K_{a-t}}{K_{a-s}}. \tag{7.121}$$

Thus, the RI_i value is equivalent to the microscopic affinity constant of the test agonist for the inactive state of the receptor divided by that of the standard agonist.

And that is the proof for the validity of RA_i and RI_i.

Toward a Facile Estimation of the Microscopic Affinity Constants of Agonists

The most common way of determining whether an agonist has activity at a particular receptor is to measure its concentration-response curve. The latter can be analyzed beyond the estimation of EC_{50} and $E_{\max}$ to define the stimulus of the receptor population and to make relative estimates of the microscopic affinity constants of the agonist for the active and inactive states of the receptor. These are useful in determining the selectivity of an agonist and in providing feedback to medicinal chemists regarding the affinity of a test compound for the two shapes of the receptor — active and inactive. Often, activity is related to the fit of the agonist within a binding pocket having a single structure. But how can the action of an agonist be explained unless both of its fits with the two distinct quantal shapes of the binding pocket are considered? The RA_i and RI_i values provide the requisite measure of the relative affinity the agonist for these active and inactive shapes of the receptor.

The methods described in this chapter for making these estimations seem complicated, but it is not that difficult to develop an algorithm that computes these parameters with little input other than the data itself. The computer algorithm could search parameter space and determine whether it is possible to estimate both τ and K_{obs} or only the product of the two. The algorithm could also be used to calculate the initial parameter estimates for regression analysis.

The RA_i and RI_i values are limited to the extent that the estimates are relative, although the RI_i value can be easily converted to K_a using Equation (7.97). When there is constitutive receptor activity, however, it is possible to estimate the actual

K_b values in units of 1.0 M^{-1}. A modification of receptor theory that enables these calculations is described in Chapter 8.

Appendix

This section gives more detailed procedures for the nonlinear regression techniques used in three of the examples covered in the text concerning the data shown in Figures 7.3a, 7.4a and 7.8. Here, I describe how to use the Prism 6 application (GraphPad Software, Inc.) to analyze the data, although the general procedures can be applied to any nonlinear regression algorithm. It is assumed that the reader is generally familiar with the Prism environment.

Example 1: Estimation of the affinity constant of a partial agonist through analysis of its concentration-response curve and that of a full agonist (Figure 7.3a).

1. The data for the concentration-response curves are entered into a Data Table for an XY plot in Prism. The log molar concentrations of the two agonists are entered into the column on the far left entitled "X". The response measurements for a single experiment or the means of several experiments can be entered, but it is recommended that the individual values of replicate experiments be entered into a Data Table. The replicate response measurements of the standard agonist are entered into subcolumns of Group A and those of the partial agonist are entered into subcolumns of Group B. A given response value is always entered into the same row as that of its corresponding log agonist concentration.
2. The Graph of the data is selected, and ultimately, the nonlinear regression (curve fit) analysis is selected.

   Before proceeding with the specific steps necessary for the analysis in Prism, let me describe the strategy. When doing nonlinear regression analysis with the equations described in this chapter, it is better to express some of the parameters as logarithms. In addition, because it is possible to estimate the individual parameters, τ and K_{obs}, for the partial agonist, but only their product (τK_{obs}) for the full agonist, we will have to enter the equation with a dummy variable for the product τK_{obs}. This will enable us to estimate the product τK_{obs} for the full agonist and the individual values, τ and K_{obs}, for the partial agonist in the Prism environment.

   Now let us return to the process of entering the regression equation into Prism. The following lines are entered into Prism to define a new equation:

```
⟨A⟩ LogR=LogTauKobsFull
```

```
⟨A⟩ LogK=LogKobsFull
⟨~A⟩ LogR=LogTauPartial + LogKobsPartial
⟨~A⟩ LogK=LogKobsPartial
Y = Msys/(1 + {[10^(X + LogK) + 1]/[10^(X + LogR)]}^M)
```

The last line represents the regression equation, which is a simplified form of the operational model (Equation (7.10)). The variable "X" denotes the log concentration of agonist, and LogK and LogR represent dummy variables.

The first two lines are preceded by the symbols "⟨A⟩", which designate that the expressions on these lines apply to the data in Group A of the Data Table (i.e., that of the full agonist). The first line designates that the dummy variable, LogR, will represent a parameter (LogTauKobsFull) corresponding to the log product of the τ and K_{obs} (log τK_{obs}) parameters of the full agonist. Similarly, the second line assigns the parameter LogKobsFull (log K_{obs} value of the full agonist) to the dummy variable, LogK.

The third and fourth lines are preceded by the symbols "⟨~A⟩ ", which instruct Prism to apply the corresponding variable assignments to all of the data except those in Group A. For the present example, we could have preceded these lines with the symbols "⟨B⟩" to refer to the single set of data for the partial agonist pilocarpine in Group B, but using the expression "⟨~A⟩" is applicable for future situations in which the data for more than one partial agonist are analyzed.

The third line designates that the dummy variable, LogR, now represents the sum of two parameters, LogTauPartial + LogKobsPartial. Finally, the fourth line assigns the parameter LogKobsPartial to the dummy variable, LogK.

3. Now that the equation has been entered, parameter constraints are applied. The log K_{obs} of the full agonist is constrained to an arbitrarily low value of zero (LogKobsFull = 0) and the parameters for the maximum response of the system, M_{sys} (Msys), and the transducer slope factor, m (M), are shared across all data sets.
4. The next step involves assigning initial parameter estimates. These are estimated using the following equations in which the parameters with a prime symbol denote the initial estimate of the parameter:

$$\text{LogTauKobsFull}' = -\log EC_{50\text{-full}}, \tag{7.122}$$

$$\text{LogTauPartial}' = \log\left(\frac{1}{\frac{E_{\max\text{-full}}}{E_{\max\text{-partial}}} - 1}\right), \tag{7.123}$$

$$\texttt{LogKobsPartial}' = -\log EC_{50\text{-partial}} \tag{7.124}$$

$$\texttt{Msys}' = E_{\text{max-full}}. \tag{7.125}$$

In these equations $EC_{50\text{-full}}$ and $EC_{50\text{-partial}}$ denote the EC_{50} values of the full and partial agonists, and $E_{\text{max-full}}$ and $E_{\text{max-partial}}$ denote the corresponding E_{max} values. The transducer slope value can be assigned an initial estimate of 1:

$$\texttt{M}' = 1, \tag{7.126}$$

or better yet, it can be assigned the value of the Hill slope of the concentration-response curve of the full agonist (H_{full}):

$$\texttt{M}' = H_{\text{full}}. \tag{7.127}$$

5. Now that the data, regression equation, parameter constraints and initial parameter estimates have all been entered, regression analysis is initiated. For this example, the following parameter estimates were obtained: Msys, 97.6 ± 1.62; m, 1.75 ± 0.10; the LogTauKobs value of the full agonist, carbachol, 7.01 ± 0.02; the LogKobs value of the partial agonist, pilocarpine, 5.81 ± 0.04 and the LogTau value of pilocarpine, 0.28 ± 0.028.

   If it were uncertain that carbachol were a full agonist, then the regression would have to be done allowing the estimate of LogKobsFull to be unconstrained. When this is done the following parameters were obtained: M, 1.78 ± 0.25; the LogTauKobs value of the full agonist, carbachol, 7.02 ± 0.07; the LogKobs value of the partial agonist, pilocarpine, 5.83 ± 0.17 and the LogTau value of pilocarpine, 0.27 ± 0.10. In this situation, it is impossible to obtain a reliable value of Msys.

Example 2: Estimation of the affinity constant by the method of partial receptor inactivation (Figure 7.4a).

1. The data for the concentration-response curves of oxotremorine-M (Figure 7.4a) are entered into a Data Table for an XY plot as described in step 1 of Example 1 above. The control data and those measured under conditions of partial receptor inactivation are entered into subcolumns of Group A and Group B, respectively.
2. The Graph of the data is selected, and ultimately, the nonlinear regression (curve fit) analysis is selected.

   The following lines are entered into Prism to define a new equation:

```
⟨A⟩Q = 1,
⟨~A⟩Q = Fraction,
Y = Msys/(1 + {[10^(X + LogKobs) + 1]/
      [Q*[10^(X + LogTau + LogKobs)]]}^M).
```

The last line represents the regression equation, which is a form of the operational model (Equation (7.10)). In this equation, `Msys` represents the maximum response of the system (M_{sys}), `X`, the log molar concentration of agonist, `LogKobs`, the log K_{obs} value of the agonist, `Q`, the fraction of the receptor population not inactivated by 4-DAMP mustard, `LogTau`, the log τ value of the agonist, and `M`, the transducer slope factor in the operational model (m).

The first line assigns the value of 1 to the dummy variable, `Q`, when the control data in Group A are being analyzed. The second line assigns the parameter "`Fraction`" to the dummy variable `Q` when the data in Group B are being analyzed. The parameter "`Fraction`" represents the fraction of the receptor population not inactivated. The significance of the "⟨A⟩" and "⟨~A⟩" entries are described above under step 2 of Example 1.

3. Next, parameter constraints are applied. This involves sharing the estimates of the parameters for the maximum response of the system (`Msys`), the transducer slope factor (M), log τ (`LogTau`) and log K_{obs} (`LogKobs`) across both sets of data.
4. The initial parameter estimates are calculated using the following equations in which the parameters with a prime symbol denote the initial estimate of the parameter:

$$\texttt{LogKobs}' = -\log EC_{50\text{-inact}}, \tag{7.128}$$

$$\texttt{LogTau}' = \log\left(\frac{EC_{50\text{-inact}}}{EC_{50\text{-control}}}\right), \tag{7.129}$$

$$\texttt{Q}' = \frac{EC_{\text{max-inact}} * EC_{50\text{-control}}}{EC_{\text{max-control}} * EC_{50\text{-inact}}}, \tag{7.130}$$

$$\texttt{Msys}' = E_{\text{max-control}}. \tag{7.131}$$

In these equations $EC_{50\text{-control}}$ and $EC_{50\text{-inact}}$ denote the EC_{50} values measured under control conditions and after partial receptor inactivation, and $E_{\text{max-control}}$ and $E_{\text{max-inact}}$ denote the corresponding E_{max} values.

As described in Example 1, the transducer slope factor can be assigned an initial estimate of 1:

$$\text{M}' = 1 \tag{7.132}$$

or the value of the Hill slope of the concentration-response curve of the full agonist (H_{full}):

$$\text{M}' = H_{\text{full}}. \tag{7.133}$$

5. After initiating the repression analysis, the following parameters are obtained: `LogKobs`, 5.49 ± 0.30; `Fraction`, 0.12 ± 0.05; `LogTau`, 0.95 ± 0.29; `Msys`, 288 ± 0.23 and `M`, 1.17 ± 0.22.

Example 3: Estimation of the RA_i values of agonists relative to that of a full agonist (Figure 7.8).

1. The data for the concentration-response curves of the agonists (Figure 7.8) are entered into a Data Table for an XY plot following the general guidelines given under step 1, Example 1. The data for the agonist with the largest E_{max} (i.e., oxotremorine-M) is entered under Group A in the Data Table. This agonist is designated as the standard agonist. The data for the other agonists (carbachol, arecoline, pilocarpine and McN-A343) are entered into the Data Table under Groups B–D. These agonists are designated as test agonists, and their group ordering within the Data Table is unimportant.
2. The Graph of the data is selected, and the nonlinear regression (curve fit) analysis is selected.

   The following lines are entered into Prism to define a new equation:

   ```
   <A> Q = LogR
   <A> LogK = LogKStand
   <~A>Q = LogRA + LogR
   <~A>LogK = LogKTest
   Y = Msys/(1 + {[10^(X + LogK) + 1]/[10^(X + Q)]}^M)
   ```

   The last line describes the regression equation, which is a form of the operational model. In this equation, `Msys` represents the maximum response of the system (M_{sys}), `X`, the log molar concentration of agonist, and `M`, the transducer slope factor in the operational model (m).

   The first two lines designate which parameters of the standard agonist are assigned to the two dummy variables. The first line assigns the parameter, `LogR` (log τK_{obs}), to the dummy variable, `Q`, and the second line assigns the parameter, `LogKStand` (log K_{obs}), to the dummy variable `LogK`.

   The second two lines assign parameters of the test agonists to the dummy variables. Line 3 assigns the parameter, `LogRA` ($LogRA_i$) to the dummy variable, `Q`, and the fourth line assigns the parameter, `LogKTest` (log K_{obs}) to the dummy variable, `LogK`.

   This approach enables one to work within the Prism environment and constrain the log K_{obs} value of the standard agonist while estimating unique values for each test agonist. It also enables one to estimate the log τK_{obs} value of standard agonist while sharing this parameter across all the datasets and obtaining unique estimates of log RA_i for each test agonist.

3. Next, parameter constraints are applied. This involves sharing the estimates of the parameters for the maximum response of the system, M_{sys} (`Msys`), the transducer slope factor, m (M), and the log τK_{obs} value for the standard agonist (`LogR`). The log K_{obs} value of the standard agonist (`LogKstand`) is constrained to zero.
4. The initial parameter estimates are calculated using the following equations in which the parameters with a prime symbol denote the initial estimate of the parameter:

$$\texttt{LogR}' = -\log EC_{50-s} \quad (7.134)$$

$$\texttt{LogKTest} = -\log EC_{50-t} \quad (7.135)$$

$$\texttt{LogRA}' = \log\left(\frac{E_{\max-t} * EC_{50-s}}{E_{\max-s} * EC_{50-t}}\right) \quad (7.136)$$

$$\texttt{Msys}' = E_{\max-s} \quad (7.137)$$

In these equations EC_{50-s} and EC_{50-t} denote the EC_{50} values of the standard and test agonists, and $E_{\max-s}$ and $E_{\max-t}$ denote the corresponding $E_{\max}$ values.

As described for Example 1, the transducer slope factor can be assigned an initial estimate of 1:

$$\mathrm{M}' = 1 \quad (7.138)$$

or the value of the Hill slope of the concentration-response curve of the full agonist (H_s):

$$\mathrm{M}' = H_s. \quad (7.139)$$

5. After initiating the regression analysis, the following parameters are obtained: oxotremorine-M, `LogTauKobs`, 6.48 ± 0.04; carbachol, `LogKobs`, 5.19 ± 0.14; `LogRA`$_i$, −0.56±.06; arecoline, `LogKobs`, 5.49±0.12; `LogRA`, −0.61±0.07; pilocarpine, `LogKobs`, 5.58 ± 0.16; `LogRA`, −1.20±0.15 and McN-A343, `LogKobs`, 5.023 ± 0.16 and `LogRA`, −1.76 ± 0.15. The estimates of `Msys` and `M` are: 53.92 ± 1.3 and 1.15 ± 0.09, respectively.

If it were uncertain that oxotremorine-M were a full agonist, then the regression would have to be done allowing the estimate of the log affinity constant of oxotremorine-M (`LogKStand`) to be unconstrained. This would have led to an unreliable estimate of `Msys`, but `LogRA`$_i$ and `LogKobs` values of all of the other agonists would be unaffected.

References

1. Leff, P. Analysis of agonist action using the operational model. *Trends Pharmacol Sci* **9**, 395–398 (1988).
2. Black, J. W. and Leff, P. Operational models of pharmacological agonism. *Proc R Soc Lond Ser B: Biol Sci* **220**, 141–162 (1983).
3. Black, J. W., Leff, P., Shankley, N. P. and Wood, J. An operational model of pharmacological agonism: the effect of $E/[A]$ curve shape on agonist dissociation constant estimation. *Br J Pharmacol* **84**, 561–571 (1985).
4. Stephenson, R. P. A modification of receptor theory. *Br J Pharmacol* **11**, 379–393 (1956).
5. Furchgott, R. F. The use of β-haloalkylamines in the differentiation of receptors and in the determination of dissociation constants of receptor-agonist complexes. *Adv Drug Res* **3**, 21–55 (1966).
6. Furchgott, R. F. and Bursztyn, P. Comparison of dissociation constants and of relative efficacies of selected agonists acting on parasympathetic receptors. *Ann NY Acad Sci* **144**, 882–899 (1967).
7. Ehlert, F. J., Griffin, M. T., Sawyer, G. W. and Bailon, R. A simple method for estimation of agonist activity at receptor subtypes: comparison of native and cloned M_3 muscarinic receptors in guinea pig ileum and transfected cells. *J Pharmacol Exp Ther* **289**, 981–992 (1999).
8. Ehlert, F. J. On the analysis of ligand directed signaling at *G* protein coupled receptors. *Naunyn Schmied Archive Pharmacol* **377**, 549–577 (2008).
9. Ehlert, F. J., Griffin, M. T. and Suga, H. Analysis of functional responses at *G* protein coupled receptors: estimation of relative affinity constants for the inactive receptor state. *J Pharmacol Exp Ther* (2011).
10. Ehlert, F. J., Suga, H. and Griffin, M. T. Quantifying agonist activity at *G* protein-coupled receptors. *J Vis Exp*, e3179 (2011).
11. Figueroa, K. W., Griffin, M. T. and Ehlert, F. J. Selectivity of agonists for the active state of M1–M4 muscarinic receptor subtypes. *J Pharmacol Exp Ther* **328**, 331–342 (2008).
12. Griffin, M. T., Figueroa, K. W., Liller, S. and Ehlert, F. J. Estimation of agonist activity at *G* protein-coupled receptors: analysis of M2 muscarinic receptor signaling through $G_{i/o}$, G_s and G_{15}. *J Pharmacol Exp Ther* **321**, 1193–1207 (2007).
13. Tran, J. A., Chang, A., Matsui, M. and Ehlert, F. J. Estimation of relative microscopic affinity constants of agonists for the active state of the receptor in functional studies on M_2 and M_3 muscarinic receptors. *Mol Pharmacol* **75**, 381–396 (2009).
14. Marsh, J. D. and Smith, T. W. Receptors for beta-adrenergic agonists in cultured chick ventricular cells. Relationship between agonist binding and physiologic effect. *Mol Pharmacol* **27**, 10–18 (1985).
15. Ehlert, F. J. and Rathbun, B. E. Signaling through the muscarinic receptor-adenylate cyclase system of the heart is buffered against GTP over a range of concentrations. *Mol Pharmacol* **38**, 148–158 (1990).

Analysis of Agonism and Inverse Agonism in Signaling Pathways that Exhibit Constitutive Activity

8

Most G protein-coupled receptors are silent in their native environment unless activated by an agonist. But when expressed at high density in heterologous cells, they often elicit small signaling responses in the absence of agonist, like changes in adenylate cyclase activity, stimulation in phosphoinositide hydrolysis, or stimulation in [^{35}S]GTPγS binding, depending upon their preference for G proteins. Given that these proteins evolved to function as molecular switches that flip between active and inactive states, it is not surprising that, at any given instant, a very small fraction of the receptor population might briefly switch into the active state in the absence of agonist because of thermal activation. In fact, it would be impossible for a ligand to induce a conformational change in a receptor unless the receptor has some tendency, however small, to undergo the conformational change spontaneously (see below). These spontaneously active receptors could generate a measurable response if the receptor population is sufficiently large or if the downstream signaling pathway is highly amplified. In this chapter, I describe how to measure the activity of orthosteric ligands at receptors exhibiting spontaneous receptor activity (constitutive activity).

One advantage of measuring a response elicited through a signaling pathway that reports the constitutive activity of the receptor is that it is possible to estimate the microscopic affinity constant of a ligand for the active (K_b) state of a receptor in absolute units, like M^{-1}, for example. In the absence of constitutive activity, it is only possible to estimate a relative measure of K_b (RA_i) as described in the preceding chapter.

The K_b and K_a values are measures of the affinity of an orthosteric ligand for the active and inactive states of the receptor, whereas the K_{obs} value represents a weighted average of the K_b and K_a values. As described in Chapter 6 under "Three types of affinity constants", there is no receptor structure that exhibits an affinity of K_{obs} with regard to a highly efficacious agonist. If there is a change in the density of G protein or concentration of GTP and GDP in the cytosol, this will affect the K_{obs} value of an efficacious agonist and the activation state of the receptor (ε). This change occurs not because of a change in the K_b and K_a values, but because of a shift in the equilibrium between active and inactive states as predicted by Le Chatelier's principle. Thus, the K_b and K_a values of an agonist are invariant, whereas the K_{obs} and relative efficacy values change depending upon the nature of the signaling pathway. There may be different active receptor states that signal through different G proteins. In this case, having a method for estimating the effector-selective states of the receptor is useful in the characterization of agonist bias as described in upcoming Chapter 9.

The estimates of K_b and K_a also provide feedback to the medicinal chemist about the activity of drug molecules. Too often, the structure of a compound is correlated with its EC_{50} or K_{obs} value. As just described, there is no receptor structure that exhibits these affinities, and the quantal nature of receptors dictates that they have at least two affinities, K_b and K_a. How the functional activity of a compound relates to its affinity for ground and active states of the receptor should be considered in drug discovery programs. Thus, screening compounds in functional assays with a defined receptor-signaling pathway that exhibits constitutive activity is a powerful approach for determining the pharmacological properties of new compounds.

Constitutive Receptor Activity

When a receptor exhibits constitutive activity, orthosteric ligands can display three types of behavior when eliciting a downstream response at a G protein-coupled receptor — agonism, inverse agonism and neutral antagonism (Figure 8.1). Agonists increase receptor activation, inverse agonists inhibit it and neutral antagonists have no effect. This tripartite classification correlates with selectivity for the active and inactive states of the receptor. Agonists exhibit selectivity for the active state ($K_b/K_a > 1$), inverse agonists the opposite selectivity ($K_b/K_a < 1$), and neutral antagonists lack selectivity for either state ($K_b/K_a = 1$). By selecting the inactive state of the receptor, inverse agonists inhibit constitutive receptor activity. In the absence of spontaneous activity, inverse agonists and neutral antagonists are without effect because an inverse agonist selects for a conformation of the receptor that it has already assumed, and a neutral antagonist does not induce conformational changes.

Many highly potent antagonists behave as inverse agonists when the receptor exhibits constitutive activity because a high-affinity ligand must have high affinity

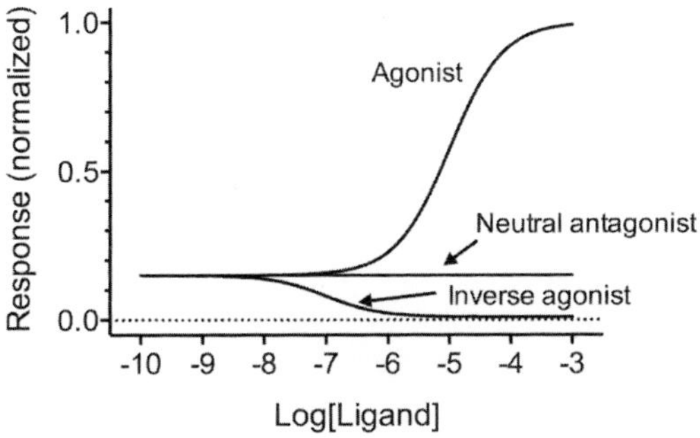

Figure 8.1: Concentration-responses curves of an agonist, neutral antagonist and inverse agonist for eliciting a response downstream in the signaling pathway of a *G* protein-coupled receptor exhibiting constitutive activity.

for at least one state of the receptor. If the ligand is an antagonist, then it must have high affinity for the inactive state. If so, there is a good chance that this high-affinity interaction will be lost when the receptor assumes a different shape, like that of the active state (see Prologue). It follows that most highly potent antagonists of low molecular weight are likely to exhibit selectivity for the inactive state and behave as inverse agonists. This prediction may not hold for an antagonist with high molecular weight when the *G* protein-coupled receptor has a much smaller endogenous ligand because very large ligands derive considerable binding affinity by interacting with loci beyond the binding pocket for the endogenous ligand. Whether this additional binding topography changes with the activation state of the receptor is open to chance because it is not an adaptation for ligand-induced activation like that of the natural ligand's binding pocket.

Different receptors exhibit different levels of constitutive activity. Rhodopsin (retinal-opsin), for example, exhibits practically no activity. The frequency that a single bovine opsin protein conjugated with 11-*cis*-retinal undergoes spontaneous activation is about once every 4,000 years.[1] This is due, in part, to its covalently attached ligand, 11-*cis*-retinal, which behaves as an inverse agonist.[2] Opsin itself exhibits low basal activity. Similarly, other *G* protein-coupled receptors often exhibit some basal activity, particularly when over expressed or when their respective *G* proteins are over expressed.[3,4] Inverse agonists inhibit basal activity under these conditions. One cannot help but wonder if there are endogenous inverse agonists that inhibit constitutively active *G* protein-coupled receptors.

One example may be related to the MC4 subtype of melanocortin receptors. At this receptor α-melanocyte stimulating hormone acts as an agonist, whereas agouti or agouti related peptide act as antagonists.[5,6] When expressed in heterologous cells,

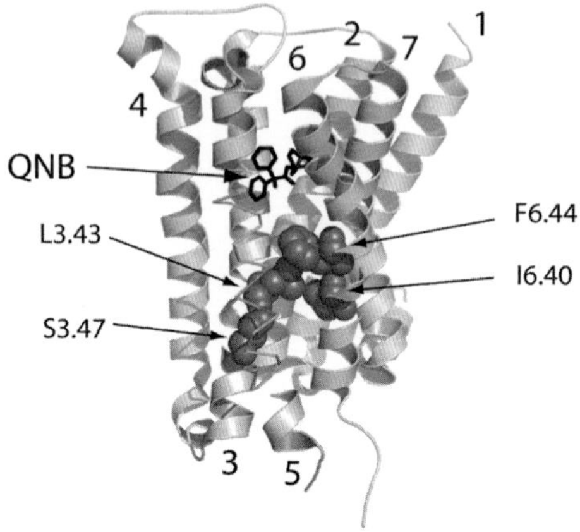

Figure 8.2: Crystal structure of the helices and extracellular loops of the human M_2 muscarinic receptor bound with 3-quinuclidinyl benzilate (QNB).[32] The secondary structure (helices 1–4, 6 and 7) and backbone (helix 5) are shown. Leucine 3.43, serine 3.47, isoleucine 6.40 and phenylalanine 6.44 (L114, S118, I392 and F396) are indicated on helices two (L114 and S118) and six (I392 and F396) (PDB ID: 3uon).

the MC4 receptor exhibits spontaneous basal activation of adenylate cyclase, and this activity is inhibited by agouti related peptide or smaller fragments of it.

While most receptors exhibit little constitutive activity, some diseases are caused by point mutations in *G* protein-coupled receptors that cause constitutive activity in the absence of agonist.[7,8] A potentially useful therapeutic drug for this condition would be an inverse agonist.

There are several conserved residues in the transmembrane spanning domains of many *G* protein-coupled receptors that when mutated to alanine cause constitutive activity. Four of these that are involved in maintaining the inactive state of the receptor (L3.43, S3.47, I370 and F3.64)[7,9–11] are shown in Figure 8.2. This type of gain-of-function mutation shifts the equilibrium between the active and inactive states of the receptor in the direction of the active state.[3,12] This change has the potential to increase the observed affinity and efficacy of the agonist because agonists exhibit selectivity for the active state. For a receptor functioning in the presence of physiological amounts of intracellular GTP, the increase in observed affinity might be negligible. But if agonist-binding properties are measured in cellular homogenates in the absence of GTP, constitutively active receptor mutants typically exhibit increased affinity for agonists.

An increase in the expression of a *G* protein-coupled receptor or its *G* protein usually increases the basal activities of receptors, including those that have imperceptible constitutive activity under more physiological levels of expression. This is consistent with how Le Chatelier's principle applies to the interaction between receptor (R) and G protein (G) (see Chapter 6):

$$R + G \overset{K_2}{\longleftrightarrow} RG. \tag{8.1}$$

An increase in the receptor or G protein will shift the equilibrium to the right and increase the amount of the receptor-G protein complex.[4] In most cells, the G protein would be loaded with GDP, and its collision with the unoccupied receptor would increase the probability of the exchange of GTP for GDP and the generation of free $G\alpha$–GTP subunits and a quantum of cellular signaling. The unoccupied receptor-G protein complex, however, is not as active as the agonist-receptor-G protein complex. Thus, like ligand-gated ion channels, G protein-coupled receptors also require the binding of two ligands to cause maximal receptor activation — the agonist and the G protein. This fact explains why the crystal structure of rhodopsin (*cis*-retinal-opsin) is not much different from that of photoactivated rhodopsin (trans-retinal-opsin),[13,14] whereas the structure of the agonist-β_2-adrenoceptor-G_s complex is much different from the antagonist-β_2-adrenoceptor complex.[15,16] Nonetheless, some activation of G protein-coupled receptors occurs during spontaneous collisions with G proteins in the absence of agonist.

The existence of constitutive receptor activity was predictable on the basis of the hypotheses of Monod Wyman and Changeaux.[17] They postulated that enzymes exist in two quantal states, active and inactive, and that the binding of ligands shifted the equilibrium depending upon their relative affinities for the two states. It seemed likely that ion channels and other receptors might behave similarly, and that a low level of spontaneous receptor activation may occur in the absence of agonist or endogenous ligand. This idea was well accepted in the field of electrophysiology where the ingenious methods of fluctuation analysis and power spectral analysis were devised to count ion channels and to determine the microscopic constants of single channels through measurement of the random spontaneous openings within a population of ion channels.[18–21] Yet in the early 1980s the concept of constitutive activity was considered novel within much of the community of pharmacologists working on G protein-coupled receptors.

Part of the reason is that the requisite technology for measuring basal receptor activity had not been applied to G protein-coupled receptors. While electrophysiological techniques were exquisitely sensitive and capable of measuring small conductances in excitable cells, the typical cellular or tissue responses to G protein-coupled receptor activation were unsuitable for detecting spontaneous receptor activity. Many responses, like secretion or smooth muscle contraction, depend on membrane depolarization, yet a low level of activation of many G protein-coupled receptors is insufficient to depolarize the membrane to threshold. Other biochemical responses were also unsuitable for detecting spontaneous receptor activity. How does one know, for example, if the resting level of cAMP in a tissue is influence by the spontaneous activity of a G_s-linked receptor? It may be that G_s, as well as adenylate cyclase, have basal activity. Or perhaps the control measure of cAMP represents a pool that is unaffected by receptor activity. The problem is more difficult when isotopic methods are used to measure cAMP, like labeling its precursor, ATP. But when molecular biological techniques were used to study receptors, it became clear that the response normally elicited by the agonist-receptor complex could sometimes be replicated without the agonist by over expressing the receptor. Being able to regulate the expression of the receptor was the key to measuring constitutive activity at G protein-coupled receptors.

It can even be said that constitutive activity is an essential property of any receptor, for there are only two ways that a drug can switch an inactive receptor (R) into an active one (R^*), and both depend on constitutive activity. A drug could wait for the receptor to switch into the active state spontaneously and then stabilize the active state by binding to it selectively (conformational selection):

$$D + R_s \overset{K_q}{\longleftrightarrow} D + R_s^* \overset{K_b}{\longleftrightarrow} DR_s^*. \tag{8.2}$$

In this scheme, the value of K_q is proportional to constitutive receptor activity, and K_b denotes the affinity constant of the active state of the receptor. Obviously, receptor activation cannot occur unless the receptor has some propensity, however small, to form the active state spontaneously ($K_q > 0$). The drug could also bind to the inactive state first and induce the active state by a process known as conformation induction:

$$D + R_s \overset{K_a}{\longleftrightarrow} DR_s^* \overset{K_b K_q / K_a}{\longleftrightarrow} DR_s^*. \tag{8.3}$$

The induction process is proportional to $K_b K_q / K_a$, in which K_a denotes the affinity constant of the inactive state of the receptor. Thus, once again, receptor activation cannot occur unless the value of K_q is greater than zero. Thus, we can conclude that constitutive activity is an inevitable property of any ligand-sensitive molecular switch.

For most wild type G protein-coupled receptors, the isomerization constant of the unoccupied receptor is probably very low ($K_q \leq 10^{-4}$). It follows that very little of the unoccupied receptor population would be in the active state, and hence, there would be few opportunities for the agonist to select the active state by the mechanism outlined in scheme 8.2. Thus, the main route for agonist-induced receptor activation is through conformational induction (scheme 8.3).

Constitutive receptor activity is frequently considered to be inconsistent with the occupancy assumption, which posits that receptor activation is proportional to receptor occupancy. If the receptor *is* active in the absence of agonist, would not this violate the theory? Not really because the theory was specifically intended to explain the activity of the agonist-receptor complex, not that of the unoccupied receptor. When there is constitutive activity, the activity of the orthosteric ligand-receptor complex is still proportional to occupancy. We just need to account for the activity of the unoccupied receptor as well. Otherwise, there will be errors in the estimates of affinity and efficacy if only the agonist–induced-response above basal activity is analyzed using the methods described in Chapter 6.

Modifying Receptor Theory to Account for Constitutive Receptor Activity

As mentioned above, the concentration-response curve of an agonist at a constitutively active receptor might look something like that shown in Figure 8.1. In the absence of agonist, a modest response occurs (15% in this example), and as the concentration of agonist increases, the response increases until a maximal effect is elicited. If the signal generated by the receptor was simply equivalent to basal receptor activation of in the absence of agonist plus that generated by the agonist-receptor complex, then the former could be subtracted from the latter and the difference (agonist concentration-response curve) could be analyzed in the normal manner. But as agonist occupancy increases, fewer receptors are unoccupied. If constitutive activity is defined as the spontaneous activity of unoccupied receptors, then constitutive activity decreases with an increase in agonist occupancy. At 100% receptor occupancy, the agonist-receptor stimulus is maximal, and constitutive receptor activity is absent because there are no unoccupied receptors. An accurate model for the receptor signal needs to account for the decline in constitutive activity coincident with the increase in the agonist-receptor signal.[22]

To begin, we need to define the behavior of the receptor population in the presence of the three different types of orthosteric ligands — agonist, inverse agonist and neutral antagonist. Figure 8.3a shows a simple model illustrating how the receptor population behaves. The level of constitutive activity is determined by the constant K_q, which is equivalent to the ratio of active (R^*) to inactive (R) states of the unoccupied receptor. For most receptors, the value of K_q is very low, but for the example shown in Figure 8.3b, it has been set to an arbitrarily high value (0.33) so

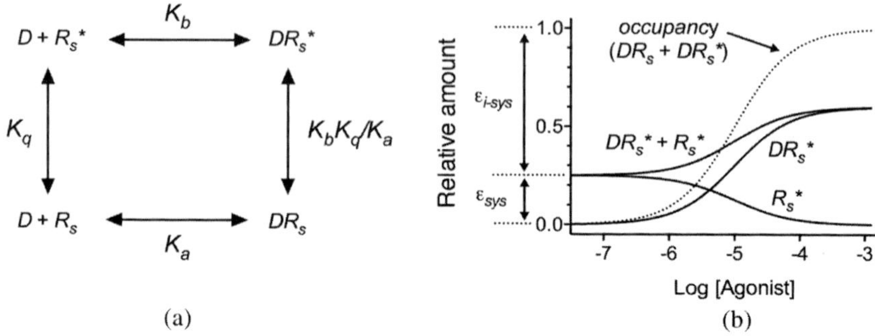

Figure 8.3: The consequences of a two-state model for the interaction of an agonist with a G protein-coupled receptor exhibiting constitutive activity. (a) The scheme shows the equilibrium binding of an agonist (D) with a receptor having two states, active (R_s*) and inactive (R_s). The parameter K_q denotes the isomerization constant of the unoccupied receptor, and K_b and K_a denote the affinity constants of the agonist for the active and inactive states of the receptor, respectively. (b) The plot shows how receptor occupancy ($DR_s + DR_s$*) and the occupied (DR_s^*), unoccupied (R_s^*) and total ($DR_s^* + R_s^*$) fractions of the receptor population in the active state change with the agonist concentration. In this plot, ε denotes the fraction of the population of occupied receptors in the active state, ε_{sys}, the fraction of the population of unoccupied receptors in the active state, ε_i, the fraction of occupied receptor population in the inactive state, and $\varepsilon_{i\text{-sys}}$, the fraction of the unoccupied receptor population in the inactive state.

that it will be easier to track the various receptor species. In Figure 8.3a, the affinity constants of the agonist (D) for the active and inactive states of the receptor are denoted by K_b and K_a, respectively.

Figure 8.3b tracks the various receptor species as a function of the agonist concentration. The fraction of the unoccupied receptor population in the active state is defined by the constant ε_{sys} (efficacy of the system). Its value is equivalent to the fraction of the total receptor population in the active state in the absence of ligand (0.25 for this example). As the concentration of agonist increases, occupancy of the active (DR_s^*) and inactive (DR_s) states of the receptor increases. The fraction of the occupied receptor population in the active state is denoted as ε (efficacy), and its value is independent of receptor occupancy for a receptor having a single orthosteric site. The value of efficacy is also equivalent to the fraction of the total receptor population in the active state when receptor occupancy by the agonist is equivalent to 100%.

As receptor occupancy increases, constitutive activity (R_s^*) decreases because the latter is defined as the fraction of the unoccupied receptor population in the active state and the amount of the unoccupied receptor complex decreases as occupancy increases. The total amount of receptor in the active state is shown as the summation of R_s^* and DR_s^* in Figure 8.3b.

There are two terms related to the fraction of the receptor population in the inactive state that will be useful later in the estimation of the affinity constant of the inactive state. The first is $\varepsilon_{i\text{-sys}}$ (inactivity of the system). This parameter is equivalent to the fraction of the unoccupied receptor population in the inactive state, and its value is equivalent to the fraction of the total receptor population in the inactive state in the absence of ligand. The second term is ε_i (inactivity), which is equivalent to the fraction of the population of ligand-receptor complexes in the inactive state. Its value is equivalent to the fraction of the total receptor population in the inactive state when receptor occupancy by the agonist is 100%. The relationships between the efficacy and inactivity terms are:

$$\varepsilon_i = 1 - \varepsilon. \tag{8.4}$$

$$\varepsilon_{i\text{-sys}} = 1 - \varepsilon_{\text{sys}}. \tag{8.5}$$

Figure 8.4 shows how the model behaves when the orthosteric ligand is a neutral antagonist (panel a) or inverse agonist (panel b). As receptor occupancy by the neutral antagonist increases, constitutive receptor activity decreases because there is less unoccupied receptor. With the decrease in constitutive activity, however, there is a coincident increase in the amount of neutral antagonist-receptor complex in the active state such that there is no change in the level of total receptor activation ($R_s^* + DR_s^*$). Thus, a neutral antagonist *has* intrinsic efficacy, and its value is equivalent to that of the system ($\varepsilon = \varepsilon_{\text{sys}}$ and $\varepsilon_i = \varepsilon_{i\text{-sys}}$). It can be seen that an inverse agonist also has intrinsic efficacy, but its value is less than that of the system ($\varepsilon < \varepsilon_{\text{sys}}$), and its intrinsic inactivity is greater than that of the system ($\varepsilon_i > \varepsilon_{i\text{-sys}}$).

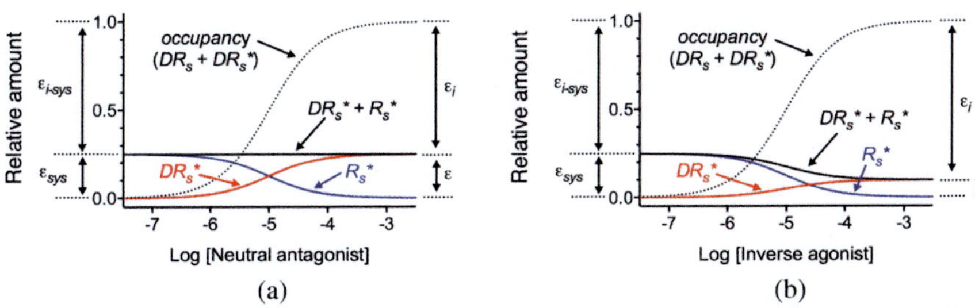

Figure 8.4: The consequences of a two-state model for the interaction of a neutral antagonist (a) and inverse agonist (b) with a *G* protein-coupled receptor exhibiting constitutive activity. The symbols are defined as described in the legend to Figure 3.

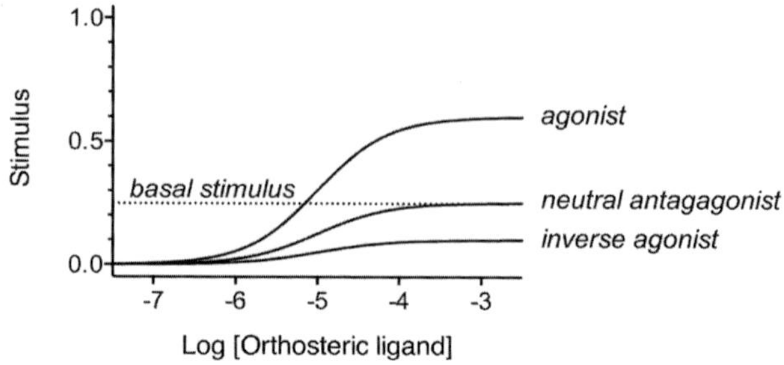

Figure 8.5: Stimulus functions for an agonist, neutral antagonist and an inverse agonist for a receptor exhibiting constitutive activity. The stimulus is defined as the fraction of the receptor population in the active state and bound with ligand. For receptors that exhibit simple one-site occupancy functions, the stimulus is equivalent to the product of receptor occupancy and efficacy.

Figure 8.5 shows the fraction of the population of orthosteric ligand-receptor complexes in the active state plotted against the concentration of the orthosteric ligand. This function is known as the stimulus, and its maximum is equivalent to ε. The intrinsic efficacy of the agonist is greater than that of the system (ε_{sys}); that of the neutral antagonist is equivalent to that of the system; and that of the inverse agonist is less than that of the system. Conversely, the intrinsic inactivity (ε_i) of the agonist is less than that of the system ($\varepsilon_{i\text{-}sys}$); that of a neutral antagonist is equivalent to that of the system; and that of an inverse agonist is greater than that of the system.

While it may seem counterintuitive that a neutral antagonist has intrinsic efficacy in a constitutively active system, this conclusion follows directly from the definition of efficacy established by Stevenson[23] and Furchgott,[24] which defines ε as the fraction of the population of ligand-receptor complexes in the active state. Because a neutral antagonist has no effect on the response, then it must replace constitutive receptor activity with an equivalent amount of ligand-receptor complex in the active state. Therefore, neutral antagonists *have* efficacy at constitutively active receptors. It may not be much, but in a highly sensitive signaling pathway its consequences are measureable. This theory also leads to the more intuitive conclusions that the intrinsic efficacy of a neutral antagonist is equivalent to that of the system and that the intrinsic inactivity of an inverse agonist is greater than that of the system.

Operational Model for Constitutive Receptor Activity

If we could measure receptor activation directly in a manner analogous to how an electrophysiologist can measure the ensemble average of a ligand-gated ion channel, then we could simply fit the model in Figure 8.3a to the ligand-receptor activation curve and estimate the relevant parameters (efficacy and the observed affinity constant). At *G* protein-coupled receptors, however, it is usually difficult to measure receptor activation directly, and it is often of great interest to measure how the orthosteric ligand affects a specific response, like changes in cAMP accumulation, mobilization of Ca^{2+} or the opening of a cyclic nucleotide-gated ion channel. To analyze these types of responses, we need to apply the reverse-engineering or null method approaches described in Chapter 7. The reverse engineering approach involves first deriving an equation that describes the total stimulus ($R_s^* + DR_s^*$) as a function of relevant parameters, like affinity and efficacy. This total-stimulus function can be substituted into the operational model for the stimulus. The resulting equation can be simplified and used to analyze concentration-response curves by nonlinear regression analysis.

First, let us develop an equation that describes the total stimulus (T) elicited by a receptor. This stimulus represents the summation of the stimuli elicited by the occupied (DR) and unoccupied (R) receptor complexes:

$$T = \frac{(\varepsilon_{\text{sys}} R + \varepsilon DR)R_T}{R + DR}. \tag{8.6}$$

In this equation, ε_{sys} represents the fraction of the unoccupied receptor population in the active state, ε, the fraction of the occupied receptor population in the active state, and R_T, the total receptor density. The observed affinity constant (K_{obs}) of the receptor is defined by:

$$K_{\text{obs}} = \frac{[DR]}{[R][DR]}. \tag{8.7}$$

As described in previous chapters (e.g., Chapter 2), K_{obs} is a population term, and it describes occupancy of the receptor. The reciprocal of K_{obs} is equivalent to the concentration of ligand that yields half-maximal receptor occupancy. It follows that the concentration of ligand required for half-maximal occupancy of both active and inactive states is equivalent to $1/K_{\text{obs}}$. Solving Equation (8.7) for DR yields:

$$[D]K_{\text{obs}}[R] = [DR]. \tag{8.8}$$

Substituting Equation (8.8) for DR in Equation (8.6) yields the following after simplification:

$$T = \frac{(\varepsilon_{\text{sys}} + \varepsilon DK_{\text{obs}})R_T}{1 + DK_{\text{obs}}}. \tag{8.9}$$

To derive an operational model for the analysis of ligand responses mediated by a receptor with constitutive activity, we simply need to substitute Equation (8.9) for the stimulus in the transducer function of the operational model. This function is described by Equation (7.5) in Chapter 7, and I reproduce it next in a form that addresses the stimulus generated by both the ligand-occupied and unoccupied receptor complexes. This simply involves replacing the variable S (stimulus of the ligand-receptor complex) with T (sum of the stimuli of the ligand-receptor complex and the unoccupied receptor):

$$\text{response} = \frac{T^m M_{\text{sys}}}{T^m + K_E^m}. \tag{8.10}$$

Substituting Equation (8.9) for T in Equation (8.10) yields:

$$\text{response} = \frac{M_{\text{sys}}}{1 + \left(\frac{1+DK_{\text{obs}}}{(\tau_{\text{sys}}+\tau DK_{\text{obs}})}\right)^m}, \tag{8.11}$$

in which,

$$\tau = \frac{\varepsilon R_T}{K_E}, \tag{8.12}$$

$$\tau_{\text{sys}} = \frac{\varepsilon_{\text{sys}} R_T}{K_E}. \tag{8.13}$$

Equation (8.11) is our goal. It can be used to analyze the concentration-response curves of agonists and inverse agonists in systems with constitutive receptor activity.

There are two additional parameters that are useful in understanding the estimation of the affinity constant of the ligand for the inactive state of the receptor as described in more detail below. These relationships are:

$$\tau_i = \frac{\varepsilon_i R_T}{K_E}, \tag{8.14}$$

$$\tau_{i\text{-sys}} = \frac{\varepsilon_{i\text{-sys}} R_T}{K_E}. \tag{8.15}$$

A significant distinction between the consequences of the form of the operational model in Equation (8.11) and that described by Equation (7.10) in Chapter 7 is that when there is constitutive receptor activity, a neutral antagonist has measurable efficacy, and hence a measurable τ value even though it does not select or induce the active state of the receptor. This is because efficacy is defined as the fraction of the population of ligand-receptor complexes in the active state regardless of whether the ligand induced the active state or simply bound to a receptor already in the active state.

By this point, it may have occurred to the reader that it should be possible to estimate the τ values of inverse agonists and neutral antagonists by analyzing their concentration-response curves with Equation (8.11). This would enable the estimation of both RA_i and RI_i values. Recall that these two parameters are relative estimates of the affinity constants of the orthosteric ligand for the active and inactive states of the receptor, respectively. The RA_i and RI_i values of a group of orthosteric ligands could be normalized relative to those of a neutral antagonist. If the K_{obs} value of the neutral antagonist is known, then it should be possible to estimate the actual K_b values of orthosteric ligands in units of M^{-1} because the K_{obs} value of a neutral antagonist is equivalent to both its K_b and K_a. In other words, for a constitutively active receptor, there is sufficient information in the concentration-response curves of orthosteric ligands to estimate the actual K_b value, not just its relative value (RA_i). The method for doing so is described in the upcoming sections.

Estimation of the Observed Affinity Constants of Partial and Inverse Agonists

In Chapter 7, it was shown that it is impossible to obtain unique estimates of the parameters of the operational model unless the observed affinity constant of the ligand or the maximal response of the system is already known. The same is true regarding the use of the operational model for constitutive receptor activity (Equation (8.11)). It is possible, however, to estimate the affinity constant of a partial or inverse agonist if the latter's concentration-response curve is analyzed simultaneously with that of a full agonist.

One way to describe this technique is to apply it to the analysis of theoretical concentration-response curves and to show that the parameters that were used to generate the data can be estimated. The theoretical data are simulated using the form of the operational model that describes the response as a function of the total stimulus (T) (Equation (8.11)). The total stimulus represents the summation of the agonist-bound and unbound forms of the active state of the receptor in a complex bound with G protein and guanine nucleotide (DR_s^*GX and R_s^*GX) (see Chapter 6, Equation (6.69)). These latter complexes are defined in terms of the microscopic affinity constants of the receptor states using the various equations that describe the level 2 parameters of the ternary complex model with guanine nucleotide in terms of the level 3 microscopic constants. This method of simulating functional responses is used throughout this chapter.

The results of such simulations are illustrated in Figure 8.6a, which shows the theoretical concentration-response curves of a full agonist (A), a partial agonist (B)

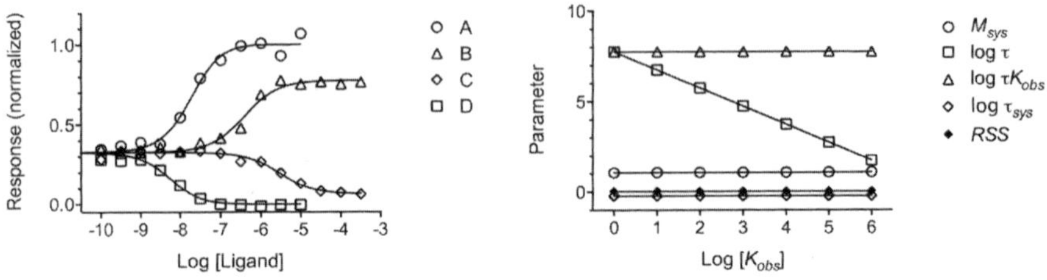

(a) (b)

Figure 8.6: Analysis of the concentration-response curves of an agonist (A), a partial agonist (B), and two inverse agonists capable of causing partial (C) and near complete (D) inhibition of constitutive receptor activity. (a) The theoretical responses were generated using the quaternary complex scheme from Chapter 6 to generate the total stimulus (T). The stimulus was then substituted into the transducer function of the operational model (Equation (8.10)) to generate response values. A random error was added to stimulate experimental error. The microscopic constants of the ligands for this simulation were: A, log K_b, 8.0; log K_a, 4.0; B, log K_b, 6.6; log K_a, 6.0; C, log K_b, 5.0; log K_a, 5.5 and D, log K_b, 6.0; log K_a, 8.0. The parameters of the transducer function were: M_{sys}, 1.0; $K_{E\text{-obs}}$, 0.005 and m, 1.5. The theoretical curves represent the global least-squares fit of Equation (8.11) to the data. (b) Summary of nonlinear regression analysis of the concentration response curve of agonist A to Equation (8.11).

and two inverse agonists (C and D). A random error has been added to the simulated data in this figure as well as those of other simulated data in this chapter so that it will be possible to determine how feasible it is to estimate specific parameters. All of the theoretical measurements represent the summation of constitutive receptor activity plus the activity of the ligand-receptor complex. Before continuing with the analysis, let us consider how real measurements of this type are actually made.

For example, suppose that the concentration-response curves of various agonists and inverse agonists for affecting cAMP accumulation were measured in a cell over expressing the β_2-adrenoceptor. The accumulation of at least some of the cAMP in the cell was probably not caused by activated β_2-adrenoceptors. Adenylate cyclase may have some basal activity in the absence of receptor stimulation. Other naturally occurring receptors in the cell may contribute to a low level of constitutive activity, and finally, the analytical technique used to measure cAMP may have a characteristic residual blank or background value even if no cAMP is present in the sample. Therefore, the measurement of cAMP in the nontransfected cell must be subtracted from the measurements made in the cell over expressing the β_2-adrenoceptor before the data can be analyzed using Equation (8.11).

This potential problem in the estimation of constitutive receptor activity places limitations on the types of preparations that can be examined. It seems likely that it might be impossible to estimate the basal activity of a naturally occurring constitutively active receptor in native tissues accurately unless there is means of preventing the expression of the receptor. The basal response measured in tissue from a knockout mouse lacking the receptor of interest could be measured and subtracted from the measurements made in wild type tissue. Compensatory changes might occur in the knockout mouse that could change this basal activity measurement, however. If a number of structurally diverse inverse agonists inhibit constitutive receptor activity to the same extent at high concentrations, it may be reasonable to assume that these agents are full-inverse agonists that completely inhibit the constitutive activity of the receptor. If so, then the residual activity measured at high concentrations of these agents could be subtracted from the response measurements to yield only the receptor-induced response.

Now let us return to the analysis of the data in Figure 8.6a. For the estimation of orthosteric ligand affinity constants, Equation (8.11) is fitted to the four curves in Figure 8.6a simultaneously using global nonlinear regression analysis sharing the estimates of the system parameters among the data (M_{sys}, τ_{sys} and m) and obtaining unique estimates of K_{obs} and τ for each agonist. When this was done, a good fit of Equation (8.11) to the data was obtained, however, the estimates of K_{obs} and τ for agonist A had large standard errors. Depending upon the data and the nonlinear regression algorithm used, it may be impossible to obtain a least-squares fit unless the τ or K_{obs} value of agonist A is constrained.

This is not a problem because it is possible to search a parameter space and identify the unique product of the K_{obs} and τ values of agonist A that yields a least-squares fit as described in Chapter 7 in relation to Figure 7.3. The plot in Figure 8.6b shows the parameter estimates for agonist A when the value of log K_{obs} is constrained to specific values over the range of 0–6. These estimates are plotted as points corresponding to the constrained value of log K_{obs} on the abscissa. A smooth curve has been drawn through the points for each parameter. The residual sum of the squared deviations of the regression equation from the theoretical data in Figure 8.6a is also plotted (*RSS*). When log K_{obs} is constrained to values between 0–6, *RSS* is at a constant minimum, indicating that the parameter estimates yield a least-squares fit. When the log K_{obs} value of agonist A is constrained to values greater than 6.0, *RSS* increases, indicating that a least-squares fit is no longer attained. Over the range of constrained K_{obs} values that yield a minimum *RSS*, the estimates of M_{sys}, log τ_{sys}, m and the log K_{obs} and log τ values of ligands B and C are also constant. As the constrained value of the log K_{obs} value of agonist A decreases, however, the best estimate of τ value of A increases proportionately, such that the product, τK_{obs}, is constant. Thus, to obtain the global least-squares fit, the K_{obs} value of agonist A is constrained to an arbitrarily low constant (e.g., 0), and the estimates of the other parameters that minimize *RSS* are obtained by global nonlinear regression analysis. The best estimates of the ligand parameters are: agonist A, log τK_{obs}, 7.75 ± 0.12; agonist *B*, log K_{obs}, 6.05 ± 0.12; log τ, 0.39 ± 0.10; inverse agonist C, log K_{obs}, 5.56 ± 0.16; log τ, −0.85 ± 0.23; and inverse agonist D, log K_{obs}, 8.22 ± 0.16. It is impossible to estimate an accurate log τ value of inverse agonist D because the simulated responses in the presence of high concentrations of this ligand are too small. The estimates of the parameters of the system are: M_{sys}, 1.07 ± 0.19; log τ_{sys}, −0.032 ± 0.074 and *m*, 1.07 ± 0.22. With the exception of the τ value of the full-inverse agonist D, these values are nearly equivalent to those used to generate the data in the first place: agonist A, log τK_{obs}, 7.78; agonist *B*, log K_{obs}, 6.01; log τ, 0.37; inverse agonist C, log K_{obs}, 5.50; log τ, −0.72; and inverse agonist D, log K_{obs}, 8.00 and log τ, −2.22. The difference being the theoretical and measured parameters is attributable to the random error in the simulated data. Example 1 in the appendix of this chapter gives the details of how this regression analysis is done.

The analysis of parameter space shown in Figure 8.6b can also be done for inverse agonists, and similar results are obtained in that the parameter estimates are constant over the range of constrained K_{obs} values of agonist A that yield a least-squares fit. It is unfeasible, however, to estimate the τ value of a full-inverse agonist accurately because the latter parameter is proportional to the difference between the activity measured in the presence of a high concentration of the inverse agonist and the background or blank measurement of the system. The latter represents the activity in the absence of the receptor. The error in the estimation of this small difference is huge, not withstanding the problems associated with determining what amount of the measurement constitutes the blank or background response in the absence of the receptor. It may seem fastidious to be concerned about whether a very small τ value differs from zero, but knowing, for example, that the observed dissociation constant of a ligand is 0.000001 M (1 μM; log K_{obs} = 6) and not 0.001 M (1 mM; log K_{obs} = 3) has a huge impact on our knowledge of how the drug behaves. Similarly knowing the log τ value of an inverse agonist is necessary for estimating its K_b value as described below.

Why not just subtract the constitutive response of the receptor from all of the response measurement and analyze the agonist-induced response above basal activity in the standard way using Equation (7.10) as described in Chapter 7 you might ask? A simple way to address this question is to analyze simulated data by the two different methods and determine if there are any differences in the results. Figure 8.7a shows a simulation of the concentration-response curves of a series of agonists for a system that exhibits substantial constitutive activity. All of the agonists

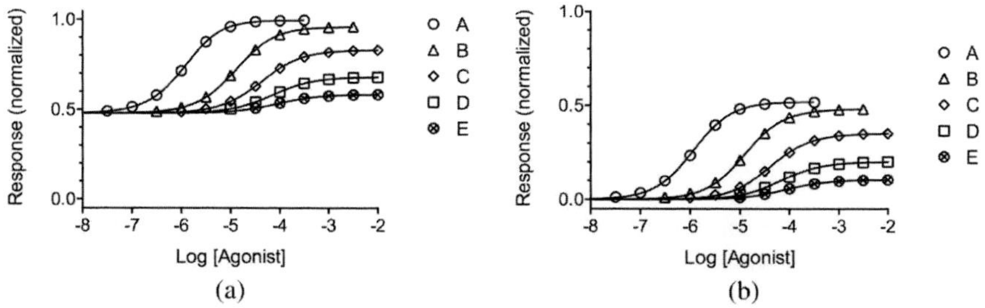

Figure 8.7: How the estimates of affinity and efficacy are affected by subtracting the constitutive stimulus from agonist responses before analysis. (a) Series of agonist concentration-response curves at a receptor exhibiting constitutive activity. The theoretical responses were generated without random error in a manner analogous to that described for the data in Figure 8.6. The microscopic constants of the ligands for this simulation were: A, log K_b, 6.0; log K_a, 4.0; B, log K_b, 5.0; log K_a, 4.0; C, log K_b, 4.5; log K_a, 4.0; D, log K_b, 4.25; log K_a, 4.0 and E, log K_b, 4.125, log K_a, 4.0. The parameters of the transducer function were: M_{sys}, 1.0; $K_{E\text{-obs}}$, 0.01 and m, 1.5. The theoretical curves represent the global least-squares fit of Equation (8.11) to the data points. (b) Same as (a) except that the constitutive response has been subtracted from all of the agonist responses. The theoretical curves represent the global least-squares fit of Equation (7.10) to the data.

have the same affinity for the inactive state of the receptor (log $K_a = 10^4\ \text{M}^{-1}$), but differ in their affinities for the active state (log K_b, 6.0, 5.0, 4.5, 4.25 and 4.125 for agonists A, B, C, D and E, respectively). No random error has been added to the data so that precise parameter estimates can be made. Figure 8.7b shows the same concentration-response curves, but with the constitutive response subtracted from the total response for each concentration of each agonist.

Both sets of data were analyzed by nonlinear regression analysis using Equation (8.11) for the data in Figure 8.7a and Equation (7.10) from Chapter 7 for the data Figure 8.7b. The corresponding parameter estimates are shown in solid and open bars, respectively, in Figure 8.8. Both approaches yield similar estimates of log τK_{obs} for the most efficacious agonists (A–C), but there is a systematic under estimation of this parameter as the efficacy of the agonist declines (Figure 8.8a). The log difference is approximately 0.6 (antilog, four-fold) for the least efficacious agonist (E). There is little error in the estimation of log K_{obs} (Figure 8.8b), showing that the error in the estimation of log τK_{obs} can be attributed to an error in the estimation of log τ (Figure 8.8c). With real experimental data having experimental error, it would be impossible to obtain individual estimates of log K_{obs} and log τ for the most efficacious agonist (agonist A) unless additional data were available (e.g., results from experiments employing partial receptor inactivation), so these estimates are not shown.

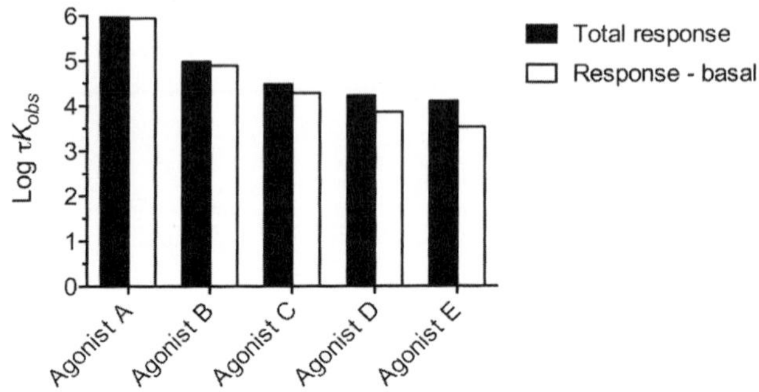

(a)

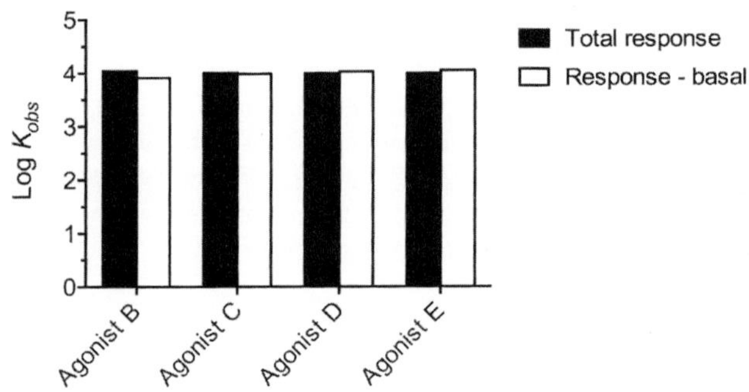

(b)

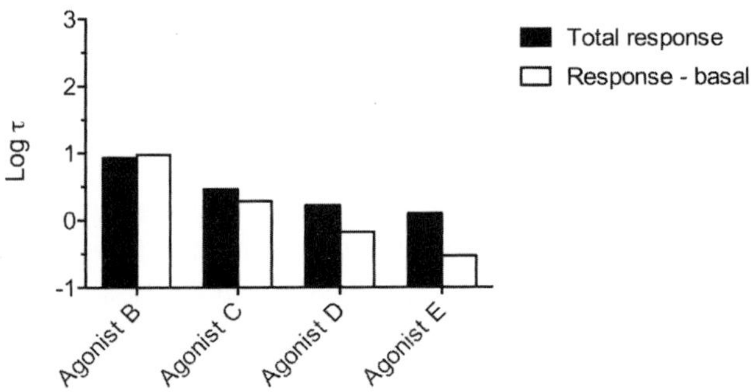

(c)

Figure 8.8: Summary of nonlinear regression analysis of the data in Figure 8.7a and 8.7b using Equations (8.11) and (7.10), respectively. The histograms show the estimates of log τK_{obs} (a), log K_{obs} (b) and log τ (c) for the data presented in panel a (open bars) and panel b (closed bars) of Figure 8.7.

There is another way to transform concentration-response curves with constitutive activity that has intuitive appeal as shown in Figure 8.9. This example is essentially the same as that shown in Figure 8.6a, but the random error has been eliminated. Figure 8.9a shows the concentration-response curves of the two agonists and inverse agonists and also defines the $E_{\max}$ values. These are equivalent to the response elicited by a receptor population when fully occupied by the orthosteric ligand. Under this condition, there is no constitutive response.

The response values in Figure 8.9b were derived by subtracting the constitutive response observed in the absence of agonist (0.31) from all of the response values in Figure 8.9a as indicated by the text beneath Figure 8.9b. Figure 8.9b also defines the *range* parameter, which represents the difference between the $E_{\max}$ value and the constitutive response. The sign of the value of the *range* is negative for the inverse agonists (C and D) and positive for the agonists (A and B).

If each concentration-response curve in Figure 8.9b is divided by its respective *range* the result would be a series of concentration-response curves each having an $E_{\max}$ value of 1 and a unique EC_{50} value. Now if each of these normalized concentration-response curves is multiplied by its respective $E_{\max}$ value as defined in Figure 8.9a, the result is the series of curves shown in Figure 8.9c. These two transformations of the data in Figure 8.9b (multiply by $E_{\max}$ and divide by range) are indicated by the text in Figure 8.9c ($\times E_{\max}/\text{range}$). The complete transformation of the response values (response) in Figure 8.9a is given by the following equation:

$$\text{response}_{\text{Trans}} = \frac{(\text{response} - \text{basal})E_{\max}}{\text{range}}. \tag{8.16}$$

in which, $\text{response}_{\text{Trans}}$ denotes the transformed response in Figure 8.9c, and basal, the constitutive response.

On the surface, the transformed responses in Figure 8.9c seem to represent only the response elicited by the agonist with the constitutive component eliminated. However, analysis of the data with standard form of the operational model (Equation (7.10)) reveals systematic deviations in the parameter estimates from the values used to simulate the data. If the transducer slope factor in the operational model is equivalent to 1, however, there is no error.

Thus, when there is substantial constitutive activity, it is necessary to analyze the concentration-response curves of partial agonists using a method that accounts for constitutive receptor activity (e.g., nonlinear regression analysis with Equation (8.11)). The standard form of the operational model could be used in cases where the Hill slopes of the concentration-response curves are equivalent to 1 if the data are transformed by Equation (8.16). This is not recommended, however, because it is usually better to analyze the untransformed data.

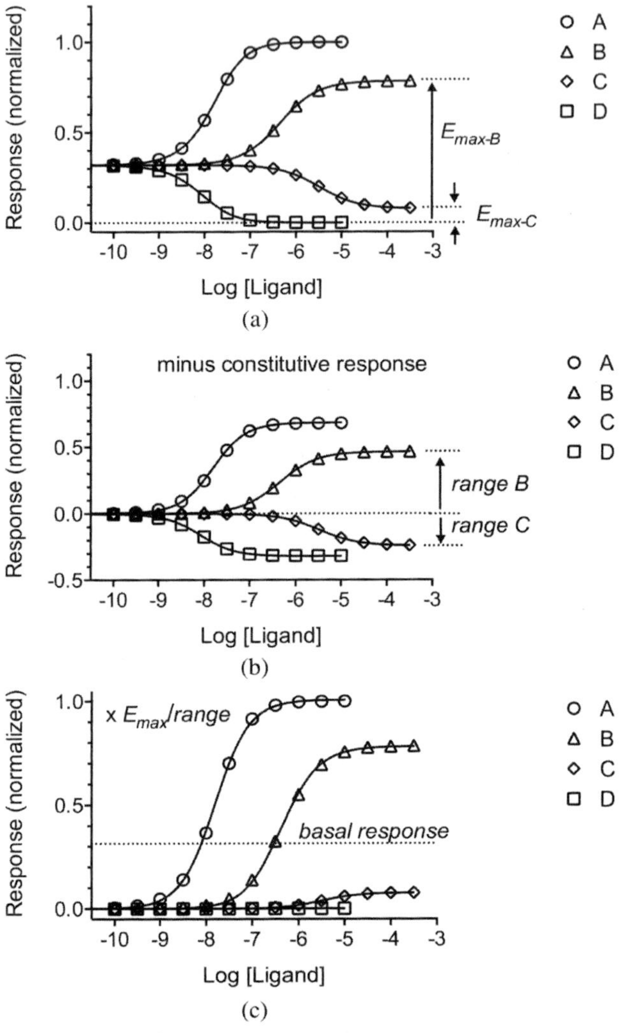

Figure 8.9: Graphical method for transforming ligand concentration-response curves to eliminate the contribution of constitutive activity to the response. (a) The simulated concentration-response curves are the same as those in Figure 8.6a but without the random error. (b) The response values were derived from those shown in panel a by subtracting the constitutive response. The range is equivalent to the response measured at a maximally effective concentration of ligand. (c) The response values in panel b have been divided by their respective ranges and multiplied by their respective $E_{\max}$ values in panel a to yield the concentration-response curves in panel c.

Estimation of the Observed Affinity Constants Orthosteric Ligands by Means of Partial Receptor Inactivation

If an irreversible antagonist is available, it should be possible to estimate both the K_{obs} and τ values of a full agonist by measuring responses under control conditions and also after partial inactivation of the receptor population as discussed in Chapter 6. The method can also be applied to partial agonists and inverse agonists, but this is unnecessary if a full agonist is available as shown by the prior simulation in Figure 8.6.

Depending upon the receptor and the goals of the experiment, it may be unnecessary to measure the K_{obs} value of a full agonist under conditions of constitutive activity because the fraction of the receptor population in the active state is probably unaffected by the level of receptor expression. So it should be possible to express the receptor at lower levels and estimate the K_{obs} value when there is no constitutive activity. If constitutive activity is induced by over expressing the G protein, however, then this could shift the equilibrium between active and inactive states in the active direction as shown by scheme 8.1.[4] This may cause an increase in the K_{obs} value of the agonist, but the increase could be very small in an intact cell with GTP in the cytosol. There may be some cases, however, where the receptor population cannot be manipulated, or the receptor exhibits constitutive activity regardless of its expression level. It is because of these latter possibilities that I describe a method for estimating the K_{obs} values of full agonists under conditions of constitutive receptor activity. The method can also be applied to partial agonists and inverse agonists.

The estimation of the K_{obs} value of a full agonist involves the alkylation of a fraction of the receptor population with an irreversible antagonist so that the available number of agonist-activated receptors is reduced. The irreversible receptor complex could have no effect on basal receptor activity, or it could display varying degrees of inverse agonism. If treatment with a receptor-saturating concentration of the irreversible antagonist for a time sufficient to cause near maximal receptor alkylation (four or five half-times for receptor alkylation) eliminates constitutive activity, then it can be assumed that the irreversible antagonist behaves like a full-inverse agonist. Irreversible neutral antagonists and full-inverse agonists are useful in the estimation of the K_{obs} values of full agonists in systems with constitutive activity, and the associated methods of analysis are described next.

As just mentioned, the technique involves measuring the concentration-response curve of the agonist under control conditions and after a fraction of the receptor population has had their orthosteric binding pockets covalently occluded with an irreversible antagonist. The degree of receptor activation should be such that the maximal effect of the agonist has been reduced to only 30–70% of its initial value. The two concentration-response curves are subsequently analyzed by global nonlinear regression analysis using either the equation for the irreversible neutral antagonist (Equation (8.18)) or the irreversible inverse agonist (Equation (8.20)) (see below).

First, I derive the equations for regression analysis, beginning with the equation for responses measured after treatment with an irreversible neutral antagonist. After such treatment, the receptor population behaves as two subpopulations — one having its orthosteric binding sites occluded and its constitutive activation unaffected and the other being unaffected by the irreversible neutral antagonist. The total stimulus (T) generated by these two subpopulations is given by:

$$T = \frac{q(\varepsilon_{\text{sys}} + \varepsilon D K_{\text{obs}}) R_T}{1 + D K_{\text{obs}}} + (1 - q)\varepsilon_{\text{sys}} R_T. \tag{8.17}$$

In this equation q denotes the fraction of the receptor population not inactivated by the irreversible neutral antagonist. The large fraction on the right side of the equation represents the stimulus elicited by the residual receptors, and it was derived by multiplying the equation for the total stimulus (Equation (8.9)) by q. The expression on the far right of the equation represents the constitutive stimulus elicited by the fraction $(1 - q)$ of the receptor population occluded by the irreversible neutral antagonist. This expression was derived by setting D equal to zero in the equation for the total stimulus (Equation (8.9)) and simplifying.

To generate an equation for the response to an agonist following partial receptor inactivation, the stimulus function described by Equation (8.17) is substituted into the transducer function of the operational model (Equation (8.10)) and simplified:

$$\text{response} = \frac{M_{\text{sys}}}{1 + \left(\frac{DK_{\text{obs}}+1}{qD\tau K_{\text{obs}}+\tau_{\text{sys}}(1+DK_{\text{obs}}(1-q))}\right)^m}. \tag{8.18}$$

Thus, the foregoing represents the equation describing the response to the orthosteric ligand following partial receptor inactivation with an irreversible, neutral antagonist in a system with constitutive activity.

The regression equation for responses after partial receptor inactivation with an irreversible inverse agonist is derived as follows. This type of irreversible ligand will not only prevent receptor activation by an orthosteric ligand, but it will also turn off the constitutive stimulus because of its full inverse agonist activity. Thus, the equation for the total stimulus after receptor inactivation can be derived by multiplying the total stimulus equation (Equation (8.9)) by the fraction of the receptor population not inactivated (q):

$$T = \frac{q(\varepsilon_{\text{sys}} + \varepsilon D K_{\text{obs}}) R_T}{1 + D K_{\text{obs}}}. \tag{8.19}$$

Substituting this equation into the transducer function of the operational model (Equation (8.10)) and simplifying yields the equation for responses measured after partial receptor inactivation with an irreversible inverse agonist:

$$\text{response} = \frac{M_{\text{sys}}}{1 + \left(\frac{1 + DK_{\text{obs}}}{q(D\tau K_{\text{obs}} + \tau_{\text{sys}})}\right)^m}. \tag{8.20}$$

If irreversible ligands are unavailable, it is possible to estimate the K_{obs} and τ values of a full agonist by analyzing responses measured under control conditions and at a reduced level of receptor expression. In this case, Equation (8.20) can be used in the analysis with the variable q representing the fractional level of receptor expression.

The theoretical data in Figure 8.10 will be used to explain the use of Equation (8.18) in the estimation of K_{obs}. Figure 8.10a shows the theoretical concentration-response curve of a full agonist in a system with constitutive receptor activity, before and after the treatment with an irreversible neutral antagonist. The theoretical curves were generated in a manner analogous to that described in Figure 8.6a. A random error has been added to all of the simulated data. Under control conditions, the agonist elicits a maximal response equivalent to the maximal response of the system. Following inactivation of a fraction of the receptor population with the

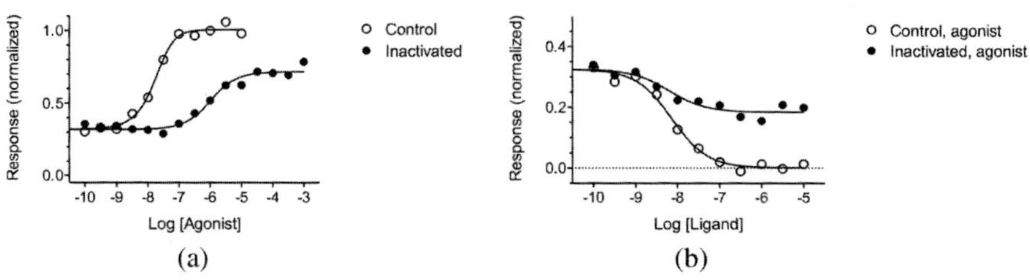

Figure 8.10: Effects of partial receptor inactivation with an irreversible neutral antagonist on the concentration-response curves of an agonist (a) and inverse agonist (b) in a system exhibiting constitutive activity. The theoretical responses were generated in a manner analogous to that described for the data in Figure 8.6, and the theoretical curves represent the least-squares fit of Equation (8.18) to the simulated response values. (a) The microscopic constants of the agonist were log K_b, 8.0 and log K_a, 4.0. The parameters of the transducer function were: M_{sys}, 1.0; $K_{E\text{-obs}}$, 0.005, q, 0.01 and m, 1.5. (b) The microscopic constants of the inverse agonist were log K_b, 6.0 and log K_a, 8.0. The parameters of the transducer function were: M_{sys}, 1.0; $K_{E\text{-obs}}$, 0.005, q, 0.40 and m, 1.5.

irreversible neutral antagonist, there is no change in the basal level of response, but the agonist concentration-response curve exhibits a shift to the right and a decrease in maximal response.

To estimate the dissociation constant of the full agonist, global nonlinear regression analysis is used to fit Equation (8.18) to the data, sharing the estimates of M_{sys}, τ_{sys}, m, K_{obs} and τ between the two curves and obtaining a unique estimate of q for the data simulated with receptor alkylation. The value of q is constrained to 1 for the control curve. Regression analysis yields estimates of 5.83 ± 0.18 and 1.94 ± 0.37 for log K_{obs} and log τ, respectively. The other parameter estimates were: M_{sys}, 0.98 ± 0.022; log τ_{sys}, -0.19 ± 0.11; m, 1.75 ± 0.90; and q, 0.013 ± 0.0049. These parameter values are generally similar those used to stimulate the data: log K_{obs}, 5.65; log τ, 2.13; M_{sys}, 1.0; log τ_{sys}, -0.22; m, 1.50 and q, 0.01. One exception is with regard to the transducer slope factor m, which exhibited a large standard error. This variation can be reduced with replicates of the simulated data.

The details of the analysis illustrated in Figure 8.10 is described in Example 2 of the appendix to this chapter.

It is also possible to estimate the dissociation constant of an inverse agonist using this method, but it is necessary to have a good estimate of M_{sys} for this analysis. This would require analyzing the inverse agonist data together with that of a full agonist. Yet as shown in the analysis associated with Figure 8.6, it is unnecessary to inactivate a fraction of the receptor population to obtain an estimate of the K_{obs} value of an inverse agonist when the data are analyzed simultaneously with those of a full agonist. So there is no benefit of using the method of partial receptor inactivation for inverse agonists.

Nonetheless, it is interesting to compare the effects of partial receptor inactivation on the concentration-response curves of agonists and inverse agonists. Figure 8.10b shows the effects of partial receptor inactivation on the concentration-response curve of an inverse agonist. The receptor system was the same as that shown in Figure 8.10a. Global nonlinear regression analysis with Equation (8.17) was used as described in Figure 8.10a, with the parameter q constrained to 1 for the control curve and M_{sys} constrained to 1 for both curves. This analysis yielded the following parameter estimates: log K_{obs}, 8.28 ± 0.19; log τ, -2.71 ± 3.05; q, 0.52 ± 0.096; log τ_{sys}, -0.30 ± 0.08 and m, 1.05 ± 0.31. With the exception of log τ, these estimates are similar to those used to simulate the data: log K_{obs}, 8.00; log τ, -2.21; log τ_{sys}, -0.22; m, 1.5 and q, 0.40. As mentioned above it is difficult to estimate the τ value of a near full inverse agonist accurately because the estimation depends on an accurate measurement of the small response measured at receptor saturating concentrations of the ligand.

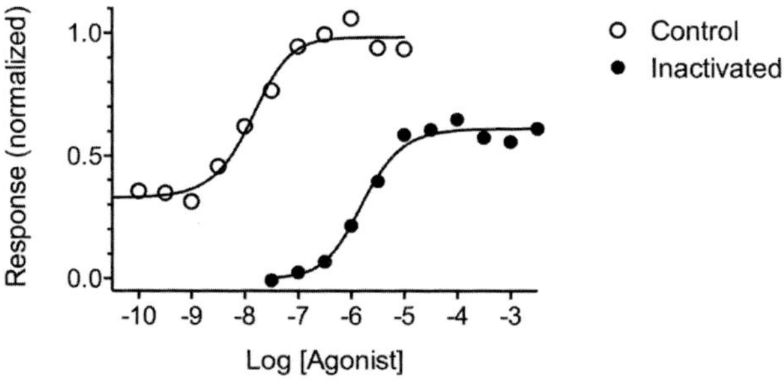

Figure 8.11: Effects of reduced receptor expression or partial receptor inactivation with a full inverse agonist on the concentration-response curves of an agonist. The theoretical responses were generated in a manner analogous to that described for the data in Figure 8.6, and the theoretical curves represent the least-squares fit of Equation (8.20) to the simulated response values. The microscopic constants of the agonist were log K_b, 8.0 and log K_a, 4.0. The parameters of the transducer function were: M_{sys}, 1.0; $K_{E\text{-obs}}$, 0.005, q, 0.01 and m, 1.5.

An interesting property of the data in Figure 8.10 is that while receptor inactivation reduces the potency of an agonist, it causes a very small increase in the potency of an inverse agonist.

The effect of an irreversible, full inverse agonist on the concentration-response curve to a full agonist in a system with constitutive activity is shown in Figure 8.11. The parameters used to simulate these data are given in the figure legend. In this case, the irreversible ligand causes a decrease in constitutive receptor activity. Global nonlinear regression analysis was used to fit Equation (8.20) to the data sharing the estimates of M_{sys}, m, K_{obs}, τ_{sys} and τ between the two curves and obtaining a unique estimate of q for the data simulated with receptor alkylation while constraining the value of q to 1 for the control curve. Regression analysis yielded estimates of 5.62 ± 0.24 and 2.24 ± 0.39 for log K_{obs} and log τ, respectively. The other parameter estimates were: M_{sys}, 0.98 ± 0.020; log τ_{sys}, -0.21 ± 0.082; m, 1.44 ± 0.45 and q, 0.0081 ± 0.0064. These parameter values are generally similar those used to stimulate the data: log K_{obs}, 5.65; log τ, 2.13; M_{sys}, 1.0; log τ_{sys}, -0.22; m, 1.50 and q, 0.01, with the minor differences being attributable to the random error in the simulated measurements.

The K_{obs} values of partial agonists and inverse agonists can also be estimated using the method of partial receptor inactivation described above. In the case of inverse agonists, however, it may be difficult to measure a reliable concentration-response curve after partial receptor inactivation with an irreversible full-inverse agonist, because the latter may reduce constitutive receptor activation to a very low level. Estimation of K_{obs} for partial and inverse agonists requires sufficient data for a reasonable estimate of M_{sys}, which can be achieved by analyzing the data simultaneously those of a full agonist. But if a full agonist is investigated, then one can estimate the K_{obs} of a partial agonist without inactivating the receptor population as described by the example illustrated in Figure 8.6. So there is no great advantage in using the method of partial receptor inactivation with partial and inverse agonists.

Estimation of the Observed Intrinsic Efficacies of Orthosteric Ligands

Observed efficacy is a measure of the fraction of the population of orthosteric ligand-receptor complexes in the active state, and it is easy to estimate a relative measure of this parameter from the analysis described so far. Equation (8.12) shows that the parameter τ is proportional to observed efficacy. Thus, if the τ value of one agonist (τ_A) is divided by that of another (τ_B), the result is the efficacy of agonist A expressed relative to that of B ($\varepsilon_A/\varepsilon_B$):

$$\frac{\tau_A}{\tau_B} = \frac{\frac{\varepsilon_A R_T}{K_E}}{\frac{\varepsilon_B R_T}{K_E}} = \frac{\varepsilon_A}{\varepsilon_B}. \tag{8.21}$$

This equation is the same as Equation (7.23) in Chapter 7. To estimate the relative efficacy values of a series of ligands for eliciting a particular response that exhibits constitutive receptor activity, their corresponding τ values are estimated using the analysis described in Figures 8.6, 8.10 or 8.11. One ligand (e.g., the most efficacious agonist) is designated as the standard to which the others are compared, and the τ value of each ligand is divided by that of the standard. The resulting ratios are the relative efficacy values of the ligands. As mentioned above, the analysis illustrated in Figure 8.6a did not provide an estimate of the τ value of the most efficacious agonist (A), but this can be estimated with additional experiments involving partial receptor inactivation.

If there is a lack of constitutive activity, then the efficacy values of a series of ligands can be appreciated by comparing their occupancy-response relationships and determining the level of receptor occupancy of each agonist that elicits a specified level of response. For example, if agonist A elicits a 30% response at one-tenth the level of receptor occupancy required by agonist B to elicit the same response, then the efficacy of agonist A is ten-fold greater than that of agonist B. An example of this logic is shown in Figure 7.6 of Chapter 7. When there is constitutive receptor activity, however, it is not a simple matter to estimate equi-active responses from a plot of the total response (constitutive plus drug-induced) vs. the ligand concentration.

This point is illustrated in Figure 8.12. Figure 8.12a shows the concentration-response curves of a full agonist, a neutral antagonist and a partial inverse agonist for a receptor signaling pathway that exhibits constitutive activity. Any horizontal line through the plot only intersects but one concentration-response curve, indicating that none of the total responses elicited by the orthosteric ligands is equivalent. It is impossible, therefore, to compare equi-active responses.

But if the component of the total response elicited by the ligand-receptor complex is plotted, the resulting curves resemble those of three agonists with different maximal responses as shown in Figure 8.12b. This can be done for the agonist and inverse agonist using Equation (8.11) with the value of τ_{sys} constrained to zero and the values of the other parameters (M_{sys}, τ and K_{obs}) constrained to those that yielded a best fit to the data in Figure 8.12a. We would need more information for the neutral antagonist (K_{obs}, see Chapter 10) to make the transformation shown in Figure 8.12b. The E_{max} values of the concentration-response curves in Figure 8.12b are the same as those of the total responses plotted in Figure 8.12a. Note that the E_{max} value of the neutral antagonist in Figure 8.12b is equivalent to the constitutive response of the unoccupied receptor, whereas that of the inverse agonist is less than that of the constitutive response. Relative efficacy can be conceptualized by plotting the responses elicited by the ligand-receptor complex from Figure 8.12b against receptor occupancy by the ligand. Knowing K_{obs}, receptor occupancy is calculated as:

$$\text{occupancy} = \frac{DK_{obs}}{DK_{obs} + 1}. \tag{8.22}$$

Because the range of occupancy values is huge, the ligand-receptor responses have been plotted against occupancy on a log scale in Figure 8.12c.

If a horizontal line is drawn across the plots in Figure 8.12c at a level as close as possible to the half-maximal response yet low enough to intersect the three curves, one can interpolate the levels of receptor occupancy by the three ligands that yield equivalent responses. The vertical dotted arrows indicate the receptor occupancy levels by the agonist (0.002), neutral antagonist (0.30) and inverse agonist (1) that yield an equivalent response (0.23). Because it requires 500-fold greater receptor occupancy by the inverse agonist to produce a response equivalent to that of the agonist, then the efficacy of the inverse agonist is only one five-hundredth that of the agonist. Similarly, the efficacy of the neutral antagonist is only one one-hundred-fiftieth that of the agonist because it requires 150-fold more occupancy by the former to produce a response equivalent to that of the latter. Thus, relative to the agonist, the efficacies of the neutral antagonist and inverse agonist are 0.0067 and 0.002, respectively.

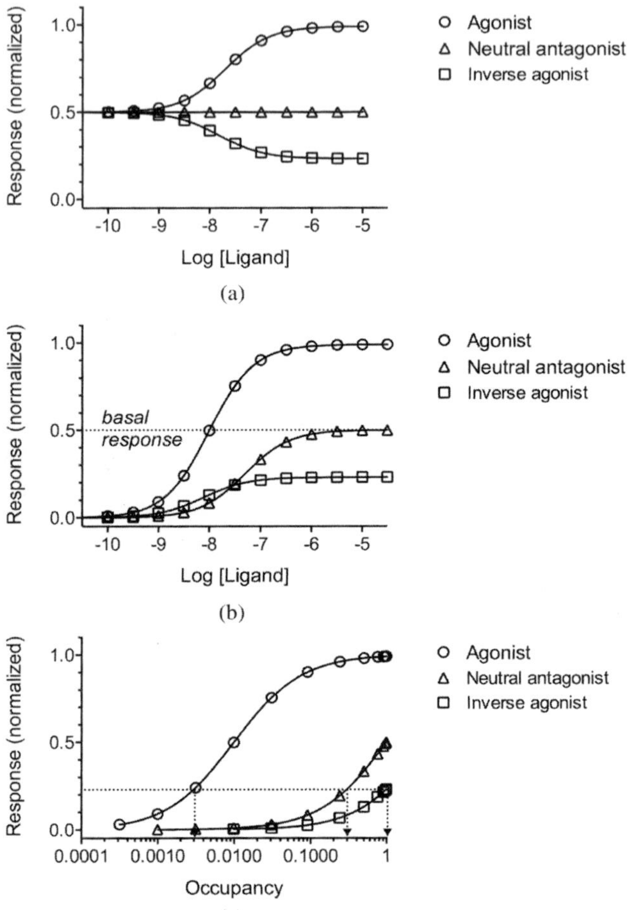

Figure 8.12: Graphical approach for determining the relationship between receptor occupancy and the response elicited by the ligand-receptor complex a system exhibiting constitutive activity. (a) Simulated concentration-response curves of an agonist (log K_{obs}, 6.0; ε, 0.5), neutral antagonist (log K_{obs}, 7.0; $\varepsilon = \varepsilon_{sys}$) and inverse agonist (log K_{obs}, 8.0; ε, 0.0015). The observed affinity constants and efficacy values of the ligands were substituted into Equation (8.9) with $\varepsilon_{sys} = 0.005$ to generate total stimulus values at various concentrations of the ligand. These estimates were then substituted into transducer function, Equation (8.10), to generate the concentration-response curves. (b) Responses elicited by the ligand-receptor complex. The responses were generated as described for panel a, but with ε_{sys} set to 0. (c) Occupancy-response relationships. The ligand-receptor response values in panel b are plotted against receptor occupancy by the ligand on a log scale (see Equation (8.22) for the calculation of receptor occupancy). The horizontal dotted line corresponds to a fractional response of 0.23. The vertical dotted lines indicate the level of receptor occupancy required by each ligand to generate a response equivalent to 0.23.

As mentioned above, it is unnecessary to make the complicated plot in Figure 8.12 to estimate the efficacy of orthosteric ligands at constitutively active receptors. It is presented here to help explain the meaning of efficacy as it applies to neutral antagonists and inverse agonists at the population level of analysis. One can simply apply Equation (8.21) after analyzing the functional data with the operational model. It is also possible to estimate the relative efficacies of orthosteric ligands using Equation (8.45) if their K_{obs} and K_b values are known as described below.

You might wonder why I am considering the amount of neutral antagonist- or inverse agonist-receptor complex required to generate a particular level of response. After all, does not a neutral antagonist lack an effect when bound to a receptor population and does not an inverse agonist inhibit the effect of a constitutively active receptor population? While the latter statements are correct, the definition of efficacy is the fraction of the occupied receptor population in the active state. Whenever a receptor system exhibits constitutive activity, then a significant fraction of the population of neutral antagonist- or inverse agonist-receptor complexes will be in the active state. If so, then these types of ligands will have significant efficacy values when defined at the population level of analysis.

A more informative way to consider efficacy is at a deeper level of analysis relating to receptor states as discussed in the next section. But to address this issue, we need to define the population parameters consistently across different systems regardless of the level of constitutive activity, and this leads to significant efficacy values for antagonists and inverse agonists whenever there is constitutive activity.

Estimation of the Microscopic Affinity Constants of the Active and Inactive States of the Receptor

At the beginning of this chapter, it was mentioned that functional responses mediated through a constitutively active receptor could be analyzed to estimate the microscopic affinity constants of an orthosteric ligand for the active (K_b) and inactive (K_a) states of the receptor. In this section, the requisite receptor theory for making these estimations is developed and examples of how to apply this theory are illustrated.

Receptor theory: In Chapter 7, it was demonstrated that the product of observed affinity (K_{obs}) and intrinsic efficacy (ε) is proportional to the microscopic affinity constant of an agonist for the active state of the receptor (K_b). If we knew the proportionality constant for this relationship, we could simply divide the product, $K_{\text{obs}} \times \varepsilon$, by this constant and obtain an estimate of K_b.

One way to estimate the proportionality constant is to use the model shown in Figure 3a to derive equations for K_{obs} and ε in terms of the microscopic constants of the model (i.e., K_a, K_b and K_q). The two equations can then be multiplied together to yield the product, $K_{\text{obs}} \times \varepsilon$. This product can then be divided by K_b to yield the proportionality constant.

For this strategy, we can use the equations for K_{obs} and ε that were derived in Chapter 7 (i.e., Equations (7.111) and (7.116)). Setting the product of K_{obs} and ε equal to the product of Equations (7.111) and (7.116) yields:

$$K_{\text{obs}}\varepsilon = \frac{K_a + K_b K_q}{1 + K_q} \times \frac{1}{1 + \frac{K_a}{K_b K_q}}. \tag{8.23}$$

Simplification yields:

$$K_{\text{obs}}\varepsilon = \frac{K_b K_q}{1 + K_q}. \tag{8.24}$$

Dividing both sides by K_b yields:

$$\frac{K_{\text{obs}}\varepsilon}{K_b} = \frac{K_q}{1 + K_q}. \tag{8.25}$$

Substituting in the ratio R_s^*/R_s for K_q and simplifying yields:

$$\frac{K_{\text{obs}}\varepsilon}{K_b} = \frac{R_s^*}{R_s + R_s^*}. \tag{8.26}$$

The expression on the right-hand side of Equation (8.26) represents the fraction of the free receptor complex in the active state, which is equivalent to ε_{sys}. Thus, Equation (8.26) reduces to:

$$\frac{K_{\text{obs}}\varepsilon}{K_b} = \varepsilon_{\text{sys}}, \tag{8.27}$$

in which,

$$\varepsilon_{\text{sys}} = \frac{K_q}{1 + K_q}. \tag{8.28}$$

Thus, ε_{sys} is the proportionality constant for the relationship between the microscopic affinity constant of the active state of the receptor and the product of observed affinity and intrinsic efficacy. In signaling pathways that lack measureable constitutive receptor activity, it is usually impossible to estimate τ_{sys}, and hence, only RA_i can be estimated. But in constitutively active receptor systems, it is possible to measure τ_{sys}, and therefore, to estimate K_b by dividing the product, $K_{obs} \times \varepsilon$, by ε_{sys}:

$$\frac{K_{obs}\varepsilon}{\varepsilon_{sys}} = K_b. \tag{8.29}$$

There is an analogous relationship for the estimation of the microscopic affinity constant for the inactive state of the receptor (K_a):

$$\frac{K_{obs}\varepsilon_i}{\varepsilon_{i\text{-sys}}} = K_a. \tag{8.30}$$

To prove this relationship, we can 1) derive equations for K_{obs}, ε_i and $\varepsilon_{i\text{-sys}}$ in terms of the microscopic constants, 2) substitute these equations for the appropriate variables in Equation (7.30) and 3) determine if the resulting equation yields the equality $K_a = K_a$.

The equation for $\varepsilon_{i\text{-sys}}$ is derived by first rearranging Equation (8.5) into the following form:

$$\varepsilon_{sys} = 1 - \varepsilon_{i\text{-sys}}. \tag{8.31}$$

Substituting in Equation (8.28) for ε_{sys} in Equation (7.31) followed by rearrangement yields:

$$\varepsilon_{i\text{-sys}} = \frac{1}{1 + K_q}. \tag{8.32}$$

The equations for K_{obs} and ε_i have already been derived in Chapter 7 (Equations (7.111) and (7.117)). Substituting in Equations (7.111), (7.117) and (8.32) for K_{obs}, ε_i and $\varepsilon_{i\text{-sys}}$ in Equation (8.30) yields:

$$\frac{\frac{K_a + K_b K_q}{1+K_q} \times \frac{1}{1+\frac{K_b K_q}{K_a}}}{\frac{1}{1+K_q}} = K_a. \tag{8.33}$$

This equation reduces to the equality $K_a = K_a$, which proves Equation (8.30) from which it was derived.

To take advantage of the theory just described, it is necessary to relate it to the operational model (Equation (8.11)). This will enable the estimation of K_b and K_a whenever a downstream response at a G protein-coupled receptor is measured provided that the response exhibits constitutive receptor activity.

To begin, Equations (8.12) and (8.13) can be rearranged in the following form:

$$\varepsilon = \frac{\tau K_E}{R_T}, \tag{8.34}$$

$$\varepsilon_{\text{sys}} = \frac{\tau_{\text{sys}} K_E}{R_T}. \tag{8.35}$$

Substituting Equations (8.34) and (8.35) for ε and ε_{sys} in Equation (8.29) yields:

$$\frac{K_{\text{obs}}\tau}{\tau_{\text{sys}}} = K_b. \tag{8.36}$$

Because both τ_{sys} and the product $K_{\text{obs}}\tau$ can always be estimated from the concentration-response curve of an orthosteric ligand at a constitutively active receptor, it follows that it is always possible to estimate K_b using Equation (8.36).

An analogous approach can be used to estimate K_a from the parameters of the operational model. Rearranging Equations (8.14) and (8.15) yields:

$$\varepsilon_i = \frac{\tau_i K_E}{R_T}, \tag{8.37}$$

$$\varepsilon_{i\text{-sys}} = \frac{\tau_{i\text{-sys}} K_E}{R_T}. \tag{8.38}$$

Substituting Equations (8.37) and (8.38) for ε_i and $\varepsilon_{i\text{-sys}}$ in Equation (8.30) yields:

$$\frac{K_{\text{obs}}\tau_i}{\tau_{i\text{-sys}}} = K_a. \tag{8.39}$$

In the next section, we will apply this theory in the estimation of K_b and K_a from the concentration-response curves of orthosteric ligands in receptor systems that exhibit constitutive activity.

Estimating K_b values from functional data

I will describe how to use Equation (8.39) to estimate K_b by applying it to the theoretical data already analyzed in Figures 8.6a and 8.10. Recall that the concentration-response curves of the full agonist A, the partial agonist B and the two inverse agonists C and D shown in Figure 8.6a were analyzed by global nonlinear regression analysis using Equation (8.11). This analysis yielded estimates the K_{obs} and τ values of the partial agonist and one of the inverse agonists as well as the τ_{sys} value for this response. At this point, all that one has to do is divide the product of K_{obs} and τ by τ_{sys} to yield the corresponding K_b values of orthosteric ligands B and C (see Equation (7.48)). These estimates are illustrated in Figure 8.13 and

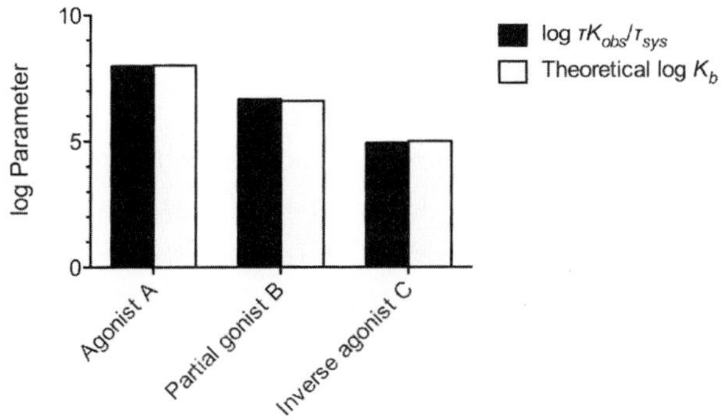

Figure 8.13: Relationship between the population parameters, τ, $K_{\rm obs}$ and $\tau_{\rm sys}$, and the microscopic affinity constant of a ligand for the active state of the receptor, K_b. The histogram shows the product, $\tau K_{\rm obs}$, divided by $\tau_{\rm sys}$ for ligands A, B and C from Figure 8.6a (solid bars). The values of these parameters were estimated from the simulated data in Figure 8.6a as described above. The open bars illustrate the theoretical values of the microscopic affinity constants of ligands for the active state of the receptor.

labeled $K_{\rm obs}\tau/\tau_{\rm sys}$. It can be seen that these estimates are in close agreement with the theoretical K_b values listed in the table that were used to simulate the data in the first place. It was impossible to estimate the K_b value of the full inverse agonist because of the large error associated with the estimate of its τ value as discussed above. In the case of the full agonist A, it was impossible to obtain distinct estimates of $K_{\rm obs}$ and τ, yet nonetheless, it was possible to estimate their product using the approach illustrated in Figure 7.6b. Dividing this product by $\tau_{\rm sys}$ yields an accurate estimate of K_b for agonist A as shown by the close agreement between the estimate of $K_{\rm obs}\tau/\tau_{\rm sys}$ and the theoretical value of K_b used to simulate the data (Figure 8.13).

Estimating the K_b values of the agonist (Figure 8.10a) and inverse agonist (Figure 8.10b) is a simple matter after analyzing the concentration-response curves measured before and after partial receptor inactivation with the appropriate form of the operational model (Equation (8.18)). Again, the product of the estimates of $K_{\rm obs}$ and τ is divided by $\tau_{\rm sys}$ to yield an estimate of K_b. These calculations yielded log K_b values of 7.96 and 5.87 for the agonist and inverse agonist, which are similar to the theoretical values used to simulate the data (8.0 and 6.0, respectively).

Another approach for estimating the K_b values of orthosteric ligands is to analyze their concentration-response curves with a form of the operational model that has the K_b value as one of the parameters. Note that the various forms of the operational model (Equations (8.11), (8.18) and (8.20)) that were used to analyze the data in Figures 8.6, 8.10 and 8.11 all have the occurrence of the product $\tau K_{\rm obs}$

in the equation. This product can be replaced with the expression:

$$\tau_{sys} K_b = \tau K_{obs}, \tag{8.40}$$

which was derived by multiplying both sides of Equation 8.36 by τ_{sys}. Substituting $\tau_{sys} K_b$ for τK_{obs} in Equations (8.11), (8.18) and (8.20) yields the corresponding equations:

$$\text{response} = \frac{M_{sys}}{1 + \left(\frac{DK_{obs}+1}{\tau_{sys}(DK_b+1)}\right)^m}, \tag{8.41}$$

$$\text{response} = \frac{M_{sys}}{1 + \left(\frac{DK_{obs}+1}{\tau_{sys}(qDK_b+(1+DK_{obs}(1-q)))}\right)^m}, \tag{8.42}$$

$$\text{response} = \frac{M_{sys}}{1 + \left(\frac{1+DK_{obs}}{\tau_{sys}(qDK_b+1)}\right)^m}. \tag{8.43}$$

These Equations (8.41)–(8.43) can be used in the same manner as the original Equations (8.11), (8.18) and (8.20) from which they were derived except that K_b values are estimated instead of τ values.

A brief review of how these equations are used follows. Equation (8.41) is fitted to the concentration-response curves of a series of orthosteric ligands, like those shown in Figure 8.6a, to estimate the K_b values of the ligands. The curves are analyzed simultaneously by global nonlinear regression analysis sharing the estimates of M_{sys}, τ_{sys} and m among the curves and estimating unique values of K_{obs} and K_b for each ligand. In the case of any full agonist, it might be necessary to constrain the K_{obs} value to an arbitrary constant equal to or less than the true K_{obs} value.

Equation (8.42) is fitted to the concentration-response of an orthosteric ligand measured before and after inactivation of a fraction of the receptor population with an irreversible neutral antagonist. The two curves are analyzed simultaneously by global nonlinear regression analysis sharing the estimates of M_{sys}, τ_{sys}, m, K_{obs} and K_b between the two curves. The value of q is constrained to 1 for the control curve and a unique estimate of q is obtained for the curve measured after partial receptor inactivation.

Equation (8.43) is used in the estimation of K_b by the method of reduced receptor expression or partial receptor inactivation using an irreversible, full, inverse agonist. In this case, the curves measured before and after reduced receptor expression or partial receptor inactivation are analyzed simultaneously by global nonlinear regression analysis, sharing the estimates of M_{sys}, m, K_{obs} and K_b between the two curves. The value of q is constrained to 1 for the control curve and a unique value of q is estimated for the curve measured after reduced expression or partial receptor inactivation.

Most nonlinear regression applications provide estimates of the asymptotic standard error of the parameter estimates. An advantage of using Equations (8.41)–(8.43) with such an application is that the standard error value of K_b will be provided.

It is also possible to estimate the relative efficacy values of orthosteric ligands from the K_b and K_{obs} values. As described in Chapter 6, the product of observed affinity and intrinsic efficacy of one orthosteric ligand expressed relative to that of another is equivalent to the corresponding ratio of affinity constants for the active state of the receptor:

$$\frac{\varepsilon K_{\text{obs}}}{\varepsilon' K'_{\text{obs}}} = \frac{K_b}{K'_b}. \tag{8.44}$$

In this equation, the apostrophes denote the parameters of the test orthosteric ligand. Rearranging this equation yields:

$$\frac{\varepsilon}{\varepsilon'} = \frac{K_b K'_{\text{obs}}}{K'_b K_{\text{obs}}}. \tag{8.45}$$

Thus, the relative efficacy values of agonist can be estimated from the K_b and K_{obs} values using Equation (8.45). Again, one can also estimate relative efficacy by calculating the ratio of the corresponding τ values of the ligands as described above.

Estimating K_a values from functional data

A method for estimating the K_a values of agonist has been described previously.[25] Here, I will summarize the steps of the process and use this technique to analyze theoretical data.

The data needed for this analysis are the concentration-response curves of the ligands. In addition, the concentration-response curve of the most efficacious agonist also needs to be measured after the receptor population has been partially inactivated. Using this information, the experimenter can estimate the K_a values of all the agonists, except for the most efficacious agonist. As mentioned above, the same information can be used to estimate the K_b values of all the ligands, except that of a full inverse agonist. To measure the K_a value of the most efficacious full agonist and the K_b of a full inverse agonist, additional concentration-response curves need to be measured under conditions of allosteric modulation or after constitutively activating point mutations have been introduced into the receptor. These approaches are not addressed in this chapter.

Before describing the derivation of the equation for this analysis, let me explain its basis. It has been shown that if the K_{obs} value of the most efficacious agonist is known, then it is possible to obtain a good estimate the K_a value of any other ligand whose efficacy is equal to less than one-half that of the most efficacious agonist.[25] The method involves constraining the value of K_E to the largest value that can still yield a least-squares fit to the data for the most efficacious agonist. Equation (8.12) shows that for given values of τ and R_T, the value of K_E is largest when the value of ε is at its maximum value of 1. This maximum ε value corresponds to the efficacy value of an agonist that completely activates the population of occupied receptors. In the method described below, the value of ε for the most efficacious agonist is constrained to 1 and the parameter K_E is left unconstrained.

This technique may seem inappropriate because the parameter K_E is embedded in the parameter τ in the various forms of the operational model that have been described so far in this chapter, and it is impossible to estimate K_E from the types of data we have considered. More importantly, it seems inappropriate to arbitrarily assign a value of 1 to ε even if the agonist is a full agonist because high intracellular concentrations of GTP are likely to reduce the fraction of the agonist-occupied receptor population in the active state to a value less than 1. The procedure yields accurate K_a values, nonetheless, so long as the efficacy of the test ligand is less than one-half that of the most efficacious agonist. Our inability to know the true value of K_E is what limits the analysis to ligands having efficacies that are less than half that of most efficacious agonist.

With a provisional value of K_E, it is possible to estimate provisional values of ε and ε_{sys}, and therefore, ε_i and $\varepsilon_{i\text{-sys}}$. Ultimately, K_a can be calculated by substituting the values of K_{obs}, ε_i and $\varepsilon_{i\text{-sys}}$ into Equation (8.30).

In the method described below, a regression equation is developed with the parameter ε_i constrained to 0 for the most efficacious agonist. This is equivalent to constraining ε to 1. Regression analysis yields provisional values of K_E and the ε_i values of the less efficacious ligands, from which the corresponding K_a values can be estimated.

As mentioned, three types of concentration-response curves need to be measured for this analysis: 1) that of the most efficacious agonist under control conditions and 2) after partial receptor inactivation, and 3) those of the other ligands. A different regression equation will be used for each type of data. Below I describe the development of a single general equation and show how to constrain specific parameters for each type of data. Using this approach, a single equation can be globally fit to all of the data.

I begin by illustrating how to modify the operational model for constitutive activity (Equation (8.11)) so that it can be used in the analysis of all but for the most efficacious ligand. First, Equations (8.12) and (8.13) are substituted for τ and τ_{sys}, respectively, in Equation (8.11). Then the efficacy terms (ε and ε_{sys}) in the resulting equation are replaced by the expressions $1 - \varepsilon_i$ and $1 - \varepsilon_{i\text{-sys}}$ (see Equations (8.4) and (8.5)), respectively, to yield the following form of Equation (8.11) after simplification:

$$\text{response} = \frac{M_{sys}}{1 + \left(\frac{K_{E\text{-obs}}(1+DK_{obs})}{(1-\varepsilon_{i\text{-sys}})+(1-\varepsilon_i)DK_{obs}}\right)^m}, \tag{8.46}$$

in which,

$$K_{E\text{-obs}} = \frac{K_E}{R_T}. \tag{8.47}$$

This form of the operational model is appropriate for analysis of the concentration-response curves of all of the ligands except the most efficacious.

To derive an equation for the analysis of the concentration-response curve of the most efficacious agonist for the condition of partial receptor inactivation, the expressions $\varepsilon/K_{E\text{-obs}}$ and $(1 - \varepsilon_{i\text{-sys}})/K_{E\text{-obs}}$ are substituted for τ and τ_{sys} in Equation (8.18). Then, a value of 1 is substituted for ε, which yields the following equation upon simplification:

$$\text{response} = \frac{M_{sys}}{1 + \left(\frac{K_{E\text{-obs}}(DK_{obs}+1)}{(1-\varepsilon_{i\text{-sys}})(1+DK_{obs}(1-q))+qDK_{obs}}\right)^m}. \tag{8.48}$$

To derive an equation for analysis of the control data for the most efficacious agonist, Equation (8.48) is solved for the condition of $q = 1$:

$$\text{response} = \frac{M_{sys}}{1 + \left(\frac{K_{E\text{-obs}}(DK_{obs}+1)}{(1-\varepsilon_{i\text{-sys}})+DK_{obs}}\right)^m}. \tag{8.49}$$

The three Equations (8.46), (8.48) and (8.49) for the estimation of K_a values can be represented by the single equation:

$$\text{response} = \frac{M_{sys}}{1 + \left(\frac{K_{E\text{-obs}}(DK_{obs}+1)}{(1-\varepsilon_{i\text{-sys}})(1+DK_{obs}(1-q))+q(1-\varepsilon_i)DK_{obs}}\right)^m}, \tag{8.50}$$

provided that the following parameter constraints are in place. First, the system parameters (M_{sys}, m and $\varepsilon_{i\text{-sys}}$) are shared across all data sets. For the analysis of the data for all of the ligands except the most efficacious agonist, the parameter q is constrained to 1, which reduces Equation (8.50) to Equation (8.46). For the analysis of the data for the most efficacious agonist under the condition of partial receptor inactivation, the parameter ε_i in Equation (8.50) is constrained to 0, which yields Equation (8.48). Finally, for the analysis of the control data for the most efficacious agonist, the parameters q and ε_i are constrained to 1 and 0, respectively, which yields Equation (8.49).

Now let us use Equation (8.50) to analyze simulated concentration-response curves to determine if it is possible to estimate reasonable K_a values.

Once again, to establish the validity of the method, a series of theoretical concentration-response curves can be analyzed to determine if it is possible to estimate the K_a values that were used to simulate the data. Theses curves are shown in Figure 8.14. They were simulated under the condition of a high concentration of guanine nucleotide (1 mM). The K_a values of the agonists were set to 10^4 M^{-1}, whereas those of the inverse agonists were set to 10^5 M^{-1}. The K_b values were adjusted so that the ratio of K_b/K_a varied from 10^3–10^{-1} as shown in the legend to the figure. The theoretical curves were generated using the same procedure described for the theoretical data in Figure 8.5.

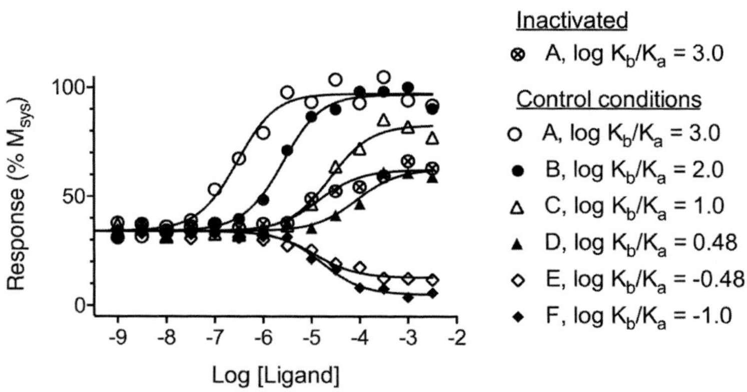

Figure 8.14: Estimation of the affinity constants of ligands for the inactive state of the receptor (K_a) through analysis of concentration-response curves exhibiting constitutive receptor activity. The theoretical responses were generated in a manner analogous to that described for the data in Figure 8.6, and the theoretical curves represent the global least-squares fit of Equation (8.50) to the simulated response values. The log K_a value of each ligand was set to 4 for the simulations, and the relative value of K_b is given in the legend. The parameters of the transducer function were: M_{sys}, 100; $K_{E\text{-obs}}$, 0.01 and m, 1.5. The concentration-response curve of the most efficacious agonist (A) was simulated for conditions of control and partial receptor inactivation with an irreversible neutral antagonist (q, 0.01).

Equation (8.50) was fitted to the theoretical data shown in Figure 8.14, and a good fit of the regression equation to the data was obtained. The estimates of the system parameters are M_{sys}, 97.6 ± 1.2 and m, 1.18 ± 0.23. The provisional estimates of $\log K_{E-obs}$ and $\log \varepsilon_{i-sys}$ are -2.02 ± 0.20 and -0.0025 ± 0.00134, respectively. The estimates of M_{sys} and m are similar to the values used to simulate the data (100 and 1, respectively). The estimate of q (0.0099 ± 0.0031) was also similar to that (0.01) used in the simulation. The estimates of the log affinity constants of ligands A–G were 4.63 ± 0.15, 3.94 ± 0.74, 4.06 ± 0.14, 3.84 ± 0.16, 5.02 ± 0.17 and 4.88 ± 0.14, respectively. The corresponding provisional estimates of $\log \varepsilon_i$ were $-\infty$, -0.35 ± 0.84, -0.018 ± 0.0065, -0.0069 ± 0.0029, -0.0084 ± 0.00061 and -0.00035 ± 0.00033. Using Equation (8.30) and the estimated values of $\log K_{obs}$, $\log \varepsilon_i$ and $\log \varepsilon_{i-sys}$, it is possible to estimate the $\log K_a$ values of ligands B–G: 3.59, 4.04, 3.83, 5.02 and 4.88. These values are similar to the corresponding values used to simulate the data: 4.0, 4.0, 4.0, 5.0 and 5.0. The third example in the appendix give detailed instructions on how to do the nonlinear regression part of this analysis.

The provisional efficacy values of ligands A–G can be calculated from their respective ε_i values using Equation (8.4): 1, 0.55, 0.04, 0.016, 0.0019 and 0.00081. Although the ε and ε_i values are unlikely to be accurate because they were estimated by constraining the ε_i value of the most efficacious agonist to 0 (i.e., $\varepsilon = 1$), the efficacy of one agonist relative to that of another is accurate. This relationship holds because although the provisional K_{E-obs} and ε values are inaccurate, the ratio of the two (ε / K_{E-obs}) is accurate and is equivalent to the τ value. Thus, the efficacy of one ligand, expressed relative to that of another, can be estimated from the ratio of the corresponding τ values. This latter ratio is also equivalent to the corresponding ratio of provisional efficacy values. Finally, because the efficacy of the most efficacious agonist was assigned an ε value of 1, the provisional ε values of ligands B–G are actually equivalent to their respective efficacy values, expressed relative to that of ligand A.

Recall that the method for estimation of K_a is valid provided that the efficacy of the ligand is less than one-half that of the most efficacious ligand. This draws attention to the efficacy of ligand B, which exceeds 0.5 when expressed relative to the most efficacious agonist (ligand A). This value could be in error because ligand B is essentially a full agonist, and it is difficult to estimate reliable τ and K_{obs} values of a full agonist unless additional information is collected (e.g., concentrations response curve after partial receptor inactivation). Thus, the discrepancy between the estimated value of $\log K_a$ for agonist B (3.59) and the value used in the simulation (4) could be due to error in its estimation. Based on the parameters used to simulate the data, it can be shown that the actual efficacy value of agonist B is only 0.34 when expressed relative to that of agonist A. Thus, the error in the estimation of the log K_a value of agonist B is caused by its full agonist behavior.

The K_b values of the ligands can be estimated using the procedure described in the prior section or by using Equation (8.29) and some of the parameters estimated in this section. We still need to estimate ε_{sys}, which can be done by subtracting the value of ε_{i-sys} from 1 (see Equation (8.5)) to yield a ε_{sys} value of -2.248 when expressed as a logarithm. Substitution of ε_{sys} and the appropriate values of ε and K_{obs} for each ligand into Equation (8.29) yields the following estimates of log K_b: A, 6.87; B, 5.93; C, 4.92; D, 4.28; E, 4.55 and F, 4.04. These are nearly the same as those used to simulate the data: A, 7.0; B, 6.0; C, 5.0; D, 4.48; E, 4.52 and F, 4.0. Finally, the estimates of log K_b/K_a for all but the most efficacious agonist are: B, 2.33; C, 0.88; D, 0.45; E, -0.47 and -0.85. These values are in reasonable agreement with those used to simulate the data: B, 2.0; C, 1.0; D, 0.48; E, -0.48 and F, -1.0.

In many analyses, there may be more than one full agonist, and in such cases, it would be necessary to measure the effect of partial receptor inactivation or reduced receptor expression on the concentration-response curves of all full agonists if there K_a values are to be estimated. In this case, Equation (8.50) can be used with the constraints mentioned above. In the case of any full agonist that is not the most efficacious agonist, it concentration-response curve measured under conditions of partial receptor inactivation or reduced receptor expression can be analyzed with Equation (8.50) without constraining any of the parameters except the system parameters. Using this approach, the K_a values of all the ligands except the most efficacious agonist can be estimated.

Estimation of K_b, K_{obs} and relative efficacy when the transducer slope factor is equivalent to one ($m = 1$)

When the transducer slope factor in the operational model is equivalent to 1, the Hill slope of the concentration-response curve is also equal to 1, and there are simple relationships among empirical parameters of the concentration-response curve of an orthosteric ligand at a constitutively active receptor and the K_b, K_{obs} and relative efficacy values. For this analysis, the EC_{50} value is defined as the concentration of orthosteric ligand that causes a half-maximal change in the total response, and the maximal effect (T_{sat}) is defined as the total response that occurs at receptor saturating concentrations of the orthosteric ligand. A schematic representation of EC_{50} and T_{sat} are illustrated in Figure 8.15 for an agonist (Figure 8.15a) and an inverse agonist (Figure 8.15b) in a receptor signaling pathway that exhibits constitutive receptor activity.

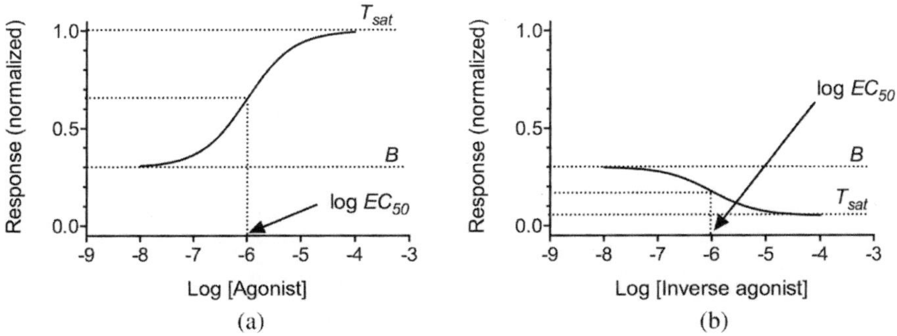

Figure 8.15: Graphical parameters of ligand concentration-response curves for a system exhibiting constitutive activity. The theoretical plots are shown for an agonist (a) and inverse agonist (b).

When the transducer slope factor in the operational model is equivalent to 1, there is a simple relationship between the ratio of K_b values of two agonists and their corresponding T_{sat} and EC_{50} values for a response:

$$RA_i = \frac{K_b}{K_b} = \frac{EC_{50}T'_{\mathrm{sat}}}{EC_{50}T'_{\mathrm{sat}}}, \tag{8.51}$$

in which the parameters of the reference orthosteric ligand are denoted with apostrophes. The K_b and K_{obs} values of an orthosteric ligand can be estimated from the following equations:

$$K_b = \frac{T_{\mathrm{sat}}}{EC_{50}B}, \tag{8.52}$$

$$K_{\mathrm{obs}} = \frac{(M_{\mathrm{sys}} - T_{\mathrm{sat}})}{EC_{50}(M_{\mathrm{sys}} - B)}. \tag{8.53}$$

Finally, the intrinsic efficacy of one agonist expressed relative to that of another is described by:

$$\frac{\varepsilon}{\varepsilon'} = \frac{T_{\mathrm{sat}}\left(M_{\mathrm{sys}} - T'_{\mathrm{sat}}\right)}{T'_{\mathrm{sat}}(M_{\mathrm{sys}} - T_{\mathrm{sat}})}. \tag{8.54}$$

The equations for the estimation of K_{obs} and relative efficacy can only be applied to partial agonists and inverse agonists, because there is little difference between the T_{sat} of a full agonist and M_{sys}. In fact, the T_{sat} of a full agonist could be used as an estimate of M_{sys} in these calculations.

To illustrate the validity of these equations, they can be applied to the analysis of the data shown in Figure 8.14, which were generated with the transducer slope

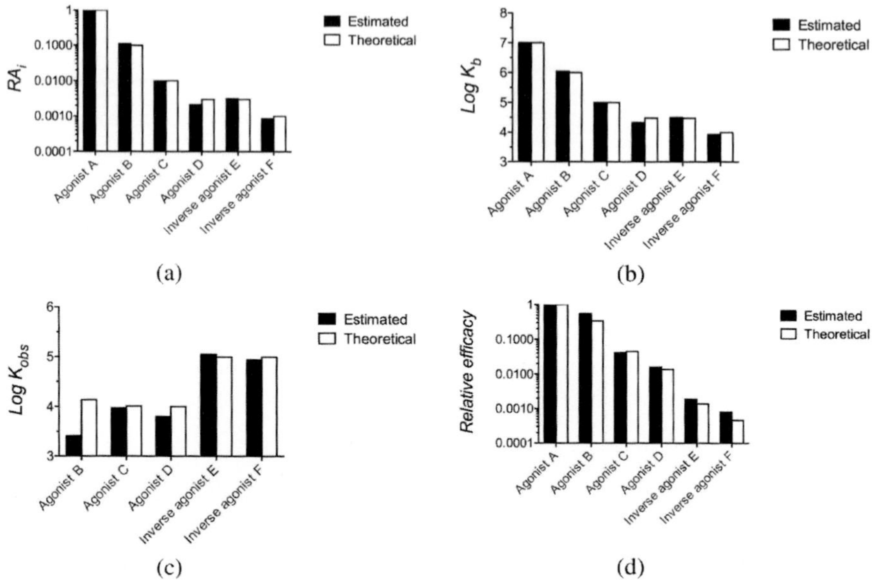

Figure 8.16: Simple method for estimating RA_i, log K_b, log K_{obs} and relative efficacy from the graphical parameters of ligand concentration-response curves for a system exhibiting constitutive activity and a transducer slope factor of 1. RA_i, log K_b, log K_{obs} and relative efficacy were estimated for the ligand concentration-response curves shown in Figure 8.14 using Equations (8.51)–(8.54), respectively. The histograms represent the theoretical (white bars) and the estimated (black bars) values of RA_i (a), log K_b (b), log K_{obs} (c) and relative efficacy (d).

factor in the operational model set to one. The T_{sat} and EC_{50} values of the various orthosteric ligands were estimated from these data, and these values were then used to estimate RA_i, K_b, K_{obs} and relative efficacy values of the ligands (see Figure 8.16). In these calculations, it was assumed the T_{sat} value of full agonist A was equivalent to M_{sys}. It can be seen that there is usually good agreement between the theoretical values and the values estimated from the T_{sat} and EC_{50} values. One exception is the estimate of the K_{obs} value of full agonist B. As described above, the estimation relies on an accurate measure of the small difference between its T_{sat} value and M_{sys} (T_{sat} value of the more efficacious agonist A (97.6%)), which is difficult to achieve. Using the true estimate of M_{sys} (100%) in the calculation, yielded a much more accurate K_{obs} value for agonist B.

To prove these equations, the operational model (Equation (7.18)) can be solved to derive equations for the empirical parameters of the concentration response curve

(T_{sat} and EC_{50}). These equations can then be substituted into Equations (7.61)–(7.64) to determine if the latter reduce to equalities.

The equation for T_{sat} is derived by taking the limit of Equation (7.18) as D approaches infinity:

$$T_{sat} = \lim_{D\to\infty} \frac{M_{sys}}{1 + \left(\frac{DK_{obs}+1}{D\tau K_{obs}+\tau_{sys}}\right)^m}. \tag{8.55}$$

This limit is:

$$T_{sat} = \frac{\tau^m M_{sys}}{\tau^m + 1}. \tag{8.56}$$

When $m = 1$, Equation (7.66) reduces to:

$$T_{sat} = \frac{\tau M_{sys}}{\tau + 1}. \tag{8.57}$$

The equation for the basal response (B) can be derived by solving the operational model for constitutive activity (Equation (7.18)) when $D = 0$:

$$B = \frac{\tau_{sys}^m M_{sys}}{\tau_{sys}^m + 1}. \tag{8.58}$$

When $m = 1$, this equation reduces to:

$$B = \frac{\tau_{sys} M_{sys}}{\tau_{sys} + 1}. \tag{8.59}$$

The equation for EC_{50} value can be derived by first considering the range of responses elicited by an orthosteric ligand. For an orthosteric agonist, this range is equivalent to the T_{sat} minus the basal response. A half-maximal response occurs when the total response is equivalent to the basal response plus half of the range of the response:

$$\text{response}_{50} = B + 0.5(T_{sat} - B). \tag{8.60}$$

In this equation, response_{50} denotes the total response at an EC_{50} concentration of the orthosteric agonist. For an orthosteric inverse agonist, the range of responses is equivalent to the basal response minus T_{sat}. A half-maximal response occurs when the total response is equivalent to the basal response minus half of the range of the response:

$$\text{response}_{50} = B - 0.5(B - T_{\text{sat}}). \tag{8.61}$$

Equations (8.60) and (8.61) both reduce to the same form:

$$\text{response}_{50} = 0.5(T_{\text{sat}} + B). \tag{8.62}$$

Replacing the variable response$_{50}$ in Equation (8.62) with the operational model for constitutive activity (Equation (7.18)) with D replaced with EC_{50} yields:

$$\frac{M_{\text{sys}}}{1 + \left(\frac{EC_{50}K_{\text{obs}}+1}{EC_{50}\tau K_{\text{obs}}+\tau_{\text{sys}}}\right)^{m}}. \tag{8.63}$$

Substituting in Equations (8.57) and (8.59) for T_{sat} and B and solving for EC_{50} yields:

$$EC_{50} = \frac{(2\tau^m\tau_{\text{sys}}^m + \tau_{\text{sys}}^m + \tau^m)^{1/m} - \tau_{\text{sys}}(\tau_{\text{sys}}^m + \tau^m + 2)^{1/m}}{K_{\text{obs}}(\tau(\tau_{\text{sys}}^m + \tau^m + 2)^{1/m} - (2\tau^m\tau_{\text{sys}}^m + \tau_{\text{sys}}^m + \tau^m)^{1/m})}. \tag{8.64}$$

When $m = 1$, this equation reduces to:

$$EC_{50} = \frac{\tau_{\text{sys}} + 1}{K_{\text{obs}}(\tau + 1)}. \tag{8.65}$$

Now, we are in a position to substitute Equations (8.57), (8.59) and (8.65) for T_{sat}, B and EC_{50} in Equations (8.51)–(8.54) to determine if the latter are equalities. Beginning with Equation (8.51), making the appropriate substitutions yields:

$$\frac{K_b}{K_b'} = \frac{\frac{\tau_{\text{sys}}+1}{(\tau+1)K_{\text{obs}}'} \times \frac{\tau M_{\text{sys}}}{\tau+1}}{\frac{\tau_{\text{sys}}+1}{(\tau+1)K_{\text{obs}}} \times \frac{\tau' M_{\text{sys}}}{\tau'+1}}. \tag{8.66}$$

Simplification yields:

$$\frac{K_b}{K_b'} = \frac{\frac{K_{\text{obs}}\tau}{\tau_{\text{sys}}}}{\frac{K_{\text{obs}}'\tau'}{\tau_{\text{sys}}}}. \tag{8.67}$$

This equation is an equality because Equation (8.36) shows that $K_{\text{obs}}\tau/\tau_{\text{sys}} = K_b$.

Substituting the appropriate equations for T_{sat}, B and EC_{50} into Equation (8.52) yields:

$$K_b = \frac{\frac{\tau M_{sys}}{\tau+1}}{\frac{\tau_{sys}+1}{K_{obs}(\tau+1)} \times \frac{\tau_{sys}M_{sys}}{\tau_{sys}+1}}. \tag{8.68}$$

This equation reduces to Equation (8.36), illustrating that Equation (8.52) is an equality.

Substituting the appropriate equations for T_{sat}, B and EC_{50} into Equation (8.53) yields:

$$K_{obs} = \frac{M_{sys} - \frac{\tau M_{sys}}{\tau+1}}{\frac{\tau_{sys}+1}{K_{obs}(\tau+1)} \times \left(M_{sys} - \frac{\tau_{sys}M_{sys}}{\tau_{sys}+1}\right)}. \tag{8.69}$$

This equation reduces to the equality, $K_{obs} = K_{obs}$, which proves Equation (8.53) from which it was derived.

Finally, to prove Equation (8.54), Equation (8.51) can be rearranged into the following form by substituting Equation (8.44) for K_b/K_b':

$$\frac{\varepsilon K_{obs}}{\varepsilon' K'_{obs}} = \frac{EC'_{50}T_{sat}}{EC_{50}T'_{sat}}. \tag{8.70}$$

Substituting Equation (8.53) for K_{obs} yields:

$$\frac{\varepsilon \frac{(M_{sys}-T_{sat})}{EC_{50}(M_{50}T_{sat})}}{\varepsilon \frac{(M_{sys}-T'_{sat})}{EC'_{50}(M_{sys}-B')}} = \frac{EC_{50}\,'T_{sat}}{EC_{50}T'_{sat}}. \tag{8.71}$$

This equation reduces to Equation (8.54), which proves the latter.

And that is the proof of Equations (8.51)–(8.54).

Constitutive Activity and the Threshold for Generating a Response

One of the advantages of a highly sensitive signaling pathway is that it will amplify a low level of receptor activation. This will enable the experimenter to detect a response to an agonist having very low intrinsic efficacy. If the sensitivity of the system is great enough, it will also enable one to detect the response elicited by constitutive receptor activation. Such systems are useful in the estimation of the microscopic affinity constants of orthosteric ligands as described earlier in this chapter.

While many signaling pathways in isolated tissues are highly sensitive, the downstream responses elicited by them do not necessarily report constitutive receptor activity even though it is often possible to measure the response to agonists with very low efficacy. A useful example is the isolated guinea pig ileum. This preparation does not appear to exhibit a basal contractile response caused by constitutive receptor activation. The highly efficacious muscarinic agonist, oxotremorine-M elicits powerful contractions of the guinea pig ileum with an EC_{50} value of approximately 10^{-8} M [26]. Its K_{obs} value is about 10^{5} M, and its $E_{\max}$ is equivalent to the maximal response of the system. Thus, the guinea pig ileum is highly sensitive because it only requires that oxotremorine-M occupy less than a fraction of 1% of the M_3 receptor population to elicit a 50% response. In contrast, the partial agonist McN-A343 elicits contraction with an EC_{50} value of approximately 10^{-5} M, and its maximal response is approximately 40% that of the maximal response of the system.[27] The K_b and K_a values of these two agonists have been estimated based on their ability to elicit phosphoinositide hydrolysis in HEK 293 cells over expressing $G_{\alpha 15}$ and the human M_3 receptor. Oxotremorine-M exhibits a K_b/K_a ratio of about 10^4 at the M_3 receptor, whereas the corresponding ratio for McN-A343 is only 3.[22,28] The latter ratio indicates that McN-A343 increases the activation level of the receptor population to a value that is only three-fold that of constitutive receptor activation. How is it that McN-A343 can elicit a maximal response equivalent to 40% that of the system while its K_b/K_a ratio is only 3 and the signaling pathway appears incapable of registering a response to constitutive receptor activation?

The explanation may be that the signal elicited by the constitutive activity of the M_3 receptor population is amplified, but the final measured response (contraction) exhibits a threshold. That is, a low level of receptor activation will not depolarize the plasma membrane sufficiently to activate the voltage-sensitive Ca^{2+} channels that allow the requisite influx of Ca^{2+} for contraction.[29,30] This could simply be caused by the expression of Ca^{2+}-activated potassium channels that act to clamp membrane potential close to the resting level.[31] The maximum level of receptor activation that can occur without triggering a measureable level of the response of interest is termed "threshold". If a response exhibits a threshold that is almost satisfied by constitutive receptor activation then an agonist with a very low K_b/K_a ratio could elicit substantial responses.

Consider the example shown in Figure 8.17. Figure 8.17a illustrates the stimulus function (receptor activation) elicited by two agonists — one with a large K_b/K_a ratio of 10^3 (agonist A) and another with a K_b/K_a ratio of only 2 (agonist B). The receptor exhibits constitutive activity such that 1% of the receptor population is active in the absence of orthosteric ligands ($K_q = 10^{-2}$). Agonist A exhibits a

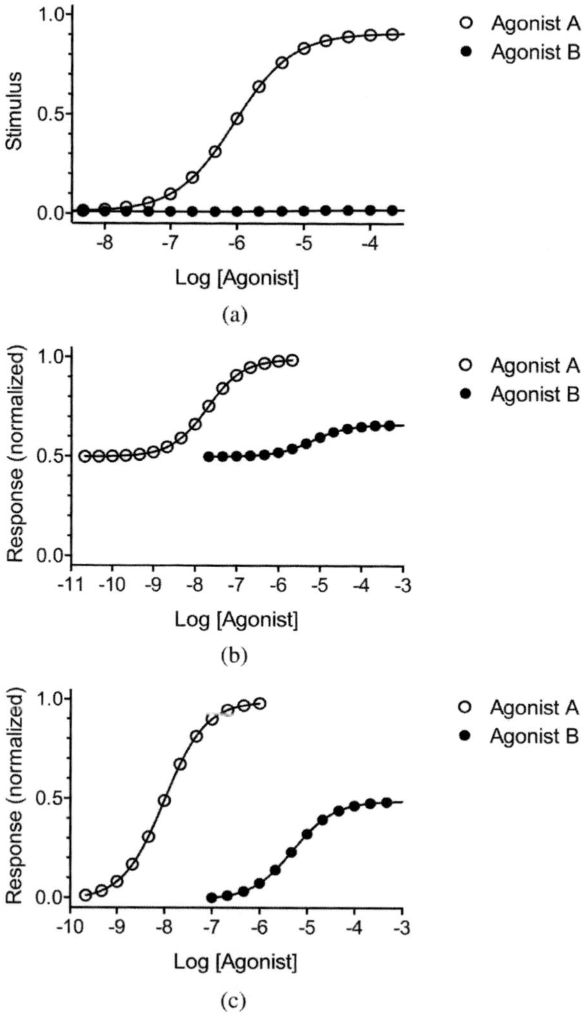

Figure 8.17: The influence of constitutive activity and the combination of constitutive activity and threshold on the concentration-response curves of a highly efficacious agonist (A, $K_b/K_a = 10^3$) and a partial agonist (B, $K_b/K_a = 2.0$). (a) The theoretical total stimulus (total receptor activation) functions were generated for the two agonists using a Monod Wyman and Changeux model[17] for total receptor activation ($T = 1/(1+(DK_a+1)/K_q(DK_b+1))$). The isomerization constant of the free receptor (K_q) was set to 10^{-2}. The microscopic affinity constants of the agonists were: A, log K_b, 8.0; log K_a, 5.0 and B, log K_b, 5.3; log K_a, 5.0. (b) The stimuli shown in panel (a) were substituted into the transducer function of the operational model (Equation (8.10)) to generate the response values. (c) Same as (b) except that a threshold stimulus of 0.01 was subtracted form the total stimulus values before substituting into the transducer function.

stimulus function that yields a maximal 90% activation of the receptor population, whereas agonist B only activates about 2% of the receptor population. Figure 8.17b shows the final response when the stimuli are amplified through a system having a $K_{E-\text{obs}}$ value of 10^{-2}. This sensitive signaling pathway amplifies constitutive receptor activity such that there is a 50% response in the absence of orthosteric ligands. Both agonists elicit substantial responses above the basal response caused by constitutive receptor activation.

Figure 8.17c shows the same signaling system but with the added requirement that 1% receptor activation is needed to satisfy the threshold for the response. In other words, only when receptor activation exceeds 1% does a measureable response occur. In this case, the threshold requirement eliminates the basal response caused by constitutive receptor activation. Now the less efficacious agonist B is able to elicit a substantial response even though its K_b/K_a ratio is only 2.

If there were no constitutive receptor activity, then it would have been impossible for agonist B with its low K_b/K_a ratio to elicit such a large response. The example shown in Figure 8.18 illustrates how much selectivity for the active state is required. Figure 8.18a shows the stimuli elicited by two agonists, C and D. Their corresponding K_b/K_a ratios are 10^5 and 10^2. This receptor elicits very little constitutive activation (0.01%; $K_q = 10^{-4}$) in the absence of orthosteric ligands. Even though the K_b/K_a ratios of agonists C and D are 100-fold greater than those of A and B, respectively, the efficacies of C and D for activating a receptor with an isomerization constant of 10^{-4} (i.e., $K_q = 10^{-4}$) (Figure 8.18a) are equivalent to those of agonists A and B for activating a receptor with a larger isomerization constant of 10^{-2} (Figure 7.17a). Figure 8.18b shows their concentration-response curves in a system having the same sensitivity ($K_{E\text{-obs}} = 10^{-2}$) as that shown in Figure 7.17b. The concentration-response curves of agonists C and D are practically the same as those of A and B in the system with a threshold for the response (Figure 7.17c).

Thus, on the surface, the systems in Figures 7.17c and 7.18b seem to lack evidence of constitutive receptor activity. In one case (Figure 7.17c) agonist B elicits a 50% response with only two-fold selectivity for the active state, whereas agonist D requires 100-fold selectivity for the active state to generate the same response (Figure 7.18b). The explanation, of course, is that there is substantial constitutive receptor activity for the response shown in Figure 7.18b; it is just hidden by the threshold for the response, which could be caused by constitutive or Ca^{2+}-activated potassium channels, for example. Thus, even though a complex system may lack a basal response, one cannot conclude that the receptor lacks constitutive receptor activity.

One final point is that the methods for estimating microscopic constants described in this chapter are valid even if there is a threshold for the response

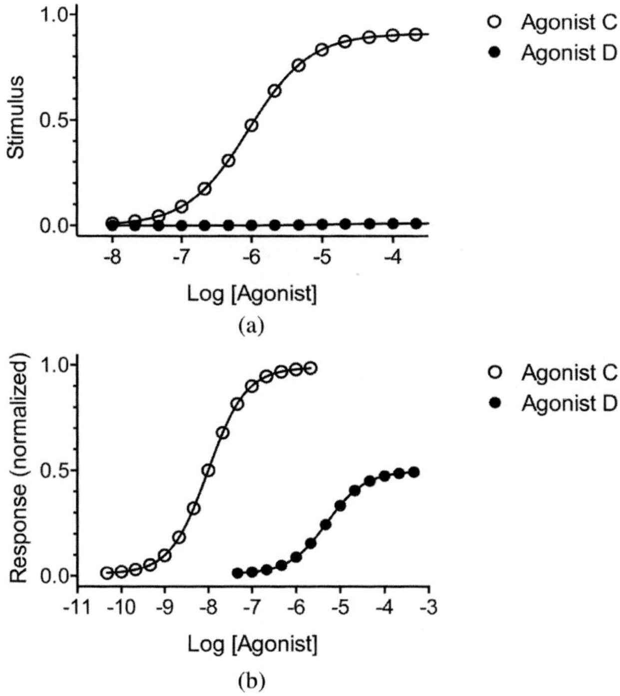

Figure 8.18: Relationship between the stimulus (a) and response (b) for a highly efficacious agonist (C, $K_b/K_a = 10^5$) and a partial agonist (D, $K_b/K_a = 10^2$) in a system lacking constitutive activity. (a) The theoretical total stimulus (total receptor activation) functions were generated for the two agonists using a Monod Wyman and Changeux model[17] for total receptor activation ($T = 1/(1+(DK_a+1)/K_q(DK_b+1))$). The isomerization constant of the free receptor (K_q) was set to 10^{-4}. The microscopic affinity constants of the agonists were: A, log K_b, 10.0; log K_a, 5.0 and B, log K_b, 7.0; log K_a, 5.0. (b) The stimuli shown in panel a were substituted into the transducer function of the operational model (Equation (8.10)) to generate the response values.

that masks some or all of the basal response to constitutive receptor activation. If the basal constitutive response is completely masked, then it is only possible to estimate a relative measure of K_b (i.e., RA_i).

Constitutive Activity and Drug Discovery

Signaling pathways that report the constitutive activity of the receptor are useful in drug discovery because they enable the estimation of the microscopic affinity constants of the active state of the receptor. This estimate is a pure, receptor-dependent parameter of the orthosteric ligand that ultimately determines how it elicits all of the various responses to receptor activation. Having a cell line in which the expression of the receptor and the downstream signaling proteins can be induced to sufficiently high levels should be useful in the development of assays for measuring these constants.

The level of receptor expression influences the type of signaling that occurs, and the overexpression of receptors and the elements of the signaling pathway can lead to a change in the nature and magnitude of signaling and even to the emergence of non-physiological signaling. Not surprisingly, it is often stated that the nature of *G* protein signaling and the properties of downstream responses measured in such a system may bear little relevance to normal physiology, and therefore, that the measurements of the affinity and efficacy of orthosteric ligands in such systems may be of little relevance. This conclusion focuses on population thinking and is expressed in terms of population vocabulary. The quantal behavior of the receptor is not addressed. The observed affinity of an agonist (K_{obs}) may change with the expression of its preferred *G* protein, and the magnitude of the receptor-*G* protein output (ε) may change accordingly, but this does not change the quantal properties of individual receptors and *G* proteins — only the probability of their unique interactions. Therefore, it should always be possible to use these systems to estimate the fundamental quantal parameters of the orthosteric ligand-receptor interaction (i.e., K_b and K_a). The high level of expression of a *G* protein may increase the number of collisions it has with a receptor that it normally does not encounter at a physiological receptosome, which could cause atypical signaling. But this should be obvious, and it is possible to compare the microscopic constants of an agonist for eliciting a response through its preferred *G* protein relative to a less-preferred, overexpressed *G* protein. It is also possible to measure the sensitivity of receptor signaling through one pathway relative to the other. All of these measurements are useful parameters that define the orthosteric ligand-receptor complex in terms of its microscopic constants for receptor states.

Remarkably one of the easiest parameters to estimate from a concentration-response curve is the affinity constant of the orthosteric ligand for the active state of the receptor. A relative measure can always be estimated for agonists from their concentration-response curves, and in the case of a signaling pathway that reports the response of constitutive receptor activation, it is possible to estimate the K_b in absolute units for any orthosteric ligand. The estimation is accurate regardless of which step in the signaling pathway is measured or how the affinity and efficacy of the receptor population changes depending on the concentration of the *G* protein or GTP. Thus, the quantum behavior of receptor activation leaves its unique signature on each response in the signaling pathway, even if the measured response is simple and several steps down the signaling pathway.

There is now substantial evidence that different active states of the receptor are involved in mediating signaling through different effector proteins. In the next chapter, I describe how different *G* proteins provide a window for measuring these different effector-selective states of the receptor.

Appendix

This section gives more detailed procedures for the nonlinear regression techniques used in three of the examples covered in the text concerning the data shown in Figures 8.6a, 8.10a and 8.14. Here, I describe how to use the Prism 6 application (GraphPad Software, Inc.) to analyze the data, although the general procedures can be applied to any nonlinear regression algorithm. It is assumed that the reader is generally familiar with the Prism environment. The method for the first example (Figure 8.6a) is essentially the same as that described in the first example of the appendix of Chapter 7, but here the variables are renamed to account for partial and inverse agonists.

Example 1: Estimation of the affinity constants of partial and inverse agonists through analysis of their concentration-response curves with that of a full agonist (Figure 8.6a).

1. The data for the concentration-response curves are entered into a Data Table for an XY plot in Prism. The log molar concentrations of the agonists and inverse agonists are entered into the column on the far left entitled "X". The response measurements for a single experiment or the means of several experiments can be entered, but it is recommended that the individual values of replicate experiments be entered into a Data Table. The replicate response measurements of the most efficacious agonist (agonist A) are entered into subcolumns of Group A and those of the other ligands (B–D) are entered into subcolumns of Groups B–D. A given response value is always entered into the same row as that of its corresponding log ligand concentration.
2. The Graph of the data is selected, and ultimately, the nonlinear regression (curve fit) analysis is selected.

   The regression equation used for this analysis is equivalent to Equation (8.11), but some of the parameters will be expressed as logarithms for this analysis. In addition, because it is possible to estimate the individual parameters, τ and K_{obs}, for the partial and inverse agonists, but only their product (τK_{obs}) for the full agonist, we will have to enter the equation with a dummy variable for the product τK_{obs}. This will enable us to estimate the product τK_{obs} for the full agonist and the individual values, τ and K_{obs}, for the less efficacious ligands in the Prism environment.

Now let us return to the process of entering the regression equation into Prism. The following lines are entered into Prism to define a new equation:

⟨ A ⟩ `LogR` = `LogTauKobsFull`
⟨ A ⟩ `LogK` = `LogKobsFull`
⟨~ A ⟩ `LogR` = `LogTau + LogKobs`
⟨~ A ⟩ `LogK` = `LogKobs`
Y = `Msys/(1 + {[10 ∧ (X + LogK) + 1]/[10 ∧ (X + LogR)]} ∧ M)`

The last line represents the regression equation, which is a simplified form of the operational model (Equation (8.11)). The variable "X" denotes the log concentration of agonist, and `LogK` and `LogR` represent dummy variables.

The first two lines are preceded by the symbols "⟨A⟩", which designate that the expressions on these lines apply to the data in Group A of the Data Table (i.e., that of the full agonist). The first line designates that the dummy variable, `LogR`, will represent a parameter (`LogTauKobsFull`) corresponding to the log product of the τ and K_{obs} (log τK_{obs}) parameters of the full agonist. Similarly, the second line assigns the parameter `LogKobsFull` (log K_{obs} value of the full agonist) to the dummy variable, `LogK`.

The third and fourth lines are preceded by the symbols "⟨~ A ⟩ ", which instruct Prism to apply the corresponding variable assignments to all of the data except those in Group A.

The third line designates that the dummy variable, `LogR`, now represents the sum of two parameters, `LogTauPartial + LogKobsPartial`. Finally, the fourth line assigns the parameter `LogKobsPartial` to the dummy variable, `LogK`.

3. Now that the equation has been entered, parameter constraints are applied. The log K_{obs} of the full agonist and the parameters for the maximum response of the system, M_{sys} (`Msys`), and the transducer slope factor, m (M), are shared across all data sets.
4. The next step involves assigning initial parameter estimates. These are estimated using the following equations in which the parameters with a prime symbol denote the initial estimate of the parameter:

$$\texttt{LogKobsFull}' = 0 \tag{8.72}$$

$$\texttt{LogTauKobsFull}' = -\log EC_{50\text{-full}} \tag{8.73}$$

$$\texttt{LogTau}' = \log\left(\frac{1}{\frac{E_{\text{max-full}}}{E_{\text{max}}} - 1}\right) \tag{8.74}$$

$$\texttt{LogKobs}' = -\log EC_{50} \tag{8.75}$$

$$\texttt{Msys}' = E_{\text{max-full}}. \tag{8.76}$$

In these equations $EC_{50\text{-full}}$ denotes the EC_{50} value of the full agonists and EC_{50} denotes that of the other ligands, and $E_{\text{max-full}}$ and E_{max} denote the corresponding E_{max} values. The transducer slope value can be assigned an initial estimate of 1:

$$\texttt{M}' = 1, \tag{8.77}$$

or better yet, it can be assigned the value of the Hill slope of the concentration-response curve of the full agonist (H_{full}):

$$\texttt{M}' = H_{\text{full}} \tag{8.78}$$

5. Now that the data, regression equation, parameter constraints and initial parameter estimates have all been entered, regression analysis is initiated. For this example, the following parameter estimates were obtained: Agonist A, LogTauKobsFull, 7.748 ± 0.12; partial agonist B, LogTau, 0.3892 ± 0.1029; LogKobs, 6.052±0.1185; inverse agonist C, LogTau, −0.8461±0.236; LogKobs, 5.557 ± 0.1636 and inverse agonist D, LogKobs, 8.219 ± 0.1858. As described in the text, it was impossible to obtain accurate estimates of the Kobs value of the full agonist and the LogTau value of the full inverse agonist.

Example 2: Estimation of the affinity constant by the method of partial receptor inactivation (Figure 8.10a).

1. The data for the concentration-response curves of oxotremorine-M (Figure 8.10a) are entered into a Data Table for an XY plot as described in step 1 of Example 1 above. The control data and those measured under conditions of partial receptor inactivation are entered into subcolumns of Group A and Group B, respectively.
2. The Graph of the data is selected, and ultimately, the nonlinear regression (curve fit) analysis is selected.
   The following lines are entered into Prism to define a new equation:

   $\langle \text{A} \rangle\ \texttt{Q} = 1$

   $\langle \sim \text{A} \rangle\ \texttt{Q} = \text{Fraction}$

   $Y = \texttt{Msys}/(1 + \{[10 \wedge (\texttt{X} + \texttt{LogKobs}) + 1]/[(10 \wedge \texttt{LogTsys}) * (1 + (1 - \texttt{Q}) * (10 \wedge (\texttt{X} + \texttt{LogKobs}))) + \texttt{Q} * (10 \wedge (\texttt{X} + \texttt{LogTau} + \texttt{LogKobs}))]\} \wedge \texttt{M})$.

The last line represents the regression equation, which is a form of the operational model (Equation (8.18)). In this equation, `Msys` represents the maximum response of the system (M_{sys}), X, the log molar concentration of agonist, `LogKobs`, the log K_{obs} value of the agonist, Q, the fraction of the receptor population not inactivated by the irreversible neutral antagonist, `LogTau`, the log τ value of the agonist, `LogTsys`, the log τ_{sys} value, and `M`, the transducer slope factor in the operational model (m).

The first line assigns the value 1 to the dummy variable, Q, when the control data in Group A are being analyzed. The second line assigns the parameter "Fraction" to the dummy variable Q when the data in Group B are being analyzed. The parameter "Fraction" represents the fraction of the receptor population not inactivated. The significance of the "⟨ A ⟩ " and "⟨ ~ A ⟩ " entries are described above under step 2 of Example 1.

3. Next, parameter constraints are applied. This involves sharing the estimates of the parameters for the maximum response of the system (`Msys`), the transducer slope factor (`M`), log τ (`LogTau`) and log K_{obs} (`LogKobs`) across both sets of data.
4. The initial parameter estimates (denoted with the prime symbol) are calculated using the following equations in which the parameters with a prime symbol denote the initial estimate of the parameter:

$$\texttt{LogKobs}' = -\log EC_{50\text{-inact}}, \tag{8.79}$$

$$\texttt{LogTau}' = \log\left(\frac{EC_{50\text{-inact}}}{EC_{50\text{-inact}}}\right), \tag{8.80}$$

$$\texttt{LogTsys}' = \log\left(\frac{B}{E_{\text{max-control}} - B}\right), \tag{8.80}$$

$$\texttt{Q}' = \frac{E_{\text{max-inact}} * EC_{50\text{-control}}}{E_{\text{max-control}} * EC_{50\text{-inact}}}, \tag{8.81}$$

$$\texttt{Msys}' = E_{\text{max-control}}. \tag{8.82}$$

In these equations $EC_{50\text{-control}}$ and $EC_{50\text{-inact}}$ denote the EC_{50} values measured under control conditions and after partial receptor inactivation, and $E_{\text{max-control}}$ and $E_{\text{max-inact}}$ denote the corresponding E_{max} values.

The initial estimate of the transducer slope factor is assigned as described in Example 1.

5. After initiating the repression analysis, the following parameters are obtained: `LogKobs`, 5.788±0.1296; Fraction, 0.01323±0.003774; `LogTau`, 1.836±0.2821; `Msys`, 1.007 ± 0.01656 and `M`, 1.988 ± 0.8358.

Example 3: Estimation of the log K_a values of ligands in a system with constitutive activity (Figure 8.14).

1. The data for the concentration-response curves of the agonists (Figure 8.14) are entered into a Data Table for an XY plot following the general guidelines given under step 1, Example 1. The data for the agonist with the largest E_{max} (i.e., agonist A) and for the condition of partial receptor inactivation with an irreversible neutral antagonist is entered under Group A in the Data Table. The control concentration-response curve of agonist A is entered under Group B in the Data Table. The data for the other ligands (B–F) are entered into the Data Table under Groups C–G.
2. The Graph of the data is selected, and the nonlinear regression (curve fit) analysis is selected.

   The following lines are entered into Prism to define a new equation:

   ⟨ A ⟩ Q= Fraction

   ⟨ ~ A ⟩ Q= 1

   ⟨ A ⟩ Logei = −20

   ⟨ A ⟩ LogKobs = LogKStand

   ⟨ B ⟩ Logei = −20

   ⟨ B ⟩ LogKobs = LogKStand

   ⟨ C ⟩ LogKobs = LogKTest

   ⟨ C ⟩ Logei = LogeiTest

   ⟨ D ⟩ LogKobs = LogKTest

   ⟨ D ⟩ Logei = LogeiTest

   ⟨ E ⟩ LogKobs = LogKTest

   ⟨ E ⟩ Logei = LogeiTest

   ⟨ F ⟩ LogKobs = LogKTest

   ⟨ F ⟩ Logei = LogeiTest

   ⟨ G ⟩ LogKobs = LogKTest

   ⟨ G ⟩ Logei = LogeiTest

   Y = Msys/(1+{[10^(X+LogKobs)+1]/[((1−10^(Logeisys))/(10^LogKE))*(1+(1−Q)*(10^(X+LogKobs)))+Q*(10^(X+LogKobs))*(1−10^Logei)/(10^LogKE)]^M)

The last two lines describe the regression equation, which is a form of the operational model. In this equation, `Msys` represents the maximum response of the system (M_{sys}), X, the log molar concentration of agonist, `LogKobs`, the log K_{obs} of the ligand, `Logeisys`, the log $\varepsilon_{i\text{-sys}}$ of the system, `LogKE`, the log $K_{E\text{-obs}}$ of the system, `Logei`, the log ε_i of the ligand, and `M`, the transducer slope factor in the operational model (m).

The first line assigns the parameter, Fraction, to the dummy variable Q for the data in Group A and the second line assigns the value 1 to the dummy variable Q for the rest of the data. The next four lines assign the value of -20 to the dummy variable, `Logei`, and the parameter, `LogKStand`, to the dummy variable, `LogKobs`, for the data in Groups A and B. The remaining lines assign the parameters, `LogKTest` and `LogeiTest`, to the dummy variables, `LogKobs` and `Logei`, respectively, for the remainder of the data.

3. Next, parameter constraints are applied. This involves sharing the estimates of the parameters, `LogKStand`, `Msys`, `Logeisys`, `LogKE` and M.
4. The initial parameter estimates (denoted with the prime symbol) are calculated using the following equations in which the parameters with a prime symbol denote the initial estimate of the parameter:

$$\texttt{Msys}' = E_{\text{max-control}}. \tag{8.83}$$

$$\text{Fraction}' = \frac{E_{\text{max-inact}} * EC_{50-\text{control}}}{E_{\text{max}-\text{control}} * EC_{50-\text{inact}}} \tag{8.84}$$

$$\texttt{Logeisys}' = \log\left(\frac{(E_{\text{max}-\text{control}} - B)EC_{50-\text{inact}}}{B * EC_{50-\text{inact}}}\right), \tag{8.85}$$

$$\texttt{LogKStand}' = -\log EC_{50\text{-inact}}, \tag{8.86}$$

$$\texttt{LogKTest}' = -\log EC_{50\text{-Test}}, \tag{8.87}$$

$$\texttt{LogeiTest}' = \log\left(\frac{(E_{\text{max-control}} - E_{\text{max-test}})EC_{50-\text{inact}}}{E_{\text{max}-\text{test}}EC_{50\text{-control}}}\right), \tag{8.88}$$

$$\texttt{LogKE}' = \log\left(\frac{EC_{50\text{-inact}}}{EC_{50\text{-control}}}\right). \tag{8.89}$$

In these equations $EC_{50\text{-control}}$ and $EC_{50\text{-Test}}$ denote the EC_{50} values of agonist A and the various test ligands, $\text{EC}_{50-inact}$, denotes the EC_{50} value of agonist A for the condition of partial receptor inactivation, `Logeisys'`, the initial estimate of log $\varepsilon_{i\text{-sys}}$, Fraction, the residual fraction of receptor not inactivated by the irreversible neutral antagonist (q), and LogKE, the log sensitivity constant of the signaling pathway ($\log K_{E\text{-}obs}$).

The initial estimate of the transducer slope factor is assigned as described in Example 1.

5. After initiating the repression analysis, the following parameters are obtained: Fraction, 0.009930 ± 0.003208; `LogKStand`, 4.629 ± 0.1486; `Msys`, 97.63 ± 1.268; `Logeisys`, -0.002451 ± 0.001366; `LogKE`, -2.02 ± 0.2061 and M, 1.176 ± 0.2347. The parameters of the other ligands were: B, `LogKTest`, 3.940 ± 0.7507; `LogeiTest` -0.3513 ± 0.8697; Agonist C, `LogKTest`, 4.059 ± 0.1387; `LogeiTest` -0.01835 ± 0.006662; Agonist D, `LogKTest`, 3.840 ± 0.1674; `Loge iTest` -0.006873 ± 0.002927; Agonist E, `LogKTest`, 5.018 ± 0.1759; `LogeiTest` -0.0008304 ± 0.0006196 and Agonist F, `LogKTest`, 4.880 ± 0.1450; `LogeiTest`, -0.0003465 ± 0.0003292.

References

1. Luo, D. G., Yue, W. W., Ala-Laurila, P. and Yau, K. W., Activation of visual pigments by light and heat. *Science* **332**, 1307–1312 (2011).
2. Surya, A., Foster, K. W. and Knox, B. E., Transducin activation by the bovine opsin apoprotein. *J Biol Chem* **270**, 5024–5031 (1995).
3. Seifert, R. and Wenzel-Seifert, K., Constitutive activity of G-protein coupled receptors: cause of disease and common property of wild-type receptors. *Naunyn Schmiedebergs Arch Pharmacol* **366**, 381–416 (2002).
4. Burstein, E. S., Spalding, T. A. and Brann, M. R., Pharmacology of muscarinic receptor subtypes constitutively activated by G proteins. *Mol Pharmacol* **51**, 312–319 (1997).
5. Haskell-Luevano, C. and Monck, E. K., Agouti-related protein functions as an inverse agonist at a constitutively active brain melanocortin-4 receptor. *Regul Pept* **99**, 1–7 (2001).
6. Nijenhuis, W. A., Oosterom, J. and Adan, R. A., AgRP(83-132) acts as an inverse agonist on the human-melanocortin-4 receptor. *Mol Endocrinol* **15**, 164–171 (2001).
7. Tao, Y. X., Constitutive activation of G protein-coupled receptors and diseases: insights into mechanisms of activation and therapeutics. *Pharmacol Ther* **120**, 129–148 (2008).
8. Seifert, R., Wenzel-Seifert, K., Gether, U. and Kobilka, B. K., Functional differences between full and partial agonists: evidence for ligand-specific receptor conformations. *J Pharmacol Exp Ther* **297**, 1218–1226 (2001).
9. Lu, Z. L., Curtis, C. A., Jones, P. G., Pavia, J. and Hulme, E. C., The role of the aspartate-arginine-tyrosine triad in the m1 muscarinic receptor: mutations of aspartate 122 and tyrosine 124 decrease receptor expression but do not abolish signaling. *Mol Pharmacol* **51**, 234–241 (1997).
10. Spalding, T. A., Burstein, E. S., Henderson, S. C., Ducote, K. R. and Brann, M. R., Identification of a ligand-dependent switch within a muscarinic receptor. *J Biol Chem* **273**, 21563–21568 (1998).

11. Wess, J., Han, S. J., Kim, S. K., Jacobson, K. A. and Li, J. H., Conformational changes involved in G-protein-coupled-receptor activation. *Trends in Pharmacol Sci* **29**, 616–625 (2008).
12. Hulme, E. C., GPCR activation: a mutagenic spotlight on crystal structures. *Trends in Pharmacol Sci* **34**, 67–84 (2013).
13. Palczewski, K., Kumasaka, T., Hori, T., Behnke, C. A., Motoshima, H., Fox, B. A., Le Trong, I., Teller, D. C., Okada, T., Stenkamp, R. E., Yamamoto, M. and Miyano, M., Crystal structure of rhodopsin: A G protein-coupled receptor. *Science* **289**, 739–745 (2000).
14. Salom, D., Lodowski, D. T., Stenkamp, R. E., Le Trong, I., Golczak, M., Jastrzebska, B., Harris, T., Ballesteros, J. A. and Palczewski, K., Crystal structure of a photoactivated deprotonated intermediate of rhodopsin. *Proc Natl Acad Sci USA* **103**, 16123–16128 (2006).
15. Rasmussen, S. G., DeVree, B. T., Zou, Y., Kruse, A. C., Chung, K. Y., Kobilka, T. S., Thian, F. S., Chae, P. S., Pardon, E., Calinski, D., Mathiesen, J. M., Shah, S. T., Lyons, J. A., Caffrey, M., Gellman, S. H., Steyaert, J., Skiniotis, G., Weis, W. I., Sunahara, R. K. and Kobilka, B. K., Crystal structure of the beta2 adrenergic receptor-G_s protein complex. *Nature* **477**, 549–555 (2011).
16. Rasmussen, S. G., Choi, H. J., Rosenbaum, D. M., Kobilka, T. S., Thian, F. S., Edwards, P. C., Burghammer, M., Ratnala, V. R., Sanishvili, R., Fischetti, R. F., Schertler, G. F., Weis, W. I. and Kobilka, B. K., Crystal structure of the human beta2 adrenergic G-protein-coupled receptor. *Nature* **450**, 383–387 (2007).
17. Monod, J., Wyman, J. and Changeux, J. P., On the Nature of allosteric transitions: A Plausible Model. *J Mol Biol* **12**, 88–118 (1965).
18. Katz, B. and Miledi, R., Membrane noise produced by acetylcholine. *Nature* **226**, 962–963 (1970).
19. Katz, B. and Miledi, R., Further observations on acetylcholine noise. *Nat New Biol* **232**, 124–126 (1971).
20. Derksen, H. E., Axon membrane voltage fluctuations. *Acta Physiol Pharmacol Neerl* **13**, 373–466 (1965).
21. Lindemann, B. and DeFelice, L. J., On the use of general network functions in the evaluation of noise spectra obtained from epithelia. *Soc Gen Physiol Ser* **36**, 1–13 (1981).
22. Ehlert, F. J., Suga, H. and Griffin, M. T., Analysis of agonism and inverse agonism in functional assays with constitutive activity: estimation of orthosteric ligand affinity constants for active and inactive receptor states. *J Pharmacol Exp Ther* **338**, 671-686 (2011).
23. Stephenson, R. P., A modification of receptor theory. *Br J Pharmacol* **11**, 379–393 (1956).
24. Furchgott, R. F., The use of β-haloalkylamines in the differentiation of receptors and in the determination of dissociation constants of receptor-agonist complexes. *Adv Drug Res* **3**, 21–55 (1966).
25. Ehlert, F. J., Griffin, M. T. and Suga, H., Analysis of functional responses at *G* protein coupled receptors: Estimation of relative affinity constants for the inactive receptor state. *J Pharmacol Exp Ther* (2011).
26. Ringdahl, B., Structural requirements for muscarinic receptor occupation and receptor activation by oxotremorine analogs in the guinea-pig ileum. *J Pharmacol Exp Ther* **232**, 67–73. (1985).

27. Ehlert, F. J., Griffin, M. T., Sawyer, G. W. and Bailon, R., A simple method for estimation of agonist activity at receptor subtypes: comparison of native and cloned M_3 muscarinic receptors in guinea pig ileum and transfected cells. *J Pharmacol Exp Ther* **289**, 981–992 (1999).
28. Ehlert, F. J., Suga, H. and Griffin, M. T., Quantifying agonist activity at *G* protein-coupled receptors. *J Vis Exp*, e3179 (2011).
29. Chang, K. J. and Triggle, D. J., Quantitative aspects of drug-receptor interactions. I. Ca2+ and cholinergic receptor activation in smooth muscle: a basic model for drug-receptor interactions. *J Theor Biol* **40**, 125–154 (1973).
30. Rosenberger, L. B., Ticku, M. K. and Triggle, D. J., The effect of Ca2+ antagonists on mechanical responses and Ca2+ movements in guinea pig ileal longitudinal smooth muscle. *Can J Physiol Pharmacol* **57**, 333–347 (1979).
31. Hong, S. J., Roan, Y. F. and Chang, C. C., Spontaneous activity of guinea pig ileum longitudinal muscle regulated by Ca(2+)-activated K+ channel. *Am J Physiol* **272**, G962–G971 (1997).
32. Haga, K., Kruse, A. C., Asada, H., Yurugi-Kobayashi, T., Shiroishi, M., Zhang, C., Weis, W. I., Okada, T., Kobilka, B. K., Haga, T. and Kobayashi, T., Structure of the human M_2 muscarinic acetylcholine receptor bound to an antagonist. *Nature* **482**, 547–551 (2012).

9

Analysis of Ligand Bias in Receptor Signaling through Different *G* Protein Pathways

How a drug interacts with the active and inactive states of a receptor ultimately determines its effects the body. An agonist, for example, increases the probability that the receptor will assume the active conformation or remain in this conformation for a longer period of time. At a ligand-gated ion channel, this corresponds to an increase in the frequency and duration of random unitary conductance events. Even though each event is probabilistic, the summation of all of the channel openings in the cell yields an average current that is a highly predictable function of how many receptors are occupied by the agonist. The size of this response is correlated with the size of the receptor population, and its consequences can vary depending upon where the receptor is expressed in the body.

Activation of neuronal nicotinic receptors in sympathetic ganglia and the adrenal medulla, for example, causes an increase in blood pressure and the release of catecholamines into the bloodstream. In both cases, we can track the chain of causation back to the unitary conductance of the ligand-gated ion channel and the microscopic affinity constants of the agonist for the open and closed states. Ultimately, these determine the frequency and magnitude of the digital information generated by the agonist-receptor complex.

An isolated *G* protein-coupled receptor, however, is incapable of eliciting a response unless it interacts with a *G* protein or other signaling partner (e.g., *G* protein-coupled receptor kinase). The active state of the receptor catalyzes the formation of free GTP–$G\alpha$ subunits that elicit downstream responses. The amount of receptor, *G* protein, *RGS* protein and effector enzymes in the signaling cascade can greatly influence the nature and magnitude of the response and the activation state of the receptor population, but these signaling elements do not affect the structures of the active and inactive states, only their probabilities of undergoing a conformational change. If the same active state of the receptor is involved in mediating a set of different responses, then the value of the affinity constant of an agonist for the active state (K_b) should be the same when estimated from its concentration-response curve for eliciting the various responses.

In some instances, however, it appears that different active states of the receptor are involved in eliciting responses through different *G* proteins. An often-cited example concerns the effects of two truncated forms of pituitary adenylate cyclase activating peptide (PACAP) on responses mediated by its receptor (see Figure 9.1).[1] The PACAP receptor is a member of the secretin family of *G* protein-coupled receptors (see Figure 5.7). Both truncated peptides stimulate cAMP accumulation with similar potencies and maximal responses, but over the range of peptide concentrations shown in Figure 8.1b, $PACAP_{1-38}$ stimulates phosphoinositide hydrolysis with about 10-fold greater potency than $PACAP_{1-27}$. Both peptides exhibit lower potency in the phosphoinositide assay as do many agonists for other *G* protein-coupled receptors.

There is sufficient data to estimate the RA_i values of the peptides for both assays using the techniques describe in Chapter 7. Recall that RA_i is a relative estimate of the microscopic affinity constant of a ligand for the active state of the receptor. When expressed relative to that estimated for PACAP 27, the log RA_i values of PACAP 38 for eliciting cAMP and inositol phosphate accumulation are −0.44 and 1.41, respectively. This means that the active state of the receptor that engages G_s has about three-fold higher affinity for PACAP 27 than PACAP 38, whereas the active state that engages G_{15} has about 40-fold higher affinity for PACAP 38 than PACAP

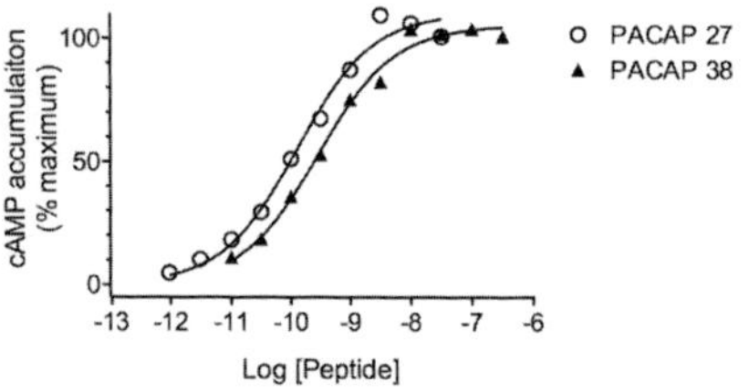

(a)

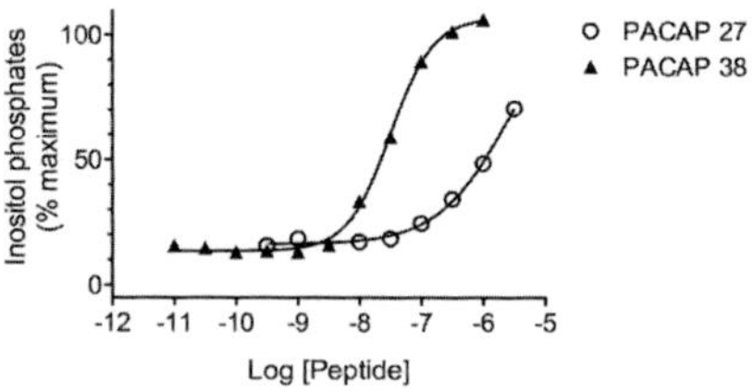

(b)

Figure 9.1: Effects of truncated peptides of pituitary adenylate cyclase activating peptide (PACAP) on PACAP receptor-mediated cAMP (a) and $[^3H]$inositolphosphate (b) accumulation. The plots show the effects of various concentrations of the peptides on these second messenger responses in LLC PK1 cells transiently expressing a short splice variant (PACAP-R) of the PACAP receptor. The plot has been constructed from data published by Spengler *et al.*[1] The curves represent the best fit of Equation (7.47) from Chapter 7 to the data.

27. There are numerous possibilities that could explain these relative estimates. For example, it could be that PACAP 38 has the same affinity for the two active states and that PACAP 27 exhibits low affinity for the active state that signals through G_q. Or perhaps PACAP 27 has similar affinities for the two states, but PACAP 38 exhibits selectivity for the state signaling through G_q. Despite the various possibilities, the data do show that different active states of the PACAP receptor are likely to be involved in activating G_s and G_q.

Another example of agonist bias concerns the effects of the muscarinic agonists, carbachol and McN-A-34, on M2 receptor-mediated inhibition of adenylate cyclase and stimulation of phosphoinositide hydrolysis in HEK 293 cells expressing $G\alpha_{15}$ (Figure 9.2).[2] The murine G protein, $G\alpha_{15}$ and its human homolog, $G\alpha_{16}$, are members of the G_q family of G proteins, and they activate phospholipase Cβ, which hydrolyzes phosphatidylinositol-4,5-*bis*phosphate.[3] These G proteins are normally

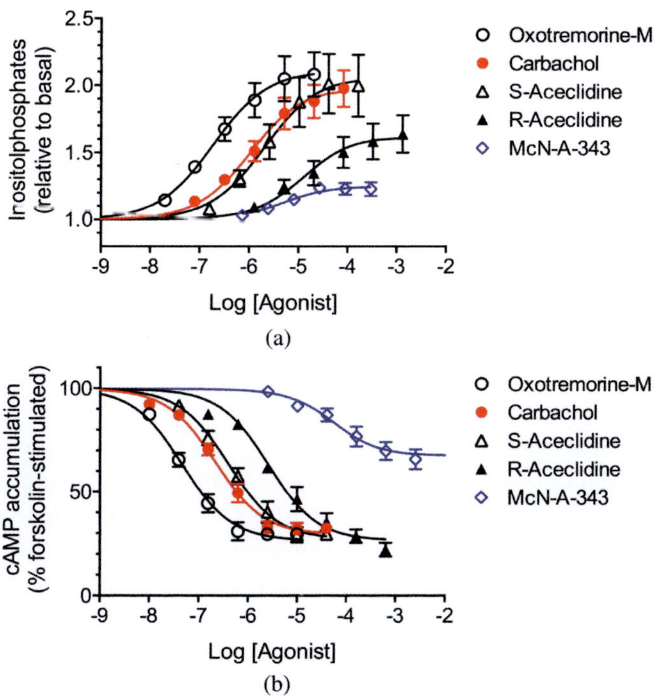

Figure 9.2: Effects of muscarinic agonists on M_2 receptor-mediated stimulation of phosphoinositide hydrolysis (a) and inhibition of cAMP accumulation (b) in HEK 293 cells stably expressing $G\alpha_{15}$ and transiently expressing the human M_2 muscarinic receptor. The data represent previously unpublished exploratory experiments associated with a prior study by Griffin *et al.*[2]

confined to cells of haematopeitic origin, like those of the spleen, thymus and bone marrow,[4,5] but when expressed in cells with G protein-coupled receptors, they readily transduce signals elicited by receptors that normally signal through G_q in other cellular contexts. $G\alpha$ 15 and 16 also mediate phosphoinositide hydrolysis elicited by some G_s-and G_i-linked receptors that do not signal through G_q very well.[6] Normally, the M_2 receptor does not signal through G_q effectively in tissues where both are expressed. But when the M_2 receptor and $G\alpha_{15}$ are expressed in HEK 293 cells, the former elicits a substantial phosphoinositide response through $G\alpha_{15}$. Even though the M_2 receptor is unlikely to signal through $G\alpha_{15}$ in the body, the experiment provides an opportunity to test that hypothesis that an agonist may select for a unique active conformation of a the M_2 receptor that exhibits selectivity for one G protein over another. The experiment also has relevance to drug screening, because $G\alpha$ 15 and 16 are frequently used to transduce signals from G_i-and G_s-linked receptors into Ca^{2+} responses, which can be easily measured at a distance with a fluorometer in a high throughput screen. Thus, it is useful to know whether the pharmacological profile that is characteristic of a receptor in its native context is changed when the receptor couples to $G\alpha_{15}$.

Carbachol elicits a potent inhibition of forskolin-stimulated cAMP accumulation (pEC_{50}, 6.67 $\pm$ 0.07) and a somewhat less potent stimulation of phosphoinositide hydrolysis (pEC_{50}, 5.96 $\pm$ 0.13) (Figure 9.2). When the log RA_i values of oxotremorine-M and the enantiomers of aceclidine are expressed relative to that of carbachol, little difference in this measure is noted for each agonist across the two assays. For example, the log RA_i values of oxotremorine-M, S-aceclidine and R-aceclidine in the phosphoinositide assay are 0.79 ± 0.21, -0.21 ± 20 and -1.47 ± 0.34, respectively, and those for the cAMP assay are 0.69 ± 0.09, -0.24 ± 0.12 and -1.05 ± 0.11. In contrast, the log RA_i values of McN-A343 in the phosphoinositide and cAMP assays are -1.71 ± 0.38 and -2.84 ± 0.19, respectively. Thus, if carbachol exhibits similar affinity for the active state of the M_2 receptor that couples to G_{15} and G_i, then McN-A343 must exhibit about 10-fold higher affinity for a state that couples to G_{15} compared to that coupling to G_i, because its RA_i value is about 10-fold greater for the phosphoinositide assay. This phenomenon has been given several names including agonist bias and agonist-directed signaling.

Differences in the sensitivity of signaling pathways can yield data that appear as though an agonist exhibits selectivity for one pathway relative another even though the same active state of the receptor might mediate both responses. The data in Figure 9.3 serve as an example of how this issue complicates our analysis of the effects of two opiate drugs and the encephalin analog, DAMGO, on μ opiate receptor-mediated inhibition of cAMP accumulation (Figure 9.3a) and recruitment of arrestin

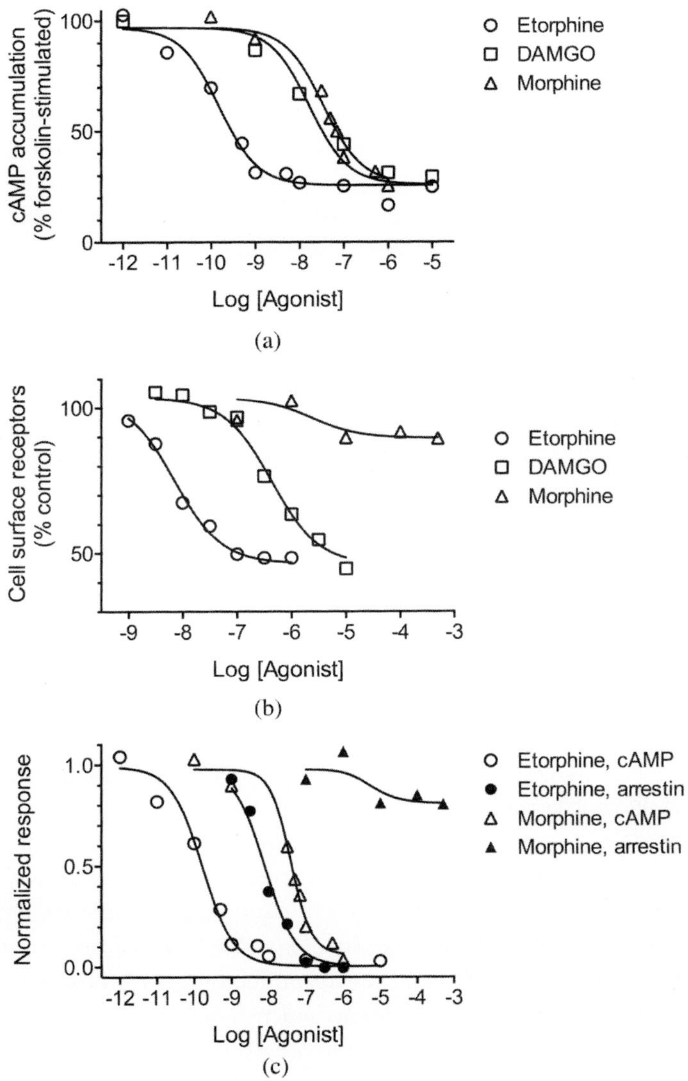

Figure 9.3: Effects of μ opioid receptor agonists on cAMP accumulation (a) and arrestin recruitment (b) in HEK 293 cells stably expressing the murine μ opiate receptor with a FLAG epitope on its amino terminus. The plot in panel c shows the concentration-response curves of etorphine and morpine from a and b plotted again, but normalized relative to the $E_{\max}$ of etorphine. The theoretical curves represent the least squares fits of Equations (7.47) (a and b) and 7.10 (c) to the data. The plot has been constructed from data published by Keith *et al.*[7] and reproduced with permission.

(Figure 9.3b) in HEK 293 cells expressing the μ-opiate receptor.[7] DAMGO is the peptide, tyrosine-*D*-alanine-glycine-*N*-methylphenylalanine, with an amide linkage to 2-aminoethanol on its C-terminus. The first response is mediated by coupling of the active state of the receptor to G_i (inhibition of cAMP accumulation) and the second by phosphorylation of the active state by a *G* protein receptor kinase (recruitment of arrestin). The resulting phosphoserine and phosphothreonine groups provide docking sites for arrestin, which rapidly forms a complex with the receptor. Ultimately, the arrestin-receptor complex is removed from the plasma membrane by endocytosis (see Chapter 5). The plot in Figure 9.3b illustrates receptor expression on the plasma membrane after incubation with different concentrations of agonist. The cyclase response is highly sensitive, and the agonists, etorphine and DAMGO, inhibit cAMP accumulation at very low levels of receptor occupancy (Figure 9.3a). Their pEC_{50} values in the cAMP assay are about 40-fold greater (9.80 and 7.89, respectively) than their corresponding potencies in the arrestin-recruitment assay (8.14 and 6.26) (Figure 9.3b).

We know that arrestin recruitment *is* proportional to receptor occupancy because this response is defined as the association of the phosphorylated receptor with arrestin. The amount of phosphorylated receptor is proportional to the fraction of the agonist-receptor complex in the active state (see Chapter 5) because the *G* protein receptor kinase only phosphorylates the active state of the receptor. It has been suggested that morphine exhibits agonist bias because it mediates a potent inhibition of cAMP accumulation ($pEC_{50} = 7.52$), but does not recruit arrestin.

Let me first suggest a hypothesis that is incorrect, but nonetheless, useful in illustrating the problem of assessing agonist bias. Perhaps morphine selects the same active state of the μ-opiate receptor when signaling through G_i to inhibit adenylate cyclase as it does when attempting to recruit arrestin. Its lack of effect in the latter assay might be caused by its low efficacy. For example, if the efficacy of morphine is, say only 10% that of DAMGO, then its $E_{\max}$ value for recruiting arrestin would be only 10% that of DAMGO because arrestin recruitment is proportional to the fraction of the agonist-receptor complex in the active state. Such a small effect would be difficult to detect and might explain the behavior in Figure 9.3b. Even with an efficacy value of only 10% that of DAMGO, morphine should mediate a maximal inhibition of cAMP accumulation, because we know that DAMGO can elicit this response at very low receptor occupancy.

To test this postulate, the approach described in Chapter 7 under the section entitled, "Is the inactivity of a partial agonist caused by the insensitivity of the system?" can be used. Briefly, the analysis begins with the assumption that the affinity and efficacy of a given agonist in the two assays is the same and that the difference in activity is attributed to changes in the sensitivity constant of the signaling pathway and the receptor density (i.e., R_T/K_E). After normalizing the data relative to the E_{max} value of the more efficacious agonist, the data are analyzed using the operational model assuming that the affinity constant of the agonist is the same for both responses (see Figure 9.3c). This assumption implies that the difference in the τ value of a given agonist across the two assays implies a difference in R_T/K_E because τ is equivalent to $\varepsilon R_T/K_E$ (see Chapter 7, Equation (7.11)) and our provisional assumption is that agonist efficacy (ε) is the same for the two signaling pathways.

This analysis showed that the log τ value of etorphine in the arrestin-recruitment assay is only about 2% that of the cAMP assay. For morphine, however, the corresponding τ value is only about 0.4% that of the cAMP assay. Thus, inactivity of morphine in the arrestin recruitment assay cannot be attributed simply to the reduced sensitivity of the assay because such a mechanism would have resulted in similar reductions in the τ values of both agonists.

This conclusion is consistent with prior data showing that there is little difference between the efficacy of DAMGO and morphine for inhibiting cAMP accumulation through the μ-opiate receptor and that the efficacy of etorphine may be substantially lower than morphine.[8] The magnitude of the inhibitory effects of sodium and guanine nucleotides on the binding affinity of these agonists for μ opiate receptors is also consistent with this postulate.[9] It follows that the hypothesis mentioned above, regarding the potential low efficacy of morphine in the cAMP assay, is untenable, and that morphine exhibits a bias for mediating an inhibition of ademylate cyclase relative to recruitment of arrestin.

Another way to address the question of agonist bias is to estimate the log RA_i values of the agonists for the two different assays. For the cAMP assay, the estimates for DAMGO and morphine, expressed relative to that for etorphine, are −2.45 and −2.36, respectively. The corresponding estimates for the arrestin recruitment assay are −1.78 and −3.14. The relatively large decline in the log RA_i value of morphine in the arrestin assay illustrates its low activity.

So how do we define the term agonist bias? In this book, the expression applies to the condition where an agonist selects for different active conformations of a given receptor when signaling through different pathways. If the agonist exhibits a difference in its affinity constants for the two states, then it exhibits ligand bias or ligand-directed signaling. It may even behave as an antagonist in one of the assays.

It follows from the definition of agonist bias that it is necessary to estimate the K_b values of agonists to determine if an agonist exhibits bias across different functional assays. This can be done using the methods described in the preceding chapter if the assay reports the constitutive activity of the receptor, otherwise the RA_i values can be estimated using the methods described in Chapter 7. Those methods were verified using theoretical examples that only considered a single active state of the receptor. The purpose of this chapter is to illustrate that these methods are appropriate when there are different active states of the receptor that signal through different pathways. Thus, I develop a model for agonist bias based on two active states of the receptor that signal through different G proteins. I use this model to simulate theoretical concentration-response curves for different responses mediated by different active states of the receptor. The resulting concentration-response curves are analyzed to show how it is possible to estimate the microscopic affinity constants of orthosteric ligands for receptor states.

Model for Ligand Bias

I begin with a simple model for a receptor having two active states and describe the relationship between the population parameters (observed affinity and efficacy) and the microscopic affinity constants of the receptor states. This model is subsequently expanded into a more complex one that includes two G proteins that elicit different signaling pathways. Then, in the following section, it is used as input to the operational model so that theoretical concentration-response curves can be simulated and analyzed to verify our methods of analysis.

Simple model for two active receptor states

The simplest model for the behavior of a receptor with multiple active states is that shown in Figure 9.4, which considers two active states (R_s^* and R_s^{**}) and one inactive state (R_s). Each active state triggers a different signaling pathway, and ultimately, a different downstream response. Thus, the model describes a receptor capable of eliciting two different sets of responses, each set mediated by a different active state that could potentially signal through a different G protein or other signaling protein. In the absence of orthosteric ligands, the inactive state of the receptor predominates. This means that the equilibrium constants describing the spontaneous isomerization of the receptor into R_s^* or the R_s^{**} are very small (i.e., $K_q << 1$ and $K_r << 1$). Thus, in most cases it would probably be difficult to measure the response mediated by constitutive activity of the receptor unless a highly sensitive response were measured.

Using this simple model, equations for the observed affinity constant (K_{obs}) and intrinsic efficacies of any orthosteric ligand for generating the two active states of the receptor can be derived using a strategy analogous to that applied in Chapter 2 for the derivation of equations for the activation of a ligand-gated ion channel having only one orthosteric binding site. Here, I list the equations for the various population parameters and show how they can be manipulated to estimate the microscopic constants of the model. The observed affinity constant (K_{obs}) of an orthosteric ligand is given by:

$$K_{\text{obs}} = \frac{K_a + K_{b1}K_q + K_{b2}K_r}{1 + K_q + K_r}. \tag{9.1}$$

In this equation K_a denotes the affinity constant of the orthosteric ligand for the inactive state and K_{b1} and K_{b2} denote the affinity constants for the two active states (R_s^* and R_s^{**}, respectively). The fraction of the orthosteric ligand-receptor complex in the R_s^* state is equivalent to the intrinsic efficacy of the orthosteric ligand for generating the R_s^* state (ε_1):

$$\varepsilon_1 = \frac{K_{b1}K_q}{K_a + K_{b1}K_q + K_{b2}K_r}. \tag{9.2}$$

An analogous equation describes the intrinsic efficacy of the orthosteric ligand for the R_s^{**} state

$$\varepsilon_2 = \frac{K_{b2}K_r}{K_a + K_{b1}K_q + K_{b2}K_r}. \tag{9.3}$$

The fractional amount of free receptor complex in the R_s^* state is defined as the intrinsic efficacy of the system for the R_s^* state ($\varepsilon_{1\text{-sys}}$):

$$\varepsilon_{1\text{-sys}} = \frac{K_q}{1 + K_q + K_r}. \tag{9.4}$$

An analogous equation describes the intrinsic efficacy of the system for ligand for the R_s^{**} state ($\varepsilon_{2\text{-sys}}$):

$$\varepsilon_{2\text{-}sys} = \frac{K_r}{1 + K_q + K_r}. \tag{9.5}$$

It is possible to derive equations for the microscopic constants for the receptor states using the latter equations for the population parameters (e.g., K_{obs}, ε_1 and $\varepsilon_{1\text{-sys}}$). Accordingly, the microscopic affinity constants of the orthosteric ligand for the R_s^*

(K_{b1}) and R_s^{**} (K_{b2}) states are given by:

$$K_{b1} = \frac{\varepsilon_1 K_{\text{obs}}}{\varepsilon_{1\text{-sys}}}, \tag{9.6}$$

$$K_{b2} = \frac{\varepsilon_2 K_{\text{obs}}}{\varepsilon_{2\text{-sys}}}. \tag{9.7}$$

If the equations for the population parameter are substituted into Equations (9.6) and (9.7), the latter reduce to the equalities, $K_{b1} = K_{b1}$ and $K_{b2} = K_{b2}$, which demonstrates their validity.

It follows from Equations (9.6) and (9.7) that it should be possible to estimate the affinity constant of an agonist for a given active state of the receptor relative to that of another agonist using the following equations:

$$RA_{i1} = \frac{K_{b1}}{K'_{b1}} = \frac{\varepsilon_1 K_{\text{obs}}}{\varepsilon'_1 K'_{\text{obs}}}, \tag{9.8}$$

$$RA_{i2} = \frac{K_{b2}}{K'_{b2}} = \frac{\varepsilon_2 K_{\text{obs}}}{\varepsilon'_2 K'_{\text{obs}}}. \tag{9.9}$$

In these equations, the apostrophes denote the parameters of the agonist to which the microscopic affinity constants are normalized. The RA_i terms represent the intrinsic relative activity of the orthosteric ligand, and it represents a relative measure of the affinity constant of the active state of the receptor.

The relevant equations for the inactive state of the receptor are described next. The fraction of the orthosteric ligand-receptor complex in the inactive state (DR_s) is defined as the intrinsic inactivity of the orthosteric ligand (ε_i):

$$\varepsilon_i = \frac{K_a}{K_a + K_{b1}K_q + K_{b2}K_r}. \tag{9.10}$$

The fractional amount of the free receptor complex in the inactive state (R_s) is defined as the intrinsic inactivity of the system ($\varepsilon_{i\text{-sys}}$):

$$\varepsilon_{i\text{-sys}} = \frac{1}{1 + K_q + K_r}. \tag{9.11}$$

Using the equations for K_{obs}, ε_i and $\varepsilon_{i\text{-sys}}$, it is possible to show that the microscopic affinity constant of the orthosteric ligand for the inactive state (K_a) is equivalent to:

$$K_a = \frac{\varepsilon_i K_{\text{obs}}}{\varepsilon_{i\text{-sys}}}. \tag{9.12}$$

It follows from Equations (9.12) that it should be possible to estimate the affinity constant of an agonist for the inactive state of the receptor expressed relative to that of another agonist using the following equation:

$$RI_i = \frac{K_a}{K_a'} = \frac{\varepsilon_i K_{\text{obs}}}{\varepsilon_i' K_{\text{obs}}'}. \tag{9.13}$$

The apostrophes denote the parameters of the agonist to which microscopic affinity constant is normalized. The term RI_i denotes the intrinsic relative inactivity of the agonist and represents a relative measure of the affinity constant of the orthosteric ligand for the inactive state.

Finally, the microscopic constants describing the amount of free receptor complex in the two active states are given by:

$$K_q = \frac{\varepsilon_{1\text{-sys}}}{\varepsilon_{i\text{-sys}}}, \tag{9.14}$$

$$K_r = \frac{\varepsilon_{2\text{-sys}}}{\varepsilon_{i\text{-sys}}}. \tag{9.15}$$

Note that the equations for the affinity constants of the two active states (Equations (9.6) and (9.7)), the relative affinity constants for the two active states (Equations (9.8) and (9.9)), the affinity constant for the inactive state (9.12), the relative affinity for the inactive state (9.13), and the two microscopic constants for the spontaneous conversion of the free receptor into the active states (Equations (9.14) and (9.15) are essentially equivalent to corresponding equations described in Chapters 6 and 7 for a receptor with only one active state and one inactive state. These results suggest, therefore, that the methods used in the analysis responses mediated by a single active state of the receptor should be applicable to the analysis of a receptor with more than one active state.

Let us examine this conclusion within the context of the operational model, which is used to analyze downstream responses at *G* protein-coupled receptors. This model provides an estimate of τ as well as τ_{sys} if the system exhibits constitutive activity. These parameters are proportional to the ε value of an orthosteric ligand and the fraction of the free receptor complex in the active state (ε_{sys}), respectively. Specifically, τ and τ_{sys} are equivalent to $\varepsilon/K_{E\text{-obs}}$ and $\varepsilon_{\text{sys}}/K_{E\text{-obs}}$, respectively. All of the equations for microscopic constants contain an efficacy term as a scalar or single parameter in both the numerator and denominator. Thus, each efficacy term can be replaced with its corresponding τ value without invalidating the equality because the $K_{E\text{-obs}}$ parameter cancels out of the relationship. It follows that the equations for microscopic constants described above can be rewritten in terms of the parameters of the operational model:

$$K_{b1} = \frac{\tau_1 K_{\text{obs}}}{\tau_{1\text{-sys}}}, \tag{9.16}$$

$$K_{b2} = \frac{\tau_2 K_{\text{obs}}}{\tau_{2\text{-sys}}}, \tag{9.17}$$

$$RA_{i1} = \frac{K_{b1}}{K'_{b1}} = \frac{\tau_1 K_{\text{obs}}}{\tau'_1 K'_{\text{obs}}}, \tag{9.18}$$

$$RA_{i2} = \frac{K_{b2}}{K'_{b2}} = \frac{\tau_2 K_{\text{obs}}}{\tau'_2 K'_{\text{obs}}}, \tag{9.19}$$

$$K_a = \frac{\tau_i K_{\text{obs}}}{\tau_{i\text{-sys}}}, \tag{9.20}$$

$$RI_i = \frac{K_a}{K'_a} = \frac{\tau_i K_{\text{obs}}}{\tau'_i K'_{\text{obs}}}, \tag{9.21}$$

$$K_q = \frac{\tau_{1\text{-sys}}}{\tau_{i\text{-sys}}} \tag{9.22}$$

$$K_r = \frac{\tau_{2\text{-sys}}}{\tau_{i\text{-sys}}}. \tag{9.23}$$

Thus, we can restate the conclusion mentioned at the end of the prior paragraph in terms of the operational model. That is, the procedures described in Chapters 6 and 7 for using the operational model to analyze responses elicited by a single active state can also be used to analyze responses mediated by different active states of the receptor. The analysis should provide estimates of the microscopic affinity constants of ligands for the different active states of the receptor.

Model of a receptor with two active states that exhibit selectivity for different G proteins

The simple model just described (Figure 9.4) is useful because it yields a ligand-receptor activation function that is appropriate for a *G* protein coupled receptor, and it is easy to solve analytically. This latter property enabled us to show how the receptor population parameters (e.g., K_{obs}, ε_1, ε_{sys}, etc.) are related to the microscopic constants of the receptor states. But we know that a more realistic model includes the interaction of the receptor with *G* proteins in the presence of guanine nucleotide. In this section, therefore, I expand the simple model shown in Figure 9.4 to include

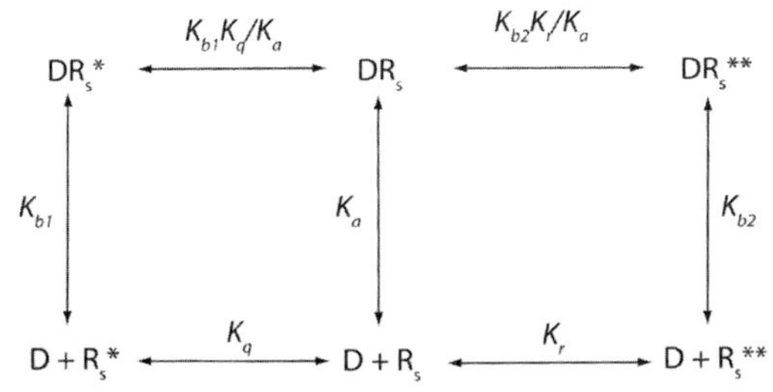

Figure 9.4: Simple model for the interaction of a drug with a receptor that exhibits two active states and a single inactive state. The microscopic affinity constants of the ligand for the inactive (R_s, K_a) and two active states (R_s^*, K_{b1}; R_s^{**}, K_{b2}) are indicated. The isomerization constants of the unoccupied active states are denoted as K_q and K_r for R_s^* and R_s^{**}, respectively.

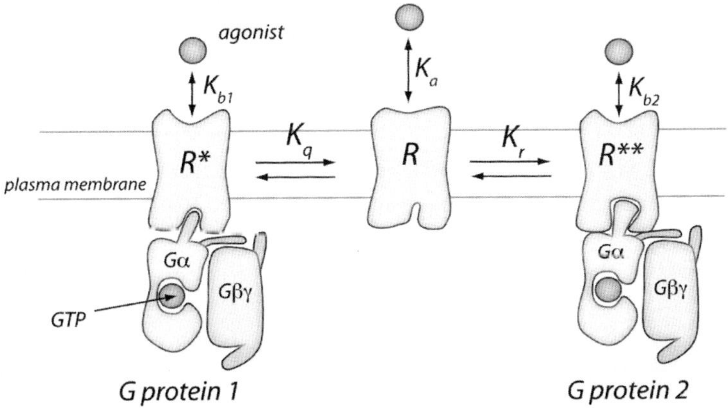

Figure 9.5: Schematic diagram for the interaction of orthosteric ligands with a single receptor having two active states, each interacting with a distinct *G* protein. The microscopic constants are defined in the text and the legend to Figure 8.4.

the interaction of each active state of the receptor with a different *G* protein. The model is used in the next section for the analysis of simulated data.

A schematic description of the model used to simulate agonist bias is shown in Figure 9.5. It is based on the assumption that there are two active states of the receptor (R_s^* and R_s^{**}) that exhibit selectivity for different *G* proteins (*G* protein 1 and *G* protein 2, respectively). The coupling protein could be something other than a *G* protein (e.g., *G* protein-coupled receptor kinase, GRK), but for this example, two *G* proteins are used. An agonist with selectivity for the pathway leading to activation of *G* protein 1 would exhibit a higher affinity for R_s^* relative to R_s^{**} ($K_{b1}/K_{b2} > 1$), whereas an agonist with the opposite signaling bias would exhibit higher affinity for R_s^{**} relative to R_s^* ($K_{b2}/K_{b1} > 1$).

To describe the behavior of the scheme shown in Figure 9.5, the ternary complex model with guanine nucleotide (see Chapter 6, Figure 6.29a) can be modified to include an additional G protein. It is easier to describe the model at the level of receptor complexes first, and then it can be expanded to define the model at the level of receptor states. The scheme shown in Figure 9.6 illustrates the different receptor complexes. The model represents two ternary complex models with guanine nucleotide, with each having a different G protein. The light gray arrows connect the two sub-models at nodes where the receptor is dissociated from the G protein. This enables the receptor to exchange G proteins as we track receptor complexes from one sub-model to the other.

To expand the two-dimensional model to include receptor states, a third dimension is needed. Imagine that the all of the receptor complexes in Figure 9.6 are in the inactive state (R_s), and then imagine an additional set of equilibriums for the active state lying above that shown in Figure 9.6. The resulting three-dimensional set of equilibriums on the left would represent the interaction of the inactive (R_s) and active states of the receptor (R_s^*) with G_1, and the set of equilibriums on the right would illustrate the corresponding interaction of R_s and R_s^{**} with G_2. A three-dimensional diagram of the complete model is shown in Figure 9.7.

This model is equivalent to the combination of two quaternary complex models defined at the level of receptor states (see Chapter 6, Figure 6.29b). The orthosteric

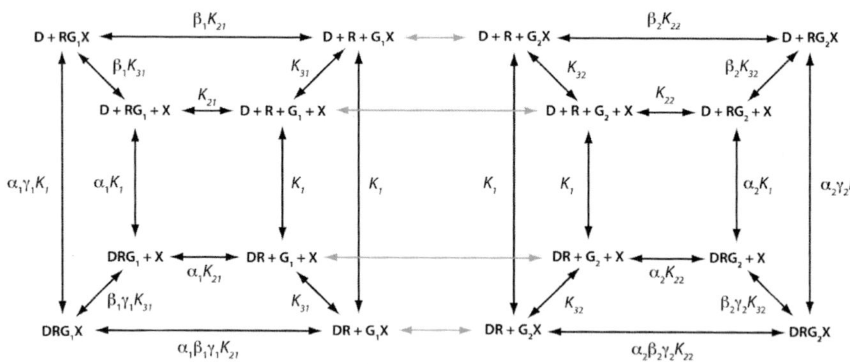

Figure 9.6: The simple model in Figure 9.4 has been expanded to show the interaction of each active state of the receptor with a different G protein (G_1 and G_2), each in equilibrium with a guanine nucleotide (X). The various population parameters have been defined in Chapter 6. An additional subscript has been added to each parameter to denote the G protein (G_1 or G_2) involved in the equilibrium.

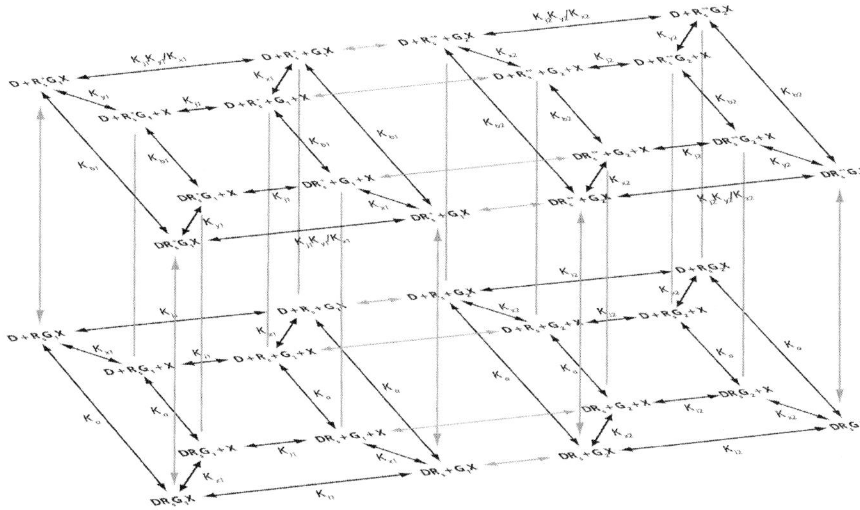

Figure 9.7: The model in Figure 9.6 has been expanded to show the inactive (R_s) and two active (R_s^* and R_s^{**}) states that interact specifically with different G protein (G_1 and G_2). The equilibriums involving the actives state are illustrated on top and those for the inactive state on the bottom. The microscopic constants are defined in the appendix.

ligand-induced stimulus elicited by the R_s^* state of the receptor is proportional to amount of the orthosteric ligand bound in a quaternary complex with the G protein and guanine nucleotide (DR_s^*GX), whereas the constitutive activity of the R_s^* state is proportional to the amount of active state in a ternary complex with G protein and guanine nucleotide (R_s^*GX). Similarly, the orthosteric ligand-induced and constitutive components of the signal elicited by the R_s^{**} state of the receptor are proportional to $DR_s^{**}GX$ and $R_s^{**}G_s$, respectively. Thus, the model shown in Figure 9.7 provides a realistic representation of how a single receptor with two active states can elicit signaling through two different G proteins.

The analytical solution to the model shown in Figure 9.7 is given in the appendix of this chapter. Equations (9.53) and (9.60) give the amount of DR_s^*GX and R_s^*GX complexes, respectively, as a function of the orthosteric ligand concentration and various parameters of the model including the microscopic affinity constants of the orthosteric ligand for the inactive (K_a) and two active (K_{b1} and K_{b2}) states of the receptor. Other parameters are related to how the guanine nucleotide (X) binds to the two G proteins (G_1 and G_2) and how the latter interact with the active and inactive states of the receptor. Finally, additional parameters are the concentration of guanine nucleotide and the relative amount of receptor and G protein.

Analysis of Simulated Data for Ligand Bias

Now that we have a realistic model for the activation of different G proteins through different active states of the receptor, it can be used to generate theoretical concentration-response curves. These can be analyzed to determine if it is possible to estimate the affinity constants of the active and inactive receptor states as well as the population parameters, K_{obs} and relative efficacy. The approach for simulating concentration-response curves involves using the model in Figure 9.7 to generate the amount of quaternary (DR_s^*GX, drug-receptor stimulus) and ternary (R_s^*GX, constitutive stimulus) complexes of an active state of the receptor (e.g., R_s^*) as a function of the orthosteric ligand concentration. The summation of these complexes represents the total stimulus (T) for the R_s^* state (Equation 8.17, Chapter 8) and is used as input to the operational model. The transducer function of the operational model is listed here for convenience:

$$\text{response} = \frac{T^m M_{\text{sys}}}{T^m + K_{E\text{-obs}}^m}. \tag{9.24}$$

As described in earlier chapters, M_{sys} denotes the maximum response of the system, m the transducer slope factor and $K_{E\text{-obs}}$ a parameter related to the sensitivity of the signaling pathway. This process yields a concentration-response curve mediated by the R_s^* state of the receptor. If amount of receptor complex in the form of $DR_s^{**}GX$ and $R_s^{**}GX$ is used as input for the total stimulus in the operational model, then a concentration-response curve mediated by the R_s^{**} state of the receptor can be generated. In the generation of both types of responses, it is assumed that only one state of the receptor contributes to each response. A random error was added to the simulated response values to insure that the estimation of parameter values was feasible.

The model for the total stimulus can be evaluated under two conditions corresponding to a dynamic equilibrium or segregated equilibriums. In the first case (Dynamic equilibrium), the receptor is in equilibrium with the two types of G protein at the same time. This means that the gray arrows between the two sets of equilibriums in Figures 9.6 and 9.7 is functional. In the second case (Segregated equilibriums), the receptor only interacts with one G protein at a time, and thus, the gray arrows between the two sets of equilibriums in Figures 9.6 and 9.7 would not exist for this condition.

Estimating K_{obs} and K_b by the method of partial receptor inactivation

As described in Chapter 7, it is impossible to estimate the observed affinity constant of an agonist from a single concentration-response curve. But if the responses are

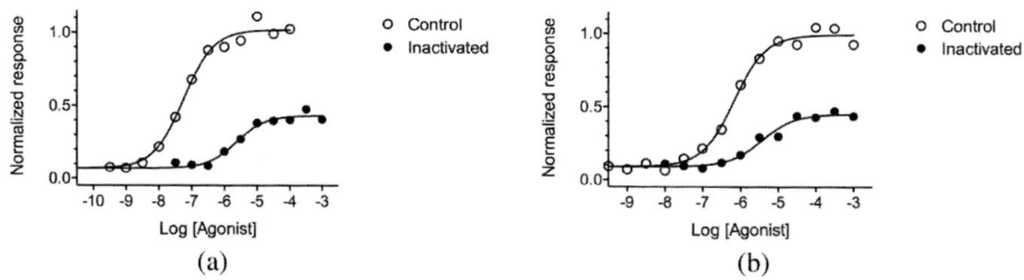

Figure 9.8: Effects of an irreversible neutral antagonist on the concentration-response curves of an agonist for eliciting responses mediated by different active states of the same receptor. The plots in the two panels show simulated concentration-response curves of two agonists for control conditions and after partial receptor inactivation. The affinity constants of the two agonists for the inactive state of the receptor are the same (log K_a, 5.0), but they differ in their affinity for the active state ((*a*) log K_{b1}, 8.0; (b) log K_{b2}, 7.0). The theoretical curves were generated using Equations 9.53 (a) and 9.60 (b) assuming a dynamic equilibrium with the two G proteins. The values of the microscopic constants are: $\log K_{i1} = \log K_{i2}, -2.55$; $\log K_{j1} = \log K_{j2}$, 2.845; $\log K_{x1} = \log K_{x2}$, 8.0; $\log K_{y1} = \log K_{y2}$, 4.602 and $\log K_q = \log K_r, -4.00$. The ratio of G_1 and G_2 to the total receptor concentration was 10, and the concentration of guanine nucleotide (X) was 10^{-3} M. The values of q were 0.01 (a) and 0.1 (b). The parameters of the operational model were: M_{sys}, 1.0; m, 1.5 and log $K_{E\text{-obs}}$, −2.0.

measured under control conditions and after some of the receptors have had their orthosteric binding sites occluded with an irreversible antagonist, then there is sufficient information to estimate K_{obs}. The theoretical data shown in Figure 9.8 represent an example of this type of experiment for a receptor exhibiting constitutive activity and two active states that mediate different responses through different G proteins (G_1 and G_2) as shown in Figures 9.8a and 9.8b, respectively. The data were simulated as described above and represent a dynamic equilibrium of the receptor with the two G proteins. The agonist mediating the responses in Figures 9.8a and 9.8b is the same. Its log affinity constants for the active states coupling to G_1 (log K_{b1}) and G_2 (log K_{b2}) are 8 and 7, respectively, and its log affinity constant for the inactive state of the receptor is 5 (i.e., $\log K_a = 5$). The percentage of receptors inactivated in Figures 9.8a and 9.8b were 99 and 90%, respectively, and the data were simulated for the case of an irreversible ligand that behaves as a neutral antagonist when bound irreversibly. The other parameters for the simulation are listed in the legend to Figure 9.8.

The agonist exhibits a bias for signaling though G_1 relative to G_2 ($K_{b1} > K_{b2}$), and its pEC_{50} value is approximately 10-fold greater for the G_1 response (7.25 ± 0.06 vs. 6.17 ± 0.05). This behavior is expected given the agonist's bias and the similarity in the value of the sensitivity constant ($K_{E\text{-obs}}$) for each pathway (0.01). To estimate the observed dissociation constant (K_{obs}) of the agonist, the concentration-response curves in each panel are analyzed by nonlinear regression analysis using

Equation (8.18) following the procedure described in Chapter 8 (see analysis of the data in Figure 8.10a). Regarding the data in Figure 9.8a, this analysis yielded the following parameter estimates: log K_{obs}, 5.38 ± 0.16; log τ, 1.90 ± 0.17; M_{sys}, 1.04 ± 0.026; log τ_{sys}, −1.24 ± 0.25; m, 0.92 ± 0.15 and q, 0.0081 ± 0.003. Except for minor differences related to random error in the simulated data, these estimates are nearly the same as those predicted from the microscopic constants used to simulate the data in the first place: log K_{obs}, 5.51; log τ, 1.79; M_{sys}, 1.0; log τ_{sys}, −0.70; m, 1.50; and q, 0.01. With regard to the data in Figure 9.8b, the following parameter values were estimated: log K_{obs}, 5.27 ± 0.16; log τ, 0.89 ± 0.17; M_{sys}, 1.10 ± 0.07; log τ_{sys}, −0.97 ± 0.06; m, 1.09 ± 0.25 and q, 0.079 ± 0.027. Again, these estimates agreed with those predicted from the microscopic constants used to simulate the data: log K_{obs}, −5.52; log τ, 0.79; M_{sys}, 1.0; log τ_{sys}, −0.70; m, 1.5; and q, 0.1.

So far, we can see that there is little or no difference in the K_{obs} value in the two different assays even though the agonist exhibits 10-fold higher affinity for the state of the receptor that signals through G_1. This is expected because the model used to simulate the data is a dynamic equilibrium, which represents a collection of receptosomes in which the receptor has equal access to both G_1 and G_2. If the receptosomes were segregated in two categories, one having G_1 and another having G_2, then a difference in K_{obs} would have been measured as described below in connection with Figure 9.10.

The affinity constants of the active states of the receptor that mediate the two responses can be calculated from the parameters estimated so far by using Equations (9.16) and (9.17) described above or by the equivalent equation (8.36) in Chapter 8. That is, if the product of K_{obs} and τ is divided by τ_{sys}, then an estimate of K_b is obtained. More simply, log K_{obs}+ log τ − log τ_{sys} = log K_b. These calculations yield estimates of 8.51 and 7.13, for K_{b1} and K_{b2}, respectively. Thus, the agonist exhibits 20-fold higher affinity for the active state of the receptor that couples to G_1 as compared to that coupling to G_2. In other words, the agonist exhibits a bias for signaling through G_1 compared to G_2. The difference between the measured selectivity for G_1 relative to G_2 (20-fold) and the actual value used for the simulation (10-fold) is due to random error in the simulated measurements.

It is also possible to estimate the K_b values of the agonist in the two different assays by nonlinear regression analysis using Equation (8.42) as described in Chapter 8. With this equation, it is possible to estimate log K_b directly from the regression analysis without having to estimate it from the K_{obs}, τ and τ_{sys} values. Although this advantage is trivial, computer applications for the nonlinear regression analysis often provide standard error values for the parameters in the regression equation. When regression analysis was done using Equation (8.42), log K_{b1} and K_{b2} values of 8.51 ± 0.26 and 7.13 ± 0.24 were estimated for the data in Figures 9.8a

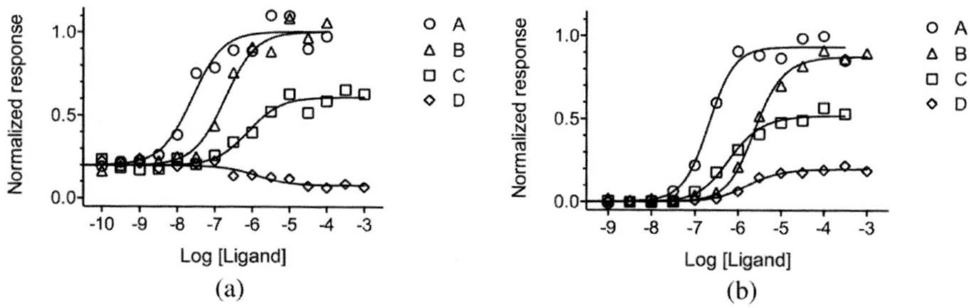

Figure 9.9: The concentration-response curves of a series of orthosteric ligands for eliciting responses through a receptor in a dynamic equilibrium with two G proteins, G_1 (a) and G_2 (b). The plots in panels a and b were simulated using Equation (9.53) (a) and (9.60) (b) assuming a dynamic equilibrium. The log microscopic affinity constants of the ligands for active (log K_{b1}) and inactive states (log K_a) of the receptor are indicated in parenthesis for each ligand in panel a: A (8.0 and 5.0), B (7.0 and 4.5), C (6.51 and 6.0) and D (5.70 and 6.0). The corresponding values for the ligands in panel b are the same except for ligand D (7.0 and 6.0). The values of the other parameters and microscopic constants are the same as those used in the simulations in Figure 9.8 except for the following: q, 1.0 and log $K_{E\text{-obs}}$, -2.30 and -1.30 for (a) and (b), respectively.

and 9.8b, respectively. These values are essentially the same as those estimated from the corresponding log K_{obs}, log τ and log τ_{sys} values as just described.

The analysis described above could also be carried out for an inverse agonist, although in this situation, it would be necessary to analyze the behavior of a full agonist as well because the latter is required in the estimation of M_{sys}. An example of the analysis of an inverse agonist is given below in connection with Figure 9.9.

If the two responses lacked constitutive activity, it would have been impossible to estimate the K_b values of the agonist in the two functional assays. In this case, it would have been necessary to measure the concentration-response curves of a series of orthosteric ligands and to estimate the K_b value of a given agonist relative to that of another (RA_i). In the next section, examples this situation are described.

Analysis of a series of orthosteric ligand concentration-response curves for ligand bias

The most common way of detecting agonist bias is to compare the concentration-response-curves of a series of orthosteric ligands in different functional assays for the same receptor. Figure 9.9 shows a theoretical example of the concentration-response curves of three agonists and an inverse agonist in two functional assays.

Table 9.1: Estimates of the EC_{50} and E_{max} (or T_{sat}) values for the simulated data in Figure 9.9.

	G_1 Response		G_2 Response	
Ligand	log EC_{50}	T_{sat}	log EC_{50}	E_{max}
A	-7.27 ± 0.14	0.99 ± 0.04	-6.68 ± 0.03	0.93 ± 0.01
B	-6.71 ± 0.08	1.01 ± 0.03	-5.57 ± 0.05	0.88 ± 0.02
C	-6.09 ± 0.15	0.62 ± 0.02	-6.12 ± 0.08	0.53 ± 0.08
D	-5.78 ± 0.36	0.20 ± 0.02	-5.78 ± 0.16	0.20 ± 0.16

The response mediated by G_1 exhibits constitutive activity (Figure 9.9a), whereas that elicited by G_2 does not. The data were simulated for a dynamic equilibrium. A summary of the EC_{50} and T_{sat} (analogous to E_{max}, see Chapter 8) values of the agonists in for eliciting responses through G_1 and G_2 are listed in Table 9.1.

The series of concentration-response curves in Figure 9.9a can be analyzed by global nonlinear regression analysis with Equation (8.41) as described in Chapter 8. This analysis yields estimates of the K_{obs} values the partial agonist C (log K_{obs} = 5.81 ± 0.18) and inverse agonist D (log $K_{obs} = 5.91 \pm 0.54$). These estimates are nearly the same as the theoretical values. It is impossible to estimate the K_{obs} values of the full agonist from only the data in Figure 8.9a, and their K_{obs} values must be constrained to constants equal to or less than their true values during regression analysis. This analysis yields the following estimates of the K_{b1} values of orthosteric ligands A–D: $8.25 \pm 0.20, 7.36 \pm 0.20, 6.56 \pm 0.18$ and 5.48 ± 0.61, respectively, which are nearly equivalent to their respective theoretical values: 8.0, 7.0, 6.5 and 5.7. Finally, regression analysis yields estimates of M_{sys} 1.00 ± 0.02, m 1.07 ± 0.21 and log τ_{sys} -0.57 ± 0.13, which are nearly the same as the theoretical values (1.0, 1.5 and −0.41, respectively). Knowing K_{obs} and K_{b1}, relative efficacy can be calculated for ligands C and D using Equation (8.45) in Chapter 8.

Because the concentration-response curves in Figure 9.9b lack constitutive receptor activity, they can be analyzed with Equations (7.47) and (7.49) as described for the analysis of the data in Figure 7.8 in Chapter 7. This analysis yields estimates of the log RA_i values of the agonists expressed relative to that of the full agonist A. The RA_i value is a relative estimate of K_{b2}, and the corresponding log values for orthosteric ligands B–D, expressed relative to that of A are -1.05 ± 0.05, -0.52 ± 0.08 and -1.18 ± 0.19, respectively. These are nearly equivalent to the respective theoretical values: −1.0, −0.5 and −1.0. The RA_i values for the orthosteric ligands in Figure 8.9a can be estimated by dividing the K_{b1} value of each ligand by that of agonist A, which yields the following log RA_i values for B–D: −0.88, −1.69

and −2.78, respectively. The log RA_i value of agonist B is similar across the two responses, where those of ligands C and D are about one and a half log units greater in the G_2 assay (Figure 9.9b). Thus, agonists C and ligand D exhibit a bias for signaling through G_2 because their affinity constants for the active state of the receptor that couples with G_2 are much greater than those that couple to G_1.

The analysis of the data in Figure 8.9b also yields estimates of the log K_{obs} values of agonist B 4.81 ± 0.14, C 6.08 ± 0.09 and D 5.88 ± 0.19, and these agreed with the theoretical values; −4.86, −6.03 and −6.01, respectively. The K_{obs} values of orthosteric ligands C and D do not differ across the two assays, which is expected for a dynamic equilibrium. Similarly, the estimates of M_{sys} (0.93 ± 0.013) and m (1.45 ± 0.11) are also similar to their theoretical values: 1.0 and 1.5, respectively.

Let us review the relationship between the estimated population parameters and the simulated data. Table 9.2 lists the K_{obs} and E_{max} values of the agonists for generating the two types of quaternary complexes, DR_s^*GX and $DR_s^{**}GX$. The K_{obs} value is equivalent to the EC_{50} value of the agonist for eliciting half-maximal formation of the quaternary complex, and the maximal amount of the complex, expressed relative to the total receptor density, is equivalent to the efficacy of the agonist. For

Table 9.2: Comparison of the estimates of the observed affinity (K_{obs}) and efficacy (ε) with the EC_{50} and E_{max} values of the ligands for generating the quaternary complexes, $DR_s^*G_1X$ and $DR_s^{**}G_2X$, for the simulations illustrated in panels a and b, respectively, of Figure 9.9.

	G_1 Response				G_2 Response			
	$DR_s^*G_1X$		Estimated		$DR_s^{**}G_2X$		Estimated	
Ligand	pEC_{50}	E_{max}[a]	log K_{obs}	ε[b]	pEC_{50}	E_{max}[c]	log K_{obs}	ε[d]
A	5.70	0.39			5.70	0.39		
B	4.86	0.27			4.86	0.27 (1.00)	4.81± 0.14	1.00
C	6.03	0.0060 (1.00)	5.81± 0.18	1.00	6.03	0.060 (0.22)	6.08± 0.09	0.18
D	6.01	0.00098 (0.16)	5.91± 0.54	0.07	6.01	0.020 (0.071)	5.88± 0.19	0.063 0.063

[a] The values in parentheses denote efficacy expressed relative to that of agonist C.
[b] Denotes efficacy expressed relative to that of agonist C. The value was calculated from the estimates of K_b and K_{obs} using Equation (8.45) in Chapter 8.
[c] The values in parentheses denote efficacy expressed relative to that of agonist B.
[d] Denotes efficacy expressed relative to that of agonist B. The value was calculated from the estimates of the RA_i and K_{obs} estimates using Equation (7.69) in Chapter 7.

those orthosteric ligands whose K_{obs} values could be calculated, there is good agreement between the K_{obs} value of the quaternary complex and the K_{obs} value estimated from the concentration-response curve for the downstream response. There is also reasonable agreement between the theoretical and estimated relative efficacy values provided that the theoretical values are expressed relative to the same agonist that was used in the estimation process. The two-fold error in the estimate of the relative efficacy of inverse agonist D for the G_1 response can be attributed to the relatively large error in the estimation of its K_{obs} value. Taking the product of the theoretical affinity and efficacy values of each agonist and dividing it by that of agonist A yields the theoretical RA_i values. There is agreement between these theoretical values and those estimated from the concentration-response curves as summarized in Table 8.2 and as just described above.

The final example of a series of concentration-response curves that illustrate the concept of agonist bias is shown in Figure 8.10. These are for the condition of segregated equilibriums. That is, for each response, the receptor is in equilibrium with only one G protein. Thus, the responses in panel a are mediated by G_1 in the absence of G_2, whereas those in panel b are mediated by G_2 in the absence of G_1. The curves have been simulated for a series of agonists having varying affinities and efficacies in two systems that lack constitutive activity. The theoretical EC_{50} and E_{max} values for these data are listed in Table 9.3.

Using the procedures described in Chapter 7, it is possible to estimate the log K_{obs} values of the partial agonists and the log RA_i value of each agonist expressed relative to that of full agonist A for each response. This analysis yielded the following K_{obs} values for agonists B–D in panel a: 4.40 ± 0.16, 6.22 ± 0.31 and 5.94 ± 0.14, respectively. The corresponding values for panel b are 4.40 ± 0.11, 5.83 ± 0.15 and 5.81 ± 0.15. The difference in the estimate of log K_{obs} for agonist C across two responses is due to the segregation of the two G proteins and to the bias for signaling through G_2 relative to G_1. The log RA_i values of the agonists (A–D) in panel a are

Table 9.3: Estimates of the EC_{50} and E_{max} values for the simulated data in Figure 9.10.

	G_1 Response		G_2 Response	
Ligand	log EC_{50}	E_{max}	log EC_{50}	E_{max}
A	-6.67 ± 0.06	0.93 ± 0.03	-6.20 ± 0.07	0.91 ± 0.02
B	-4.96 ± 0.07	0.85 ± 0.02	-4.79 ± 0.08	0.65 ± 0.02
C	-6.03 ± 0.17	0.19 ± 0.01	-6.77 ± 0.07	0.82 ± 0.02
D	-6.07 ± 0.09	0.56 ± 0.02	-5.94 ± 0.09	0.30 ± 0.01

-1.50 ± 0.01, -1.03 ± 0.04 and -0.52 ± 0.02, respectively, and the corresponding values for panel b are $-1.49 \pm 0.01, 0.52 \pm 0.005$ and -0.52 ± 0.01. The log RA_i value of each agonist is the same across the two assays with the exception of agonist C, which exhibits a 1.5-fold larger value for the state of the receptor that couples to G_2 compared to that coupling to G_1. Thus, agonist C exhibits a bias for signaling through the active state of the receptor that signals through G_1.

Knowing the K_{obs} and RA_i values of agonists B–D, it is possible to estimate their relative efficacies in the two assays using Equation (7.67) from Chapter 7. When normalized relative to agonist B, the log efficacies of C and D were -1.10 and -0.54 for the response mediated by G_1 and 0.57 and -0.53 for the response mediated by G_2, respectively. Thus, agonist C exhibited 47-fold greater efficacy for eliciting responses through G_2 compared to G_1.

All of the parameter estimates are nearly the same as the theoretical values: Figure 9.10a, log K_{obs}: A, 5.71; B, 4.67; C, 6.18 and D, 6.10; log RA_i: B, -1.5; C, -1.0 and D -0.5; Figure 9.10b, log K_{obs}: A, 5.71; B, 4.67; C, 6.37 and D, 6.10; log RA_i: B, -1.5; C, 0.5 and D, -0.5.

Table 9.4 summarizes the relationship between the theoretical and estimated population parameters (K_{obs} and ε). Again, K_{obs} represents the EC_{50} value of the agonist for eliciting half-maximal formation of the quaternary complex, and the

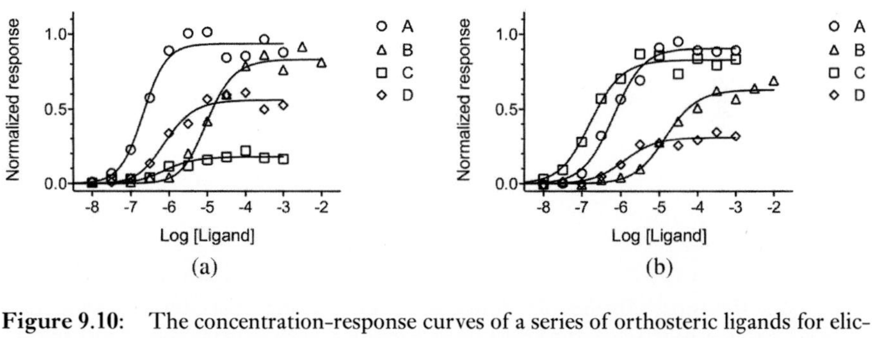

Figure 9.10: The concentration-response curves of a series of orthosteric ligands for eliciting responses through a receptor in two segregated equilibriums, each involving a different G protein, G_1 (a) *and* G_2 (b). The plots in panels a and b were simulated using Equation (9.53) (a) and (9.60) (b) assuming a segregated equilibrium, which involves assigning a value of 0 to the concentration of G protein not eliciting a stimulus. The log microscopic affinity constants of the ligands for active (log K_{b1}) and inactive states (log K_a) of the receptor are indicated in parenthesis for each ligand in panel a: A (8.0 and 5.0), B (6.5 and 4.5), C (7.0 and 6.0) and D (7.5 and 6.0). The corresponding values for the ligands in panel b are the same except for ligand C (8.5 and 6.0). The values of the other parameters and microscopic constants are the same as those used in the simulations in Figure 9.8 except for the following: q, 1.0 and log $K_{E\text{-obs}}$, -1.0 and -0.7 for a and b, respectively.

Table 9.4: Comparison of the estimates of the observed affinity (K_{obs}) and efficacy (ε) with the EC_{50} and E_{max} values of the ligands for generating the quaternary complexes, $DR_s^*G_1X$ and $DR_s^{**}G_2X$, for the simulations illustrated in panels a and b, respectively, of Figure 9.10.

	G_1 Response				G_2 Response			
	$DR_s^*G_1X$		Estimated		$DR_s^{**}G_2X$		Estimated	
Ligand	pEC_{50}	E_{max}[a]	$\log K_{obs}$	ε[b]	pEC_{50}	E_{max}[a]	$\log K_{obs}$	ε[b]
A	5.71	0.78			5.71	0.78		
B	4.67	0.28 (1.00)	4.40 ± 0.16	1.00	4.67	0.28 (1.00)	4.40 ± 0.11	1.00
C	6.18	0.038 (0.13)	6.22 ± 0.32	0.085	6.37	0.55 (1.95)	5.83 ± 0.15	3.99
D	6.10	0.11 (0.40)	5.94 ± 0.15	0.31	6.10	0.11 (0.40)	5.81 ± 0.15	0.26

[a] The values in parentheses denote efficacy expressed relative to that of agonist B.
[b] Denotes efficacy expressed relative to that of agonist B. The value was calculated from the estimates of the RA_i and K_{obs} estimates using Equation (7.69) in Chapter 7.

relative amount of the quaternary complex is a measure of efficacy. There is reasonable agreement between the theoretical and estimated parameters for those agonists whose K_{obs} and relative efficacy values could be calculated, provided that the latter are normalized to the same agonist.

Estimation K_a and RI_i values when there is more than one active state

When there are two active states of the receptor, K_a or RI_i values can be calculated for all but the most efficacious agonist in a series using procedures described in Chapters 7 and 8, but the method requires further explanation to account for the complexities associated with multiple active states.

The estimation of K_b and RA_i is easier than the analogous measurements for the inactive state of the receptor (K_a or RI_i) because accurate estimates of τ and τ_{sys} can be estimated, and these are proportional to the intrinsic efficacy of the ligand (ε) and the intrinsic efficacy of the system (ε_{sys}):

$$K_b = \frac{\varepsilon K_{obs}}{\varepsilon_{sys}} = \frac{\tau K_{obs}}{\tau_{sys}}. \quad (9.25)$$

Thus, K_b can be estimated from the parameters of the operational model (i.e., τ, K_{obs} and τ_{sys}) even though it is impossible to estimate precise values of ε and ε_{sys}. The analogous relationship for K_a is:

$$K_a = \frac{\varepsilon_i K_{obs}}{\varepsilon_{i\text{-sys}}} = \frac{\tau_i K_{obs}}{\tau_{i\text{-sys}}}. \tag{9.26}$$

There is no simple way to estimate precise values of τ_i and $\tau_{i\text{-sys}}$, however. At best, only approximate values of ε_i and $\varepsilon_{i\text{-sys}}$ can be estimated from the approximate values of ε and ε_{sys}. These are calculated using rearranged forms of Equations (8.4) and (8.5) from Chapter 8, which are shown below for convenience:

$$\varepsilon_i = 1 - \varepsilon, \tag{9.27}$$

$$\varepsilon_{i\text{-sys}} = 1 - \varepsilon_{sys}. \tag{9.28}$$

It is the error in these estimations that make it impossible to estimate the K_a value of the most efficacious agonist in a series accurately. Fortunately, this error is minor for a less efficacious agonists in a series provided that its efficacy is less than one-half that of the most efficacious agonist. This error would seem to be compounded in the case of two active states of the receptor because the intrinsic efficacies and inactivities of an agonist must sum to one, and there are three such values:

$$1 = \varepsilon_1 + \varepsilon_2 + \varepsilon_i. \tag{9.29}$$

Similarly, the corresponding parameters for the efficacies and inactivities of the free receptor complex must also sum to one:

$$1 = \varepsilon_{1\text{-sys}} + \varepsilon_{2\text{-sys}} + \varepsilon_{i\text{-sys}}. \tag{9.30}$$

Nonetheless, it is shown below that the calculations described in Equations (9.27) and (9.28) provide reasonable estimates of K_a for the less efficacious agonists in a series.

To illustrate this point, the K_a or RI_i (relative estimate of K_a) values can be estimated for the various orthosteric ligands shown in Figures 9.9 and 9.10 using the procedures described in Chapters 7 and 8. I will describe the process again to show that the method of estimating ε_i and $\varepsilon_{i\text{-sys}}$ only introduces error in the estimate of the K_a of the most efficacious agonist in a series. Recall that it is necessary to have an estimate of the K_{obs} values of all of the orthosteric ligands, including those of the full agonists. Thus, additional experiments involving the method of partial receptor inactivation would be needed to estimate the K_{obs} values of the full agonists in Figures 9.9 and 9.10, essentially as described in Chapter 8 in connection with Figure 8.14. In the analysis below, the K_{obs} value will be constrained to its theoretical value.

To estimate K_a values of the ligands illustrated in Figure 9.9a, the data are analyzed by nonlinear regression analysis using Equation (8.50) in Chapter 8. This equation is used when the response reports the constitutive activity of the receptor. The value of q is constrained to 1 and the ε_i value of the most efficacious agonist is constrained to 0. The value of log K_{obs} of the most efficacious agonist is constrained to its theoretical value (5.704). This analysis yielded reasonable estimates of log ε_i-sys (-0.00131 ± 0.00065) and the log ε_i values of agonists C (-0.00696 ± 0.0018) and D (-0.0005 ± 0.0007). Substituting these estimates into Equation (9.20) yields estimates of log K_a for agonists C (5.82) and D (5.91) that are approximately the same as the theoretical value used to simulate the data (6.0). It was impossible to estimate a reliable value of log K_a for agonist B because it behaved as a full agonist.

For the estimation of the log K_a values of the ligands in Figure 9.9b, the data are analysis using Equation (7.95) in Chapter 7. This analysis yields estimates of the log K_{obs} values of agonists B (4.99 ± 0.10), C (6.14 ± 0.09) and D (5.93 ± 0.20), and their corresponding log ε_i values (-0.26 ± 0.06, -0.050 ± 0.004 and -0.018 ± 0.002). Dividing the product of the K_{obs} and ε_i values of each agonist by that of the least efficacious agonist (D) yields the RI_i value. When expressed as logarithms, the corresponding values for agonists B–D are -1.18, 0.18 and 0.00. These values represent the log K_a value expressed relative to that of agonist D. The log K_a value can be estimated from RI_i using Equation (7.97) in Chapter 7. These calculations yielded the following estimates of log K_a for agonists B–D: 4.74, 6.10 and 5.93, respectively. These are nearly the same as those used to simulate the data (4.5, 6 and 6, respectively).

The same analysis was used to estimate the log K_a values of the agonists in Figure 9.10. The following log K_a values were estimated for agonists B–D in panel a: 4.40, 6.30 and 6.0, respectively. These are nearly the same as those used to simulate the data (4.5, 6.0 and 6.0). Finally, the estimates of log K_a for agonists B and D in panel b were 4.43 and 5.89, respectively, which are nearly the same as those used to simulate the data (4.5 and 6.0). Again, it is difficult to estimate log K_a for the efficacious agonists A and C in Figure 9.10b.

Estimation of K_b, RA_i, K_{obs} and Relative Efficacy When the Hill Slope is One

When the Hill slope of the orthosteric ligand concentration-response curve is equal to one and there are multiple active states of the receptor, the parameters for the active state of the receptor and the K_{obs} and relative efficacy values of partial agonists can be estimated from the empirical parameters of the concentration-response curve (i.e., EC_{50}, E_{max} and T_{sat}) as described in Chapters 7 and 8. For concentration-response curves exhibiting constitutive activity, K_b can be estimated using Equation (8.52), and the K_{obs} and relative efficacy values of partial agonists can be calculated using Equations (8.53) and (8.54), respectively. For concentration-response curves lacking constitutive activity, RA_i can be calculated using Equation (7.56), and the relative affinity and relative efficacy values of partial agonists can be calculated using Equations (7.60) and (7.70), respectively.

Fundamental Corollary for Microscopic Affinity Constants

The fundamental property of drug receptor interactions that provides the basis for the analyses describe in this chapter as well as the preceding two chapters can be stated in a single sentence: For any ligand that binds reversibly to a homogeneous population of receptors that exits in several conformations, the microscopic affinity constant of a ligand for the jth conformation of the receptor (K_j) is equivalent to the product of its observed affinity (K_{obs}) and the fraction of the ligand-receptor complex in that state (ε_j) divided by the fraction of the free receptor complex in the same state $\varepsilon_{j\text{-sys}}$:

$$K_j = \frac{K_{obs}\varepsilon_j}{\varepsilon_{j\text{-sys}}}. \tag{9.31}$$

This corollary enables the estimation of relative or actual estimates of K_j provided at least proportional estimates of ε_j or both ε_j and $\varepsilon_{j\text{-sys}}$ can be made, respectively.

Why Multiple Active States

When bound with its endogenous ligand, an efficient receptor assumes an optimum structure complementary to the conformation of the G protein that has the guanine nucleotide-binding pocket open so that GTP can readily displace GDP. The longer the ligand-receptor complex remains in this optimal conformation, the more efficient the signaling. For this to occur, the binding pocket on the receptor must have high affinity for the endogenous ligand when the G protein-coupling part of the receptor is in its optimum structure. If there were multiple conformations of the binding pocket allosterically linked to multiple conformations of the G protein-coupling domain, then some of the ligand-binding energy would be wasted on unproductive conformational changes in the receptor. Thus, a unique high-affinity, active conformation of the ligand-binding pocket allosterically linked to a unique, highly efficient conformation of the *G* protein-coupling domain would appear to be an adaptation for efficient signaling. A gene that encoded such a receptor would thrive at the expense of other genes for less efficient receptors.

So why are there different active states of the receptor that mediate agonist bias? One possibility is based on the observation that most instances of agonist bias are attributed to the actions of xenobiotics. Some synthetic ligands are likely to select for rare conformations of the receptor, perhaps by interacting with regions adjacent to the binding pocket of the natural ligand. This conformation may be less complementary with one type of *G* protein relative to another, ultimately resulting in agonist bias. The overall signaling efficiency of the synthetic ligand-receptor complex at the receptor's preferred *G* protein is likely to be less than that of the natural ligand, and this is often observed in cases of agonist bias.

Some receptors have more than one natural ligand. The β_1 adrenoceptor is activated by both epinephrine and norepinephrine, and a given opioid receptor is activated by more than one opioid peptide. Perhaps these different natural ligands select for different conformational states of their receptors.

Ligand Bias and Drug Discovery

Agonist bias implies that there are different active states of the receptor that exhibit selectivity for different *G* proteins or other signaling partners resulting in a variation in the population parameters (observed affinity, K_{obs}, and intrinsic efficacy, ε) of a ligand when measured at different responses. This seems like the activity of the agonist depends on the response being measured, but the *G* protein is not so much a shaper of drug action so much as a window onto the different active states of the receptor. Rather, different *G* proteins report the activity of different receptor states. The ability of a drug to select for these different states can be determined in assays for different signaling pathways (e.g., recruitment of arrestin) including those elicited by different *G* proteins. Not only can an orthosteric ligand exhibit bias for one signaling pathway relative to another, but an allosteric ligand can potentially direct the endogenous ligand to signal through a subset of it normal signaling pathways (see Chapter 10). Thus, drugs exhibiting a bias in signaling have the potential to be more selective therapeutic agents.

Appendix

This appendix describes the solution to the model shown in Figure 9.7. The goal in solving the model is to generate an equation that describes the fractional amount of receptor signaling through G proteins 1 (G_1) and 2 (G_2) as a function of the ligand concentration and microscopic constants. The stimulus for the G_1 pathway is equivalent to the summation of the constitutive ($R_s^* G_1 X$) and ligand-induced ($DR_s^* G_1 X$) components of the receptor activation function. The analogous components contribute to signaling through G_2 ($R_s^{**} G_1 X$ and $DR_s^{**} G_2 X$). In addition, the function is also solved for the condition of partial receptor inactivation by an irreversible ligand that behaves as a neutral antagonist when bound irreversibly.

To begin, the various microscopic constants are defined beginning with the affinity constants of the orthosteric ligand for the inactive (R_s) and two active states (R_s^* and R_s^{**}) of the receptor:

$$K_a = \frac{[DR_s]}{[D][R_s]}, \tag{9.32}$$

$$K_{b1} = \frac{[DR_s^*]}{[D][R_s^*]}, \tag{9.33}$$

$$K_{b2} = \frac{[DR_s^{**}]}{[D][R_s^{**}]}. \tag{9.35}$$

Those for the binding of guanine nucleotide (X) to the inactive state of the receptor-G_1 complex are:

$$K_{x1} = \frac{[G_1X]}{[X][G_1]}, \tag{9.36}$$

$$K_{x1} = \frac{[R_sG_1X]}{[X][R_sG_1]}, \tag{9.37}$$

$$K_{x1} = \frac{[DR_sG_1X]}{[X][DR_sG_1]}. \tag{9.38}$$

The corresponding constants for G_2 are:

$$K_{x2} = \frac{[G_2X]}{[X][G_2]}, \tag{9.39}$$

$$K_{x2} = \frac{[R_sG_2X]}{[X][R_sG_2]}, \tag{9.40}$$

$$K_{x2} = \frac{[DR_sG_2X]}{[X][DR_sG_2]}. \tag{9.41}$$

The affinity constants of guanine nucleotide (X) for the active state of the receptor-G_1 complex are defined as:

$$K_{y1} = \frac{[R_s^* G_1 X]}{[X][R_s^* G_1]}, \tag{9.42}$$

$$K_{y1} = \frac{[DR_s^* G_1 X]}{[X][DR_s^* G_1]}. \tag{9.43}$$

The corresponding constants for G_2 are:

$$K_{y2} = \frac{[R_s^{**} G_2 X]}{[X][R_s^{**} G_2]}, \tag{9.44}$$

$$K_{y2} = \frac{[DR_s^{**} G_2 X]}{[X][DR_s^{**} G_2]}. \tag{9.45}$$

The affinity constants of the two G proteins for the inactive state of the receptor are:

$$K_{x1} = \frac{[R_s G_1]}{[R_s][G_1]}, \tag{9.46}$$

$$K_{x2} = \frac{[R_s G_2]}{[R_s][G_2]}. \tag{9.47}$$

The corresponding constants for the active state of the receptor are:

$$K_{y1} = \frac{[R_s^* G_1]}{[R_s^*][G_1]}, \tag{9.48}$$

$$K_{y2} = \frac{[R_s^{**} G_2]}{[R_s^{**}][G_2]}. \tag{9.49}$$

The isomerization constants for the two active states of the receptor are defined as:

$$K_q = \frac{[R_s^*]}{[R_s]}, \tag{9.50}$$

$$K_r = \frac{[R_s^{**}]}{[R_s]}. \tag{9.51}$$

The total stimulus of the G_1 signaling pathway (T_1) is defined as:

$$T_1 = \frac{[R_s^* G_1 X] + [DR_s^* G_1 X]}{[R_T]}. \tag{9.52}$$

In this equation, $[R_T]$ denotes the total concentration of all of the various receptor complexes. Substitutions are then made for the receptor complexes in the numerator and denominator using Equations (9.32)–(9.51). After simplification, the following expression can be derived for T_1:

$$T_1 = \frac{q\frac{[G_{1-T}]}{[R_T]}XK_{j1}K_qK_{y1}(1+DK_{b1})}{A+B+C} + \frac{(1-q)\frac{[G_{1-T}]}{[R_T]}XK_{j1}K_qK_{y1}}{E+F+G}. \tag{9.53}$$

In this equation, $[G_{1-T}]$ denotes the total concentration of G_1, and q denotes the fraction of the receptor population not inactivated by the irreversible ligand. The other variables are defined as:

$$A = \left(1 + K_q + K_r + D(K_a + K_{b1}K_q + K_{b2}K_r)\right), \tag{9.54}$$

$$B = \frac{[G_{1-T}]}{[R_T]}(K_{i1}(1+DK_a)(1+XK_{x1}) + K_{j1}K_q(1+DK_{b1})(1+XK_{y1})), \tag{9.55}$$

$$C = \frac{[G_{2-T}]}{[R_T]}(K_{i2}(1+DK_a)(1+XK_{x2}) + K_{j2}K_r(1+DK_{b2})(1+XK_{y2})), \tag{9.56}$$

$$E = 1 + K_q + K_r, \tag{9.57}$$

$$F = \frac{[G_{1-T}]}{[R_T]}\left(K_{i1}(1+XK_{x1}) + K_{j1}K_q(1+XK_{y1})\right), \tag{9.58}$$

$$G = \frac{[G_{2-T}]}{[R_T]}\left(K_{i2}(1+XK_{x2}) + K_{j2}K_r(1+XK_{y2})\right). \tag{9.59}$$

In these latter equations, $[G_{2-T}]$ denotes the total concentration of G_2. Using a similar strategy, the equation for the total stimulus of the G_2 pathway (T_2) can be derived:

$$T_2 = \frac{q\frac{[G_{2-T}]}{[R_T]}XK_{j2}K_rK_{y2}(1+DK_{b2})}{A+B+C} + \frac{(1-q)\frac{[G_{2-T}]}{[R_T]}XK_{j2}K_rK_{y2}}{E+F+G}. \tag{9.60}$$

References

1. Spengler, D., Waeber, C., Pantaloni, C., Holsboer, F., Bockaert, J., Seeburg, P. H. and Journot, L. Differential signal transduction by five splice variants of the PACAP receptor. *Nature* **365**, 170–175 (1993).

2. Griffin, M. T., Figueroa, K. W., Liller, S. and Ehlert, F. J. Estimation of Agonist Activity at G Protein-Coupled Receptors: Analysis of M2 Muscarinic Receptor Signaling through Gi/o,Gs, and G15. *J Pharmacol Exp Ther* **321**, 1193–1207 (2007).
3. Wettschureck, N. and Offermanns, S. Mammalian *G* proteins and their cell type specific functions. *Physiol Rev* **85**, 1159–1204 (2005).
4. Wilkie, T. M., Scherle, P. A., Strathmann, M. P., Slepak, V. Z. and Simon, M. I. Characterization of *G*-protein alpha subunits in the G_q class: expression in murine tissues and in stromal and hematopoietic cell lines. *Proc Natl Acad Sci USA* **88**, 10049–10053 (1991).
5. Davignon, I., Catalina, M. D., Smith, D., Montgomery, J., Swantek, J., Croy, J., Siegelman, M. and Wilkie, T. M. Normal hematopoiesis and inflammatory responses despite discrete signaling defects in Galpha15 knockout mice. *Mol Cell Biol* **20**, 797–804 (2000).
6. Offermanns, S. and Simon, M. I. $G\alpha$ 15 and $G\alpha$ 16 couple a wide variety of receptors to phospholipase C. *J Biol Chem* **270**, 15175–15180 (1995).
7. Keith, D. E., Anton, B., Murray, S. R., Zaki, P. A., Chu, P. C., Lissin, D. V., Monteillet-Agius, G., Stewart, P. L., Evans, C. J. and von Zastrow, M. mu-Opioid receptor internalization: opiate drugs have differential effects on a conserved endocytic mechanism in vitro and in the mammalian brain. *Mol Pharmacol* **53**, 377–384 (1998).
8. Selley, D. E., Liu, Q. and Childers, S. R. Signal transduction correlates of mu opioid agonist intrinsic efficacy: receptor-stimulated [35S]GTP gamma S binding in mMOR-CHO cells and rat thalamus. *J Pharmacol Exp Ther* **285**, 496–505 (1998).
9. Lee, K. O., Akil, H., Woods, J. H. and Traynor, J. R. Differential binding properties of oripavines at cloned mu-and delta-opioid receptors. *Eur J Pharmacol* **378**, 323–330 (1999).

10

Competitive Interactions Between Orthosteric Ligands in Functional Assays on *G* Protein-Coupled Receptors

So far we have considered how to analyze the responses that individual drugs elicit through *G* protein-coupled receptors. In this chapter, I describe how to analyze the response resulting from the interaction between two drugs that compete for the same binding site. This site is usually the locus where the natural neurotransmitter binds (orthosteric site), but the analysis is applicable to competitive interactions at any site on the receptor. When an allosteric agonist binds to a receptor and turns it on, for example, the inhibitory behavior of an allosteric antagonist is often indistinguishable from that of a conventional orthosteric antagonist inhibiting the effects of an endogenous neurotransmitter.

The proximate goals of a competitive inhibition experiment in the context of functional assays are usually two fold — to estimate the affinity constant of a competitive ligand for its binding site and to determine if the competing ligands bind to the same site. Because many antagonists have no effect by themselves in functional assays, competition with an agonist provides a means of determining if they interact with the receptor, and if so, how potently (first goal). This can be accomplished with a cellular assay that reports the activity of a specific receptor whose DNA has been expressed in the cell or with a well-characterized assay on an isolated tissue that is known to elicit a unique response when treated with a selective agonist. If the antagonist is highly potent and selective and its properties are well-known, it can be used as a diagnostic probe to determine if the actions of a less characterized agonist are mediated through the same receptor in a native tissue. The decision is based on whether the estimate of the affinity constant of the antagonist for antagonizing the response is equivalent to that expected for binding to its preferred receptor. Complications can arise if more than one receptor mediates the response to the agonist, and these are addressed in this chapter.

This brings attention to the second goal of determining where ligands interact on a receptor. The definition of competitive inhibition includes the requirement that the ligands bind to the same site. This could mean different things to different scientists, depending upon whether they are interested in knowing the amino acid residues on the receptor that interact with the ligand or whether they are a physician treating an opiate-overdose patient with an opiate antagonist. For our analysis, the definition is entirely functional. That is, if the binding of two ligands is mutually exclusive, then the two ligands bind competitively. Competition occurs because the three-dimensional space that each ligand occupies when it binds to the receptor by itself overlaps or intersects with that of the other ligand so that both ligands cannot bind at the same time. The explanation for this type of mechanism lends itself nicely to Venn diagrams as shown in Figure 10.1. Thus, the conclusions

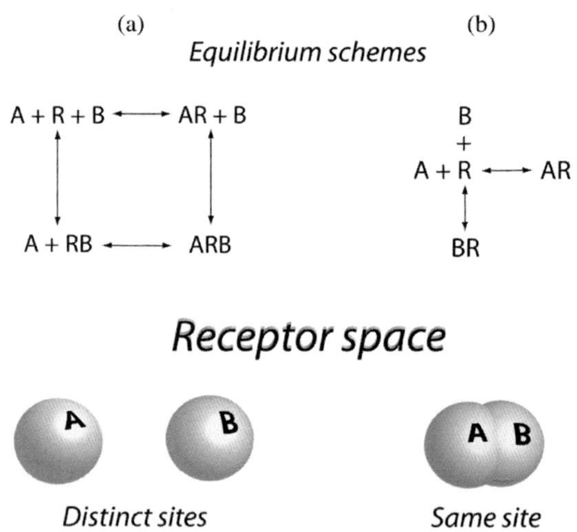

Figure 10.1: Definition of a competitive interaction between two ligands at a receptor. (a) The upper left side of the figure shows an equilibrium scheme for the binding of two ligands, A and B, to distinct sites on the same receptor. This mechanism allows both drugs to bind simultaneously to yield the ternary complex, ARB. The lower part of panel (a) shows the three-dimensional space occupied by each ligand when bound to the receptor. Because these two sets of receptor space are distinct, the ligands do not interact competitively. (b) The upper right side of the figure shows an equilibrium scheme for the binding of two ligands, A and B, to the same site on the receptor, which limits receptor binding to only one drug at a time. This competitive interaction implies that the space occupied by each ligand when bound to the receptor intersects with that of the other ligand.

drawn from competitive inhibition studies are relative. We cannot determine where a ligand binds specifically, but only whether it binds to the same site as another ligand. If the binding site of one ligand has been identified with certainty, such as in the determination of the crystal structure of the receptor bound with ligand, then we can assume that a competitor acts at the same site.

This functional definition might seem unambiguous, but it does not take much scrutiny to discover its limitations. As mentioned above, if the interactions of an endogenous ligand with amino acid residues in the binding pocket have been determined, then it is seems likely that a competitive inhibitor of comparable size may interact with at least some of these, but not necessarily all. An obvious example relates to many common antagonists of the amine class within the rhodopsin family of *G* protein-coupled receptors, like the muscarinic receptors. The prototypical antagonists, atropine, scopolamine and many others, have large aromatic and aliphatic rings on the acyl part of the molecule corresponding to additions onto the acetyl group of acetylcholine. These extra substituents must interact with residues that acetylcholine does not. If we extend the analogy to consider how a peptide toxin of considerable molecular weight occludes the orthosteric site of an endogenous ligand, then there may be little intersection of the two binding sites at all. The toxin could occlude the orthosteric site and prevent access to it without deriving any binding affinity from the residues that the endogenous ligand interacts with.

Some antagonists are bipartite and resemble two drug molecules connected by a carbon chain of sorts. If one end of the molecule interacts with the orthosteric site, the compound can behave as a competitive antagonist. This type of compound is termed bitopic because of its dual interaction with two loci on the receptor. If the compound derives considerable binding affinity by its interaction with a locus adjacent to the orthosteric site, then it could still remain bound to the receptor after the competing orthosteric ligand binds. But there is usually a range of concentrations over which a bitopic antagonist will compete with orthosteric ligands. This type of behavior does not fit cleanly into competitive or allosteric. It is a useful mode of drug action, nonetheless, because receptor subtypes often have highly conserved orthosteric-binding pockets with divergence in peripheral areas that could be exploited to develop subtype selective antagonists.

The differences between the binding characteristics of two competing ligands can also be subtle. Consider the agonist oxotremorine, which incorporates an ionizable pyrrolidine ring analogous to the trimethylammonium group of acetylcholine, and at the other end of the molecule, a pyrrolidone ring analogous to the carbonyl group of acetylcholine. Oxotremorine is highly potent and efficacious although not as efficacious as acetylcholine. The addition of a single methyl group at position 4 on the butyne chain converts the compound into a competitive antagonist. Oxotremorine and its methylated antagonist derivative (BOK1) are competitive inhibitors of each other because they occupy much of the same three-dimensional space when bound to the receptor, but BOK1 clearly interacts with the orthosteric site of the M_3 muscarinic receptor differently from that of oxotremorine.

Apart from mutually exclusive binding, the other component of the definition of a competitive interaction is reversibility. Competitive inhibitors interact reversibly with their target receptors. This seems like an unambiguous property, but reversibility is a relative concept and one that has created considerable confusion. A problem occurs because the standard analysis of competitive inhibition is based on the assumption that the competing ligands are in equilibrium with the receptor when the response is measured. This condition is difficult to achieve when a highly potent ligand with a slow dissociation rate from the receptor is involved because the time required for two competing ligands to reach equilibrium depends on the half-time for dissociation of the more slowly dissociating ligand (see Chapter 12).

Consider a pharmacologist measuring the effect of a highly potent antagonist on agonist-mediated stimulation of $[^{35}S]GTP\gamma S$ binding in a broken cell preparation. The assay can be done with a 30 min pre-incubation of agonist and antagonist with the cellular homogenate before $[^{35}S]GTP\gamma S$ is added for another 30 min incubation as described by Lazareno and Birdsall.[1] The amount of $[^{35}S]GTP\gamma S$-labeled $G\alpha$ subunits that accumulate during the subsequent 30 min is likely to do so under equilibrium conditions because of the lengthy pre-incubation with both competing ligands.

Now consider an electrophysiologist studying the effect of an antagonist on the whole-cell current response of a ligand-gated ion channel under voltage-clamp conditions. To insure that the antagonist reaches equilibrium, the preparation is usually incubated with the antagonist for a period of time (e.g., 10–30 min) before adding the agonist and measuring the current response over a brief time (10 s) to avoid desensitization. The measurements can be repeated with different agonist concentrations. The problem is that if it took 10–30 min for the antagonist to equilibrate with the receptor in the absence of agonist, it will take another 10–30 min for it to re-equilibrate after the agonist is added. It is unfeasible, if not impossible, to measure the response under equilibrium conditions because the current response is likely to desensitize within a few seconds. Under the condition just described, the so-called competitive antagonist is actually behaving like an irreversible antagonist because it binds nearly irreversibly over the brief time that the response is measured (i.e., 10 s).

Scientists often use the established terminology for specific mechanisms of enzyme inhibition as a convenience for describing the empirical effect of an inhibitor on the concentration-response curve of an agonist without concern for the mechanism. For example, many of us know that a competitive inhibitor causes a parallel shift to the right in the concentration-response curve of an agonist. Some scientists use the term competitive inhibition to describe the behavior of any inhibitor that does this. The problem is that any type of reversible inhibitor as well as irreversible inhibitor may cause a parallel shift to the right in the agonist concentration-response curve. The oxymoron, competitive-allosteric, has even been used to describe known allosteric inhibitors because of the tradition of using the phrase "competitive inhibition" as short hand for "something that shifts the curve to the right without reducing the E_{max}."

To make matters worse, enzymologists adopted the exasperating custom of defining specific inhibitory mechanisms in terms of what they are not. For example, if you were to explain the definition of competitive inhibition to a layperson and then ask them to define noncompetitive inhibition, they would probably give a very rational English definition of, any mechanism that is not competitive. But they would be wrong because the term noncompetitive refers to a specific type of inhibition; namely, an inhibitor that binds to the enzyme and inhibits its activity without affecting the binding of substrate or ligand. If the three primary colors were so named we might have red, nonred and blue. Once initiated, this tradition is contagious as in the case of the term, uncompetitive, which refers to an allosteric modulator that increases the affinity of the substrate for its site on the enzyme while reducing the activity of the enzyme. The analogous trichromatic color-naming scheme would now be red, nonred and un–red. Fortunately, we have yet to be burdened with incompetitive and nonincompetitive.

Throughout this book, the term competitive inhibition is used to designate a mechanism. That is, competitive inhibition is a consequence of the interaction of two drugs that bind reversibly in a mutually exclusive manner to the same site on the receptor. As a result, each ligand competitively inhibits the binding of the other. The nature of their effects on the concentration-response curve may or may not be prototypical, depending upon the conditions of the assay and the intrinsic efficacies of the competing ligands.

In this chapter, I describe the analysis of competitive interactions in functional assays involving various orthosteric ligands including agonists, partial agonists, competitive antagonists and inverse agonists. The discussion includes assays that exhibit constitutive activity as well as those that do not. Because many potent competitive antagonists dissociate very slowly from receptors, I also cover the analysis of irreversible inhibitors that bind to the same site on the receptor as the reversible agonist. We have already discussed part of this analysis in Chapters 7 and 8 in connection with the estimation of the observed affinity constants of agonists using the method of partial receptor inactivation. I also describe how to analyze the total inhibitory effect of a drug when only part of its inhibition is caused by competition with the agonist (resultant analysis). Finally, I conclude with a discussion of the analysis of the competitive inhibition of responses elicited by more than one type of receptor, including the condition where one or both of the interacting receptors behaves as a coincidence detector.

Competitive Inhibition of the Stimulus

The effect of a competitive antagonist on the response to an agonist is caused by the inhibition of the binding of the agonist to the receptor. This effect reduces agonist occupancy, which reduces the stimulus of the agonist-receptor complex (product of agonist occupancy and intrinsic efficacy), which in turn reduces the response. Thus, if we can predict how a competitive inhibitor reduces the binding of the agonist, we should be able to predict how it inhibits the response to an agonist.

We begin by defining the observed affinity constants of the agonist (K_{obs}) and competitive antagonist (K_I):

$$K_{obs} = \frac{[DR]}{[D]^*[R]}, \tag{10.1}$$

$$K_I = \frac{[IR]}{[I]^*[R]}. \tag{10.2}$$

Solving these equations for $[DR]$ and $[IR]$ yields:

$$[D]K_{obs}[R] = [DR], \tag{10.3}$$

$$[D]K_I[R] = [IR]. \tag{10.4}$$

At equilibrium, the fractional amount of agonist bound in the presence of the competitive antagonist is given by:

$$\frac{[DR]}{R_T} = \frac{[DR]}{[DR] + [IR] + [R]}. \tag{10.5}$$

Substituting Equations (10.3) and (10.4) for $[DR]$ and $[IR]$, respectively, followed by simplification yields:

$$[DR] = \frac{[D]R_T}{[D] + \frac{1}{K'_{\text{obs}}}}, \tag{10.6}$$

in which:

$$K'_{\text{obs}} = \frac{K_{\text{obs}}}{1 + [I]K_I}. \tag{10.7}$$

The occupancy functions of an agonist in the presence (Equation (10.6)) and absence (see Equation (7.1)) of a competitive inhibitor are of the same except that K_{obs} is replaced with the apparent observed affinity constant, K'_{obs}, when a competitive antagonist is present. Thus, the competitive inhibitor has the effect of causing the agonist to bind as if it has lower affinity. More precisely, the agonist binds to the receptor as though its affinity constant has been reduced by the factor $1/(1 + [I]K_I)$. This effect is often described as an antagonist-induced reduction in agonist affinity, but strictly speaking, the antagonist does not affect K_{obs}, only K'_{obs}.

The stimulus function (S) of an agonist in the presence of a competitive inhibitor is simply equivalent to the product of occupancy and the efficacy (ε) of the agonist:

$$S = \frac{[D]\varepsilon R_T}{[D] + \frac{1+IK_I}{K_{\text{obs}}}}. \tag{10.8}$$

This stimulus function is an accurate representation of the activation state of the receptor population in the presence of a competitive inhibitor. This function provides the basis for using reverse engineering (operational model) or a null method for estimating the affinity constant of a competitive antagonist in functional studies.

Estimation of antagonist affinity in functional assays lacking constitutive activity

The goal of this section is to develop methods for estimating the affinity constant of a competitive antagonist in a functional assay. The analysis focuses on the condition of no constitutive receptor activity and is based on the assumption that the agonist and competitive antagonist have reached equilibrium with the receptor when the response is being measured. The consequences of deviation from this condition are described in another section below.

Null method

The problem in analyzing agonist responses is that we are uncertain of the relationship between receptor occupancy and response (see Chapter 7). For example, we cannot assume that the response is proportional to $[DR]$ in Equation (10.6). But we can assume that equivalent responses in the absence and presence of a competitive inhibitor imply equal levels of receptor occupancy. Schild[2,3] determined how to solve this problem in the case of competitive antagonism, and I will describe his solution using a theoretical example. Recall from Chapter 7 that the response can be expressed as a function of the stimulus ($g(S)$). In the absence of a competitive inhibitor this function is:

$$\text{response} = g\left(\frac{[D]\varepsilon R_T}{[D] + \frac{1}{K_{\text{obs}}}}\right) \tag{10.9}$$

Imagine that a specific concentration of agonist ($[D]$) is added to the functional assay such that the response elicited is equivalent to 32% of $E_{\max}$. Suppose that the preparation is washed, and a competitive inhibitor is added at a specific concentration ($[I]$). After allowing the system to reach equilibrium, a higher concentration of agonist is added ($[D']$). Now the response in the presence of both agonist and antagonist can be described by (see Equation (10.8)):

$$\text{response} = g\left(\frac{[D']\varepsilon R_T}{[D'] + \frac{1+IK_I}{K_{\text{obs}}}}\right) \tag{10.10}$$

For the sake of argument, let us assume that this higher concentration of agonist ($[D']$) just so happened to elicit a response equivalent to that caused by the concentration ($[D]$) used in the absence of antagonist (32%). If so, we can write an equality in which the responses in the absence and presence of competitive inhibitor are the same:

$$g\left(\frac{[D]\varepsilon R_T}{[D] + \frac{1}{K_{\text{obs}}}}\right) = g\left(\frac{[D']\varepsilon R_T}{[D'] + \frac{1+IK_I}{K_{\text{obs}}}}\right) \tag{10.11}$$

Because the function describing the relationship between the stimulus and response ($g(S)$) occurs on both sides of the equation, it can be eliminated. Additional simplification yields:

$$\frac{[D]}{[D] + \frac{1}{K_{\text{obs}}}} = \frac{[D']}{[D'] + \frac{1+IK_I}{K_{\text{obs}}}}. \tag{10.12}$$

This equation describes the relationship between equiactive agonist concentrations in the absence ($[D]$) and presence ($[D']$) of a competitive antagonist. Further simplification yields:

$$K_I = \frac{\frac{[D']}{[D]} - 1}{[I]}. \tag{10.13}$$

The expression, $[D']/[D]$, represents the ratio of equiactive agonist concentrations and is sometimes referred to as the concentration ratio (CR). Thus, Equation (10.13) can be rewritten as:

$$K_I = \frac{CR - 1}{[I]}. \tag{10.14}$$

This equation shows that the ratio of equiactive agonist concentrations in the presence and absence of a competitive antagonist is a predictable function of the concentration of the antagonist and its observed affinity constant. This concept is more easily understood from Equation (10.14) when it is rearranged into the following form:

$$CR = [I]K_I + 1. \tag{10.15}$$

To apply the null method for estimation of K_I, it is unnecessary to use trial and error to identify equiactive concentrations as in the example described above. One can simply measure the concentration-response curve of the agonist in the absence and presence of a given concentration of the competitive antagonist. Each response on the curve measured in the presence of antagonist is elicited by the agonist at a concentration that is a constant factor (CR) greater than that required to elicit the same response in the absence of agonist. This means that when the concentration-response curves are plotted on a conventional log agonist-concentration scale, the curve in the presence of antagonist should be shifted to the right in a parallel fashion by log CR. All that is necessary is to subtract 1 from this shift and divide the result by the antagonist concentration to estimate the affinity constant of the antagonist in units of M^{-1}. More commonly, the molar concentration of antagonist is divided by $CR - 1$ to yield the dissociation constant of the antagonist in molar units, which is often designated as K_B.

An example of this calculation is shown in Figure 10.2. The shift is usually estimated as the ratio of the EC_{50} value of the agonist measured in the presence of antagonist (EC'_{50}) divided by that measured in its absence (EC_{50}). The calculation assumes that there is no change in E_{max}. If the competitive antagonist causes a

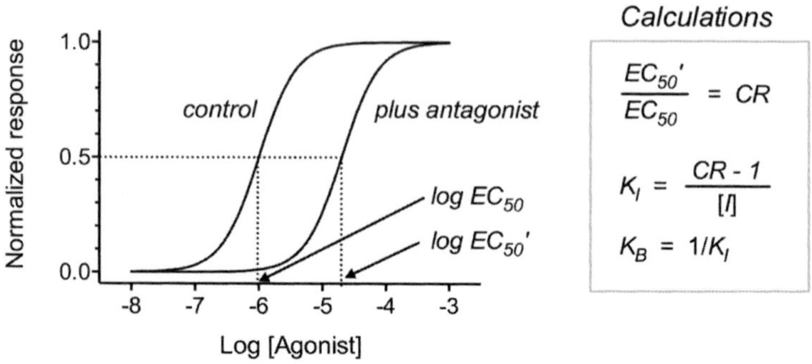

Figure 10.2: Estimation of the affinity constant of a competitive antagonist (K_I) in functional assays. The plot show the concentration-response curve of an agonist measured in the absence (control) and presence of a fixed concentration of antagonist ([I]). The box on the right, entitled "Calculations", shows the analytical relationship between the shift in the curve caused by the competitive antagonist and [I] and K_I.

change in the E_{max} because of its slow dissociation from the receptor upon agonist addition, then it is better to estimate K_I using a different approach as describe below.

Operational model

The reverse engineering strategy described in Chapter 7 is another way to get around the seemingly unknown relation between occupancy and response in functional assays. The advantage is that the operational model can be fitted directly to the measured responses to estimate the relevant pharmacological parameters by regression analysis rather than by making calculations on interpolated data. For the sake of completeness, I mention that a form of the operational model for competitive inhibition can be fitted to the data to estimate K_I. For this analysis, the K_{obs} value of the agonist is constrained to an arbitrarily low value (e.g., 0). Regression analysis provides an accurate estimate of K_I. If the agonist is a full agonist, then the estimates M_{sys}, m and the product, τK_{obs}, are accurate, but not those of τ or K_{obs} as explained in Chapter 7. If the agonist is a partial agonist, then there may be error in the estimates of M_{sys}, m and to a lesser extend, τK_{obs}, but not K_I. This method of analysis is described in the appendix in connection with the simulated data in Figure 10.7 below.

A simpler approach to estimate K_I is outlined in Figure 10.2. Nonlinear regression analysis is used to fit a logistic equation to the agonist concentration-response curves measured in the absence and presence of the antagonist. The estimate of E_{max} is shared between the two concentration-response curves, and K_I is estimated

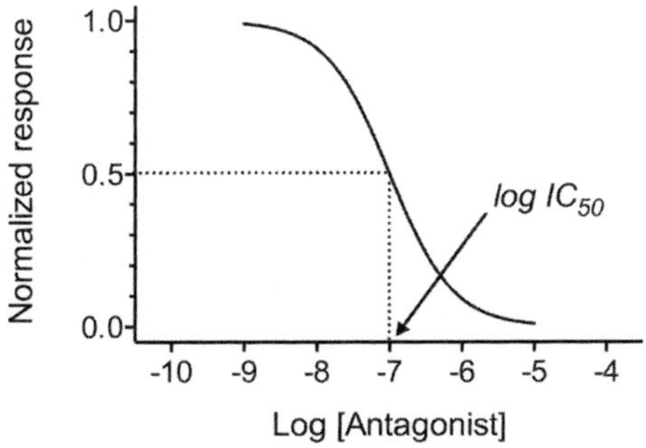

Figure 10.3: Effects of various concentrations of a competitive inhibitor on the response to an agonist at a fixed concentration. The IC_{50} value represents the concentration of antagonist that causes a 50% inhibition of the response elicited by the agonist.

from the EC_{50} values as described in the figure. A further development of this method is described below under the section "Testing the hypothesis of competitive inhibition" in connection with Equations (10.25) and (10.26).

There is another common assay format in which knowledge of the consequences of the operational model is useful for the estimation of the K_I of a competitive inhibitor. This format involves measuring the response to a single concentration of agonist in the absence and presence of increasing concentrations of competitive antagonist as illustrated in Figure 10.3. This format is commonly used in biochemical assays in which samples of a homogenous preparation of cells are incubated with the same concentration of agonist in the presence of different concentrations of competitive inhibitor. If the agonist and antagonist equilibrate rapidly with the receptor, the data can be collected as a cumulative concentration-inhibition curve for the antagonist in real time, but this approach is uncommon because of slow equilibration and possible desensitization of the response to the agonist. The K_I value of the antagonist can be estimated by algebraically by correcting its IC_{50} value, but how is this done?

The method for correction the IC_{50} value can be determined with the operational model for competitive inhibition. This model is derived by substituting the function for the stimulus of an agonist in the presence of a competitive inhibitor (Equation (10.8)) into the logistic transducer function of the operational model (Equation (7.5), Chapter 7). Simplification yields:

$$\text{response} = \frac{M_{\text{sys}}}{1 + \left(\frac{1+DK_{\text{obs}}+IK_I}{D\tau K_{\text{obs}}}\right)^m}. \tag{10.16}$$

This equation is used to derive an equality for estimating K_I from the IC_{50} value of the inhibitor. The approach described next is essentially equivalent to that described by Lazareno and Birdsall in an appendix to their manuscript on estimation of competitive antagonist affinity from functional inhibition curves.[4]

When the competitive inhibitor is present at its IC_{50} concentration (i.e., $I = IC_{50}$), the level of the response is equal to half that observed in the presence of the same concentration of agonist but without the antagonist. The corresponding equality for this concept is:

$$\frac{1}{2} \times \frac{M_{\text{sys}}}{1 + \left(\frac{1+DK_{\text{obs}}}{D\tau K_{\text{obs}}}\right)^m} = \frac{M_{\text{sys}}}{1 + \left(\frac{1+DK_{\text{obs}}+IC_{50}K_I}{D\tau K_{\text{obs}}}\right)^m}. \tag{10.17}$$

The left-hand side of this equation simply represents the standard form of the operational model described in Chapter 7 multiplied by one-half. Rearrangement of Equation (10.17) yields:

$$\frac{1}{K_I} = \frac{IC_{50}}{\left((D\tau K_{\text{obs}})^m + 2\,(1 + DK_{\text{obs}})^m\right)^{1/m} - 1 - DK_{\text{obs}}}. \tag{10.18}$$

Because this equation is complicated, I will consider the simple case when the transducer slope factor is equivalent to one (i.e., $m = 1$). This simplification of Equation (10.18) yields:

$$\frac{1}{K_I} = \frac{IC_{50}}{D\tau K_{\text{obs}} + DK_{\text{obs}} + 1}. \tag{10.19}$$

This equation can be rearranged into the form:

$$\frac{1}{K_I} = \frac{IC_{50}}{1 + \frac{D}{EC_{50}}}, \tag{10.20}$$

in which

$$EC_{50} = \frac{1}{K_{\text{obs}}\,(1 + \tau)}. \tag{10.21}$$

Equation (10.21) can also be derived from the operational model as described in Chapter 7 (see Equation (7.55) with the appropriate substitution for τ, Equation (7.11)). Equation (10.20) shows that the reciprocal of the observed affinity constant of the competitive inhibitor (observed dissociation constant, K_B) is equivalent to its IC_{50} value divided by one plus the concentration of agonist used in the experiment divided by its EC_{50} value. This equation is analogous to the Cheng and

Prusoff equation[5] for the correction of the IC_{50} value of a competitive inhibitor in competitive enzyme assays:

$$K_i = \frac{IC_{50}}{1 + \frac{[S]}{K_M}}. \qquad (10.22)$$

Note that the reciprocal of the affinity constant ($1/K_I$) in Equation (10.20) is equivalent to the dissociation constant of the competitive inhibitor (K_i) in Equation (10.22).

Why does Equation (10.20) work when the concentration of substrate ([S]) in the Cheng and Prusoff equation is divided by the dissociation constant of the substrate (K_M), whereas the analogous expression in Equation (10.20) has the concentration of agonist divided by its EC_{50} value? This question is particularly salient because the EC_{50} value can be much less than the reciprocal of K_{obs}, particularly when the agonist elicits a half-maximal response at a concentration 1/100 that of $1/K_{obs}$ (see Figure 7.13, Chapter 7).

The answer is related to the difference in receptor occupancy by the antagonist at its IC_{50} concentration for inhibiting an enzyme reaction and for antagonizing a functional response at a G protein-coupled receptor. When the EC_{50} value of an agonist is much less that the reciprocal of K_{obs}, then only a very small fraction of the receptor population need be occupied by the agonist to elicit a half-maximal response. If the agonist elicits a 50% response at 1% receptor occupancy, for example, then the antagonist would have to occupy about 99% of the receptors to reduce the occupancy of the agonist at its EC_{50} concentration to only 0.5%. In contrast, a competitive inhibitor in an enzyme assay only needs to occupy 50% of the substrate binding sites to reduce the velocity of the reaction to 50%, because the velocity measurement is proportional to the activity of the enzyme population. It is the greater occupancy requirement of the competitive antagonist in the functional assay that makes it necessary to overcorrect the IC_{50} value by an expression involving D/EC_{50} instead the analogous expression involving the dissociation constant of the agonist (reciprocal of K_{obs}). Remarkably, the correction factor is strikingly similar to a competitive inhibition equation even though EC_{50} is not equal to the dissociation constant of the agonist.

At this point, it is worth repeating that the validity of Equation (10.20) rests on the assumption that the agonist concentration-response curve exhibits a transducer slope factor of 1. Deviations from this condition can yield substantial error in the estimation of K_I as shown by Lazareno and Birdsall.[4] The error is most prominent at agonist concentrations equal to or less than EC_{50}.

Testing the hypothesis of competitive inhibition

An antagonist-induced, parallel, dextral shift in the agonist concentration-response curve is insufficient evidence for a competitive mechanism because many types of inhibitors have this effect. Negative allosteric modulators can behave this way, particularly when the signaling pathway is highly sensitive. In this situation, the response at E_{max} occurs at a low level of receptor occupancy so that even an irreversible orthosteric antagonist could cause a dextral shift in the agonist-concentration response curve without reducing E_{max} (see discussion associated with Figure 7.16, Chapter 7). Thus, to determine if the inhibitory mechanism is competitive, the behavior of the antagonist over a range of concentrations needs to be investigated.

Before proceeding, let me explain that the approach for testing whether a mechanism is competitive is not necessarily the best approach for measuring the affinity constant of the competitive antagonist in a functional assay. Drugs can interfere with the response at many points in the signaling pathway. A competitive inhibitor with moderate selectivity for a given receptor may, at high concentrations, interfere with the functional response by interacting with other proteins in the signaling pathway. Thus, if an estimate of K_I is the goal, it is best to test the inhibitor at a concentration that induces about a 10-fold shift in the agonist concentration-response curve. This will provide an accurate shift in the concentration-response curve for the estimation of K_I using the approach illustrated in Figure 10.2.

The most unequivocal method for testing for competition is to use a reliable competitive binding assay as described in Chapter 12. If the ligand causes a complete displacement of the specific binding of the radioligand over a range of concentrations, then it is likely to be a competitive inhibitor. The strength of the conclusion rests on the breadth of the concentration range, and in some instances, this range may be small because of a practical limit in how high the concentration of radioligand can be increased relative to its dissociation constant, or else there may be no suitable radioligand available. Thus, testing for competition in functional assays may be essential for a variety of reasons.

It is best to characterize the receptor in a functional assay with a defined recombinant receptor with as few signaling steps as possible (e.g., second messenger assay) to minimize the probability that the test inhibitor will interfere with a signaling protein mediating the response. It is possible that when the receptor is expressed in its native context, it may associate with a binding partner (chaperone or other receptor monomer) that alters its observed affinity for the antagonist, although this affinity change is unlikely for many G protein-coupled receptors. It may be that the intent of the experiment is to examine a specific downstream function of the receptor in

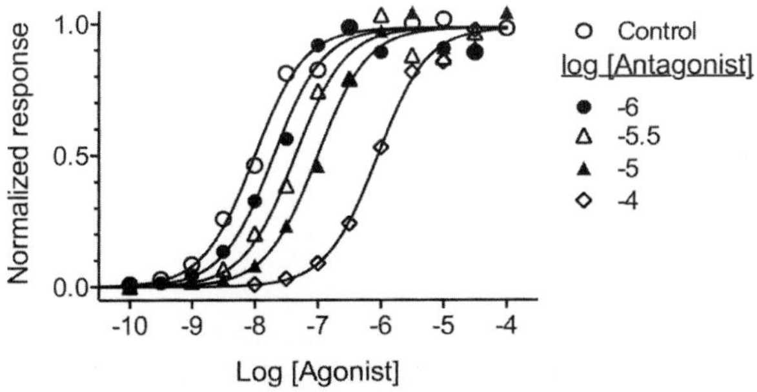

Figure 10.4: Simulation of the effects of various concentrations of a competitive antagonist on the concentration-response curve of an agonist. The curves were simulated using Equation (10.16) for the condition of no antagonist (Control) and increasing concentrations of a competitive antagonist as indicated in the legend. The parameters of the operational model used in the simulation were, M_{sys}, 1; m, 1; log K_{obs}, 6; log τ, 2 and log K_I, 6.

its native context, in which case the abbreviated functional assay is of no use except, perhaps, for comparison purposes.

Simply stated, the method for testing whether an inhibitor is competitive involves estimating the K_I value at various concentrations of the antagonist using the method described above in Figure 10.2. If the estimate of K_I shows insignificant variation, then there is no reason to reject the competitive hypothesis. Also, the K_I estimate of the antagonist should be the same regardless of the agonist that is used to elicit the response provided that the agonists all elicit the response through the same receptor.

A simulation of the type of experiment that is commonly done to test for competitive inhibition is shown in Figure 10.4. The experiment involves measuring a control concentration-response curve for the agonist in the absence and presence of various concentrations of the antagonist. Each curve can be fitted with a logistic equation to estimate the EC_{50} value. The curves can be fitted simultaneously sharing the estimate of E_{max} and the Hill slope as described below. From the shift in the concentration-response curve caused by each concentration of antagonist, it is possible to calculate K_I using Equation (10.14). The experiment can be repeated, and the log means of several estimates can be made. For the data in Figure 10.4, the estimate of log K_I for the log antagonist concentrations of -6, -5.5, -5 and -4 were 6.05, 5.95, 5.97 and 6.04, respectively. In this case there is little difference among the estimates so there is no reason to reject the hypothesis of competitive inhibition. The parameters for this simulation are given in the legend to Figure 10.4.

In some cases, usually involving intact tissues, it may be possible to make replicate measurements of the concentration-response curve on the same tissue. One potential problem with this assay format is that any error in the estimation of the control EC_{50} value is translated into an error in the estimate of all of the antagonist-induced shifts in the concentration-response curves. Another type of problem is that the sensitivity and the maximum response of the preparation can change with time as additional concentration-response curves are measured. This can lead to a time-dependent error in the estimation of EC_{50}. A simple solution is to run additional control tissues throughout the duration of the experiment so that it is possible to estimate the antagonist-induced shift in the concentration-response curve relative to a control EC_{50} value measured at the same time. Many of these problems can be eliminated in cellular assays in which a second messenger, Ca^{2+}, or $[^{35}S]GTP\gamma S$ binding is measured noncumulatively. Essentially, each well in a cell culture plate or assay tube represents a different, noncumulative measurement of the response. Assuming that the preparation is homogeneous, there should be little problem associated with time-dependent changes in sensitivity. The control responses can be measured in sufficient replication to minimize the translation of error into all of the measured antagonist-induced shifts in the concentration-response curve.

Figure 10.5 illustrates a simple way to analyze the results of the competitive inhibition experiment shown in Figure 10.4 provided that there is no evidence of changes in E_{max}. The EC_{50} value is simply plotted against the concentration of antagonist. The line through the mean log EC_{50} values represents the least-squares

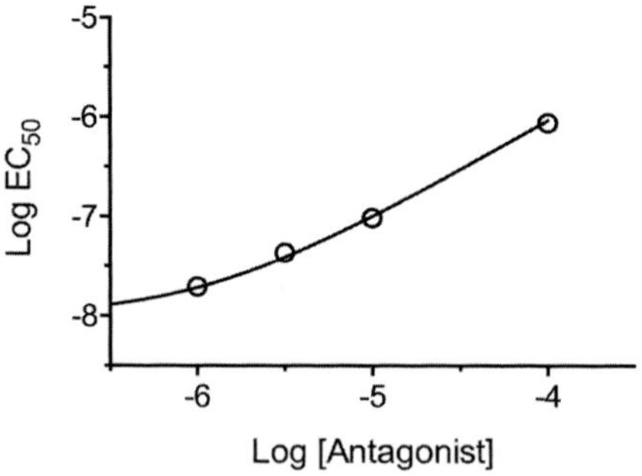

Figure 10.5: Simple method of analysis of competitive inhibition in functional assays. The plot shows an analysis of the data illustrated in Figure 10.4. The EC_{50} value of the agonist is plotted against the log concentration of antagonist. The theoretical curve represents the least-squares fit of Equation (10.23) to the data, including the log of the control EC_{50} value of the agonist, which is not shown on the plot. The best fitting parameters are, log $EC_{50\text{-con}}$, -8.00 ± 0.03 and log K_I, 5.95 ± 0.04.

fit of the following equation to the data:

$$\log EC_{50} = \log \left(10^{EC50\text{-con}} \left(10^{\log I + \log KI} + 1\right)\right). \qquad (10.23)$$

In this equation, $EC_{50\text{-con}}$ denotes the control EC_{50} value, and log EC_{50} denotes that measured in the presence of different concentrations of the antagonist. This equation was derived by rearranging Equation (10.13). Regression analysis yields an estimate of 5.95 ± 0.04 for log K_I and a control log EC_{50} value of -8.00 ± 0.03. Although not shown on the plot, the log control EC_{50} value was included in the regression and was plotted against at an arbitrarily large negative abscissa value (−20), equivalent to a log antagonist concentration of essentially zero (i.e., $-\infty$). An advantage of this analysis is that all of the log EC_{50} estimates are given equivalent weight. This reduces error in the estimate of the control EC_{50} and the probability that error in its estimation will translate into errors in the estimates of each of the antagonist-induced shifts in the concentration-response curves. The plot is consistent with competitive inhibition because the EC_{50} value of the agonist increases proportionately with the antagonist concentration when the latter is approximately 10-fold greater than the reciprocal of K_I. As described in Chapter 11, it is possible to discriminate between competitive and allosteric mechanisms with this analysis.

A more traditional method of analyzing the effect of an antagonist involves constructing a Schild plot, which is based on a linear form of Equation (10.15):

$$\log (CR - 1) = \log I + \log K_I. \qquad (10.24)$$

This approach was a sensible way to analyze data at a time when computing power for nonlinear regression analysis was not readily available because the estimate of log K_I can be obtained by linear regression. The usual convention is to use the negative log dissociation constant of the antagonist (−log K_B) rather the log of the corresponding affinity constant (log K_I). Regardless, if log $(CR - 1)$ is plotted against log I, the resulting graph is known as a Schild plot. Figure 10.6 illustrates a Schild plot of the data in Figure 10.4. If the mechanism is competitive, the points should define a straight line having a slope of one that intersects the X-axis at log K_B or negative log K_I. In the original plot described by Arunlakshana and Schild[6], log $CR - 1$ was plotted against the −log of the antagonist concentration. The X-intercept was designated as the pA_2 value because it is equivalent to the negative log of the antagonist concentration that causes a two-fold shift in the agonist concentration-response curve. Note that when CR is equivalent to 2 in Equation (9.24), then $\log I = -\log K_I$. Thus, pA_2 is equivalent to log K_I provided that the mechanism of antagonism is competitive. The linear behavior of the Schild plot illustrates a proportionality between $CR-1$ and the concentration of antagonist.

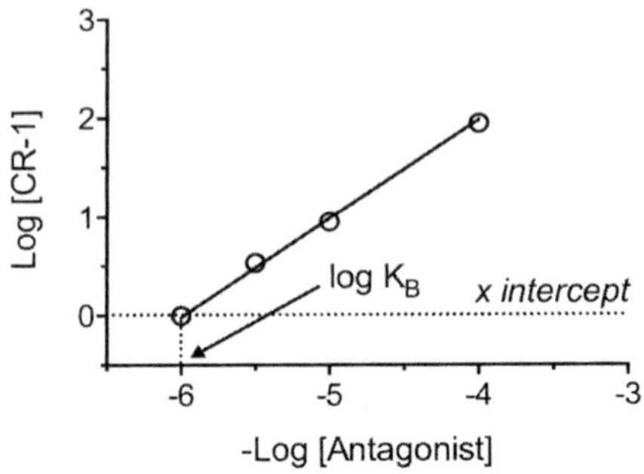

Figure 10.6: Schild plot for the analysis of competitive inhibition: The log of $CR - 1$ is plotted against the log concentration of antagonist used in the assay. The theoretical curve represents the least squares fit of the data to the Equation (10.24). The estimate of log K_I is 5.81 ± 0.02.

At high concentrations of antagonist, CR–1 is approximately equal to CR, and hence, the shift is proportional to the antagonist concentration.

While the Schild plot is commonly used to illustrate and analyze the effect of an inhibitor on the agonist concentration-response curve, it is not the best way to analyze data because there can be substantial error in the estimation of $CR - 1$ at low antagonist concentrations. One common error is to assume that the pA_2 value is an appropriate estimate of log K_I when the slope of the plot is substantially less than one. Because error in the estimation of CR is amplified in the transformation, log $CR - 1$, substantial error in the estimate of log K_I can occur with Schild analysis particularly with a plot having low CR estimates. It is better to estimate the log K_I using nonlinear regression analysis as described in Figure 10.5, provided that there is no evidence of changes in E_{max}.

Perhaps the most thorough method to analyze the inhibition data shown in Figure 10.4 is to fit an appropriate equation to the concentration-response curves directly and not simply to the EC_{50} values estimated from them. This can be done by global nonlinear regression analysis using the following logistic equation:

$$\text{response} = \frac{D^n E_{max}}{D^n + (EC_{50\text{-con}}\,(1 + IK_I))^n}. \tag{10.25}$$

Regression analysis is done by sharing the estimate of the maximal response (E_{max}), $EC_{50\text{-con}}$ value, Hill slope (n) and affinity constant of the competitive antagonist (K_I) among the curves obtained at different concentrations of the competitive inhibitor (I). Analysis of the data in Figure 10.4 using the log form of Equation (9.25) (see Equation (9.26) below) yielded a log K_I value of 5.95 ± 0.08. The other parameters were: E_{max}, 0.99 ± 0.01; log $EC_{50\text{-con}}$, -8.00 ± 0.06; and n, 1.08 ± 0.08. The appendix to the next chapter (Chapter 11) illustrates how to do this regression analysis in connection with discriminating between competitive and allosteric models. As mentioned, Equation (10.25) can be rearranged into a form that expresses the independent variables D and I and the parameters EC_{50} and K_I as logarithms (log$D$, log$I$, log$EC_{50\text{-con}}$ and logK_I):

$$\text{response} = \frac{E_{\max}}{1 + \left(\left(10^{\log EC_{50\text{-con}} - \log D}\right) \times \left(1 + 10^{\log I + \log K_I}\right)\right)^n}. \tag{10.26}$$

While this method of analysis is the more rigorous, it is also more sensitive to nonequilibrium effects, changes in the sensitivity of the preparation, nonreceptor-mediated inhibitory effects of the antagonist, and procedural errors in the execution of the experiment. Provided that the potential problem of nonequilibrium can be addressed, a good assay format for this method of analysis is an experiment involving the noncumulative measurements of a signaling response (second messenger) in a homogenous cellular preparation.

It is also possible to test the mechanism of competitive inhibition in the assay format described in Figure 10.3. In this experiment, the response is measured in the presence of a fixed concentration of agonist and in the absence and presence of increasing concentrations of the antagonist. The K_I value of the antagonist is estimated from the IC_{50} value of the inhibition curve using Equation (10.22), which is based on a transducer slope factor of one ($m = 1$) and knowledge of the EC_{50} value of the agonist. This experiment can be repeated using higher concentrations of the agonist. If the K_I value estimated from each experiment is not significantly different, then there is no reason to reject the hypothesis of competitive inhibition.

Thus, the process of testing competitive inhibition involves measuring the response to various concentrations of agonist in the presence of various concentrations of an antagonist. We can define the requisite data for the test as a two-dimensional matrix consisting of the response values measured in the presence of difference combinations of agonist and antagonist concentrations as shown in Table 10.1. Each row in the matrix corresponds to a different agonist concentration, whereas each column corresponds to a different antagonist concentration. If the responses in a column are plotted against the log agonist concentration, the result is a standard agonist concentration-response curve. If several columns are plotted, the result is an illustration of the effect of different concentrations of antagonist on the agonist concentration-response curve as shown in Figure 10.7a for the data in Table 10.1. In contrast, if a single row were plotted, the result would be an antagonist-inhibition curve like that shown in Figure 10.3. If all of the rows are plotted, the result is a series of antagonist inhibition curves, each measured

Table 10.1: Simulated responses to an agonist at various concentrations (rows) and in the presence of various concentrations of antagonist (columns). The data were simulated using Equation (10.16). The parameters of the model were M_{sys}, 1; m, 1; log τ, 2; log K_{obs}, 6 and log K_I, 6. A random error of approximately $\pm$ 7.5% of the simulated value plus 0.2 was added to the data.

	log [Antagonist]						
log [Agonist]	$-\infty$[a]	−6	−5.5	−5	−4	−3	−2
−9	0.099	0.045	0.025	0.0092	0.0010	0.00010	0.000010
−8	0.47	0.31	0.21	0.088	0.0091	0.0010	0.000093
−7	0.94	0.81	0.72	0.51	0.087	0.011	0.0010
−6	1.01	0.96	0.91	0.82	0.46	0.90	0.0091
−5	1.03	1.05	0.91	0.93	0.88	0.48	0.090
−4	0.92	1.01	0.99	1.06	1.05	0.95	0.50
−3	0.97	1.01	1.07	0.91	1.02	1.05	0.98
−2	1.00	1.01	1.03	0.96	0.96	0.91	1.06

[a] Represents the log concentration of 0.

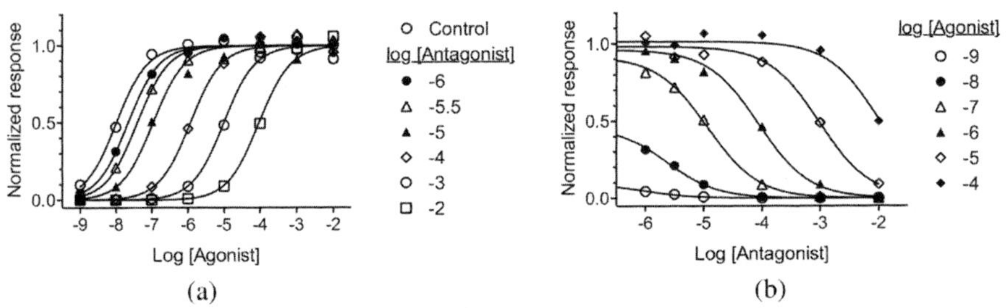

Figure 10.7: Simulation of the effects of various concentrations of a competitive antagonist on the responses elicited by an agonist at various concentrations. (a) The simulated data are plotted as a series of agonist concentration-response curves for the condition of no antagonist (control) and various concentrations of antagonist as indicated in the legend. (b) Here, the same data are plotted as a series of antagonist inhibition curves, each curve representing a simulation with a different concentration of agonist. The numerical data for these plots are given in Table 10.1. The theoretical curves represent the least-squares fit of Equation (10.16) to the data. In both panels (a) and (b), the parameter estimates were, M_{sys}, 1.00 $\pm$ 0.01; m, 1.01 $\pm$ 0.05; log τK_{obs}, 7.88 $\pm$ 0.04 and log K_I, 5.98 $\pm$ 0.06. The theoretical parameters used to simulate the data were: M_{sys}, 1; m, 1, log K_{obs}, 6; log τ, 2 and log K_I, 6.

at a different concentration of agonist, as shown in Figure 10.7b for the data in Table 10.1.

By now it should be apparent that there is no advantage to collecting data in either format — a series of agonist concentration-response curves or a series of antagonist-inhibition curves. Both represent different ways of looking at the same data. The choice of format is based on how the data are collected (cumulative or noncumulative) and issues related to desensitization and equilibrium. To analyze a series of inhibition curves, global nonlinear regression analysis with the operational model for competitive inhibition (Equation (10.16)) can be employed. As mentioned above in the section entitled "Operational model", this analysis requires that K_{obs} be constrained to an arbitrarily low value (e.g., 0). This method also applies to a series of agonist concentration-response curves. Example 1 in the appendix describes how to do this analysis for the data in Table 10.1 (i.e., data illustrated in Figure 10.7).

Before leaving this section I would like to return to the issue raised above regarding the estimation of K_I from a series of inhibition curves when the transducer slope factor differs from one. In this case, there is significant error in the estimation of K_I if the Cheng and Prusoff equation (Equation (10.20)) is used. To illustrate this point, we can analyze the simulated data in Figure 10.8, which lacks random error and was generated for the case of a transducer slope factor of two ($m = 2$). The data in panel a show a series of agonist concentration-response curves simulated for the condition of no antagonist (control) and increasing concentrations of antagonist

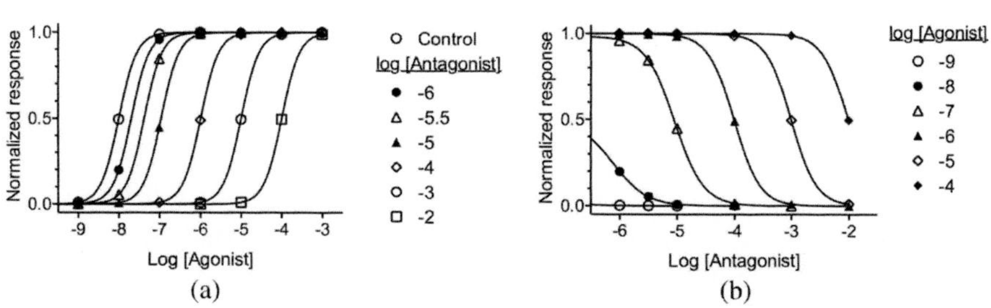

Figure 10.8: Simulation of the effects of various concentrations of a competitive antagonist on the responses elicited by an agonist at various concentrations. (a) The simulated data without error are plotted as a series of agonist concentration-response curves for the condition of no antagonist (control) and various concentrations of antagonist as indicated in the legend. (b) Here, the same data are plotted as a series of antagonist inhibition curves, each curve representing a simulation with a different concentration of agonist. The theoretical curves represent the least-squares fit of Equation (10.16) to the data. In both panels (a) and (b), the parameter estimates were, M_{sys}, 1; m, 2; log τK_{obs}, 8 and log K_I, 6. The theoretical parameters used to simulate the data were: M_{sys}, 1; m, 2, log K_{obs}, 6; log τ, 2 and log K_I, 6.

having a log K_I value of 6. For each concentration of antagonist, the same value of log K_I (6) can be calculated from the estimate of CR using Equation (10.14). Panel b shows the same data, but plotted as series of antagonist inhibition curves, each measured at a different concentration of agonist. When Equation (10.20) is used to estimate the log K_I values from the corresponding IC_{50} values, the following estimates are obtained at log agonist concentrations of −9, −8, −7, −6 and −4: 6.35, 6.42, 6.08, 6 and 6, respectively. Thus, there is moderate deviation (about 0.4 log units) from the theoretical value of log K_I (6) when Equation (10.20) is used for the low agonist concentrations (log concentrations of −9 and −8). As mentioned above, however, if global nonlinear regression analysis is used with Equation (10.26), there is no error in the estimation of K_I from the data in Figure 10.8b.

Although the question of the validity of Equation (10.20) (Cheng and Prusoff calculation or Equation (10.22)) is important, the issue can always be circumvented as Lazareno and Birdsall[4] have reminded us with the following reasoning: To apply the Cheng and Prusoff equation, an estimate of the EC_{50} value is needed. To estimate an EC_{50}, one needs the agonist concentration-response curve, and if one has such a curve, then there is sufficient information to estimate K_I using the standard concentration-ratio method (i.e., Equation (10.13) or (10.14)).

This logic is illustrated in Figure 10.9. The upper part of the figure shows a simulated agonist concentration-response curve (left) and a simulated antagonist inhibition curve at a fixed concentration of agonist (10^{-8} M) (right). The standard competitive inhibition equation for estimating K_I from the antagonist-induced shift in the agonist concentration-response curve can be used with these data (Equation (10.13)). This equation contains the variables D, D', I and K_I. As described earlier, the variables D and D' represent equiactive concentrations of agonist in the absence and presence of the antagonist, respectively. The concentration and affinity constant of the antagonist are denoted by I and K_I.

These variables can also be defined in terms of the experimental format shown in the figure without invalidating Equation (10.13). That is, when the antagonist is present at a concentration equivalent to its IC_{50} ([I], 1.28 μM) the fractional response elicited by the agonist (0.23) at a concentration of 10 nM (D') is equivalent to that elicited by the agonist at a lower concentration of 4.8 nM (D) in the absence of antagonist (see interpolated agonist concentration for response = 0.23 on the upper left side of figure). If these values are substituted for the appropriate variables in Equation (10.13), the value of K_I can be calculated as shown in the lower box in the figure. The estimated value of log K_I (6.07) is essentially the same as that used in the simulation (6).

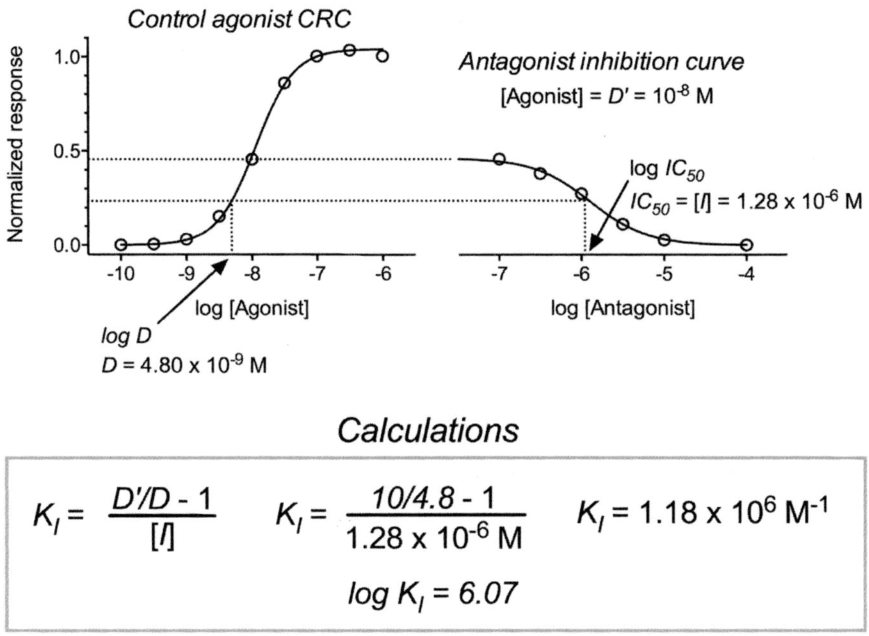

Figure 10.9: Use of the standard competitive inhibition Equation (10.14) to estimate K_I from data often analyzed with the Cheng and Prusoff equation. The plots on the upper left and right show a simulated agonist concentration-response curve and an antagonist inhibition curve for the condition of a fixed concentration of agonist (10^{-8} M), respectively. The upper horizontal dotted line indicates the level of response elicited by the agonist at a concentration of 10^{-8} M, and the lower horizontal dotted line indicates the level of response equivalent to one-half this level. The vertical dotted line on the right intersects the abscissa at the IC_{50} concentration of antagonist. The vertical dotted line on the left intersects the abscissa at an interpolated log agonist concentration (D) that elicits a response equivalent to that elicited by the agonist (10^{-8} M, D') in the presence of antagonist at its IC_{50} concentration. The data were simulated with Equation (10.16) using the following parameter values: M_{sys}, 1; m, 1.5; log τ, 2; log K_{obs}, 6 and log K_I, 6. A random error equivalent to $\pm$ 7.5% of the simulated response value plus 0.2 was added to the data. The lower portion of the figure illustrates how to apply Equation (10.14) to estimate the K_I value of the antagonist from the data.

To sum up, Equation (10.13) can always be used to estimate the K_I of an antagonist from its IC_{50} value for inhibiting the response to a fixed concentration of agonist. This option avoids potential error in the use of the Cheng and Prusoff equation (Equation (10.20)) when the concentration-response curve of the agonist exhibits a transducer slope factor different from 1.[4]

Interpretation of Deviations from Competitive Inhibition

If data deviate significantly from the competitive inhibition equations described in the prior section, then the mechanism of inhibition could be different from competitive or else there could be some other factor influencing the data. One common explanation is that the assay has not reached equilibrium, and a discussion of this condition begins in the next paragraph. With regard to mechanisms that are not competitive, the analysis of irreversible inhibitors is described in the next section, and various forms of allosterism are described in Chapter 11. The inhibitor could have a competitive effect on the receptor as well as another inhibitory or stimulatory effect on the signaling pathway, and this situation is described below under, "Resultant analysis."

Depending on the response being measured and the potency of the antagonist, it may be impossible to measure competitive inhibition under equilibrium conditions. Consider a response that desensitizes quickly, like the Ca^{2+} response of many G_q-linked receptors expressed in HEK 293 cells loaded with the fluorescent Ca^{2+} dye, fura-2. This response is often measured in drug-screening efforts because the increase in fluorescence can be readily measured in a high throughput format. The activity of compounds at non-G_q-linked, G protein-coupled receptors can be studied by expressing a chimeric $G_{\alpha q}$ protein that has the last few residues in its C-terminus replaced with those from $G_{\alpha s}$ or $G_{\alpha i}$ (see Chapter 5). This Ca^{2+} response is readily measured, but it is not highly potent because many efficacious agonists have to occupy a substantial fraction of the receptor population to elicit a maximal Ca^{2+} response. Depending upon the agonist, receptor and cell type, the increase in fluorescence rises to a peak within seconds of agonist addition and gradually decays over the course of a minute or two.

Suppose that the antagonist being investigated is highly potent and has a log K_I of about 9.0 and a half-time for dissociation from the receptor of about 10 min. When the agonist is added to HEK 293 cells that have equilibrated with the antagonist for 40 min, the agonist may not produce a Ca^{2+} response because it may take several minutes for it to displace enough antagonist from the receptor so that it can elicit a response. This is not a situation that can be solved by recording the response at a later time when equilibrium is achieved, however, because the response in the absence of antagonist desensitizes within a minute or two. As time passes the agonist will slowly displace the antagonist and slowly desensitize the response so that little or no response will ever be recorded under this condition. High concentrations of agonist may produce a response depending upon how much antagonist has equilibrated with the receptor. Nonetheless, the inhibition caused by the antagonist could greatly exceed that expected at equilibrium, and the suppression in the agonist response will probably include a decrease in E_{max}, which clearly deviates from a competitive mechanism at equilibrium. With this assay format, it is simply impossible to measure the response of the agonist in the presence of some antagonists at equilibrium.

Clues to the mechanism can be gleaned by determining if the inhibitory effect of the antagonist persists after it has been washed from the cells. If subsequent agonist responses are still inhibited after washout, the results may indicate that the antagonist dissociates slowly from the receptor. Another useful experiment would be to add the agonist and antagonist simultaneously. If high concentrations of both are added, they should reach equilibrium with the receptor population rapidly, and previously effective concentrations of antagonist may be unable to suppress the response to high concentrations of agonist. Again, this would suggest that the deviation from competitive inhibition in the prior assay format may be caused by the slow dissociation of the antagonist from the receptor.

This quasi-irreversible effect of a slowly dissociating competitive antagonist might not be noticed if the response is highly sensitive and can be elicited by a highly efficacious agonist at a very low level of receptor occupancy. Consider the muscarinic contractile response of the guinea pig ileum or the cAMP response of many G_s or G_i-linked G protein-coupled receptors expressed in HEK 293 cells. Often only a fraction of one percent of the receptor population needs to be occupied by a highly efficacious agonist for a half-maximal response to occur. Normally, these responses are recorded within tens of seconds to a few minutes after agonist addition. If the guinea pig ileum is pre-incubated with atropine (10 nM) for 20 min, it will cause a subsequent 10-fold shift in the concentration-response curve of a highly efficacious agonist without causing a decrease in the E_{max} value. Based on this effect, a log molar K_I value of about 9 can be calculated for scopolamine assuming competitive inhibition. This value agrees with its binding affinity at the M_3 receptor, which is the subtype of the muscarinic receptor responsible for intitiating direct contraction of the ileum. In contrast, the same concentration of atropine almost completely eliminates the contractile response to the partial agonist pilocarpine.

There is no need to postulate a deviation from competitive inhibition even though the E_{max} of pilocarpine is greatly suppressed and the effect of atropine differs depending upon the agonist that is used to elicit the response. Both sets of data can be explained by the hypothesis that atropine is a competitive inhibitor of the M_3 receptor with slow dissociation kinetics. Over the time period that the concentration-response curve is measured, atropine behaves as an irreversible inhibitor because of its slow dissociation from the M_3 receptor. Irreversible inhibition of 90% of the receptor population will greatly suppress the maximal response to a partial agonist, like pilocarpine. In contrast, the same effect will shift the concentration-response curve of a highly efficacious agonist to the right, 10-fold without decreasing the E_{max} as predicted by the consequences of Equation (10.28) below. These ideas were first developed by Paton.[7] In the next section, the methods for analyzing this type of inhibition are described.

Analysis of Irreversible and Slowly Dissociating Orthosteric Ligands

It should now be apparent that two mechanisms can explain why an orthosteric antagonist might behave like an irreversible inhibitor. It may indeed bind irreversibly, or it could bind reversibly but not dissociate from the receptor to a significant extent during the brief time that the response is measured.

If the antagonist binds irreversibly and reduces the E_{max} of the agonist, the relative amount of receptors inactivated can be estimated as described in Chapter 7 in connection with Figure 7.4a. That is, Equation (7.10) is fitted to the agonist concentration-response curves measured before and after treatment with the irreversible antagonist. Regression analysis provides estimates of the affinity constant of the agonist and the ratio of τ values, which yields and estimate of the fraction residual free receptors (q). The relative amount of inactivated receptors is equal to $1 - q$.

A different method is used to calculate receptor inactivation when the irreversible inhibitor shifts the agonist concentration-response curve to the right without affecting E_{max}. Recall from Chapter 7 that if two agonists, A and B, are both full agonists and have the same affinity, then the efficacy of B expressed relative to that of A is:

$$\frac{\varepsilon_B}{\varepsilon_A} = \frac{EC_{50\text{-}A}}{EC_{50\text{-}B}}. \tag{10.27}$$

Receptor inactivation has an effect equivalent to reducing the efficacy of an agonist without affecting its affinity. The apparent reduction in efficacy is proportional to receptor inactivation ($1 - q$). Thus, it follows that the fractional amount of free receptors after treatment with an irreversible antagonist is equivalent to:

$$q = \frac{EC_{50}}{EC'_{50}}. \tag{10.28}$$

In this equation, EC'_{50} denotes the EC_{50} value measured after treatment with the irreversible antagonist. Paton[7] developed an identical equation using a different rationale.

It is also possible to estimate the affinity constant of the irreversible antagonist, although the method is tedious and seldom done. The analysis is based on the assumption that the antagonist first forms a reversible complex with the receptor (IR) that subsequently converts to a covalent complex (I-R):

$$I + R \overset{K_I}{\longleftrightarrow} IR \overset{k}{\longrightarrow} I - R. \tag{10.29}$$

Essentially, the amount free receptors at various times after incubation with various concentrations of the irreversible antagonist is determined. The estimates of q at each concentration of antagonist are fitted to an exponential equation (i.e., $q = e^{-kt}$) to estimate the rate constant for inactivation (k). The concentration of antagonist that yields a half-maximal increase in the rate constant is equivalent to the reciprocal of K_I, provided that the rate constant for alkylation is slower than the rate of dissociation of the reversible antagonist-receptor complex.[8] If not, then the analysis described by Gill and Rang.[9] can be used to estimate K_I. A problem with this analysis is that it is never possible to inactivate the receptor population completely because of recycling of new receptors from an intracellular pool. Functional responses on cellular homogenates lack this problem.

The affinity constant of a competitive antagonist that behaves like an irreversible antagonist during the short time that the response is measured can also be estimated. If the antagonist is incubated with the receptor preparation for a sufficient time before the response is measured (perhaps 20 min — 1 h), it can often be assumed that it has reached equilibrium and that its level of receptor occupancy should be maintained when the agonist is added provided that the response is measured over a short time interval. This assumption is used to estimate the K_I value the antagonist.

Let us return to the example described at the end of the prior section. Recall that incubation of ileal smooth muscle with atropine at a concentration of 10 nM caused a subsequent 10-fold shift in the concentration-response curve of a highly efficacious agonist. Assuming that atropine acts irreversibly during the time that the response is measured, its fractional occupancy $(1 - q)$ can be estimated to be 0.9 using Equation (10.28). If the standard occupancy equation for a competitive antagonist at equilibrium,

$$1 - q = \frac{IR}{R_T} = \frac{[I]}{[I] + \frac{1}{K_I}}, \tag{10.30}$$

is solved for K_I, the result is:

$$\frac{1 - q}{[I]q} = K_I. \tag{10.31}$$

This equation can be used to estimate the K_I of atropine assuming that occupancy when the response is measured is equivalent to that at equilibrium immediately before the response was measured. This analysis yields a K_I value of 9×10^8 ($\log K_I = 8.95$).

The K_I value can also be estimated using the standard competitive inhibition relationship based on the shift in the agonist concentration-response curve caused by atropine (Equation (10.14)). This analysis also yields a K_I value of 9×10^8 ($\log K_I = 8.95$). Thus, if a competitive antagonist acts "irreversibly" during the brief time that the response is measured and shifts the agonist concentration-response curve to the right without affecting $E_{\max}$, its K_I can be estimated using a competitive relationship (Equation (10.14)) or the irreversible inhibitor assumption (Equations (10.28) and (10.31)). Both methods yield the same estimate of K_I. In fact if Equation (10.28) is substituted for q in Equation (10.31), the result is competitive inhibition Equation (10.14).

With regard to the partial agonist pilocarpine, atropine affects its concentration-response curve in a manner characteristic of an irreversible antagonist. That is, it reduces the $E_{\max}$ of the curve. Receptor occupancy by atropine $(1 - q)$ can be calculated using Equation (7.10) as described at the beginning of this section. The K_I value of atropine can then be calculated using Equation (10.31). When this is done, there is usually little difference in the estimate of the K_I value of atropine when measured in competition experiments with pilocarpine or a highly efficacious agonist.

So far, irreversible and competitive antagonists that behave irreversibly have been discussed. But what about a competitive antagonist that slowly dissociates from the receptor as the response is being measured, but does not reach equilibrium? This issue is relevant when the product of a reaction (e.g., second messenger) is measured over a time interval during which receptor occupancy by the antagonist is slowly changing as the system strives to equilibrate. There is no simple way to analyze this type of data unless kinetics are considered. Nevertheless, there are some situations where the assay can be easily modified to avoid the problem.

If an accumulated product is measured as the response, the assay conditions might possibly be modified so that it accumulates under equilibrium conditions. For example, the response of agonists that either stimulate G_s- or, to a lesser extent, G_i-linked receptors can sometimes be measured as a change in adenylate cyclase activity in broken cell preparations. Because the integrity of the cytosol is lost, receptor desensitization does not occur. The agonist and antagonist can be added to the incubation mixture in the presence of ATP and a nucleotide regenerating system for a while to allow equilibrium to occur. The labeled substrate (labeled ATP) can be added subsequently, and the amount of product (labeled cAMP) that accumulates over time can then be measured under equilibrium conditions. A similar strategy can be used when agonist-stimulated [^{35}S]GTPγS binding is measured in tissue or cellular homogenate.[1] The agonist and antagonist can be incubated together to allow equilibrium to occur before [^{35}S]GTPγS is added. The amount of [^{35}S]GTPγS binding that occurs can be subsequently measured under equilibrium conditions.

A related strategy can be used in many instances to measure agonist-stimulated phosphoinositide hydrolysis in intact cells. The products of the reaction (inositolphosphates) can be measured using a mass technique.[10] or by quantifying the radioactivity of an inositolphosphate fraction in cells or tissue prelabeled with [^{3}H]inositol.[11] The inositolphosphates usually do not accumulate to substantial levels because phosphatases in the cell continually hydrolyze inositolphosphates and allow the free inositol to recycle into new phosphoinositide. Usually the assay does not desensitize provided that inositol is continuously recycled. If the assay is carried out in the presence of lithium, a substantial quantity of inositol-1-phosphate accumulates because lithium inhibits myoinositol-1-phosphatase. Thus, the agonist and antagonist can be added to the assay, and after equilibrium is achieved, lithium is added, and the amount of inositol-1-phosphate that accumulates over time can be measured at equilibrium.

Competitive Interactions Between Orthosteric Ligands with Different Efficacies in Assays that Exhibit Constitutive Activity

The theory of competitive antagonism predicts that an inhibitor shifts the agonist concentration-response curve to the right in a parallel fashion. It follows that for every level of response on the control concentration-response curve, an equivalent response exists on the curve measured in the presence of the antagonist. This enables a null equation to be derived that has a simple solution (see Equation (10.14)). Not so when the competition involves an agonist and a partial agonist or when an inverse agonist competes with an agonist in a system that reports the constitutive activity. In this section, therefore, I develop the requisite methods for examining these types of interactions.

Operational model for competitive interactions at constitutively active receptors

To develop a robust model that accounts for all possible situations, each receptor term in the competitive inhibition model (Figure 10.1b) needs to be assigned a value for efficacy. The corresponding terms for the agonist-receptor complex (*DR*), the inhibitor-receptor complex (IR) and the free receptor (R) are $\varepsilon, \varepsilon_I$ and ε_{sys}, respectively. The ε_I term represents the intrinsic efficacy of the competitor (partial agonist, neutral antagonist or inverse agonist) and ε_{sys} the intrinsic efficacy of the unoccupied receptor. These terms can be used to develop an equation for the total stimulus (T), which can then be substituted into the logistic-transducer function of the operational model (Equation 7.5 of Chapter 7) to derive an equation that expresses the response as a function of the concentrations of agonist and competitor. This equation is used to estimate the affinity constant of the competitor by nonlinear regression analysis of the agonist concentration-response curves measured in the absence and presence of competitor.

First, the equation for the total stimulus (T) is derived. The fractional total stimulus represents the summation of the product of each receptor term and its corresponding efficacy value divided by the total amount of receptor:

$$T = \frac{\varepsilon_{\text{sys}}[R] + \varepsilon[DR] + \varepsilon_I[IR]}{[R] + [DR] + [IR]}. \tag{10.32}$$

Substituting Equations (10.3) and (10.4) for $[DR]$ and $[IR]$ yields:

$$T = \frac{\varepsilon_{\text{sys}}[R] + \varepsilon DK_{\text{obs}}[R] + \varepsilon_I IK_I[R]}{[R] + DK_{\text{obs}}[R] + IK_I[R]}. \tag{10.34}$$

Simplification yields the equation for the total stimulus:

$$T = \frac{\varepsilon_{\text{sys}} + \varepsilon DK_{\text{obs}} + \varepsilon_I IK_I}{1 + DK_{\text{obs}} + IK_I}. \tag{10.35}$$

This total stimulus function is substituted into the logistic transducer function of the operational model for the total stimulus (Equation (7.5), Chapter 7) to yield an equation for the response as a function of the population parameters. After simplification the result is:

$$\text{response} = \frac{M_{\text{sys}}}{1 + \left(\frac{1+DK_{\text{obs}}+IK_I}{\tau_{\text{sys}}+\tau DK_{\text{obs}}+\tau_I IK_I}\right)^m}, \tag{10.36}$$

in which

$$\tau_{\text{sys}} = \frac{\varepsilon_{\text{sys}}}{K_{E\text{-obs}}}, \tag{10.37}$$

$$\tau = \frac{\varepsilon}{K_{E\text{-obs}}}, \tag{10.38}$$

$$\tau_I = \frac{\varepsilon_I}{K_{E\text{-obs}}}. \tag{10.39}$$

Equation (10.36) is our goal. It expresses the response as a function of the affinity constants and τ values of the free receptor and ligand-receptor complexes. Accordingly, M_{sys} denotes the maximum response of the system, m, the transducer slope factor, K_{obs}, the affinity constant of the agonist-receptor complex, and K_I, the affinity constant of the inhibitor-receptor complex.

Analysis of simulated competitive inhibition data

Figure 10.10 shows a simulation of the inhibitory effect of an inverse agonist on the response to a full agonist in a system that exhibits constitutive receptor activity. The data were simulated using Equation (10.36), and the parameter estimates are listed in the figure legend. A random error of $\pm$10% was added to each data point. Because the interaction only involves one agonist, additional information is required before the operational model (Equation (10.36)) can be fitted to the data. Recall from Chapter 7 that it is impossible to estimate the M_{sys} of a response unless more than one agonist concentration-response curve is measured. For the sake of argument, let's assume that we know from prior experiments that the agonist is a full agonist.

Equation (9.36) is fitted to all of the concentration-response curves simultaneously using global nonlinear regression analysis. The parameters M_{sys}, m, τ_{sys}, τ_I, K_I

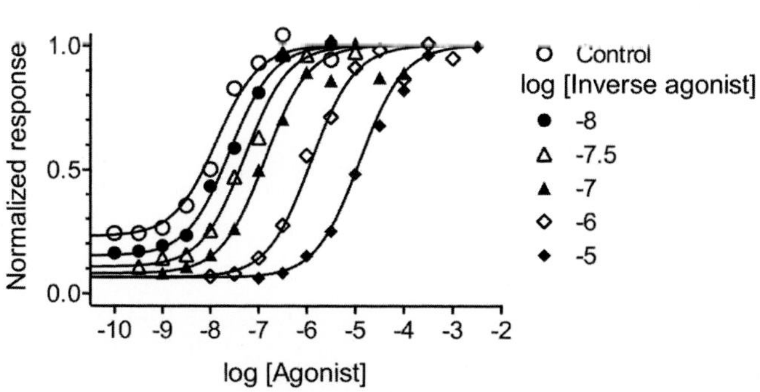

Figure 10.10: Analysis of the competitive inhibition of the response to an agonist by an inverse agonist at a receptor signaling pathway exhibiting constitutive activity. The plot shows simulated agonist concentration-response curves for the conditions of no inverse agonists (control) and increasing concentrations of inverse agonist as indicated in the legend. The theoretical curves represent the least-squares fit of Equation 10.36 to data. The parameter estimates were M_{sys}, 1.01 $\pm$ 0.01; m, 0.95 $\pm$ 0.09; log τ_{sys}, -0.55 ± 0.08; log τK_{obs}, 8 $\pm$ 0.08; log τ_I, -1.22 ± 0.19 and log K_I, 8.05 $\pm$ 0.08. The data were simulated using Equation (10.36) and the following parameter values: M_{sys}, 1.0; m, 1.0; log τ_{sys}, -0.50; log τ, 2.0; K_{obs}, 6.0; log τ_I, -1.2 and log K_I, 8. A random error equivalent to $\pm$ 7.5% of the simulated response value plus 0.2 was added to the data.

and τ are shared and the value of K_{obs} is constrained to an arbitrarily high value (e.g., log $K_{\text{obs}} = 0$). The details of the regression analysis are described in the appendix to this chapter (see Example 2). Regression analysis yielded the following parameter estimates: M_{sys}, 1.01 ± 0.01; m, 0.95 ± 0.09; log τ_{sys}, -0.55 ± 0.08; log τ_I, -1.22 ± 0.20 and log K_I, 8.05 ± 0.08. It is impossible to obtain accurate estimates of K_{obs} and τ from the data, only their product (log τK_{obs}, 8.00 ± 0.07). These parameter estimates are similar to those used to simulate that data (M_{sys}, 1.0; m, 1.0; log τ_{sys}, -0.5; log τ_I, -1.2; log K_I, 8 and log τK_{obs}, 8), which illustrates the feasibility of the analysis.

If the separately measured concentration-response curves of the inverse agonist and the full agonist were analyzed simultaneously as described in Chapter 7, it would be possible to estimate the K_{obs} and τ values of the inverse agonist (in this example, τ_I and K_I) and the product of τK_{obs} for the full agonist. These estimates should agree with those described above for the simulated experiment shown in Figure 10.10. One would not have known from the individual concentration-response curves, however, if the inverse agonist and full agonist were interacting competitively at the same locus on the receptor. The inverse agonist could have been acting at an allosteric site to inhibit constitutive receptor activity, for example. But the competitive interaction shown in Figure 10.10 argues against the allosteric hypothesis. The characteristic feature of competition is the proportional relationship between the inhibitor concentration and the shift in the agonist concentration-response curve. There is no indication that the curve-shifting effect of the inverse agonist diminishes at high concentrations. In contrast, there is a limit to the effect of an allosteric modulator as described in the next Chapter.

Another example of a competitive interaction is shown in Figure 10.11 where the effects of a partial agonist on the response to a full agonist are illustrated for a system that lacks constitutive activity. For this analysis, Equation (10.36) is simplified to account for the lack of constitutive activity (i.e., $\tau_{\text{sys}} = 0$):

$$\text{response} = \frac{M_{\text{sys}}}{1 + \left(\frac{1+DK_{\text{obs}}+IK_I}{\tau DK_{\text{obs}}+\tau_I IK_I}\right)^m}. \tag{10.40}$$

As described in Chapter 7, it is possible to estimate the observed affinity constant and τ value of of a partial agonist when its concentration-response curve is analyzed simultaneously with that of a full agonist. The same applies to the situation where various concentrations of full agonist and partial agonist are added to the assay at the same time. Once again, however, there is insufficient information to estimate the K_{obs} value of the full agonist, and the regression analysis is done with K_{obs} constrained to an arbitrarily low value (e.g., log $K_{\text{obs}} = 0$). The parameters M_{sys}, m, τ, τ_I and K_I

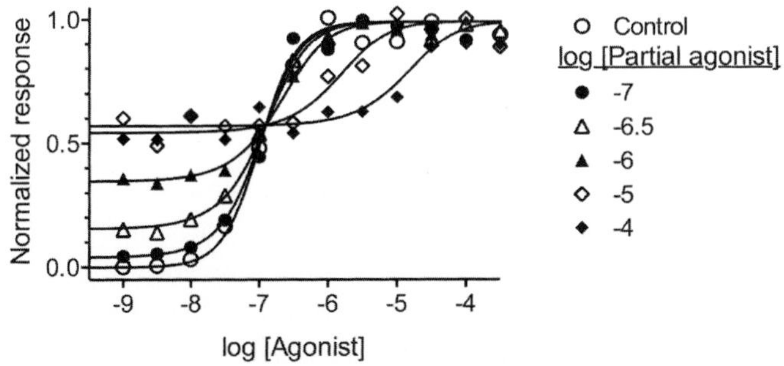

Figure 10.11: Analysis of the competitive inhibition of the response to a full agonist by a partial agonist. The plot shows simulated agonist concentration-response curves for the conditions of no partial agonists (control) and increasing concentrations of a partial agonist as indicated in the legend. The theoretical curves represent the least-squares fit of Equation (10.36) to data with log τ_{sys} constrained to -20. The parameter estimates were M_{sys}, 0.99 ± 0.01; m, 1.60 ± 0.17; log τK_{obs}, 6.99 ± 0.03; log τ_I, 0.086 ± 0.02 and log K_I, 6.10 ± 0.09. The data were simulated using Equation (10.36) and the following parameter values: M_{sys}, 1; m, 1; log τ_{sys}, -20; log τ, 2.0; K_{obs}, 5.0; log τ_I, 0.1 and log K_I, 6. A random error equivalent to $\pm 7.5\%$ of the simulated response value plus 0.2 was added to the data.

are shared during the analysis. Regression analysis yielded the following parameter estimates: M_{sys}, 0.99 ± 0.01; m, 1.60 ± 0.16; log τ_I, $0.086 \pm 0.0.18$; log K_I, 6.10 ± 0.09 and τK_{obs}, 6.99 ± 0.03. These parameter estimates are similar to those used to simulate that data (M_{sys}, 1.0; m, 1.5; log τ_I, 0.1; log K_I, 6 and log τK_{obs}, 7).

Again, the estimated values of the parameters of the partial agonist (log τ_I and log K_I) and full agonist (log τK_{obs}) should be the same as those estimated from their individual concentration-response curves. The experiment shown in Figure 10.10 tests for their competitive interaction, which is apparent in how increasing concentrations of the partial agonist continue to shift the concentration-response curve of the full agonist to the right with no sign of limit to this effect.

Can the standard null equation be used to analyze data with inverse and partial agonists?

There is an obvious similarity in the effect of the antagonist on the concentration-response curve of the agonist in Figure 10.2 and the effects of the inverse agonist and partial agonist shown in Figures 10.10 and 10.11. In each interaction, the drug with the lower efficacy causes a dextral shift in the concentration-response curve of the more efficacious drug. The differences lie in how the baseline response changes with increasing concentrations of the inhibitory drug. Using the operational model to analyze the data is useful for testing deviations from competitive inhibition, but is it really necessary for estimating the K_I value of the competitor if the data are consistent with competitive inhibition? Would it be possible to analyze the data using the simple relationship described by Equation (10.14)? This worked so well with competitive antagonists when the response lacked constitutive receptor activity.

The answer is yes and no, but let me first define what is meant by a shift in the concentration-response curve under these conditions because different shifts can be estimated depending upon the level of response considered. Suppose the shift (CR) is calculated as the ratio of the EC_{50} value of the agonist measured in the presence of the less efficacious competitor divided by that measured in its absence. By EC_{50} I mean the concentration of agonist causing a half-maximal increase in response over the span of responses elicited by the agonist. Thus, for the control agonist concentration-response curve in Figure 10.10, the response at EC_{50} is approximately 60% because constitutive receptor activity generates about a 20% response in the absence of agonist. As the concentration of inverse agonist increases, the response at EC_{50} approaches the 50% response of the system. With this definition, a mathematical expression for EC_{50} can be derived using the operational model (Equation (10.36)), and the validity of the simple calculation (Equation (10.14)) can be addressed. This strategy can show that Equation (10.14) is invalid in this situation, but the analysis is complicated, so I will describe the results of simulations.

For the data illustrated in Figure 10.10, which was simulated with a transducer slope factor of 1 (i.e., $m = 1$), there was little difference in log K_I when estimated from each shift in the agonist concentration-response curve using Equation (10.14) (average log K_I, 7.97) or when regression Equation (10.36) was globally fit to all of the data simultaneously (log K_I, 8.05). In the case the antagonism between full and partial agonists (Figure 10.11), there was also good agreement between the estimates of log K_I for the two methods for partial agonist concentrations from log 6 to log 4. At lower concentrations of partial agonist (log -7 and -6.5), there was moderate error in the estimate when Equation (10.14) was used to calculate log K_I from each shift (5.81 and 5.29, respectively). Part of the error is due to the simulated random error, but when data with no error were simulated, Equation (10.14) yielded log K_I estimates of 5.37 and 5.76 for partial agonists concentrations of log -7 and log -6.5. At the higher concentrations, there was practically no error. Thus, it seems that there is little error in using this approach provided that the partial agonist is used at a concentration that causes at least a moderate shift in the agonist concentration-response curve (at least three- to 10-fold).

If the competitor is a partial agonist with a substantial maximal response (e.g., 70% that of the full agonist) and the system exhibits substantial cooperativity ($m = 2 - 3$), then low concentrations of the partial agonist (e.g., $I = 1/K_I$)

can actually reduce the EC_{50} value of the full agonist, which precludes the use of Equation (10.14). In all of these instances, however, regression analysis with Equation (10.36) yields correct estimates of log K_I.

Resultant Analysis

It is surprising how often experts seem unaware of the possibility that an unusual effect of a drug in a functional assay might be caused by something other than its interaction with the receptor that normally mediates the response. This issue is especially relevant when a putative antagonist is allowed to incubate with the assay preparation for a while to determine if it can antagonize the response to a subsequently added agonist. If the compound can penetrate the plasma membrane, it might have an inhibitory effect inside the cell that is more potent than its effect on the receptor.

The title of this section is an expression first introduced by Black and coworkers.[12] to denote the analysis of a drug effect mediated by two or more specific actions. Thus, in this section, I describe how to analyze the action of a compound whose net effect is only partly mediated by competitive inhibition. The proximate goal is to describe how to estimate the affinity constant of such an antagonist for the orthosteric site on the receptor.

Theory of resultant analysis

Recall that the basis of measuring the affinity constant of a competitive antagonist in a functional assay is that equivalent responses in the absence and presence of a competitive inhibitor imply equal levels of receptor occupancy by the agonist. Simplification of the corresponding mathematical equality illustrates that the antagonist-induced shift (*CR*) in the agonist concentration-response curve at equilibrium is solely dependent on the concentration of antagonist and its affinity constant (see Equation (10.14)).

Consider the corresponding equality for the situation where the competitive inhibitor (*J*) not only competes with the agonist, but also affects the transducer function in some way:

$$g\left(\frac{[D]\varepsilon R_T}{[D]+\frac{1}{K_{\text{obs}}}}\right)=g^*\left(\frac{[D']\varepsilon R_T}{[D']+\frac{1+JK_J}{K_{\text{obs}}}}\right) \tag{10.41}$$

The derivation of this equality is analogous to that described for Equation (10.11). It illustrates the relationship between equiactive agonist concentrations in the absence (D) and presence (D') of an antagonist. The expression on the left communicates the idea that the response in the absence of the competitive inhibitor is a function ($G(x)$) of the stimulus and that the stimulus is equivalent to the product of intrinsic efficacy and receptor occupancy. An equivalent response can be elicited by a different concentration of the agonist (D') in the presence of the antagonist (J), and the value of this response is defined by the expression on the right. In this instance, however, the function that transduces the stimulus into the response ($G^*(x)$) is not the same as that on the left as indicated by the asterisk. This is because part of the effect of the antagonist includes an action distinct from competitive inhibition, and this auxiliary effect can be attributed to a change in the transducer function. Because the transducer function is different on the two sides of the equation, it cannot be eliminated.

A clever solution to this problem was described by Black and coworkers.[12] They realized that it should be possible to derive an equality based on the interactions among an agonist, a standard competitive antagonist (I), and an unknown antagonist (J) whose action might be only partly competitive. Before presenting this ingenious solution, I need to derive the receptor occupancy function for an agonist in equilibrium with two competitive inhibitors.

The affinity constants of the agonist (D) and competitive inhibitors (I and J) are defined in the normal manner as described by Equations (10.1) and (10.2) for D and I, respectively. The fractional receptor occupancy by the agonist in the presence of two competitors is given by:

$$\frac{[DR]}{R_T} = \frac{[DR]}{[DR] + [IR] + [JR] + [R]}. \tag{10.42}$$

Replacing each receptor complex with an equivalent expression in terms of $[R]$ and the appropriate affinity constants and drug concentrations yields:

$$\frac{[DR]}{R_T} = \frac{DK_{\text{obs}}[R]}{DK_{\text{obs}}[R] + IK_I[R] + JK_J[R] + [R]}. \tag{10.43}$$

Simplification yields:

$$[DR] = \frac{D[R_T]}{D + \frac{\left(1+IK_I+JK_J\right)}{K_{\text{obs}}}}. \tag{10.44}$$

Now that we have an equation for receptor occupancy by the agonist in the presence of two competitive antagonists, let us return to the null equation of Black and coworkers.[12] These investigators reasoned that it should be possible to derive an equality based on equiactive responses to an agonist in the presence of J and in the presence of both I and J. Recall that I is a standard competitive inhibitor and J is an unknown competitive inhibitor that also affects the transducer function. Thus, the corresponding null equation is given by:

$$g^*\left(\frac{[D]\varepsilon R_T}{[D]+\frac{1+JK_J}{K_{\text{obs}}}}\right)=g^*\left(\frac{[D']\varepsilon R_T}{[D']+\frac{1+IK_I+JK_J}{K_{\text{obs}}}}\right). \tag{10.45}$$

Notice that the perturbed transducer function ($G^*(x)$) occurs on both sides of the equation so that it can be eliminated:

$$\frac{[D]\varepsilon R_T}{[D]+\frac{1+JK_J}{K_{\text{obs}}}}=\frac{[D']\varepsilon R_T}{[D']+\frac{1+IK_I+JK_J}{K_{\text{obs}}}}. \tag{10.46}$$

Further simplification yields:

$$\frac{[D']}{[D]}=\frac{1+IK_I+JK_J}{1+JK_J}. \tag{10.47}$$

If CR^* is substituted for $[D']/[D]$ and the resulting expression simplified, the result is:

$$K_J=\frac{1+IK_I-CR^*}{J(CR^*-1)}. \tag{10.48}$$

The variable CR^* represents the shift in the concentration-response curve of the agonist caused by the standard competitive antagonist I in the presence of J. The behavior of the standard competitive antagonist can be characterized in separate experiments to determine K_I. CR^* can be measured in experiments with both I and J, and from this new data K_J can be estimated using Equation (10.48) and the prior estimate of K_I.

Equation (10.48) can be rearranged into a form that throws light on how the competitive component of the resultant effect J behaves:

$$CR^*=[I]K_I'+1, \tag{10.49}$$

in which

$$K_I'=\frac{K_I}{([J]K_J+1)}. \tag{10.50}$$

Equation (10.49) shows that if the functional assay is always carried out in the presence of a fixed concentration of J, then the shift in the agonist concentration-response curve caused by the competitive inhibitor I (CR^*) will resemble a competitive relationship except that the apparent estimate of the affinity constant of I (K_I')

will differ from its true value by the factor, $1/([J]K_J + 1)$, as shown by Equation (10.50). Thus, the competitive component of J influences the estimate of the affinity constant of I.

Let me explain how this interaction behaves with a theoretical example. Imagine an unknown inhibitor J that shifts an agonist's concentration-response curve to the right 11-fold at a concentration of 10^{-7} M (see Figure 10.12a). The effect resembles competitive inhibition in the sense that there is no change in E_{max}. If we assumed that the mechanism is competitive, an estimate of 10^8 M^{-1} could be estimated for K_J using the competitive inhibition Equation (10.14). Of course, this estimate is only as good as the assumption of competition.

Now imagine a well-characterized inhibitor, I, whose mechanism is known to be competitive. It shifts the agonist concentration-response curve to the right 11-fold at a concentration of 10^{-8} M (see Figure 10.11b). This effect yields an estimate of 10^9 M^{-1} for the affinity constant of I. Imagine that the experiment with I is repeated in the presence of J (10^{-7} M). That is, the agonist concentration-response curves in both the absence and presence of I (10^{-8} M) are measured in the presence of J (see Figure 10.11c). Suppose that antagonist I now only causes a two-fold shift. Under this condition its apparent affinity constant (K'_I) could be estimated by rearrangement of Equation (10.49):

$$K'_I = \frac{CR^* - 1}{[I]}. \tag{10.51}$$

Using this equation, the apparent affinity constant of antagonist I in the presence of J is now only 10^8 M^{-1}. The competitive effect of J has reduced the apparent affinity of I to one-tenth its true value. Finally, we can rearrange Equation (9.50) into a form that will enable the estimation of the affinity constant of J from this information:

$$\frac{\frac{K_I}{K'_I} - 1}{[J]} = K_J. \tag{10.52}$$

Substituting in a value of 10 for K_I/K'_I ($10^9/10^8$) and a value of 10^{-7} M for $[J]$ yields an estimate of 0.9×10^8 M^{-1} for K_J. This value is approximately the same as that estimated for J when it was used by itself (10^8 M^{-1}). Thus, our competitive hypothesis for the action of J in this case appears valid. In other words, the inhibitory effect of J can be attributed entirely to a competitive mechanism.

But suppose the results of the last experiment came out differently, and antagonist I still caused an 11-fold shift in the concentration-response curve when J was present at a concentration of 10^{-7} M (see Figure 10.11d). Using this information and Equation (10.52), the estimate of K_J is now zero. This means that J

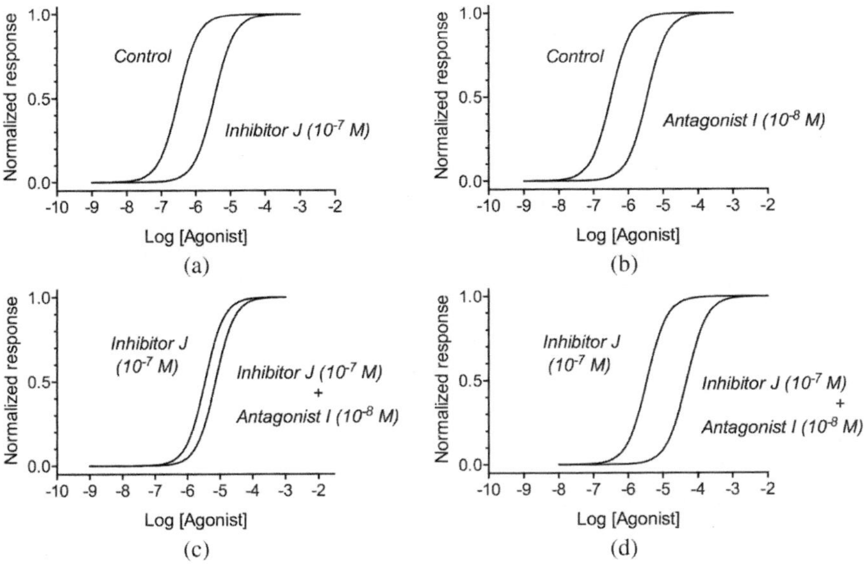

Figure 10.12: Resultant analysis of the interactions among an agonist, a known competitive inhibitor (I) and an inhibitor with an unknown mechanism (J). Theoretical concentration-response curves of an agonist in the presence of I and J are shown. (a) The plot of the concentration-response curve of an agonist in the absence (*control*) and presence of a fixed concentration of inhibitor J (10^{-7} M). (b) Plot of the concentration-response curve of an agonist in the absence (*control*) and presence of a fixed concentration of competitive inhibitor I (10^{-8} M). (c) Plot of the concentration response curves of an agonist in the presence of inhibitor J (10^{-7} M) and in the presence of inhibitor J (10^{-7} M) and competitive inhibitor I (10^{-8} M). The plot is consistent with a competitive mechanism for J. (d) Plot of the concentration-response curves of an agonist in the presence of inhibitor J (10^{-7} M) and in the presence of inhibitor J (10^{-7} M) and competitive inhibitor I (10^{-8} M). The plot is consistent with a non-receptor inhibitory mechanism for J.

does not bind to the orthosteric site to which the agonist and antagonist I bind, or more likely, J has very low affinity for the orthosteric site. Thus, in this case, J is not a competitive inhibitor of the receptor, and its large inhibitory effect on the concentration-response curve of the agonist shown in Figure 10.12a is through some other mechanism involving an effect on the transducer function ($G(S)$).

There is no need to divide the calculation of K_J into the two steps using Equations (10.51) and (10.52). This was done simply to explain how the resultant effect of J behaves. After estimating the K_I value of antagonist I from the experiment shown in Figure 10.12b, one can apply Equation (10.48) to estimate K_J using the value of J and the estimate of CR^*. In the next section, I apply this method to explain the unusual inhibitory effect of the muscarinic antagonist, *p*-FHHSiD, on gastrointestinal smooth muscle. But first, a little background information.

Estimation of the affinity constant p-FHHSiD for muscarinic receptors in smooth muscle

Acetylcholine is an essential neurotransmitter in the gastrointestinal tract. It is released from enteric nerves[13,14] and activates M_2 and M_3 muscarinic receptors on the smooth muscle to cause a synchronized contraction resulting in peristalsis.[15] Activation of the M_3 receptor elicits contraction of intestinal smooth muscle through a mechanism ultimately triggered by G_q. The M_2 receptor also affects contraction, but the signaling mechanisms are contingent upon M_3 receptor activation, and the M_2 receptor signaling pathway is less potent than that of the M_3. For these reasons, M_3-selective muscarinic antagonists potently inhibit contraction, whereas M_2 selective antagonists are less effective. This situation applies to smooth muscle from various gastrointestinal tissues as well as the trachea[16,17] and urinary bladder.[18] Some M_3-selective muscarinic antagonists inhibit contraction of intestinal smooth muscle more potently than that of the trachea and urinary bladder, even though the same type of M_3 receptor causes contraction in these tissues. An example of such a compound is *p*-FHHSiD. It antagonizes muscarinic agonist-induced contraction of the guinea pig ileum with ten-fold greater potency than that of the trachea.[19,20]

Figure 10.13a shows a concentration-response curve for the highly efficacious muscarinic agonist, oxotremorine-M, in the guinea pig ileum. It elicits contraction with high potency (pEC_{50}, 7.75 ± 0.02). When the experiment is repeated in the presence of *p*-FHHSiD, the concentration-response curve shifts to the right, depending on the concentration of the *p*-FHHSiD. This antagonist appears to increase the $E_{\max}$ of oxotremorine-M, but this effect can be attributed to time because control experiments reveal a time-dependent increase in the $E_{\max}$.

The log affinity constant of *p*-FHHSiD for the receptor mediating contraction in the ileum can be estimated from the shifts that it causes at concentrations of 0.03 and 0.10 μM assuming a competitive mechanism (7.8 and 8.1, respectively, see Equation (10.14)). The two estimates of log K_J are approximately the same, which would be expected for competition. The prototypic muscarinic antagonist, atropine, causes a 6.4-fold shift in the concentration-response curve, and this effect yields a log K_I estimate of 8.7. If the inhibitory effect of *p*-FHHSiD is mediated entirely through competitive inhibition, then the estimate of its affinity constant should be the same when measured from its ability to inhibit the contraction to oxotremorine-M (Figure 10.13a) or to inhibit the competitive effect of atropine

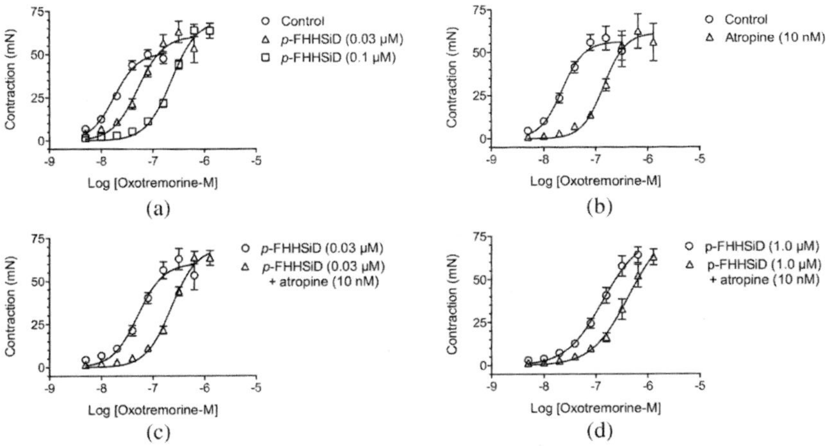

Figure 10.13: Resultant analysis of the interactions among the highly efficacious agonist, oxotremorine-M, atropine and *p*-FHHSiD at muscarinic receptors eliciting contraction in the guinea pig ileum. (a), The concentration-response curves of oxotremorine-M in the absence (*control*) and presence of two concentrations of *p*-FHHSiD (0.03 and 0.1 μM) are shown. (b) Plot of the concentration-response curve of oxotremorine-M in the absence (*control*) and presence of atropine (10 nM) are shown. (c) Plot of the concentration-response curves of oxotremorine-M in the presence of *p*-FHHSiD (0.03 μM) and in the presence of *p*-FHHSiD (0.03 μM) and atropine (10 nM) are shown. (d) Same as *c* except that the concentration of *p*-FHHSiD is 1 μM of the concentration response curves of an agonist in the presence of atropine J (10^{-8} M). The data are from Ehlert *et al.*[20] and have been reproduced with permission.

on the contraction to oxotremorine-M. Examples of the latter type of experiment are shown in Figures 10.13c and 10.13d. When these atropine induced shifts in the concentration-response curve (CR^*) are analyzed with Equation (10.48), the estimate of the log K_J value of *p*-FHHSiD at concentrations of 0.03 and 0.1 μM are 7.17 and 7.25, respectively. Thus, the average estimate of the log K_J from the latter experiment is 0.75 log units less than that estimated when *p*-FHHSiD was used by itself. The data suggest that the most potent inhibitory effect of *p*-FHHSiD is mediated not through competition at the M_3 receptor, but through inhibition of the signaling pathway that elicits contraction.

If the same type of experiment is done with the competitive antagonists, 4-DAMP and atropine, there is no difference in the log K_J value of 4-DAMP when estimated from its competitive effect on the contraction to oxotremorine-M (9.13) or from its competitive effect on the ability of atropine to inhibit the contraction to oxotremorine-M (9.08). Thus, the two known competitive inhibitors act competitively under both conditions.[20] When the experiment with *p*-FHHSiD and atropine was done in the trachea, the two experimental paradigms yielded similar estimates of the log affinity constant of *p*-FHHSiD (about 6.77), and the average value was similar to that estimate in the ileum using resultant analysis. Finally, the log K_I value of *p*-FHHSiD was estimated from its ability to inhibit oxotremorine-M-mediated phosphoinositide hydrolysis in the ileum (7.08) and trachea (7.06), and approximately the same values were estimated in both tissues. Phosphoinositide hydrolysis is mediated by the M_3 receptor in ileum and trachea.

These data are consistent with the postulate that *p*-FHHSiD binds competitively to the M_3 muscarinic receptor with a log K_I of approximately 7 to inhibit muscarinic responses in ileum and trachea. Its more potent inhibitory effect on contraction of the ileum exceeds its binding affinity for any known muscarinic receptor by one log unit. Presumably, this inhibitory effect is downstream from phosphoinositide hydrolysis, because its log K_I value for antagonizing this response (7) is consistent with its ability to competitively antagonize the M_3 receptor. Perhaps other tissue selective antagonists, like xamifenacin, have a similar action.

Using Antagonists to Identify Receptor Function

Antagonists with selectivity for specific receptor subtypes can be used to identify the receptor mediating a specific function provided that the function of interest can be measured in an isolated tissue so that the concentrations of agonist and antagonist are known. The method involves estimating the affinity constant of the antagonist in two ways. First, by applying Equation (10.14) to estimate K_I from the antagonist-induced shift in the agonist concentration-response curve, and second, by using a competitive binding assay to measure K_I in experiments on different cells, each expressing a subtype of the receptor thought to mediate the response. The latter estimates are unequivocal because the binding affinity is measured in an experiment on a homogeneous population of a known receptor. If the estimate of binding affinity for one of the receptor subtypes agrees with that measured in the functional response, then it would seem likely that this is the subtype that mediates the response. If the experiment is repeated with a handful of subtype selective antagonist whose actions are entirely competitive and all of the binding affinity estimates point to a single subtype, then the conclusion is nearly unequivocal.

Of course if the response of interest can be measured in tissues from mutant mice lacking the subtypes of interest, then the question can be easily addressed using these animals. But there are often species differences in the receptors that contribute to a specific response so experiments on nonmouse tissues can be highly relevant.

There are some common errors and pitfalls in using antagonists to identify the receptor mediating function. The buffer used in the binding assay should be the same or similar to that of the functional assay because ionic strength can have a large effect on the affinity of antagonists.[21] Binding assays run in a hypotonic buffer of 30 mM Tris/HCl, pH 7.4 and 10 mM $MgCl_2$, for example, are not useful for this type of determination. The major ion species in the binding buffer (NaCl, $MgCl_2$ and $NaHCO_3$) should be the same as that of the physiological buffer used for the functional assay. The receptor preparation for the binding assay needs to be a defined receptor. This can be achieved by expressing a known recombinant receptor in a cell line. The species from which the receptor was derived should be the same as that of the tissue because some antagonists exhibit marked species differences in their binding affinity. A potential problem is that other proteins that naturally occur with the receptor in its native receptosome might influence the affinity of the antagonist for the receptor. An example is how the pharmacological profile of the calcitonin-like-receptor receptor changes depending upon the chaperone with which it dimerizes (see Chapter 5).[22] The occurrence of dimerization-induced changes in antagonist affinity seems uncommon, and if it does occur, one would have to express the receptor in a system that reproduced the naturally occurring dimerization forms of the receptor or else use antagonists whose affinity is not so affected. Another problem is that more than one receptor might contribute to the response in the isolated tissue, and this issue is discussed below.

Because the strength of the conclusion from this method of analysis involves a comparison of antagonist affinities between two assays, the validity of the conclusion would seem to be proportional to the number of antagonists investigated. But before considering additional subtype selective antagonists, their mechanism of action should be established as competitive. With regard to muscarinic receptors, the compound *p*-FHHSiD and xamiphenacin, should probably be avoided because their most potent mechanism of inhibition in some tissues appears unrelated to antagonism of muscarinic receptors. Thus, it is better to use a few well-characterized antagonists rather than a long list of agents whose mechanisms have not been established.

This approach of comparing two estimates of antagonist affinity (binding and function) is far superior to the common method of adding an antagonist at a single blocking concentration and determining if the response to a single concentration of agonist is inhibited *in vivo* or *in vitro*. The same can be said of single blocking concentrations of antagonists used to inhibit the action of an endogenous neurotransmitter. One can never be certain that the antagonist does not have additional inhibitory effects unrelated to antagonism of the specific receptor of interest. Also the amount of antagonist required to block the response depends on several issues including the amount of endogenous neurotransmitter being released, the amount of agonist administered *in vivo* or *in vitro*, and the antagonist concentration at the receptor site following *in vivo* administration. There are simply too many variables that prevent an unequivocal determination of the receptor that mediates the response. In contrast the approach of comparing two estimates of K_I (binding an function) from *in vitro* assays involves determining quantitative estimates of K_I, not just an all-or-none answer to the question of whether the antagonist blocks the response. If the estimates of the K_I values of several subtype-selective antagonists agree across the two assays, then it is highly likely that the response is mediated by the receptor for which there is this agreement.

Examples of using antagonists to identify receptor function

My first example is an old one, and at the time, its purpose was to confirm that the radioligand [^{3}H]N-methylscopolamine ([^{3}H]NMS) bound to the muscarinic receptor and not that the muscarinic receptor mediated a response. The results in Figure 10.14 show the binding affinities (log K_I) of several antagonists plotted against the corresponding estimates measured in antagonism studies on the guinea pig ileum. The binding experiment was done on homogenates of the rat cerebral

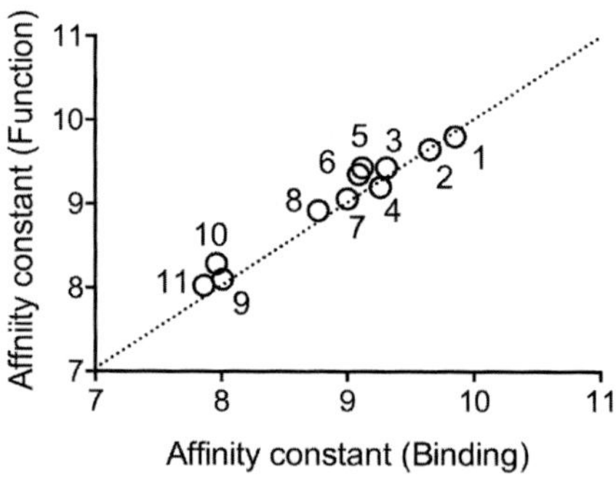

Figure 10.14: Comparison of the affinity constants of selected antagonists for binding to muscarinic receptors labeled with [^{3}H] N-methylscopolamine in rat forebrain with the affinity constants measured by the competitive antagonism of agonist-induced contraction of the guinea pig ileum. The source of the data is described in the text. 1, (−) 3-quinuclidinyl benzilate; 2, N-methylatropine; 3, (−) atropine; 4, (−) scopolamine; 5, (+) N-methyl-3-quinuclidinyl benzilate; 6, (−) N-methyl-3-quinuclidinyl benzilate; 7, N,N-dimethyl-4-piperidinyl diphenylacetate; 8, tricyclamol; 9, secergan and 10, propylbenzilylcholine.

cortex.[23,24] The log K_I values were estimated from the concentration of antagonist that caused a 50% inhibition of the binding of [^{3}H]NMS after correcting for the competitive effect of [^{3}H]NMS as described in Chapter 12. The functional assay involved measuring muscarinic agonist-induced contraction of the guinea pig ileum in the absence and presence of different antagonists.[25–29] The log K_I values were subsequently estimated using the competitive inhibition theory described earlier in this chapter. Given that there are five subtypes of the muscarinic receptor having a differential distribution in brain and smooth muscle, one might wonder how such a good agreement between the two estimates could be obtained. For example the ileum expresses mainly M_2 and some M_3 receptors,[30,31] whereas the cerebral cortex expresses mainly the M_1, M_4 and M_2 subtypes and smaller amount of the M_3 and M_5.[32,33] The explanation is simple. In the physiological buffer that was used for both assays, [^{3}H]NMS binds with similar affinity for the different receptor subtypes, none of the antagonists discriminate among subtypes, nor do they exhibit difference in their affinity for rat and guinea pig receptor. Thus, given these additional points, the data are consistent with the idea that [^{3}H]NMS specifically binds to muscarinic receptors and that contraction is elicited by a muscarinic receptor.

Figure 10.15a shows the K_I values of muscarinic antagonists, measured by antagonism of agonist-mediated inhibition of adenylate cyclase in the rat ileum, plotted against the binding affinities of the same antagonists for the major muscarinic receptor expressed in the ileum.[34] There is reasonable agreement between

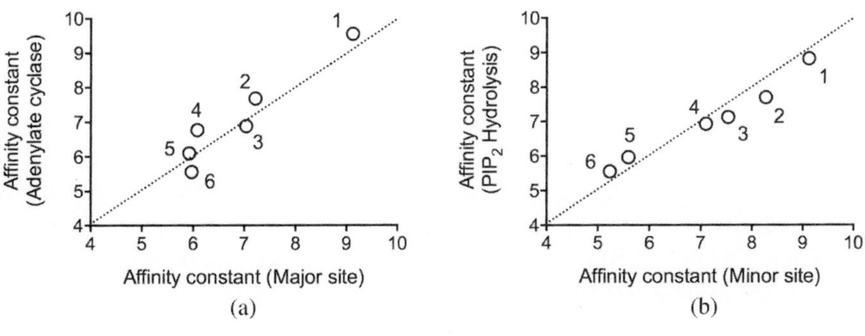

Figure 10.15: Comparison of the relationship between the binding affinity of selected antagonists for major (a) and minor (b) muscarinic sites and their antagonism of agonist-mediated inhibition of adenylate cyclase activity (a) and stimulation of phosphoinositide hydrolysis in the longitudinal muscle of the rat ileum (b). The data have been reproduced from Candell *et al.* with permission. 1, (−) N-methylscopolamine; 2, methoctramine; 3, AF-DX 116 (11[[2-[(diethylamino)methyl]-1-piperidinyl]acetyl]-5,11-dihydro-6H-pyrido[2,3-b][1,4]benzodiazepine-6-one); 4, hexahydrosiladifendidol; 5, *p*-fluorohexahydrosila-difendidol and 6, pirenzepine.

the functional and binding estimates of K_I, indicating that the major muscarinic receptor in the ileum mediates an inhibition of adenylate cyclase. Figure 10.15b shows the K_I values of the same antagonists, measured by antagonism of muscarinic agonist-stimulated phosphoinositide hydrolysis in the ileum, plotted against the binding affinity estimates for the minor muscarinic receptor in the ileum.[34] In this case, the data indicate that the minor site mediates phosphoinositide hydrolysis.

Simulation of antagonism of responses mediated by two receptors

When more than one type of receptor mediates a response, the estimate of the K_I value of the competitive antagonist can vary depending upon the signaling pathways of the receptors. In this section I describe the behavior of this antagonism using the operational model and a stimulus function that incorporates an interaction between two receptors.

My first example involves a response having two independent components, each mediated by a different receptor and both readily activated by the agonist. The receptors behave independently in the sense that the total response is equal to the sum of the responses elicited by receptor 1 (R_1) and receptor 2 (R_2). This could happen, for example, if the response were cAMP accumulation in a tissue composed of two cell types, one type expressing R_1 and the other R_2. Figure 10.16 shows an example of such a tissue where 40% of the response is mediated by R_1 and 60% by R_2. The concentration-response curves of an agonist in the absence and presence of an R_1 selective (panel a), an R_2 selective (panel b) and a nonselective (panel c) antagonist are shown. The log K_I values of the antagonists for R_1 and R_2 are: A_1 (9.0 and 7.5), A_2 (7.5 and 9.0) and A_3 (9.0 and 9.0). The concentration of each antagonist is 10^{-8} M. Because of the independent behavior of the two responses, the selective antagonists A_1 and A_2 cause a nonparallel shift in the concentration-response curve, whereas the nonselective antagonist A_3 causes a parallel shift. All of the data could be fitted simultaneously to a model composed of two independent components, and accurate estimates of log K_I for R_1 and R_2 could be estimated for each antagonist. In the case of A_3, a single estimate of log K_I for both R_1 and R_2 could be calculated using the simple Equation (9.14). With real data, it is usually uncommon for the responses to be completely independent.

The theoretical data in Figure 10.17 represent the response to an agonist elicited by two receptors (R_1 and R_2) that converge on the same signaling pathway. The signal elicited by each receptor is highly amplified so that a maximal response can be elicited

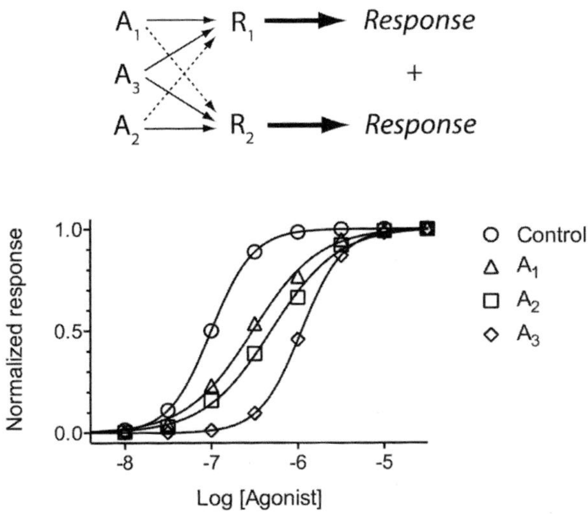

Figure 10.16: Simulation of the effects of receptor-selective antagonists on a response elicited by an agonist acting on two populations of distinct receptors (R_1 and R_2) expressed on different cells. The relative abundances of R_1 and R_2 are 0.4 and 0.6, respectively. The scheme above the concentration-response curves illustrates the nature of the signaling pathway and the selectivity of the antagonists (A_1, A_2 and A_3) for R_1 and R_2. The solid arrows emanating from R_1 and R_2 and pointing towards their separate responses indicate that each receptor contributes independently to the total response. The arrows emanating from the antagonists and pointing toward the receptors indicate high (log K_I, 9; solid arrow) and low (log K_I, 7.5; dotted arrow) affinity. Thus, A_1 has selectivity for R_1, A_2 selectivity for R_2, and A_3 equivalent high affinity for R_1 and R_2. In the lower half of the figure the effects of the antagonists (10^{-8} M) on the concentration-response curve of an agonist are illustrated.

by either receptor. The affinity constants of the antagonists and their concentrations are the same as that in Figure 10.16. In this situation, the shifts in the concentration-response curve caused by A_1 (panel a) or A_2 (panel b) are both small. That is, A_1 has an effect suggesting that R_2 mediates the response and A_2 has an effect suggesting that R_1 mediates the response. This is because each receptor converges on the same signaling pathway, and if one receptor is antagonized the response can be rescued by the other receptor. If both receptors are blocked simultaneously, then a substantial shift in the concentration response curve occurs as shown in the case of A_3 (panel c). Estimation of the log K_I value of A_3 using the simple Equation (9.14) would yield an accurate estimate for the affinity of A_3 for R_1 and R_2.

In the foregoing examples, each receptor elicited a signaling pathway that was capable of generating the measured response or output of the assay. Some receptors

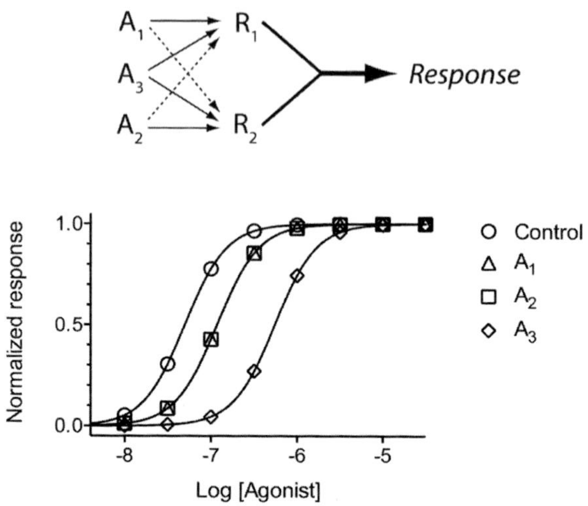

Figure 10.17: Simulation of the effects of receptor-selective antagonists on a response elicited by an agonist acting on two populations of distinct receptors (R_1 and R_2) expressed on the same cells. The scheme above the concentration-response curves illustrates the nature of the signaling pathway and the selectivity of the antagonists (A_1, A_2 and A_3) for R_1 and R_2. The solid arrows emanating from R_1 and R_2 converge to form a single arrow, indicating that each receptor converges on the same signaling pathway to elicit the response. The arrows emanating from the antagonists and pointing toward the receptors indicate high (log K_I, 9; solid arrow) and low (log K_I, 7.5; dotted arrow) affinity. Thus, A_1 has selectivity for R_1, A_2 selectivity for R_2, and A_3 equivalent high affinity for R_1 and R_2. In the lower half of the figure the effects of the antagonists (10^{-8} M) on the concentration-response curve of an agonist are illustrated.

mediate signaling pathways that do not elicit a response, but amplify the response mediated by another receptor. An example is the M_2 muscarinic receptor in smooth muscle. It mediates an inhibition of Ca^{2+}-activated K^+ channels (BK_{Ca}).[15,35–37] These channels elicit hyperpolarizing currents whenever the intracellular Ca^{2+} concentration increases near the plasma membrane. Any Ca^{2+} mobilizing receptor that elicits contraction of smooth muscle must overcome this inhibitory effect of BK_{Ca} channels. It follows that activation of the M_2 receptor will prevent the inhibitory effect of BK_{Ca} channels on the contraction elicited by a G_q-linked, Ca^{2+}-mobilizing receptor, like the M_3 muscarinic receptor. By itself, M_2 receptor activation has little effect because little Ca^{2+} is mobilized under resting conditions, and thus, there is little BK_{Ca} current for the M_2 receptor to inhibit.

Thus, in the latter example, the M_2 receptor is defined as conditionally acting because its response is contingent upon activation of the M_3 receptor. The M_3 receptor is defined as directly acting because it can elicit a response when activated by itself. In the following example the antagonism of responses elicited through an interaction between a conditionally and directly acting receptor is simulated. The maximal effect of R_1 receptor activation is to cause a five-fold amplification in the signal elicited by the R_2 receptor. The behavior of this interaction is dependent on the relative sensitivities of the two signaling pathways, and the behavior of interest is the shift in the agonist concentration-response curve caused by the antagonist. This depends on whether the antagonist (A) has selectivity for the conditionally acting (R_1) or directly acting (R_2) receptor (A_1 and A_2, respectively). Thus, in the following simulation (Figure 10.18), the shift in the agonist concentration-response curve caused by A_1 or A_2 is plotted for three theoretical conditions relating to the relative sensitivities of the two signaling pathways. The affinity constants of A_1 and A_2 for R_1 and R_2 are as described in the prior examples.

Figure 10.18a shows how a single concentration of A_1 (10^{-8} M) shifts the concentration-response curve. When the sensitivity of the R_1 signaling pathway

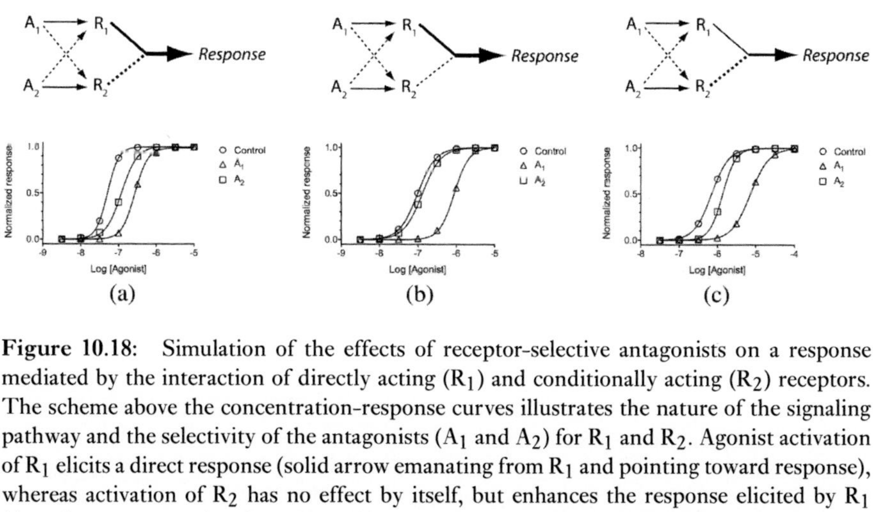

Figure 10.18: Simulation of the effects of receptor-selective antagonists on a response mediated by the interaction of directly acting (R_1) and conditionally acting (R_2) receptors. The scheme above the concentration-response curves illustrates the nature of the signaling pathway and the selectivity of the antagonists (A_1 and A_2) for R_1 and R_2. Agonist activation of R_1 elicits a direct response (solid arrow emanating from R_1 and pointing toward response), whereas activation of R_2 has no effect by itself, but enhances the response elicited by R_1 (dotted arrow emanating from R_2 and intersecting the solid arrow from R_1). The thickness of the solid and dotted arrows emanating from the receptors is proportional to the sensitivity of their signaling pathway. The arrows emanating from A_1 and A_2 and pointing toward the receptors indicate high (solid line, log K_I, 9.0) and low (dotted line, log K_I, 7.5) affinity. Accordingly, A_1 has selectivity for R_1 and A_2 selectivity for R_2. In the lower half of the figure the effects of the antagonists (10^{-8} M) on the concentration-response curve of an agonist are illustrated. The relative sensitivities of the signaling pathways for R_1 and R_2 differ for each panel. (a) The sensitivities are the same. (b) The sensitivity of the signaling pathway of R_1 is greater than that of R_2. (c) The sensitivity of the signaling pathway of R_2 is greater than that of R_1.

is the same as that of R_2 (Figure 18a), the response is more potently antagonized by A_2, but with a K_I estimate (see equation 10.14) a little less than that expected for a response mediated entirely by R_2. When the sensitivity of the R_2 pathway is greater (panel b) or less (panel c) than that of R_1, the response is still more potently antagonized by A_2. The results show that the antagonism this type of receptor interaction has a tendency to follow the profile of the directly acting receptor.

My final example concerns a response elicited by two conditionally acting receptors (R_1 and R_2). That is, activation of either receptor by itself elicits no response, but the simultaneous activation of both receptors elicits the response. The behavior of this interaction is shown by the simulation in Figure 10.19. The affinity constants of the antagonist (A_1 and A_2) for the two receptors (R_1 and R_2, respectively) are as described in the prior examples. When the sensitivities of the two signaling pathways are the same (panel a) the response is potently antagonized by both antagonists (10^{-8} M). When the sensitivity of the R_1 pathway is greater than that of R_2 (panel b), the response is potently antagonized by A_2. In contrast,

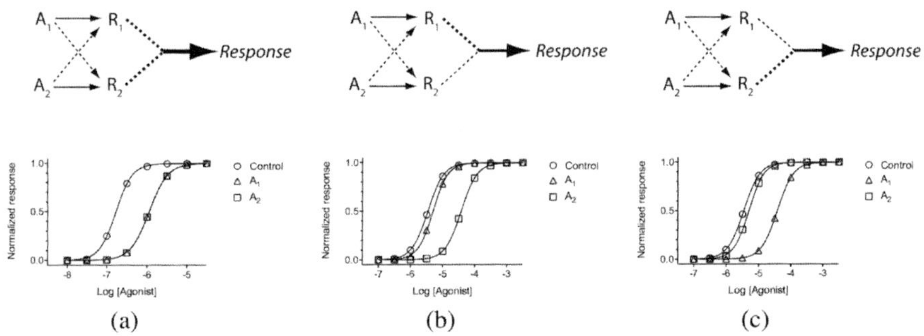

Figure 10.19: Simulation of the effects of receptor-selective antagonists on a response elicited by an agonist at two conditionally acting receptors (R_1 and R_2). The scheme above the concentration-response curves illustrates the nature of the signaling pathway and the selectivity of the antagonists (A_1 and A_2) for R_1 and R_2. Activation of either receptor by itself elicits no effect (dotted arrows emanating from each receptor), but the simultaneous activation of both receptors elicits a response (solid arrow emanating from the convergence of the two dotted lines from R_1 and R_2). The thickness of the solid and dotted arrows emanating from the receptors is proportional to the sensitivity of the signaling pathway. The arrows emanating from A_1 and A_2 and pointing toward the receptors indicate high (solid line, log K_I, 9) and low (dotted line, log K_I, 7.5) affinity. Accordingly, A_1 has selectivity for R_1 and A_2 selectivity for R_2. In the lower half of the figure the effects of the antagonists (10^{-8} M) on the concentration-response curve of an agonist are illustrated. The relative sensitivities of the signaling pathways for R_1 and R_2 differ for each panel. (a) The sensitivity are the same. (b) The sensitivity of the signaling pathway of R_1 is greater than that of R_2. (c) The sensitivity of the signaling pathway of R_2 is greater than that of R_1.

when the sensitivity of the R_2 pathway is greater than that of R_1, the response is potently antagonized by A_1. The results shows that the antagonism resembles the pharmacological profile of the least sensitive signaling pathway.

This type of interaction can explain the antagonism of muscarinic agonist-induced contraction of smooth muscle. Many prominent investigators assume that the M_3 receptor is the most important receptor for eliciting contraction because M_3 antagonists inhibit contraction with high potency. While it may be reasonable to conclude that the M_3 is most important, the rationale does not address the role of the M_2 receptor. An M_3 profile for pharmacological antagonism is entirely consistent with the known conditional signaling mechanisms of the M_2 receptor as shown by the examples in Figures 10.18a–10.18c. The observation that muscarinic agonist-induced contraction is almost completely absent in urinary bladder from the M_3 muscarinic receptor knockout (M_3 KO) mouse is also construed as evidence for the near exclusive role of the M_3 receptor in mediating contraction. But if the contractile effect of the M_2 receptor depends on that of the M_3, then would not it be expected that the action of the former would be lost in tissue lacking the M_3 receptor?

In intestinal smooth muscle, it has been shown that moderate concentrations of acetylcholine elicit heterologous desensitization of contraction.[18,38] That is, prior treatment with acetylcholine (1 μM) for 20 min causes a subsequent reduction in the sensitivity of the ileum to histamine and other agonists like $PGF_{2\alpha}$. This desensitization is prevented by inactivation of either the M_2 or M_3 receptor. Thus, activation of both M_2 and M_3 receptors is required for heterologous desensitization. M_2 selective antagonists specifically inhibit acetylcholine-induced heterologous desensitization, indicating that the M_2 signaling pathway is less sensitive than that of the M_3. This conclusion follows from the principles outlined in Figure 10.19c. If heterologous desensitization is an unavoidable consequence of a powerful contractile stimulus, then the results are consistent with the postulate that the M_2 receptor causes a conditional enhancement of M_3 receptor mediated contraction and that the potency of the M_2 signaling pathway is less than that of the M_3 receptor.

Clinical Uses of Antagonists

This chapter is concerned with the competitive antagonism of the response to an agonist, and the two main goals are to describe how to determine if the action of a drug is competitive and, if so, how to estimate its affinity constant for its binding site on the receptor. We have seen that it is possible to make these determinations in many instances provided that the response can be measured in an isolated cell or tissue format. These types of assays can be used to determine how potent and selective a drug is for one receptor subtype relative to another. An antagonist or partial agonist with a highly selective action could very well be a useful therapeutic drug. Its effects *in vivo* may be highly selective, even though *in vivo* assays are not suitable for estimating the affinity constant of a drug or determining whether its mechanism is competitive.

In the next chapter, the effects of drugs that alter receptor function by binding to a site different from that of the endogenous ligand is described. This mode of interaction is defined as allosteric, and it is a useful mechanism for drug action.

Appendix

This section gives more detailed procedures for the nonlinear regression techniques used in two of the examples covered in the text concerning the data shown in Figures 10.7 and 10.10. Here I describe how to use the Prism 6 application (GraphPad Software, Inc.) to analyze the data, although the general procedures can be applied to any nonlinear regression algorithm. It is assumed that the reader is generally familiar with the Prism environment.

Example 1: Analysis of the effect of various concentrations of an antagonist on the concentration response curve of a full agonist (Figure 10.7a).

The data for this example are given in Table 10.1. Because the data in Figures 10.7a and 10.7b are the same, but simply plotted in different ways, the method described below applies to both figures.

1. The data for the concentration-response curves are entered into a Data Table for an XY plot in Prism. The log molar agonist concentrations in the column farthest to the left in Table 10.1 (row labels) are entered into the Prism Data Table under the column on the left entitled "X". The columns of response measurements in Table 10.1 identified by the column labels $-\infty$, -6, -5.5, -5, -4, -3 and -2 are entered into the columns labeled Groups A–G, respectively, in the Prism Data Table.

   In cases with replicate measurements, these can be added into subcolumns of a given group. Subcolumns can be added to the Data Table using the Format menu.
2. The log antagonist concentrations are entered as column titles for the response measurements of each concentration-response curve. Under Group A, an arbitrarily large negative number (e.g., -20) is entered to indicate the antilog of a very small number of essentially zero. This value corresponds to the log antagonist

concentration for the control concentration-response curve. Under Groups B–G log antagonist concentrations of −6, −5.5, −5, −4, −3 and −2 are entered for the column titles, respectively. These entries correspond to the column titles in Table 10.1.

3. The Graph of the data is selected, followed by the icon, "= Analyze", and ultimately, the nonlinear regression (curve fit) analysis is selected.

   The regression equation used for this analysis is equivalent to Equation (10.16), but some of the parameters will be expressed as logarithms for this analysis. In addition, because it is only possible to estimate the product of τ and K_{obs} (τK_{obs}), but not the individual parameters, we will have to enter the equation with a single variable for the product τK_{obs}.

   The following lines are entered into Prism to define a new equation:

   $$\mathrm{A} = 1 + 10\hat{}(\mathrm{X} + \texttt{LogKobs}) + 10\hat{}(\mathrm{I} + \texttt{LogKI})$$

   $$\mathrm{B} = 10\hat{}(\mathrm{X} + \texttt{LogTauKobs})$$

   $$\mathrm{Y} = \texttt{Msys}/(1 + (\mathrm{A}/\mathrm{B})\hat{}\mathrm{M})$$

   The first two lines define the variables A and B. The last line is the regression equation, which is equivalent to Equation (10.16) after the substitutions for A and B are made. The variable "X" denotes the log concentration of agonist, and I, `LogKobs`, `LogKI`, `LogTauKobs`, `Msys` and M represent log I, log K_{obs}, log K_I, the log of the product, τK_{obs}, M_{sys} and m.

4. Now that the equation has been entered, parameter constraints are applied. The log K_{obs} of the agonist (`LogKobs`) is constrained to an arbitrarily low value of zero and log I (I) is constrained to the "Data set constant (= column title)" option. All of the parameters (`Msys`, `M`, `LogKI` and `LogTauKobs`) are shared across all data sets.

5. The next step involves assigning initial parameter estimates. These are estimated using the following equations in which the parameters with a prime symbol denote the initial estimate of the parameter:

$$\texttt{LogTauKobs}' = -\log EC_{50\text{-control}}, \tag{10.53}$$

$$\texttt{LogKI}' = pA_2, \tag{10.54}$$

$$\texttt{Msys}' = E_{\max\text{-control}}. \tag{10.55}$$

In these equations $EC_{50\text{-control}}$ and $E_{\max\text{-control}}$ denote the EC_{50} and $E_{\max}$ values of the control agonist concentration-response curve and pA_2 denotes a rough estimate of the negative log concentration of antagonist the causes a twofold shift in the agonist concentration-response curve. For the data in Figure 10.10, we will take the negative log of one of the antagonist concentrations used in the simulated experiment that cause a small shift of about two-fold. This gives us a value of 8 for pA_2. The transducer slope value can be assigned an initial estimate of 1:

$$M' = 1, \tag{10.56}$$

or better yet, it can be assigned the value of the Hill slope of the concentration-response curve of the control agonist concentration response curve (H):

$$M' = H. \tag{10.57}$$

6. Now that the data, regression equation, parameter constraints and initial parameter estimates have all been entered, regression analysis is initiated. For this example, the following parameter estimates were obtained: `Msys`, 0.998 ± 0.0076; `M`, 1.005 ± 0.0518; `LogTauKobs`, 7.975 ± 0.0521 and `LogKI`, 5.976 ± 0.0629.

Example 2: Analysis of the effect of various concentrations of an inverse agonist on the concentration-response curve of a full agonist in a signaling pathway that exhibits constitutive activity (Figure 10.10).

The data for this example are illustrated in Figure 10.10. This analysis is similar to that described above in Example 1 except that a different regression equation is used.

1. The data for the concentration-response curves are entered into a Data Table for an XY plot in Prism as described above in Example 1. The log molar agonist concentrations are entered into the left most column entitled "X". The measurements for the concentration-response curves measured in the absence and presence of various concentrations of inverse agonist are entered into the columns labeled Groups A–F.
2. The log antagonist concentrations are entered as column titles for the response measurements of each concentration-response curve as described above in Example 1.
3. The Graph of the data is selected, followed by the icon, "= Analyze", and ultimately, the nonlinear regression (curve fit) analysis is selected.

   The regression equation used for this analysis is equivalent to Equation (10.36). Some of the parameters will be expressed as logarithms for this analysis, and a single variable will be used for the product, τK_{obs}, because it is impossible to estimate the individual parameters (τ and K_{obs}) from the data in Figure 10.10.

The following lines are entered into Prism to define a new equation:

A = 1 + 10^(X + LogKobs) + 10^(I + LogKI)

B = 10^LogTsys + 10^(X + LogTauKobs) + 10^(I + LogTauI + LogKI)

Y = Msys/(1 + (A/B)^M)

The first two lines define the variables A and B. The last line is the regression equation, which is equivalent to Equation (10.36) after the substitutions for A and B are made. The variable "X" denotes the log concentration of agonist, and I, LogKobs, LogKI, LogTauKobs, LogTauI, Msys and M represent log I, log K_{obs}, log K_I, the log of the product, τK_{obs}, log τ_I, M_{sys} and m.

4. Now that the equation has been entered, parameter constraints are applied. The log K_{obs} of the agonist (LogKobs) is constrained to an arbitrarily low value of zero and log I (I) is constrained to the "Data set constant (= column title)" option. All of the parameters are shared across all data sets.
5. The next step involves assigning initial parameter estimates. The initial estimates of LogTauKobs, LogKI and Msys are derived using Equations (10.53)–(10.57). The other parameters are estimated using the following equations in which the parameters with a prime symbol denote the initial estimate of the parameter:

$$\text{LogTsys}' = \log\left(\frac{B}{E_{\text{max-control}} - B}\right), \qquad (10.58)$$

$$\text{LogTauI}' = \log\left(\frac{E_{\text{max-ia}}}{E_{\text{max-control}} - E_{\text{max-ia}}}\right). \qquad (10.59)$$

In these equations $EC_{50\text{-control}}$ and $E_{\text{max-control}}$ denote the EC_{50} and E_{max} values of the control agonist concentration-response curve, B denotes the response elicited by constitutive receptor activity and $E_{\text{max-ia}}$ denotes the maximal response elicited by the inverse agonist. The latter may be difficult, if not impossible, to estimate if the inverse agonist completely inhibits constitutive activity. In such cases, LogTau can be constrained to an arbitrarily large negative integer (e.g., −20).
6. Now that the data, regression equation, parameter constraints and initial parameter estimates have all been entered, regression analysis is initiated. For this example, the following parameter estimates: Msys, 1.01 ± 0.01; M, 0.95 ± 0.09; LogTauKobs, 8 ± 0.07, LogTsys, −0.55 ± 0.08; LogTauI, −1.22 ± 0.20 and LogKI, 8.05 ± 0.08.

References

1. Lazareno, S. and Birdsall, N. J. Pharmacological characterization of acetylcholine-stimulated [^{35}S]-GTP gamma S binding mediated by human muscarinic m1-m4 receptors: antagonist studies. *Br J Pharmacol* **109**, 1120–1127 (1993).
2. Schild, H. O. pAx and competitive drug antagonism. *Br J Pharmacol Chem* **4**, 277–280 (1949).
3. Schild, H. O. Drug antagonism and pAx. *Pharmacol Rev* **9**, 242–246 (1957).
4. Lazareno, S. and Birdsall, N. J. Estimation of competitive antagonist affinity from functional inhibition curves using the Gaddum, Schild and Cheng–Prusoff equations. *Br J Pharmacol* **109**, 1110–1119 (1993).
5. Cheng, Y. and Prusoff, W. H. Relationship between the inhibition constant (K1) and the concentration of inhibitor which causes 50% inhibition (I_{50}) of an enzymatic reaction. *Biochem Pharmacol* **22**, 3099–3108 (1973).
6. Arunlakshana, O. and Schild, H. O. Some quantitative uses of drug antagonists. *Br J Pharmacol* **14**, 48–58 (1959).
7. Paton, W. D. M. A theory of drug action based on the rate of drug-receptor combination. *Proc R Soc Lond, Ser B* **154**, 21–69 (1961).
8. Ehlert, F. J. and Jenden, D. J. The binding of a 2-chloroethylamine derivative of oxotremorine (BM 123) to muscarinic receptors in the rat cerebral cortex. *Mol Pharmacol* **28**, 107–119 (1985).
9. Gill, E. W. and Rang, H. P. An alkylating derivative of benzilylcholine with specific and long-lasting parasympatholytic activity. *Mol Pharmacol* **2**, 284–297 (1966).
10. Kendall, D. A. and Hill, S. J. in *Methods in Neurotransmitter Receptor Analysis* (ed. H. I. Yamamura) (Raven Press, 1990), pp. 68–87.
11. Berridge, M. J., Downes, C. P. and Hanley, M. R. Lithium amplifies agonist-dependent phosphatidylinositol responses in brain and salivary glands. *Biochem J* **206**, 587–595 (1982).
12. Black, J. W., Gerskowitch, V. P., Leff, P. and Shankley, N. P. Analysis of competitive antagonism when this property occurs as part of a pharmacological resultant. *Br J Pharmacol* **89**, 547–555. (1986).
13. Kilbinger, H. and Wagner, P. Inhibition by oxotremorine of acetylcholine resting release from guinea pig-ileum longitudinal muscle strips. *Naunyn Schmiedebergs Archives of Pharmacology* **287**, 47–60 (1975).
14. Kilbinger, H. Modulation by oxotremorine and atropine of acetylcholine release evoked by electrical stimulation of the myenteric plexus of the guinea-pig ileum. *Naunyn Schmiedebergs Archives of Pharmacology* **300**, 145–151 (1977).
15. Ehlert, F. J., Pak, K. J. and Griffin, M. T. Muscarinic agonists and antagonists: effects on gastrointestinal function. *Handbook of Experimental Pharmacology*, 343–374 (2012).
16. Ostrom, R. S. and Ehlert, F. J. M2 muscarinic receptors inhibit forskolin- but not isoproterenol-mediated relaxation in bovine tracheal smooth muscle. *J Pharmacol Exp Ther* **286**, 234–242 (1998).
17. Ostrom, R. S. and Ehlert, F. J. Comparison of functional antagonism between isoproterenol and M2 muscarinic receptors in guinea pig ileum and trachea. *J Pharmacol Exp Ther* **288**, 969–976 (1999).

18. Ehlert, F. J., Griffin, M. T., Abe, D. M., Vo, T. H., Taketo, M. M., Manabe, T. and Matsui, M. The M2 muscarinic receptor mediates contraction through indirect mechanisms in mouse urinary bladder. *J Pharmacol Exp Ther* **313**, 368–378 (2005).
19. Eglen, R. M., Michel, A. D., Montgomery, W. W., Kunysz, E. A., Machado, C. A. and Whiting, R. L. The interaction of parafluorohexahydrosiladiphenidol at muscarinic receptors in vitro. *Br J Pharmacol* **99**, 637–642 (1990).
20. Ehlert, F. J., Hsu, C.-H. J. and Shehnaz, D. Effects of *p*-FHHSiD on muscarinic receptor mediated contraction and phosphoinositide hydrolysis in ileum and trachea. *Proceedings of the XIV World Congress of Pharmacology* (2002).
21. Pedder, E. K., Eveleigh, P., Poyner, D., Hulme, E. C. and Birdsall, N. J. Modulation of the structure-binding relationships of antagonists for muscarinic acetylcholine receptor subtypes. *Br J Pharmacol* **103**, 1561–1567 (1991).
22. Parameswaran, N. and Spielman, W. S. RAMPs: The past, present and future. *Trends Biochem Sci* **31**, 631–638 (2006).
23. Birdsall, N. J. M., Burgen, A. S. V. and Hulme, E. C. in *Cholinergic Mechanisms and Psychopharmacology* (ed. D. J. Jenden) (Plenum Press, 1977), pp. 25–33.
24. Hulme, E. C., Birdsall, N. J., Burgen, A. S. and Mehta, P. The binding of antagonists to brain muscarinic receptors. *Mol Pharmacol* **14**, 737–750 (1978).
25. Yamamura, H. I. and Snyder, S. H. Muscarinic cholinergic receptor binding on the longitudinal muscle of the guinea pig ileum with [^{3}H]quinuclidinyl benzilate. *Mol Pharmacol* **10**, 861–867 (1974).
26. Paton, W. D. M. and Rang, H. P. The uptake of atropine and related drugs by intestinal smooth muscle of the guinea-pig in relation to acetylcholine receptors. *Proc R Soc* **B 163**, 1–44 (1965).
27. Wiedling, S. Pharmacological and toxicological studies on beta-diethylaminoethyl phenothiazine-10-carboxylate hydrochloride, transergan, a new spasmolytic and anti-parkinsonism agent. *Acta Pharmacol Toxicol* (*Copenh*) **13**, 59–79 (1957).
28. Abramson, F. B., Barlow, R. B., Franks, F. M. and Pearson, J. D. Relationships between chemical structure and affinity for postganglionic acetylcholine receptors of the guinea-pig ileum. *Br J Pharmacol* **51**, 81–93 (1974).
29. Wiedling, S. Secergan, a quaternary phenothiazine compound having anticholinergic and ganglion-blocking actions. *Acta Pharmacol Toxicol* (*Copenh*) **14**, 112–129 (1958).
30. Giraldo, E., Monferini, E., Ladinsky, H. and Hammer, R. Muscarinic receptor heterogeneity in guinea pig intestinal smooth muscle: binding studies with AF-DX 116. *Eur J Pharmacol* **141**, 475–477 (1987).
31. Michel, A. D. and Whiting, R. L. Direct binding studies on ileal and cardiac muscarinic receptors. *Br J Pharmacol* **92**, 755–767 (1987).
32. Maeda, A., Kubo, T., Mishina, M. and Numa, S. Tissue distribution of mRNAs encoding muscarinic acetylcholine receptor subtypes. *FEBS Lett* **239**, 339–342 (1988).
33. Buckley, N. J., Bonner, T. I. and Brann, M. R. Localization of a family of muscarinic receptor mRNAs in rat brain. *J Neurosci* **8**, 4646–4652 (1988).
34. Candell, L. M., Yun, S. H., Tran, L. L. and Ehlert, F. J. Differential coupling of subtypes of the muscarinic receptor to adenylate cyclase and phosphoinositide hydrolysis in the longitudinal muscle of the rat ileum. *Mol Pharmacol* **38**, 689–697 (1990).
35. Cole, W. C., Carl, A. and Sanders, K. M. Muscarinic suppression of Ca2+-dependent K current in colonic smooth muscle. *Am J Physiol* **257**, C481–487 (1989).

36. Wade, G. R. and Sims, S. M. Muscarinic stimulation of tracheal smooth muscle cells activates large-conductance Ca(2+)-dependent K+ channel. *Am J Physiol* **265**, C658–665 (1993).
37. Nakamura, T., Kimura, J. and Yamaguchi, O. Muscarinic M2 receptors inhibit Ca2+-activated K+ channels in rat bladder smooth muscle. *Int J Urol* **9**, 689–696 (2002).
38. Shehnaz, D., Ansari, K. Z. and Ehlert, F. J. Acetylcholine-induced desensitization of the contractile response to histamine in guinea pig ileum is prevented by either pertussis toxin-treatment or by selective inactivation of muscarinic M3 receptors. *J Pharmacol Exp Ther* **297**, 1152–1159 (2001).

11

Analysis of Allosterism in Functional Assays on *G* Protein-Coupled Receptors

When an endogenous ligand binds to a *G* protein-coupled receptor, it initiates a signaling cascade that ultimately causes a physiological response. An allosteric drug that binds to a secondary site on the receptor may alter the output of the pathway by modifying the affinity and efficacy components of the orthosteric ligand-receptor complex. In this chapter, I describe how to measure this modulation.

We have already discussed a number of allosteric interactions at *G* protein-coupled receptors including, how an agonist increases the affinity of the receptor for the *G* protein, how it also reduces the affinity of GDP for the *G* protein, and how GDP and GTP have opposite allosteric effects on the binding of the $G\beta\gamma$ subunits to the $G\alpha$ subunit (Chapters 5 and 6). The allosteric interactions described in this chapter follow the same rules, but are examined from the perspective of an exogenous allosteric modulator whose concentration can be manipulated and whose cooperative interactions vary depending on chemical structure.

As discussed throughout this book, drug-receptor interactions can be analyzed at different hierarchical levels. At the deepest level, the affinity of a drug for the active and inactive states of single receptors is considered in terms of microscopic affinity constants. The summation of thousands of single-receptor events gives rise to the average behavior of a receptor population, and at this level of analysis, drug action is characterized in terms of observed affinity and efficacy. Finally, we can analyze drug action at a superficial level relating to any number of responses downstream in the signaling pathway. In Chapter 4, we considered how to analyze the whole-cell current responses of ligand-gated ion channels (population behavior) to estimate the microscopic constants of allosteric modulators. In this chapter, I describe how to infer the observed affinity and efficacy components of the receptor population from the behavior of a downstream response using the null and reverse engineering approaches described in Chapter 7. I also discuss how observed affinity and efficacy are related to the more fundamental microscopic constants that determine single receptor behavior. This chapter focuses on the analysis of allosterism in functional assays; the next chapter includes the analysis allosterism in ligand-binding assays.

I use the term allosteric to refer to the ability of a drug to bind to a specific site on a receptor and alter a distant region of the receptor. This use is consistent with the two specific meanings of allosterism given in Chapter 4: (1) pertaining to a change in the activity of a receptor by the binding of a ligand and (2) pertaining to the modification in the binding of one ligand by the binding of another ligand to the receptor. I will also use the terminology described in Chapter 4 to define allosteric ligands. That is, an allosteric modulator has no effect by itself but modifies the observed affinity and efficacy of orthosteric ligands. Positive modulators increase affinity and efficacy, whereas negative modulators have the opposite effect. Many allosteric modulators exhibit striking probe dependence, and hence, the nature and sign (positive or negative) of the modulation often depends on the interacting orthosteric ligands.[1] An allosteric agonist is defined as an allosteric ligand that elicits a response in the absence of orthosteric ligand.

Some ligands may bind to a site on the receptor without allosterically affecting the binding of the orthosteric ligand, but affect the activity of the receptor. Another type of ligand could bind in the same way and lack allosteric effects on both receptor activity and the binding of the orthosteric ligand, but inhibit receptor activity by competitively inhibiting the receptor-*G* protein interaction. The analysis developed in this chapter is suitable for measuring these effects even though their mechanisms may not be fully allosteric. Hence, I will cover the analysis of these interactions as well.

Our discussion of allosterism is based on the assumption that the interacting ligands bind to distinct sites on the receptor. The definition of distinct sites in this chapter is based on functional criteria as described in the prior chapter on competitive inhibition. That is, two ligands bind to distinct sites if the three-dimensional space occupied by each ligand when bound to the receptor does not intersect with that of the other, thereby excluding a competitive interaction. While this definition may seem unambiguous, it does not take much scrutiny to discover its limitations, some of which I have described in the beginning of the Chapter 10. One issue worth repeating is related to drugs of sufficient size, which interact simultaneously with both allosteric and orthosteric sites. Often, such compounds are large and analogous to two ligands connected by a carbon chain of sorts. Even though these agents may bind to a previously defined allosteric site, they do not fit the functional definition of allosteric. There is good reason to avoid using the adjective allosteric to describe bitopic ligands because these agents interact competitively with the endogenous ligand, and hence, they lack the prototypical beneficial effects of an allosteric mechanism, including protection against overdose and the preservation of the timing of endogenous signaling (see Chapter 4).[1–3] A bitopic ligand may have the advantage of recognizing differences in receptor subtypes based on variation in the less conserved allosteric site, but this would be a good reason to use the modifier, selective orthosteric, instead of allosteric.

The analysis of this chapter is based on behavior under equilibrium conditions. This does not appear to introduce complications because most allosteric modulators of *G* protein-coupled receptors have observed affinity constants that are much less than 10^8 M^{-1}, and hence, their rate constants for association and dissociation are likely to be very fast. The same argument applies to the orthosteric agonists with which they interact. Hence, assuming that drug diffusion to the receptor is rapid, responses to agonists in the presence of allosteric modulator should be easy to measure under equilibrium conditions before desensitization occurs. Even if the latter does occur, the methods of analysis described here are still valid for those situations where the response represents the amount of an accumulated product whose breakdown has been prevented (e.g., second messenger).

In thinking about how receptors work, one is struck by the evolutionary adaptation that allows a small endogenous ligand to bind to a site on the receptor and induce the active state. It seems unlikely that a given drug could act at other sites on the receptor to cause a comparable effect unless the site evolved for that purpose. We also might expect that receptors evolved to be stealth and immune to perturbation by other ligands, although it is possible that some xenobiotics and natural toxins may interfere with the molecular workings of the receptor. Allosteric agonism, however, would seem to require an evolutionary adaptation in most instances.

Consider the allosteric modulation of Cys-loop receptors. These ligand-gated ion channels are composed of five homologous subunits and often have two orthosteric sites, each between the amino terminal domains of two subunits. This stoichiometry implies that there are often three additional binding pockets on a given Cys-loop receptor, which are homologous to the orthosteric sites and may function as an allosteric site for a compound with the appropriate structure. Something like this probably accounts for the benzodiazepine site on the $GABA_A$ receptor, which resides within the γ-α subunit interface (see Chapter 4). The latter may represent a vestigial orthosteric site that still retains its ability to affect the channel pore allosterically if occupied by the appropriate ligand. This would explain how a small allosteric ligand could act at a distance to cause a chloride flux by inducing a conformational change in the pore-lining M2 domain of the channel. The actions of allosteric agonists, like etomidate and anesthetics, could be explained by their disruption of the highly ordered closed structure of the channel pore, which would not seem to require an evolutionary adaptation.

Many *G* protein-coupled receptors are capable of acting as single units, even though they may coexist in oligomeric complexes in receptosomes,[4,5] which presumably increases their efficiency. This has been elegantly shown by reconstitution of receptor-*G* protein complexes in in lipoprotein discs that can only accommodate a single receptor and *G* protein.[6] These complexes exhibit a large guanine nucleotide-induced reduction in agonist affinity. Given that allosteric interactions are reciprocal, the latter implies that single receptor-*G* protein complexes can support an agonist-mediated reduction in the affinity of GDP for the *G* protein, and hence, guanine nucleotide exchange.

An additional point is that allosteric ligands clearly modulate some *G* protein-coupled receptors when only one type of receptor protein is expressed. For example, the ability of allosteric ligands to modify the M_2 muscarinic receptor depends only on expression of the M_2 receptor and not an additional homolog of the receptor that has lost its ability to be activated by acetylcholine, but can be modulated through an allosteric site that is homologous to the orthosteric site on the M_2 receptor. In other words, abundant allosteric interactions occur at the M_2 receptor that depend on only one type of receptor protein, and not a hetero-oligomeric complex of the type characteristic of the allosterically diverse Cys-loop receptors.

So why do *G* protein-coupled receptors have allosteric sites? A quick look at some *G* protein-coupled receptors whose allosteric sites have been reasonably well localized by site-directed mutagenesis suggests that the vestigial-orthosteric-site hypothesis is operating on the single receptor subunit scale. For example, some members of the rhodopsin family have endogenous peptide ligands that bind with the extracellular loops of the receptor.[7] Similarly, although members of the secretin family of *G* protein-coupled receptors have large amino terminal domains that are essential for ligand binding, receptor activation often requires that the ligand or an exposed ligand sequence in the amino terminal domain to interact with the extracellular loops of the receptor.[8] In contrast, the M_2 muscarinic receptor has an orthosteric-binding pocket located within the transmembrane spanning helices,[9] whereas the allosteric site is located more superficially in the extracellular loops,[10,11] like the orthosteric site for some peptide receptors. With regard to mGluRs, the orthosteric site resides in the amino terminal Venus flytrap domains,[12,13] whereas in some cases, the allosteric site is located in the transmembrane spanning helices,[14] like the orthosteric site of the M_2 muscarinic receptor and other members of the amine class within the rhodopsin family. Thus, the locus of the orthosteric site of some *G* protein-coupled receptors is homologous to the allosteric site for other receptors, and the converse also seems to occur.

All of this speculation does not rule out the possibility that there can be found allosteric ligands which interact with loci that are not homologous to orthosteric sites. It simply suggests that positive modulation is more likely to require the vestige of an evolutionary adaptation for the binding site of an orthosteric ligand or a hot spot on the receptor suited for adaption into an orthosteric site in homologous receptors.

Let me give a brief summary of how the remainder of this chapter is organized. I begin by presenting models for allosterism at both the level of the receptor population and receptor states. The population model is used to derive an operational model that can be used to analyze functional data and to determine changes in the affinity and efficacy components of the orthosteric ligand-receptor complex. Examples of how to do this are first given using theoretical data and then using real data from the literature. The results are discussed within the context of probe dependence, biased allosterism, the existence of allosterically regulated sites on the receptor that relay the agonist to the orthosteric activation site, and how allosteric modulation of population parameters relate to the microscopic constants of receptor states.

Those uninterested in the development of the mathematical models for allosteric interactions may want to skip ahead to the section "Simulation and analysis of allosteric interactions."

Models for Allosteric Interactions

First we begin by developing a model that describes the activation state of a receptor population in terms of the population parameters (observed affinity and efficacy). This model is then used to develop an operational model that expresses response as a function of the concentrations of orthosteric agonist and allosteric ligand. Finally, equivalent models are developed that are defined in terms of the microscopic constants of the receptor states, including a model with two active states. These models are used to simulate data and to derive the relationship between the population parameters and the microscopic constants.

General population model

The general population model used to describe the activation state of a receptor population (stimulus) undergoing allosteric interactions is shown in Figure 11.1.

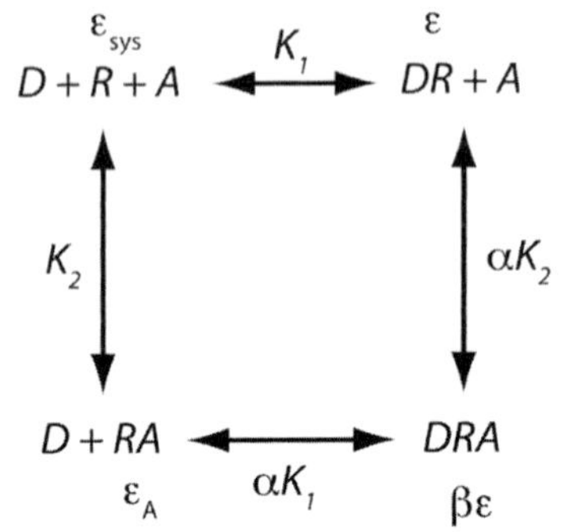

Figure 11.1: Allosteric ternary complex scheme: The scheme shows the equilibrium for the binding of orthosteric (D) and allosteric (A) ligand to a receptor having a distinct binding site for each ligand. The parameters K_1 and K_2 represent the affinity constants (units of M^{-1}) for the binding of D and A to the unliganded receptor. The parameter α denotes the cooperativity constant for the cooperative interaction between the binding of D and A to R. Hence, αK_1 denotes the affinity constant for the binding of D to RA and αK_2, the affinity constant for the binding of A to DR. The efficacy terms (ε_{sys}, ε, ε_A and $\beta\varepsilon$) denote the fraction of the population of the receptor species (R, DR, RA and DRA) in the active state.

This model is known as the allosteric ternary complex model, and it describes the interaction between an orthosteric ligand (D) and an allosteric ligand (A) with a receptor (R) having one orthosteric site and one allosteric site. The observed affinity constants of the orthosteric (K_1) and allosteric (K_2) ligands are defined in the usual way:

$$K_1 = \frac{[DR]}{[D][R]}, \tag{11.1}$$

$$K_2 = \frac{[RA]}{[A][R]}. \tag{11.2}$$

These represent the observed affinity constants for the binding of each ligand in the absence of the other, and they have units of inverse molarity (M^{-1}). The reciprocal of each constant is equivalent to the concentration of ligand required for half-maximal receptor occupancy in the absence of the other ligand. If the binding of A allosterically affects the binding of D, then D will affect the binding of A to the same extent. The proof for this relationship is described in Chapter 4 within the context of ligand-gated ion channels. The amount of reciprocal modulation in binding affinity is known as the cooperativity, and it is denoted by the Greek symbol α in the model.

Thus, when the receptor is already occupied by one of the interacting ligands, the observed affinity constant of the other ligand is given by one of the following:

$$\alpha K_1 = \frac{[DRA]}{[D][RA]}, \tag{11.3}$$

$$\alpha K_2 = \frac{[DRA]}{[A][DR]}. \tag{11.4}$$

In these equations, αK_1 and αK_2 denote the observed affinity constants of the orthosteric and allosteric ligands, respectively, when the receptor is already occupied by the other ligand.

The value of α determines the magnitude and sign of the allosterism. For example, a cooperativity value of 10 (i.e., $\alpha = 10$) denotes positive cooperativity and that the interacting ligands each cause a reciprocal tenfold increase in the affinity of the other ligand. Conversely, an α value of 0.1 indicates negative cooperativity equivalent to a reduction in affinity to one-tenth the original value. The domain of α (possible range of α values) is any number greater than zero, although practically speaking, there is a physical limit to how much a small ligand can allosterically affect the binding of another ligand. A cooperativity value of zero indicates that the binding is mutually exclusive (i.e., only one ligand can bind to the receptor at a time). Under this condition, there is no allosterism, and the resulting model describes competitive inhibition.

Even though there are only two states of the receptor (active and inactive), each receptor complex in the model can exhibit a different level of activity. This is because each complex in the model represents a population of complexes, and the activity of the receptor population depends on the average activity of the individual receptors, which are either active or inactive. When a ligand binds to the receptor population, it can increase or decrease the probability of single receptor activation, and therefore, cause a graded change in the activation state of the population of occupied receptors. Efficacy is the parameter used to describe the fraction of a population of ligand-receptor complexes, which is in the active state, and therefore, we will use the symbols ε, ε_A and $\beta\varepsilon$ to denote the efficacies of the *DR*, *RA* and *DRA* complexes, respectively. One way to think about β is that it is the scalar value by which the allosteric ligand affects the efficacy of the orthosteric ligand. Usually, the free receptor complex has little or no activity, but the signaling pathway of a given *G* protein-coupled receptor may be highly amplified enabling the constitutive activity of the receptor to be detected. Thus, we need to account for the activity of the free receptor complex, and this is denoted as the efficacy of the system (ε_{sys}).

Now that we have a model describing the activation state of a receptor undergoing allosteric interactions, equations can be derived that express receptor activation as a function of the concentrations of interacting ligands. The fraction of the receptor population in the active state is referred to as the stimulus, but because our model includes constitutive activity, we will use the term total stimulus (T) for this purpose. It is defined as:

$$T = \frac{\varepsilon_{\rm sys}R + \varepsilon DR + \beta\varepsilon DRA + \varepsilon_A ARA}{R + DR + DRA + RA}. \qquad (11.5)$$

This equation communicates the idea that the summation of the fractions of each population of receptor complexes in the active state is equal to the fraction of the total receptor population in the active state. This equation is solved in the usual manner by replacing each receptor complex with an expression derived from the equations for the definitions of the observed affinity constants (Equations (11.1)–(11.4)). For example, rearrangement of Equation (1) yields, $[DR] = K_1[D][R]$. After making these substitutions and simplifying the equation, the following result can be obtained.

$$T = \frac{\varepsilon_{\rm sys} + \varepsilon_A AK_2 + \varepsilon DK_1\,(1 + \alpha\beta AK_2)}{1 + AK_2 + DK_1\,(1 + A\alpha K_2)}. \qquad (11.6)$$

When the receptor exhibits no measureable constitutive activity ($\varepsilon_{\rm sys} \approx 0$), Equation (11.6) can be simplified as:

$$S = \frac{\varepsilon_A AK_2 + \varepsilon DK_1\,(1 + A\alpha\beta K_2)}{1 + AK_2 + DK_1\,(1 + A\alpha K_2)}. \qquad (11.7)$$

In this equation, S is used to denote the stimulus instead of T because there is no constitutive activity. When the allosteric ligand lacks efficacy ($\varepsilon_A \approx 0$), but nonetheless modulates the activity of the orthosteric ligand, the following equation can be used to describe receptor activation:

$$S = \frac{\varepsilon DK_1\,(1 + A\alpha\beta K_2)}{1 + AK_2 + DK_1\,(1 + A\alpha K_2)}. \qquad (11.8)$$

This equation can be rearranged in a way that gives us a better appreciation of how the modulator affects the observed affinity and efficacy of the orthosteric ligand:

$$S = \frac{q\varepsilon DK_1}{1 + DpK_1}, \qquad (11.9)$$

in which,

$$p = \frac{1 + A\alpha K_2}{1 + AK_2}, \quad (11.10)$$

$$q = \frac{1 + A\alpha\beta K_2}{1 + A\alpha K_2}. \quad (11.11)$$

Equation (11.9) is consistent with the simple one-site model introduced in Chapter 7 to describe the stimulus function, except that the affinity and efficacy terms have been multiplied by the expressions, p and q. Thus, the modulator has the affect of modifying the observed affinity and efficacy of the agonist. Equations (11.10) and (11.11) show that, as the concentration of allosteric ligand increases, p and q approach a plateau equivalent to α and β, respectively. Thus, α and β represent the maximal effect of the allosteric ligand on the observed affinity and efficacy of the orthosteric ligand.

In many instances, it is only possible to estimate the product of α and β, and not their individual values. Hence, for the purposes of nonlinear regression analysis, it is useful to express Equation (11.11) in the following form:

$$q = \frac{1 + A\gamma K_2}{1 + A\alpha K_2}. \quad (11.12)$$

in which,

$$\gamma = \alpha\beta. \quad (11.13)$$

Now let us use these equations to develop an operational model for analyzing the effect of an allosteric ligand on a downstream response.

Operational model for allosterism

If it were possible to measure the activation state of a GPCR easily, we could analyze the corresponding agonist concentration-response curve with one of the appropriate equations (Equations (11.6)–(11.8)), and estimate the population parameters for allosteric regulation. Measuring the activation of a GPCR is difficult, however, and a much simpler approach is to deduce the population parameters from the behavior of a downstream response. Because the downstream response is amplified and the sensitivity of the agonist for eliciting the response is often high relative to its ability to activate the receptor population, we need a transducer function that accurately represents the conversion of the stimulus into the final response. We will use the reverse engineering approach of Black and Leff[15] described in Chapter 7. These investigators showed that if the input function to a signaling pathway is consistent with a simple one-site model of the type shown by Equation (11.9) and the output consistent with a logistic function of the sort commonly used to analyze agonist concentration-response curves, then the transducer function can be described by Equation (7.5) in Chapter 7. This equation is given below for convenience:

$$\text{response} = \frac{T^m M_{\text{sys}}}{T_m + K_E^m} \tag{11.14}$$

In this equation, M_{sys} represents the maximum possible response that the system can elicit, K_E denotes the sensitivity constant of the signaling pathway, and m denotes the transducer slope factor.

Simply stated, Equation (11.14) predicts how the stimulus function is converted into the measured downstream response. Thus, if we substitute 11.6 for the total stimulus (T) in Equation (11.14) we will have an equation that describes the response elicited by an agonist in the presence of an allosteric ligand at a receptor having constitutive activity. Further simplification yields:

$$\text{response} = \frac{M_{\text{sys}}}{1 + \left(\frac{1+AK_2+DK_1(1+A\alpha K_2)}{\tau_{\text{sys}}+\tau_A AK_2+\tau DK_1(1+A\gamma K_2)}\right)^m}, \tag{11.15}$$

in which,

$$\tau_{\text{sys}} = \frac{\varepsilon_{\text{sys}}}{K_E}, \tag{11.16}$$

$$\tau_A = \frac{\varepsilon_A}{K_E}, \tag{11.17}$$

$$\tau = \frac{\varepsilon}{K_E}. \tag{11.18}$$

Equation (11.15) can be modified to account for the condition of no measureable constitutive activity[16]:

$$\text{response} = \frac{M_{\text{sys}}}{1 + \left(\frac{1+AK_2+DK_1(1+A\alpha K_2)}{\tau_A AK_2+\tau DK_1(1+A\gamma K_2)}\right)^m}, \tag{11.19}$$

or the case where there is no constitutive activity and the allosteric modulator lacks efficacy:

$$\text{response} = \frac{M_{\text{sys}}}{1 + \left(\frac{1+AK_2+DK_1(1+A\alpha K_2)}{\tau DK_1(1+A\gamma K_2)}\right)^m}. \tag{11.20}$$

This latter equation can also be arranged in the following for using the parameters p and q defined above:

$$\text{response} = \frac{M_{\text{sys}}}{1 + \left(\frac{D + \frac{1}{pK_1}}{q\tau D} \right)^m}. \tag{11.21}$$

We will use Equations (11.15) and (11.19)–(11.21) in the analysis of allosterism in functional assays. In the next section, the relationship between the observed constants of the receptor population and the microscopic constants that define the receptor states is described.

Simple two-state model

Derivation of the simple two-state model: Figure 11.2 illustrates the allosteric ternary complex model at the level of receptor states. This model is known as the simple two-state model, and it was derived from the population model (Figure 11.1) by expanding each receptor complex into its active (R_s^*) and inactive (R_s) components. At this level of analysis, the affinity of a ligand for a receptor state is determined by the state of the receptor and not by whether another ligand is bound to it. Thus,

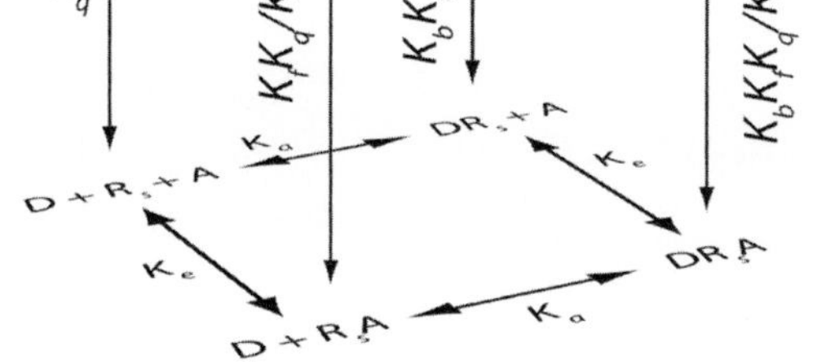

Figure 11.2: Two-state scheme for allosteric interactions. The model is equivalent to the Monod Wyman and Changeaux model,[21] but with the addition of conformational induction steps. The microscopic affinity constants of the orthosteric ligand for the active (R_s^*) and inactive (R_s) states of the receptor are denoted as K_b and K_a, respectively. The corresponding constants for the allosteric ligand are K_f and K_e. The isomerization constant for the unoccupied receptor is denoted as K_q. The other isomerization steps are assigned with expressions consistent with the law of microscopic reversibility as described in Chapters 2 and 4.

the microscopic affinity constants of the orthosteric ligand for the active and inactive receptor states are given by:

$$K_a = \frac{[DR_s]}{[D][R_s]} = \frac{[DR_sA]}{[D][R_sA]}, \tag{11.22}$$

$$K_b = \frac{[DR_s^*]}{[D][R_s^*]} = \frac{[DR_s^*A]}{[D][R_s^*A]}. \tag{11.23}$$

The corresponding constants for the allosteric ligand are given by:

$$K_e = \frac{[R_sA]}{[A][R_s]} = \frac{[DR_sA]}{[A][DR_s]}, \tag{11.24}$$

$$K_f = \frac{[R_s^*A]}{[A][R_s^*]} = \frac{[DR_s^*A]}{[A][DR_s^*]}. \tag{11.25}$$

Finally, K_q describes the isomerization constant of the free receptor:

$$K_q = \frac{[R_s^*]}{[R_s]}. \tag{11.26}$$

The other isomerization constants are given on the vertical edges of the model and were assigned in a manner consistent with the law of microscopic reversibility as described in Chapter 4.

To derive an equation for fractional activation of the receptor population (total stimulus, T), we begin with the following relationship:

$$T = \frac{[R_{\text{active}}]}{[R_T]} = \frac{[R_{\text{active}}]}{[R_{\text{active}}] + [R_{\text{inactive}}]}, \tag{11.27}$$

in which,

$$[R_{\text{active}}] = [R_s^*] + [DR_s^*] + [DR_s^*A] + [R_s^*A], \tag{11.28}$$

$$[R_{\text{inactive}}] = [R_s] + [DR_s] + [DR_sA] + [R_sA]. \tag{11.29}$$

In Equation (11.27), $[R_T]$ denotes the total concentration of receptors. Each receptor complex can be replaced with an equation derived from the definitions of the microscopic constants (Equations (11.22)–(11.26)). Following simplification, the following result is obtained:

$$T = \frac{1}{1 + \frac{(1+DK_a)(1+AK_e)}{K_q(1+DK_b)(1+AK_f)}}. \tag{11.30}$$

To express the downstream response as a function of the microscopic constants, Equation (11.30) is substituted into the transducer function of the operational model (Equation (11.14)) to obtain the following equation after simplification:

$$\text{response} = \frac{M_{\text{sys}}}{1 + \left(K_E\left(1 + \frac{(1+DK_a)(1+AK_e)}{K_q(1+DK_b)(1+AK_f)}\right)\right)^m}. \tag{11.31}$$

Relation between population parameters of the simple two-state model and the microscopic constants: Using the strategy outlined in Chapter 4, it is possible to express the cooperativity constant α and the observed affinity constants of the interacting ligands (K_1 and K_2) in terms of the microscopic constants:

$$K_1 = \frac{K_a + K_b K_q}{1 + K_q}, \tag{11.32}$$

$$K_2 = \frac{K_e + K_f K_q}{1 + K_q}, \tag{11.33}$$

$$\alpha K_1 = \frac{K_a K_e + K_b K_f K_q}{K_e + K_f K_q}. \tag{11.34}$$

By expressing the ratio of $\alpha K_1/K_1$ in terms of microscopic constants, the resulting equation for α is obtained following simplification:

$$\alpha = \frac{\left(K_a K_e + K_b K_f K_q\right)\left(1 + K_q\right)}{\left(K_a + K_b K_q\right)\left(K_e + K_f K_q\right)}. \tag{11.35}$$

Before deriving an equation for β, let me first give the overall strategy. First, equations for the efficacy of the orthosteric agonist in the presence ($\beta\varepsilon$) and absence (ε) of the allosteric modulator are derived in terms of the microscopic constants. Then, the ratio of the equation for $\beta\varepsilon$ is divided by that for ε to obtain an equation for β.

The efficacy of the *DR* complex (ε) is defined as the fraction of population of *DR* complexes in the active state at any given level of receptor occupancy, including 100%. Thus, a simple way to derive an equation for ε, is to take limit of the receptor activation function (Equation (11.30)) as D approaches infinity (i.e., 100% receptor occupancy) when $A = 0$:

$$\varepsilon = \lim_{D\to\infty} \frac{1}{1 + \frac{(1+DK_a)}{K_q(1+DK_b)}} = \frac{1}{1 + \frac{K_a}{K_q K_b}}. \tag{11.36}$$

To estimate $\beta\varepsilon$, the limit of the same function is evaluated when both A and D approach infinity:

$$\beta\varepsilon = \lim_{D\to\infty, A\to\infty} \frac{1}{1+\frac{(1+DK_a)(1+AK_e)}{K_q(1+DK_b)(1+AK_f)}} = \frac{1}{1+\frac{K_aK_e}{K_qK_bK_f}}. \tag{11.37}$$

The parameter β is equivalent to Equation (11.37) divided by Equation (11.36):

$$\beta = \frac{K_bK_fK_q + K_aK_f}{K_bK_fK_q + K_aK_e}. \tag{11.38}$$

The equation for efficacy of the allosteric ligand ε_A can be derived using a procedure analogous to that used for the derivation of ε. The limit of the receptor activation function (Equation (11.30)) is taken as A approaches infinity when $D = 0$:

$$\varepsilon_A = \lim_{A\to\infty} \frac{1}{1+\frac{(1+AK_e)}{K_q(1+AK_f)}} = \frac{1}{1+\frac{K_e}{K_qK_f}}. \tag{11.39}$$

Finally, the equation for the parameter γ is derived by taking the product of the equations for α (Equation (11.36)) and β (Equation (11.35)):

$$\gamma = \frac{K_f + K_fK_q}{K_e + K_fK_q}. \tag{11.40}$$

In the case of allosteric modulators, which lack sufficient selectivity to activate the receptor, their ratio of K_f/K_e is much less than the reciprocal of the isomerization constant of the free receptor ($1/K_q$). For this condition, Equation (11.40) predicts that:

$$\gamma \approx \frac{K_f}{K_e}. \tag{11.41}$$

Complex two-state model

Derivation of the complex two-state model: The model shown in Figure 11.1 can also be used to explore the behavior of allosterism when there are multiple active and inactive states of the receptor. In this section, we will consider the case of one inactive state and two active states (R_s^* and R_s^{**}). This model is known as the complex two-state model, and it is derived from the simple population model in Figure 11.1 by expanding each receptor complex into one inactive (R_s) and two active (R_s^* and R_s^{**}) states as shown in Figure 11.3. The microscopic affinity constants of the inactive

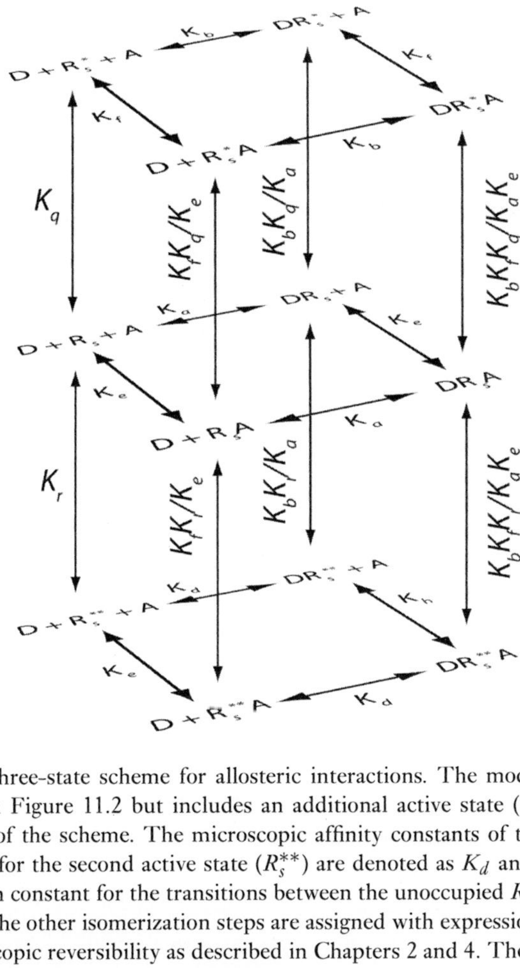

Figure 11.3: Three-state scheme for allosteric interactions. The model is analogous to that illustrated in Figure 11.2 but includes an additional active state (R_s^{**}) on the lower horizontal plane of the scheme. The microscopic affinity constants of the orthosteric and allosteric ligands for the second active state (R_s^{**}) are denoted as K_d and K_d, respectively. The isomerization constant for the transitions between the unoccupied R_s and R_s^{**} states is denoted by K_r. The other isomerization steps are assigned with expressions consistent with the law of microscopic reversibility as described in Chapters 2 and 4. The definitions of K_a, K_b and K_q are given in the legend to Figure 11.2.

(R_s) and first active state (R_s^*) are described by Equations (11.22)–(11.26). Those for the second active state are given by:

$$K_d = \frac{[DR_s^{**}]}{[D][R_s^{**}]} = \frac{[DR_s^{**}A]}{[D][R_s^{**}A]}, \tag{11.42}$$

$$K_h = \frac{[DR_s^{**}]}{[D][R_s^{**}]} = \frac{[DR_s^{**}A]}{[D][R_s^{**}A]}. \tag{11.43}$$

The isomerization constant for the transition of the free receptor into the second active state is denoted by K_r:

$$K_r = \frac{[R_s^{**}]}{[R_s]}. \tag{11.44}$$

The other isomerization steps of the complex two-state model are shown on the vertical edges of the scheme in Figure 11.3, and these are derived by applying the law of microscopic reversibility as described in Chapter 4.

The complex two-state model shown in Figure 11.3 does not show the one-step transitions between the two active states of the various ligand-receptor complexes (i.e., $R_s^* \leftrightarrow R_s^{**}$, $DR_s^* \leftrightarrow DR_s^{**}$ and $DR_s^*A \leftrightarrow DR_s^{**}A$). There is no reason to eliminate these possibilities from an accurate model of population kinetics or single receptor behavior. But with regard to equilibrium behavior, the law of microscopic reversibility shows that two different routes to the same destination are equivalent at equilibrium. For example, the equilibrium constant for the transition $DR_s^* \leftrightarrow DR_s^{**}$ is equivalent to that (K_dK_r/K_bK_q) for the alternate route to the same destination $(DR_s^* \leftrightarrow DR_s \leftrightarrow DR_s^{**})$. Thus, the model shown in Figure 11.3 is sufficient to describe the amounts of all of the ligand-receptor complexes at equilibrium, even though all possible transitions are not shown in the model.

The complex two-state model can be used to explore two main conditions, 1) both active states couple to the same G protein and elicit the same response and 2) each active state couples with a different G protein and elicits a different response. For the condition of two active states coupling to the same G protein, it seems likely that there could be a difference in how well each active state activates the G protein. We will use a simple scalar denoted by the Greek letter δ to express the activity of R_s^{**} relative to R_s^*. These two versions of the model are described below in the sections, "*Two active states mediate the same response*" and "*Allosteric modulation of pathway selectivity*." Before addressing these two conditions, I describe how the observed binding affinity is related to the microscopic constants of the complex two-state model.

Relation between population parameters of the complex two-state model and the microscopic constants: Using the strategy outlined in Chapter 4, it is possible to express the cooperativity constant α and the observed affinity constants of the interacting ligands (K_1 and K_2) in terms of the microscopic constants of the complex two-state model:

$$K_1 = \frac{K_a + K_bK_q + K_dK_r}{1 + K_q + K_r}, \tag{11.45}$$

$$K_2 = \frac{K_e + K_fK_q + K_hK_r}{1 + K_q + K_r}, \tag{11.46}$$

$$\alpha K_1 = \frac{K_a K_e + K_b K_f K_q + K_d K_h K_r}{K_e + K_f K_q + K_h K_r}, \tag{11.47}$$

$$\alpha K_2 = \frac{K_a K_e + K_b K_f K_q + K_d K_h K_r}{K_a + K_b K_q + K_d K_r}. \tag{11.48}$$

By expressing the ratio of $\alpha K_1/K_1$ in terms of microscopic constants and simplifying the resulting equation for α is obtained:

$$\alpha = \frac{(K_a K_e + K_b K_f K_q + K_d K_h K_r)(1 + K_q + K_r)}{(K_e + K_f K_q + K_h K_r)(K_a + K_b K_q + K_d K_r)}. \tag{11.49}$$

The equation describing the effect of the allosteric modulator on the observed efficacy of the orthosteric ligand (β) depends on whether the two active states of the receptor couple with the same or different G proteins. Hence, these derivations are given in the appropriate sections.

Two active states mediate the same response: In this section, we will use the complex two-state model to derive an equation for receptor activation assuming that both active states of the receptor contribute to the response. We will also use this function to derive (1) an equation for the allosteric modulation in observed efficacy (β) and (2) an operational model for the analysis of the allosteric regulation of a response generated by two active states of a receptor.

To derive an equation for fractional activation of the receptor population (total stimulus, T), we begin with:

$$T = \frac{[R^*_{\text{active}}] + \delta[R^{**}_{\text{active}}]}{[R_T]} = \frac{[R^*_{\text{active}}] + \delta[R^{**}_{\text{active}}]}{[R^*_{\text{active}}] + [R^{**}_{\text{active}}] + [R_{\text{inactive}}]}, \tag{11.50}$$

in which,

$$[R^*_{\text{active}}] = [R^*_s] + [DR^*_s] + [DR^*_s A] + [R^*_s A], \tag{11.51}$$

$$[R^{**}_{\text{active}}] = [R^{**}_s] + [DR^{**}_s] + [DR^{**}_s A] + [R^{**}_s A], \tag{11.52}$$

$$[R_{\text{inactive}}] = [R_s] + [DR_s] + [DR_s A] + [R_s A]. \tag{11.53}$$

Note that because the level of activity of the two states may differ, all complexes involving R^{**}_s in the numerator have been multiplied by the scalar δ to account for this potential difference. If there is a difference in the activities of the two actives states, the less active state is assigned to R^{**}_s so that $\delta < 1.0$, and the fractional activation of the receptor population does not exceed 1.0. Each receptor complex can be replaced with an equation derived from the definitions of the microscopic constants (Equations (11.22)–(11.26) and (11.32)–(11.44)). Following simplification, the following result is obtained:

$$T=\frac{K_q(1+DK_b)(1+AK_f)+\delta K_r(1+DK_d)(1+AK_h)}{K_q(1+DK_b)(1+AK_f)+K_r(1+DK_d)(1+AK_h)+(1+DK_a)(1+AK_e)}. \tag{11.54}$$

Equation (11.54) can be used to derive an equation for the observed efficacy of the orthosteric ligand in the absence (ε) and presence ($\beta\varepsilon$) of the allosteric modulator. Taking the limit of Equation (11.54) as D approaches infinity when $A=0$ yields the equation for ε:

$$\begin{aligned}\varepsilon &= \lim_{D\to\infty}\frac{K_q(1+DK_b)+\delta K_r(1+DK_d)}{K_q(1+DK_b)+K_r(1+DK_d)+(1+DK_a)}\\ &= \frac{K_bK_q+\delta K_dK_r}{K_a+K_bK_q+K_dK_r}.\end{aligned} \tag{11.55}$$

To estimate $\beta\varepsilon$, the limit of the same function is evaluated when both A and D approach infinity:

$$\begin{aligned}\beta\varepsilon &= \lim_{D\to\infty,A\to\infty}\frac{K_q(1+DK_b)(1+AK_f)+\delta K_r(1+DK_d)(1+AK_h)}{K_q(1+DK_b)(1+AK_f)+K_r(1+DK_d)(1+AK_h)+(1+DK_a)(1+AK_e)}\\ &= \frac{(K_a+K_bK_q+K_dK_r)(K_bK_fK_q+\delta K_dK_hK_r)}{(K_aK_e+K_bK_fK_q+K_dK_hK_r)(K_bK_q+\delta K_dK_r)}.\end{aligned} \tag{11.56}$$

The parameter β is equivalent to Equation (11.56) divided by Equation (11.55):

$$\beta=\frac{(K_bK_fK_q+\delta K_dK_hK_r)(K_a+K_bK_q+K_dK_r)}{(K_aK_e+K_bK_fK_q+K_dK_hK_r)(K_bK_q+\delta K_dK_r)}. \tag{11.57}$$

The equation for the parameter γ, which represents the product of α and β, is derived by multiplying the equation for α (Equation (11.49)) by that for β (Equation (11.57)):

$$\gamma=\frac{(1+K_q+K_r)(K_bK_fK_q+\delta K_dK_hK_r)}{(K_e+K_fK_q+K_hK_r)(K_bK_q+\delta K_dK_r)}. \tag{11.58}$$

The equation for the efficacy of the allosteric ligand (ε_A) is derived by taking the limit of the receptor activation function as A approaches infinity when $D = 0$:

$$\varepsilon_A = \lim_{A\to\infty} \frac{K_q\left(1 + AK_f\right) + \delta K_r\left(1 + AK_h\right)}{K_q\left(1 + AK_f\right) + K_r\left(1 + AK_h\right) + \left(1 + AK_e\right)}$$
$$= \frac{K_f K_q + \delta K_h K_r}{K_e + K_f K_q + K_h K_r}. \quad (11.59)$$

To express the downstream response as a function of the microscopic constants, Equation (11.50) is substituted into the transducer function of the operational model (Equation (11.14)) to obtain the following equation after simplification:

$$\text{response} = \frac{M_{\text{sys}}}{1 + \left(K_E\left(\frac{K_q(1+DK_b)(1+AK_f)+\delta K_r(1+DK_d)(1+AK_h)}{K_q\left(1+DK_b\right)\left(1+AK_f\right) + K_r\left(1+DK_d\right) \times \left(1+AK_h\right) + \left(1+DK_a\right)\left(1+AK_e\right)}\right)\right)^m}. \quad (11.60)$$

Allosteric modulation of pathway selectivity: In this section, it is assumed that the two active states of the complex two-state model elicit different responses. Therefore, a receptor activation function will be derived for each active state. These functions will be used to derive (1) equations for the allosteric modulation in observed efficacy (β) for each active state and (2) operational models for the analysis of the allosteric regulation of the responses generated by the two active states.

To derive an equation for the fraction of the receptor population in the R_s^* state (total stimulus, T^*), we begin with Equation (11.27), but with the active ($[R_{\text{active}}]$) and inactive ($[R_{\text{inactive}}]$) components defined by:

$$[R_{\text{active}}] = [R_s^*] + [DR_s^*] + [DR_s^*A] + [R_s^*A], \quad (11.61)$$
$$[R_{\text{inactive}}] = [R_s] + [DR_s] + [DR_sA] + [R_sA] + [R_s^{**}] + [DR_s^{**}]$$
$$+ [DR_s^{**}A] + [R_s^{**}A]. \quad (11.62)$$

Each receptor complex is replaced with an equation derived from the definitions of the microscopic constants, and the following result is obtained after simplification:

$$T^* = \frac{1}{1 + \frac{(1+DK_a)(1+AK_e)+K_r(1+DK_d)(1+AK_h)}{K_q(1+DK_b)(1+AK_f)}}. \quad (11.63)$$

Equation (11.63) can be used to derive an equation for the observed efficacy of the orthosteric ligand with regard to the R_s^* state (ε^*). Taking the limit of Equation (11.63) as D approaches infinity when $A = 0$ yields the equation for ε^*:

$$\varepsilon^* = \lim_{D\to\infty} \frac{1}{1 + \frac{(1+DK_a)+K_r(1+DK_d)}{K_q(1+DK_b)}} = \frac{1}{1 + \frac{K_a+K_dK_r}{K_bK_q}}. \quad (11.64)$$

To estimate $\beta\varepsilon$ for the R_s^* state, the limit of the same function is evaluated when both A and D approach infinity:

$$\beta\varepsilon^* = \lim_{D\to\infty, A\to\infty} \frac{1}{1 + \frac{(1+DK_a)(1+AK_e)+K_r(1+DK_d)(1+AK_h)}{K_q(1+DK_b)(1+AK_f)}}$$

$$= \frac{1}{1 + \frac{K_aK_e+K_dK_hK_r}{K_bK_fK_q}}. \quad (11.65)$$

The parameter β is equivalent to Equation (11.65) divided by Equation (11.64):

$$\beta^* = \frac{K_f\left(K_a + K_bK_q + K_dK_r\right)}{\left(K_aK_e + K_bK_fK_q + K_dK_hK_r\right)}. \quad (11.66)$$

The parameter γ^*, which represents the product of α and β^*, is derived by multiplying the equation for α (Equation (11.49)) by that for β^* (Equation (11.66)):

$$\gamma^* = \frac{K_f\left(1 + K_q + K_r\right)}{\left(K_e + K_fK_q + K_hK_r\right)}. \quad (11.67)$$

In the case of allosteric modulators, which lack sufficient selectivity to induce the R_s^* state, their ratio of K_f/K_e is much less than the reciprocal of the isomerization constant of the free receptor ($1/K_q$). For this condition, Equation (11.67) predicts the following when there is little or no constitutive activity (i.e., both K_q and $K_r \ll 1$):

$$\gamma^* \approx \frac{K_f}{K_e}. \quad (11.68)$$

The equation for the efficacy of the allosteric ligand (ε_A^*) is derived by taking the limit of the receptor activation function (Equation (11.63)) as A approaches infinity when $D = 0$:

$$\varepsilon_A^* = \lim_{D\to\infty} \frac{1}{1 + \frac{(1+AK_e)+K_r(1+AK_h)}{K_q(1+AK_f)}} = \frac{1}{1 + \frac{K_e+K_hK_r}{K_fK_q}}. \quad (11.69)$$

To express the downstream response for active state R_s^* as a function of the microscopic constants, Equation (11.63) is substituted into the transducer function of the operational model (Equation (11.14)) to obtain the following equation after simplification:

$$\text{response}^* = \frac{M_{\text{sys}}}{1 + \left(K_E\left(1 + \frac{(1+DK_a)(1+AK_e)+K_r(1+DK_d)(1+AK_h)}{K_q(1+DK_b)(1+AK_f)}\right)\right)^m}. \tag{11.70}$$

In this equation, response* denotes the response elicited by the R_s^* state.

Now that we have completed the equations for the response elicited through the R_s^* state, we need to derive the corresponding equations for the R_s^{**} state. This is accomplished by repeating the steps associated with Equations (11.61)–(11.70), but with the receptor complexes containing the R_s^{**} state being active and those of the R_s^* state inactive.

Thus, to derive an equation for the fraction of the receptor population in the R_s^{**} state (total stimulus, T^{**}), we begin with Equation (11.27), but with the active ($[R_{\text{active}}]$) and inactive ($[R_{\text{inactive}}]$) components defined by:

$$[R_{\text{active}}] = [R_s^{**}] + [DR_s^{**}] + [DR_s^{**}A] + [R_s^{**}A], \tag{11.71}$$

$$[R_{\text{inactive}}] = [R_s] + [DR_s] + [DR_sA] + [R_sA] + [R_s^*] + [DR_s^*]$$
$$+[DR_s^*A] + [R_s^*A]. \tag{11.72}$$

Each receptor complex is replaced with an equation derived from the definitions of the microscopic constants, and the following result is obtained after simplification:

$$T^{**} = \frac{1}{1 + \frac{(1+DK_a)(1+AK_e)+K_q(1+DK_b)(1+AK_f)}{K_r(1+DK_d)(1+AK_h)}}. \tag{11.73}$$

Equation (11.73) can be used to derive an equation for the observed efficacy of the orthosteric ligand with regard to the R_s^{**} state (ε^{**}). Taking the limit of Equation (11.73) as D approaches infinity when $A = 0$ yields the equation for ε^{**}:

$$\varepsilon^{**} = \lim_{D\to\infty} \frac{1}{1 + \frac{(1+DK_a)+K_q(1+DK_b)}{K_r(1+DK_d)}} = \frac{1}{1 + \frac{K_a+K_bK_q}{K_dK_r}}. \tag{11.74}$$

To estimate $\beta\varepsilon^{**}$ for the R_s^{**} state, the limit of the same function is evaluated when both A and D approach infinity:

$$\beta\varepsilon^{**} = \lim_{D\to\infty, A\to\infty} \frac{1}{1+\frac{(1+DK_a)(1+AK_e)+K_q(1+DK_b)(1+AK_f)}{K_r(1+DK_d)(1+AK_h)}}$$

$$= \frac{1}{1+\frac{K_aK_e+K_bK_fK_q}{K_dK_hK_r}}. \qquad (11.75)$$

The parameter β is equivalent to Equation (11.75) divided by Equation (11.74):

$$\beta^{**} = \frac{K_h\left(K_a + K_bK_q + K_dK_r\right)}{\left(K_aK_e + K_bK_fK_q + K_dK_hK_r\right)}. \qquad (11.76)$$

The parameter γ^{**}, which represents the product of α and β^{**}, is derived by multiplying the equation for α (Equation (11.49)) by that for β^{**} (Equation (11.76)):

$$\gamma^{**} = \frac{K_h\left(1 + K_q + K_r\right)}{\left(K_e + K_fK_q + K_hK_r\right)}. \qquad (11.77)$$

In the case of allosteric modulators which lack sufficient selectivity to induce the R_s^{**} state, their ratio of K_h/K_e is much less than the reciprocal of the isomerization constant of the free receptor ($1/K_r$). For this condition, Equation (11.77) predicts the following when there is little or no constitutive activity (i.e., both K_q and $K_r \ll 1$):

$$\gamma^{**} \approx \frac{K_h}{K_e}. \qquad (11.78)$$

The equation for the efficacy of the allosteric ligand (ε_A^{**}) is derived by taking the limit of the receptor activation function (Equation (11.73)) as A approaches infinity when $D = 0$:

$$\varepsilon_A^{**} = \lim_{D\to\infty} \frac{1}{1+\frac{(1+AK_e)+K_q(1+AK_f)}{K_r(1+AK_h)}} = \frac{1}{1+\frac{K_e+K_fK_r}{K_hK_r}}. \qquad (11.79)$$

To express the downstream response for active state R_s^{**} as a function of the microscopic constants, Equation (11.73) is substituted into the transducer function of the operational model (Equation (11.14)) to obtain the following equation after simplification:

$$\text{response}^{**} = \frac{M_{\text{sys}}}{1 + \left(K_E\left(1 + \frac{(1+DK_a)(1+AK_e)+K_q(1+DK_b)(1+AK_f)}{K_r(1+DK_d)(1+AK_h)}\right)\right)^{m}}, \quad (11.80)$$

in which response** denotes the response elicited by the R_s^{**} state.

This concludes the derivations of the equations describing the various states of the allosteric ternary complex model. In the next two sections, these are used to analyze both simulated and real data from the literature.

Simulation and Analysis of Allosteric Interactions

One way to test a method for analyzing data is to apply the technique to theoretical data and see if it yields good estimates of the parameters that were used to simulate the data in the first place. This will be our strategy in this section.

Let me give a brief description of the overall approach. First, theoretical data are generated using Equations (11.31), (11.60), (11.70) and (11.80). These define the response in terms of the microscopic constants of the interacting ligands and the parameters of the signaling pathway. A random error is added so that ultimately it will be possible to determine if reliable parameters estimates can be estimated from real data exhibiting experimental error. Then I describe how to use global nonlinear regression analysis with the operational model to estimate some of the observed parameters of the receptor population and its signaling pathway (K_1, K_2, τ, τ_A, τ_{sys}, α, γ and M_{sys}). Finally, the relationship between the microscopic constants and the population parameters is explained.

It might seem more straightforward to estimate the microscopic constants directly by fitting to the data to an equation defined in terms of microscopic constants. With real data, however, it is often unclear what the underlying model is at the level of receptor states. Therefore, I describe the more general population analysis, which is consistent on a deeper level with a wide range of state models of varying degrees of complexity.

Perhaps the most rigorous simulation to validate the analysis is to generate the theoretical data using the quaternary complex model defined at the level of receptor states as described in Chapter 6. This can be done with a modified form of the model that includes an allosteric site on the receptor. Whether the data are simulated using the latter approach or with the equations mentioned above, the method of analysis described in this chapter yields the same parameter estimates for the allosteric modulator and orthosteric ligand. The less complicated approach is used here simply because it is easier to describe.

Practically any response elicited by an agonist can be used as an assay to investigate the effect of potential allosteric ligands. A useful format is to measure complete concentration-response curves for the agonist in the absence and presence of increasing concentrations of the allosteric ligand. A hallmark feature of allosterism is that the effect of the allosteric ligand reaches a limit at high concentrations, corresponding to the product (γ) of its modulatory effects on the affinity (α) and efficacy (β) of the agonist. Therefore, it is necessary to test the allosteric ligand over a range of concentrations, which capture the ceiling effect. Otherwise, it will be impossible to estimate an accurate value for γ. In many instances, negative allosteric modulators cause a parallel shift to the right in the agonist concentration-response curve in a manner that resembles competitive inhibition. The relationship between the shift in the curve and the modulator concentration differs from competition, however, in the concentration range where the ceiling effect occurs. Thus, to discriminate competitive inhibition from negative allosteric modulation, it is necessary to measure the full range of allosteric effects.

Changes in the sensitivity of the signaling pathway affect the behavior of allosterism but not the estimates of K_2 and γ

The simulation in Figure 11.4 shows how the sensitivity of the signaling pathway affects the allosteric modulation of the responses to a highly efficacious (panel *a* and *c*) and a less efficacious (panel *b* and *d*) agonist. The data were simulated using Equation (11.31) with the K_E value set to 0.01 and 0.1 for the more sensitive (panels *a* and *b*) and less sensitive (panels *c* and *d*) signaling pathways, respectively. This form of the operational model is based on receptor activation being consistent with a simple two-state model. The more efficacious agonist exhibited lower potency in the less sensitive signaling pathway, whereas the less efficacious agonist exhibited both lower potency and a reduced maximal response. Increasing concentrations of the positive allosteric modulator ($K_f/K_e = 10$) increased the potency of both agonists, and also, the maximal response of the less efficacious agonist in the less sensitive signaling pathway. The various microscopic constants and system parameters used to simulate the data are given in the figure legend.

To estimate the population parameters, all of the data in a given panel are analyzed simultaneously by global nonlinear regression analysis. Let me begin by giving a detailed description of how to do this using the Prism application and the

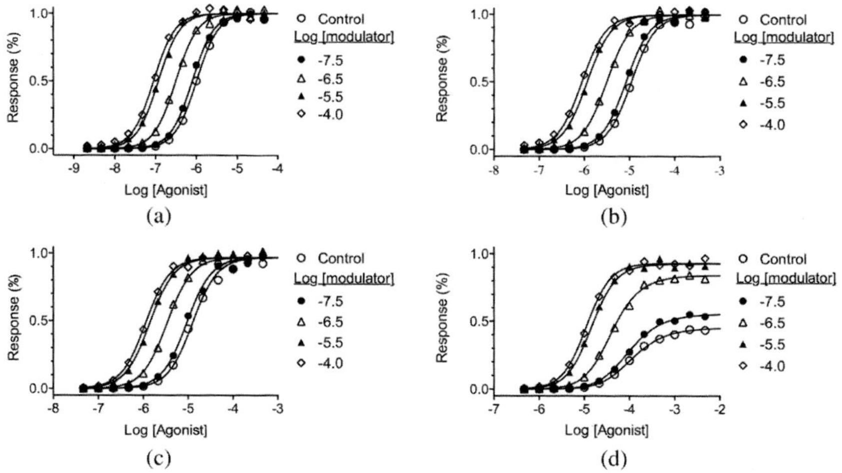

Figure 11.4: Simulation and analysis of the influence of a positive allosteric modulator on the response to highly efficacious (a, c) and less efficacious (b, d) agonists in signaling pathways exhibiting high (a, b) and low (c, d) sensitivity. The data were simulated using Equation (11.31) for the conditions of no modulator (control) and the indicated log molar concentrations of modulator. The microscopic constants used in the simulation were $\log K_b$, 8.0; $\log K_a$, 4.0; $\log K_e$, 6.0, $\log K_f$, 7.0; $\log K_q$, -4.0, $\log K_E$, -2.0 (panels *a* and *b*) and -1.0 (panels *c* and *d*) and *m*, 1.8.

data in panel *a* as an example. Then I will give a quick overview of how to analyze the subsequent data.

The data in panel *a* are entered into a Data Table in Prism. The log molar concentrations of agonist are entered into the first column on the left entitled "X", and the measured responses of each concentration-response curve are entered into a separate lettered column. If there are replicate measurements of a given curve, these data are entered into subcolumns of a given column. Each column of data is entitled with the log molar concentration for the modulator used in the experiment. For the control condition (no modulator), a large negative number (e.g., -20) is entered because the log of zero is negative infinity. Collectively the data represent a matrix of response values. The position of each measurement in the matrix is determined by the agonist (row) and modulator (column) concentrations.

Next, a user-defined equation is entered into Prism. This is a modified form of Equation (11.20) with all of the parameters except M_{sys} and m expressed as logarithms:

$$\text{response} = \frac{M_{\text{sys}}}{1 + \left(\dfrac{1+10^{\log A+\log K_2}+10^{X+\log K1}\left(1+10^{\log A+\log\alpha+\log K_2}\right)}{10^{X+\log\tau+\log K1}\left(1+10^{\log A+\log\gamma+\log K_2}\right)}\right)^{m}}. \tag{11.81}$$

The independent variables are the log agonist (X) and log modulator ($\log A$) concentrations.

The next step entails calculating the initial parameter estimates. Because all of the curves have the same $E_{\max}$ value, it is impossible to estimate the observed affinity constant (K_1) and τ value of the agonist unless additional data are collected. Nonetheless, it should be possible to estimate the product of τK_1 and most of the other parameters by constraining K_1 to an arbitrarily low value for the regression ($\log K_1 = 0$). The rationale for this approach is described in Chapter 7.

The initial estimate of $\log \tau$ ($\log \tau^*$) is calculated as:

$$\log \tau^* = pEC_{50\text{-control}} - \log K_1^*. \tag{11.82}$$

In this equation, $\mathrm{pEC}_{50\text{-control}}$ denotes the negative logarithm of the control concentration-response curve and $\log K_1^*$ denotes the initial estimate of $\log K_1$ (constrained value of 0 in this case).

Assuming that the agonist is a full agonist, the initial estimate of the maximal response of the system (M_{sys}^*) is assigned the $E_{\max}$ value of the agonist:

$$M_{\text{sys}}^* = E_{\max}. \tag{11.83}$$

The initial estimate of the transducer slope factor (m^*) is assigned a value of 1:

$$m^* = 1. \tag{11.84}$$

The initial parameter estimates of the allosteric ligand can be accurately calculated from the EC_{50} and $E_{\max}$ values of the agonist concentration-response curves. All of the $E_{\max}$ values are approximately the same and are given normalized values of 1.0. The log EC_{50} values of the agonist in the absence and presence of the modulator at log concentrations of −7.5, −6.5, −5.5 and −4 are −6.00, −6.12, −6.48, −6.94 and −7.04, respectively. These are used to estimate the product of the change in the observed affinity and efficacy of the agonist caused by the allosteric modulator. This change is known as the relative activity (RA) value. When the Hill slope of the agonist concentration-response curve is equal to 1.0, or if all of the curves have the same $E_{\max}$ value, the RA value can be accurately calculated from Equation (11.85). We will designate this value as the observed RA (RA_{obs}) value because the Hill slopes in panel a are greater than one:

$$RA_{\text{obs}} = \frac{E_{\max} EC_{50\text{-control}}}{E_{\max\text{-control}} EC_{50}}. \tag{11.85}$$

In this equation, EC_{50} and $E_{\max}$ denote the parameters measured in the presence of the modulator and $EC_{50\text{-control}}$ and $E_{\max\text{-control}}$ denote the parameters of the control curve. Using this equation, log RA_{obs} values of 0.13, 0.48, 0.94 and 1.04 can be calculated for data simulated at log modulator concentrations of −7.5, −6.5, −5.5, and −4, respectively.

Now these RA_{obs} values are useful in calculating initial parameter estimates for the modulator. The maximum value of log RA_{obs} ($RA_{obs\text{-}max}$) is a good estimate of $\log\gamma$. Hence, the initial estimate for this parameter ($\log\gamma^*$) is calculated as:

$$\log\gamma^* = \log RA_{obs\text{-}max}. \tag{11.86}$$

This equation yields an estimate of $\log\gamma$ of 1.04 ($\gamma \approx 10$). For the condition of positive cooperativity, the observed affinity constant of the modulator (K_2) is approximately equal to the reciprocal of the modulator concentration generating an RA_{obs} value of 2 ($\log RA_{obs} = 0.32$) divided by γ. Hence, using the variable A_2 to designate this concentration of modulator, the equation for the initial estimate of $\log K_2$ for the condition of positive cooperativity can be described as:

$$\log K_2^* = -\log A_2 - \log RA_{obs\text{-}max}. \tag{11.87}$$

From the values given above, I estimate a log value of about −7.0 for concentration of modulator causing a two-fold increase in EC_{50}. Taking the negative of this value and subtracting $\log RA_{obs\text{-}max}$ from it yields an estimate of 6.0 for $\log K_2$. For the condition of negative cooperativity, the initial estimate of $\log K_2$ is estimated as:

$$\log K_2^* = -\log A_2. \tag{11.88}$$

Having assigned the initial parameter estimates, it is necessary to apply the appropriate parameter constraints. As described above, $\log K_1$ is constrained to 0, and the second independent variable, $\log A$, is assigned values equal to the appropriate column headings. Without having sufficient data to estimate K_1, it is impossible to estimate α. Consequently, the parameter, $\log\alpha$, is constrained to 0. All of the other parameters are shared among the curves.

Now that preparation for global nonlinear regression analysis is complete, the regression can be initiated. When this is done, the following parameter values are obtained $\log K_2$, 5.93 ± 0.04; $\log\gamma$, 1.06 ± 0.02; M_{sys}, 1.00 ± 0.005 and m, 1.64 ± 0.05. The regression analysis also provides an estimate of $\log\tau$, however, this estimate is inaccurate because $\log K_1$ was constrained during the analysis. Nonetheless, the product of the constrained K_1 value and the inaccurate estimate of τ *is* accurate (i.e., $\log\tau K_1$, 5.99 ± 0.01).

The theoretical microscopic constants that were used to simulate the data are given in the legend to Figure 11.4. From these microscopic constants, it is possible to estimate the values of the population parameters using equations described previously in this chapter: $\log \tau K_1$, 5.00; $\log K_2$, 6.00 and $\log \gamma$, 1.00. The parameters of the system were: M_{sys}, 1.0 and m, 1.8. Thus, there was very good agreement between the estimated parameters and those used to simulate the data. Note that the estimate of $\log \gamma$ (1.06) is a close estimate of the log ratio of the microscopic affinity constants of the modulator for the active ($\log K_f = 7.0$) and inactive ($\log K_e = 6.0$) states of the receptor (see Equations (11.36) and (11.37)).

The simulated data in panels *b* and *c* are analyzed in the same way. When this was done for the data in panel *b*, the following parameter estimate were obtained: $\log \tau K_1$, 4.96 ± 0.015; $\log K_2$, 5.95 ± 0.04; $\log \gamma$, 1.10 ± 0.02; M_{sys}, 1.00 ± 0.005 and m, 1.65 ± 0.05. These agree nicely with the system parameters and the population parameters calculated from the microscopic constants used to simulate the data: $\log \tau K_1$, 5.00; $\log K_2$, 6.00 and $\log \gamma$, 1.00; M_{sys}, 1.0 and m, 1.8. For the simulated data in panel *c*, the following parameters were estimated: $\log \tau K_1$, 4.90 ± 0.02; $\log K_2$, 6.04 ± 0.05; $\log \gamma$, 1.05 ± 0.02; M_{sys}, 0.97 ± 0.005 and m, 1.60 ± 0.05. Again, these are consistent with the theoretical value of $\log \tau K_1$ (5.00) and the theoretical parameters of the system and allosteric modulator. The system (M_{sys} and m) and modulator parameters (K_2 and γ) are the same for all of the simulations in Figure 11.4.

The simulated data with the less efficacious agonist (panel *d*) show that the control E_{max} value is substantially less than that obtained with the more efficacious agonist in panel *c* and that high concentrations of the modulator increase the E_{max} to a plateau level similar to the maximum response of the system (M_{sys}). Recall from Chapter 7 that it is possible to estimate the K_{obs} value of a partial agonist provided that there is sufficient information to estimate M_{sys}. Thus, we should be able to estimate $\log K_1$. One approach is to analyze the data in panels *c* and *d* simultaneously, so that it would be possible to obtain a more reliable estimate of M_{sys}. Regardless, I will use the approach applied for the other data in Figure 11.4, except that the parameter $\log K_1$ is shared during the regression. Its initial estimate is assigned a value equivalent to the $-\log EC_{50}$ value of the control concentration-response curve. The initial estimate of $\log \tau$ is calculated as the log of the E_{max} of the control curve divided by M_{sys}. With these constraints and initial parameter values, the following estimates are obtained: $\log K_1$, 4.03 ± 0.04; $\log \tau$, -0.025 ± 0.007; $\log K_2$, 6.00 ± 0.02; $\log \gamma$, 0.99 ± 0.03; M_{sys}, 0.95 ± 0.006, and m, 1.82 ± 0.07. The estimates of $\log K_1$ and $\log \tau$ are similar to the predicted theoretical values of 4.04 and -0.041, respectively. The other parameters for the allosteric modulator and system are similar to the theoretical values given above.

It is still difficult to obtain a reliable estimate of $\log \alpha$ for these data because errors in the estimates of α and β are highly correlated. That is, a good fit to the data can be obtained with an erroneous low value of α and an erroneously high value of β. For this reason, the parameters $\log \alpha$ is constrained to 0 for the regression. If there is sufficient data to estimate M_{sys} with great precision, then it would be possible to obtain a reasonable estimate of α for the data in panel *d*.

In summary, the analysis shows that it is possible to estimate reliable values of K_2 and γ for the allosteric modulator regardless of changes in the sensitivity of the system and the efficacy of the agonist. It is usually, if not impossible in most instances, to separate γ into its α and β components unless additional data is collected. If the underlying mechanism is consistent with the simple two-state model, then the estimate of γ is constant and independent of the orthosteric ligand. Even though the modulator affects of concentration-response curves of the agonists differently in the less sensitive system (panels *c* and *d*), the estimates of K_2 and γ are the same.

Estimating the cooperativity constant α using the method of partial receptor inactivation

In the previous section, it was shown that a series of agonist concentration-response curves measured in the presence of various concentrations of allosteric modulator represents insufficient information to estimate the cooperativity constant of the allosteric ligand (α) and the observed affinity constant of the orthosteric agonist (K_{obs}). Part of the problem rests on being unable to estimate the observed affinity constant of the orthosteric ligand. In this section, it is shown that inactivating a fraction of the receptor population enables the estimation of both K_{obs} and α.

Figure 11.5 shows a simulation of how partial receptor inactivation affects the concentration-response curve of the agonist and the modulatory effect of the allosteric ligand. In panel *a*, the concentration-response curve of an agonist is simulated in the absence and presence of various concentrations of a negative allosteric modulator. In addition, the condition of reducing the receptor population to only 1% of its original level is simulated for the concentration-response curve with no modulator (Inactivated). In panel *b*, a control concentration-response curve was simulated in addition to a set of curves illustrating the effect of the allosteric modulator under the condition of receptor inactivation. In this case, the residual receptor population was 2%.

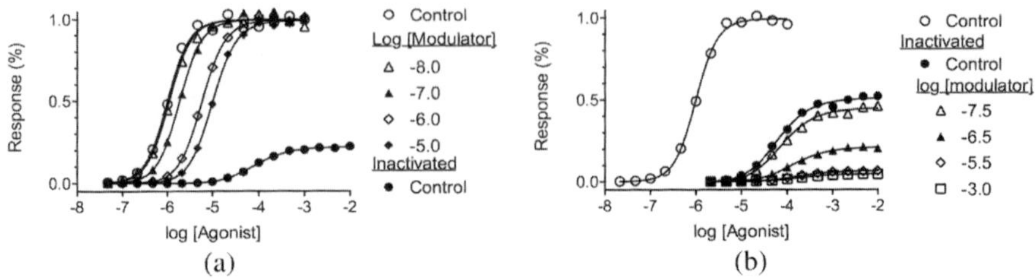

Figure 11.5: Effects of partial receptor inactivation on negative allosteric modulation of the response to an agonist. (a) The concentration-response curve to an agonist is simulated for the conditions of no modulator (control) and the indicated concentrations of negative allosteric modulator. The control curve is also simulated for the condition of partial receptor inactivation with an irreversible antagonist ($q = 0.01$). (b) Concentration-response curves to an agonist are simulated for conditions of partial receptor inactivation ($q = 0.02$) and no modulator (control) or the indicated concentrations of negative allosteric modulator. The control curve is also simulated without receptor inactivation. The microscopic constants used in the simulation were $\log K_b$, 8.0; $\log K_a$, 4.0; $\log K_e$, 7.0, $\log K_f$, 6.0; $\log K_q$, -4.0, $\log K_E$, -2.0 and m, 1.8. The data were simulated using a variant of Equation (11.31) derived by multiplying Equation (11.30) by q and substituting the result into the transducer function (Equation (11.14)).

I begin by describing how the data in panel *a* are analyzed within the context of the Prism application. The data for each concentration-response curve are entered into a separate column on a data sheet. Each column is entitled with the log concentration of modulator as described in the previous section. For the sake of discussion, let us assume that the measurements for the condition of no modulator and the four concentrations of modulator are entered into columns A–E, respectively, whereas the measurements for the condition of receptor inactivation with no modulator are entered into column F. The regression equation used for this analysis is the same as Equation (11.81), but with an additional parameter representing the fraction of residual active receptors. Because this parameter is estimated for the data in column F only and constrained to 1.0 for the other data, we need to introduce a dummy variable (P) and assign it a value of 1.0 for the data in columns A–E and allow it to be an estimated parameter (q) for the data in column F. This can be accomplished by entering three lines into Prism for the user-defined equation:

$$\langle F \rangle P = q \tag{11.89}$$

$$\langle {\sim} F \rangle P = 1.0 \tag{11.90}$$

$$\text{response} = \frac{M_{\text{sys}}}{1+\left(\frac{1+10^{\log A+\log K_2}+10^{X+\log K_1}\left(1+10^{\log A+\log \alpha+\log K_2}\right)}{P\left(10^{X+\log \tau+\log K_1}\right)\left(1+10^{\log A+\log \gamma+\log K_2}\right)}\right)^{m}}. \tag{11.91}$$

The independent variables are the log agonist (X) and log modulator ($\log A$) concentrations. Equation (11.91) is essentially the same as Equation (11.81) but with the denominator of the fraction within the large set of parentheses multiplied by the dummy variable P. The two lines before Equation (11.91) instruct Prism to estimate the value of P for the data in column F only and assign the estimate to parameter q. For the data in the other columns, P is treated as part of the equation and is assigned a value of one. Regression analysis is done as described above, but with all of the parameters ($\log K_1$, $\log \tau$, $\log K_2$, $\log \gamma$, $\log \alpha$, M_{sys}, m and q) shared among all the data. The dummy variable P does not appear as an estimated parameter, only q.

When this is done, the following parameter estimates are obtained: $\log K_1$, 4.32 ± 0.10; $\log \tau$, 1.69 ± 0.10; $\log K_2$, 7.02 ± 0.03; $\log \alpha$, -0.52 ± 0.25; $\log \gamma$, -1.02 ± 0.03; M_{sys}, 1.00 ± 0.004, m, 1.82 ± 0.05 and q, 0.010 ± 0.002. With the exception of $\log \alpha$, these estimates accurately reflect the theoretical population parameters that can be calculated from the microscopic constants used to simulate the data: $\log K_1$, 4.301; $\log \tau$, 1.70; $\log K_2$, 7.00; $\log \alpha$, -0.26; $\log \gamma$, -1.00; M_{sys}, 1.00; m, 1.82 and q, 0.01. The estimate of $\log \alpha$ is inaccurate, and its 95% confidence interval spans about one log unit.

A better estimate of α can be obtained from the data in panel b. Let us assume that the measurements for the control concentration-response curve are entered into a data sheet in Prism in column A, whereas the condition of receptor inactivation with various concentrations of modulator, including no modulator (Control), are entered into columns B–F. These data are also globally fitted to Equation (11.91), but with the dummy variable P controlled by the following two lines, which precede the equation when entered as a user-defined equation in Prism:

$$\langle A \rangle P = 1 \tag{11.92}$$

$$\langle \sim A \rangle P = q. \tag{11.93}$$

This sequence instructs Prism to estimate the value of the dummy variable P for the data in columns B–F and assign the estimate to parameter q. For the data in column A, P is treated as part of the equation and is assigned a value of 1.0. Again, global nonlinear regression analysis is done with all of the parameters shared.

When this is done, the following parameter estimates are obtained: $\log K_1$, 4.32 ± 0.02; $\log \tau$, 1.70 ± 0.02; $\log K_2$, 7.00 ± 0.05; $\log \alpha$, -0.26 ± 0.05; $\log \gamma$, -0.95 ± 0.05; M_{sys}, 0.99 ± 0.004, m, 1.94 ± 0.05 and q, 0.021 ± 0.001. All of these estimates accurately reflect the theoretical population parameters of the simulated data as described in connection with panel a. For this simulation, the residual fraction of active receptors was 0.02, however. Knowing γ and α, the efficacy modulation parameter β can be calculated by rearrangement of Equation (11.13) (i.e., $\log \beta = \log \gamma - \log \alpha$). The calculated value of $\log \beta$ (−.69) agrees reasonably with the theoretical value (−0.76) used for the simulation.

In summary, the simulations show that the receptor-inactivation paradigm outlined in Figure 11.5b enables the estimation of all of the population parameters of the interacting ligands. Reducing the population of active receptors is usually accomplished by treating the preparation with an irreversible antagonist. If the response is measured in cultured cells, the same effect can be achieved by expressing the receptor at a lower level.

Analyzing the effect of an allosteric agonist

So far we have only considered allosteric modulators — compounds that have no demonstrable effect by themselves. In this section, I describe how to analyze the modulatory effect of an allosteric agonist.

Figure 11.6 shows the effect of an allosteric agonist on the concentration-response curve of a highly efficacious agonist. The data were simulated using the two-state model (Equation (11.31)) and the microscopic constants given in the figure legend. These data are analyzed using Equation (11.19) described above, but with each parameter, except M_{sys} and m, expressed in log form:

$$\text{response} = \frac{M_{sys}}{1 + \left(\frac{1+10^{\log A+\log K_2}+10^{\log D+\log K_1}\left(1+10^{\log A+\log \alpha+\log K_2}\right)}{10^{\log \tau_A+\log A+\log K_2}+10^{\log \tau+X+\log K_1}\left(1+10^{\log A+\log \gamma+\log K_2}\right)} \right)^m}. \tag{11.94}$$

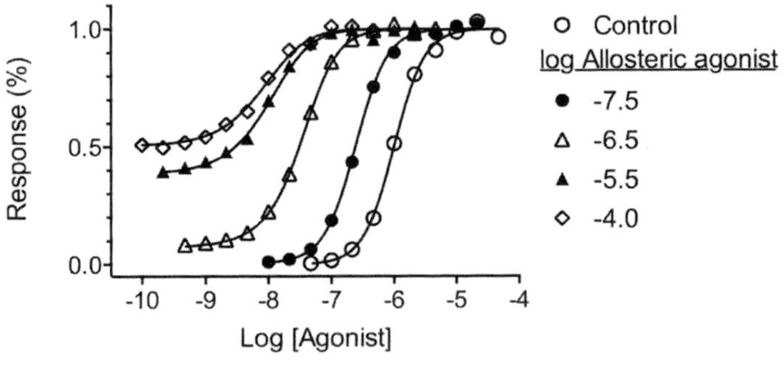

Figure 11.6: Effects of an allosteric agonist on the concentration-response curve to an agonist. The data were simulated using Equation (11.31) for the conditions of no modulator (control) and the indicated log molar concentrations of modulator. The microscopic constants used in the simulation were $\log K_b$, 8.0; $\log K_a$, 4.0; $\log K_e$, 6.0, $\log K_f$, 8.0; $\log K_q$, −4.0, $\log K_E$, −2.0 and m, 1.8.

The independent variables are the log agonist (X) and modulator ($\log A$) concentrations. The orthosteric agonist appears to be a full agonist, because the allosteric agonist does not cause an increase in its E_{max}. It is impossible, therefore, to estimate $\log K_1$ and $\log \alpha$ from these data alone, and these parameters can be constrained during the regression although this is not essential. $\text{Log}\, K_1$ is constrained to an arbitrarily low value (e.g., 0), and it is usually a safe bet to constrain $\log \alpha$ to 0. The initial estimate of parameter $\log \tau$ is calculated using Equation (11.82).

The initial estimates of $\log \gamma$ and $\log K_2$ are estimated using Equations (11.86) and (11.87), but there may be some confusion with regard to estimating the requisite RA_{obs} values for this calculation because the concentration-response curves exhibit a response in the absence of the orthosteric agonist. First, the curves are analyzed by a four-parameter logistic equation of the following form to estimate EC_{50} and E_{max}:

$$\text{response} = \text{basal} + \frac{E_{max} - \text{basal}}{1 + 10^{n(\log EC_{50} - \log D)}}. \tag{11.95}$$

In this equation, basal denotes the response measured in the absence of the orthosteric agonist (D), and n denotes the Hill slope. The EC_{50} value represents the concentration of agonist that elicits a response midway between the basal response and the E_{max}. It follows that the response at EC_{50} is equivalent to the average of the basal and E_{max} values. Nonlinear regression analysis of the data with Equation (11.95) yields estimates of the log EC_{50} values (−6.0, −6.6, −7.5, −8.0 and −8.1) of the agonist measured in the absence and presence of the modulator at log concentrations of −7.5, −6.5, −5.5 and −4.0, respectively. Alternatively, reasonable values can be estimated by interpolating a value from a plot of the data. The E_{max} values are nearly the same for all of the curves and are equivalent to a normalized value of 1.0. Using Equation (11.85) and the estimates of E_{max} and EC_{50}, the following log RA_{obs} values can be estimated for the modulator at log molar concentrations of −7.5, −6.5, −5.5 and −4.0, respectively: 0.63, 1.47, 2.0 and 2.11. The maximum RA_{obs} value ($RA_{obs\text{-}max}$) represents a good estimate of $\log \gamma$ (2.11). I estimate that the log concentration of modulator causing a two-fold increase in RA_{obs} is approximately −8.0. Using the equation for calculating the initial estimate of $\log K_2$ for the condition of positive cooperativity (i.e., Equation (11.87)), a value of 6.0 can be calculated for initial estimate of $\log K_2$.

The initial estimates of the other parameters are calculated as described above. Equation (11.94) is fitted to the data by global nonlinear regression analysis, sharing the estimates of $\log \tau$, $\log \tau_A$, $\log K_2$, $\log \gamma$, M_{sys} and m among the data. Regression analysis yields the following parameter estimates: $\log \tau K_1$, 6.00 ± 0.01; $\log K_2$, 5.97 ± 0.03; $\log \tau_A$, 0.008 ± 0.007; $\log \gamma$, 2.02 ± 0.04; M_{sys}, 1.00 ± 0.004;

m, 1.78 ± 0.06 and q, 0.021 ± 0.001. These values are nearly the same as the theoretical population parameters calculated from the microscopic constants used to simulate the data: $\log K_1$, 6.00; $\log K_2$, 6.004; $\log \tau_A$, −0.004; $\log \gamma$, 1.996; M_{sys}, 1.00, and m, 1.8. Again, the regression will yield an estimate of τ for the orthosteric ligand, but the estimate is inaccurate. This is because errors in the estimation of $\log K_1$ and $\log \tau$ are inversely correlated and constraining $\log K_1$ to an erroneously low value will cause the estimate of $\log \tau$ to be erroneously high. Nonetheless, the product of K_1 and τ is accurate (see Chapter 7).

Some of the parameters can be estimated by analyzing the individual concentration-response curves of the two agonists simultaneously using the procedures described in Chapter 7 as if both agonists were orthosteric agonists. This analysis yields estimates of $\log \tau K_1$, $\log K_2$, and $\log \tau_A$ that should be the same as those estimated above, notwithstanding minor differences caused by the omission of data measured in the presence of both agonists.

Analyzing allosterism at constitutively active receptors

Accurate estimates of the population parameters can also be measured when there is measurable constitutive activity. To illustrate the procedure, the simulated data in Figure 11.7 will be analyzed. This plot shows the effect of positive (panel *a*) and negative (panel *b*) allosteric ligands on the response to an agonist elicited by a receptor exhibiting constitutive activity. The microscopic constants used to simulate the data are given in the figure legend.

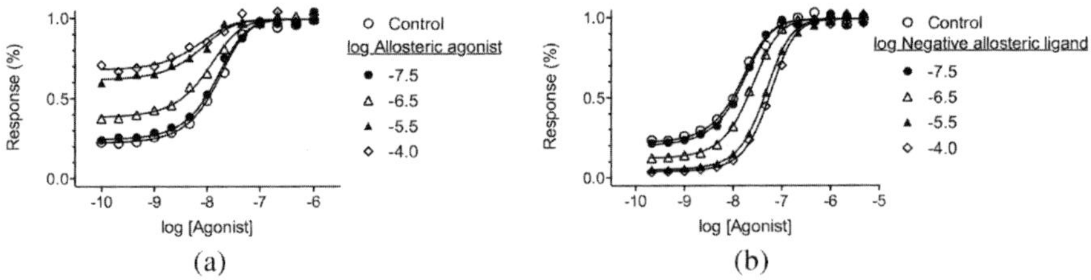

Figure 11.7: Effects of positive (a) and negative (a) allosteric ligands on the response to an agonist in a signaling pathway that exhibits constitutive receptor activity. The data were simulated using Equation (11.31) for the conditions of no modulator (control) and the indicated log molar concentrations of modulator. The log affinity constants of the allosteric ligands for the active and inactive state of the receptor were: *a*, $\log K_f$, 6.5 and $\log K_e$, 6.0; b, $\log K_f$, 6.0, and $\log K_e$, 6.5. The other microscopic constants used in the simulation were $\log K_b$, 8.0; $\log K_a$, 4.0; $\log K_q$, −3.0; $\log K_E$, −2.7 and m, 1.8.

Let us begin with the data in panel *a*. These are analyzed by global nonlinear regression analysis using Equation (11.15) with most of the parameters expressed as logarithms:

$$\text{response} = \frac{M_{\text{sys}}}{1 + \left(\frac{1+10^{\log A+\log K_2}+10^{\log D+\log K_1}\left(1+10^{\log A+\log\alpha+\log K_2}\right)}{10^{\log\tau_{\text{sys}}}+10^{\log\tau_A+\log A+\log K_2}+10^{\log\tau+X+\log K_1}\left(1+10^{\log A+\log\gamma+\log K_2}\right)}\right)^m}. \tag{11.96}$$

The independent variables are the log agonist (X) and modulator ($\log A$) concentrations. Because the orthosteric agonist appears to be a full agonist, it is impossible to estimate K_1, τ and the cooperativity constant, α, from the data in panel *a*. Additional data involving partial receptor inactivation, for example, would be needed to estimate all of the parameters. Thus, when the data are analyzed by global nonlinear regression, the parameters $\log K_1$ and $\log\alpha$ are constrained to 0, and the other parameters are shared among the data sets.

The initial estimates of the parameters, $\log\gamma$ and $\log K_2$, are calculated using Equations (11.86) and (11.87), respectively. The RA_{obs} values for these calculations are estimated with Equation (11.85), and the requisite EC_{50} and $E_{\max}$ values for this latter equation are estimated in accordance with Equation (11.95). The initial estimate of $\log\tau_{\text{sys}}$ ($\log\tau^*_{\text{sys}}$) is calculated using a simplification of Equation (8.20) in Chapter 8:

$$\tau^*_{\text{sys}} = \left(\frac{B}{M_{\text{sys}} - B}\right). \tag{11.97}$$

The other initial parameter estimates are calculated as described above.

With these parameter constraints and initial estimates, global nonlinear regression analysis can be initiated using Equation (11.96): When this is done, the following parameter estimates are obtained: $\log\tau K_1$, 7.63 ± 0.07; $\log K_2$, 6.09 ± 0.05; $\log\tau_A$, 0.16 ± 0.023; $\log\gamma$, 0.40 ± 0.06; M_{sys}, 0.99 ± 0.006; $\log\tau_{\text{sys}}$, -0.25 ± 0.04; and m, 2.18 ± 0.33. These values agree with the system parameters and the theoretical population parameters calculated from the microscopic constants used to simulate the data: $\log\tau K_1$, 7.70; $\log K_2$, 6.00; $\log\tau_A$, 0.20; $\log\gamma$, 0.50; M_{sys}, 1.00; $\log\tau_{\text{sys}}$, -0.54; and m, 1.8.

Whenever there is constitutive activity, it is possible to estimate some of the microscopic constants from the population parameters. Using the relationship described by Equation (8.36) in Chapter 8 ($K_b = \tau K_{\text{obs}}/\tau_{\text{sys}}$), the following estimates can be calculated for the microscopic affinity constants of the orthosteric ($K_b = \tau K_1/\tau_{\text{sys}}$) and allosteric ($K_f = \tau_A K_2/\tau_{\text{sys}}$) ligands for the active state of the receptor: $\log K_b$, 8.24 and $\log K_f$, 6.74. These are nearly the same as the theoretical values used for the simulation: $\log K_b$, 8.0 and $\log K_f$, 6.5. Another way to estimate K_f is to take the product of K_2 and γ:

$$K_f = \gamma K_2. \tag{11.98}$$

Substituting Equations (11.33) and (11.40) for K_2 and γ, respectively, and simplifying establishes the proof of this equation. This calculation yields an estimate of 6.49 for K_f, which is in close agreement with the theoretical value (6.50) and more accurate than that estimated using Equation (8.36) from Chapter 8.

Because the efficacy of the allosteric ligand is clearly less than one-third that of the orthosteric agonist, its $\log K_2$ value (6.09) is a reasonable estimate of K_e (see Chapter 8). Again, this value ($\log K_2 = 6.09$) agrees with the $\log K_e$ value used in the simulation (6.0). Also note that the estimate of $\log K_f/K_e$ (0.65) is approximately equal to $\log \gamma$. With regard to the microscopic affinity constant of the orthosteric agonist for the inactive state of the receptor (K_a) all we can say from the present data is that it is less than K_1.

The data for the negative allosteric modulator in panel *b* are analyzed in the same way, taking care to use Equation (11.88) for calculating the initial estimate of $\log K_2$ for this negative allosteric modulator. When the data are analyzed by global nonlinear regression analysis with Equation (11.96), the following parameter estimates are obtained: $\log \tau K_1$, 7.71 ± 0.03; $\log K_2$, 6.49 ± 0.05; $\log \tau_A$, -0.73 ± 0.08; $\log \gamma$, -0.54 ± 0.04; M_{sys}, 0.99 ± 0.005; $\log \tau_{sys}$, -0.29 ± 0.03; and m, 1.91 ± 0.13. These values agree with the system parameters and the theoretical population parameters calculated from the microscopic constants used to simulate the data: $\log \tau K_1$, 7.70; $\log K_2$, 6.50; $\log \tau_A$, -0.80; $\log \gamma$, -0.50; M_{sys}, 1.00; $\log \tau_{sys}$, -0.54; and m, 1.8.

The microscopic affinity constants of the ligands for the active state of the receptor can be estimated using the relationships described above ($K_b = \tau K_1/\tau_{sys}$ and $K_f = \gamma K_2$). These calculations yield estimates of $\log K_b$ (8.25) and $\log K_f$ (6.0) nearly the same as those used in the simulation (8.00 and 6.00, respectively). With a negative allosteric modulator, the estimate of K_{obs} (i.e., K_2) is a good approximation of K_e as shown by the close agreement between the estimate of $\log K_2$ (6.49) and the theoretical value (6.50) calculated from the microscopic constants used to simulate the data. Again, the estimated ratio of $\log K_f/K_e$ (-0.49) is a good approximation of $\log \gamma$ (-0.50).

Multiple active states of the receptor give rise to biased allosteric effects

When an allosteric interaction is consistent with the simple two-state model, the estimates of K_2 and γ are independent of the orthosteric ligand. This is because

K_2 is a measure of the observed affinity constant of the allosteric ligand for the receptor when the orthosteric site is unoccupied, and γ is a close approximation of the ratio of the microscopic affinity constants of the modulator for the active and inactive states of the receptor (K_f/K_e). The latter microscopic constants are fundamental properties of the receptor, which are unique to the allosteric ligand and are unaffected by the orthosteric ligand. There may be differences in affinity and efficacy modulation depending on the efficacy of the agonist and the sensitivity of the signaling pathway, but K_2 and γ are invariant (see Figure 11.4). When there are more than two active states of the receptor, however, the estimate of γ can vary depending upon the orthosteric ligand. In this section, an example of this type of biased allosterism is analyzed.

Figure 11.8 shows the effect of a positive allosteric modulator on the responses elicited by two agonists at a receptor, having two active states of equivalent activity that couple to the same signaling pathway. The microscopic affinity constants of the allosteric modulator for the inactive state (K_e) and the first (K_f) and second (K_h) active states are 10^6, 10^7 and 10^8 M^{-1}, respectively. The corresponding values for the orthosteric agonist in panel *a* (agonist A) are 10^4, 10^8 and 10^6 M^{-1} (K_a, K_b, and K_d, respectively), whereas the corresponding values for the agonist in panel *b* (agonist B) are 10^4, 10^6 and 10^8 M^{-1}. Thus, both the allosteric modulator and agonist B exhibit selectivity for the second active state, whereas agonist A exhibits selectivity for the first active state. Consequently, the positive allosteric modulator causes about a 100-fold increase in the potency of the agonist B and only about a ten-fold increase in the potency of agonist A.

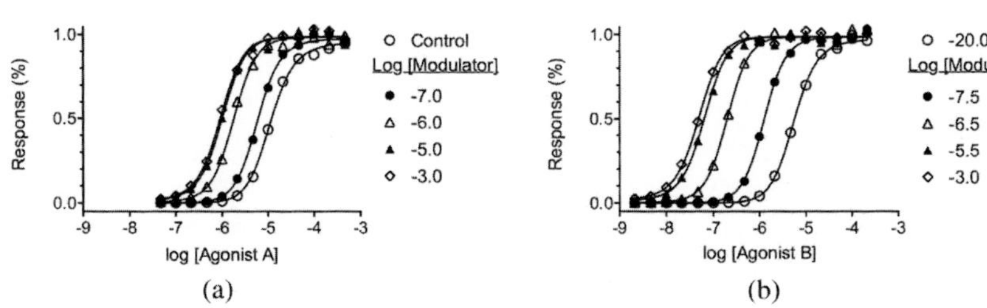

Figure 11.8: Biased allosteric modulation based on two active receptor states, each having equivalent activity but different selectivities for the agonists. The data were simulated using Equation (11.60) for the conditions of no modulator (control) and the indicated log molar concentrations of modulator. The log affinity constants of the allosteric ligand for the inactive and two active states of the receptor were: $\log K_e$, 6.0; $\log K_f$, 7.0 and $\log K_h$, 8.0. The corresponding values for agonist A in panel *a* were: K_a, 4.0; $\log K_b$, 8.0 and $\log K_d$, 6.0, and for agonist B in panel *b*, K_a, 4.0; $\log K_b$, 6.0 and $\log K_d$, 8.0. The other constants were $\log K_q$, -4.0; $\log K_r$, -2.5; $\log \delta$, 0; $\log K_E$, -1.0 and m, 2.0.

These data can be analyzed by global nonlinear regression analysis with Equation (11.81), following the procedure outlined above for the condition where the orthosteric agonist is a full agonist, the allosteric modulator lacks effect on E_{max}, and there is neither constitutive activity nor does the modulator exhibit an effect by itself. Recall that under these conditions, Equation (11.81) is used with the $\log K_1$ constrained to a low value (e.g., 0) and $\log \alpha$ constrained to 0. When the data in panel *a* are analyzed in this manner, the following parameter estimates are obtained: $\log \tau K_1$, 4.92 ± 0.03; $\log K_2$, 5.93 ± 0.04; $\log \gamma$, 1.14 ± 0.02; M_{sys}, 0.98 ± 0.006; and m, 1.69 ± 0.06. These agree with the system parameters and the population parameters calculated from the microscopic constants used to simulate the data: $\log \tau K_1$, 5.04; $\log K_2$, 6.00 and $\log \gamma$, 1.03; M_{sys}, 1.0 and m, 1.8. Analyzing the data in panel *b* yields the following parameter estimates: $\log \tau K_1$, 4.92 ± 0.02; $\log K_2$, 5.97 ± 0.02; $\log \gamma$, 2.10 ± 0.02; M_{sys}, 0.97 ± 0.004 and m, 1.71 ± 0.04. These also agree with the system parameters and the population parameters calculated from the microscopic constants used to simulate the data: $\log \tau K_1$, 5.00; $\log K_2$, 6.00; and $\log \gamma$, 1.99; M_{sys}, 1.0 and m, 1.8. The theoretical parameters used in the simulation are listed in the legend to Figure 11.8.

To summarize, the simulated data in Figure 11.8 show that if the allosteric modulator discriminates between different active states of the receptor, it can have different allosteric effects on the response to an agonist depending upon the selectivity of the agonist for the different active states. In this situation, the estimate of γ represents a weighted average of the two ratios of microscopic affinity constants, K_f/K_e and K_h/K_e. Agonist A, for example, exhibits much greater selectivity for active state 1 ($K_b/K_a = 10^4$) compared to active state 2 ($K_d/K_a = 10^2$). Consequently, it generates the response primarily through active state 1. Because the allosteric modulator has low selectivity for this state ($K_f/K_e = 10$), the estimate of γ for the interaction is about equal to K_f/K_e (10). In contrast, agonist B exhibits selectivity for active state 2 ($K_d/K_a = 10^4$) relative to active state 1 ($K_b/K_a = 10^2$) and so does the allosteric modulator ($K_h/K_e = 10^2$). Consequently, the γ parameter for the interaction is approximately equal to K_h/K_e (100). Despite these agonist-dependent differences in the estimate of γ, the estimate of K_2 is independent of the agonist because it represents the affinity of the allosteric ligand for the agonist-unoccupied receptor complex.

It seems likely that different active states of the receptor would exhibit different activities. Figure 11.9 shows a simulation for a receptor having two active states with differing activities. In this case the agonist exhibits the same selectivity for both active states ($K_b/K_a = K_d/K_a = 10^5$), whereas the allosteric ligand exhibits selectivity

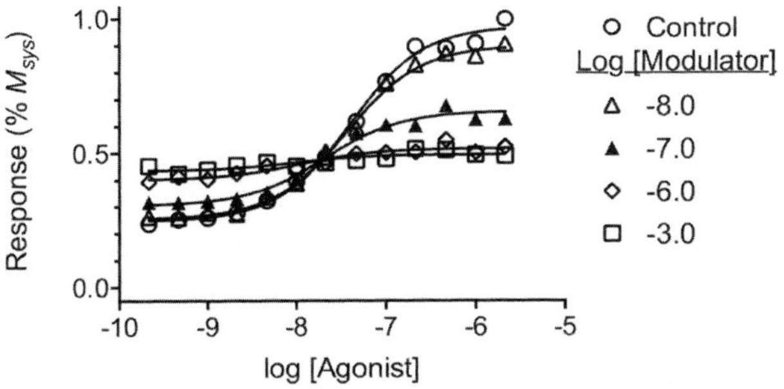

Figure 11.9: Allosteric modulation of the response to an agonist elicited by a receptor having two active states with strikingly different activities. The data were simulated using Equation (11.60) for the conditions of no modulator (control) and the indicated log molar concentrations of modulator. The log affinity constants of the allosteric ligand for the inactive and two active states of the receptor were: $\log K_e$, 6.0; $\log K_f$, 7.0 and $\log K_h$, 9.0. The corresponding values for agonist A were: K_a, 4.0; $\log K_b$, 8.0 and $\log K_d$, 8.0. The other constants were $\log K_q$, −2.5; $\log K_r$, −2.5; $\log \delta$, −3.5; $\log K_E$, −2.0 and m, 2.0.

for active state 2 ($K_h/K_e = 10^3$) compared to active state 1 ($K_f/K_e = 10$). Also, the activity of active state 2 is only about 0.0003 that of active state 1.

When viewed from the perspective of a simple two-state model, the simulated data in Figure 11.9 seem baffling. The allosteric ligand causes an increase in constitutive activity, which is consistent with a positive allosteric effect, but it also reduces the E_{max} of the agonist, which is consistent with a negative allosteric effect. These seemingly conflicting observations can be rationalized by a two-active-state hypothesis. The allosteric ligand induces both active states to cause a small response by itself at the constitutively active receptor. But it has much greater selectivity for the less-active, active state, and hence, it reduces the response to the agonist.

The data can be analyzed using global nonlinear regression analysis with Equation (11.94), which includes parameters for constitutive activity (τ_{sys}) and the response elicited by the allosteric ligand (τ_A) in the absence of agonist. The initial parameter estimates are calculated using the approach described in connection with Figure 11.7. Without additional information, it is impossible to estimate the maximum response of the system even though the orthosteric agonist is a full agonist. Let us assume that we have other data confirming that the agonist is a full agonist, and that the responses have been normalized relative to E_{max}, which is a good estimate of M_{sys}. Therefore, regression analysis is done with M_{sys} constrained to 1.0, and all of the other parameters shared. When this is done, the following parameter estimates are obtained: $\log K_1$, 6.06 ± 0.33; $\log \tau$, 1.41 ± 0.36; $\log K_2$, 6.58 ± 0.11; $\log \tau_A$, -0.10 ± 0.02; $\log \alpha$, 1.86 ± 0.33; $\log \gamma$, 0.45 ± 0.12; $\log \tau_{\text{sys}}$, -0.44 ± 0.07; and m, 1.11 ± 0.17. $\log \beta$ can be estimated from the estimates of $\log \alpha$ and $\log \gamma$ ($\log \beta = \log \gamma - \log \alpha = -1.69$). Even though the estimates of $\log K_1$, $\log \tau$, and $\log \alpha$ have large standard errors, all of the estimated parameter values agree reasonably with the system parameters and the theoretical population parameters calculated from the microscopic constants (see figure legend) used to simulate the data: $\log K_1$, 5.80; $\log \tau$, 1.70; $\log K_2$, 6.62; $\log \tau_A$, -0.11; $\log \alpha$, 2.08; $\log \gamma$, -0.39; $\log \tau_{\text{sys}}$, -0.50; and m, 1.0.

As mentioned above, the errors in the estimation of K_1 and τ are inversely correlated so the estimate of product τK_1 typically has much less error than that of either K_1 or τ ($\log \tau K_1$, 7.47 ± 0.04). The same applies to K_2 and τ_A ($\log \tau_A K_2$, 6.48 ± 0.11), and to α and β ($\log \alpha\beta = \log \gamma$, 0.45 ± 0.12).

Allosteric modulation of pathway selectivity

If there are different active states of the receptor, these may couple differentially to different G proteins, and ultimately, elicit different responses. This raises the possibility that the allosteric modulator might exhibit selectivity for a specific pathway relative to another. Figure 11.10 shows such an example. In this case, the agonist behaves as a full agonist at response 1 (panel a) and a competitive antagonist

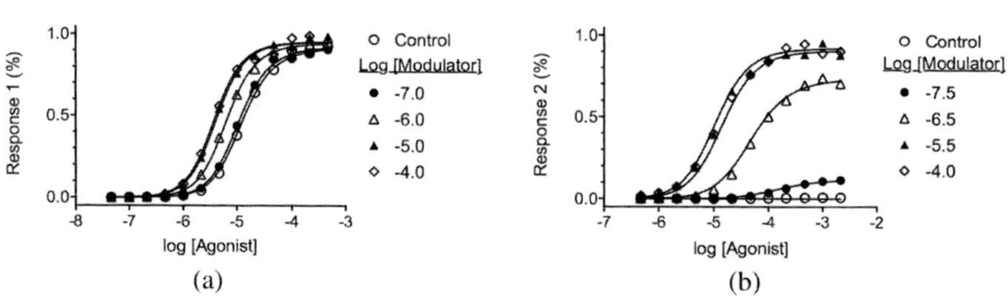

Figure 11.10: Allosteric modulation of pathway selectivity. The plots show the allosteric modulation of different responses ((a) response 1 and (b) response 2) to an agonist elicited through different active states of the same receptor. The data were simulated using Equation (11.60) for the conditions of no modulator (control) and the indicated log molar concentrations of modulator. The log microscopic affinity constants of the allosteric ligand for the inactive and active states eliciting response 1 and response 2, respectively, were: $\log K_e$, 6.0; $\log K_f$, 6.8 and $\log K_h$, 8.0. The corresponding values for the agonist were: K_a, 4.0; $\log K_b$, 8.0 and $\log K_d$, 6.0. The other constants were $\log K_q$, -4.0; $\log K_r$, -4.0; $\log \delta$, 0; $\log K_E$, -1.0 and m, 2.0.

at response 2 (panel *b*). This behavior is explained by the high ($K_b/K_a = 10^4$) and low ($K_d/K_a = 10^2$) selectivity of the agonist for the active states of the receptor-mediating responses 1 and 2, respectively. The allosteric ligand causes a modest increase in the potency of the orthosteric ligand at response 1, but converts it from a competitive antagonist into a full agonist at response 2. These results are consistent with the modest selectivity of the modulator for the active state of the receptor eliciting response 1 ($K_f/K_e \approx 3$) and its much greater selectivity for the active state mediating response 2 ($K_h/K_e = 10^2$). All of the microscopic constants and system parameters are given in the legend to the figure.

Global nonlinear regression analysis is used to analyze the data in each panel separately following the procedures described above. The data in panel *a* are analyzed using Equation (11.75) with $\log K_1$ and $\log\alpha$ constrained to 0 because the allosteric modulator lacks a significant effect on E_{max}. All of the other parameters are shared. Regression analysis yields the following estimates: $\log \tau K_1$, 4.88 ± 0.02; $\log K_2$, 5.83 ± 0.08; $\log\gamma$, 0.55 ± 0.02; M_{sys}, 0.93 ± 0.006; and m, 1.58 ± 0.05. These agree nicely with the system parameters and the population parameters calculated from the microscopic constants used to simulate the data: $\log \tau K_1$, 5.00; $\log K_2$, 6.00; $\log\gamma$, 0.50; M_{sys}, 1.0; and m, 2.0. The value of $\log\gamma$ for this interaction is approximately equal to $\log K_f/K_e$ (0.50).

The data in panel *b* are analyzed in the same manner, but with the parameter $\log K_1$ shared among the data sets because the modulator causes a change in E_{max}, thereby enabling an estimate of K_1. Regression analysis yields the following parameter estimates: $\log K_1$, 4.10 ± 0.07; $\log\tau$, -1.25 ± 0.31; $\log K_2$, 5.96 ± 0.05; $\log\gamma$, 2.18 ± 0.35; M_{sys}, 0.95 ± 0.014; and m, 1.66 ± 0.10. These agree nicely with the system parameters and the population parameters calculated from the microscopic constants used to simulate the data: $\log K_1$, 4.30; $\log\tau$, -1.30; $\log K_2$, 6.00; $\log\gamma$, 3.00; M_{sys}, 1.0; and m, 1.8. Also, the value of $\log\gamma$ for this interaction is approximately equal to $\log K_h/K_e$ (2.00).

Summary of the determinants of allosteric modulation in affinity and efficacy

Observed affinity and efficacy are population parameters that describe the empirical behavior of the receptor activation function. These parameters are determined by the more fundamental microscopic constants that define the isomerization of the free receptor (K_q) and the affinity constants of the orthosteric ligand for active (K_b) and inactive (K_a) states of the receptor. The modification of affinity and efficacy by an allosteric ligand depends on the former constants as well as the affinity of the

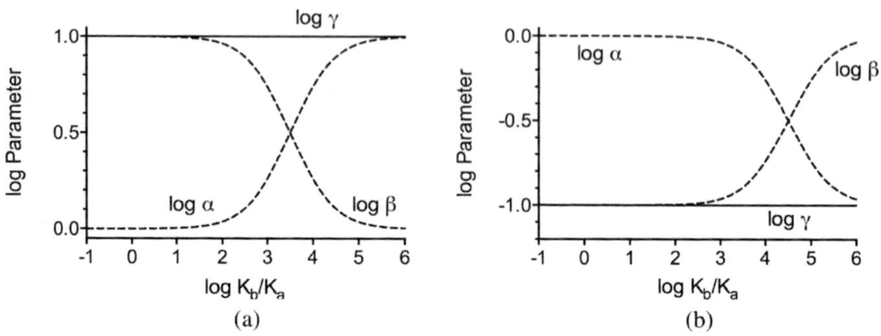

Figure 11.11: Relationships among agonist selectivity for active and inactive states (K_b/K_a) and the allosteric modulation in affinity (α), efficacy (β) and their product (γ) for positive (a) and negative (b) allosteric modulators. The plots in *a* and *b* were generated using Equations (11.35), (11.38), and (11.40) for positive (*a*, $\log K_f/K_e = 1.0$) and negative (*b*, $\log K_f/K_e = -1.0$) allosteric modulators. The other constants used in the calculations were K_a, 4.0; $\log K_b$, 8.0 and $\log K_q$, −4.0.

allosteric ligand for the active (K_f) and inactive (K_e) states of the receptor. In this section, the relationship between these parameters is described.

Figure 11.11 illustrates the relationship between the nature of the allosteric modulation and the selectivity of an agonist for the active and inactive states of a receptor that conforms to a simple two-state model. The graphs show the scalar allosteric modulation in affinity (α) and efficacy (β), plotted against the ratio of affinity constants of the orthosteric ligand for the active and inactive states of the receptor (K_b/K_a). For this simulation, the isomerization constant of the free receptor (K_q) was 10^{-4}.

Panel *a* addresses the case of a positive allosteric modulator having tenfold higher affinity for the active state of the receptor relative to the inactive state ($K_f/K_e = 10$). For inverse agonists, neutral antagonists, and agonists with low efficacy, the allosteric effect consists entirely of efficacy modulation (β). As the efficacy of an agonist increases, there is a gradual reduction in efficacy modulation and a corresponding increase in affinity modulation (α). For highly efficacious agonists that fully activate the receptor population (i.e., $\varepsilon \approx 1$), it is impossible for the modulator to cause a further increase in efficacy and so the allosteric effect consists primarily of affinity modulation. The product of affinity and efficacy modulation (γ) is constant for a given allosteric modulator, regardless of the efficacy of the agonist, provided that the interactions are consistent with a simple two-state model.

It may seem odd that a positive allosteric modulator would increase the efficacy of neutral antagonists and inverse agonists. After all, these orthosteric ligands are usually without effect in both the absence and presence of allosteric modulators. But all receptors have some finite level of receptor activation equivalent to $K_q/(1+K_q)$. Otherwise it would be impossible for an agonist to activate the receptor. Allosteric ligands have an effect that is indistinguishable from causing a scalar change in K_q equivalent to K_f/K_e. No matter how small the level of constitutive receptor activation, it changes by a factor of K_f/K_e when an allosteric ligand is present at a concentration that saturates the allosteric site. Recall from Chapter 8 that the efficacy of a neutral antagonist is equivalent to the basal activity of the receptor (ε_{sys}). It follows that the allosteric modulator has an effect indistinguishable from increasing both ε_{sys} and the efficacy of a neutral antagonist by the factor K_f/K_e. The efficacy of an inverse agonist is equivalent to the fraction of the population of inverse agonist-receptor complexes in the active state, and this low fraction also increases in the presence of a positive allosteric modulator.

Although not shown in this example, a positive allosteric modulator with a K_f/K_e ratio of a magnitude similar to or greater than $1/K_q$ reduces the affinity of inverse agonists ($\log\alpha < 0$) and increases their efficacy to an extent greater than the product of the two effects ($\beta > \gamma$).

The effects of a negative allosteric modulator having a K_f/K_e ratio of 0.1 are shown in panel *b* of Figure 11.11. Again, the modulator primarily affects the efficacy of inverse agonists, neutral antagonists, and agonists with low efficacy. As the efficacy of the orthosteric ligand increases, there is a gradual reduction in efficacy modulation and a corresponding rise in affinity modulation.

The relationship between the value of the product of affinity and efficacy modulation (γ) and the ratio of K_f/K_e is illustrated in Figure 11.12. For this simulation, the value of K_q is 10^{-4}. The value of γ is essentially equal to the ratio, K_f/K_e, for all values of K_f/K_e that are equal to or less than about one-tenth $1/K_q$. With a further increase in K_f/K_e above $0.1/K_q$, the value of γ asymptotically approaches a maximum value of $1/K_q$ (i.e., 10^4 in this example). This relationship can be derived by rearranging the equation for γ (Equation (11.40)) into the following form:

$$\gamma = \frac{\frac{K_f}{K_e} + \frac{K_f}{K_e}K_q}{1 + \frac{K_f}{K_e}K_q}, \tag{11.99}$$

and taking its limit as K_f/K_e approaches infinity:

$$\lim_{Kf/Ke \to \infty} \frac{\frac{K_f}{K_e} + \frac{K_f}{K_e}K_q}{1 + \frac{K_f}{K_e}K_q} = \frac{1 + K_q}{K_q} = \gamma. \tag{11.100}$$

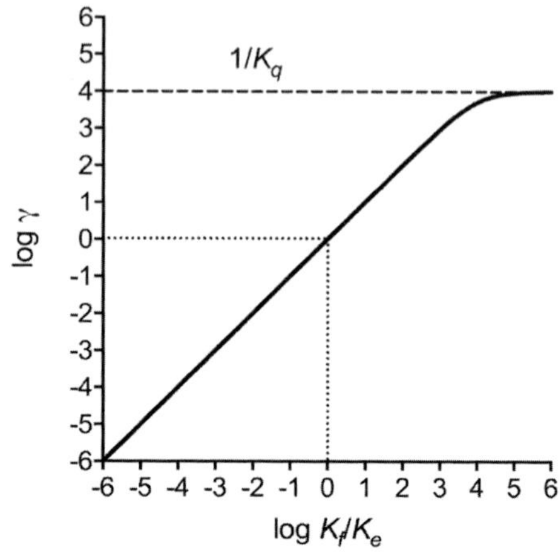

Figure 11.12: Relationship between the product of affinity and efficacy modulation (γ) and the selectivity of the allosteric ligand active and inactive receptor states (K_b/K_a). The plot were generated using Equation (11.40). The other constants used in the calculations were K_a, 4.0; log K_b, 8.0; and log K_q, -4.0.

Because K_q is usually much less than 1, Equation 11.100 shows that the maximum possible value of γ is essentially equal to $1/K_q$.

All of the relationships described in this section apply to the simple two-state model. While there are probe-dependent differences in affinity and efficacy modulation, the product of these parameters are constant for all orthosteric ligands and is equivalent to K_f/K_e except when $K_f/K_e > 0.1/K_q$. When there are multiple active states of the receptor, there can be probe-dependent changes in γ. Generally, the value of γ represents a weighted average of the orthosteric ligand's ratio of affinity constants for the different active states relative to the inactive state. An allosteric ligand with high selectivity for a given allosteric state (e.g., R^{**}) will exhibit a γ value characteristic of that state provided that the isomerization constant for the state ($K_r = R_s^{**}/R$) is not much less than that of another active state and the agonist does not lack selectivity for the R_s^{**} state.

Relationship between allosteric modulation and occupancy of the allosteric site

Allosteric ligands have the effect of causing a concentration-dependent change in the observed affinity and efficacy of the orthosteric ligand. In this section, I consider how this graded modulation is related to occupancy of the allosteric site by the modulator.

One way to think about affinity and efficacy modulation is to imagine that the concentration-response curves of an agonist, measured in the absence and presence of various concentrations of allosteric ligand, are the concentration-response curves of different agonists with different affinity and efficacy values. In this thought experiment, the responses measured in the absence of modulator are assigned to a standard agonist and those measured in the presence of different concentrations of allosteric ligand represent the responses of different agonists. The product of the observed affinity and efficacy of an agonist, expressed relative to that of the standard agonist, is designated as the RA_i value, and it can be calculated using the techniques described in Chapter 7. For our purpose of analyzing allosterism, we will simply define this parameter as the RA value (relative activity). Figure 11.13, panel *a*, shows a plot of these RA estimates for the simulated example shown in Figure 11.4, panel *a*, which describes the effects of a positive allosteric modulator.

Another way to estimate the RA value is by fitting the concentration-response curves to Equation (11.21) by global nonlinear regression analysis to estimate the scalar changes in affinity (p) and efficacy (q) for each concentration-response curve. The RA value is equal to the product of p and q (i.e., $RA = pq$). For the sake of discussion, assume that the control concentration-response curve is entered into a worksheet in Prism in column A, and the data measured in the presence of the

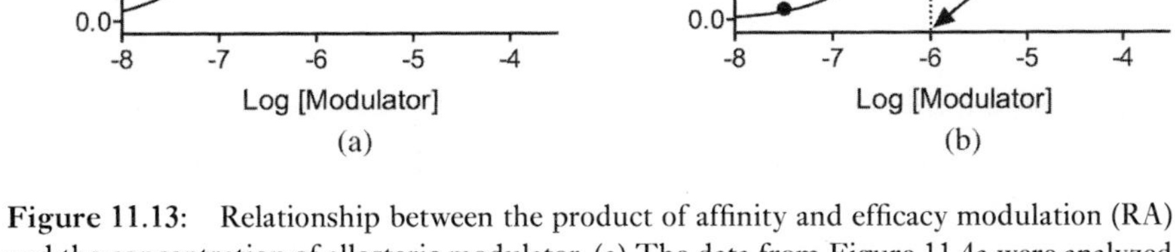

Figure 11.13: Relationship between the product of affinity and efficacy modulation (RA) and the concentration of allosteric modulator. (a) The data from Figure 11.4a were analyzed by global nonlinear regression analysis using Equation (11.21) to estimate the scalar changes in the observed affinity (p) and efficacy (q) of the agonist caused by the indicated concentrations of allosteric modulator. The product of these (pq) is equivalent to RA and is plotted against the log concentration of allosteric modulator. The theoretical curve represents the least squares fit of Equation (11.106) to the data points. (b) Same as panel *a* except that the normalized RA value (RA_{norm}, see Equation (11.107)) is plotted against the log concentration of modulator. The theoretical curve represents the least–squares fit of a simple one-site model to the data points.

four concentrations of modulator are entered into columns B–E. Because the values of p and q need to be constrained to 1.0 for the control concentration-response curve, the following five lines are entered into Prism as the User-defined equation to accomplish this:

$$\langle A \rangle P = 0, \tag{11.101}$$

$$\langle A \rangle Q = 0, \tag{11.102}$$

$$\langle \sim A \rangle P = \log p, \tag{11.103}$$

$$\langle \sim A \rangle Q = \log q, \tag{11.104}$$

$$\text{response} = \frac{M_{\text{sys}}}{1 + \left(\frac{X + \frac{1}{10^{P + \log K_1}}}{10^{Q + \log \tau + X}} \right)^{m}}. \tag{11.105}$$

The independent variable in Equation (11.105) is the log agonist (X) concentration. Equation (11.105) is equivalent to Equation (11.21) but with the parameters p, q, K_1 and τ expressed as logarithms (P, Q, $\log K_1$ and $\log \tau$, respectively). The four lines before Equation (11.105) instruct Prism to estimate values of P and Q for the data in columns B – E and assign the values to parameters $\log p$ and $\log q$. For the control, data in column A, P and Q are treated as part of the equation and are assigned values of zero. When the regression is done sharing the estimates of $\log K_1$, $\log \tau$, M_{sys}, and m among the curves, it is only possible to estimate M_{sys} (1.00 ± 0.007) and m (1.68 ± 0.06) with reasonable accuracy. All of the other parameters have huge standard errors. Nonetheless, it is possible to estimate the log product, $\log \tau K_1$ (5.99 ± 0.03), and the log product of log pq (i.e., log RA) for each concentration of modulator. These log RA values are 0.15 ± 0.03, 0.52 ± 0.04, 0.95 ± 0.03 and 1.05 ± 0.05 for log allosteric ligand concentrations of -7.5, -6.6, -5.5, and 4.0, respectively. Another way to do the regression is with $\log K_1$ constrained to a low value (e.g., 0) because the modulator is without effect on E_{max}. This approach yields the same estimates of $\log K_1$, log RA, M_{sys}, and m. The parameters estimated using Equation (11.105) are essentially the same as those calculated using the method described in the previous paragraph. Hence, the plot in Figure 11.12a accurately describes both sets of estimates.

The plot in Figure 11.13a shows that the allosteric ligand causes a concentration-dependent increase in the RA value and that the maximal log RA value of the modulator approaches an asymptote corresponding to $\log \gamma$. Nonlinear regression analysis of the log RA values according to the following equation yields estimates of the observed affinity constant of the allosteric ligand (K_2) and γ:

$$\log RA = \log\left(\frac{1 + A\gamma K_2}{1 + AK_2}\right). \tag{11.106}$$

When the regression analysis is performed, the estimates of $\log K_2$ (5.97 ± 0.04) and $\log \gamma$ (1.06 ± 0.02) are nearly the same as those estimated using the operational model as described in the discussion of the simulation in Figure 11.4a.

In the absence of modulator, the RA value is equal to one, because the product of the scalar changes in p (1.0) and q (1.0) equals 1.0 for the control curve. Thus, a normalized RA value (RA_{norm}) can be estimated by subtracting 1 from the RA estimate and dividing it by the maximum value of RA ($RA_{\max}$ or γ) minus 1.0:

$$RA_{\mathrm{norm}} = \frac{RA - 1}{RA_{\max} - 1}. \tag{11.107}$$

These normalized RA values are plotted in Figure 11.13b against the log concentration of modulator for the simulated data in Figure 11.4a. The plot shows that the RA_{norm} increases as a function of the modulator concentration in a manner that is consistent with a simple one-site model, with a half-maximal increase occurring at a modulator concentration equivalent to the reciprocal of its observed affinity constant for the receptor when unoccupied by the orthosteric ligand ($-\log K_2$, -5.99 ± 0.02). Thus, it follows that the normalized RA value is a measure of receptor occupancy by the allosteric ligand.

The graphs in Figure 11.14 show an analogous plot for the negative allosteric modulator, gallamine, on the response to the highly efficacious agonist, oxotremorine-M, at the M2 muscarinic receptor. The data for this plot are shown in Figure 11.18 and are discussed in the section, "Analysis of data from the literature." In this case, the negative allosteric modulator causes a concentration-dependent reduction in RA (panel *a*). The limiting value of log RA measured at high concentrations of gallamine is approximately -2. Regression analysis of the log RA values with Equation (11.106) yields estimate of $\log K_2$ (6.27 ± 0.08) and $\log \gamma$ (-1.92 ± 0.09). A plot of the normalized RA values is shown in panel *b*. The relationship between RA_{norm} and the gallamine concentration is not inconsistent with a one-site model having an observed log affinity constant ($\log K_2$) of 6.27, although there are insufficient data points to define the occupancy relationship completely. The data clearly show, nonetheless, that the effect of gallamine reaches a limit at high concentrations, illustrating the consistency with an allosteric mechanism.

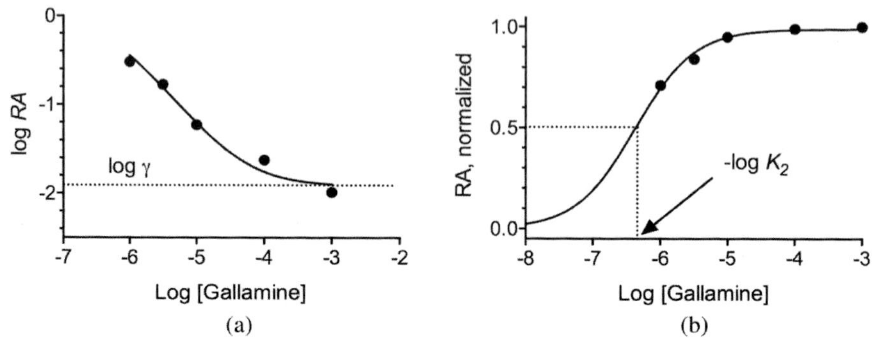

Figure 11.14: Relationship between the product of affinity and efficacy modulation (RA) of oxotremorine-M-mediated inhibition of adenylate cyclase activity in the rat myocardium and the concentration of gallamine. (a) The data from Figure 11.17d were analyzed by global nonlinear regression analysis to estimate the scalar changes in the observed affinity (p) and efficacy (q) of the agonist caused by the indicated concentrations of allosteric modulator. The product of these (pq) is equivalent to RA and is plotted against the log concentration of gallamine. The theoretical curve represents the least-squares fit of Equation (11.106) to the data points. (b) Same as panel *a* except that the normalized RA value (RA_{norm}, see Equation (11.107)) is plotted against the log concentration of modulator. The theoretical curve represents the least-squares fit of a simple one-site model to the data points.

Analysis of steric inhibitory effects on receptor function

So far, we have analyzed interactions caused by the ability of the allosteric ligand to shift the equilibrium between active and inactive states of the receptor. Sometimes a ligand may bind to an allosteric site on a receptor and inhibit activity without selecting the inactive state of the receptor. This could occur, for example, by steric interference of the receptor-G protein interaction. In this section, two examples are covered: (1) the allosteric ligand binds to a secondary site on the receptor and inhibits the activity of the receptor without altering its observed affinity and (2) the allosteric ligand binds to the active state of the receptor selectively and inhibits its activity. In the field of enzyme kinetics, these two mechanisms are known as noncompetitive and uncompetitive inhibition, respectively. The second mechanism is analogous to the open-channel block discussed in Chapter 4.

Noncompetitive inhibition: I begin with the derivation of the total stimulus function — defined at the level of receptor states — for an orthosteric ligand in the presence of a noncompetitive inhibitor. All receptor complexes bound with the allosteric ligand are defined as inactive. Thus, the active and inactive components of the receptor population are defined by:

$$[R_{\text{active}}] = [R_s^*] + [DR_s^*], \tag{11.108}$$

$$[R_{\text{inactive}}] = [R_s] + [DR_s] + [DR_sA] + [R_sA] + [DR_s^*A] + [R_s^*A]. \tag{11.109}$$

Substituting these expressions into Equation 11.27 yields the following equation for the total stimulus after simplification:

$$T = \frac{1}{\left(1 + AK_f\right) + \frac{(1+DK_a)(1+AK_e)}{K_q(1+DK_b)}}. \tag{11.110}$$

Because a noncompetitive inhibitor has no effect on the observed affinity of the orthosteric ligand, the values of its microscopic affinity constants are the same, which in turn are equivalent to its observed affinity constant (i.e., $K_2 = K_e = K_f$). Thus, a simpler equation for the total stimulus of an agonist in presence of a noncompetitive inhibitor can be derived by substituting K_2 for K_e and K_f in Equation (11.110):

$$T = \frac{1}{(1 + AK_2)\left(1 + \frac{(1+DK_a)}{K_q(1+DK_b)}\right)}. \tag{11.111}$$

Substituting Equation (11.111) for the total stimulus in the operational model (Equation (11.14)) yields the following equation for the response elicited by an agonist in the presence of a noncompetitive inhibitor:

$$\text{response} = \frac{M_{\text{sys}}}{1 + \left(K_E\,(1 + AK_2)\left(1 + \frac{(1+DK_a)}{K_q(1+DK_b)}\right)\right)^m}. \tag{11.112}$$

Figure 11.15 shows a simulation of the effect of a noncompetitive inhibitor on the response to an agonist in signaling pathways exhibiting high (panel *a*) and low (panel *b*) sensitivity. High concentrations of the inhibitor completely inhibit the response to the agonist because the allosteric ligand-receptor complex is inactive. Thus, there is no ceiling effect of the inhibitor, and the data are consistent with the hypothesis that the orthosteric ligand does not select for the inactive state, but rather, causes a direct inhibition of receptor activity when bound to it. When the sensitivity of the signaling pathway is high (panel *a*), low concentrations of the noncompetitive inhibitor shift the agonist concentration-response curve to the right, whereas higher concentration cause a depression in E_{max}. When the system exhibits low sensitivity, even low concentrations of the noncompetitive inhibitor cause a reduction in the E_{max}.

There is insufficient information in the data in panel *a* to estimate all of the allosteric parameters. Nonetheless, Equation (11.81) can be fitted to the data by

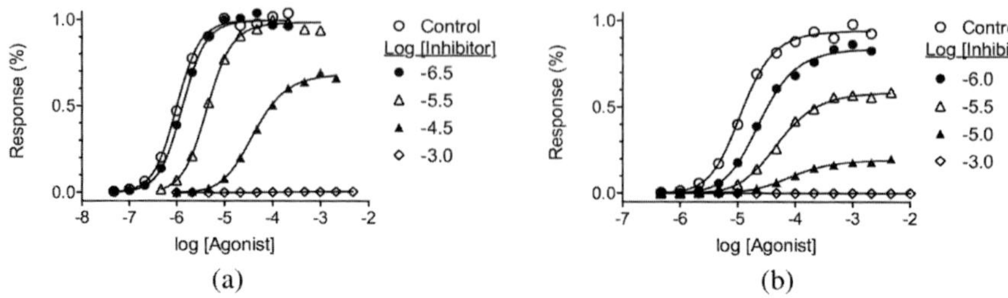

Figure 11.15: Effects of a noncompetitive inhibitor on the concentration-response curve to an agonist in signaling pathways exhibiting high (a) and low (b) sensitivity. The data were simulated using Equation (11.112) for the conditions of no noncompetitive inhibitor (control) and the indicated log molar concentrations of inhibitor. The log sensitivity constant ($\log K_E$) for the simulations in panels *a* and *b* were -2.0 and -1.0, respectively. The microscopic constants used in the simulation were $\log K_b$, 8.0; $\log K_a$, 4.0; $\log K_e$, 6.0, $\log K_f$, 6.0; $\log K_q$, -4.0 and m, 1.8.

global nonlinear regression analysis with $\log \alpha$ constrained to zero and $\log \gamma$ constrained to an arbitrarily low value (-20) to represent a γ value of zero. When this is done, the following parameter estimates are obtained: $\log \tau K_1$, 6.00 ± 0.03; $\log K_2$, 6.00 ± 0.02; M_{sys}, 1.00 ± 0.005 and m, 1.81 ± 0.05. These estimates agree with the theoretical system parameters and the observed population parameters calculated from the microscopic constants used for the simulation: $\log \tau K_1$, 6.00; $\log K_2$, 6.00; M_{sys}, 1.00 and m, 1.80. Fitting the equation with the parameter γ shared yields a very low estimate of γ (-3.96 ± 4.39) with a huge standard error and does not result in a significant reduction in residual error nor a change in the estimates of the other parameters. Thus, the data are consistent with the hypothesis that the allosteric ligand-receptor complex is inactive ($\gamma = 0$; $\log \gamma = -\infty$).

With the data in panel *a* only, it is impossible to determine if the allosteric ligand modulates the observed affinity of the orthosteric ligand because it is possible to obtain an equally good fit when global nonlinear regression analysis is done with $\log \alpha$ constrained to values greater than zero, but not much less than zero ($\infty \geq \log \alpha \geq 0$). When $\log \alpha$ is constrained to values greater than zero, the estimate of $\log K_1$ decreases and the estimate of $\log \tau$ increases such that the log product ($\log \tau K_1$) remains constant. Thus, from the data in panel *a*, all we can estimate are the parameters mentioned above and the domain of $\log \alpha$ ($-\infty \leq \log \alpha \leq 0$). To obtain estimates of $\log K_1$, $\log \tau$ and $\log \alpha$ from functional data, it is necessary to use the technique of partial receptor inactivation as described above.

The same methods and restrictions apply to the data in panel *b*. Global nonlinear regression analysis with Equation (11.81) and $\log \alpha$ constrained to zero yields the following parameter estimates: $\log \tau K_1$, 5.01 ± 0.03; $\log K_2$, 5.97 ± 0.03; M_{sys},

1.00 ± 0.01 and m, 1.86 ± 0.01. Again, these agree with the theoretical values: $\log \tau K_1$, 5.00; $\log K_2$, 6.00; M_{sys}, 1.00 and m, 1.80. Searching parameter space with $\log \alpha$ constrained to various values showed that the domain of $\log \alpha$ is essentially $\infty > \log \alpha \geq 0$.

Uncompetitive inhibition: The mechanism of uncompetitive inhibition is based on the premise that the inhibitor binds selectively to the active state of the receptor and inhibits its activity. I consider the more general mechanism where the inhibitor exhibits a preference for the active state. That is, it has much higher affinity for the active state than the inactive state (i.e., $K_f/K_e \gg 1$). The prior Equation (11.110) describes how an allosteric ligand that inhibits activity when bound to the receptor affects the total stimulus of an orthosteric ligand. This equation considers the general case where the inhibitor can exhibit a range of selectivities for the active state (i.e., $0 < K_f/K_e < \infty$). Substituting Equation (11.110) for the total stimulus (T) in the operational model (Equation (11.14)) yields an equation for the response to an agonist in the presence of this type of inhibitor:

$$\text{response} = \frac{M_{\text{sys}}}{1 + \left(K_E\left(1 + AK_f\right) + \frac{(1+DK_a)(1+AK_e)}{K_q(1+DK_b)}\right)^m}. \qquad (11.113)$$

To simulate the response to an agonist in the presence of an uncompetitive inhibitor, we will use this equation with the stipulation that the allosteric ligand has high selectivity for the active state ($K_f/K_e \gg 1$).

Figure 11.16 shows a simulation of the effect of an uncompetitive inhibitor on the concentration-response curves of an agonist in signaling pathways having high

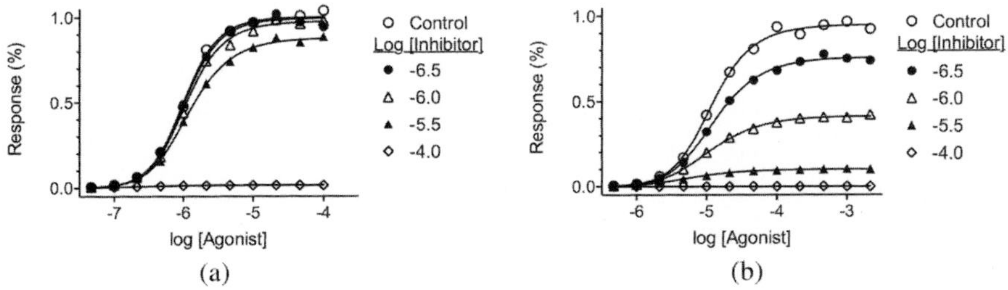

Figure 11.16: Effects of an uncompetitive inhibitor on the concentration-response curve to an agonist in signaling pathways exhibiting high (a) and low (b) sensitivity. The data were simulated using Equation (11.113) for the conditions of no uncompetitive inhibitor (control) and the indicated log molar concentrations of inhibitor. The log sensitivity constant ($\log K_E$) for the simulations in panels *a* and *b* were -2.0 and -1.0, respectively. The microscopic constants used in the simulation were $\log K_b$, 8.0; $\log K_a$, 4.0; $\log K_e$, 3.0, $\log K_f$, 7.0; $\log K_q$, -4.0 and m, 1.8.

(panel *a*) and low (panel *b*) sensitivity. High concentrations of the inhibitor cause a complete inhibition of receptor activity, which is consistent with the postulate that the allosteric ligand-receptor complex is inactive. In the system with high sensitivity, moderate to high occupancy of the allosteric site has hardly an effect on the concentration-response curve of the agonist because the reduction in agonist efficacy is opposed by a proportional increase in affinity. In the less sensitivity system, occupancy of the allosteric site causes reductions in the E_{max} and EC_{50} (increase in potency) values of the agonist.

Let me begin with the simulated data in panel *b*. These are analyzed by global nonlinear regression analysis with Equation (11.81). When this is done, only the parameters M_{sys} (1.08 ± 0.04) and m (1.97 ± 0.09) can be estimated with reasonable standard errors. Although the estimates of $\log K_1$ and $\log \tau$ are inaccurate and have substantial standard errors, it is possible to obtain a reliable estimate of the log product, $\log \tau K_1$ (5.05 ± 0.02). Similarly, while $\log \alpha$ and $\log K_2$ have colossal standard errors, the estimate of the log product, $\log \alpha K_1 K_2$, is accurate (11.00 ± 0.02). All of these estimated parameters agree with the theoretical values: $\log \tau K_1$, 5.00; $\log \alpha K_1 K_2$, 11.00; M_{sys}, 1.00 ± 0.01; and m, 1.80. By constraining $\log \alpha$ to a constant and doing the regression several times with a range of different constrained $\log \alpha$ values, it can be shown that a least-squares fit is obtained whenever $\log \alpha$ is greater than or equal to about 2.0 and $\log K_2$ is less than about 4.4. The theoretical values of $\log \alpha$ and $\log K_2$ are 3.40 and 3.3, respectively. Thus, aside from the system parameters, the only ligand parameters that we can estimate are $\log \tau K_1$ and $\log \alpha K_1 K_2$. Using the method of partial receptor inactivation, it is possible to estimate $\log K_1$, $\log \tau$ and the product, αK_2. Whenever the uncompetitive inhibitor has high selectivity for the active state, however, it is impossible to obtain separate estimates of α and K_2 from functional experiments.

Using the same approach on the data in panel *a* yields the following parameter estimates: $\log \tau K_1$, 6.04 ± 0.03; $\log \alpha K_1 K_2$, 10.98 ± 0.04; M_{sys}, 1.01 ± 0.01 and m, 2.00 ± 0.12. These are nearly the same as the theoretical values: $\log \tau K_1$, 6.00; $\log \alpha K_1 K_2$, 11.00; M_{sys}, 1.00 ± 0.01 and m, 1.80. It is also possible to establish the domains of $\log \alpha$ (greater than about 2) and $\log K_2$ (less than about 3.8). Separate estimates of $\log K_1$ and τ can be obtained using the method of partial receptor inactivation.

Analysis of Data from the Literature

The analysis of three sets of data from the literature is described in this section to gain some perspective on how the methods and theory relate to real data.

Positive allosteric modulation of the metabotropic glutamate receptor 4

The first set of data comes from Marino and coworkers[17] who investigated the effect of a positive allosteric modulator (N-phenyl-7-(hydroxyimino)cyclopropa-[*b*]chromen-1a-carboxamide; PHCCC) on the Ca^{2+} response to L-(+)-2-amino-4-phosphonobutyric acid (L-AP4) in Chinese hamster ovary (CHO) cells expressing the mGluR4 and a chimeric *G* protein composed mainly of G_i and the C terminal tail of G_q. As described in Chapter 5, this chimeric *G* protein is useful for measuring a G_q-mediated response by a *G* protein-coupled receptor that normally prefers to signal through G_i. Figure 11.17a shows that increasing concentrations of PHCCC

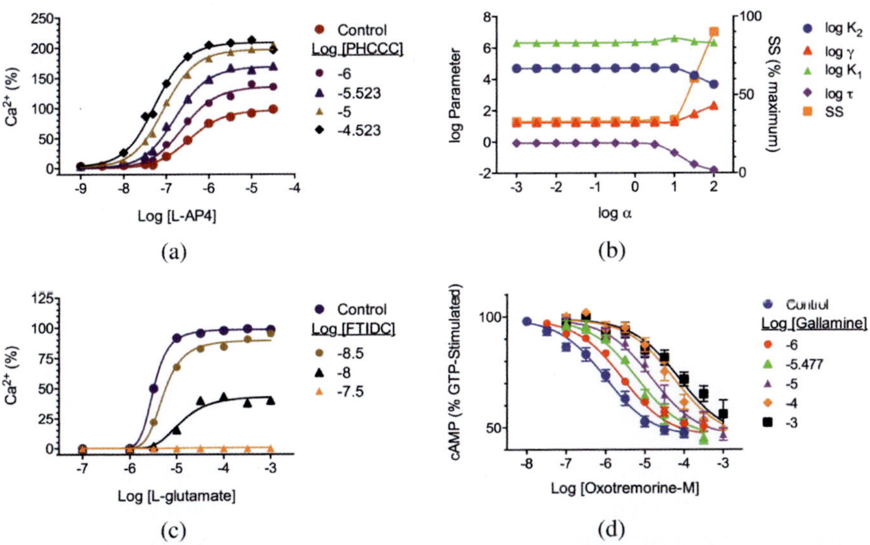

Figure 11.17: Analysis of positive allosteric modulation of the MGluR4 (a) and negative allosteric modulation of the MGluR1a (c) and M_2 muscarinic receptor (d). (a) The plot shows the concentration response curve of Log [L-glutamate] for eliciting Ca^{2+} mobilization in the absence (control) and presence of the indicated log concentrations of FTIDC. The data are from Marino *et al.*[17] and have been reproduced with permission. (b) Summary of nonlinear regression analysis of the data in panel *a* using Equation (11.81). (c) The plot shows the concentration response curve of L-glutamate for eliciting Ca^{2+} mobilization in the absence (control) and presence of the indicated log concentrations of FTIDC. The data are from Suzuki *et al.*[18] and have been reproduced with permission. (d) The plot shows the concentration-response curves of oxotremorine-M in the absence (control) and presence of the indicated log concentrations of gallamine. The data are from Ehlert,[19] and have been reproduced with permission.

cause an increase in the potency and E_{max} of L-AP4 for eliciting Ca^{2+} mobilization. The mean data points were estimated from a published figure of the data.

Because the data lack constitutive activity and the allosteric ligand lacks activity by itself, the data are analyzed by global nonlinear regression analysis with Equation (11.81) (i.e., Equation (11.20) with most of the parameters expressed as logarithms). The allosteric modulator causes an increase in the E_{max} of L-AP4, and it appears that this effect reaches a ceiling level. These observations suggest that it should be possible to estimate $\log K_1$ and M_{sys}. Hence, all of the concentration-response curves are analyzed simultaneously sharing the estimates of $\log K_1$, $\log K_2$, $\log \gamma$, $\log \tau$, M_{sys} and m among the curves and constraining $\log \alpha$ to zero. The approach is the same as that used for the analysis of the data in Figure 11.4d. When this is done, the following parameter estimates are obtained: $\log K_1$, 6.31 ± 0.09; $\log \tau$, $-.068 \pm 0.09$; $\log K_2$, 4.70 ± 0.07; $\log \gamma$, 1.24 ± 0.08; M_{sys}, 214 ± 3; and m, 1.19 ± 0.09.

Because PHCCC is an allosteric modulator that is incapable of eliciting a response by itself, then the ratio of its affinity constants for the active and inactive states of the receptor (K_f/K_e) is much less than the reciprocal of K_q. Hence, $\log K_2$ is a good estimate of $\log K_e$, and $\log \gamma$ is a good estimate of $\log K_f/K_e$. This yields estimates of 5.94 and 4.70 for $\log K_f$ and $\log K_e$, respectively.

One might ask, how we can be sure that constraining $\log \alpha$ to zero is an acceptable way to perform the regression analysis? After all, how do we know what the value of $\log \alpha$ is? One way to address these questions is to perform numerous nonlinear regressions, each time with $\log \alpha$ constrained to a different value. After each regression, the parameter estimates and residual sum of the square deviations of the data points from the regression equation is recorded. Figure 11.17b shows the results of this walk through parameter space. The results of each regression consists of a set of plotted points that describe a vertical line intersecting the abscissa at the value to which $\log \alpha$ was constrained during regression. Each point in the set represents a parameter estimate or *RSS*. The plot shows the results of 11 nonlinear regression analyses, each done with $\log \alpha$ constrained to a different value ranging from -3 to 2 in 0.5 log increments. Whenever $\log \alpha$ is constrained to a value equal to or less than 0.5, RSS is at a constant minimum, and the estimates of the other parameters are constant. In other words, the best fit of the regression equation to the data is obtained whenever $\log \alpha$ is constrained to a value less than 0.5. It does not matter to what value $\log \alpha$ is constrained so long as it is less than 0.5. Over this domain of $\log \alpha$, the parameter estimates are constant. In contrast, when $\log \alpha$ is constrained to a value greater than 0.5, a least squares fit is not obtained (i.e., RSS increases), and the parameter estimates vary depending on the value of $\log \alpha$. In summary, these results show that our strategy of constraining $\log \alpha$ to zero is valid.

Negative allosteric modulation of the metabotropic glutamate receptor 1a

The next set of data for our analysis also comes from Geoffrey Conn's lab (Suzuki *et al*[18]) and is shown in Figure 11.17c. These data illustrate the effect of the negative allosteric modulator, 4-[1-(2-fluoropyridin-3-yl)-5-methyl-1H11,2,3-triazol-4-yl]-N-isopropyl-N-methyl-3,6-dihydropyridine-1(2H)-carboxamide (FTIDC), on the Ca^{2+} response to L-glutamate in CHO cells expressing the human mGluR1a. Increasing concentrations of FTIDC cause a modest reduction in potency and a large decrease in the maximum response of L-glutamate. The data suggest a noncompetitive mechanism because the inhibition is complete.

The data can be analyzed using the approached described in relation to the simulated data in Figure 11.14. Equation (11.81) is fitted to the data by global nonlinear regression analysis with $\log \alpha$ constrained to zero. $\log \gamma$ is constrained to an arbitrarily low value (−20) to represent an antilog of zero because FTIDC completely inhibits the effect of L-glutamate. When this is done, estimates of $\log K_1$ and $\log \gamma$ are obtained, but the errors in the estimates of these parameters are inversely correlated and depend on the value to which $\log \alpha$ is constrained. Hence, only the estimate of their product is given here together with the other estimates: $\log \tau K_1$, 5.98 ± 0.17; $\log K_2$, 7.71 ± 0.18; M_{sys}, 1.03 ± 2.0 and m, 8.5 ± 2.6. Searching parameter space with $\log \alpha$ constrained to a range of values shows that a least squares fit is obtained whenever $\log \alpha$ is greater than zero, but not much less than zero. Hence the domain of $\log \alpha$ is approximately, $\infty \geq \log \alpha \geq 0$. Assuming a noncompetitive mechanism (i.e., $\log \alpha$ constrained to 0) yields estimates of $\log K_1$ (5.82 ± 0.17) and $\log \tau$ (0.16 ± 0.06).

The noncompetitive mechanism seem likely because Susuki *et al.*[18] reported that FTIDC has no effect on the binding of [^{3}H]L-glutamate to the mGluR1a. Also, the effects of site-directed mutagenesis on the ability of FTIDC to modulate the Ca^{2+} response suggests that FTIDC interacts with mGluR1a on TM6 and TM7 near intracellular loop 3. These results suggest the FTIDC may prevent the receptor-*G* protein interaction without affecting the binding of L-glutamate, which is consistent with a noncompetitive mechanism.

Negative allosteric modulation of the M_2 muscarinic receptor

My last example concerns some of my own data illustrating the effect of gallamine on M_2 muscarinic receptor-mediated inhibition of adenylate cyclase activity in the rat myocardium.[19] As shown in Figure 11.17d, the highly efficacious agonist, oxotremorine-M, causes a concentration-dependent inhibition of adenylate cyclase activity. When measured in the presence of a range of concentrations of gallamine, the concentration-response curves of oxotremorine-M shift to the right without a significant change in E_{max}. A careful look at the relation between the shift and the gallamine concentration suggest a limit to the shift at high concentrations of gallamine.

The data are analyzed using an approach analogous to that described in Figure 11.4 for the simulated data in panels $a - c$. We need to modify regression Equation (11.81) to account for an inhibitory effect, however. This can be done by taking the appropriate regression equation for analyzing an agonist-induced gain in function and subtracting it from a parameter representing the amount of activity measured in the absence of ligands (R_0):

$$\text{response} = R_0 - \frac{M_{\text{sys}}}{1 + \left(\frac{1+10^{\log A+\log K2}+10^{X+\log K1}\left(1+10^{\log A+\log\alpha+\log K2}\right)}{10^{X+\log\tau+\log K1}\left(1+10^{\log A+\log\gamma+\log K2}\right)}\right)^m}. \qquad (11.114)$$

Incidentally, the same modification can be made with the other forms of the operational model (Equations (11.94) and (11.96)) when the response to the agonist is inhibitory. For the analysis of the data in Figure 11.18, global nonlinear regression is done using Equation (11.114) with $\log K_1$ constrained to an arbitrarily low value (e.g., zero). Regression analysis yields an inaccurate estimate of $\log\tau$ because it is inversely correlated with the estimate of $\log K_1$, and we cannot estimate $\log K_1$ from the data in Figure 11.17d only. Nonetheless, the product of the estimate of $\log\tau$ and the constrained value of $\log K_1$ is accurate. The results of the analysis yield the following parameter estimates: R_0, $99.1 \pm 1.1\%$; $\log\tau K_1$, 6.00 ± 0.07; $\log K_2$, 6.26 ± 0.09; $\log\gamma$, -1.88 ± 0.08; M_{sys}, $53 \pm 2.2\%$ and m, 0.73 ± 0.06. The estimate of $\log\tau K_1$ is equivalent to the $-\log EC_{50}$ value of oxotremorine-M in the absence of gallamine.

One might be concerned that the data are consistent with a model of competitive inhibition and that there is no basis for using an allosteric model for the analysis. The competitive hypothesis can be tested by constraining γ to zero during global nonlinear regression analysis. Recall that when γ is equal to zero, the binding of the orthosteric and allosteric ligands is mutually exclusive and Equation (11.114) and its antecedent (Equation (11.81)) are consistent with a competitive interaction. When the data in Figure 11.17d are analyzed with $\log\gamma$ constrained to an arbitrarily low integer (e.g., -20) to approximate the log of zero ($-\infty$), there is a highly significant increase in residual error ($F = 154$; $P = 1.7 \times 10^{-16}$) indicating that the data are inconsistent with competitive inhibition.

The data can also be analyzed by estimating the RA values of the concentration–response curves measured in the presence of each concentration of gallamine. This is accomplished by global nonlinear regression analysis of the data using a modification of Equation (11.105) that accounts for an agonist-mediated inhibition of a response (R_0):

$$\text{response} = R_0 - \frac{M_{\text{sys}}}{1 + \left(\frac{X + \frac{1}{10^{P+\log K_1}}}{10^{Q+\log\tau+X}}\right)^m}. \tag{11.115}$$

When using Prism, the four lines mentioned above (Equations (11.101)–(11.104)) precede Equation (11.115) to instruct Prism to constrain P and Q to zero for the control data (entered in column A of the datasheet) and estimate values of P and Q for the other curves and assign the estimate to parameters $\log p$ and $\log q$, respectively. Regression analysis is done with $\log K_1$ constrained to an arbitrarily low value (e.g., zero). This analysis yields estimates of $\log p$ and $\log q$ for each curve, and although the latter are inaccurate, their sum yields accurate estimates of the $\log RA$ values (−0.52, −0.73, −1.23, −1.68 and −1.99) for the various log molar concentrations of gallamine (−6, −5.52, −5, −4.52 and −4, respectively). These values are plotted against the log gallamine concentration in Figure 11.14a. If the data were consistent with competitive inhibition, the plot should attain a slope of negative one as the modulator concentration increases above the reciprocal of K_2. It is clear, however, that the $\log RA$ value does not continue to decrease inversely proportionally with an increase in the modulator concentration, but rather, approach a limiting value corresponding to $\log\gamma$. Regression analysis of the $\log RA$ values with Equation (11.106) yields estimates of $\log K_2$ (6.27 ± 0.08) and $\log\gamma$ (-1.92 ± 0.09) as mentioned above.

A plot of the normalized RA values (RA_{norm}) is shown in panel b of the same figure. These are calculated using Equation (11.107) and represent an accurate estimate of fractional occupancy of the allosteric site by gallamine when the orthosteric site is unoccupied. The −log concentration of gallamine yielding and RA_{norm} value of 0.5 is equivalent to $\log K_2$ (6.27). Again, the latter estimate is not well-defined by the range of gallamine concentrations shown in Figure 11.14b.

What Does the Allosteric Modulation of GPCRs Tell us About Receptor States?

Now that we have analyzed a handful of data from the literature, what do they tell us about the underlying receptor states that give rise to the behavior of the receptor population? One way to address this question is to investigate how well the state models described in Figures 11.2 and 11.3 fit the data. These models, after all, determine the parameters of the allosteric ternary complex model (Figure 11.1) that were estimated from the data. The critical issue is that the parameters of the allosteric ternary complex model cannot assume any value, but rather, are constrained by the underlying state model. To address the nature of the state model directly, we can attempt to fit an operational model to the data, which has the stimulus function defined at the level of receptor states (see Figures 11.2 and 11.3) rather than at the level of the receptor population (see Figure 11.1) as we did in the preceding analyses.

The data on the mGluR4 in Figure 11.17a can be analyzed using the operational model with the stimulus defined at the level of receptor states assuming a simple two-state model (Equation (11.31). As in the analysis of open probability or the population behavior of ligand-gated ion channels (whole cell current response), it is impossible to obtain independent estimates of K_b and K_q from the data unless there is significant constitutive activity. Nonetheless, the error in the estimate of K_b is inversely correlated with that of K_q so that it is possible to obtain an estimate a parameter that is proportional to the product K_bK_q and inversely proportional to the sensitivity constant of the signaling pathway (K_E). This parameter (K_{bqs}) is defined by the following equation:

$$K_{bqs} = \frac{K_b K_q}{\left(1 + K_q\right) K_{E-\mathrm{obs}}}. \tag{11.116}$$

This complicated parameter, K_{bqs}, is actually equivalent to the product τK_1 (proof by substitution using Equation (11.32) for K_1 and Equations (7.11) and (7.116) for τ and ε). The latter product is easy to estimate as described in this and in previous chapters. Thus, regression analysis of the data for the mGluR4 can be done with Equation (11.31) rearranged to include the parameter, K_{bqs}:

$$\text{response} = \frac{M_{\mathrm{sys}}}{1 + \left(K_E\left(1 + \frac{(1+DK_a)(1+AK_e)}{K_q\left(1+\frac{DK_{bqs}\left(1+K_q\right)K_E}{K_q}\right)(1+AK_f)}\right)\right)^m}. \tag{11.117}$$

To obtain a fit, we need to constrain K_q to any arbitrarily low value so long as it is low enough ($\log K_q < -2.3$) to eliminate constitutive activity (less than 0.5% of M_{sys}). Regression analysis can be done with the forgoing constraints (i.e., $\log K_q = -4$) and with a form of the equation in which the microscopic constants are expressed as logarithms. When the data in Figure 11.16a are analyzed in this manner, a fit indistinguishable from that shown in Figure 11.16a is obtained. Thus, the data are consistent with a simple two-state model. The analysis yields the following parameter estimates: log K_{bqs}, 6.28 ± 08; log K_a, 6.37 ± 0.07; log K_f, 5.92 ± 0.06; log K_e, 4.75 ± 0.08 and log K_E, −1.48 ± 0.62. The estimates of log K_f and log K_e are essentially the same as those estimated above (approximately log K_2 (4.70), and γK_2 (5.94), respectively) and their log ratio, log K_f/K_e (1.17), provides an estimate of log γ similar to that estimated above (1.24). The estimate of log K_E is unreliable, however, as shown by its large standard error and associated confidence interval that spans about three log units.

It is not surprising that the data in Figure 11.17a are consistent with a simple two-state model for they exhibit behavior analogous to that of the simulated data in Figure 11.4d and the effects of etomidate on the current response to a partial agonist, P4S (see Chapter 4, Figure 4.21). In both cases, the underlying mechanism is a simple two-state model.

The data on the allosteric effects of FTIDC (Figure 11.17c) tell us nothing about receptor states because FTDIC appears to be a noncompetitive inhibitor. That is, it has no allosteric effect on the binding of the agonist, and it does not appear to affect the activity of the receptor allosterically. Rather, it may simply prevent the interaction of the G protein with the receptor competitively. One could argue that the effect of FTIDC is indirectly allosteric because it prevents the positively cooperative effect of the endogenous G protein.

The data in Figure 11.17d regarding the effects of gallamine are a puzzle of sorts. By itself, the collection of measurements in Figure 11.17d is consistent with a two-state model. There is additional data, however, showing that gallamine only modulates the affinity of oxotremorine-M and other muscarinic agonists.[20] Also, when the same experiment was done with a partial agonist (N-methyl-N-(4-pyrrolinyl-1-methyl-2-butynyl)carboximide, BM5) gallamine has no effect on $E_{\max}$, indicating affinity-only modulation.[19]

So why does affinity-only modulation pose a problem with state models of the type shown in Figure 11.2 and 11.3? Ultimately, both of these models predict that allosteric modulation of orthosteric ligands that do not fully activate the receptor population should be associated with efficacy modulation. But as just described, the evidence shows that gallamine only modulates the affinity of several agonists for mediating an inhibition of adenylate cyclase through the M_2 receptor.

Let me throw light on this conundrum by providing an explanation that is not entirely correct, but in theory, can at least account for affinity-only modulation between gallamine and a given orthosteric ligand, but not all orthosteric ligands.[20] The provisional explanation is that there could be multiple active and inactive states of the receptor. Let us assume that there are two active (R_s^* and R_s^{**}) and two inactive (R_{s-1} and R_{s-2}) states. For our discussion, the affinity constants of the orthosteric ligand for the two active states are designated as K_b and K_d and those for the inactive states are designated as K_a and K_c, respectively. The corresponding pairs of constants for gallamine are K_f and K_h (active states) and K_e and K_g (inactive states). We also need to define the various isomerization constants of the unoccupied receptor. These are K_q (R^*_{s-1}/R_{s-1}), K_r (R_{s-2}/R_{s-1}) and Ks (R^{**}_{s-2}/R_{s-2}).

Suppose that the isomerization constants for the transitions of the two inactive states into the two active states are the same (i.e., $K_q = R^*_{s-1}/R_{s-1} = 10^{-4}$ and $K_s = R^{**}_{s-2}/R_{s-2} = 10^{-4}$), but that the isomerization of inactive state 1 into inactive state 2 is very low ($K_r = R_{s-2}/R_{s-1} = 10^{-3}$). In the absence of ligands, therefore, the abundances of the four states of the receptor would be 0.9999 (R_{s-1}), 0.001 (R_{s-2}), 0.0001 (R^*_{s-1}) and 0.0000001 (R^{**}_{s-2}). It follows that the most probable route of receptor activation by an orthosteric ligand would involve an R_{s-1} to R^*_{s-1} transition rather than an R_{s-2} to R^{**}_{s-2} transition.

Now suppose that gallamine has equivalent high affinity for the less abundant active and inactive states (R^{**}_{s-2} and R_{s-2}) and equivalent low affinity for the more abundant active and inactive states (R_{s1} and R^*_{s-1}). Let us also suppose that the ratio of microscopic affinity constants of the agonist for the first pair of active and inactive states (K_b/K_a) is equivalent to that of the second pair (K_d/K_c), but that the agonist exhibits a lower average affinity for the second pair of states compared to that of the first pair. In the presence of gallamine, the agonist would exhibit lower observed affinity because it has a lower average affinity for the pair of active and inactive states (R_{s-2} and R^{**}_{s-2}) that gallamine selects. Nonetheless, its efficacy would be unaffected because its ratio of K_b/K_a is the same as K_d/K_c. A summary of this combination of parameters and a simulation of the stimulus function for this allosteric interaction is shown in Figure 11.18.

The problem with this hypothesis is that it cannot explain why gallamine modulates the affinity, but not the efficacy, of other agonists for inhibiting adenylate cyclase through the M_2 receptor. One would expect a random relationship between the ratios of K_b/K_a and K_d/K_c for a group of agonists because the three-dimensional shape of the orthosteric binding pocket on each of the four states of the receptor is different. One way of thinking about this situation is to consider the observed affinity constants (K_{obs}) of a random set of ligands for four different G protein-coupled receptors (e.g., M_2 muscarinic, H_1 histamine, D_2 dopamine and β_2 adrenergic). What is the likelihood that, for a given ligand, the ratio of its K_{obs} *value* for the M_2 receptor divided by that of the H_1 histamine receptor is the same as the ratio of its K_{obs} value for the D_2 dopamine receptor divided by that of the β_2 adrenoceptor? Now consider the likelihood that this relationship holds for all ligands. This seems

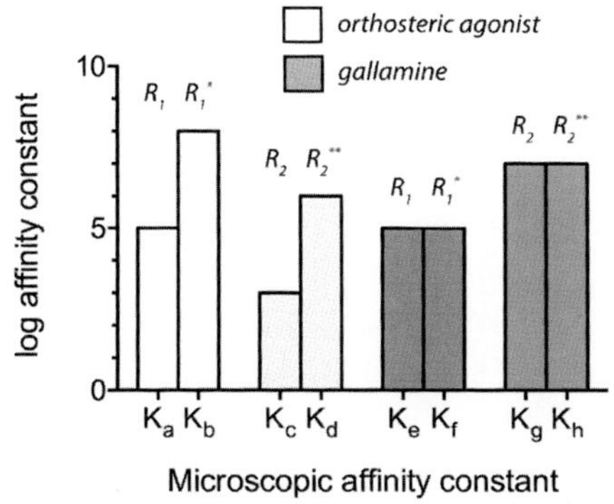

Figure 11.18: An unlikely explanation for affinity-only modulation by gallamine of the response to muscarinic agonists at the M_2 receptor. The explanation is based on the existence of two active and inactive states of the receptor. The histogram shows the affinity constants of the agonist and gallamine for the pairs (R_1 and R_2) of active and inactive states. Further details are given in the text.

impossible because one would expect a random variation between these ratios for each ligand.

The combination of microscopic constants shown in Figure 11.18 is not the only way that a model of two active and two inactive states of the receptor could yield affinity-only modulation by an allosteric ligand. The ratio of affinity constants of the orthosteric ligand for the first pair of active and inactive states (K_b/K_a) could differ from that of the second pair (K_d/K_c) provided that the allosteric ligand has the appropriate ratios of microscopic affinity constants for the two pairs of receptor states (K_f/K_e and K_h/K_g) to offset the difference. In other words, affinity-only modulation can occur whenever the following relationship among microscopic constants is satisfied between the allosteric and orthosteric ligands, assuming that the activities of the two active states are the same:

$$\frac{K_bK_q + K_dK_rK_s}{K_a + K_cK_r} = \frac{K_bK_fK_q + K_dK_hK_rK_s}{K_aK_e + K_cK_gK_r} \tag{11.118}$$

If an orthosteric ligand is found that satisfies this relationship with a given allosteric ligand, then other orthosteric ligands will not exhibit affinity-only modulation with the allosteric ligand unless their ratios of K_b/K_a and K_d/K_c are the same as those of the first orthosteric ligand. This seems highly unlikely. It is also possible that affinity-only modulation could arise with differences in the activities of the two actives states provided that the microscopic constants of the interacting ligands offset this difference. The solution sets that yield affinity-only modulation are sufficiently unique that the possibility of finding a group of orthosteric ligands that exhibit affinity-only modulation with a given allosteric ligand is vanishingly small.

The conclusion of this digression is that multistate models cannot explain how an allosteric ligand can affect the observed affinity of a group of orthosteric ligands acting on the same receptor without also affecting their efficacy.

A speculative explanation for the allosteric effect of gallamine is that when an orthosteric ligand interacts with muscarinic receptors and other *G* protein-linked receptors, it first binds to a peripheral relay site and then shuttles to the orthosteric site. The evidence for this mechanism includes site-directed mutagenesis data showing that aromatic residues and acidic residues outside the orthosteric binding pocket are critical for the high affinity binding of acetylcholine to M_1 and M_2 muscarinic receptors. Also, the crystal structure of the β_1 adrenoceptor suggests that extracellular loop 2 on the receptor is important for endogenous ligand binding and may contribute to the difference in the activity of norepinephrine at β_1 and β_2 adrenoceptors. Thus, what I have been referring to throughout this book as the microscopic affinity constants for active (K_b) and inactive (K_a) states of the receptor may represent a composite of constants ($K_{b-\text{obs}}$ and $K_{a-\text{obs}}$) having two components consisting of the affinity constants of the orthosteric site *and* the relay site. According to this interpretation, $\log K_{b-\text{obs}}$ and $\log K_{a-\text{obs}}$ can be approximated by the following equations:

$$\log K_{b-\text{obs}} = \log K_b + \log K_{\text{rel}}, \tag{11.119}$$

$$\log K_{a-\text{obs}} = \log K_a + \log K_{\text{rel}}. \tag{11.120}$$

In these equations, K_{rel} represents the observed affinity constant of the orthosteric ligand for the relay site. If the relay site can undergo conformational changes independently of the orthosteric site and if only the relay site is modulated by gallamine, then this situation could explain how gallamine could modulate the affinity of orthosteric ligands without modulating their efficacy. For a modulation in the affinity of K_{rel} would cause equivalent changes in $K_{b-\text{obs}}$ and $K_{a-\text{obs}}$, and hence, the ratio of $K_{b-\text{obs}}/K_{a-\text{obs}}$ would be unaffected. If the ratio of $K_{b-\text{obs}}/K_{a-\text{obs}}$ is unaffected, then the modulator will have no effect on the efficacy of the orthosteric ligand.

The relay site model has important consequences that relate to our understanding of how *G* protein-coupled receptors work. First, the model gives rise to an unlimited number of observed microscopic affinity constants for receptor states, based on allosteric modulation of the relay site. This diversity comes about without variation in the microscopic constants of the orthosteric site (K_b and K_a), which is reassuring because receptors that exhibit multiple active and inactive states with varying activity do not function as efficient molecular switches. We would expect that a receptor is an evolutionary adaptation that exhibits a quantal change from low to high activity coincident with the binding of the endogenous ligand. This can only occur if the increase in activation is associated with a rapid change in conformation to a state with much higher affinity for the endogenous ligand ($K_b/K_a \approx 10^4$).

The second consequence of this model is that it distributes the observed affinity over two different binding pockets having different structure-activity relationships. An effective orthosteric ligand must have the requisite structure to satisfy the requirements of both sites. One way to think about receptor selectivity is in probabilistic terms. Imagine the set of all ligands within a size range that is appropriate for a given binding site. The number of ligands that bind to the receptor site with a minimum level of affinity (e.g., $K_{\mathrm{obs}} = 10^3\ \mathrm{M}^{-1}$) divided by the set of all ligands within the size range represents the probability that a randomly selected ligand of appropriate size will bind to the receptor. The probability that a ligand will bind to a pair of different receptors is equal to the product of the individual probabilities for each site. This analogy shows that the combination of a relay site and an orthosteric-binding site greatly increases the selectivity of a receptor because it greatly reduces the probability that a randomly selected ligand will interact with the receptor effectively. For example, some ligands could have high affinity and efficacy at the orthosteric site and have low potency because of a poor fit with the relay site, while other ligands could have high affinity for the relay site, but have low potency and efficacy because of their low affinity for the active state of the orthosteric site. Only ligands that can satisfy the requirements of both sites will be active. The consequences of a relay model indicate that structure–activity relationships are determined, in part, by the relay site.

Analyzing Allosterism Using a Combination of Functional and Binding Methods

This chapter has focused on the analysis of functional data, and it has shown that in many instances, it is impossible to discriminate between affinity (α) and efficacy (β) components of allosteric regulation when the concentration response curve of an agonist is measured in the presence of various concentrations of allosteric modulator. It is always possible to estimate the product of affinity and efficacy modulation (γ) in this type of experiment. If additional data are collected after partial receptor inactivation, then it is often possible to estimate affinity and efficacy modulation. Another approach is to estimate the affinity modulation using a radioligand binding assay. Using a combination of binding and functional methods, it would then be possible to separate out the affinity and efficacy components of allosteric regulation. In the next chapter, the analysis of both competitive and allosteric interactions in radioligand binding assays is described.

References

1. Christopoulos, A. Allosteric binding sites on cell-surface receptors: novel targets for drug discovery. *Nat Rev Drug Discov* **1**, 198–210 (2002).
2. Burgen, A. S. Conformational changes and drug action. *Federation Proceedings* **40**, 2723–2728 (1981).
3. Ehlert, F. J., Roeske, W. R., Gee, K. W. and Yamamura, H. I. An allosteric model for benzodiazepine receptor function. *Biochem Pharmacol* **32**, 2375–2383 (1983).
4. Fung, J. J., Deupi, X., Pardo, L., Yao, X. J., Velez-Ruiz, G. A., Devree, B. T., Sunahara, R. K. and Kobilka, B. K. Ligand-regulated oligomerization of beta(2)-adrenoceptors in a model lipid bilayer. *Embo J* **28**, 3315–3328 (2009).
5. Angers, S., Salahpour, A., Joly, E., Hilairet, S., Chelsky, D., Dennis, M. and Bouvier, M. Detection of beta 2-adrenergic receptor dimerization in living cells using bioluminescence resonance energy transfer (BRET). *Proc Natl Acad Sci USA* **97**, 3684–3689 (2000).
6. Whorton, M. R., Bokoch, M. P., Rasmussen, S. G., Huang, B., Zare, R. N., Kobilka, B. and Sunahara, R. K. A monomeric *G* protein-coupled receptor isolated in a high-density lipoprotein particle efficiently activates its *G* protein. *Proc Natl Acad Sci USA* **104**, 7682–7687 (2007).
7. Wu, B., Chien, E. Y., Mol, C. D., Fenalti, G., Liu, W., Katritch, V., Abagyan, R., Brooun, A., Wells, P., Bi, F. C., Hamel, D. J., Kuhn, P., Handel, T. M., Cherezov, V. and Stevens, R. C. Structures of the CXCR4 chemokine GPCR with small-molecule and cyclic peptide antagonists. *Science* **330**, 1066–1071 (2010).
8. Hoare, S. R. Mechanisms of peptide and nonpeptide ligand binding to Class B G-protein-coupled receptors. *Drug Discov Today* **10**, 417–427 (2005).
9. Haga, K., Kruse, A. C., Asada, H., Yurugi-Kobayashi, T., Shiroishi, M., Zhang, C., Weis, W. I., Okada, T., Kobilka, B. K., Haga, T. and Kobayashi, T. Structure of the human M2 muscarinic acetylcholine receptor bound to an antagonist. *Nature* **482**, 547–551 (2012).
10. Matsui, H., Lazareno, S. and Birdsall, N. J. Probing of the location of the allosteric site on m1 muscarinic receptors by site-directed mutagenesis. *Mol Pharmacol* **47**, 88–98 (1995).
11. Trankle, C., Dittmann, A., Schulz, U., Weyand, O., Buller, S., Johren, K., Heller, E., Birdsall, N. J., Holzgrabe, U., Ellis, J., Holtje, H. D. and Mohr, K. Atypical muscarinic allosteric modulation: cooperativity between modulators and their atypical binding topology in muscarinic M2 and M2/M5 chimeric receptors. *Mol Pharmacol* **68**, 1597–1610 (2005).
12. Kunishima, N., Shimada, Y., Tsuji, Y., Sato, T., Yamamoto, M., Kumasaka, T., Nakanishi, S., Jingami, H. and Morikawa, K. Structural basis of glutamate recognition by a dimeric metabotropic glutamate receptor. *Nature* **407**, 971–977 (2000).

13. Tsuchiya, D., Kunishima, N., Kamiya, N., Jingami, H. and Morikawa, K. Structural views of the ligand-binding cores of a metabotropic glutamate receptor complexed with an antagonist and both glutamate and Gd3+. *Proc Natl Acad Sci USA* **99**, 2660–2665 (2002).
14. Suzuki, Y., Moriyoshi, E., Tsuchiya, D. and Jingami, H. Negative cooperativity of glutamate binding in the dimeric metabotropic glutamate receptor subtype 1. *The Journal of Biological Chemistry* **279**, 35526–35534 (2004).
15. Black, J. W. and Leff, P. Operational models of pharmacological agonism. *Proceedings of the Royal Society of London. Series B: Biological Sciences* **220**, 141–162 (1983).
16. Gregory, K. J., Hall, N. E., Tobin, A. B., Sexton, P. M. and Christopoulos, A. Identification of orthosteric and allosteric site mutations in M2 muscarinic acetylcholine receptors that contribute to ligand-selective signaling bias. *J Biol Chem* **285**, 7459–7474 (2010).
17. Marino, M. J., Williams, D. L., Jr., O'Brien, J. A., Valenti, O., McDonald, T. P., Clements, M. K., Wang, R., DiLella, A. G., Hess, J. F., Kinney, G. G. and Conn, P. J. Allosteric modulation of group III metabotropic glutamate receptor 4: a potential approach to Parkinson's disease treatment. *Proc Natl Acad Sci USA* **100**, 13668–13673 (2003).
18. Suzuki, G., Kimura, T., Satow, A., Kaneko, N., Fukuda, J., Hikichi, H., Sakai, N., Maehara, S., Kawagoe-Takaki, H., Hata, M., Azuma, T., Ito, S., Kawamoto, H. and Ohta, H. Pharmacological characterization of a new, orally active and potent allosteric metabotropic glutamate receptor 1 antagonist, 4-[1-(2-fluoropyridin-3-yl)-5-methyl-1H-1,2,3-triazol-4-yl]-N-isopropyl-N- methyl-3,6-dihydropyridine-1(2H)-carboxamide (FTIDC). *J Pharmacol Exp Ther* **321**, 1144–1153 (2007).
19. Ehlert, F. J. Gallamine allosterically antagonizes muscarinic receptor-mediated inhibition of adenylate cyclase activity in the rat myocardium. *Journal of Pharmacology and Experimental Therapeutics* **247**, 596–602 (1988).
20. Ehlert, F. J. and Griffin, M. T. Two-state Models and the Analysis of the Allosteric Effect of Gallamine at the M2 Muscarinic Receptor. *J Pharmacol Exp Ther* **325**, 1039–1060 (2008).
21. Monod, J., Wyman, J. and Changeux, J. P. On the Nature of Allosteric Transitions: a Plausible Model. *J Mol Biol* **12**, 88–118 (1965).

PART 3
Radioligand Binding

Analysis of Drug-Receptor Interactions Using Radioligand Binding Assays on *G* Protein-Coupled Receptors

12

In prior chapters, the interaction of drugs with receptors has been discussed exclusively from the perspective of functional assays. In this chapter, I describe how radioligand binding methods can be used to throw light on this topic. Having an alternative means of measuring drug-receptor interactions can often strengthen or extend conclusions drawn from functional studies.

For example, in Chapter 10 methods for estimating the affinity constant of an antagonist in functional assays were presented. If several such measurements are made for different antagonists based on their ability to antagonize a particular response to an agonist, one can characterize the pharmacological profile of the receptor mediating the response. This profile can then be compared with that measured using radioligand binding assays on cells expressing a known receptor subtype. By comparing the functional profile with the binding profiles of different known receptor subtypes, it should be possible to identify the receptor mediating the response.

Figure 12.1 shows an example of this type of comparison. Here, the affinity constants of a series of muscarinic antagonists, estimated by antagonism of the phosphoinositide response in the rat ileum, are plotted against the binding affinities measured on homogenates of cells expressing recombinant M_1–M_5 receptors (Figures 12.1a–12.1e, respectively). While a given antagonist may exhibit a functional estimate of affinity that agrees with binding affinity at more than one receptor subtype, it is only at the M_3 receptor where there is reasonable agreement for all of the antagonists. This same type of analysis could include a comparison of the

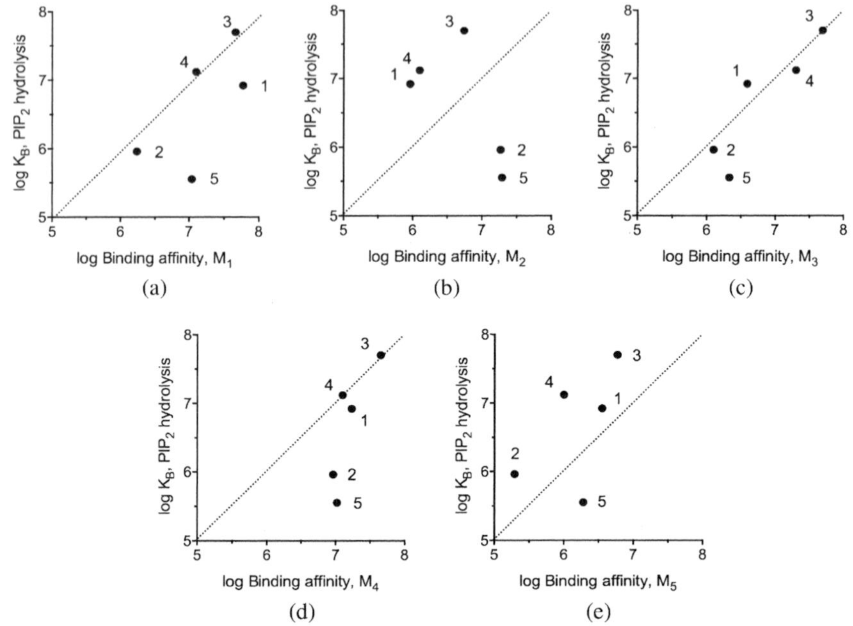

Figure 12.1: Relationship between the affinity constants of antagonists estimated by antagonism of the phosphoinositide response in gastrointestinal smooth muscle and by competitive inhibition of [^{3}H]N-methylscopolamine binding to M_1 (a), M_2 (b), M_3 (c), M_4 (d) and M_5 (e) subtypes of the muscarinic receptor. The antagonists are pirenzepine, 1; AF-DX 116, 2; hexahydrosiladifenidol, 3; *p*-fluorohexahydrosiladifenidol, 4 and methoctramine, 5. The K_B values, estimated by antagonism of the muscarinic phosphoinositide response in the longitudinal muscle of the rat ileum, are from Candell *et al.*[13] The estimates of binding affinity for muscarinic receptor subtypes are from Esqueda *et al.*[14] and Ehlert.[15]

affinities of allosteric ligands and the magnitude of the cooperativity (α) between the binding of the allosteric ligand and the agonist used in the functional assay.

Binding assays also provide a means of determining the nature of drug-receptor interactions. For example, an inhibitor might interfere with a response to an agonist in a manner that resembles competitive inhibition but through a mechanism that involves interrupting a step downstream in the signaling pathway. In this case, the inhibitor blocks the response to the agonist by a mechanism that does not involve competitive inhibition at the orthosteric site. Therefore, if the functional data were analyzed assuming a competitive mechanism, the corresponding estimate of the affinity constant (K_I) would be inaccurate. This situation could be addressed by measuring the binding properties of the inhibitor to the receptor that mediates the response. If the inhibitor's binding affinity is substantially less than its functional estimate of affinity, then this result would indicate that it causes an inhibition of function through a target other than the receptor. A more accurate functional estimate of the antagonist's affinity for the receptor could be measured using resultant analysis as described in Chapter 10 in relation to Figure 10.12.

Another example involves comparing allosteric and competitive mechanisms of inhibition. Some negative allosteric modulators with substantial negative cooperativity have effects that resemble competitive inhibitors. Whether the drug interacts with the receptor at the orthosteric site or a distinct allosteric site could be investigated using radioligand binding techniques. Ultimately, the observed affinity and cooperativity of an allosteric modulator determined in a binding assay should agree with estimates made in functional assays if the same receptor mediates the response in both cases.

This chapter is written from the perspective of comparing the results of radioligand-binding and functional studies. Obviously, there are many other important uses of receptor radioligands, but this chapter focuses on the latter topic. Several studies have shown that ionic strength can have substantial effects on the interaction of agonists and antagonist with GPCRs (e.g., see Pedder *et al.*[1]). Hence, in any comparison of functional and binding data, only the results of experiments run in a buffer of similar composition should be considered.

As described in the Prologue and throughout this book, drug-receptor interactions can be addressed at different hierarchical levels. In considering ligand binding data, two levels apply. We can consider whether the drug-receptor complex is in the active or inactive state (level 3) or whether the receptor is occupied regardless of what state it is in (level 2, population analysis). We can also consider the behavior of single receptors or a population of receptors. In the following discussion, I consider occupancy of a population of receptors without concern for the state of the receptor. That is, drug-receptor complexes in either the active or inactive state both count as occupied receptors.

From this perspective, consider occupancy of a population of a single type of receptor in a uniform environment. As the concentration of drug increases, occupancy of the receptor increases. The relationship between the drug concentration and occupancy resembles a simple one-site model having a maximum equivalent to the total amount of receptors in the population. The free concentration of drug that yields half-maximal receptor occupancy is equivalent to the observed dissociation constant (K_D). The reciprocal of this parameter is known as the observed affinity constant (K_{obs}). The observed affinity constant of a ligand measured in a functional assay as described throughout Chapters 7–12, for example, should agree with that measured in a binding assay, provided that the same receptor is being addressed in both assays.

In the case of antagonists, the latter relationship is often easy to demonstrate in experiments on real receptors in cellular homogenates, for example.[2,3] When measured in buffer of physiological ionic strength many GPCRs are primarily in the inactive state dissociated from the *G* protein. Alternatively, they may form a reversible complex with the holoprotein form of the *G* protein bound with GDP. In either case, the receptor population is almost entirely in the inactive state except for rare spontaneous activations (very low constitutive activity). This condition is likely to hold even if only a part of the receptor population communicates with *G* proteins. Thus, the binding properties of antagonists for several GPCRs often conform to a simple one-site model[3] and the observed affinity constant measured in a binding assay often agrees with that measured in a functional assay using the procedures described in Chapter 10 for example.

In the case agonists, however, there are complications as described in detail in Chapter 6. The results of binding studies on washed homogenates of cells expressing G_s- and G_i-linked receptors often exhibit high- and low-affinity agonist binding sites.[4,5] This type of data is often analyzed using a two-site binding equation, and the two sites are often designated as high and low affinity states of the receptor. The associated affinity constants (K_H and K_L, respectively) describe occupancy of the two receptor populations and do not represent those of the active and inactive states of the receptor, however. Thus, they are not receptor states in the same sense as how the term "state" is applied throughout this book. The most likely explanation is that the high-affinity site represents a population of receptors that communicates with a *G* protein whereas the low-affinity represents another population that does not. The *G* protein allosterically increases the observed affinity of the *G* protein-coupled population but not that of the population with lower observed affinity. The observed affinity constant of an agonist (K_{obs}) measured using a functional assay (see Chapter 7) is not expected to agree with either observed binding affinity of the high and low affinity receptor populations (i.e., K_H and K_L).

The factor that causes a discrepancy between the functional estimate of K_{obs} and the observed affinity of the receptor population that communicates with the *G* protein (K_H) is GTP. For a receptor to signal through *G* proteins, GTP must be present, and it is the observed affinity of the agonist for the receptor-*G* protein-GDP complex (quaternary complex) that should agree with the functional estimate of K_{obs} (see Chapter 6).

Measuring the affinity constant of the quaternary complex using ligand-binding methods is not trivial. It is often assumed that the concentration of GTP is at *G* protein-saturating concentrations in the intact cell.[6,7] If so, binding properties could be measured in the presence of high concentrations of GTP or GDP. There are additional complications, however. The observed affinity of the quaternary complex might be different from that of the receptor population that does not interact with G protein, so it might be difficult to accurately discriminate between these populations. Also, there is some circumstantial evidence that the concentration of GTP is not sufficiently high to saturate G proteins in some cells. Finally, there may be different G proteins that communicate with the receptor and the affinities of the corresponding agonist quaternary complexes may be different. For all of these reasons it may be difficult to determine the observed affinity of an agonist using binding methods under the conditions that the receptor normally functions. Thus, the approach of comparing binding affinity with functional estimates of affinity is potentially ambiguous when applied to the analysis of agonists.

The remainder of this chapter describes methods used to analyze binding data under equilibrium conditions. The first section describes how to analyze the binding of a radioligand. Having a high affinity radioligand with low nonspecific binding enables one to investigate how nonradioactive ligands interact with the receptor by examining how they modify the binding of the radioactive probe. This latter topic is subsequently covered in two sections on competitive inhibition and allosteric modulation.

Analysis of the Binding of Radioligand

Different methods can be used to measure the binding of a radioligand to a receptor preparation. My intent here is not to describe the details of these assays, but rather, to describe the nature of reversible binding interactions so that the appropriate experimental conditions can be achieved for obtaining optimal estimates of binding affinity.

Nature of reversible binding interactions

When incubated with homogenized cells expressing a receptor of interest, a radioligand not only binds in a reversible manner to the receptor, but also to other nonreceptor entities. The ligand-receptor interaction is usually of high affinity and low capacity, whereas those of nonreceptor interactions are of low affinity and high capacity. The former is designated as specific binding and the latter as nonspecific.

The specific component of binding represents the reversible binding of the radioligand (L) to the receptor (R), and it is described by the following scheme for the condition of equilibrium:

$$L + R \overset{K_L}{\longleftrightarrow} LR. \tag{12.1}$$

The equilibrium affinity constant of the radioligand (K_L) is defined as:

$$K_L = \frac{[LR]}{[L][R]}. \tag{12.2}$$

The constant K_L is equivalent to the observed affinity constant (K_{obs}). The reciprocal of K_L is equivalent to the observed dissociation constant (K_D).

To derive an equation for occupancy of the receptor by the radioligand at equilibrium, expressed as a function of the ligand concentration, we begin with the following equation:

$$\text{Occupancy} = \frac{[LR]}{[LR] + [R]}. \tag{12.3}$$

Equation (12.2) can be rearranged to solve for $[LR]$ and the result substituted into Equation (12.3). With simplification, the following equation is obtained:

$$\text{Occupancy} = \frac{[L]}{[L] + \frac{1}{K_L}}. \tag{12.4}$$

Multiplying both sides of this equation by a constant representing receptor density (B_{max}) yields an equation for specific binding (Bound_{spec}), expressed in the same units as B_{max}.

$$\text{Bound}_{spec} = \frac{[L]B_{max}}{[L] + \frac{1}{K_L}}. \tag{12.5}$$

In this equation, Bound_{spec} is equivalent to the product of Occupancy and B_{max}. In contrast, nonspecific binding ($\text{Bound}_{nonspec}$) usually increases in a linear fashion with the concentration of ligand. Therefore, it can be described by the following equation:

$$\text{Bound}_{nonspec} = [L]C_{nonspec}, \tag{12.6}$$

in which, $C_{nonspec}$ represents the coefficient of nonspecific binding.

Figure 12.2a illustrates the relationship between the specific and nonspecific binding of a radioligand ($L1$) at equilibrium plotted against its free concentration on a linear scale. The same functions are plotted in Figure 12.2b against the log concentration of radioligand. Specific binding asymptotically reaches a plateau equivalent to B_{max} as the concentration of radioligand increases. The free concentration or radioligand required to achieve half-maximal binding is equivalent to the reciprocal of K_L.

In contrast, nonspecific binding increases linearly and is proportional to the concentration of radioligand. In this example, the nonspecific binding of radioligand

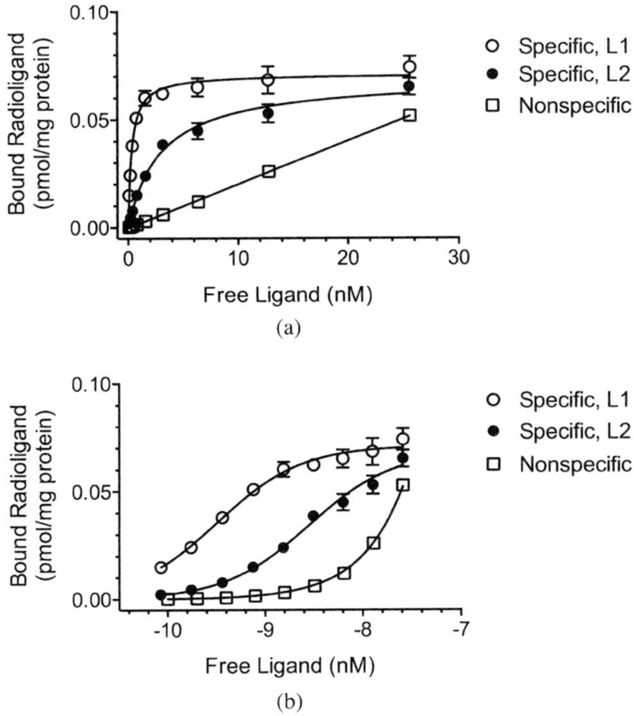

Figure 12.2: Theoretical relationship between the specific and nonspecific binding of high (L1) and low (L2) affinity radioligands having the same coefficient of nonspecific binding. (a) Specific and nonspecific binding are plotted against the free concentration of radioligand. (b) Specific and nonspecific binding are plotted against the log free concentration of radioligand.

L1 only represents about 1% that of specific binding when the radioligand concentration is sufficient for half-maximal specific binding. Thus, there is an adequate signal to noise ratio, and it should be possible to estimate the specific component of binding reasonably well for this hypothetical ligand.

Figure 12.2a also shows the binding profile of radioligand *L2*, which has the same coefficient of nonspecific binding as radioligand *L1*, but an affinity constant only one-tenth that of *L1*. In the case of radioligand *L2*, nonspecific binding is approximately 10% that of specific binding at a concentration of radioligand that yields half-maximal binding. At a radioligand concentration that yields near maximal binding (25 nM), nonspecific binding is approximately the same as specific binding. Estimates of the specific binding of radioligand *L2* and its associated binding parameters (log K_L and $B_{\max}$) are expected to exhibit much greater variance than those of *L1*. When the data are plotted against the log concentration of radioligand (Figure 12.2b), it can be seen that nonspecific binding increases dramatically relative to B_{max} at concentrations of *L2* exceeding 10^{-8} M. Not surprisingly, many commercially available radioligands have affinity constants that are greater the 10^8 M^{-1} (i.e., $K_D < 10^{-8}$ M).

In most cases, it is impossible to measure specific binding directly. Rather, total binding (specific plus nonspecific) and nonspecific binding are measured, and specific binding is calculated as their difference. Nonspecific binding can be measured directly by taking advantage of the properties of specific binding. Again, specific binding represents the reversible binding of the radioligand to a receptor target in very low abundance, often only 0.001–0.10 pmol/mg wet weight of tissue. Such binding can be competitively displaced by a ligand that binds to the same site. If a high affinity nonlabeled competitive inhibitor is incubated with the receptor preparation and radioligand, it will nearly completely displace the binding of radioligand from the receptor without affecting nonspecific binding. Nonspecific binding resembles the manner in which compounds partition across an interface of aqueous and organic solvents. The extent to which a given compound occupies the organic phase is independent of other compounds. Thus, by measuring the binding of radioligand by itself (total binding, $\text{Bound}_{\text{total}}$) and in the presence of a high concentration of competitive inhibitor (nonspecific binding, $\text{Bound}_{\text{nonspec}}$), one can obtain the necessary information to calculate specific binding ($\text{Bound}_{\text{spec}}$):

$$\text{Bound}_{\text{spec}} = \text{Bound}_{\text{total}} - \text{Bound}_{\text{nonspec}}. \tag{12.7}$$

A common strategy is to measure nonspecific binding for every concentration of radioligand used in the experiment so that the calculation described by Equation (12.7) can be made. In other words, for every measurement of total binding an additional set of nonspecific binding measurements are made under the same conditions except for the inclusion of a high concentration of nonlabeled competitive inhibitor in the assay.

With regard to the concentration of noncompetitive inhibitor, a useful approach is to include it in the assay at a concentration about 1000-fold greater than its dissociation constant (reciprocal of its affinity constant). For example, consider a radioligand and nonlabeled competitor with affinity constants of 10^9 and 10^8 M^{-1}, respectively. When the radioligand is used at concentrations equivalent to its dissociation constant (1 nM) and 10-fold greater (10 nM), the specific binding of the radioligand will be reduced to values only 0.2% and 1.1% of those measured in the absence of competitor. Generally, it is a good idea, but not essential, to use a competitive inhibitor that exhibits high affinity for the receptor so that its concentration is never very high. This strategy will reduce the already unlikely possibility that it will displace nonspecific binding.

An example of the types of measurements made in the estimation of the affinity constant and maximal binding capacity of a radioligand-receptor complex is illustrated in Figure 12.3. In Figure 12.3a, the total and nonspecific binding measurements of the radioligand are shown for various concentrations of the radioligand. In this hypothetical example, let us assume that a centrifugation assay is used to measure binding. Often, this approach is unfeasible for lipophilic ligands because it

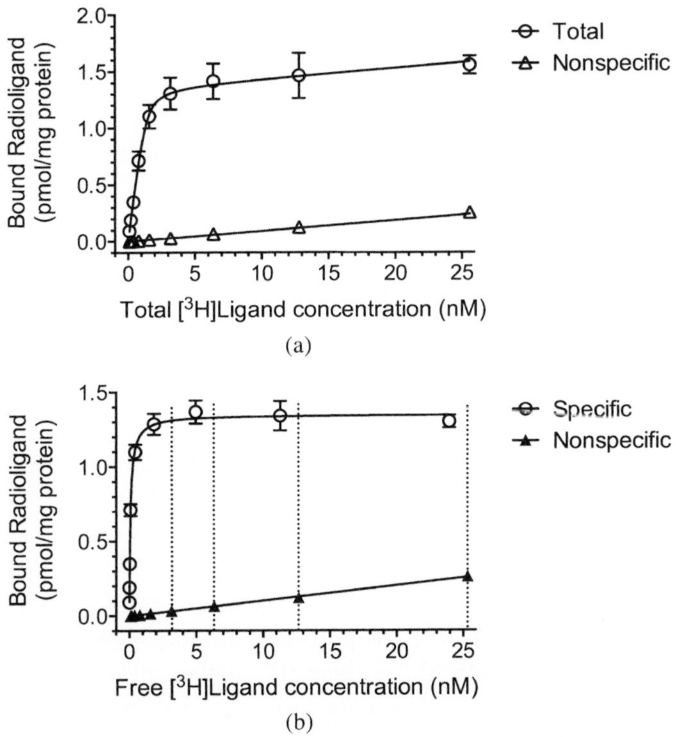

Figure 12.3: Simulation of binding curves plotted against the total (a) and free (b) radioligand concentrations. (a) Theoretical values for total and nonspecific binding are plotted against the total radioligand concentration. The theoretical curves represent the least-squares fit of Equations (12.20) (total binding) and (12.24) (nonspecific binding) to the data. (b) Theoretical values for specific and nonspecific binding are plotted against the free radioligand concentration. The theoretical curves represent the least-squares fit of Equations (12.5) (specific binding) and (12.6) (nonspecific binding) to the data. The vertical dotted lines intersect the abscissa at the corresponding free radioligand concentration of the nonspecific binding measurement.

yields high levels of nonspecific binding. For hydrophilic ligands, like those often used for the muscarinic receptor, it yields very low nonspecific binding.

Cellular homogenate is first incubated in small centrifuge tubes with radioligand for a period of time until equilibrium is achieved. An analogous set of measurement is made under the same conditions except for the inclusion of the nonlabeled competitor for estimation of nonspecific binding. At the end of the incubation, the cellular particles are pelleted by centrifugation. The supernatant is aspirated, and the pellets are superficially washed twice. After inverting the tubes to drain residual wash buffer, the pellets are dissolved in strong base and subsequently neutralized. The contents of the tubes are quantitatively transferred to scintillation vials for measurement of binding (i.e., the radioactivity associated with the cellular homogenate). The results of such an experiment are shown in Figure 12.3a.

Although measurements of total and nonspecific binding were made at the same total concentration of radioligand, the free concentration of radioligand for the measurement of nonspecific binding is greater than that of the total binding determination. This situation occurs because the assay was carried out in a small volume, and the amount of binding in the total binding case is greater than that of the nonspecific binding case. Any radioligand that is bound is no longer free in solution, and consequently, binding causes depletion in the free concentration of radioligand. This depletion is proportionately greater at the lower concentrations of radioligand.

If total and nonspecific binding are plotted against the free concentration of radioligand, the plot shown in Figure 12.3b is obtained. This plot shows that for a pair of total and nonspecific binding measurements made at the same total concentration of radioligand, there is a small difference between the measured value of nonspecific binding and the predicted value for the total binding measurement (see discrepancy between the free radioligand concentration for the specific binding measurements and the dotted vertical lines denoting the free radioligand concentrations for the associated nonspecific binding measurements). This is because the free concentration of radioligand is lower for the total binding measurement. Therefore, the level of nonspecific binding must be interpolated at the free radioligand concentration of the total binding measurement if the calculation described by Equation (12.7) is to be used precisely. Using this approach, the specific binding values shown in Figure 12.3b were calculated.

This example was specifically described for a centrifugation assay because it is easy to understand how the depletion in the free ligand concentration caused by total and nonspecific binding can be estimated. One simply subtracts the mass of radioligand bound from the total mass of drug in solution and determines the free concentration based on the difference.

A much more common method for measuring binding employs a filtration technique. This approach often reduces the nonspecific binding measurement because the cellular homogenate is trapped on filters after the incubation with radioligand and washed. The washing reduces nonspecific homogenate binding. If the specific binding of the radioligand is of sufficiently high affinity, little or no radioligand will dissociate from the receptor during the washing process. This method introduces an additional component to the nonspecific binding measurement — nonspecific filter binding. This is usually not a problem in terms of obtaining a nice signal to noise ratio with regard to specific and nonspecific binding. With this method, however, it is impossible to estimate what level of nonspecific binding to cellular homogenate occurred during the experiment. Much of the nonspecific cellular binding has been washed away and the nonspecific filter binding component of the nonspecific binding measurement does not deplete the free radioligand concentration because this binding occurred after the incubation. Therefore, when the filtration assay is used, there is usually no choice but to ignore the potential depletion in the free radioligand concentration caused by nonspecific binding. This is usually not a problem with ligands that have low nonspecific binding.

The focus of this last chapter is the measurement of binding parameters at equilibrium because this is an ideal condition for estimating the observed affinity constants of ligands. Therefore, it is essential that data should be collected at equilibrium if the intent is to analyze the data with an equilibrium model.

The time required to achieve equilibrium depends on the observed rate constants for association and dissociation, which are described below:

$$[L] + [R] \xrightarrow{k_{+1}} [DL], \tag{12.8}$$

$$[DL] \xrightarrow{k_{-1}} [L] + [R]. \tag{12.9}$$

These rate constants are observed constants and are not necessarily the same as the rate constants for association and dissociation of the active (k_{b+} and k_{b-}) or the inactive (k_{a+} and k_{a-}) states of the receptor that were described in association with Markov analysis of single channel events in Chapter 2. We are using observed constants because our analysis focuses on ligand binding without concern for the state of the receptor as described above. It can be shown that the function describing the specific binding as a function of time (t) in the presence of a fixed concentration of radioligand is:

$$\text{Bound}_{\text{spec}} = (1 - e^{-t/\tau})\frac{[L]B_{\max}}{[L] + \frac{1}{K_L}}, \tag{12.10}$$

in which,

$$\frac{1}{\tau} = k_{-1} + k_{+1}[L]. \tag{12.11}$$

In these expressions, τ, denotes the time constant.

Equation (12.10) shows that binding as a function of time is predicted by the product of an increasing exponential $(1 - e^{-t/\tau})$ and equilibrium binding $([L]B_{\max}/([L] + 1/K_L)$. Thus, the rate of the process is governed by the time constant, τ.

Figure 12.4 illustrates the binding of a hypothetical radioligand at various concentrations and incubation times. The plot shows that the time required to reach equilibrium depends on the concentration of radioligand. At a high concentration, the radioligand reaches equilibrium quickly, whereas at a low concentration, equilibrium is achieved more slowly. Thus, it is important to insure that equilibrium is achieved with the lowest concentration of radioligand used in the experiment. Equation (12.11) shows that the time constant for binding achieves a maximum value equivalent to the reciprocal of k_{-1} when the concentration of radioligand approaches zero. Thus, a simple strategy for estimating an incubation period for the binding assay is to first measure the rate constant for dissociation (k_{-1}). This can be done by incubating receptor with radioligand for a period of time, and then measuring specific binding at various times after addition of a high concentration of competitive inhibitor. The competitive inhibitor should prevent the association of radioligand with the receptor. Consequently, specific binding should decay exponentially following the addition of competitive inhibitor. The binding values, expressed relative to that measured before addition of the competitor, should be consistent

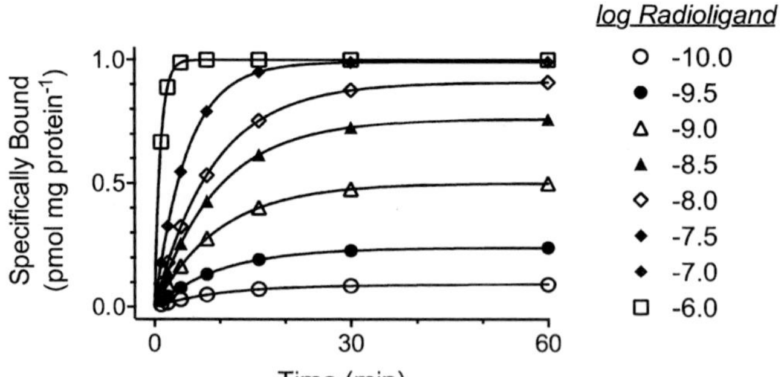

Figure 12.4: Simulation of the kinetics of radioligand binding. Equation (12.10) was used to simulate the binding of a radioligand at various concentrations and after various incubation times. The rate constants for the simulations were k_{+1}, 10^6 M^{-1} min^{-1} and k_{-1}, 0.1 min^{-1}.

with the following exponential decay equation:

$$\text{Bound}_{\text{spec}} = e^{-tk_{-1}}. \tag{12.12}$$

Thus, one can fit the binding data for the dissociation experiment to an exponential decay equation to estimate k_{-1}. If the reciprocal of k_{-1} is multiplied by the negative of the natural log of 0.5 (−0.69) the result is the half time for dissociation. A useful incubation time for insuring the equilibrium is achieved is equivalent to four or five half times for dissociation.

In some instances, it has been shown that the dissociation rate of a radioligand depends on the time that the radioligand is first incubated with the receptor. Slower rates have been observed for longer equilibration periods, suggesting that a relay site is involved in binding[8] or that the ligand-receptor complex undergoes a slow isomerization to a higher affinity complex.[9,10] Thus, it is prudent to use a long pre-incubation time when estimating the rate constant for dissociation if the goal is to determine an appropriate incubation time for equilibrium binding.

Estimation of the equilibrium binding parameters of the radioligand

In this section, examples of how to estimate the binding parameters of a radioligand (K_{obs} and B_{max}) are described. These examples differ in terms of how nonspecific binding is estimated and whether the free or total radioligand concentration is used as the independent variable.

Let us begin with a theoretical example like that shown in Figure 12.3. Recall that this simulation represents the results of an equilibrium binding experiment employing the centrifugation technique described above. The amount of depletion in the free ligand concentration by both nonspecific and specific binding can be calculated as well as the difference in the free concentration of radioligand for the total and nonspecific binding measurements.

The most accurate approach is to fit a regression equation to the measured data with the total ligand concentration being the independent variable. We will need two equations, one to describe the nonspecific binding data and the other to describe the total binding data. We will begin with the derivation of the total binding equation. This equation needs to describe the summation of the specific and nonspecific components of binding as a function of the total ligand concentration ($[L_T]$). In the sequence below, I begin with the derivation of an equation that expresses the free ligand concentration as a function of the total ligand concentration. Then this equation is substituted for $[L]$ in the equation for total binding. Those uninterested in

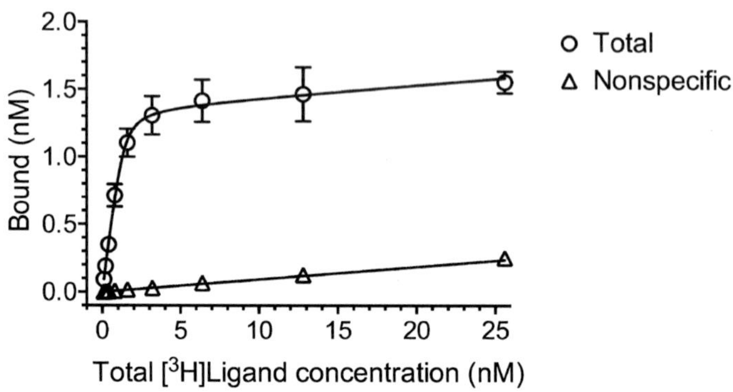

(a)

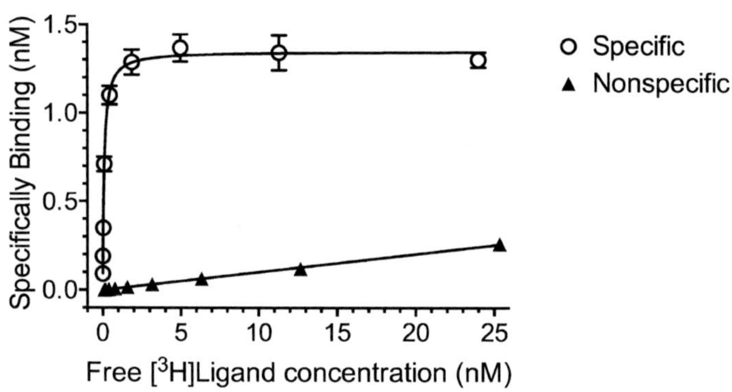

(b)

Figure 12.5: Simulation of binding curves plotted against the total (a) and free (b) radioligand concentration. The data were simulated for the condition of a centrifugation assay in which both total and nonspecific binding cause depletion in the free concentration of radioligand. (a) Theoretical values for total and nonspecific binding are plotted against the total radioligand concentration. The theoretical curves represent the least-squares fit of Equations (12.20), with Equation (12.19) substituted for $[L]$ (total binding), and Equation (12.24) (nonspecific binding) to the data. (b) Theoretical values for specific and nonspecific binding are plotted against the free radioligand concentration. The theoretical curves represent the least-squares fit of Equations (12.5) (specific binding) and (12.6) (nonspecific binding) to the data.

this derivation may want to skip ahead after Equation (12.24), where an explanation of how to analyze the theoretical data in Figure 12.5 is given.

For the total binding condition, the total ligand concentration is equivalent to the free ligand concentration ($[L]$) plus the amount bound to the receptor

($\text{Bound}_{\text{spec}}$) and the amount bound nonspecifically ($\text{Bound}_{\text{nonspec}}$).

$$[L_T] = [L] + [\text{Bound}_{\text{spec}}] + [\text{Bound}_{\text{nonspec}}]. \tag{12.13}$$

Substituting Equation (12.5) and (12.6) for specific and nonspecific binding yields:

$$[L_T] = [L] + \frac{[L]B_{\max}}{[L] + \frac{1}{K_L}} + [L]C_{\text{nonspec}}. \tag{12.14}$$

This equation is rearranged to yield a quadratic equation:

$$0 = A[L]^2 + B[L] + C, \tag{12.15}$$

in which

$$A = K_L(1 + C_{\text{nonspec}}), \tag{12.16}$$

$$B = (1 + C_{\text{nonspec}} + K_L(B_{\max} - [L_T])), \tag{12.17}$$

$$C = -[L_T]. \tag{12.18}$$

The solution is:

$$[L] = \frac{-B + \sqrt{B^2 - 4AC}}{2A}. \tag{12.19}$$

Equation (12.18) expresses the free ligand concentration as a function of the variables A, B and C, which are expressed in terms of the total ligand concentration and binding parameters (K_L, $B_{\max}$ and C_{nonspec}) (Equations (12.15)–(12.17)).

The equation for total binding is simply equal to the summation of the equations for specific (Equation (12.5)) and nonspecific binding (Equation (12.6)):

$$\text{Bound}_{\text{total}} = \frac{[L]B_{\max}}{[L] + \frac{1}{K_L}} + [L]C_{\text{nonspec}}. \tag{12.20}$$

Equation (12.20) is our goal even though it expresses total binding as a function of the free ligand concentration. We can simply replace $[L]$ in Equation (12.20) with the equation for $[L]$ expressed in terms of $[L_T]$ (Equation (12.19)).

The next step is to derive an equation for nonspecific binding as a function of the total ligand concentration. Again we begin with an equation expressing the total ligand concentration as the summation of free and bound ligand. In this case, there is only nonspecific binding:

$$[L_T] = [L] + [\text{Bound}_{\text{nonspec}}]. \tag{12.21}$$

Substituting Equation (12.6) for nonspecific binding yields:

$$[L_T] = [L] + [L]C_{\text{nonspec}}. \tag{12.22}$$

Solving for $[L]$ yields:

$$[L] = \frac{[L_T]}{1 + C_{\text{nonspec}}}. \tag{12.23}$$

This equation for the free ligand concentration is substituted for $[L]$ in the equation for nonspecific binding (Equation (12.6)) to yield:

$$\text{Bound}_{\text{nonspec}} = \frac{[L_T]C_{\text{nonspec}}}{1 + C_{\text{nonspec}}}. \tag{12.24}$$

Let us now use Equations (12.20) and (12.24) to analyze the simulated binding data in Figure 12.5a. Nonlinear regression analysis is used to fit Equations (12.20) and (12.4) to the total and nonspecific binding data, respectively, sharing the estimate of the coefficient of nonspecific binding (C_{nonspec}) between the data and obtaining estimates of the log affinity constant (log K_L) and binding capacity (B_{max}) from the total binding curve. Below I describe an approach using the Prism application (GraphPad Software Inc., San Diego).

Replicate estimates of total binding are entered into a Data Table under sub-columns of column A and the corresponding measurements of nonspecific binding are entered into subcolumns of column B. For this type of analysis, measurements of both total and nonspecific binding are expressed in units of concentration, the same as those used for the total ligand concentration. The total concentration of radioligand for each measurement of total and nonspecific binding is entered into the column X on the far left.

The following lines are entered into Prism for the regression:

```
<A>P = 1

<B>P = 0

A = ((10^LogKL)*(Cnonspec + 1)

B = Cnonspec + 1 + (10^LogKL)*(Bmax-X)

C = -X

Free = (-B + (B^2 - 4*A*C)^0.5)/(2*A)
```

Y = P*(Free*Bmax/(Free + 1/(10^LogKL)) + Free*Cnonspec) + (1 − P)*X/
(1 + Cnonspec).

The last line represents the regression equation, and it is equivalent to the summation Equations (12.20) and (12.24) except that the total binding component is multiplied by the dummy variable P and the nonspecific binding component is multiplied by $1 - P$. Note that X denotes the total radioligand concentration and *Free*, the free concentration of radioligand. The other variables are recognizable based on the resemblance to the variables used in the equations. When the total binding data are analyzed, P is assigned a value 1 so that only the total binding component of the equation is used. When the nonspecific component of binding is analyzed, P is assigned a value 0 so that only the nonspecific binding component of the equation is used.

The first two lines assign the dummy variable a value of 1 or 0, depending whether the data in the columns labeled A (total binding) or B (nonspecific binding) are being analyzed. The next three lines define the variables A, B and C of the quadratic equation, and these are equivalent to Equations (12.16)–(12.18). Finally, the equation for the free concentration of radioligand for the total binding case (*Free*) is given, which is equivalent to Equation (12.19).

When this equation is used to analyze the simulated data in Figure 12.5a the following parameter estimates were obtained: B_{max}, 1.36 ± 0.042 nM; log K_L, $0.97 \pm 0.042\,\text{nM}^{-1}$ and $C_{nonspec}$, 0.0093. In this simulated example, the volume of the assay was 1 ml, and 1 mg of protein was added to each incubation. Thus, the B_{max} value can be converted to 1.36 ± 0.04 pmol/mg protein. The log K_L is expressed in units of inverse nM. This estimate can be converted to inverse molar units: log K_L, $9.97 \pm 0.04\,\text{M}^{-1}$. The estimates of $C_{nonspec}$, B_{max} and log K_L are nearly the same as those used to simulate the data: 0.01, 1.40 pmol/mg protein and 10, respectively.

The plot in Figure 12.5b shows specific and nonspecific binding plotted against the free concentration of radioligand under the conditions of the total binding and nonspecific binding measurements, respectively. In this example, there was substantial depletion of the free ligand concentration at low total concentrations for the total binding case. Inspection of Figures 12.5a and 12.5b show that half maximal binding occurred when the total and free radioligand concentrations were approximately 1 and 0.1 nM, respectively. Given the high specific activity of many commercially available radioligands, the incubation could have easily been designed so that the receptor concentration was only one-tenth (0.1 mg protein) or one-twentieth (0.05 mg protein) that used in the simulation.

The filtration assay is commonly used to measure radioligand binding. Figure 12.6a shows a simulation of binding data obtained using the filtration

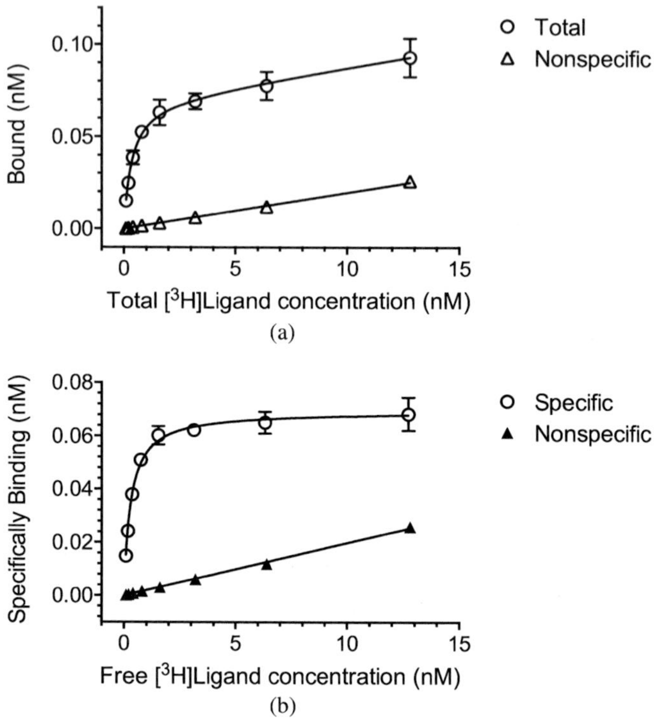

Figure 12.6: Simulation of binding curves plotted against the total (a) and free (b) radioligand concentration. The data were simulated for the conditions of a filtration assay in which only specific binding causes depletion in the free concentration of radioligand. (a) Theoretical values for total and nonspecific binding are plotted against the total radioligand concentration. The theoretical curves represent the least-squares fit of Equations (12.20), with Equation (12.31) substituted for $[L]$ (total binding), and (12.32) (nonspecific binding) to the data. (b) Theoretical values for specific and nonspecific binding are plotted against the free radioligand concentration. The theoretical curves represent the least-squares fit of Equations (12.5) (specific binding) and (12.6) (nonspecific binding) to the data.

method. Both total and nonspecific binding are plotted against the total concentration of radioligand. As described above, it is only feasible to account for the depletion in the radioligand concentration caused by specific binding in the filtration assay. Thus, the equations used for nonlinear regression analysis are modified accordingly. For purposes of deriving an equation for the free ligand concentration, the total ligand concentration is calculated as the summation of the specifically bound and free radioligand concentrations:

$$[L_T] = [L] + [\text{Bound}_{\text{nonspec}}]. \tag{12.25}$$

Substituting Equation (12.5) for specific binding yields:

$$[L_T] = [L] + \frac{[L]B_{\max}}{[L] + \frac{1}{K_L}}. \tag{12.26}$$

This equation is rearranged to yield a quadratic equation:

$$0 = A[L]^2 + B[L] + C, \tag{12.27}$$

in which

$$A = K_L, \tag{12.28}$$

$$B = (1 + K_L(B_{\max} - [L_T])), \tag{12.29}$$

$$C = -[L_T]. \tag{12.30}$$

The solution is:

$$[L] = \frac{-B + \sqrt{B^2 - 4AC}}{2A}. \tag{12.31}$$

In the analysis of the simulated data in Figure 12.6a, Equation (12.31) is substituted into Equation (12.20) for the free radioligand concentration.

For the nonspecific binding data, the following equation is fitted to the data:

$$\text{Bound}_{\text{nonspec}} = [L_T]C_{\text{nonspec}}. \tag{12.32}$$

With these equations, the best fit can be obtained to the data in Figure 12.6a using a strategy like that described above for the data in Figure 12.5a. When this was done, the following parameter estimates were obtained: $B_{\max}$, 0.070 ± 0.0007 nM; log K_L, $0.52 \pm 0.018\,\text{nM}^{-1}$ and C_{nonspec}, 0.0020. These are very close to those used to simulate the data: $B_{\max}$, 0.07 nM; log K_L, $0.52\,\text{nM}^{-1}$ and C_{nonspec}, 0.002. In this example, the assay volume was 1 ml, and 0.05 mg of protein was added to each incubation. Thus, the $B_{\max}$ value can be converted to 1.40 ± 0.014 pmol/mg protein. The log K_L is expressed in units of inverse nM. This estimate can be converted to inverse molar units: log K_L, $9.52 \pm 0.02\,\text{M}^{-1}$.

Figure 12.6b shows specific and nonspecific binding plotted against the free concentration of radioligand for the two measurements. In the case of the nonspecific binding measurement, the free radioligand concentration is assumed to be equal to the total radioligand concentration because of the unfeasibility of estimating this value. Compared to the data in Figure 12.5, there was much less depletion in the free radioligand concentration in this example because of the much lower concentration of receptor.

In the types of analyses illustrated in Figures 12.5 and 12.6, all of the binding parameters can be estimated from the total binding measurement alone. When this was done, very similar estimates of the binding parameters were obtained. Nonetheless, it is advisable to make separate estimates of nonspecific binding to improve the analysis of the parameters and to insure that the assay is working properly. It is unnecessary to measure nonspecific binding at the same total radioligand concentrations used to estimate total binding.

Many radioligands have a low coefficient of nonspecific binding, and hence the slope of the nonspecific binding plot is often low. The change in the level of nonspecific binding associated with the change in the free ligand concentration caused by specific binding is often very low. Consequently, the correction of nonspecific binding from the measured value to that estimated for the free concentration of radioligand during the total binding measurement is often trivial. A simpler way to analyze data in this situation, therefore, is to subtract the measured nonspecific binding from the measured total binding and do the regression analysis on the calculated specific binding data. This can be done using Equation (12.5) for specific binding with the quadratic solution (Equation (12.31)) substituted for the free radioligand concentration and with Equations (12.28)–(12.30) substituted for A, B and C in Equation (12.31).

Figure 12.7a illustrates an analysis of this type using the data from Figure 12.6a. Regression analysis of the data using Equation (12.5) and the substitution for L (Equation (12.31)) yields the following parameter estimates: $B_{\max}$, 0.069 ± 0.002 nM and log K_L, $0.59 \pm 0.06\ \text{nM}^{-1}$. These are very close to the parameters used to simulate the data: $B_{\max}$, 0.07 nM and log K_L, $0.52\ \text{nM}^{-1}$. The former estimates from the simulated data can be converted to more standard units as described for the data in Figure 12.6: $B_{\max}$, 1.38 ± 0.04 pmol/mg protein and log K_L, $9.59 \pm 0.06\ \text{M}^{-1}$.

Perhaps the simplest way to analyze the data is to calculate specific binding directly as described in the prior example and to subtract each specific binding value, expressed in concentration units, from the appropriate total ligand concentration in the same units to estimate the free concentration of radioligand for each specific binding estimate. The estimated specific binding values are plotted against the appropriate estimated free radioligand concentrations as shown in Figure 12.7b. Nonlinear regression analysis is done using Equation (12.5) directly, without the substitution for the quadratic solution for the free radioligand concentration. This analysis yields the following estimates: $B_{\max}$, 1.38 ± 0.04 pmol/mg protein and log K_L, 9.59 ± 0.06. These are very close to the parameters used to simulate the data: $B_{\max}$, 1.4 pmol/mg protein and log K_L, 9.52 and are also similar to those estimated in the prior analysis. Generally, when there is more variation in the

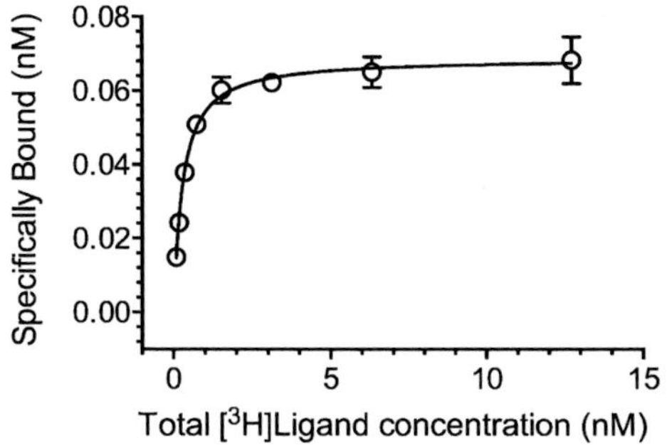

(a)

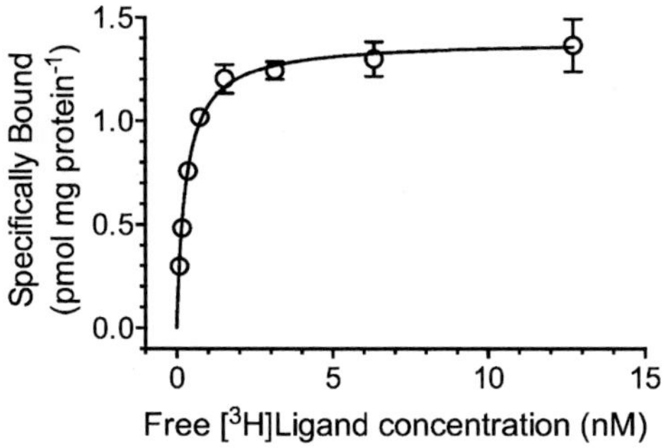

(b)

Figure 12.7: Simulation of total (a) and specific (b) binding curves plotted against the total and free radioligand concentrations, respectively. The data were simulated for the conditions of a filtration assay in which only specific binding causes depletion in the free concentration of radioligand. (a) Theoretical values for total binding are plotted against the total radioligand concentration. The theoretical curve represents the least-squares fit of Equation (12.5) to the data (with Equation (12.31) substituted for $[L]$). (b) Theoretical values for specific binding are plotted against the free radioligand concentration. The theoretical curve represents the least-squares fit of Equation (12.5) to the data.

data, it is better to do the regression using the true independent variable, the total radioligand concentration. The differences between the two analyses are often very small.

Competitive Inhibition of Radioligand Binding

Having a suitable radioligand that binds with high affinity to a receptor subtype or class of receptor subtypes enables one to examine how other nonlabeled drugs bind to the same target. This is accomplished by measuring their ability to inhibit the binding of the radioligand. In this section, I consider the case of competitive displacement of radioligand binding.

There are usually two goals in this type of experiment, 1) estimation of the affinity constant of the nonlabeled drug and 2) determination of the mechanism of inhibition. Both can be addressed by measuring the binding of the radioligand at a fixed concentration and in the presence of various concentrations of the nonlabeled inhibitor. The resulting data are referred to as a competition curve. Another approach is to measure the binding of the radioligand at different concentrations in the absence and presence of a fixed concentration of nonlabeled drug. Although the former approach is used more commonly used it is helpful to understand the relationship between the two methods of analysis.

I begin by addressing the two approaches for estimating the affinity of nonlabeled drugs for a target receptor. Then, I discuss the method for determining the mechanism of the interaction between the nonlabeled drug and radioligand.

Estimating the affinity constants of nonlabeled competitors

The first method for estimating the affinity constant of a nonlabeled drug involves measuring the binding of the radioligand at various concentrations as described in the preceding section. This type of data is often referred to as a saturation curve. The saturation curve is measured again in the presence of a fixed concentration of nonlabeled drug. There is sufficient information in the two sets of data to estimate the affinity constant of the nonlabeled drug.

Figure 12.8 shows an example of a simulation of this type of experiment. To simplify the analysis, specific binding is plotted against the free concentration of radioligand. The plot shows the saturation curve of the radioligand in the absence and presence of a competitive inhibitor. The control curve is the same as that shown in Figure 12.7b. To estimate the affinity constant of the inhibitor (K_I), we need to develop the appropriate equation for nonlinear regression analysis. This is described next. Those not interested in this derivation may want to skip ahead to the section on nonlinear regression analysis immediately after the derivation (Equation (12.38)).

The definition of the affinity constant of the competitive inhibitor is given by:

$$K_I = \frac{[IR]}{[I][R]}. \tag{12.33}$$

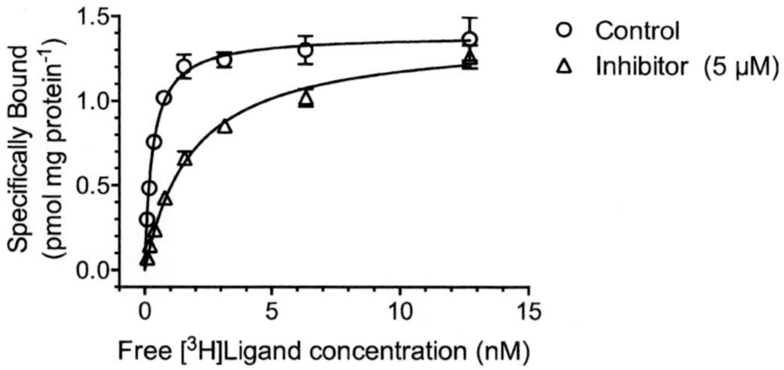

Figure 12.8: Simulation of the effect of a competitive inhibitor on the binding of a radioligand. Theoretical values for specific binding in the absence (control) and presence of a competitive inhibitor are illustrated. The theoretical curve represents the least-squares fit of Equation (12.37) to the data (with Equation (12.38) substituted for $K_{L\text{-app}}$).

To derive an equation for the binding of radioligand in the presence of competitive inhibitor, we begin with the following expression:

$$\frac{[LR]}{B_{\max}} = \frac{[LR]}{[LR]+[IR]+[R]}. \tag{12.34}$$

The concentrations of receptor complexes on the right side of the equation can be substituted with expressions based on the definitions of the equilibrium constants for L and I. For example, rearrangement of Equation (12.33) yields:

$$[IR] = [I]K_I[R]. \tag{12.35}$$

Making these substitutions and canceling out the variable $[R]$ yields:

$$[LR] = \frac{[L]K_L B_{\max}}{[L]K_L + [I]K_I + 1}. \tag{12.36}$$

Further rearrangement yields the following expression:

$$[LR] = \frac{[L]K_{L\text{-app}}B_{\max}}{[L]K_{L\text{-app}} + 1}, \tag{12.37}$$

in which,

$$K_{L\text{-app}} = \frac{K_L}{[I]K_I + 1}. \tag{12.38}$$

Equation (12.37) can be fitted to both the control saturation curve and that measured in the presence of the competitive inhibitor by global nonlinear regression analysis sharing the estimate of K_L between the curves and obtaining an estimate of K_I for the curve measured in the presence of I. A convenient way to do this is to constrain the variable I to zero for the control curve and to the appropriate value for the saturation curve measured in the presence of I (5 μM). When this was done for the analysis of the simulated data in Figure 12.8, the estimate of the log affinity constant of the competitive inhibitor was 6.01 ± 0.05, essentially the same as that used in the simulation (log K_I, 6). The estimates of the affinity constant and binding capacity of the control saturation curve were very similar to that estimated in Figure 12.7b (log K_L, 9.52 ± 0.04; B_{max}, 1.39 ± 0.03).

Another more common approach for estimating the affinity constant of a nonlabeled competitor is to measure the specific binding of radioligand at a fixed concentration and in the presence of different concentrations of the nonlabeled inhibitor. An example of this assay format is shown in Figure 12.9a. An advantage of this type of experiment is that it is unnecessary to use the radioligand at high concentrations to saturate the receptor. A reasonable approach is to use the radioligand at a concentration near its dissociation constant. It is also useful if depletion in the free concentration of radioligand is kept to a minimum (<10%), although this is not essential because an appropriate equation can be derived to account for the depletion. In the analysis below, I consider the case of little depletion in the free concentration of radioligand.

There are different forms of competitive regression equations that can be used to analyze the type of data illustrated in Figure 12.9a. A useful equation is one that expresses binding normalized to that measured in the absence of inhibitor. This equation can be derived by dividing Equation (12.37) (with Equation (12.38) substituted for $K_{L\text{-app}}$) by Equation (12.5). After simplification, the following result is obtained:

$$f = \frac{1}{1 + \frac{I}{IC_{50}}}, \tag{12.39}$$

in which

$$IC_{50} = \frac{1}{K_I}(1 + [L]K_L). \tag{12.40}$$

In Equation (12.39), f denotes the fractional amount of radioligand bound in the presence of the competitive inhibitor divided by that measured in the absence, and IC_{50}, the concentration of inhibitor that causes half-maximal displacement of radioligand binding. Because one never has an error-free estimate of the binding measured in the absence of competitive inhibitor, it is useful to introduce the parameter P_1, which is an estimate of the normalized value of binding in the absence of

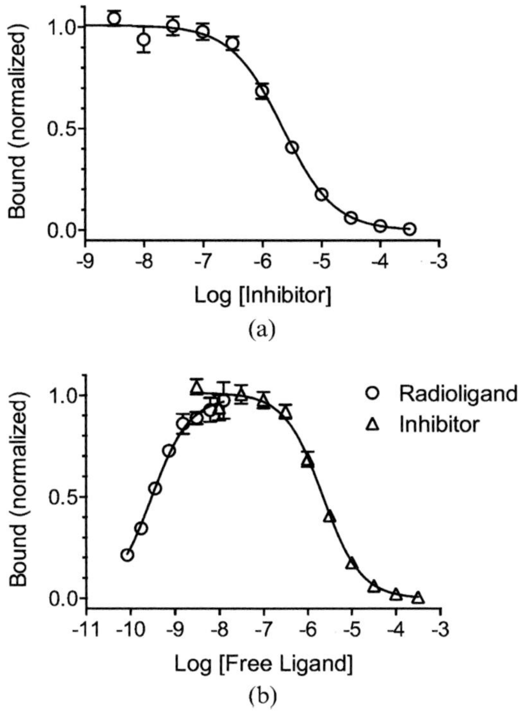

Figure 12.9: Analysis of the competitive inhibition of the binding of a radioligand at a fixed concentration. (a) The plot shows a simulation of the competition curve of an inhibitor of radioligand binding. The theoretical curve represents the least squares fit of Equation (12.41) to the data (with Equation (12.40) substituted for IC_{50}). (b) Simultaneous analysis of the saturation curve of a radioligand and the competition curve of a nonlabeled inhibitor. The theoretical curves represent the global least-squares fit of Equations (12.5) and (12.41) to the saturation and competition curves, respectively.

inhibitor:

$$f = \frac{P_1}{1 + \frac{I}{IC_{50}}}. \tag{12.41}$$

The simulated data in Figure 12.9a were analyzed with Equation (12.41) and the following parameter estimates were obtained: P_1, 1.01 ± 0.015; log IC_{50}, -5.66 ± 0.042. This IC_{50} value can be corrected to yield K_I by using a rearranged form of Equation (12.40):

$$K_I = \frac{(1 + [L]K_L)}{IC_{50}}. \tag{12.42}$$

Applying this correction yields an estimate of 5.97 for log K_I, which is in close agreement with the value used in the simulation (6). One problem with this correction factor is that there is no simple way to apply the appropriate SEM to the corrected value because there is usually error in the estimation of K_L, and hence, the error in the estimate of K_I should be greater than that of IC_{50}. There are two relatively easy ways to obtain the correct estimate of the SEM of log K_I.

The first is to analyze the data used to estimate K_L simultaneously with the competition curve. Equations (12.5) and (12.41) are fitted to the saturation curve and competition curve simultaneously sharing the estimate of K_L between the two curves and obtaining estimates of B_{max} from the saturation data and log K_I from the competition curve. When Equation (12.41) is used in this analysis, IC_{50} is substituted with Equation (12.40).

An example of this type of analysis is shown in Figure 12.9b. The binding values for the saturation curve have been normalized relative to B_{max} so that these data will fit on the same graph with those of the competition curve. The data can be analyzed in Prism using a strategy related to that described above in the analysis of the total binding and nonspecific binding curves. The equations for the saturation (Equation (12.5)) and competition curves (Equation (12.41) with IC_{50} substituted with Equation (12.40)) are multiplied by dummy variables, P and $1-P$, respectively, and the resulting equations are added together. The values of the dummy variable P is set equal to 1 or 0, depending upon whether the saturation or competition curve is being analyzed, respectively. Thus, the following lines can be entered in Prism for the regression equation:

$\langle A \rangle P = 1$

$\langle B \rangle P = 0$

IC50 = (1 + 10^(LogL + LogKL))*(10^ − LogKI)

Y = P*((10^X)*Bmax/((10^X) + 1/(10^LogKL))) + (1 − P)*P1/
(1 + (10^X)/IC50).

In this equation, X denotes the log free concentration of radioligand used in the saturation experiment or the concentration of competitive inhibitor used in the competition assay, depending upon which data set is being analyzed. The variable Log$L$ represents the fixed concentration of radioligand used in the competition experiment, and it is constrained to a constant in the analysis. The other variables are recognizable based on their similarity to the variables in Equations (12.5), (12.40) and (12.41).

The data for the experiment are entered into a Data Table in Prism. Replicates of the saturation experiment are entered into subcolumns of column A and those of the competition curves into subcolumns of column B, taking care to align the entries with the independent variables in the far left column. These are the log free concentration of radioligand in the saturation experiment (X) and the log concentration of inhibitor in the competition experiment (also denoted as X).

Regression analysis of the data in Figure 12.9b yielded the following parameter estimates: B_{max}, 0.99 ± 0.02, log K_L, 9.52 ± 0.05; P_1, 1.01 ± 0.02; log K_I, 5.96 ± 0.053. The estimate of log K_I is practically the same as that estimated above (Figure 12.9a, 5.97), but with this analysis, we now have accurate estimates of its asymptotic standard error. Note that the latter log value (0.053) is a little greater than that of the corresponding IC_{50} value estimated above (0.042). While this difference is trivial, it could be important with data having larger error, depending on the hypothesis being tested.

The second approach for obtaining valid estimates of the SEM of log K_I involves measuring a competition curve for the nonlabeled form of the radioligand at the same time the competition curve is measured for the inhibitor. The simulated data in Figure 12.10 illustrate this type of experiment. The regression equation for this analysis is Equation (12.41) with Equation (12.40) substituted for IC_{50}. When the competition curve of the nonlabeled form of the radioligand is being analyzed it is assumed that $K_L = K_I$, otherwise, K_L and K_I are distinct variables. To set up this equation in Prism, the following lines are entered.

$\langle\text{A}\rangle\texttt{LogK} = \texttt{LogKL},$

$\langle\sim\text{A}\rangle\texttt{LogK} = \texttt{LogKI},$

$\text{LogIC50} = \text{Log}((1 + 10\text{^}(\texttt{LogL} + \texttt{LogKL}))\text{*}(10\text{^} - \texttt{LogK})),$

$\text{Y} = \text{P1}/(1 + (10\text{^}\text{X})/(\texttt{LogIC50})).$

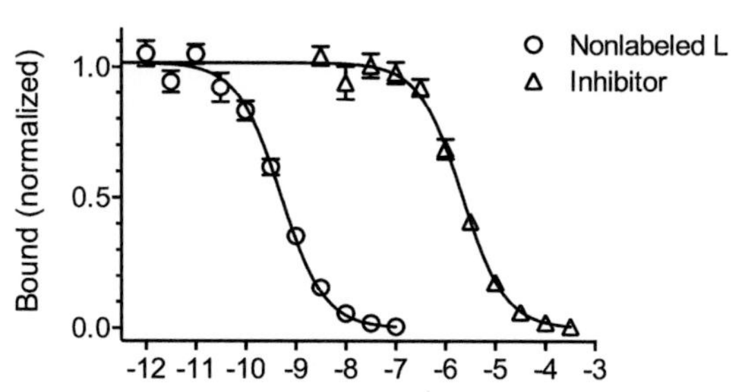

Figure 12.10: Simultaneous analysis of the competitive displacement of radioligand binding by a nonlabeled inhibitor and the nonlabeled form of the radioligand. The theoretical curves represent the least-squares fit of Equation (12.41) to the data (with Equation (12.40) substituted for IC_{50}).

When this method was used to analyze the data in Figure 12.9, the following parameter estimates were obtained: P_1, 1.02 $\pm$ 0.011; log K_L, 9.69 $\pm$ 0.10 and log K_I, 6.06 $\pm$ 0.075. These are nearly the same as those used to simulate the data: P_1, 1; log K_L, 9.56 and log K_I, 6. Similarly, the estimate of the affinity constant of the radioligand measured by competition with the nonlabeled form of the ligand (log K_L, 9.69 $\pm$ 0.10) is very similar to that estimated in the saturation experiments in Figures 12.9b (9.52 $\pm$ 0.04) and Figure 12.8 (9.52 $\pm$ 0.04).

When the nonlabeled compound in the competition experiment is the same as the radioligand (autoinhibition experiment), the relationship between IC_{50}, K_L and the concentration of radioligand used in the competition experiment ($[L]$) is given by:

$$\frac{1}{K_L} = IC_{50} - [L]. \tag{12.43}$$

This equation can be derived by substituting K_L for K_I in Equation (12.40). A simple way to correct the IC_{50} value of the nonlabeled ligand in this type of autoinhibition experiment is to subtract the radioligand concentration from the IC_{50} to yield the dissociation constant (K_D) or the reciprocal of the affinity constant (K_L). The accuracy of the estimate of K_L depends on several factors including the radioligand concentration. This can be appreciated by rearranging Equation (12.43):

$$IC_{50} = [L] + \frac{1}{K_L}. \tag{12.44}$$

This relationship shows that the IC_{50} value increases with an increase in the radioligand concentration, particularly as the latter increases above the reciprocal of the constant, K_L. When $[L]$ is very large, there is little difference between IC_{50} and $[L]$, and hence, it may be difficult to calculate K_L accurately from Equation (12.43). Thus, whenever the nonlabeled form of the radioligand is used in a competition experiment, the radioligand concentration should not be too high. Otherwise, the error in the estimate of K_L will be large.

How is Occupancy by the Inhibitor Related to Displacement of Radioligand Binding

When a sufficient concentration of competitive inhibitor is added to a binding assay, the inhibitor reduces the binding of the radioligand. This effect can be represented as the fractional binding of the radioligand as described by Equation (12.39). For example, if the fractional binding values of a radioligand in the absence and presence of the competitor are 1 and 0.8, respectively, then the competitor reduces the binding of the radioligand by 0.2 or 20%. The latter numbers represent the fractional displacement or percent displacement of radioligand binding by the competitor, respectively. If the fractional binding of the radioligand is subtracted from one on both sides of Equation (12.39), the result following simplification is an equation describing the fractional displacement of radioligand binding:

$$f_{\text{occ}} = \frac{I}{I + IC_{50}}. \tag{12.45}$$

In this equation, f_{occ} denotes the fractional occupancy by the inhibitor ($f_{\text{occ}} = (1 - f)$), and IC_{50} is defined by Equation (12.40). Note that Equation (12.45) is consistent with a simple one-site occupancy equation for the inhibitor with IC_{50} representing the apparent dissociation constant of the competitor, or alternatively, the reciprocal of its apparent affinity constant. Thus, the fractional displacement of radioligand binding by a competitor is equivalent to fractional receptor occupancy by the inhibitor in the presence of the radioligand.

The IC_{50} value of a competitive inhibitor at low radioligand concentrations can be determined by taking the limit of Equation (12.42) as the radioligand concentration approaches zero. This limit is equivalent to:

$$K_I = \frac{1}{IC_{50}}. \tag{12.46}$$

It follows that, at very low radioligand concentrations, the reciprocal of the IC_{50} value of a competitor is a good approximation of its K_I value and that the fractional displacement of radioligand binding by the competitor is equivalent to fractional receptor occupancy by the inhibitor in the absence of radioligand. In many instances, it may be unfeasible to measure binding at such low radioligand concentrations because of a low binding or excessive depletion in the free concentration of radioligand. Regardless, it is useful to understand how occupancy by the inhibitor relates to inhibition of radioligand binding.

Testing the hypothesis of competitive inhibition

In the analyses associated with Figures 12.8–12.10, a competitive model was used to simulate and analyze the data. Not surprisingly, accurate estimates of the affinity constant of the inhibitor were obtained. With real data, it may be that the inhibitor acts by a mechanism distinct from competitive inhibition. Other mechanisms could account for the behavior of inhibitor I in Figures 12.8, 12.9b and 12.10.

For example, the inhibitor could occupy the same site as the radioligand but react covalently with a residue in the binding pocket. Dividing the assay into two phases could address this mechanism. In phase one the receptor preparation is incubated with the inhibitor for a period of time and subsequently washed. In phase two, the binding of the radioligand is measured to the inhibitor-treated receptor preparation. If the inhibitor has an irreversible mechanism, then its effect should outlast its presence in the binding assay and a persistent inhibition of binding should be measured.

Alternatively, the inhibitor might bind to an allosteric site on the receptor and cause a reduction in the observed affinity of the radioligand. If the negative cooperativity is large, the inhibitory effect on binding could resemble competitive inhibition. At high concentrations, however, the radioligand might still be able to bind to the receptor even when the negative allosteric inhibitor is also bound. Thus, the procedure for determining whether the mode of inhibition is competitive involves measuring the competition curve over a range of concentrations of the radioligand. Ultimately, the inhibitor should be able to out compete the radioligand, even if the latter is present at high concentrations.

Table 12.1 lists a series of simulated measurements representing the equilibrium binding of a hypothetical radioligand at different concentrations (columns) and in the presence of different concentrations of competitive inhibitor (rows). If each column of data is plotted against the log free concentration of radioligand, the result is the set of saturation curves shown in Figure 12.11a. If each row is

Table 12.1: Simulation of the specific binding values of a radioligand (L) at different concentrations and in the presence of different concentrations of competitive inhibitor (I).[a]

	log [I]						
log [L]	−20	−7	−6	−5	−4	−3	−2
−11.0	0.032	0.029	0.016	0.003	0.000	0.000	0.000
−10.5	0.092	0.079	0.046	0.009	0.001	0.000	0.000
−10.0	0.24	0.23	0.13	0.028	0.003	0.000	0.000
−9.5	0.51	0.48	0.34	0.085	0.010	0.001	0.000
−9.0	0.79	0.74	0.63	0.23	0.030	0.003	0.000
−8.5	0.92	0.90	0.81	0.47	0.092	0.010	0.001
−8.0	0.95	1.00	0.95	0.73	0.24	0.029	0.003
−7.5	0.96	1.03	1.01	0.88	0.52	0.090	0.010
−7.0	1.02	1.01	1.02	0.97	0.74	0.25	0.031
−6.5	0.95	1.02	0.98	0.98	0.94	0.50	0.092
−6.0	1.01	1.01	1.00	0.99	1.01	0.75	0.23

[a]The binding data were simulated using Equation (12.37) with Equation (12.38) substituted for $K_{L\text{-app}}$. The values of the parameters were, $B_{\max}$, 1.0; log K_L, 9.5 and log K_I, 6. A random error of 10% was added to the simulated data.

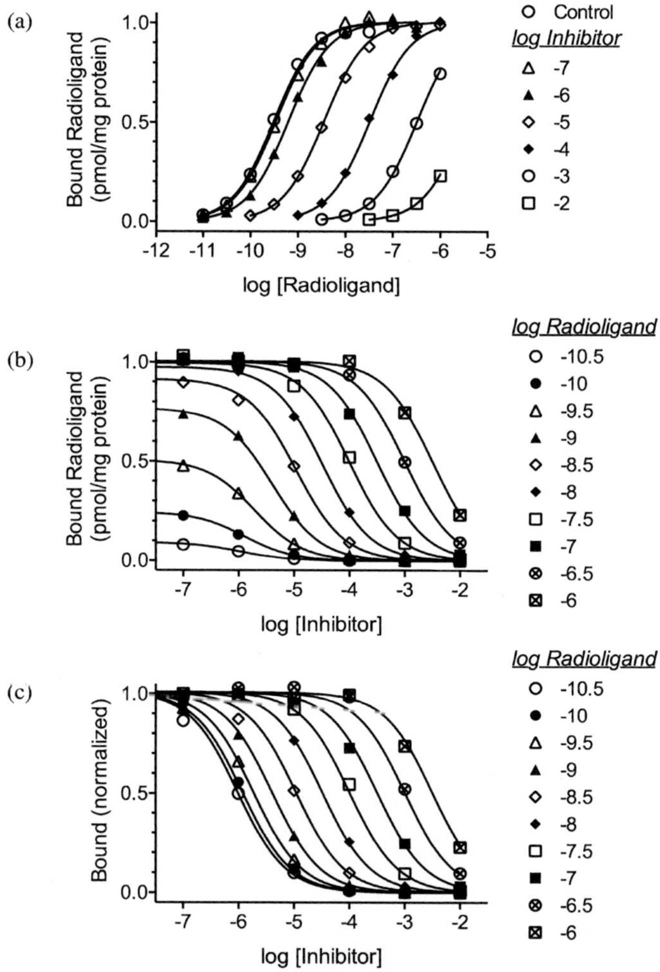

Figure 12.11: Testing the hypothesis of competitive inhibition. Simulated values of specific binding were calculated for different radioligand concentrations and in the presence of different concentrations of nonlabeled competitor. (a) Simulated binding values are plotted as a series of saturation curves for the condition of no inhibitor (control) and the indicated log concentrations of inhibitor. (b) Simulated binding values are plotted as a series of competition curves, each representing the binding of a radioligand at a fixed concentration and in the presence different concentrations of nonlabeled inhibitor. (c) The competition curves in panel (b) are plotted with binding normalized relative to that simulated in the absence of competitor. The theoretical curves in (a) and (b) represent the global least-squares fit of Equation (12.37) to the data with Equation (12.38) substituted for $K_{L\text{-app}}$. The theoretical curves in panel (c) represent the global least-squares fit of Equation (12.41) to the data with Equation (12.40) substituted for IC_{50}. The simulated data used in this analysis are given in Table 12.1.

plotted, the set of competition curves shown in Figure 12.11b are obtained. Finally, if the competition curves in Figure 12.11b are normalized relative to binding in the absence of inhibitor, the set of curves shown in Figure 12.11c is obtained. Thus, the data in Table 12.1 can be represented as a set of saturation curves, measured in the presence of different concentrations of inhibitor, or a set of competition curves, measured at different concentrations of radioligand. Both represent different ways of looking at the same data.

When viewed from the perspective of a saturation-binding curve, the inhibitor causes the curve to shift to the right on the axis representing the log concentration of radioligand. The estimate of the affinity constant of the competitor based on this shift should be the same for each concentration of inhibitor if the mechanism is competitive. For example, Equation (12.38) shows that competitive inhibitors have the effect of causing the apparent affinity of the radioligand ($K_{L\text{-app}}$) to decrease by a factor equivalent to the reciprocal of $[I]K_I + 1$. This equation can be rearranged to enable the estimation of the K_I value of the competitor from the reduction in the observed affinity of the radioligand:

$$K_I = \frac{\frac{K_L}{K_{L\text{-app}}} - 1}{[I]}. \tag{12.47}$$

As described above, K_L represents the affinity constant of the radioligand, and $K_{L\text{-app}}$, the apparent affinity of the radioligand in the presence of inhibitor. This value is equivalent to the concentration of radioligand required for half-maximal receptor occupancy in the presence of the inhibitor. Thus, the expression $K_L/K_{L\text{-app}}$ is equivalent to the scalar shift in the saturation curve caused by the inhibitor. Rewriting Equation (12.47) with the variable *shift* substituted for $K_L/K_{L\text{-app}}$ yields:

$$K_I = \frac{\text{shift} - 1}{[I]}. \tag{12.48}$$

Equation (12.48) shows that when the inhibitor is present at a concentration that causes a two-fold shift in the saturation curve, the reciprocal of the inhibitor concentration is equivalent to the affinity constant of the competitor. If this equation is used to estimate the log K_I value of the inhibitor at log inhibitor concentration (I) of −6—−3, the following log K_I values are obtained: 6.04, 6.03, 6.01 and 6.03. These values are essentially equivalent to one another and nearly equal to that used to simulate the data (6). Hence, the data are consistent with a competitive mechanism.

Equation (12.48) is written using affinity constants instead of dissociation constants. When written using dissociation constants, the following more conventional equation is obtained:

$$K_i = \frac{[I]}{\frac{K_{D\text{-app}}}{K_D} - 1}. \tag{12.49}$$

In this equation, K_D denotes the dissociation constant of the radioligand (reciprocal of K_L), $K_{D\text{-app}}$, the apparent dissociation constant of the radioligand (concentration of radioligand required for half-maximal receptor occupancy in the presence of the inhibitor), and K_i, the dissociation constant of the competitor (reciprocal of K_I). When using Equation (12.49), the variable *shift* is defined as $K_{D\text{-app}}/K_D$, and the relationships among K_i, $[I]$ and *shift* are given by:

$$K_i = \frac{[I]}{shift - 1}. \tag{12.50}$$

Another way to determine if the mechanism is competitive is to substitute Equation (12.38) for $K_{L\text{-app}}$ in Equation (12.37), and fit the resulting equation to all of the data by global nonlinear regression analysis simultaneously sharing the estimates of K_L and K_I. This can be done using Prism with a form of the equation in which the radioligand and inhibitor concentrations and the affinity constants of the radioligand and inhibitor are expressed as logarithms:

$$\text{Y} = \text{Bmax}^*(10^\wedge\text{X})/\{[1 + 10^\wedge(\text{LogI} + \text{LogKI})]/(10^\wedge\text{LogKL}) + 10^\wedge\text{X}\}.$$

In this equation, X denotes the log concentration of radioligand, and log$I$, the log concentration of inhibitor. The other variables are recognizable on the basis of their similarly to the variables in Equations (12.38) and (12.37). The log inhibitor concentrations are entered in the appropriate column headings in the Data Table. For the control saturation curve a large negative number is entered to represent the antilog of an inhibitor concentration of essentially zero (e.g., −20). For parameter constraints, Bmax, LogKL and LogKI are shared, and LogI is constrained to the value of the column title.

When this method is used to analyze the data in Figure 12.11a, the following parameter estimates are obtained: B_{max}, 1.01 ± 0.004; log K_L, 9.51 ± 0.014 and log K_I, 6.02 ± 0.02. These are nearly the same as those used to simulate the data: B_{max}, 1; log K_L, 9.55 and log K_I, 6.

The same data can be analyzed from the perspective of a series of competition curves, measured using different fixed concentrations of radioligand, as shown in Figure 12.11b. These data can also be analyzed by global nonlinear regression analysis with Equation (12.38) (with Equation (12.37) substituted for $K_{L\text{-app}}$), sharing the estimates of K_L and K_I. In this case, however, the independent variable (X) is log $[I]$ and the log values of the concentration of radioligand used in the competition curves are entered as column headings in the Data Table when Prism is used for the regression. In this case, the equation is entered into Prism in the following format:

$$\text{Y} = \text{Bmax}^*(10^\wedge\text{LogL})/\{[1 + 10^\wedge(\text{X} + \text{LogKI})]/(10^\wedge\text{LogKL}) + 10^\wedge\text{LogL}\}.$$

Note that this equation is the same as that described above for the analysis of saturation curves except that the log radioligand concentration has been changed from X to LogL and the log inhibitor concentration has been changed from LogI to X. Regression analysis of the data in Figure 12.11b yielded the following parameter estimates: B_{max}, 1.01 ± 0.004; log K_L, 9.51 ± 0.014 and log K_I, 6.02 ± 0.02. Not surprisingly, these are essentially the same as those described above for the saturation curve analysis.

Another more intuitive way to examine the competition curves is to estimate the K_I value from the IC_{50} value of each inhibition curve. If the mechanism is competitive, the K_I value of each competition curve should be the same. Using Equation (12.41) to estimate the K_I value of the inhibitor from the IC_{50} values estimated from the competition curves at log molar radioligand concentrations of -10.5 to -6 yields values of 6.07, 6.02, 6.03, 6.05, 6.06, 6, 5.95, 6.06, 5.95 and 6.06. These are all nearly the same and similar to that used to simulate the data (6).

Because the series of competition curves in Figure 12.11c were measured using radioligand concentrations that varied from a low value less than the reciprocal of the affinity constant ($1/K_L$), to a high value that saturated the receptor, the there is sufficient information to estimate both K_L and K_I, even though the data are normalized. Thus, the data can be analyzed by global nonlinear regression analysis using Equation (12.41) with IC_{50} substituted with Equation (12.40). This analysis yields the following parameter estimates: P_1, 1.01 ± 0.005; log K_L, 9.56 ± 0.04 and log K_I, 6.06 ± 0.03. Again these estimates are similar to the values used to simulate the data: P_1, 1; log K_L, 9.56 and log K_I, 6.

Allosteric Modification of Radioligand Binding

Allosteric ligands bind at a site on the receptor different from that of the orthosteric binding pocket and modify the binding of orthosteric ligands. This mechanism is illustrated by the population scheme in Figure 12.12a, which shows the binding of an allosteric ligand (A) to a receptor in the presence of orthosteric radioligand, L.[11] The affinity constant of the allosteric ligand is denoted by K_A and that of the radioligand by K_L. When the allosteric ligand is occupying all of the allosteric sites in the receptor population, the orthosteric ligand binds to its site with an observed affinity modified by the scalar, α. Conversely, when the orthosteric ligand is occupying all

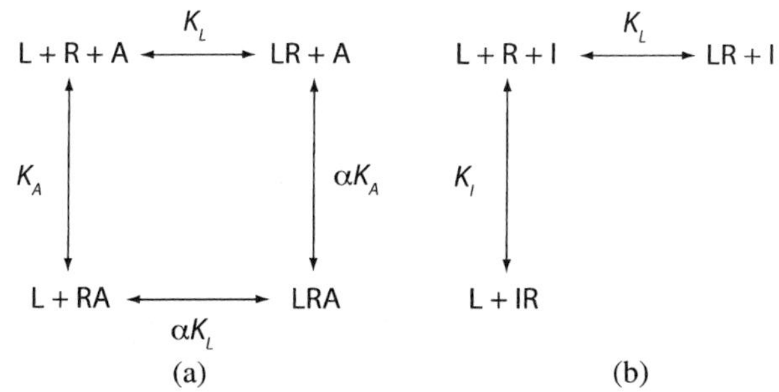

Figure 12.12: Comparison of the receptor binding schemes for allosteric (a) and competitive interactions (b). (a) The parameters K_L and K_A denote the affinity constants of the orthosteric and allosteric ligands, respectively, and α denotes the cooperativity constant. (b) The parameters K_L and K_I denote the affinity constants of the two competing orthosteric ligands.

of the allosteric sites in the receptor population, the allosteric ligand binds with an observed affinity modified by the same scalar, α. Thus, αK_L and αK_A denote the observed affinity constants of the orthosteric and allosteric ligands, respectively, when both ligands are present. The scalar, α, is known as the cooperativity constant. Values of α between 0 and 1 indicate negative cooperativity, whereas values greater than one indicate positive cooperativity.

In contrast, a competitive inhibitor binds to the same site as the radioligand, and thus, both ligands cannot bind to the receptor at the same time (see, Figure 12.12b). For example, if a high concentration of radioligand is added to a receptor preparation (i.e., $[L]K_L = 100$), it will occupy about 99% of the receptor sites. If a competitor is also added at a very high concentration (i.e., $[I]K_I = 10,000$), it will displace the radioligand and occupy about 99% of the orthosteric sites. But if the concentration of the radioligand is raised to an exceedingly high concentration (i.e., $[L]K_L = 1,000,000$), it will displace the competitor and occupy about 99% of the sites even though the competitor is present at a very high concentration (i.e., $[I]K_I = 10,000$). This behavior occurs because the interaction scheme in Figure 12.12b excludes the possibility that the two ligands can bind to the receptor at the same time.

As shown by the scheme in panel (a), however, an allosteric mechanism includes the possibility that both ligands can bind to the receptor at the same time, provided that any negative cooperativity between them is not too great. For example, let us reconsider the prior example assuming instead that the competitive inhibitor is a negative allosteric modulator having an α value of 0.01 (negative cooperativity) for its interaction with the orthosteric radioligand. If the allosteric ligand is present at a very high concentration ($[A]K_A = 10,000$), it would only cause about a 50% displacement in the binding of the orthosteric radioligand at a high concentration ($[L]K_L = 100$). If the radioligand concentration is increased to an exceedingly high level (i.e., $[L]K_L = 1,000,000$), then it would cause essentially no inhibition of the binding of the allosteric ligand. Occupancy of the receptor by the radioligand and allosteric ligand would be approximately 99.9% and 99%, respectively.

In the case of positive cooperativity ($\alpha > 1$), the allosteric ligand increases the observed affinity of the radioligand, and has the potential to increase binding provided that the radioligand is not already present at a high concentration that saturates the receptor in the absence of the allosteric ligand. Of course, positive allosteric effects are easy to discriminate from competitive inhibitors because of the opposite change in binding. Not withstanding potential artifacts influencing the data, a ligand-induced increase in radioligand binding affinity constitutes strong evidence for allosterism.

Estimating the affinity constants of allosteric ligands

A simple way to estimate the affinity constant of an allosteric ligand involves measuring the specific binding of a radioligand at a fixed concentration and in the presence of different concentrations of nonlabeled allosteric ligand. Below this set of data is referred to as a modulator curve. Figure 12.13a shows a simulated example of this type of experiment involving a negative allosteric modulator. Increasing concentrations of the negative allosteric modulator inhibit binding, but there is a limit to the inhibition. This behavior contrasts with that caused by the competitive inhibitor in Figures 12.9–12.11. The plot in Figure 12.13b shows the influence of a positive allosteric ligand on the binding of a radioligand. If the observed affinity constant of the radioligand (K_L) is known, there is sufficient information in Figures 12.13a and 12.13b to estimate the affinity constants of the two allosteric ligands and their cooperativity constants (α) for interacting with the radioligand.

To analyze the data in Figure 12.13, the appropriate equations for nonlinear regression analysis need to be derived. This derivation is described next, followed by a description of how the data are analyzed, which begins after Equation (12.62).

The affinity constant of the allosteric ligand is defined in the normal way:

$$K_A = \frac{[RA]}{[R][A]}. \tag{12.51}$$

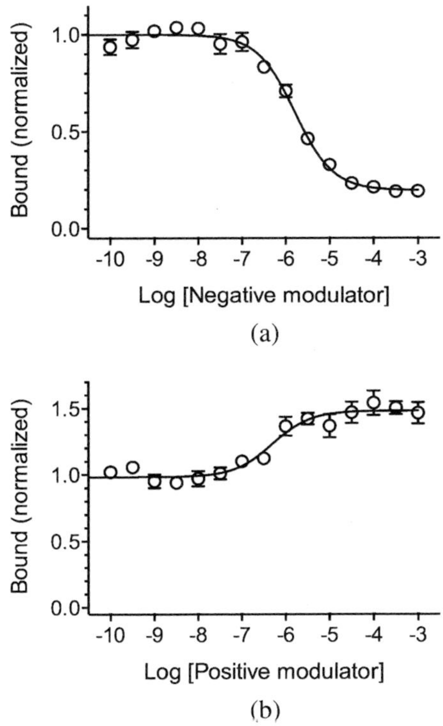

Figure 12.13: Simulation of negative (a) and positive (b) allosteric modulation of the binding of a radioligand at a fixed concentration. The theoretical curves represent the least-squares fit of Equation (12.63) to the data with KL constrained to the value used in the simulation (0.3 nM).

The two affinity constants describing the formation of the ternary complex ($[LRA]$) are defined as:

$$\alpha K_L = \frac{[LRA]}{[L][RA]}, \tag{12.52}$$

$$\alpha K_A = \frac{[LRA]}{[LR][A]}. \tag{12.53}$$

The derivation of the equation describing the binding of radioligand in the presence of allosteric modulator begins with the following equation:

$$\frac{[LR] + [LRA]}{B_{\max}} = \frac{[LR] + [LRA]}{[LR] + [LRA] + [RA] + [R]}. \tag{12.54}$$

The concentrations of the receptor complexes on the right side of the equation are substituted with expressions based on the definitions of the affinity constants (Equations (12.2) and (12.51)–(12.53)). For example, the equation for the ternary complex is given by:

$$[LRA] = [L][A]\alpha K_L K_A[R]. \tag{12.55}$$

Making these substitutions and cancelling out $[R]$ yields:

$$[LR] + [LRA] = \frac{([L][A]\alpha K_L K_A + [L]K_L)B_{\max}}{[L]K_L + [A]K_A + [L][A]\alpha K_L K_A + 1}. \tag{12.56}$$

Further rearrangement yields:

$$[LR] + [LRA] = \frac{[L]K_{L\text{-app}}B_{\max}}{[L]K_{L\text{-app}} + 1}, \tag{12.57}$$

in which:

$$K_{L\text{-app}} = K_L^* \frac{1 + [A]\alpha K_A}{1 + [A]K_A}. \tag{12.58}$$

Equation (12.57) describes the binding of radioligand in the presence of an allosteric ligand. This equation is useful for analyzing saturation curves for the radioligand in the presence of allosteric ligand.

To derive an equation for the fractional binding of the radioligand in the presence of different concentrations of allosteric ligand (modulator curve), Equation (12.57) (with Equation (12.58) substituted for $K_{L\text{-app}}$) is divided by Equation (12.5). After simplification the result is:

$$f = \frac{L + \frac{1}{K_L}}{L + \frac{1}{K_{L\text{-app}}}}. \tag{12.59}$$

In this equation, $K_{L\text{-app}}$ is defined by Equation (12.58). Equation (12.59) is consistent with the one-site modulation curves shown in Figure 12.13. These curves can be characterized by the AC_{50} value (concentration of allosteric ligand that causes a half-maximal effect on radioligand binding) and the fractional nonzero plateau in radioligand binding at high concentrations of the allosteric ligand (δ). The binding parameters of the allosteric ligand ($\log K_A$ and $\log \alpha$) can be determined from these parameters provided that the affinity constant of the radioligand is known.

To derive an equation for the nonzero plateau (δ), Equation (12.59) is substituted for $K_{L\text{-app}}$ in Equation (12.58) and the limit of the resulting equation is taken as A approaches infinity. The results is:

$$\delta = \frac{\alpha(1 + LK_L)}{1 + L\alpha K_L}. \tag{12.60}$$

The equation for the AC_{50} value of the allosteric modulation curve can be derived by solving Equation (12.59) for the value of A (AC_{50}) at the AC_{50} point of the modulator curve. This point of the binding curve is equivalent to one-half the summation of δ and fractional binding in the absence of modulator (1). Thus, the following equation is solved for AC_{50}:

$$\frac{L + \frac{1}{K_L}}{L + \frac{1}{K_{L\text{-app}}}} = \frac{1}{2}\left[1 + \frac{\alpha(1 + LK_L)}{1 + L\alpha K_L}\right]. \tag{12.61}$$

The left-hand side of this equation is equivalent to Equation (12.59), and the right is equivalent to one-half the summation of 1 and Equation (12.60). Substituting Equation (12.58) for $K_{L\text{-app}}$ and solving for A (AC_{50}) yields:

$$AC_{50} = \frac{1 + LK_L}{K_A(1 + L\alpha K_L)}. \tag{12.62}$$

Equation (12.62) shows that when the radioligand concentration approaches zero, the AC_{50} value of the allosteric modulation curve is equivalent to the reciprocal of its affinity constant ($1/K_A$). Thus, if a modulator curve is measured at a very low concentration of radioligand, the corresponding AC_{50} value is equivalent to the reciprocal of K_A.

Now let us use this math to analyze the simulated data in Figure 12.13. A simple approach is to use nonlinear regression analysis to fit a one-site inhibition equation to the data. This equation is similar to Equation (12.41), but it includes an addition parameter that defines a fractional plateau value (δ):

$$f = \frac{P_1(1 - \delta)}{1 + \frac{A}{AC_{50}}} + \delta P_1. \tag{12.63}$$

When this equation is fitted to the simulated data in Figure 12.13a, the following parameter estimates are obtained: P_1, 1 ± 0.02; δ, 0.195 ± 0.015 and log AC_{50}, -5.79 ± 0.053. From the estimates of δ and AC_{50} and the values of L and K_L used in the simulation, the values of the cooperativity constant, α, and K_A can be estimated from the following equations, which are rearrangements of

Equations (12.60) and (12.62):

$$\alpha = \frac{\delta(1 + L\alpha K_L)}{1 + LK_L}, \tag{12.64}$$

$$K_A = \frac{1 + LK_L}{AC_{50}(1 + L\alpha K_L)}. \tag{12.65}$$

These calculations yield estimate of -0.98 and 6.07 for log α and log K_A, which are close to the values used to simulate the data (-1 and 6, respectively).

Another way to estimate the parameters is to substitute Equation (12.58) for $K_{L\text{-app}}$ in Equation (12.57) and fit the resulting equation to the data in Figure 12.13a by nonlinear regression analysis constraining the values of $K_L$and L to the values used in the simulation. When this is done the following estimates of log K_L and log α were obtained: 6.06 and -0.98, respectively.

The same strategies can be used to analyze the data in Figure 12.13b. Regression analysis with Equation (12.63) yielded the following estimates of log δ and log AC_{50}: 1.51 and -6.30, respectively. When these values were substituted into Equations (12.64) and (12.65), the following values of log α and log K_A were obtained: 0.52 and 5.95, respectively. Finally, analyzing the data using Equation (12.63) with the appropriate substitutions (Equation (12.64) and (12.65)) yielded the following estimates: log α, 0.49 and log K_A, 5.93. All of these estimates of log α and log K_A are similar to those used in the simulation, 0.5 and 6, respectively.

The foregoing analyses, however, do not address the error associated with the estimates of K_L. In all of these calculations, K_L is assumed to be a constant, yet we know that there is error in the estimate of this parameter. To solve this problem the data used to estimate K_L can be analyzed simultaneously with those used to estimate K_A and α. This approach is analogous to that used in the analysis of the competition experiments illustrated in Figures 12.9b and 12.10.

Figure 12.14a shows a plot of the normalized saturation data for the radioligand and the modulator inhibition curve from Figure 12.13a. The data can be analyzed simultaneously using Equation (12.5) for the saturation data and Equation (12.59) (with Equation (12.58) substituted for $K_{L\text{-app}}$) for the modulator inhibition curve. During the analysis, the estimate of K_L is shared between the curves and estimates of K_A and α are obtained from the modulator inhibition curve. To perform this analysis in Prism, the saturation data are entered into a Data Table under subcolumns of column A and the data for the modulator inhibition curve are entered into subcolumns of column B. The following lines are entered for the regression equation:

<A>P = 1,

<~A>P = 0,

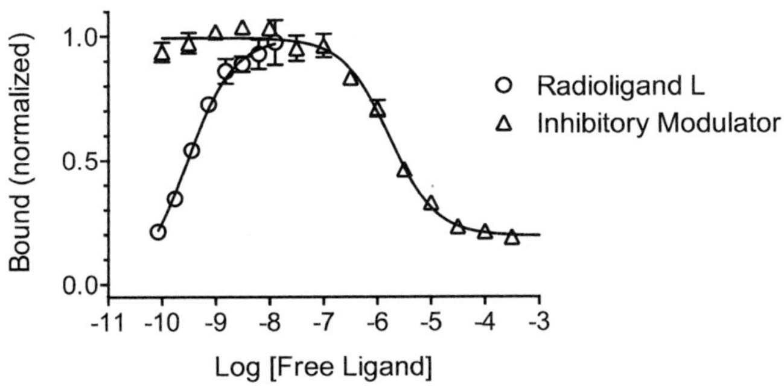

(a)

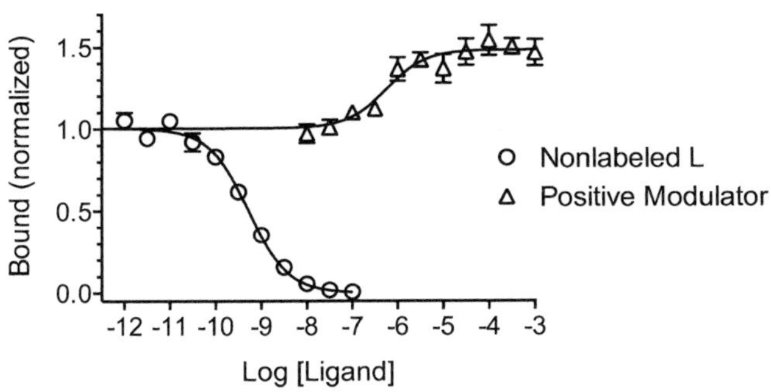

(b)

Figure 12.14: Simultaneous analysis of the binding of radioligand and the modification of its binding by an allosteric modulator. (a) The simulated data represent a saturation curve of radioligand L (specific binding plotted against the log free concentration of radioligand) and the modulation of its binding at a fixed concentration (0.3 nM) by different concentrations of a negative allosteric modulator. The theoretical curves represent the global least-squares fit of Equation (12.5) (saturation curve) and (12.59) (modulation curve) to the data with Equation (12.58) substituted for $K_{L\text{-app}}$ in Equation (12.59). (b) The simulated data represent the competitive displacement of the binding of a fixed concentration of radioligand (0.3 nM) by different concentrations of the nonlabeled form of the radioligand and the modulation of its binding at a fixed concentration (0.3 nM) by different concentrations of a positive allosteric modulator. The theoretical curves represent the global least-squares fit of Equation (12.41) (nonlabeled radioligand competition curve) and (12.59) (modulation curve) to the data with Equation (12.44) substituted for IC_{50} in Equation (12.41).

$$\text{LogKLapp} = \text{Log}((10^\wedge\text{LogKL})^*(1 + 10^\wedge(\text{X} + \text{Logalpha} + \text{LogKA}))/(1 + 10^\wedge(\text{X} + \text{LogKA}))),$$

$$\text{Y} = \text{P}^*(10^\wedge(\text{X} + \text{LogKL}))\text{B}\max/((10^\wedge(\text{X} + \text{LogKL}) + 1) + (1 - \text{P})^*\text{P1}^*(10^\wedge\text{LogL} + 10^\wedge(-\text{LogKL}))/(10^\wedge\text{LogL} + 10^\wedge(-\text{LogKLapp})).$$

The regression equation consists of the summation saturation Equation (12.5) and the equation for the fractional binding of radioligand in the presence of allosteric ligand (Equation (12.59)). The variable X denotes the log of the radioligand concentration when the saturation data are being analyzed. When the modulator data are being analyzed, X represents the log concentration of modulator. The dummy variable P is used to designate which half of the equation is used to analyze each set of data as described in connection with the analysis of the data in Figure 12.9b. The other variables should be recognizable, based on their similarity to those of the regression equations in the text.

Regression analysis of the data in Figure 12.14a yielded the following estimates: P_1, 0.99 ± 0.01; log K_L, 9.52 ± 0.043; $B_{\max}$, 0.99 ± 0.02; log α, -0.96 ± 0.05 and log K_A, 6.03 ± 0.06. These estimates are similar to the values used in the simulation: P_1, 1.0; log K_L, 9.55; $B_{\max}$, 1.0; P_1, 1.0; log α, -1.0 and log K_A, 6.

An analogous approach in which the modulator data is analyzed simultaneously with a competition curve for the nonlabeled form of the radioligand can also be used. Figure 12.14b shows an example of this type of analysis using the modulator data from Figure 12.13b and the competition data for the nonlabeled form of the radioligand from Figure 12.10. This analysis can be done using Prism in a manner to that just described for the data in Figure 12.14a. The following lines represent the regression equation.

```
<A>P = 1
<~A>P = 0
LogKLapp = Log((10^LogKL)*(1 + 10^(X + Logalpha + LogKA))/
           (1 + 10^(X + LogKA)))
Y = P*P1/(1 + (10^X)/(10^LogL + 10^( - LogKL))) + (1 - P)*
    P1*(10^LogL + 10^( - LogKL))/(10^LogL + 10^( - LogKLapp)).
```

Again, the regression equation represents the summation of two equations, each multiplied by P or 1−P to determine which one is used in the analysis. The first represents Equation (12.41) with Equation (12.44) substituted for IC_{50} and the second represents Equation (12.59) with an additional parameter for maximal binding in the absence of modulator (P1). The substitution for LogKLapp in this Equation (12.58) is given above on a separate line. The variables are defined earlier and should be recognizable based on their similarity to the corresponding variables in the equations in the text.

Regression analysis yielded the following parameter estimates: P_1, 1.00 ± 0.02; log K_L, 9.65 ± 0.14; log K_A, 5.79 ± 0.31 and log α, 0.62 ± 0.25. In this case, there is moderate variation in the estimates of log K_A and log α, because of the difficulty in resolving these parameters in the presence of modest error in the estimation of $\log K_L$.

How is Occupancy by the Modulator Related to Displacement of Radioligand Binding

Positive and negative allosteric modulators cause concentration-dependent increases or decreases, respectively, in the specific binding of the radioligand when the latter is present at fixed concentration. The maximal effect reaches a plateau, corresponding to the value of the variable δ in Figure 12.15a and 12.15b. How then do we relate

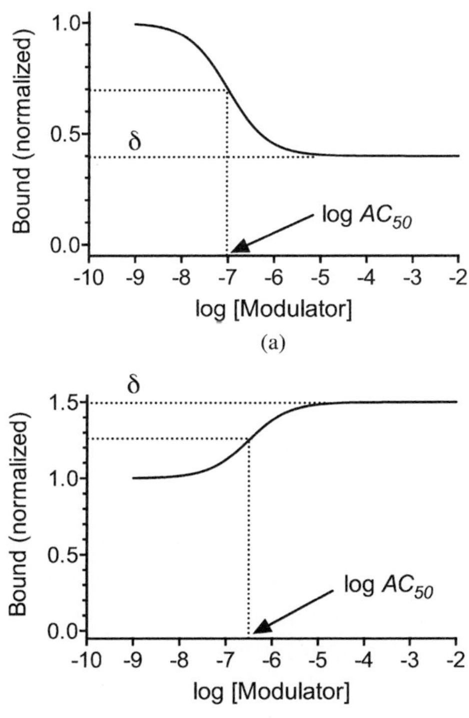

Figure 12.15: Definition of the empirical parameters of the modulation curve representing the modification in the binding of a radioligand at a fixed concentration by different concentrations of negative (a) and positive (b) allosteric modulators. The parameter δ represents the fractional binding of the radioligand in the presence of a maximally effective concentration of allosteric ligand, and AC_{50}, the concentration of allosteric ligand causing a half-maximal effect on radioligand binding.

a percent change in the binding of a radioligand to receptor occupancy by the modulator?

This can be illustrated by normalizing the modulator-induced change in binding relative to the maximal change (δ). The function describing a positive modulator-induced change in binding is equivalent to Equation (12.59) minus one. Recall that Equation (12.59) describes the binding of the radioligand in the presence of modulator normalized relative to binding in the absence of modulator. Dividing this change in binding by the maximal modulator-induced change ($\delta - 1$) yields a normalized binding curve for the positive modulator, based on its ability to modify the binding of the radioligand. This binding function is described by the following equation, based on the description just given:

$$f_{\text{pos}} = \frac{f(A) - 1}{1}. \tag{12.66}$$

In this equation f_{pos} denotes the fractional change in the binding of radioligand caused by the positive allosteric ligand, and $f(A)$, the function described by Equation (12.59). Making the substitution for $f(A)$ and δ (Equation (12.60)) yields the following equation upon simplification:

$$f_{\text{pos}} = \frac{A}{A + AC_{50}}, \tag{12.67}$$

in which AC_{50} is defined by Equation (12.62). Thus, the normalized effect of a positive modulator on the binding of a radioligand resembles a simple one-site occupancy curve with an observed dissociation constant of AC_{50}. As mentioned previously, Equation (12.62) shows that AC_{50} approaches the value of the reciprocal of K_A ($1/K_A$) as the concentration of radioligand approaches zero.

For a negative allosteric modulator, the equation describing the change in binding is equivalent to one minus Equation (12.59). Dividing this change by the maximal change induced by the negative modulator ($1 - \delta$) yields a normalized binding curve for the modulator, based on its ability to modify the binding of the radioligand. This binding function is described by the following equation:

$$f_{\text{neg}} = \frac{1 - f(A)}{1}. \tag{12.68}$$

Making the substitutions for $f(A)$ (Equation (12.59) with $K_{L\text{-app}}$ substituted by Equation (12.58)) and δ (Equation (12.60)) yields Equation (12.68) except for the replacement of f_{pos} on the left side of the equation by f_{neg}. Thus, we can rewrite Equation (12.68) using the variable f_{norm} to denote the normalized binding effect of both negative and positive modulators:

$$f_{\mathrm{norm}} = \frac{A}{A + AC_{50}}. \tag{12.69}$$

Testing the hypothesis of allosterism

In the analyses of the data illustrated in Figure 12.13a, it was assumed that the underlying mechanism for the modulator-induced change in radioligand binding was allosteric. This assumption yielded a good fit of the appropriate regression equation to the data because the data were simulated assuming an allosteric mechanism in the first place. Without prior knowledge of the underlying mechanism, other hypotheses might explain the data. Perhaps the radioligand binds to two types of sites and the nonlabeled ligand only binds to one of them and is therefore unable to displace the radioligand from all of the sites. Alternatively, there may be only one type of receptor present, but some of the sites may be trapped within membrane compartments. If the radioligand is sufficiently permeable to membrane barriers, whereas the nonlabeled ligand is not, then the inhibitor would be unable to displace a component of radioligand binding. Again, behavior like that shown in Figure 12.13a could arise from this situation. Ultimately, to verify that the mechanism is allosteric, the interaction needs to be investigated over a range of radioligand concentrations and in the presence of different concentrations of nonlabeled modulator.

The data in Table 12.2 list a series of simulated binding values for a radioligand at different concentrations (rows) and in the presence of different concentrations of a negative allosteric modulator (columns). If each column of data is plotted, the result is a series of saturation curves simulated in the presence of different concentrations of allosteric modulator as shown in Figure 12.16a. If each row of data is plotted, the result is the series of modulator curves shown in Figure 12.16b. Finally, if each modulator curve is normalized relative to maximal binding in the absence of modulator, the result is the series of modulator curves shown in Figure 12.16c.

Compared to the analogous plots of the competitive interaction shown in Figure 12.11, the results in Figure 12.16 are strikingly different. Once the concentration of the competitive inhibitor is increased above the reciprocal of its affinity constant ($1/K_I$) the shift in the saturation curve of the radioligand increases proportionately to the concentration of the competitor. In contrast, as the concentration of allosteric modulator increases, there is a clear limit to the shift as shown in Figure 12.16a. If the competitive relationship (Equation (12.48)) used to estimate the affinity constant of an inhibitor based on its ability to shift the saturation curve,

Table 12.2: Simulation of the specific binding values of a radioligand (L) at different concentrations and in the presence of different concentrations of allosteric inhibitor (A).[a]

	log [I]						
log [L]	−20	−7	−6	−5	−4	−3	−2
−11.0	0.029	0.028	0.017	0.006	0.004	0.003	0.003
−10.5	0.091	0.088	0.050	0.018	0.010	0.010	0.010
−10.0	0.23	0.22	0.14	0.055	0.033	0.031	0.032
−9.5	0.52	0.48	0.35	0.16	0.095	0.088	0.089
−9.0	0.77	0.71	0.65	0.38	0.26	0.25	0.23
−8.5	0.87	0.88	0.84	0.67	0.50	0.49	0.49
−8.0	0.96	1.00	0.90	0.88	0.76	0.77	0.75
−7.5	1.01	0.96	1.01	0.91	0.88	0.91	0.95
−7.0	0.97	1.04	0.95	1.00	1.02	0.97	0.99
−6.5	0.99	1.03	1.01	1.00	1.01	0.98	1.00
−6.0	0.97	1.04	0.97	1.01	0.96	1.03	0.96

[a]The binding data were simulated using Equation (12.57) with Equation (12.58) substituted for $K_{L\text{-app}}$. The values of the parameters were, $B_{\max}$, 1; log K_L, 9.5, log K_A, 6 and log α, −1.0. A random error of 10% was added to the simulated data.

a different estimate of K_I would be estimated for each concentration of allosteric ligand. This behavior is clearly inconsistent with competitive inhibition.

In the case of the competition curves shown in Figure 12.11b, increasing the concentration of radioligand causes the competition curve to shift to the right and at sufficiently high concentrations, the competitive inhibitor is capable of fully displacing the binding of the radioligand. In contrast, as the concentration of radioligand increases, a negative allosteric modulator has less of an inhibitory effect on radioligand binding. At sufficiently high concentrations of radioligand, the allosteric modulator has essentially no effect. This behavior is particularly apparent with the modulator curves are normalized like those shown in Figure 12.16c.

To analyze the data in Figure 12.16a, global nonlinear regression analysis is used to fit Equation (12.57) to the data with Equation (12.58) substituted for $K_{L\text{-app}}$. This can be done using Prism with a form of the equation in which the radioligand and inhibitor concentrations and the affinity constants of the radioligand and inhibitor are expressed as logarithms:

```
LogKLapp = Log((10^LogKL)*(1 + 10^(LogA + Logalpha + LogKA))/
           (1 + 10^(LogA + LogKA)))

Y = Bmax/(1 + 10^( - (X + LogKLapp))).
```

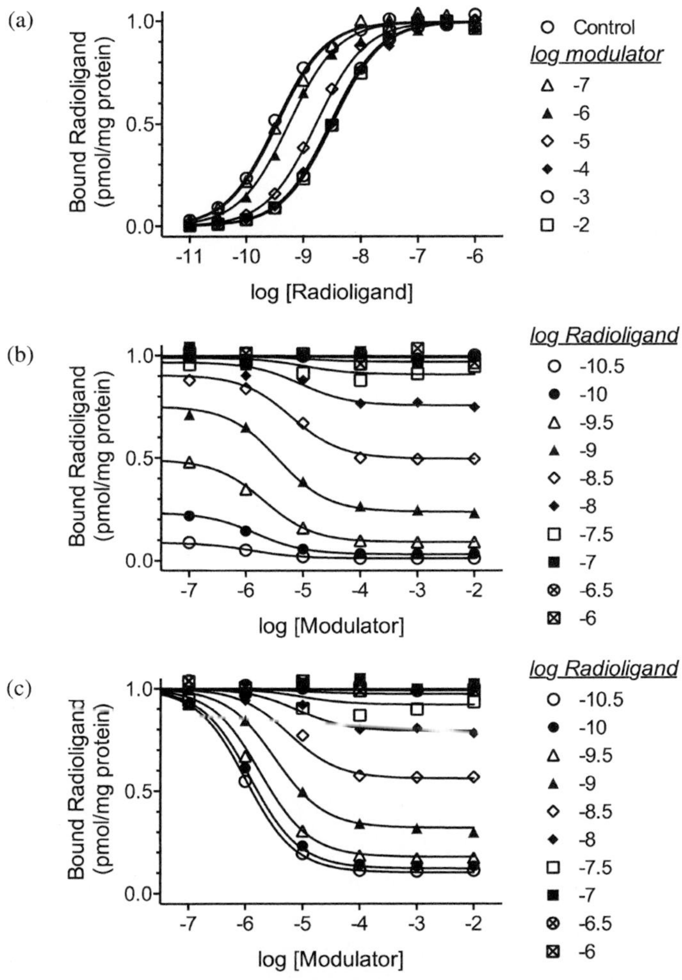

Figure 12.16: Testing the hypothesis of allosteric inhibition. Simulated values of specific binding were calculated for different concentrations of radioligand and in the presence of different concentrations of negative allosteric modulator. (a) Simulated binding values are plotted as a series of saturation curves for the condition of no modulator (control) and the indicated log concentrations of modulator. (b) Simulated binding values are plotted as a series of modulator curves, each representing the binding of radioligand at a fixed concentration and in the presence of different concentrations of nonlabeled modulator. (c) The modulator curves in panel b are plotted with binding normalized relative to that simulated in the absence of modulator. The theoretical curves in (a) and (b) represent the global least-squares fit of Equation (12.57) to the data with Equation (12.58) substituted for $K_{L\text{-app}}$. The theoretical curves in panel *c* represent the global least-squares fit of Equation (12.59) to the data with Equation (12.58) substituted for $K_{L\text{-app}}$. The simulated data used in this analysis are given in Table 12.2.

In this equation, X denotes the log concentration of radioligand, and `LogA`, the log concentration of modulator. The other variables are recognizable on the basis of their similarly to the variables in Equations (12.57) and (12.58). The log modulator concentrations are entered in the appropriate column headings in the Data Table. For the control saturation curve a large negative number is entered to represent the antilog of a modulator concentration of essentially zero (e.g., −20). For parameter constraints, `Bmax`, `LogKL`, `LogKA` and `Logalpha` are shared, and LogI is constrained to the value of the column title.

When this method is used to analyze the data in Figure 12.11a, the following parameter estimates are obtained: B_{max}, 1 ± 0.003; log K_L, 9.49 ± 0.02 and log K_A, 5.94 ± 0.05 and log α, -0.99 ± 0.02. These are nearly the same as those used to simulate the data: B_{max}, 1.0; log K_L, 9.50 and log K_A, 6.0 and log α, −1.0.

The same data can be analyzed from the perspective of a series of modulator inhibition curves, measured using different fixed concentrations of radioligand, as shown in panel b of Figure 12.16. These data can also be analyzed by global nonlinear regression analysis with Equation (12.57) (with Equation (12.58) substituted for $K_{L\text{-app}}$). In this case, however, the independent variable (X) is log [A] and the log values of the concentration of radioligand used in the competition curves are entered as column headings in the Data Table when Prism is used for the regression. These adjustment are reflected in the regression equation:

```
LogKLapp = Log((10^LogKL)*(1 + 10^(X + Logalpha + LogKA))/
                (1 + 10^(X + LogKA)))
Y = Bmax/(1 + 10^( − (LogL + LogKLapp))).
```

Note that this equation is the same as that described above for the analysis of saturation curves except that the log radioligand concentration has been changed from X to `LogL` and the log modulator concentration has been changed from `LogA` to X. Regression analysis of the data in Figure 12.16b yielded parameter estimates essentially the same as that described for the data in panel a of Figure 12.16.

The data in Figure 12.16c can be analyzed by global nonlinear regression analysis using Equation (12.59) with $K_{L\text{-app}}$ substituted with Equation (12.58). This analysis yields the following parameter estimates: log K_L, 9.56 ± 0.05; log K_A, 6.03 ± 0.03 and log α, -1.03 ± 0.04. Again these estimates are similar to the values used to simulate the data: log K_L, 9.50, log K_A, 6 and log α, −1.

As mentioned above, allosteric modulators that have a highly negatively cooperative interaction with the radioligand (log $\alpha \ll 0$) may exhibit inhibition curves resembling those of a competitive inhibitor. To discriminate between negative cooperativity and competitive inhibition, the analysis just described can be repeated but with the cooperativity constant constrained to zero (i.e., log α constrained to a large negative number (−20)). When the value of α is equivalent to zero, the equation for $K_{L\text{-app}}$ (Equation (12.58)) reduces to that for competitive inhibition (i.e., Equation (12.38)). Because the competitive form of Equation (12.58) (α constrained to 0 or $\log \alpha$ constrained to −20) has one less parameter than the allosteric form of the equation (α not constrained), the allosteric model will always fit the data better, even if the mechanism *is* competitive. There is no reason to accept the allosteric mechanism, however, unless there is a significant improvement in residual error with the more complicated model (i.e., allosteric model).

To determine if there is a significant improvement in residual error when the allosteric model is fitted to the data, the following statistic (F) is calculated:

$$F = \frac{\frac{RSS_1 - RSS_2}{df_1 - df_2}}{\frac{RSS_2}{df_2}}. \qquad (12.70)$$

In this equation, RSS_1 denotes the residual sum of squared deviations between the data points and the regression equation for competitive inhibition, RSS_2, the residual sum of squared deviations between the data points and the regression equation for allosterism, df_1 (degrees of freedom for the competitive model), the total number of data points minus the number of parameters in the competitive equation, and df_2 (degrees of freedom for the allosteric model), the total number of data points minus the number of parameters in the allosteric equation. In computing degrees of freedom, parameters that are constrained do not count as parameters. The statistic F exhibits an F distribution having degrees of freedom in the numerator and denominator of df_1 and df_2, respectively.

How is occupancy by the modulator related to its effect on the saturation curve of the radioligand?

The plots in Figure 12.16 show that an allosteric modulator causes a parallel shift in the saturation curve of the radioligand. Lazareno and Birdsall[12] have shown that a normalized measure of this shift is an estimate of receptor occupancy by the modulator in the absence of radioligand. Their rationale is described next.

The shift in the saturation curve caused by a positive allosteric modulator is equivalent to $K_{L\text{-app}}$ divided by K_L ($K_{L\text{-app}}/K_L$). In the absence of modulator, this ratio is equivalent to one. By subtracting one from the ratio $K_{L\text{-app}}/K_L$, the effect of the modulator can be calculated. If this effect is normalized relative to the maximal effect of the modulator, then the fractional effect of the modulator can be calculated. This maximal effect is equivalent to $\alpha - 1$. Thus, the normalized modification in the affinity of the radioligand caused by the modulator (M_{norm}) can be defined as:

$$M_{norm} = \frac{\frac{K_{L\text{-app}}}{K_L} - 1}{1}. \quad (12.71)$$

If Equation (12.58) is substituted for $K_{L\text{-app}}$ in Equation (12.71) and the result simplified, the following equation is obtained:

$$M_{norm} = \frac{AK_A}{AK_A + 1}. \quad (12.72)$$

This equation represents the fractional occupancy equation for the allosteric modulator in the absence of the radioligand. These mathematics also apply to the effect of a negative allosteric modulator.

Thus, the normalized change in the observed affinity of an orthosteric ligand caused by an allosteric ligand is equivalent to receptor occupancy by the allosteric ligand. As shown in Chapter 11, the normalized change in the product of the changes in affinity and efficacy of an orthosteric agonist by an allosteric modulator is also proportional to receptor occupancy by an allosteric ligand.

Measuring the interaction between the modulator and a nonlabeled orthosteric ligand

Often the nature of the allosteric interaction between the modulator and another nonlabeled orthosteric ligand is of interest. For example, one might want to measure how the allosteric modulator influences the binding of the endogenous ligand. Usually the natural ligand has low observed affinity for its orthosteric site on the receptor. Thus, it may be unfeasible to use the natural ligand as a radioactive probe. An alternative approach is to measure the competitive displacement of the binding of a high affinity antagonist radioligand by the natural ligand in the absence and presence of different concentrations of allosteric ligand.[12] In this section, the method for analyzing this type of experiment is described for the general case of a nonlabeled orthosteric ligand.

I begin by first deriving the requisite equations. Following these derivations (i.e., after Equation (12.78)), the method of data analysis is described using simulated data.

The scheme showing the competitive interaction between a radioligand (L) and another orthosteric ligand (I) in the presence of allosteric modulator is illustrated in Figure 12.17. The variable, α_1, denotes the cooperativity constant for the interaction between the binding of I and the allosteric ligand. The other variables have been described previously.

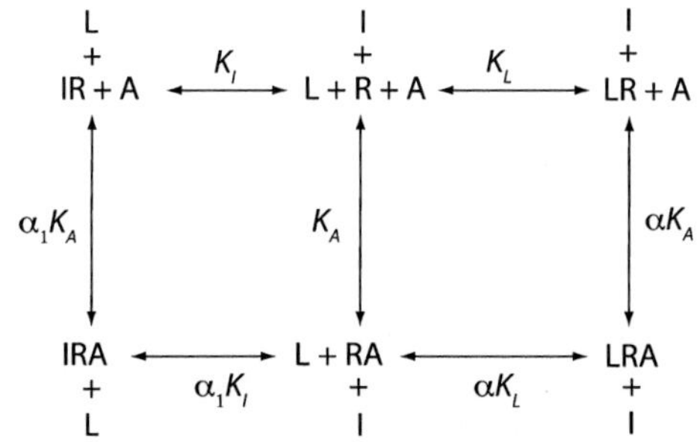

Figure 12.17: Scheme for the allosteric modulation of the competitive inhibition of the binding of a radioligand by a nonlabeled inhibitor. (a) The parameters K_L, K_I and K_A denote the affinity constants of the two orthosteric ligands and the allosteric modulator, respectively, and α and α_1 denote the cooperativity constants for the interactions between A and the two orthosteric ligands (L and I, respectively).

The two affinity constants describing the formation of the ternary complex ($[IRA]$) are defined as:

$$\alpha_1 K_D = \frac{[IRA]}{[I][RA]}, \tag{12.73}$$

$$\alpha_1 K_A = \frac{[IRA]}{[IR][A]}. \tag{12.74}$$

The derivation of the equation describing the binding of radioligand in the presence of allosteric modulator begins with the following equation:

$$\frac{[LR]+[LRA]}{B_{\max}} = \frac{[LR]+[LRA]}{[LR]+[IR]+[RA]+[LRA]+[IRA]+[R]}. \tag{12.75}$$

The concentrations of the receptor complexes on the right side of the equation are substituted with expressions based on the definitions of the affinity constants (Equations (12.2), (12.33), (12.51)–(12.53), (12.73) and (12.74)) as described in the derivations of Equations (12.37) and (12.57). Making these substitutions and simplifying yields:

$$[LR]+[LRA] = \frac{[L]K_{L\text{-app}}B_{\max}}{[L]K_{L\text{-app}}+1}, \tag{12.76}$$

in which:

$$K_{L\text{-app}} = K_L^* \frac{1+[A]\alpha K_A}{1+[I]K_I+[A]K_A(1+[I]\alpha_1 K_I)}. \tag{12.77}$$

Equation (12.76) describes the binding of the radioligand in the presence of a second orthosteric competitor (I) and an allosteric modulator (A).

To derive an equation for the fractional binding of the radioligand (f) at a fixed concentration and in the presence of different concentrations of I and A, Equation (12.76) is divided by the equation for the binding of radioligand in the absence of other ligands (Equation (12.5)). The result is Equation (12.59) but with $K_{L\text{-app}}$ defined by Equation (12.77) instead of 12.58. For convenience, Equation (12.59) is reproduced below with the assumption that the variable $K_{L\text{-app}}$ is now defined by Equation (12.77). Because the estimate of radioligand binding in the absence of other ligands has a modest error associated with it, the parameter P_1 has been introduced to represent this value:

$$f = P_1 \times \frac{L + \frac{1}{K_L}}{L + \frac{1}{K_{L\text{-app}}}}. \tag{12.78}$$

Equation (12.78) is our goal. This equation can be used to analyze the competition curves of an orthosteric ligand in the presence of different concentrations of an allosteric ligand.

The simulated data in Figure 12.18 illustrate the competitive binding curves of a nonlabeled competitor measured in the absence and presence of different

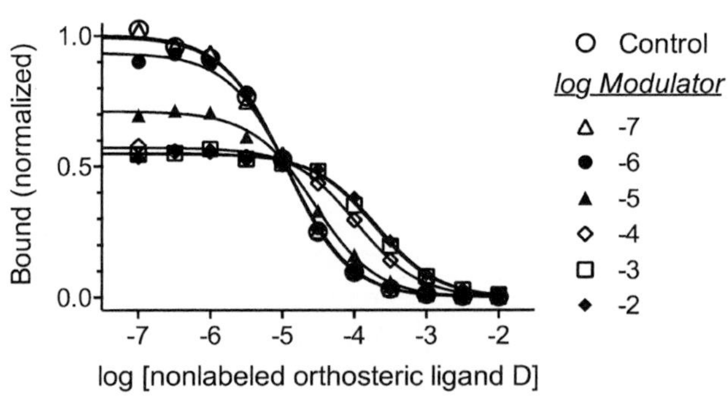

Figure 12.18: Analysis of the allosteric modulation of the displacement of the binding of radioligand by a competitive inhibitor. The plot shows a simulation of the competitive inhibition of the binding of radioligand at a fixed concentration (3.2 nM) by different concentrations of competitive inhibitor. The competition curves were simulated for the condition of no modulator (control) and the indicated log concentrations of modulator. The theoretical curves represent the global least-squares fit of Equation (12.78) to the data with Equation (12.77) substituted for $K_{L\text{-app}}$.

concentrations of allosteric ligand. The molar concentration of the radioligand is fixed at log −8.5. The affinity constant of the radioligand was set to log K_L, 9.5. The allosteric ligand causes a reduction in the binding of the radioligand, which is consistent with the log α value used in the simulation (log α, −1). In the presence of increasing concentrations of allosteric ligand, the IC_{50} value of the nonlabeled orthosteric ligand increases from a log control value of approximately −5 to a value of approximately −3.5 in the presence of a maximally effective concentration of allosteric ligand. Even though the allosteric ligand reduces the observed affinity of the radioligand about tenfold at maximal concentrations, the nonlabeled orthosteric exhibits lower observed affinity in the presence of a maximally effective concentration of modulator. Thus, it appears that the negative cooperativity between the binding of nonlabeled orthosteric ligand and the modulator (α_1) is greater than that between the modulator and the radioligand (α).

To estimate the parameters of the model, global nonlinear regression analysis is done using Equation (12.78) with $K_{L\text{-app}}$ substituted with Equation (12.77). For this regression, the equation is transformed so that the affinity constants and cooperativity factors are expressed as logarithms. The regression is done with log K_L constrained to a previously determined value. Global nonlinear regression analysis yielded the following parameter estimates: P_1, 1 ± 0.005; log K_A, 5.99 ± 0.03; log α, -1.01 ± 0.01; log K_I, 6 ± 0.01 and log α_1, -2.01 ± 0.03. These values are essentially the same as those used to simulate the data: P_1, 1.0; log K_A, 6; log α, −1.0; log K_I, 6 and log α_1, −2.0.

The asymptotic SEM values of the parameters are underestimates of the true values because log K_L was constrained during the analysis. To obtain accurate values, it is possible to analyze the data in Figure 12.17 simultaneously with those of a saturation curve for the radioligand using an approach analogous to that described in relation to Figures 12.9b and 12.14a, for example.

In the example illustrated in Figure 12.17, the allosteric ligand exhibited negative cooperativity with both the radioligand (log α, −1) and the nonlabeled competitor (log α_1, −2). For this reason, the data were simulated using a radioligand concentration tenfold larger than the reciprocal of the affinity constant, K_L, so that substantial binding would occur in the presence of high concentrations of modulator. If the modulator and the radioligand exhibit very large negative cooperativity, it may be difficult to measure competition curves in the presence of high concentrations of the modulator. This represents the main limitation of this analysis. If the radioligand and the modulator exhibit positive cooperativity, then this issue is not a problem.

References

1. Pedder, E. K., Eveleigh, P., Poyner, D., Hulme, E. C. and Birdsall, N. J. Modulation of the structure-binding relationships of antagonists for muscarinic acetylcholine receptor subtypes. *Br J Pharmacol* **103**, 1561–1567 (1991).
2. Birdsall, N. J. M., Burgen, A. S. V. and Hulme, E. C. in *Cholinergic Mechanisms and Psychopharmacology* (ed. D. J. Jenden) (Plenum Press, 1977), pp. 25–33.
3. Hulme, E. C., Birdsall, N. J., Burgen, A. S. and Mehta, P. The binding of antagonists to brain muscarinic receptors. *Mol Pharmacol* **14**, 737–750. (1978).
4. Birdsall, N. J., Burgen, A. S. and Hulme, E. C. The binding of agonists to brain muscarinic receptors. *Mol Pharmacol* **14**, 723–736 (1978).
5. De Lean, A., Stadel, J. M. and Lefkowitz, R. J. A ternary complex model explains the agonist-specific binding properties of the adenylate cyclase-coupled beta-adrenergic receptor. *J Biol Chem* **255**, 7108–7117 (1980).
6. Ehlert, F. J. in *Biomedical Applications of Computer Modeling* (ed. A. Christopoulos) (CRC Press, 2000), pp. 21–85.
7. Strange, P. G. Mechanisms underlying agonist efficacy. *Biochem Soc Trans* **35**, 733–736 (2007).
8. Jakubik, J., El-Fakahany, E. E. and Tucek, S. Evidence for a tandem two-site model of ligand binding to muscarinic acetylcholine receptors. *J Biol Chem* **275**, 18836–18844 (2000).
9. Klein, W. L. Multiple binding states of muscarinic acetylcholine receptors in membranes from neuroblastoma x glioma hybrid cells. *Biochem Biophys Res Commun* **93**, 1058–1066 (1980).
10. Jarv, J., Hedlund, B. and Bartfai, T. Isomerization of the muscarinic receptor . antagonist complex. *J Biol Chem* **254**, 5595–5598 (1979).
11. Ehlert, F. J. Estimation of the affinities of allosteric ligands using radioligand binding and pharmacological null methods. *Mol Pharmacol* **33**, 187–194. (1988).
12. Lazareno, S. and Birdsall, N. J. Detection, quantitation, and verification of allosteric interactions of agents with labeled and unlabeled ligands at *G* protein-coupled receptors: interactions of strychnine and acetylcholine at muscarinic receptors. *Mol Pharmacol* **48**, 362–378 (1995).
13. Candell, L. M., Yun, S. H., Tran, L. L. and Ehlert, F. J. Differential coupling of subtypes of the muscarinic receptor to adenylate cyclase and phosphoinositide hydrolysis in the longitudinal muscle of the rat ileum. *Mol Pharmacol* **38**, 689–697 (1990).
14. Esqueda, E. E., Gerstin, E. H., Jr., Griffin, M. T. and Ehlert, F. J. Stimulation of cyclic AMP accumulation and phosphoinositide hydrolysis by M3 muscarinic receptors in the rat peripheral lung. *Biochem Pharmacol* **52**, 643–658 (1996).
15. Ehlert, F. J., Thomas, E. A., Gerstin, E. H. and Griffin, M. T. in *Muscarinic Receptor Subtypes in Smooth Muscle* (ed. R. M. Eglen) (CRC Press, 1997), pp. 92–147.

Index